Introduction to Acoustics

INTRODUCTION TO ACOUSTICS

Robert D. Finch

Professor Emeritus
Department of Mechanical Engineering
University of Houston

Upper Saddle River, NJ 07458

Library of Congress Cataloging-in-Publication Data
Finch, Robert D.
Introduction to acoustics / Robert D. Finch.--1st ed.
p. cm
Includes bibliographical references and index.
ISBN 0-02-337570-1
1. Acoustical engineering. I. Title

TA365.F55 2004
620.2—dc22

2004049277

Vice President and Editorial Director, ECS: *Marcia J. Horton*
Editorial Assistant: *Andrea Messino*
Vice President and Director of Production and Manufacturing, ESM: *David W. Riccardi*
Executive Managing Editor: *Vince O'Brien*
Managing Editor: *David A. George*
Production Editor: *Kevin Bradley*
Director of Creative Services: *Paul Belfanti*
Art Director: *Jayne Conte*
Cover Designer: *Bruce Kenselaar*
Art Editor: *Greg Dulles*
Manufacturing Manager: *Trudy Pisciotti*
Manufacturing Buyer: *Lynda Castillo*
Senior Marketing Manager: *Holly Stark*

About the Cover: Gamma ray photograph of the bell of an English horn. Photograph by B.N. Nagarkar. Reprinted by permission from the *Journal of the Acoustical Society of America* (vol. 50, pp. 23–31, 1971). Courtesy of the American Institute of Physics.

Pearson Prentice Hall
Pearson Education, Inc.
Upper Saddle River, NJ 07458

Printed in the United States of America

10 9 8 7 6 5 4 3 2 1

ISBN 0-02-337570-1

Pearson Education Ltd., *London*
Pearson Education Australia Pty. Ltd., *Sydney*
Pearson Education Singapore, Pte. Ltd.
Pearson Education North Asia Ltd., *Hong Kong*
Pearson Education Canada, Inc., *Toronto*
Pearson Educación de Mexico, S.A. de C.V.
Pearson Education—Japan, *Tokyo*
Pearson Education Malaysia, Pte. Ltd.
Pearson Education, Inc., *Upper Saddle River, New Jersey*

Contents

Preface

The biological evolution of hearing is believed to have begun while our progenitors were still in the sea. Fish hear with an organ similar to the human cochlea, and it is believed that the external and middle ears evolved in amphibians and mammals as a means of matching the acoustic impedances of the atmosphere and the cochlea (or inner ear). This evolution would have had to occur when amphibians first emerged from the sea onto the land. On the other hand, while many animals use sound for communication, speech is a uniquely human activity, and presumably recent on the evolutionary time scale. One of the major anatomical differences between ourselves and the great apes is that we have an elaborate vocal tract above the larynx, and can articulate with great versatility, but they cannot. Because of this difference there has been speculation recently that it was the evolution of the voice as well as improvement in brain capacity that were the crucial factors in the ascendancy of humans over apes. It has even been suggested that perhaps the Neanderthals could not speak any better than the apes, and that the Cro-Magnons were the first humans to show this unique trait. Regardless of how science assesses this speculation in the future, it does make us realize just how important the acoustic system of speech and hearing is in our lives, and that to be deprived of speech or hearing is a major impediment.

Perhaps as old as our speaking abilities is our inclination to make music. Unwanted sounds can intrude on our speech and music, interfere with sleep, or cause deafness if loud enough. Noise control is therefore one of the main tasks of the acoustical engineer. The industrial revolution of the eighteenth century ushered in an age of high-power machinery, and along with it the unwanted by-product: vibration. This unwanted by-product led to the recognition of vibration as a field of serious study for mechanical engineers. In turn, the study of vibration led to the development of a discipline sometimes called "automatic" controls, which became especially important in aeronautical engineering. The pioneering work of Bell and Edison on the telephone and in broadcasting resulted in the growth of two new industries that have vastly extended our abilities to communicate. The use of submarines to destroy Allied shipping during World War I led to the invention of sonar, since electromagnetic waves will not travel very far under water. Sonar is now used for many peaceful purposes. The interest in underwater sound led to the development of practical ultrasonic transducers, which it was discovered could be used in medical diagnosis and treatment as well as in nondestructive evaluation and processing in industry. Starting in the 1920s, it became common for oil companies to use sound in seismic prospecting. We are also gaining an understanding of how to deal with earthquakes, another form of acoustic wave. Currently, significant efforts are underway to develop systems for machine recognition of speech. In addition, mechanical engineers are increasingly using machines for the acoustic and vibrational monitoring of machinery.

To some people, the word "acoustics" brings to mind orchestras and concert halls. Others think of hearing aids or noise control. Some engineers realize that mechanical oscillations may occur in discrete (lumped) systems, but fail to understand that they may also occur in continuous (distributed) systems. An example of a discrete system is the motion of a mass on a spring, and the study of such phenomena is usually termed "vibrations." An example of a continuous system is the motion of the air during the playing of music, and the term "sound" is employed in this case. Sound can also propagate at frequencies beyond the range of human hearing; at the high end, we speak of ultrasonics, and of infrasonics at the low end. Ultrasonics have become part of our lives through medical imaging and sonar, while earthquakes are an example of infrasonic waves. This wide variety of phenomena and applications is linked by a definition from the Acoustical Society of America (1965): Acoustics: "The Science of mechanical radiation in all its aspects and applications, including origin (vibration of material media, etc.), transmission through material media, at all frequencies and under the most diverse conditions, and reception."

An old adage characterizes a lecture as the process whereby the notes of the lecturer become the notes of the student. Textbooks tend to procreate in a similar manner. In defense of textbook authors, it should be noted that if our intention is to pass on to the student the most important discoveries and inventions of the past, then it is only to be expected that successive generations of textbooks should change in an evolutionary way. There is little doubt that the ancestor of all modern texts in acoustics is Rayleigh's *Theory of Sound*, first published in two volumes in 1877 and expanded in several later editions. In the United States, this book was followed by Stewart and Lindsay's *Acoustics* (1930), and then by Lindsay's *Mechanical Radiation* (1960). Another descendant in the Rayleigh tradition is Morse's *Vibration and Sound* (1936) followed by Morse and Ingard's *Theoretical Acoustics* (1968). In Britain, the most notable contributions are A.B. Wood's *A Textbook of Sound* (1930) and Stephens and Bates's *Wave Motion and Sound* (1950), the latter being used in the Physics Department at Imperial College when this author was a student. Historically, vibrations were the first branch of acoustics to enter the engineering curriculum, and Rayleigh's volume one was an early text among engineers. His influence is seen in Timoshenko's text (1928), and subsequently in that by Den Hartog (1934). Later came the well-accepted work by Thomson (1993) and many others, including the recent notable contribution by Dimarogonas (1996). Now, the most widely used general acoustics texts are probably those by Kinsler et al., *Fundamentals of Acoustics* (1982), and Pierce's *Acoustics* (1981). There are many other fine texts in acoustics influenced by Rayleigh in varying degrees, but those mentioned here are some of the better known and will suffice to make the point.

Sooner or later all educators confront this question of conscience: What it is that I am trying to accomplish? It is clear that acoustics is flourishing, but I would still argue strongly with those representatives of industry who advocate "training" young people for "jobs." Our task is to educate human beings to achieve the potential of which they are capable. Even when it comes to designing a strategy for those who must meet economic necessity, it is a poor teacher who equips the student with the tricks of only one trade. During the 30 years I have been teaching, there have been numerous shifts in employment for the acoustics graduate. Opportunities in seismic exploration are notorious for their ups and downs linked to the price of oil. The Navy's employment in underwater sound depends on national priorities for defense. Vibration and noise control and architectural acoustics tend to follow trends in the overall economy. A certain number of graduates find homes in the telephone and communications industries. Nondestructive testing of industrial components and medical ultrasonics are also important areas for employment. Very few people with backgrounds in acoustics are unemployed for long, but it is a good idea to be prepared for work in more than one industry. The task of the university should therefore be to provide the student with a broad repertoire of basic ideas. As time moves on, the content of this repertoire needs to change with the advance of science and of engineering practice. Recent years have seen the practice of acoustics revolutionized by the digital computer and digital signal processing. In the future, we can expect even greater changes in acoustical engineering as systems theorists better comprehend such functions of the human brain as speech recognition and music appreciation.

Frederick V. Hunt, the famous teacher of acoustics and professor of physics at Harvard, recognized as early as 1967 that acoustics had much to offer as a "fusion" subject, and I quote his words: "Some of the subjects now involved in this fusion process acquired their own recognized status only within the last few decades. One would cite in this connection modern information theory, the theory of stochastic processes, ... modern control theory, ... and the exploitation of nonlinearity." Today, we would recognize these topics as a part of systems theory. In this book, the intention is to stress the systems and engineering aspects of acoustics. Acoustics is a good introduction to general systems theory, both because of the heavy emphasis on mathematics and because of its interdisciplinary nature.

The book is intended for senior- or graduate-level engineers, with prerequisites in mathematics through differential equations, basic mechanics, and electromagnetism. The readership is expected to be primarily students majoring in mechanical or electrical engineering, although the book could also be of use to practicing engineers as a reference. The courses that might use this book include introductory acoustics, or vibrations, transducers, acoustic radiation, noise control, stress waves, physical acoustics, and acoustic measurements. Also included is some coverage of the effects of sound on man and on speech and hearing, insofar as these subjects define the tasks of the acoustical engineer, as well as an account of acoustic systems and a review of their design. The total length of a continuous course based on this book would be about two years.

ACKNOWLEDGMENTS

This work is dedicated to R.W.B. Stephens and Isadore Rudnick, the two great teachers of acoustics who guided my early studies. To my colleagues, Douglas Muster and Bill Cook, with thanks for wonderful years that were both informative and filled with fun. To my students, with thanks for their enthusiasm in pursuing (and occasionally evading) professorial direction. I also want to thank those who reviewed my manuscript: Singiresu Rao, University of Miami; Mike Kidner, Virginia Tech; and Nesrin Sarigul-Klijn, University of California, Davis. Finally, to my dearest wife, Sheila, and my long-suffering family, who had to endure the passage of this production.

ROBERT D. FINCH

REFERENCES AND FURTHER READING

Acoustical Society of America. 1965. *Report on Education in Acoustics*. (See 37 *J. Acoust. Soc. Am.* 359).

Den Hartog, J.P. [1934] 1974. *Mechanical Vibrations*. 4th ed. McGraw Hill.

Dimarogonas, A. 1996. *Vibration for Engineers*. 2nd ed. Prentice Hall.

Hunt, F.V. 1967. "Acoustics at the Crossroads." 87 *J. Engrg. Educ.* 693–696.

Kinsler, L.E., A.R. Frey, A.B. Coppen, and J.V. Saunders. 1982. *Fundamentals of Acoustics*. 3rd ed. Wiley.

Lindsay, R.B. 1960. *Mechanical Radiation*. McGraw Hill.

Morse, P.M. [1936] 1948. *Vibration and Sound*. 2nd ed. McGraw Hill. Republished: American Institute of Physics for Acoustical Society of America, 1976.

Morse, P.M., and K.U. Ingard. 1968. *Theoretical Acoustics*. McGraw Hill.

Pierce, A.D. 1981. *Acoustics*. McGraw Hill. Republished: American Institute of Physics, 1989.

Rayleigh, Lord. [1877] 1945. *The Theory of Sound*. Vols. 1 & 2. Dover.

Stephens, R.W.B., and A.E. Bates. [1950]. *Wave Motion and Sound*. 1966. *Acoustics and Vibrational Physics*. 2nd ed. Edward Arnold.

Stewart, G.W., and R.B. Lindsay. 1930. *Acoustics*. Van Nostrand.

Thomson, W.T. 1933. *Theory of Vibration With Applications*. 4th ed. Prentice Hall.

Timoshenko, S.P. [1928] 1990. *Vibration Problems in Engineering*. 5th ed. Wiley.

Wood, A.B. [1930] 1941. *A Textbook of Sound*. Rev. ed. MacMillan.

Chapter 1

Vibration

1.1 Phasors

1.1.1 Kinematics of Harmonic Motion

All acoustic phenomena involve vibrations, sometimes of great complexity. We can analyze these complex motions in terms of the least complex vibration, the so-called simple harmonic motion. Suppose a particle moves with uniform angular velocity ω around a circle of radius a, as in Fig. 1.1. Consider the projection of its motion on the diameter AB, the particle being presumed to start from B. An example of such a situation is the shadow thrown on a screen by parallel light illuminating a conical pendulum (Fig. 1.2).

Let the displacement of the projected motion of the shadow, (OC), be x. Then

$$x = a\cos(\omega t) \tag{1.1.1}$$

where t is the time taken for the particle to move from B to P. As is well known from elementary kinematics, the particle has a velocity $a\omega$ tangential to the circle at P (Fig. 1.3). The velocity, v, of the projected point C is thus the component of the tangential velocity in a direction parallel to AB; that is,

$$v = -a\omega\sin(\omega t) \tag{1.1.2}$$

The negative sign arises from the convention that A to B is a positive direction. Similarly, the acceleration of the particle is $a\omega^2$ and is directed radially inward, as shown in Fig. 1.4, so that the acceleration, s, of the point C is given by

$$s = -a\omega^2\cos(\omega t) \tag{1.1.3}$$

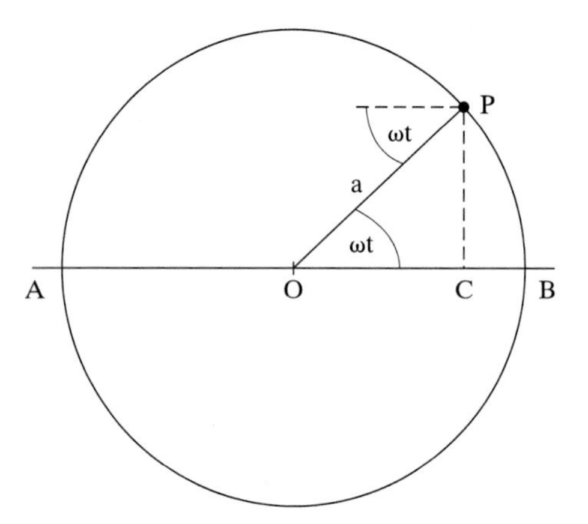

Fig. 1.1 Projection of circular motion onto a diameter results in harmonic motion of point C.

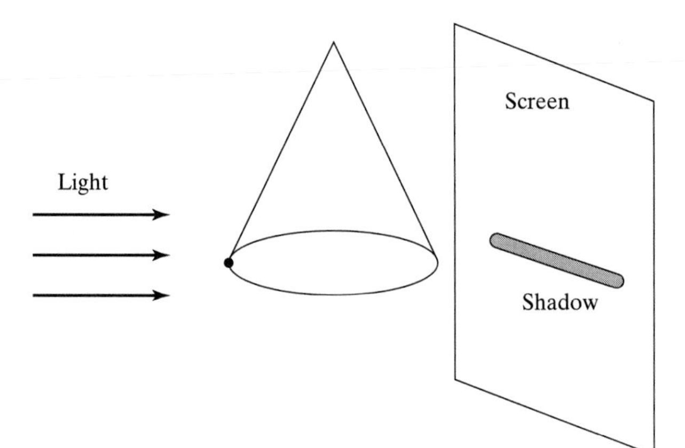

Fig. 1.2 The shadow of the bob of a conical pendulum moves in a harmonic motion.

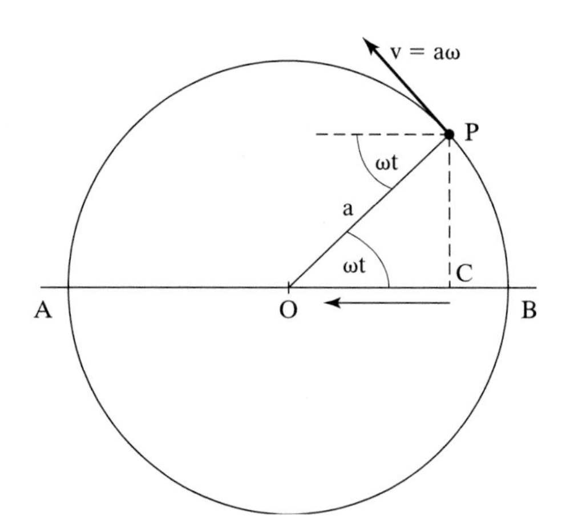

Fig. 1.3 Velocity in harmonic motion is $-v \sin \omega t = -a\omega \sin \omega t$.

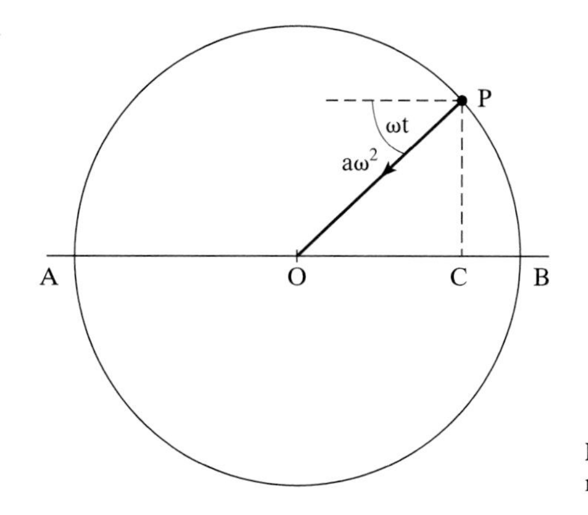

Fig. 1.4 Acceleration in harmonic motion is $-a\omega^2 \cos \omega t$.

We could have obtained these results simply by differentiating with respect to time, thus

$$v = \dot{x} = -a\omega \sin(\omega t) \tag{1.1.4}$$

and

$$s = \ddot{x} = -a\omega^2 \cos(\omega t) \tag{1.1.5}$$

Notice that

$$\ddot{x} = -\omega^2 x \tag{1.1.6}$$

Any motion that can be shown to satisfy this differential equation is called a simple harmonic motion.

We could plot the displacement as a function of time (Fig. 1.5). The maximum value, a, in this case is called the displacement amplitude. Similarly, the velocity amplitude is $a\omega$, and the acceleration amplitude is $a\omega^2$. Notice also that the projection of the

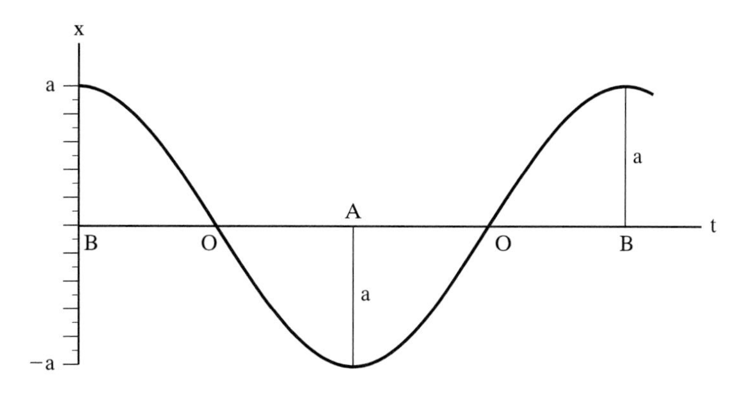

Fig. 1.5 Displacement in harmonic motion versus time.

motion moves from B through O to A and back to B, and so on. From B back to B is one cycle, and the duration of one cycle is called the period, T. Obviously,

$$T = 2\pi/\omega \tag{1.1.7}$$

Since T is the time taken to complete one cycle, the number of cycles made in unit time, called the frequency, f, is given by

$$f = 1/T = \omega/2\pi \tag{1.1.8}$$

ω is often referred to as the angular frequency. The concept of frequency is very important in acoustics. We are able to follow very low frequency vibrations by eye, but oscillations that are faster than about 10 cycles per second appear to be blurred. Audible sound involves vibrations that are much faster still: from about 30 times per second to 15,000 to 20,000 times per second. Instead of referring to a frequency as "so many times per second," we have designated a unit, the "hertz" (Hz), named for the celebrated German scientist Heinrich Hertz. The musical scale is now defined by specifying the frequencies of certain notes. Thus, middle C is 261 Hz and A is 440 Hz.

Returning to our discussion of the motion of a particle around a circle: Suppose that the particle does not start at B, but at some other point, say at D (Fig. 1.6). Then we must revise the equations as follows:

$$x = a\cos(\omega t + \delta) \tag{1.1.9}$$

$$\dot{x} = -a\omega\sin(\omega t + \delta) \tag{1.1.10}$$

$$\ddot{x} = -a\omega^2\cos(\omega t + \delta) \tag{1.1.11}$$

and still

$$\ddot{x} = -\omega^2 x \tag{1.1.12}$$

δ is called the phase angle or epoch. The harmonic motion $a\cos(\omega t + \delta)$ is said to lead the harmonic motion $a\cos(\omega t)$ by δ in phase.

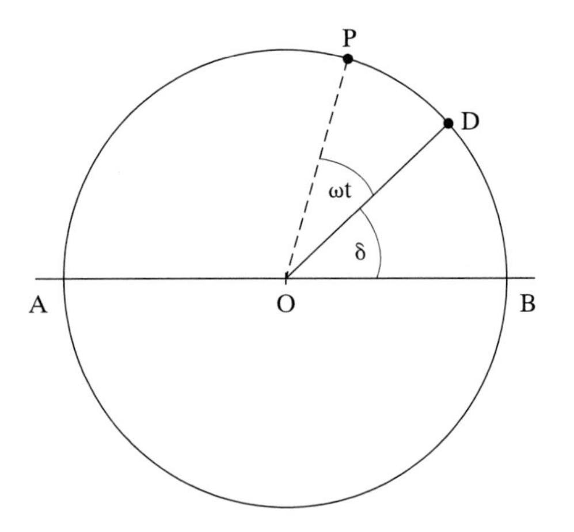

Fig. 1.6 Equivalent circular motion for a harmonic motion with initial phase angle δ.

Consider the harmonic motion a $\cos(\omega t + 3\pi/2)$. Here $\delta = 3\pi/2$. We could equally well write this harmonic motion as a $\sin(\omega t)$. So both a $\cos(\omega t)$ and a $\sin(\omega t)$ are solutions of the harmonic motion equation. In fact, the most commonly stated general solution of the equation is

$$x = A\cos(\omega t) + B\sin(\omega t) \tag{1.1.13}$$

which has two arbitrary constants, consistent with the equation being of second order. We can show that the solution given in Eq. (1.1.13) is completely equivalent to

$$x = C\cos(\omega t + \delta) \tag{1.1.14}$$

where

$$C^2 = A^2 + B^2 \tag{1.1.15}$$

and

$$\tan\delta = -B/A \tag{1.1.16}$$

C and δ are now the arbitrary constants. Alternatively, we can say that the solution is equivalent to

$$x = D\sin(\omega t + \varepsilon) \tag{1.1.17}$$

where D and ε are the arbitrary constants.

1.1.2 Phasor Representation of Oscillatory Quantities

There is an efficient and elegant method for representing oscillatory quantities (electrical currents and voltages as well as mechanical vibrations) using complex numbers. Just as the use of vectors greatly simplifies the study of mechanics, the use of complex numbers simplifies the study of oscillations. Recollect that the study of complex numbers starts by defining

$$i = \sqrt{-1} \tag{1.1.18}$$

Then any complex number, z, may be written as the sum of real and imaginary parts:

$$z = x + iy \tag{1.1.19}$$

Short-form representations of these real and imaginary parts of z are

$$x = R_e(z) \text{ and } y = I_m(z) \tag{1.1.20}$$

We may also represent z on the complex plane, in graphical form, as in Fig. 1.7. Then $|z|$ or the modulus of z is the length of the hypotenuse of the triangle. Notice that

$$|z^2| = x^2 + y^2 \tag{1.1.21}$$

$$x = |z|\cos\theta \tag{1.1.22}$$

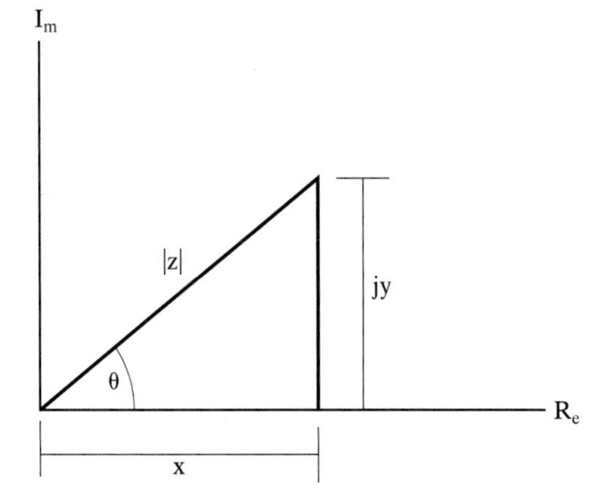

Fig. 1.7 Graphical representation of the complex number, $z = x + jy$.

and

$$y = |z|\sin\theta \qquad (1.1.23)$$

Hence

$$z = |z|\cos\theta + i|z|\sin\theta \qquad (1.1.24)$$

The angle θ is called the argument or phase of z. Note also that

$$\left(-\sqrt{-1}\right) \times \left(-\sqrt{-1}\right) = -1 \qquad (1.1.25)$$

Thus, we can define the quantity j by

$$j = -i = -\sqrt{-1} \qquad (1.1.26)$$

So, the complex number

$$x - iy = x + jy = |z|\cos\theta - i|z|\sin\theta = |z|\cos\theta + j|z|\sin\theta \qquad (1.1.27)$$

which can be represented in the fourth quadrant on the complex plane as shown in Fig. 1.8. This number is called the complex conjugate of z and is usually designated z^*. Notice that

$$zz^* = (x + iy)(x - iy) = x^2 + y^2 = |z|^2 \qquad (1.1.28)$$

So far we have represented a complex number as a point on the complex plane. The position vector from the origin to this point is called a "phasor." A useful result, known as Euler's formula, states that

$$e^{i\theta} = \cos\theta + i\sin\theta \qquad (1.1.29)$$

This theorem can be proven by using the Taylor series:

$$f(\theta) \approx f(\theta_0) + f'(\theta_0)(\theta - \theta_0) + f''(\theta_0)(\theta - \theta_0)^2/2! + \cdots \qquad (1.1.30)$$

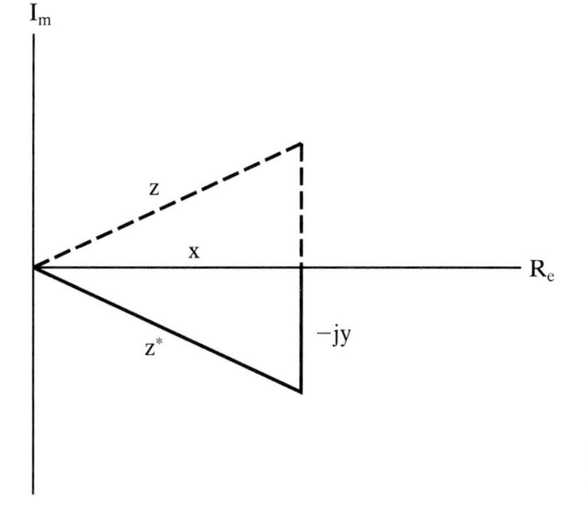

Fig. 1.8 The complex conjugate of z is $z^* = x - jy$.

Notice that $e^{i\theta}$ and $e^{j\theta}$ are "unit" phasors with phase angle $\pm\theta$. This development permits us to represent any phasor as a product of its magnitude and a unit vector:

$$z = |z|e^{i\theta} \quad \text{or} \quad z^* = |z|e^{j\theta} \tag{1.1.31}$$

We can now see the relationship between simple harmonic motion and the concept of a phasor. Let

$$\theta = \omega t + \delta \tag{1.1.32}$$

where ω is angular velocity, t is time, and δ is an arbitrary (but constant) phase angle. Then $e^{i\theta}$ will be a unit vector that rotates with angular velocity ω on the complex plane in a counterclockwise direction, and $e^{j\theta}$ is the same, except that it revolves in a clockwise direction (Fig. 1.9). Thus, we may write a simple harmonic motion as follows:

$$x = a\cos(\omega t + \delta) = R_e[a\, e^{i(\omega t+\delta)}] = R_e[a\, e^{j(\omega t+\delta)}] \tag{1.1.33}$$

Only the real part of the phasor is of significance, so it does not matter whether we use the form involving i or j. Since i is often used as a symbol for current in electrical engineering, it has become common practice to use j to symbolize the phasor. It is tedious to keep writing R_e ahead of the phasor, so it is normally omitted, as in the short-form given by the lefthand side of Eq. (1.1.20). We should therefore assume that when a physical quantity is written as a phasor, it is to be understood that only the real part has significance, unless stated otherwise. So, it would be common to write a harmonic motion as

$$x = a\, e^{j(\omega t+\delta)} \tag{1.1.34}$$

However, we must be particularly careful in performing algebraic operations to check that the presence of the imaginary component does not affect the operation.

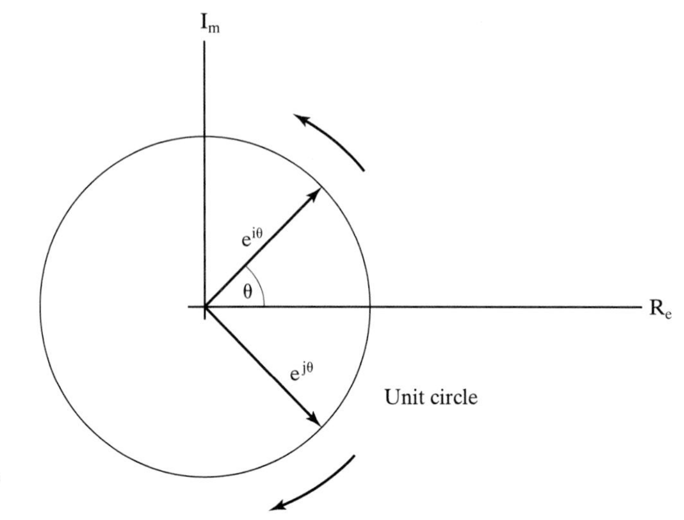

Fig. 1.9 Phasors rotate in the complex plane.

1.1.3 Superposition of Oscillations

Let two harmonic motions in the same spatial direction and of the same angular frequency, ω, be given by

$$x_1 = a \cos \omega t \tag{1.1.35}$$

and

$$x_2 = b \cos(\omega t + \delta) \tag{1.1.36}$$

Then the resultant of the superposition will be given by

$$\begin{aligned} x = x_1 + x_2 &= a \cos \omega t + b \cos \omega t \cos \delta - b \sin \omega t \sin \delta \\ &= [a + b \cos \delta] \cos \omega t - [b \sin \delta] \sin \omega t \end{aligned} \tag{1.1.37}$$

We can see that the resultant will also be a harmonic motion of amplitude A, say, and phase Φ if

$$A \cos \Phi = a + b \cos \delta \tag{1.1.38}$$

and

$$A \sin \Phi = b \sin \delta \tag{1.1.39}$$

To determine A, square and add:

$$A^2 = a^2 + b^2 + 2ab \cos \delta \tag{1.1.40}$$

Notice the similarity to addition of vectors. The situation is illustrated in Fig. 1.10. Notice also that by applying the sine rule to triangle XYZ in Fig. 1.10,

$$\frac{A}{\sin \delta} = \frac{b}{\sin \angle(YXZ)} \tag{1.1.41}$$

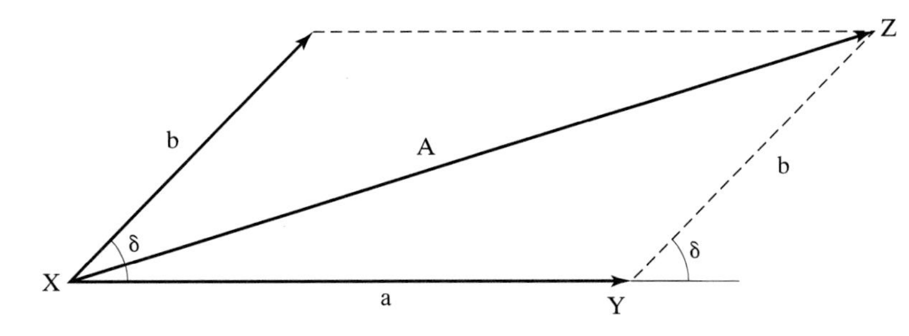

Fig. 1.10 Addition of phasors.

We can conclude then that two harmonic motions in the same direction and of the same frequency add as vectors whose magnitudes are equal to the amplitudes and whose inclination is equal to their phase difference. Note, however, that these vectors do not exist in real space, since the harmonic motions act in the same physical direction. The same mathematics applies for the addition of acoustic air pressures varying sinusoidally, although air pressure is nondirectional. The harmonic motions may be thought of only as vectors on the complex plane. Consider the complex number

$$N_1 = a\,e^{j\omega t} = a\cos\omega t + ja\sin\omega t \tag{1.1.42}$$

N_1 is a vector on the complex plane and it rotates as t changes (see Fig. 1.11). Both its real and imaginary components are harmonic motions, of which the real component is the first of the two harmonic motions we were originally considering. Now consider the complex number

$$N_2 = b\,e^{j(\omega t+\delta)} = b\cos(\omega t + \delta) + jb\sin(\omega t + \delta) \tag{1.1.43}$$

N_2 is also a vector on the complex plane, at an inclination δ to N_1. The resultant harmonic motions must thus be given by the vector addition of N_1 and N_2, as shown in Fig. 1.12.

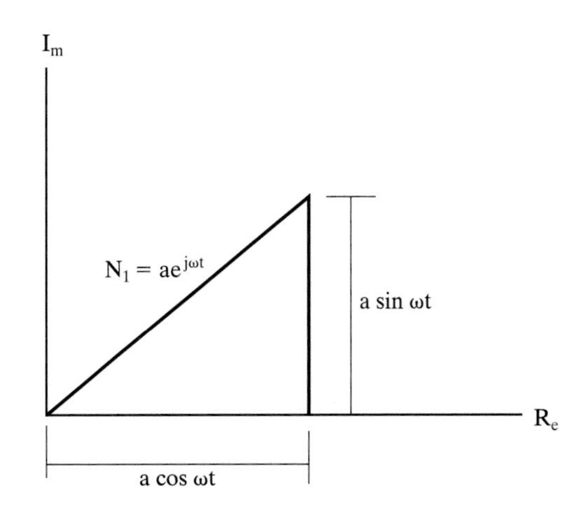

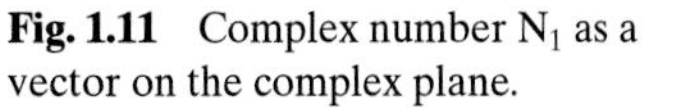
Fig. 1.11 Complex number N_1 as a vector on the complex plane.

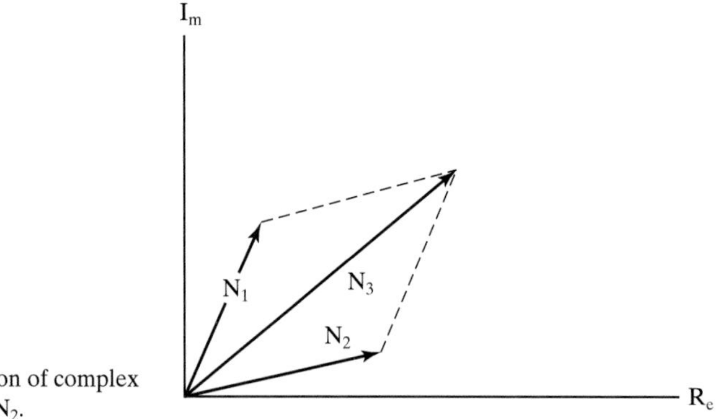

Fig. 1.12 Addition of complex numbers N_1 and N_2.

We will now return to our discussion of Eq. (1.1.40) and consider some special cases. If $\delta = 0$,

$$A^2 = (a + b)^2 \text{ and } A = a + b \tag{1.1.44}$$

The two motions are in phase and reinforce one another. If, however, $\delta = \pi$,

$$A^2 = (a - b)^2 \text{ and } A = a - b \tag{1.1.45}$$

By constructing the locus of the head of the vector A on the complex plane (Fig. 1.13), we can readily see that the above two cases give the extreme values for A. It is this variability in the resultant of adding harmonic motions of the same amplitudes but differing phases that is the basic explanation of the phenomenon of interference in optics and acoustics.

A further point regarding the parallelograms is shown in Figs. 1.10 and 1.12. Obviously, as time t changes the entire parallelograms will rotate. Consequently, to find the magnitude (amplitude) and relative phase of N_3, the time is immaterial and we arbitrarily select a reference on the real axis, as in Fig. 1.13. If it is necessary to add a number of harmonic motions, the construction on the complex plane becomes a polygon (Fig. 1.14), and the resultant is the vector that closes the polygon. In the limit of infinitesimally small changes in phase and a corresponding gradual decrease in amplitude, the polygon becomes a smooth spiral, known as the Cornu spiral, which is important in diffraction theory.

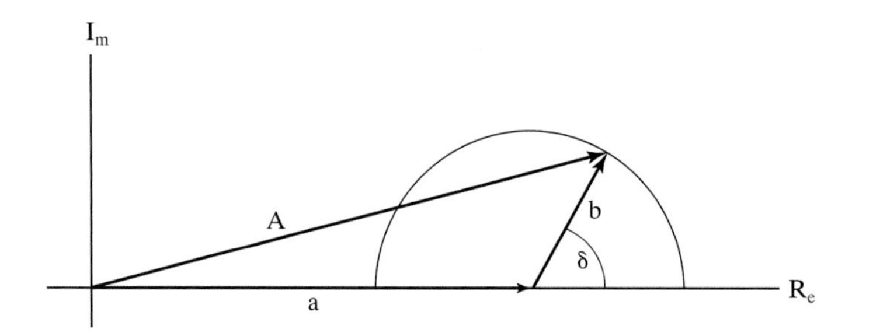

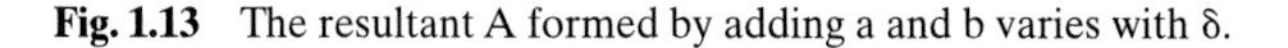

Fig. 1.13 The resultant A formed by adding a and b varies with δ.

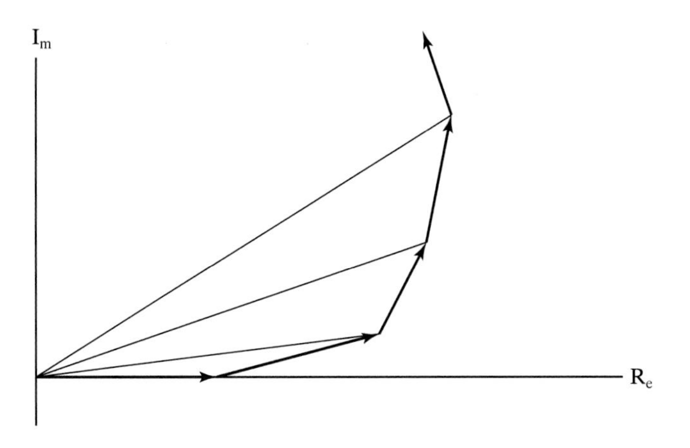

Fig. 1.14 Addition of a series of phasors with progressive phase change.

1.1.4 Dynamics and Energy of Harmonic Motion

So far we have discussed the kinematics of harmonic motion, stating that a particle undergoing such a motion has an acceleration proportional to its displacement from a certain reference point and directed towards that point (Fig. 1.15). It follows from Newton's second law of motion that it must be subject to a force F, such that

$$F = \text{mass} \times \text{acceleration} = m\ddot{x} = -m\omega^2 x \tag{1.1.46}$$

F is also always directed towards O. It follows that work must be done in moving from x to $(x + dx)$ in the amount

$$\delta W = m\omega^2 x \delta x \tag{1.1.47}$$

So the total work done in achieving a displacement x is given by

$$W = \int_0^x m\omega^2 x dx = m\omega^2 x^2/2 \tag{1.1.48}$$

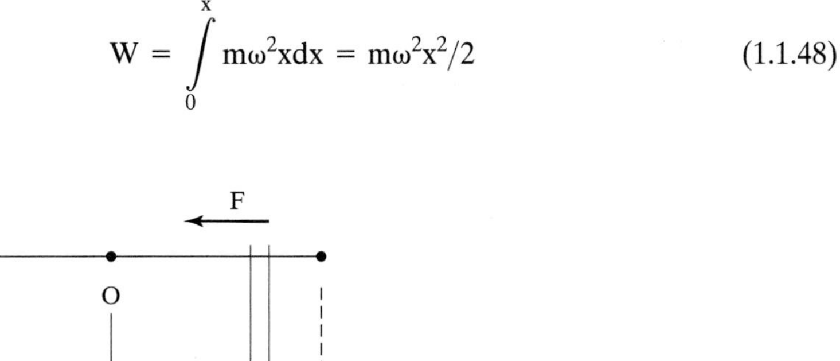

Fig. 1.15 A harmonic motion.

which must be the potential energy at displacement x (Fig. 1.16). The potential energy must be a maximum at the maximum of displacement; that is,

$$W_{max} = m\omega^2 a^2/2 \tag{1.1.49}$$

The particle has potential energy by virtue of its position, but it also has energy by virtue of motion (i.e., kinetic energy):

$$T = m\dot{x}^2/2 \tag{1.1.50}$$

Suppose, for example,

$$x = a \cos \omega t \tag{1.1.51}$$

then

$$\dot{x} = -a\omega \sin \omega t \tag{1.1.52}$$

and the total energy

$$\begin{aligned} E = W + T &= \frac{m\omega^2 x^2}{2} + \frac{m\dot{x}^2}{2} \\ &= \frac{ma^2\omega^2 \cos^2\omega t}{2} + \frac{ma^2\omega^2 \sin^2\omega t}{2} = \frac{ma^2\omega^2}{2} \end{aligned} \tag{1.1.53}$$

which is a constant, regardless of the value of displacement or time (Fig. 1.17). This constancy with time is to be expected, since there is no provision for dissipation of energy in our mathematical model. It is sometimes convenient to determine the angular frequency of a harmonic motion from the starting point of conservation of energy

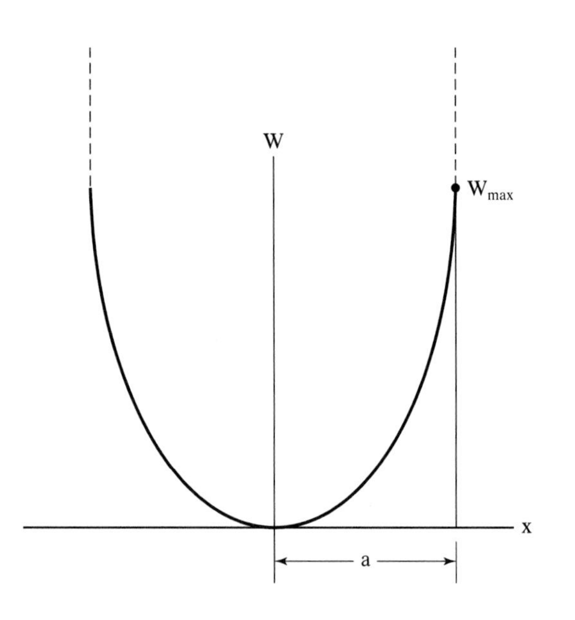

Fig. 1.16 Potential energy of a particle undergoing the motion of Fig. 1.15.

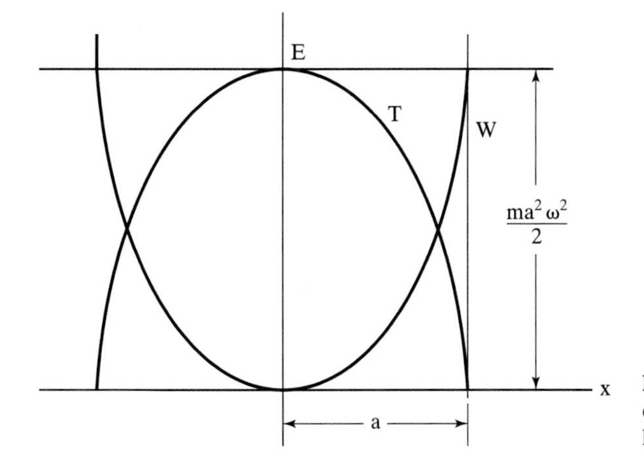

Fig. 1.17 Kinetic energy T, potential energy W, and total energy E of the harmonic motion of Fig. 1.15.

rather than from dynamic considerations. Thus, if we know that a particle has constant total energy, which at any instant is part potential and part kinetic, we may write

$$E = \frac{Sx^2}{2} + \frac{m\dot{x}^2}{2} \tag{1.1.54}$$

Then,

$$\frac{dE}{dt} = Sx\dot{x} + m\dot{x}\ddot{x} = 0 \tag{1.1.55}$$

Thus

$$\ddot{x} = -\frac{S}{m}x \tag{1.1.56}$$

which is a harmonic motion if

$$\omega = \sqrt{\frac{S}{m}} \tag{1.1.57}$$

This technique for finding the resonant frequency—in which a displacement is assumed and the principle of conservation of energy is applied—is known sometimes as Rayleigh's method, Rayleigh being the first to exploit the method in more complicated situations.

1.2 Single Degree of Freedom Oscillators

1.2.1 Mass-Loaded Spring

The most important example of harmonic motion for our purposes is the motion of a loaded spring. Consider a mass M moving on a horizontal, frictionless table and attached to a massless spring of stiffness S (Fig. 1.18). If the equilibrium position of the mass is $x = 0$, then for any displacement x

$$F = -Sx \tag{1.2.1}$$

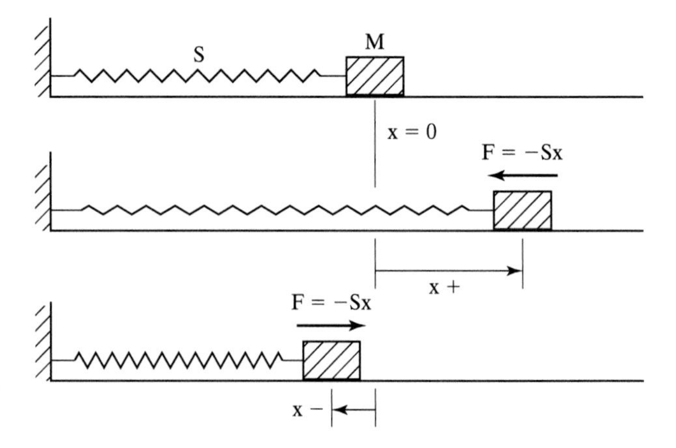

Fig. 1.18 A mass M oscillates on a spring of stiffness S.

and thus,

$$M\ddot{x} = -Sx \tag{1.2.2}$$

which is a harmonic motion with

$$\omega = \sqrt{\frac{S}{M}} \tag{1.2.3}$$

This result is not affected if the spring is hanging vertically, since the weight is balanced by a tension of the same magnitude at all times, and motion is affected only by unbalanced forces.

Of course, real springs are not massless. In the case of a linear spring, as in Fig. 1.19, it is fairly easy to take this into account. Let the spring have a mass ρ/unit length. Let the coordinate of an element (length δz, mass ρδz) on the unstretched spring be z. Let the mass M at the end of the spring be displaced by a distance y. Now we can obtain an approximate solution if we assume that the displacement is proportional to distance along

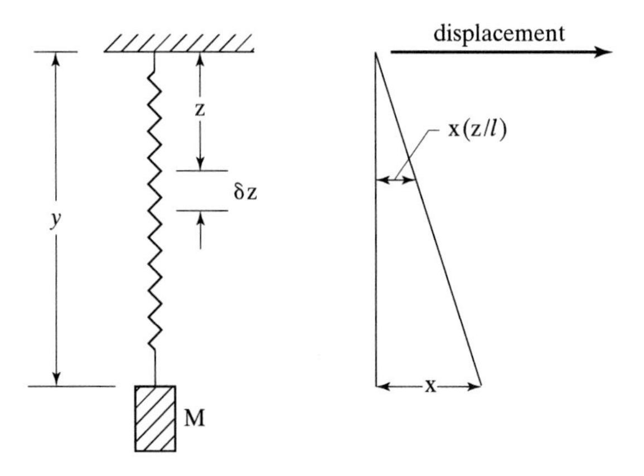

Fig. 1.19 The displacement at distance z along the spring is a fraction (z/l) of the displacement at the end with the mass.

the spring, or $\frac{z}{l}y$. Consequently, the velocity of the element must be $\frac{z}{l}\dot{y}$ and the kinetic energy of the element will be

$$\delta T = \frac{1}{2}\rho\delta z\frac{z^2}{l^2}\dot{y}^2 \tag{1.2.4}$$

Thus, the total kinetic energy of the spring at an instant of time is

$$T = \int_0^l \frac{\rho}{2}\frac{z^2}{l^2}\dot{y}^2\,dz = \frac{\rho\dot{y}^2 l}{6} \tag{1.2.5}$$

But the total mass of the spring is

$$m = \rho l \tag{1.2.6}$$

and thus the kinetic energy is $\frac{m/3}{2}\dot{y}^2$. The elastic properties of the spring are not changed by the fact that it has mass, so the potential energy remains the same. Thus, the total energy of the system is

$$\frac{1}{2}\left(M + \frac{m}{3}\right)\dot{y}^2 + \frac{Sy^2}{2} = \text{constant} \tag{1.2.7}$$

Differentiating with respect to time,

$$\left(M + \frac{m}{3}\right)\dot{y}\ddot{y} + Sy\dot{y} = 0 \tag{1.2.8}$$

and

$$\ddot{y} = \frac{-S}{\left(M + \frac{m}{3}\right)}y \tag{1.2.9}$$

and the corrected angular frequency is

$$\omega = \sqrt{\frac{S}{\left(M + \frac{m}{3}\right)}} \tag{1.2.10}$$

Note that this solution depends on our assumption of a displacement proportional to distance along the spring. This assumption is valid at low frequencies, but at high frequencies we must use a better approach by treating the spring as a distributed system.

1.2.2 Elasticity

All materials exhibit elasticity, and there is no such thing as a perfectly rigid body. Since all material objects deform, to greater or lesser extent, when subject to loads, all objects may be regarded as springs. This is the reason that the one degree of freedom oscillator

is such an important example. Most engineering materials are either fluids or solids. An enclosed fluid shows elastic behavior when compressed by an excess pressure, δP. The dilatation, or fractional volume change, is proportional to this excess pressure, and the constant of proportionality is termed the bulk modulus, B:

$$B = -\frac{\delta P}{(\delta V/V)} \tag{1.2.11}$$

The negative sign indicates that a volume decrement $(-\delta V)$ results from a pressure increment. This definition of bulk modulus applies to solids also, under circumstances of "hydrostatic" pressurization. However, there are other circumstances in which solids exhibit elastic behavior. For instance, a common situation is the case of a rod of length l under axial load, F. Then we define Young's modulus as the ratio of the stress F/A to the longitudinal strain $\delta l/l$, or

$$Y = \frac{F/A}{\delta l/l} \tag{1.2.12}$$

where A is the cross-sectional area and δl is the extension (see Fig. 1.20). In this case, the lateral dimensions of the rod usually decrease and the negative ratio of lateral strain to longitudinal strain is defined as Poisson's ratio, ν, where

$$\nu = -\frac{(\text{lateral strain})}{\delta l/l} \tag{1.2.13}$$

Finally, we recognize that it is possible to subject a solid to shear (Fig. 1.21). Then the shear modulus μ is given by

$$\mu = \frac{F/A}{\theta} \tag{1.2.14}$$

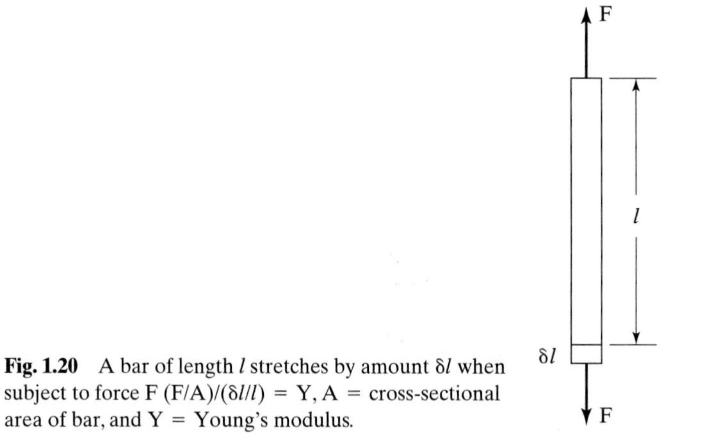

Fig. 1.20 A bar of length l stretches by amount δl when subject to force F $(F/A)/(\delta l/l) = Y$, A = cross-sectional area of bar, and Y = Young's modulus.

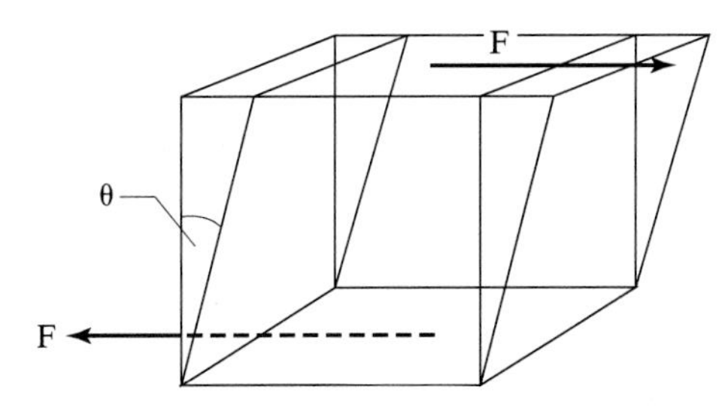

Fig. 1.21 A couple acting on an element of material causes it to shear through angle θ.

We can use these basic definitions of elastic moduli to calculate the stiffness of various objects. Take a simple example, the rod. Since Y is given by Eq. (1.2.12), it follows that

$$F = \left(\frac{YA}{l}\right)\delta l \tag{1.2.15}$$

In other words, the rod behaves as a spring of stiffness:

$$S = \frac{YA}{l} \tag{1.2.16}$$

Certainly, a solid rod would be very stiff, but we have illustrated the point that all objects may be regarded as springs.

1.2.3 Pneumatic Springs

Of considerable importance in acoustics is the case of the pneumatic spring, which is any closed volume of gas, V, subject to piston action over an area A of the enclosing surface (see Fig. 1.22). Suppose that the equilibrium pressure is P and that the expansions and contractions follow a polytropic process such that

$$PV^k = \text{constant} \tag{1.2.17}$$

and k is a constant. For a completely adiabatic process, $k = \gamma$, the ratio of specific heats. If heat is lost, then k is less than γ. For an isothermal process, $k = 1$. So, in general, $1 < k < \gamma$. On compression or expansion

$$(P + \delta P)(V + \delta V)^k = PV^k \tag{1.2.18}$$

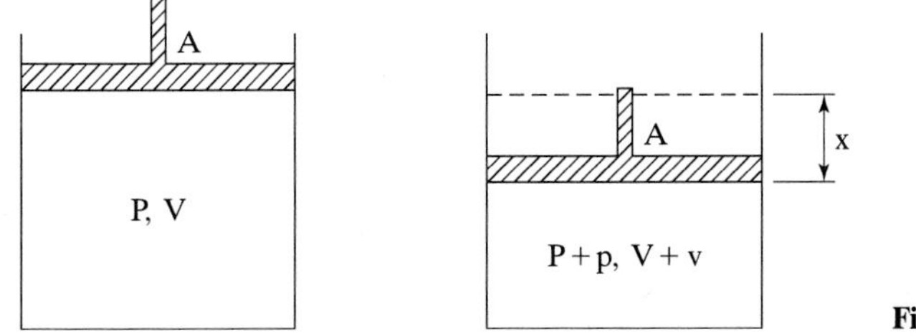

Fig. 1.22 An air spring.

and thus

$$PV^k(1 + \delta P/P)(1 + \delta V/V)^k = PV^k \tag{1.2.19}$$

Assuming that $\delta P/P, \delta V/V \ll 1$, we may use the binomial series expansion and neglect terms of higher than first order; then

$$1 + \delta P/P + k\delta V/V = 1 \tag{1.2.20}$$

or

$$\delta P = -kP\,\delta V/V \tag{1.2.21}$$

But $dV = Ax$, and the restoring force $F = \delta PA$, thus

$$F = (-kPA^2/V)x \tag{1.2.22}$$

and the stiffness is

$$S = kPA^2/V \tag{1.2.23}$$

Air springs are used to support trucks and other heavy loads. Another example of an air spring is the Helmholtz resonator. Imagine a bottle with a neck of length l, filled with a plug of density ρ, that is free to oscillate without friction (Fig. 1.23). The mass of the plug is then

$$m = \rho A l \tag{1.2.24}$$

and its resonant frequency as it vibrates on the air spring formed by the closed volume is given by

$$\omega = \sqrt{\frac{S}{m}} = \sqrt{\frac{kPA^2}{V\rho A l}} = \sqrt{\frac{kP}{\rho}}\sqrt{\frac{A}{V l}} \tag{1.2.25}$$

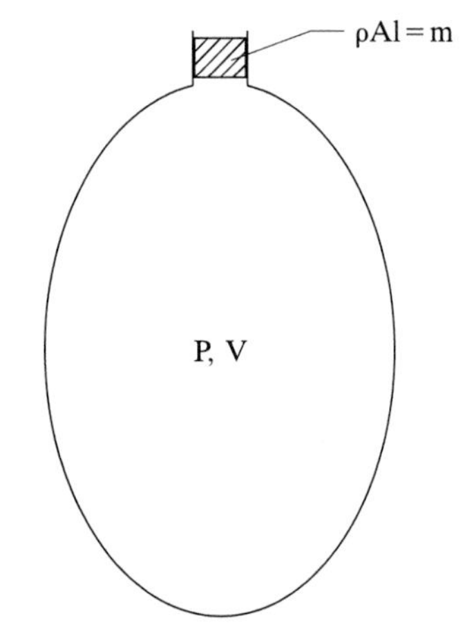

Fig. 1.23 A Helmholtz resonator.

If this plug were in fact constituted by air in the neck of an open bottle, the vibration frequency would be high and there would be no time for dissipation of heat. Then $k = \gamma$, the ratio of specific heats. Now let

$$\sqrt{\frac{\gamma P}{\rho}} = c \tag{1.2.26}$$

then the resonant frequency

$$f = \frac{c}{2\pi}\sqrt{\frac{A}{Vl}} \tag{1.2.27}$$

This formula could be tested experimentally by finding the resonant frequency of an air-filled bottle with a variable neck length. It would be found to be incorrect; but the experimental data could be fitted by adding a length correction, l', thus

$$f = \frac{c}{2\pi}\sqrt{\frac{A}{V(l + l')}} \tag{1.2.28}$$

The reason for this result is that air outside the bottle (and inside to some extent) participates in the motion of the air plug, so we underestimated its effective mass.

An interesting consequence is that even if $l \rightarrow 0$, there still will be a resonance; that is, a volume closed except by a hole that connects it to the surrounding air will give rise to vibrations that may be of audible frequency. Such resonators were first used by Helmholtz in his pioneer work in the analysis of complex tones (Fig. 1.24). These devices are sometimes called cavity resonators, a term borrowed by electrical engineers to describe an analogous electromagnetic device. Mechanical cavity resonators are used to absorb noise in automobile mufflers and in air conditioning systems.

We may generalize this treatment for any fluid volume (Fig. 1.25). Since the bulk modulus, B, is given by Eq. (1.2.11), it follows that the restoring force, F_r, is

$$F_r = -F = BA\,\delta V/V \tag{1.2.29}$$

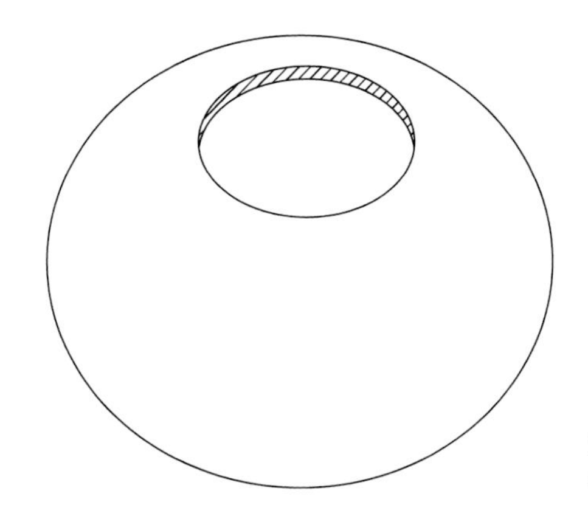

Fig. 1.24 Resonator used by Helmholtz in analyzing complex tones.

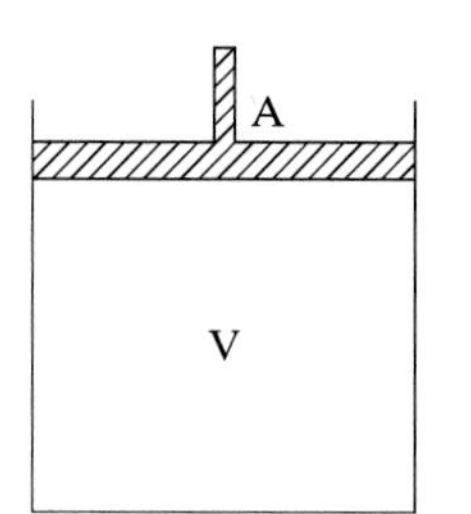

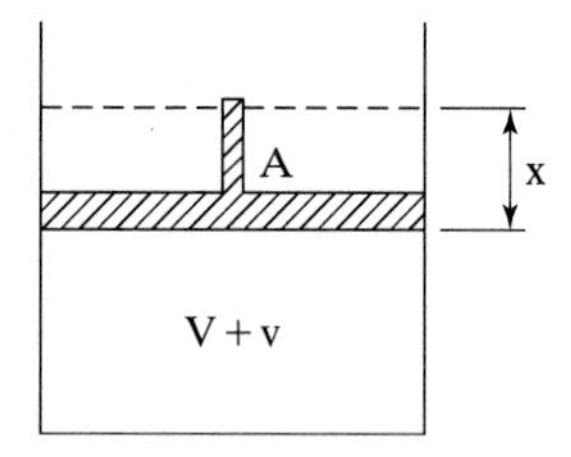

Fig. 1.25 Stiffness of a fluid volume is BA^2/V, where B is the bulk modulus.

But

$$\delta V = Ax \tag{1.2.30}$$

and thus

$$F_r = (BA^2/V)x \tag{1.2.31}$$

and the stiffness is

$$S = BA^2/V \tag{1.2.32}$$

A comparison with the previous result for a gas yields

$$B = kP \tag{1.2.33}$$

1.2.4 Stiffness of Mechanical Elements

We will now consider the stiffness of other mechanical elements. Our first example is a straight wire or shaft of length l, with a circular cross section of radius a, subject to equal opposite torques at either end (Fig. 1.26). From symmetry, it may be seen that cross sections remain plane and straight radii remain straight. Consequently, when the

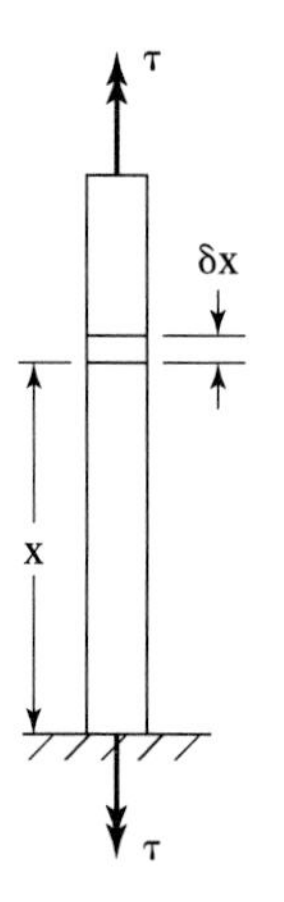

Fig. 1.26 A shaft or wire in torsion. The section δx is shown magnified in Fig. 1.27.

wire is twisted, the opposite faces of an elementary section undergo a relative shearing (see Fig. 1.27). With some manipulation, it follows that for the whole wire the torsional stiffness is given by

$$S = \frac{\mu\pi a^4}{2l} \tag{1.2.34}$$

where μ is the shear modulus.

Suppose that an object of polar moment of inertia I is suspended at the end of the wire just mentioned (Fig. 1.28). Then the equation of motion is

$$I\ddot{\theta} = -\frac{\mu\pi a^4}{2l}\theta \tag{1.2.35}$$

The result is a harmonic motion whose angular frequency is given by

$$\omega^2 = \frac{\mu\pi a^4}{2lI} \tag{1.2.36}$$

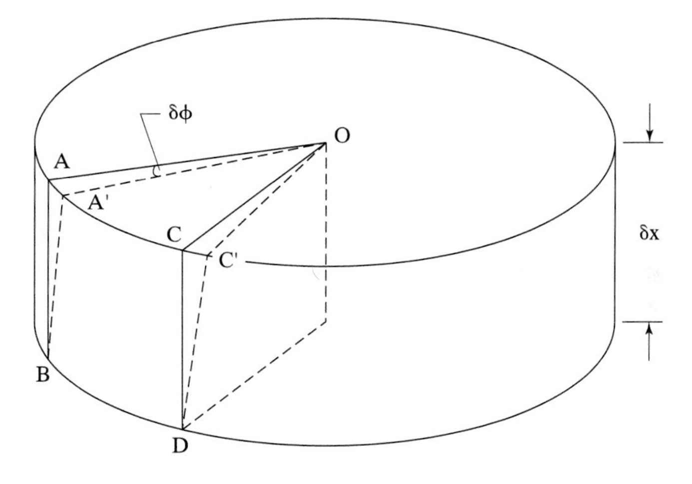

Fig. 1.27 The section δx. The top face is sheared relative to the bottom.

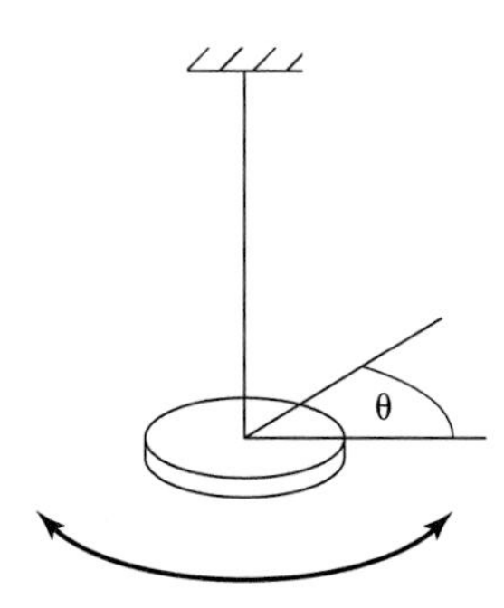

Fig. 1.28 A wire in torsional oscillation.

Thus, the resonant frequency is

$$f = \frac{1}{2\pi}\sqrt{\frac{\mu\pi a^4}{2lI}} \tag{1.2.37}$$

The extension of a helical spring (Fig. 1.29) is equivalent to a torsion. This fact may be visualized by thinking of the coil as a series of short straight sections, each progressively inclined to its neighbors. The effect of torsion on any element is to change the pitch of the coil. The cumulative effect of torsion in all the elements causes the spring to be extended or compressed along its length. It may be shown that the stiffness is given by

$$S = \frac{\mu d^4}{64nr^3} \tag{1.2.38}$$

where n = the number of turns, d is the wire diameter, and r is the coil radius.

We will take one more example: a solid ring of radius R and cross-sectional area A (Fig. 1.30). If the radius increases to $R + \delta R$, then the circumference increases to $2\pi(R + \delta R)$ and the strain will be $2\pi\delta R/2\pi R = \delta R/R = \xi/R$, where ξ is the radial displacement. We will now consider the resultant force on a segment defined by $\delta\theta$, noting first that this resultant will be directed radially inwards and so given by

$$F_R = -2F \sin \delta\theta/2 \tag{1.2.39}$$

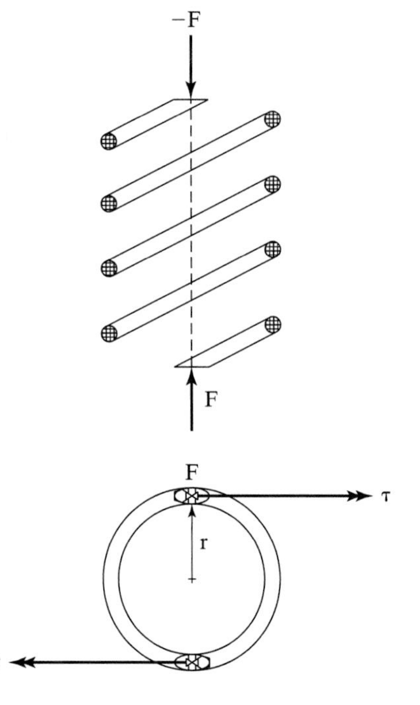

Fig. 1.29 Subjecting a helical spring to forces F and −F as shown is equivalent to putting the wire under a torsional moment τ, where τ = rF.

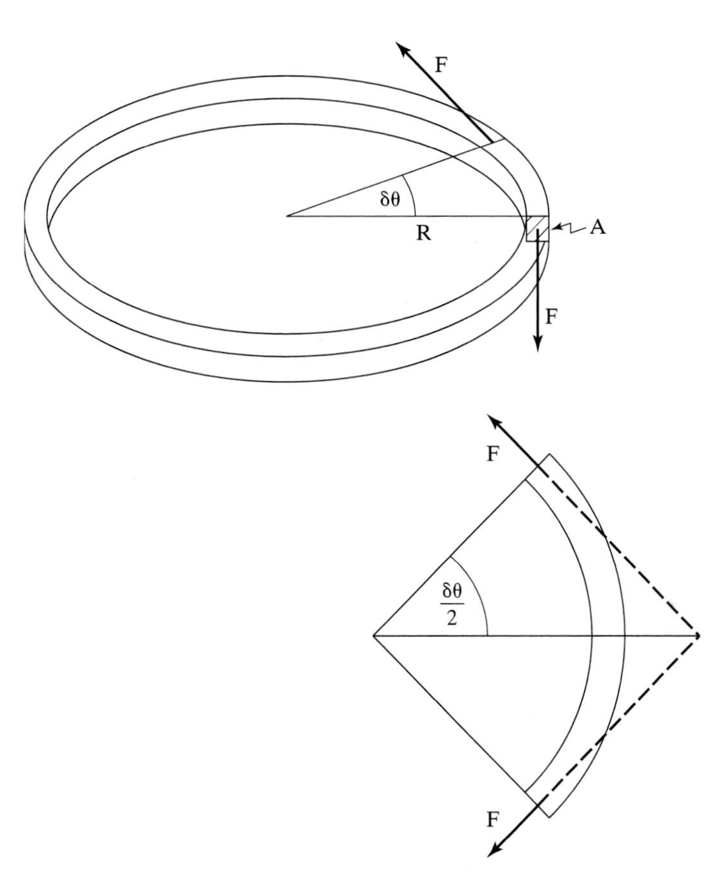

Fig. 1.30 Forces on a ring in hoop mode oscillation.

Now if $\delta\theta$ is small,

$$F_R = -F\delta\theta = -YA\xi\delta\theta/R \tag{1.2.40}$$

Applying Newton's second law,

$$-\frac{Y\xi}{R} = \rho R\ddot{\xi} \tag{1.2.41}$$

where ρ is the density. Thus, the ring's displacement is a harmonic motion of angular frequency

$$\omega = \frac{1}{R}\left(\frac{Y}{\rho}\right)^{1/2} \tag{1.2.42}$$

Recollecting Newton's formula for sound velocity, c, and setting the circumference $2\pi R = L$, we finally obtain for the resonant frequency

$$f_r = c/L \tag{1.2.43}$$

It is interesting to note that if we were to propagate a wave in a straight bar made of the same material at the frequency f_r, the wavelength would be c/f_r. The resonant frequency of a ring is, then, the frequency at which the circumference is equal to a wavelength of sound in a straight bar of the same material. Magnetostrictive scroll transducers (so called because they consist of a nickel scroll) and ferroelectric cylindrical transducers are widely used as hydrophones, and the formula in Eq. (1.2.43) may be used for a rough estimation of the frequency of maximum response. There are, however, many other modes of vibration in a cylinder (involving flexure for instance), and there will be many other "resonant" frequencies besides the basic "breathing" mode that we have just discussed. Finally, note that the stiffness $YA\ \delta\theta/R$, as in the previous cases, involves only the geometry and elasticity of the body.

Table 1.1 is a summary of spring stiffness for the examples we have discussed and for other simple mechanical elements (such as flat coils and cantilevers). Notice again that the stiffness of these various objects depends on the elastic properties of the material and on the geometry of the configuration. The stiffnesses of other configurations may be found in Roark (1989) and Blevins (1979). Finally, note that springs may be used in combination. For a set of springs in mechanical series (Fig. 1.31), the displacement of the combination is equal to the sum of the displacements of the members. Hence

$$x = x_1 + x_2 + x_3 + \cdots \tag{1.2.44}$$

and since all the springs experience the same force,

$$1/S = 1/S_1 + 1/S_2 + 1/S_3 + \cdots = \Sigma(1/S_n) \tag{1.2.45}$$

For a set of springs in mechanical parallel (Fig. 1.32), all members experience the same displacement, but transmit a different fraction of the applied force. Hence

$$F = F_1 + F_2 + \cdots + F_n = xS_1 + xS_2 + \cdots + xS_n \tag{1.2.46}$$

So,

$$F/x = S = \Sigma S_n \tag{1.2.47}$$

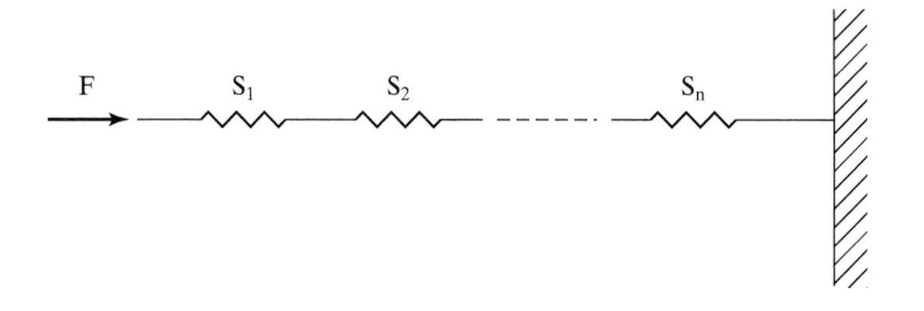

Fig. 1.31 Springs in series all transmit the same force.

Table 1.1 Table of Spring Stiffnesses

Configuration	Stiffness
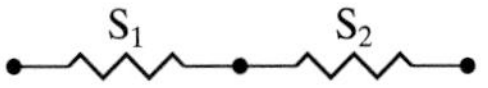 S_1, S_2 (in series)	$\frac{1}{1/S_1 + 1/S_2}$
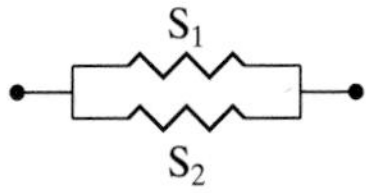 S_1, S_2 (in parallel)	$S_1 + S_2$
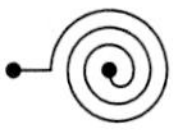 (spiral spring)	$\frac{YI}{l}$, I = moment of inertia of cross-sectional area, l = total length
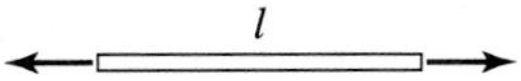 l	$\frac{YA}{l}$, A = cross-sectional area
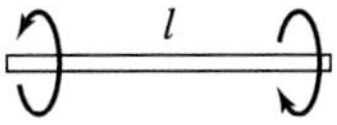 l	$\frac{\mu J}{l}$, J = polar moment of area
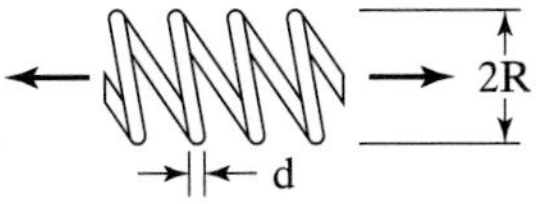 2R, d	$\frac{\mu d^4}{64nR^3}$, n = number of turns
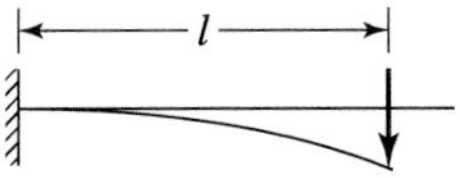 l	$\frac{3YI}{l^3}$
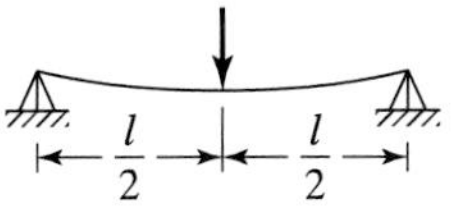 $\frac{l}{2}$, $\frac{l}{2}$	$\frac{48YI}{l^3}$
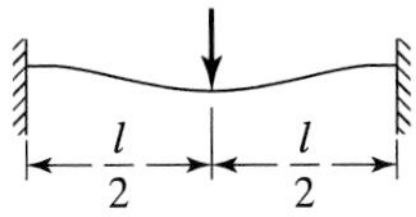 $\frac{l}{2}$, $\frac{l}{2}$	$\frac{192YI}{l^3}$
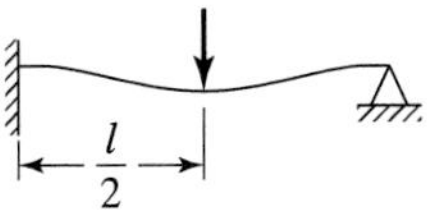 $\frac{l}{2}$	$\frac{768YI}{7l^3}$
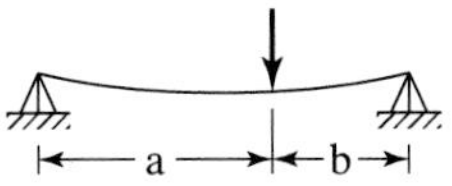 a, b	$\frac{3YIl}{a^2b^2}$
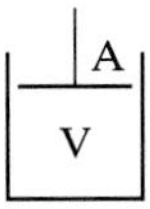 A, V	$\frac{BA^2}{V}$, for gas $B = \gamma P_0$

Source: After Thomson, 1966.

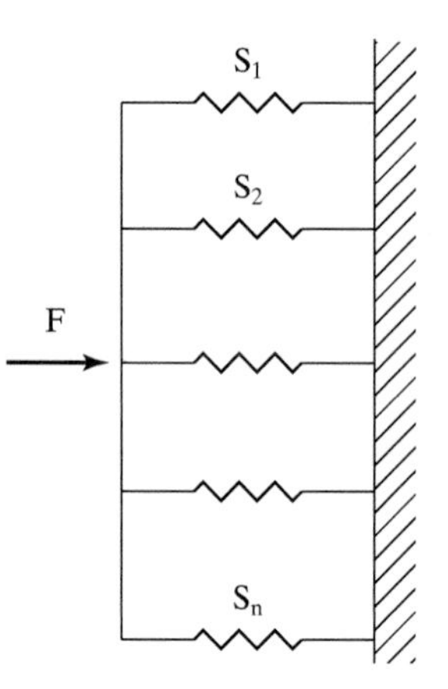

Fig. 1.32 Springs in parallel all suffer the same displacement.

1.2.5 Damping

So far, we have considered only the undamped motion of a mass on a spring. Suppose, however, that the moving mass experiences a resistive force proportional to its velocity. Such a force results if a body moves through a viscous fluid without causing non-laminar flow and is thus termed viscous damping. Coulomb, or dry friction, damping depends on the nature of the contacting surfaces and is usually taken to be independent of the relative velocities. In turbulent flow, the frictional drag is proportional to velocity squared. We will restrict ourselves to the case of linear, viscous damping for the sake of simplicity. The conventional representation for viscous damping is the dashpot, as shown in Fig. 1.33. The unbalanced forces acting on the mass are then

$$F = -Sx - R\dot{x} \tag{1.2.48}$$

where R is the damping coefficient. By Newton's second law,

$$-Sx - R\dot{x} = M\ddot{x} \tag{1.2.49}$$

which results in the well-known equation

$$M\ddot{x} + R\dot{x} + Sx = 0 \tag{1.2.50}$$

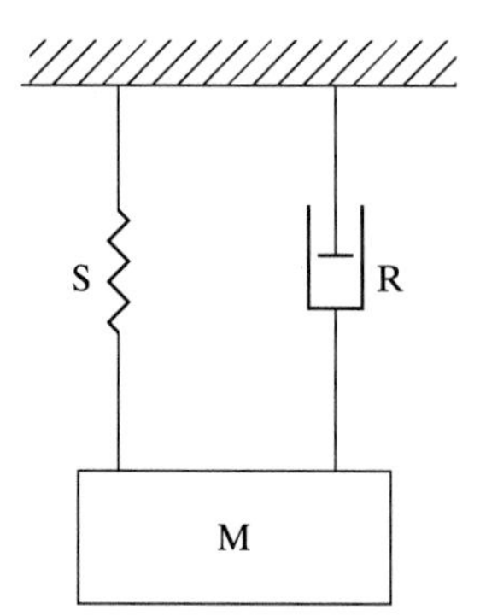

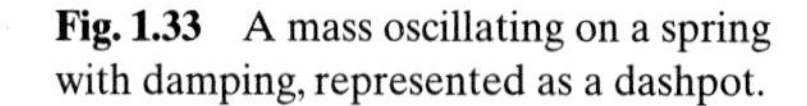

Fig. 1.33 A mass oscillating on a spring with damping, represented as a dashpot.

Compared with the undamped motion case we solved in Section 1.2.1 we have added the term $R\dot{x}$ to Eq. (1.2.2). Using physical intuition to anticipate a solution, we expect that the damping will cause the oscillations to die away with time in an exponential manner (see Fig. 1.34):

$$x = x_0 e^{-\alpha t} e^{j\omega_D t} \tag{1.2.51}$$

Substitute for x in Eq. (1.2.50) to determine α and ω, noting that

$$\dot{x} = (-\alpha + j\omega_D)x \text{ and } \ddot{x} = (\alpha^2 - 2j\alpha\omega_D - \omega_D^2)x \tag{1.2.52}$$

so that

$$M(\alpha^2 - 2j\alpha\omega_D - \omega_D^2)x + R(-\alpha + j\omega_D)x + Sx = 0 \tag{1.2.53}$$

Dividing by x and equating both real and imaginary parts to zero:

$$M(\alpha^2 - \omega_D^2) - R\alpha + S = 0 \tag{1.2.54}$$

and

$$-2M\alpha\omega_D + R\omega_D = 0 \tag{1.2.55}$$

From Eq. (1.2.55),

$$\alpha = R/2M \tag{1.2.56}$$

and thus from Eq. (1.2.54),

$$\omega_D^2 = S/M - R^2/4M^2 \tag{1.2.57}$$

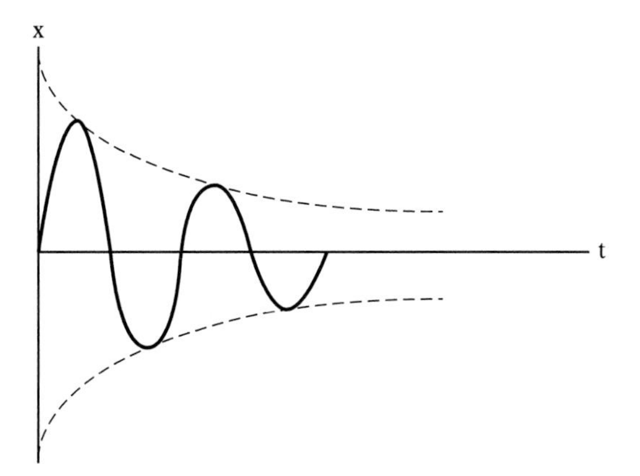

Fig. 1.34 The effect of damping is to cause the oscillations to die away exponentially with time.

Solving Eq. (1.2.50) by a more rigorous method vindicates our assumed solution, verifies the results for α and ω_D, and yields additional solutions. Thus, dividing Eq. (1.2.50) by M and setting $r = R/M$ and $s = S/M$ yield

$$\ddot{x} + r\dot{x} + sx = 0 \tag{1.2.58}$$

Introducing the differential operator notation ($D = \partial/\partial t$),

$$(D^2 + rD + s)x = 0 \tag{1.2.59}$$

Factorizing this equation, we obtain

$$\left(D + \frac{r}{2} + \sqrt{\frac{r^2}{4} - s}\right)\left(D + \frac{r}{2} - \sqrt{\frac{r^2}{4} - s}\right)x = 0 \tag{1.2.60}$$

Setting the roots

$$\lambda_1 = -r/2 - [(r/4) - s]^{1/2} \tag{1.2.61}$$

and

$$\lambda_2 = -r/2 + [(r/4) - s]^{1/2} \tag{1.2.62}$$

gives

$$(D - \lambda_1)(D - \lambda_2)x = 0 \tag{1.2.63}$$

that is, either

$$\frac{dx}{dt} = \lambda_1 x \quad \text{and} \quad x = Ae^{\lambda_1 t} \tag{1.2.64}$$

or

$$\frac{dx}{dt} = \lambda_2 x \quad \text{and} \quad x = Be^{\lambda_2 t} \tag{1.2.65}$$

Thus, the general solution is

$$x = Ae^{\lambda_1 t} + Be^{\lambda_2 t} \tag{1.2.66}$$

We will now recognize the following special cases:

1. If $R = 0 = rM$, then

$$\lambda_1 = -j\sqrt{s} = -j\sqrt{S/M} = -j\omega_0 \tag{1.2.67}$$

$$\lambda_2 = j\sqrt{S/M} = j\omega_0 \tag{1.2.68}$$

and

$$x = Ae^{-j\omega_0 t} + Be^{j\omega_0 t} = C\cos(\omega_0 t) + D\sin(\omega_0 t) \tag{1.2.69}$$

This is simply the case of harmonic motion as is to be expected in the case of zero damping.

2. If $r^2/4 > s$, then

$$x = e^{-rt/2}\left[A\exp\left(-\left\{\frac{r^2}{4} - s\right\}^{1/2} t\right) + B\exp\left(\left\{\frac{r^2}{4} - s\right\}^{1/2} t\right)\right] \tag{1.2.70}$$

Note that since

$$\left\{\frac{r^2}{4} - s\right\}^{1/2} < r/2 \tag{1.2.71}$$

both terms are nonoscillatory decaying exponentials, and the system is said to be overdamped.

3. If $r^2/4 = s$ (i.e., $R = 2\sqrt{SM}$), then

$$\lambda_1 = \lambda_2 = -r/2 \tag{1.2.72}$$

and

$$x = Ce^{-rt/2} \tag{1.2.73}$$

which contains only one arbitrary constant. To obtain the general solution in this case, add the solution $Bte^{-rt/2}$, giving

$$x = Ce^{-rt/2} + Bte^{-rt/2} \tag{1.2.74}$$

Again, there is no oscillation. This is said to be the case of critical damping, since it involves the smallest value of damping possible without throwing the system into oscillation. Some galvanometers are designed to operate at critical damping.

4. If $r^2/4 < s$, the case of small damping, then we obtain the oscillatory solution previously anticipated intuitively (Fig. 1.34):

$$\lambda_1 = -r/2 - j\{s - r^2/4\}^{1/2} \text{ and } \lambda_2 = -r/2 + j\{s - r^2/4\}^{1/2} \tag{1.2.75}$$

Substituting into Eq. (1.2.66),

$$x = e^{-rt/2}\{Ae^{-j(s-r^2/4)^{1/2}t} + Be^{j(s-r^2/4)^{1/2}t}\}^{1/2} \tag{1.2.76}$$

which is exactly the form of Eq. (1.2.51) if $B = 0$, if

$$\alpha = r/2 = R/2M \tag{1.2.77}$$

as in Eq. (1.2.56), and if

$$\omega_D{}^2 = s - r^2/4 = S/M - R^2/4M^2 \tag{1.2.78}$$

as in Eq. (1.2.57). Notice that we may write Eq. (1.2.57) as

$$\begin{aligned} x &= e^{-rt/2}[C \cos \omega_D t + D \sin \omega_D t] \\ &= Ee^{-rt/2}\sin(\omega_D t + \beta) \end{aligned} \tag{1.2.79}$$

where β = an arbitrary phase angle. $Ee^{-rt/2}$ is an exponentially decaying amplitude. The rate of decay is sometimes specified by a quantity called the logarithmic decrement, δ, defined by

$$\delta = \ln\left(\frac{x_n}{x_{n+1}}\right) \approx \ln\left(\frac{e^{-rt_n/2}}{e^{-r(t_n+T)/2}}\right) = \frac{RT}{2M} \tag{1.2.80}$$

where T = the period of the motion. This result is true only for small damping, when displacement maxima are given by

$$\sin(\omega_D t + \beta) = 1 \tag{1.2.81}$$

1.3 Forced Oscillation

1.3.1 Equation of Motion

We will now consider the case in which a harmonically varying force is applied to a system of a mass on a spring with damping at a frequency that does not in general correspond to the natural frequency (Fig. 1.35). The net unbalanced force acting on the mass is then the sum of the spring and resistance forces and the applied force. Applying Newton's second law,

$$-Sx - R\dot{x} + F_0 e^{j\omega t} = M\ddot{x} \tag{1.3.1}$$

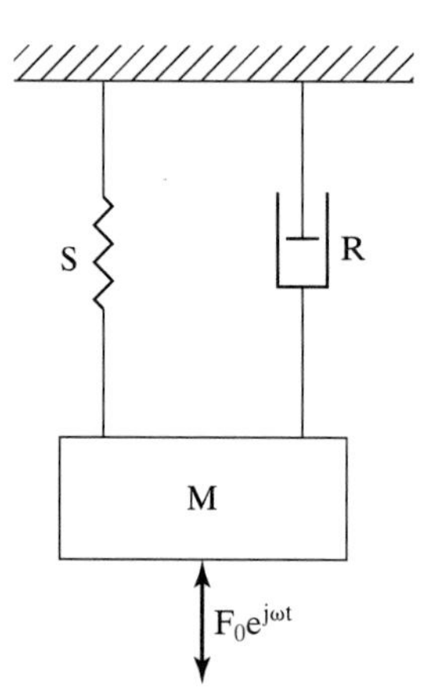

Fig. 1.35 A mass on a spring with damping is subject to forced oscillations.

or rearranging,

$$M\ddot{x} + R\dot{x} + Sx = F_0 e^{j\omega t} \tag{1.3.2}$$

Physical intuition suggests that provided the amplitude is not excessive, the mass will eventually settle down in an oscillatory motion, in constant phase relationship to the driving force. We will therefore try the solution

$$x = x_0 e^{j(\omega t - \alpha)} \tag{1.3.3}$$

Substituting Eq. (1.3.3) into Eq. (1.3.2) yields

$$[-M\omega^2 + jR\omega + S]x_0 e^{-j\alpha} = F_0 \tag{1.3.4}$$

Considering both real and imaginary parts of Eq. (1.3.4), we have

$$[-M\omega^2 + S]x_0 \cos\alpha + R\omega x_0 \sin\alpha - F_0 = 0 \tag{1.3.5}$$

and

$$R\omega x_0 \cos\alpha + [M\omega^2 - S]x_0 \sin\alpha = 0 \tag{1.3.6}$$

From Eq. (1.3.6), we identify α by

$$\tan\alpha = \frac{R}{-M\omega + S/\omega} \tag{1.3.7}$$

Now from Eq. (1.3.5),

$$x_0 = \frac{F_0}{[\{-M\omega + S/\omega\}\cos\alpha + R\sin\alpha]\omega} \tag{1.3.8}$$

But from the representation of α in the triangle (see Fig. 1.36),

$$\cos\alpha = \frac{-M\omega + S/\omega}{\{[M\omega - S/\omega]^2 + R^2\}^{1/2}} \text{ and } \sin\alpha = \frac{R}{\{[M\omega - S/\omega]^2 + R^2\}^{1/2}} \tag{1.3.9}$$

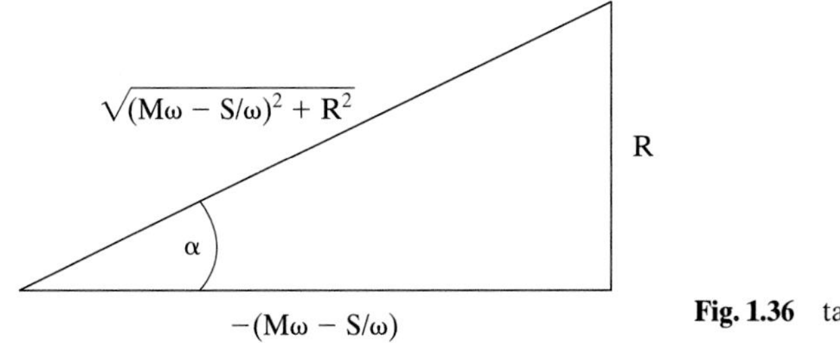

Fig. 1.36 $\tan\alpha = \dfrac{R}{-(M\omega - S/\omega)}$

Substituting for cos α and sin α in Eq. (1.3.8):

$$x_0 = \frac{F_0}{\left\{[M\omega - S/\omega]^2 + R^2\right\}^{1/2}\omega} \tag{1.3.10}$$

Finally, we obtain the steady-state solution for velocity:

$$v = \frac{jF_0 e^{j(\omega t - \alpha)}}{\left\{[M\omega - S/\omega]^2 + R^2\right\}^{1/2}} \tag{1.3.11}$$

But

$$e^{j\pi/2} = j \tag{1.3.12}$$

and thus

$$v = \frac{F_0 e^{j(\omega t - \beta)}}{\left\{[M\omega - S/\omega]^2 + R^2\right\}^{1/2}} \tag{1.3.13}$$

where

$$\beta = \alpha - \pi/2 \tag{1.3.14}$$

It follows that

$$\tan \beta = \frac{-1}{\tan \alpha} = \frac{M\omega - S/\omega}{R} \tag{1.3.15}$$

using Eq. (1.3.7). For reasons to be pointed out in the next section, it is easier to remember the results for the velocity obtained by Eqs. (1.3.13) and (1.3.15) than the displacement. Before proceeding to this discussion, however, we will derive the same results by the method of linear differential operators.

Rewriting Eq. (1.3.2) in operator notation ($D = \partial/\partial t$) and dividing by M,

$$(D^2 + rD + s)x = \frac{F_0 e^{j\omega t}}{M} \tag{1.3.16}$$

where $r = R/M$ and $s = S/M$ as before. Thus

$$x = \frac{1}{(D^2 + rD + s)} \frac{F_0 e^{j\omega t}}{M} \tag{1.3.17}$$

Employing the shift principle, we obtain, after a little algebra:

$$x = \frac{(F_0/\omega) e^{j(\omega t - \alpha)}}{\left[(M\omega - S/\omega)^2 + R^2\right]^{1/2}} \tag{1.3.18}$$

where α is given by Eq. (1.3.7).

Under the theory of differential equations, it is recognized that a complete solution of Eq. (1.3.2) is the sum of the particular integral above plus a complementary function that is the solution of the equation

$$M\ddot{x} + R\dot{x} + Sx = 0 \tag{1.3.19}$$

We have already discussed one solution of this equation for small damping (see Section 1.2.5) and found it to be a decaying exponential for the case in which the motion starts with an initial displacement. In the present case, we do not have such an initial condition, but there may still be a complementary function, whose nature depends on the initial value of acceleration. This function will be a transient that dies away exponentially with time.

1.3.2 Impedance

If we take the view of Newton's law which holds that an applied force is opposed by a force $-M\ddot{x}$, known as the inertial force, then in the case of the damped system the total applied force has to be equal and opposite to the resultant of forces $-M\ddot{x}$, $-Sx$, and $-R\dot{x}$. In the steady state, the displacement will be

$$x = x_0 e^{j(\omega t - \alpha)} \tag{1.3.20}$$

which is a phasor whose position at time $t = 0$, is as shown on the complex plane in Fig. 1.37. The elementary forces are $-Sx$, $-j\omega Rx$, and $M\omega^2 x$, and the forces that have to be applied to overcome them are Sx, $j\omega Rx$, and $-M\omega^2 x$. Now $j\omega Rx$ is $\omega Rxe^{j\pi/2}$, a phasor ahead of x in phase by $\pi/2$. If we designate the forces required to overcome stiffness, resistance, and inertia as F_S, F_R, and F_M, then F_S is Sx, F_R is $jR\omega x$, and F_M is $-M\omega^2 x$. Notice the direct proportionality between these component forces and

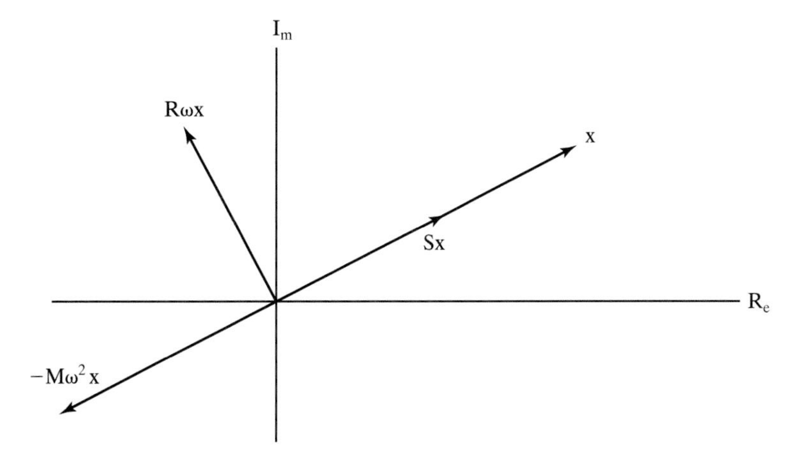

Fig. 1.37 Forces to be overcome in setting a mass into forced oscillation.

the displacement. A similar relationship exists between the forces and the velocity, v, thus

$$F_S = (S/j\omega)v = -j(S/\omega)v \tag{1.3.21}$$

$$F_R = Rv \tag{1.3.22}$$

and

$$F_M = jM\omega v \tag{1.3.23}$$

The ratio of the force applied to overcome opposition by an element to the velocity resulting in the element is known as the mechanical impedance, Z. Plotting the impedances on the complex plane results in the scheme shown in Fig. 1.38. Notice that the relative dispositions of the impedance vectors are the same as those of the forces in Fig. 1.37.

Thus, the impedance resultant Z_T is in the same ratio to the resultant force as the component impedances are to the component forces. So

$$F_T = Z_T v \tag{1.3.24}$$

or

$$\begin{aligned} v = \frac{F_T}{Z_T} &= \frac{F_0 e^{j\omega t}}{j(M\omega - S/\omega) + R} \\ &= \frac{F_0 e^{j\omega t}[-j(M\omega - S/\omega) + R]}{(M\omega - S/\omega)^2 + R^2} \\ &= \frac{F_0 e^{j(\omega t - \beta)}}{\{(M\omega - S/\omega)^2 + R^2\}^{1/2}} \end{aligned} \tag{1.3.25}$$

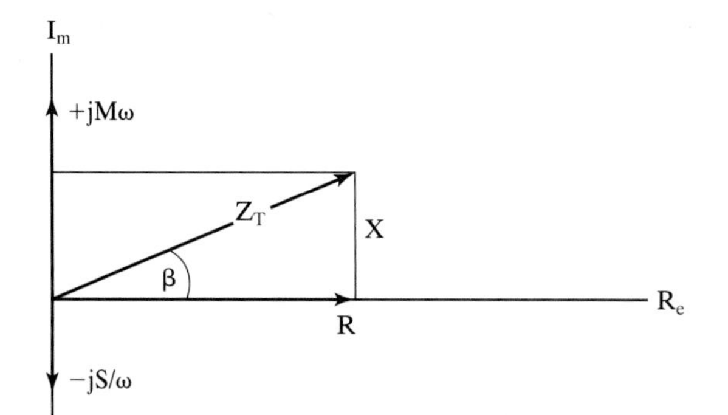

Fig. 1.38 Impedance vectors in forced oscillations, shown on complex plane.

where $\tan \beta = (M\omega - S/\omega)/R$, in agreement with Eq. (1.3.15). We see how useful the concept of impedance is in summarizing the steady-state results. The linear relationship between force and velocity is known as Ohm's Law, as in the familiar electrical circuit law. We will discuss the analogy between the electrical and mechanical cases in more detail in Section 1.3.6.

1.3.3 Resonance

The foregoing analysis explains a phenomenon known since ancient times, namely, resonance, or the dramatic increase in response at a certain driving frequency. Historically, resonance was known in connection with pendula and room acoustics, but the phenomenon is encountered in many mechanical and electrical systems and in other branches of physics, including atomic and nuclear physics. We will consider first the case of velocity resonance. The velocity amplitude is given by

$$v_0 = F_0/\left\{(M\omega - S/\omega)^2 + R^2\right\}^{1/2} = F_0/|Z_T| \tag{1.3.26}$$

This result is shown in Fig. 1.39. Obviously, v_0 will be a maximum when $|Z_T|$ is a minimum and this will occur when

$$M\omega - S/\omega = 0 \tag{1.3.27}$$

that is,

$$\omega = \omega_0 = \sqrt{S/M} \tag{1.3.28}$$

The case of velocity resonance is not of such great practical importance as that of displacement resonance. The displacement amplitude, x_0, is given by

$$x_0 = F_0/\omega|Z_T| \tag{1.3.29}$$

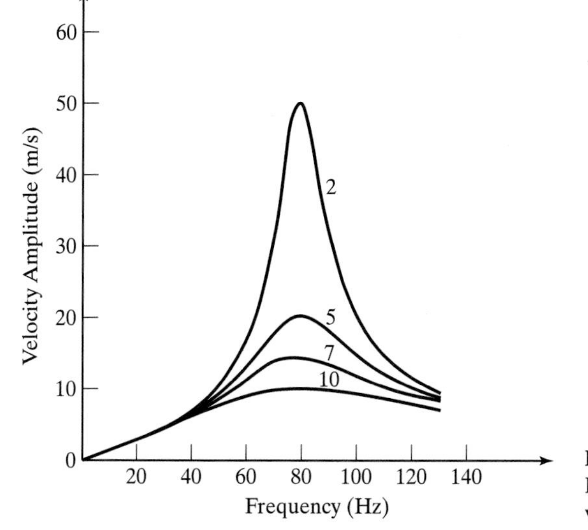

Fig. 1.39 Velocity amplitude in forced oscillation. M = 0.02 kg, S = 5000 N/m, force amplitude 100 N, various values of damping.

Now the ratio x_0/x_s—where x_s is the static displacement caused by a nonoscillatory force F_0—is called the magnification factor, G. Since x_s is F_0/S, we have

$$G = S/\omega|Z_T| \tag{1.3.30}$$

G is shown in Fig. 1.40. It follows that G will be a maximum when the displacement is a maximum (i.e., at displacement resonance), which will not occur at the same frequency as velocity resonance, except for the case of zero damping. Now it is easily shown that

$$G = \left[\left(1 - \frac{\omega^2}{\omega_0^2}\right)^2 + \frac{R^2}{SM}\frac{\omega^2}{\omega_0^2}\right]^{-1/2} \tag{1.3.31}$$

For a maximum, it may then be shown that

$$\omega^2 = \omega_0^2 - R^2/2M^2 = \omega_A^2 \tag{1.3.32}$$

It is interesting to compare this frequency with that for natural damped oscillations given in Eq. (1.2.57). The maximum magnification factor is given by substituting ω_A^2 into Eq. (1.3.31):

$$G_M = \left[\left(1 - \frac{\omega_A^2}{\omega_0^2}\right)^2 + \frac{R^2}{SM}\frac{\omega_A^2}{\omega_0^2}\right]^{-1/2} \tag{1.3.33}$$

Now, for small damping $\omega_A = \omega_0$, and

$$G_M \approx \sqrt{SM}/R \tag{1.3.34}$$

Let $\nu = \omega/\omega_0$. The sharpness of the resonance curve is also of some interest, and this may be specified by the range of frequencies over which the magnification exceeds $GM/\sqrt{2}$. These values are given by solving the equation

$$G_M/\sqrt{2} = \left[(1 - \nu^2)^2 + \frac{R^2}{SM}\nu^2\right]^{-1/2} \tag{1.3.35}$$

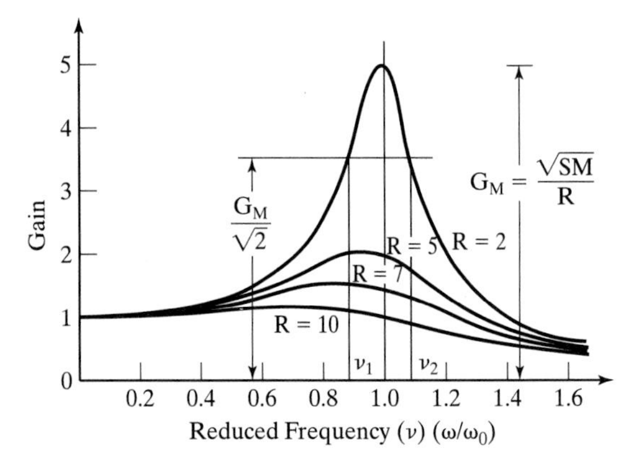

Fig. 1.40 Displacement gain versus reduced frequency. M = 0.02 kg, S = 5000 N/m, various values of damping.

which can be written as a quadratic equation in ν^2, and taking the difference of its roots we obtain for small damping

$$\nu_2^2 - \nu_1^2 = \frac{2R}{\sqrt{SM}} = \frac{2R}{\omega_0 M} = \frac{RT}{\pi M} = \frac{2\delta}{\pi} \tag{1.3.36}$$

where δ is the logarithmic decrement, as defined in Eq. (1.2.80). This result may be further simplified as follows:

$$\nu_2 - \nu_1 = \frac{2\delta}{\pi(\nu_2 + \nu_1)} = \frac{2\delta\omega_0}{\pi(\omega_2 + \omega_1)} \tag{1.3.37}$$

But the smaller the damping, the larger G_M is and the smaller $\nu_2 - \nu_1$ is (i.e., the curve is increasingly symmetrical as the damping diminishes), and thus

$$\omega_2 + \omega_1 \cong 2\omega_0 \tag{1.3.38}$$

Finally, for small damping,

$$\omega_2 - \omega_1/\omega_0 = \delta/\pi \tag{1.3.39}$$

The quality factor, Q, of the system is defined by

$$Q \doteq f_0/(f_2 - f_1) = \omega_0/(\omega_2 - \omega_1) = \pi/\delta = M\omega_0/R \tag{1.3.40}$$

A highly damped system, then, has a low-quality factor.

1.3.4 Energy and Phase in Forced Oscillations

From Eq. (1.3.15), the displacement lags behind the exciting force by the angle α where

$$\alpha = \tan^{-1}\frac{R}{(S/\omega - M\omega)} = \tan^{-1}\frac{\omega R/M}{(\omega_0^2 - \omega^2)} \tag{1.3.41}$$

when

$$\omega < \omega_0, \quad 0 < \alpha < \pi/2 \tag{1.3.42}$$

$$\omega = \omega_0, \quad \alpha = \pi/2 \tag{1.3.43}$$

$$\omega > \omega_0, \quad \pi/2 < \alpha < \pi \tag{1.3.44}$$

The variation of α with the frequency ratio is shown in Fig. 1.41. The phase change is relatively abrupt with small damping. Notice that $\alpha = \pi/2$ at velocity resonance for all values of damping. Since for small damping, from Eq. (1.3.32)

$$\omega_0 \approx \omega_A \tag{1.3.45}$$

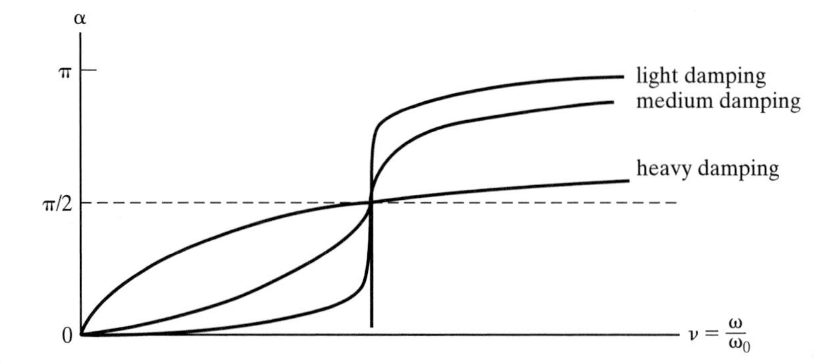

Fig. 1.41 The phase lag of the displacement behind the exciting force in forced oscillation.

it follows that

$$x_R = \frac{F}{\omega|Z_T|}e^{j(\omega t-\pi/2)} = \frac{-jFe^{j\omega t}}{\omega|Z_T|} \tag{1.3.46}$$

Hence

$$v_R = \frac{Fe^{j\omega t}}{|Z_T|} \tag{1.3.47}$$

that is, the velocity and driving force are in phase at resonance, but the displacement lags both by $\pi/2$. This result may be seen directly by inspecting the impedance diagram of Fig. 1.38. A greater significance may be attached to the result in connection with energy considerations.

The work done by the driver in causing a change dx in the displacement is

$$dW = \Re_e F_0 e^{j\omega t}\, dx = \Re_e F_0 e^{j\omega t} \Re_e \frac{dx}{dt} dx$$

$$= F_0 \cos \omega t \frac{F_0}{|Z_T|} \cos(\omega t + \beta) dt \tag{1.3.48}$$

Thus, the instantaneous power

$$P = \frac{dW}{dt} = \frac{F_0^2}{|Z_T|} \cos \omega t \cos(\omega t + \beta) \tag{1.3.49}$$

Letting $\omega t = \theta$, it is easy to see that the average power, $\overline{P}$, is given by

$$\overline{P} = \frac{1}{2\pi} = \int_0^{2\pi} \frac{F_0^2}{|Z_T|} \cos \omega t \cos(\omega t + \beta)\, d\theta = \frac{F_0^2}{2|Z_T|} \cos \beta \tag{1.3.50}$$

This important result may be expressed in a variety of ways. Since $\beta = \pi/2 - \alpha$, where α is the phase lag of displacement behind the driving force,

$$\overline{P} = \frac{F_0^2}{2|Z_T|}\sin\alpha \tag{1.3.51}$$

where

$$\sin\alpha = \frac{R}{|Z_T|} \tag{1.3.52}$$

Thus

$$\overline{P} = \frac{F_0^2 R}{2|Z_T|^2} = \frac{v_0^2 R}{2} = v_{RMS}^2 R \tag{1.3.53}$$

where v_0 = velocity amplitude and v_{RMS} = average value of velocity $= v_0/\sqrt{2}$. The above result is analogous to the expression for electrical power dissipated as Joule heating in a resistor. Similarly, we may show that this mechanical power is required to overcome frictional energy loss. Since the resistive force

$$F_R = R\,dx/dt \tag{1.3.54}$$

the power required to overcome the resistance is

$$P_R = \left(R\frac{dx}{dt}\right)\frac{dx}{dt} = Rv^2 \tag{1.3.55}$$

and averaging over one cycle

$$P_R = Rv_{RMS}^2 \tag{1.3.56}$$

These results imply that all the power goes into driving the resistance and none into driving against inertia or stiffness. The instantaneous power involved with inertia is

$$P_I = -F_I\frac{dx}{dt} = M\dot{x}\ddot{x} \tag{1.3.57}$$

but since velocity and acceleration are $\pi/2$ out of phase, the average of P_I is zero. The maximum kinetic energy stored is

$$T_{max} = \frac{1}{2}Mv_0^2 \tag{1.3.58}$$

The ratio of this quantity to the energy dissipated per cycle is

$$T_{max}/P_R T = \frac{M}{RT} = \frac{M\omega_0}{R2\pi} = \frac{Q}{2\pi} \tag{1.3.59}$$

In this way we obtain a second interpretation of the quality factor: A system with a high Q is one in which relatively little energy is dissipated compared with the energy stored.

One source of energy loss from a vibrating body is through radiation of sound. A bell struck by a hammer or an organ pipe subject to an air flow can be made to "sound" or "resonate." The force variation of the hammer blow or air blast may be Fourier analyzed, and part of the energy will be contained in the frequency range of resonant response of the sounder. This energy will be only gradually dissipated thereafter in the form of sound. Usually internal losses in the material of the sounder cause greater damping than sound radiation, and to achieve the most efficient sounding, a material with low internal losses, such as high-quality steel or glass, is required. The systems referred to here are distributed constant systems and are mentioned to illustrate the importance of the quality factor in later developments.

1.3.5 Transmissibility and Vibration Isolation

A problem frequently encountered is the transmission of vibrations to the floor from a machine with rotating or reciprocating parts (Fig. 1.42). This problem is handled by considering the ratio of the amplitude of the force felt by the floor to the amplitude of the driving force. The force felt by the floor, P, is given by

$$P = R\dot{x} + Sx = \left(R + \frac{S}{j\omega}\right)v \tag{1.3.60}$$

Hence

$$\frac{P_0}{F_0} = \frac{\sqrt{R^2 + S^2/\omega^2}}{|Z|} = \frac{\sqrt{R^2 + S^2/\omega^2}}{\sqrt{(M\omega - S/\omega)^2 + R^2}} \tag{1.3.61}$$

that is, the transmissibility

$$T = \sqrt{1 + R^2\omega^2/S^2} \Bigg/ \sqrt{\left(\frac{M\omega^2}{S} - 1\right)^2 + \frac{R^2\omega^2}{S^2}} \tag{1.3.62}$$

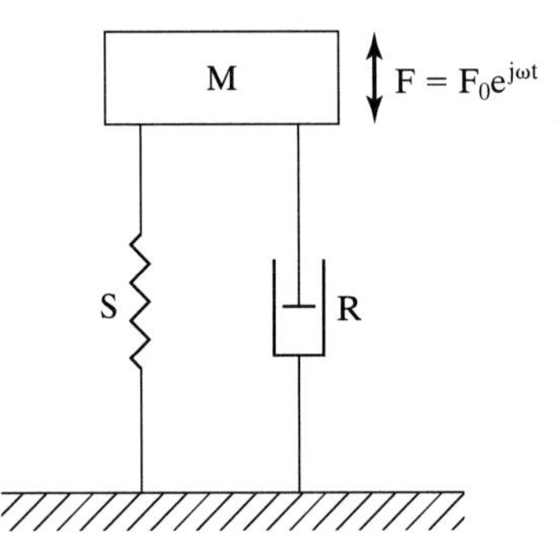

Fig. 1.42 The vibrations of the mass are transmitted to the floor.

Note that T = 1 if

$$(M\omega^2/S - 1)^2 = 1 \tag{1.3.63}$$

that is,

$$\omega = 0 \quad \text{or} \quad \omega = \sqrt{2}\sqrt{S/M} \tag{1.3.64}$$

But $\sqrt{S/M}$ is ω_0, the natural angular frequency. Consequently, the variation of the transmissibility with frequency is as shown in Fig. 1.43. The high- and low-damping curves cross when $\omega = \sqrt{2}\omega_0$. Below this frequency T is decreased by increasing R, while above this frequency T is increased by increasing R.

It is often necessary to isolate a mass from ground vibrations (Fig. 1.44). Suppose the ground has displacement

$$y = y_0 e^{j\omega t} \tag{1.3.65}$$

and the mass M has displacement x. Then the force on the mass will be

$$S(y - x) + R(\dot{y} - \dot{x}) = M\ddot{x} \tag{1.3.66}$$

Thus,

$$M\ddot{x} + R\dot{x} + Sx = (R + S/j\omega)\dot{y} \tag{1.3.67}$$

The resulting motion is a forced oscillation with equivalent forcing function

$$F = (R + S/j\omega)\dot{y} \tag{1.3.68}$$

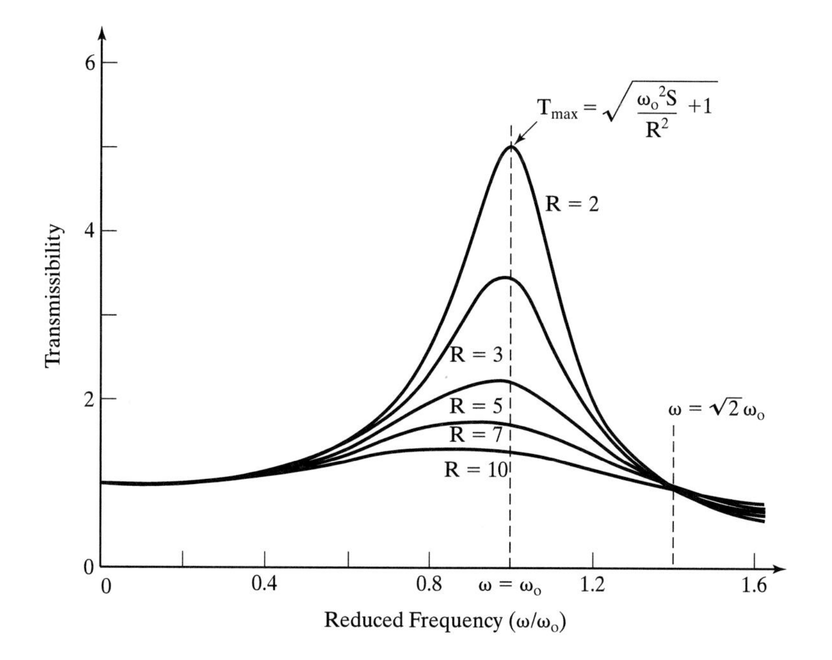

Fig. 1.43 Transmissibility of force to a floor. M = 0.02 kg, S = 5000 N, various values of damping.

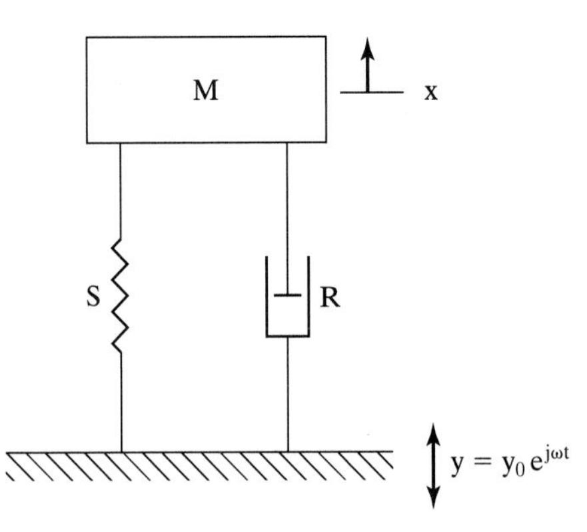

Fig. 1.44 The mass is to be isolated from floor vibrations.

where

$$|F| = (R^2 + S^2/\omega^2)^{1/2}\dot{y}_0 \tag{1.3.69}$$

Hence

$$\dot{x} = v = F/Z \tag{1.3.70}$$

and

$$v_0 = |F|/|Z| = \sqrt{(R^2 + S^2/\omega^2)}\dot{y}_0/\sqrt{(M\omega - S/\omega)^2 + R^2} \tag{1.3.71}$$

So $v_0/\dot{y}_0 = T$ as defined in Eq. (1.3.62) and the same considerations apply in the use of damping to increase isolation as in reducing transmissibility.

1.3.6 Electromechanical Analogies

Analogies may be drawn between the mechanical system in Fig. 1.45 and either the electrical series circuit in Fig. 1.46 or the electrical parallel circuit in Fig. 1.47. As we

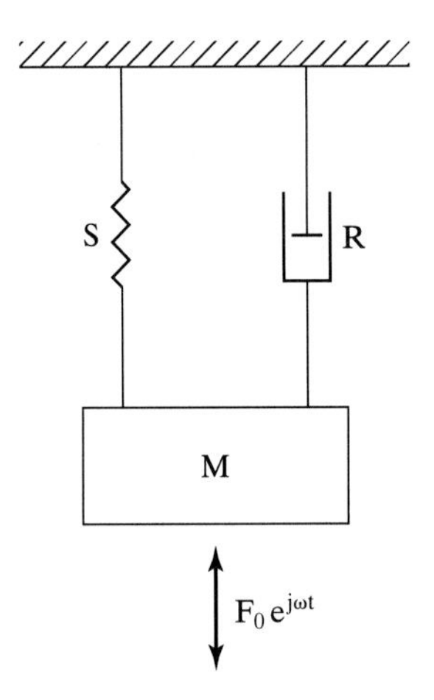

Fig. 1.45 One degree of freedom mechanical system.

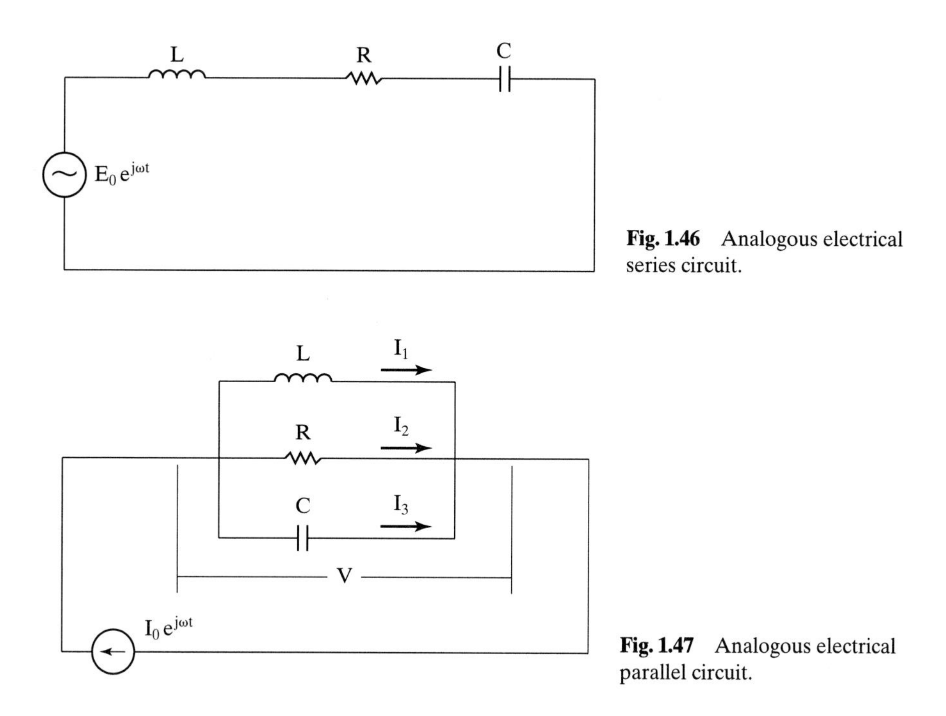

Fig. 1.46 Analogous electrical series circuit.

Fig. 1.47 Analogous electrical parallel circuit.

have seen in Section 1.3.1, the application of Newton's second law to the mechanical system results in the equation

$$M\ddot{x} + R\dot{x} + Sx = F_0 e^{j\omega t} \tag{1.3.72}$$

Kirchoff's second law states that the sum of the electromotive forces in a closed loop is equal to the sum of the potential drops across the elements in the loop. Applying this law to the circuit of Fig. 1.46 results in the equation

$$L\frac{d^2q}{dt^2} + R\frac{dq}{dt} + \frac{q}{C} = E_0 e^{j\omega t} \tag{1.3.73}$$

where q is the charge stored on the capacitor.

Kirchoff's first law states that the algebraic sum of the current flowing into any point in a circuit is zero. Applying this law to the circuit in Fig. 1.47 results in the equation

$$C\frac{dV}{dt} + \frac{V}{R} + \int \frac{V}{L} dt = I_0 e^{j\omega t} \tag{1.3.74}$$

If we now let $\int V dt = u$, then

$$C\frac{d^2u}{dt^2} + \frac{1}{R}\frac{du}{dt} + \frac{u}{L} = I_0 e^{j\omega t} \tag{1.3.75}$$

The analogies are established by comparing the coefficients and variables of these three second-order differential equations. Thus, by comparing Eqs. (1.3.72) and (1.3.73), we obtain the *series analogy*.

$\dot{x}$ (velocity)	→	$\dot{q} = I$ (current)
F (force)	→	V (voltage of electromotive force)
M (mass)	→	L (inductance)
R (mechanical resistance)	→	R (electrical resistance)
S (stiffness)	→	1/C (reciprocal capacitance)

On the other hand, by comparing Eqs. (1.3.72) and (1.3.73) we obtain the so-called *parallel or mobility analogy*.

$\dot{x}$ (velocity)	→	$du/dt = V$ (voltage or electromotive force)
F (force)	→	I (current)
M (mass)	→	C (capacitance)
R (mechanical resistance)	→	1/R (reciprocal electrical resistance)
S (stiffness)	→	1/L (reciprocal inductance)

This is called the mobility analogy, since one has to imagine the force "flowing" like a current through the mechanical elements.

In using the series analogy, it is important to realize that elements suffering the same displacement and velocity must be put in series in the electrical equivalent circuit. Consequently, on this analogy, which is the more frequently used in acoustical work, mechanical elements in parallel are represented by series electrical elements. Fig. 1.48 shows an example of the equivalent circuit of a two degree of freedom system.

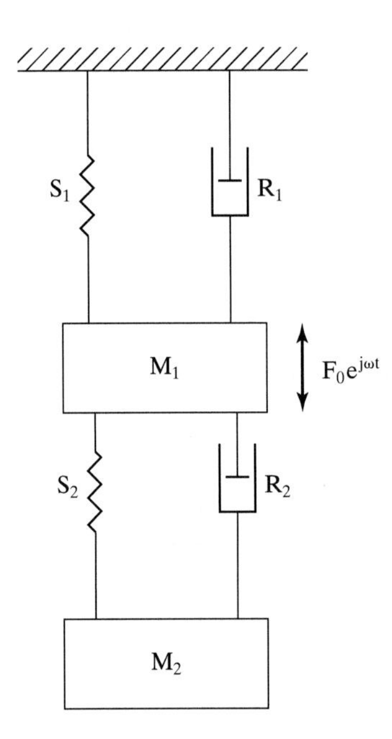

Fig. 1.48 Two degree of freedom mechanical system.

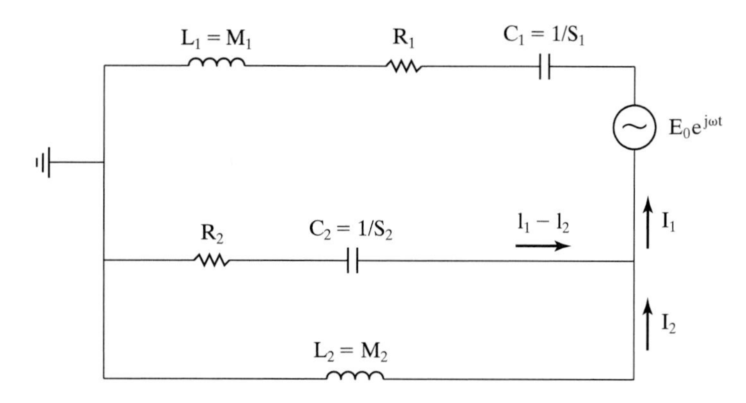

Fig. 1.49 Series analogous electrical circuit.

S_1, R_1, and M_1 all have the same velocity $\dot{x}_1$ and so must be in series. The point of application of the force also has the same displacement x_1 and must be represented in the same branch of the equivalent circuit shown in Fig. 1.49. S_2 and R_2 have displacement $(x_1 - x_2)$, while M_2 has displacement x_2; thus these elements have to be set in two separate branches joining at a node with equivalent currents adding algebraically as shown. The remaining problem is where the equivalent circuit should be grounded. It is clear the current I_1 should run to ground through C_1, R_1, and L_1, but what of the current through $L_2(M_2)$? This difficulty is resolved by imagining an additional spring of zero stiffness to be added to the mechanical system, joining to mechanical ground. This will not of course affect the mechanical system, but may be seen to be equivalent to an infinite capacitance (of zero impedance), which is essentially a short to ground.

Equivalent circuits are helpful in understanding electromechanical systems, such as transducers. In the case of very complicated or nonlinear mechanical systems, forces, displacements, and velocities resulting from various driving forces may be determined by measurements of currents and voltages in the equivalent circuit. This equivalency forms the basis of the analog computer technique in vibration analysis. A knowledge of equivalent electrical and mechanical systems is also useful for electrical and mechanical engineers having to learn each other's systems.

1.4 Two Degree of Freedom Oscillators

1.4.1 Natural Modes

When a system requires two coordinates to describe its motion, it is said to have two degrees of freedom. We will study this case in some detail because it offers a simple introduction to the behavior of systems with several degrees of freedom. A two degree of freedom system has two natural frequencies. There are two equations of motion, with each one involving both the coordinates. When vibration occurs at one of the natural frequencies, a definite relationship exists between the amplitudes of the two coordinates, and the configuration is referred to as the normal mode. The two degree of freedom system will then have two normal modes of vibration corresponding to the two natural frequencies.

Consider a system, such as the one shown in Fig. 1.50, in which damping has been ignored. Assuming that the applied forces and displacements are in the positive direction as shown, the two differential equations of motion may be written, from Newton's second law, as follows:

$$M_1\ddot{x}_1 = -S_1x_1 - S_c(x_1 - x_2) + F_1 \tag{1.4.1}$$

and

$$M_2\ddot{x}_2 = S_c(x_1 - x_2) - S_2x_2 + F_2 \tag{1.4.2}$$

Rearranging,

$$M_1\ddot{x}_1 + (S_1 + S_c)x_1 - S_cx_2 = F_1 \tag{1.4.3}$$

and

$$M_2\ddot{x}_2 + (S_2 + S_c)x_2 - S_cx_1 = F_2 \tag{1.4.4}$$

Note that the difference between these equations and the equation for a single one degree of freedom system is the presence of the "coupling" term $S_c(x_2 - x_1)$. Just as for the one-degree system, we will consider both free and forced oscillations. For free oscillations, the equations become

$$M_1\ddot{x}_1 + (S_1 + S_c)x_1 - S_cx_2 = 0 \tag{1.4.5}$$

and

$$M_2\ddot{x}_2 + (S_2 + S_c)x_2 - S_cx_1 = 0 \tag{1.4.6}$$

Physical intuition suggests the existence of a class of solution in which both masses oscillate at the same frequency; that is,

$$x_1 = A_1e^{j\omega t} \tag{1.4.7}$$

and

$$x_2 = A_2e^{j\omega t} \tag{1.4.8}$$

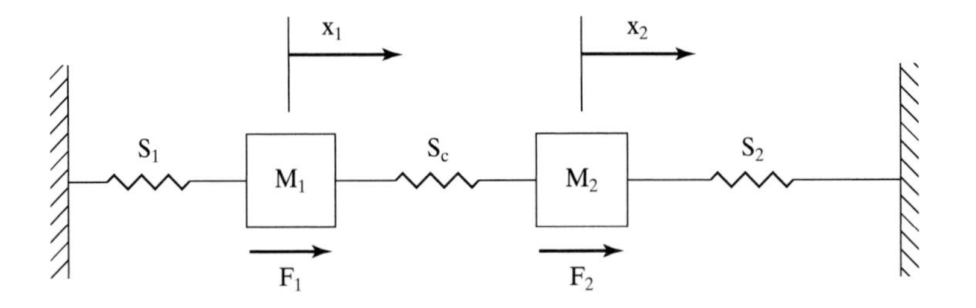

Fig. 1.50 A two degree of freedom system.

where A_1 and A_2 are arbitrary constants that include the amplitude and a phase angle. If our physical intuition is correct, substituting the assumed solutions into the equations of motion will determine the permissable values of A_1, A_2, and ω. Substituting

$$[S_1 + S_c - \omega^2 M_1]A_1 - S_c A_2 = 0 \tag{1.4.9}$$

and

$$-S_c A_1 + [S_2 + S_c - \omega^2 M_2]A_2 = 0 \tag{1.4.10}$$

Dividing Eqs. (1.4.9) and (1.4.10) gives

$$M_1 M_2 \omega^4 - \omega^2\{M_2(S_1 + S_c) + M_1(S_2 + S_c)\} - \{(S_1 + S_c)(S_2 + S_c) - S_c^2\} = 0 \tag{1.4.11}$$

which is a quadratic equation in ω^2 that leads to two physically meaningful and allowable frequencies, ω_1 and ω_2. Now the ratio of amplitudes is given by

$$\frac{A_1}{A_2} = \frac{-M_2\omega^2 + S_2 + S_c}{S_c} \tag{1.4.12}$$

There will be two possible values of the amplitude ratio depending on the value of ω^2. It is easiest to illustrate the nature of the solutions by taking specific examples, for instance the case when $M_1 = M_2 = M$ and $S_1 = S_2 = S_c = S$. Then, from Eq. (1.4.11),

$$\omega^4 - \omega^2(4S/M) + 3S^2/M^2 = 0 \tag{1.4.13}$$

Thus

$$\omega_1^2 = S/M \tag{1.4.14}$$

and

$$\omega_2^2 = 3S/M \tag{1.4.15}$$

When $\omega^2 = \omega_2^2$, we have from Eq. (1.4.12)

$$A_1/A_2 = -1 \tag{1.4.16}$$

and when $\omega^2 = \omega_1^2$,

$$A_1/A_2 = +1 \tag{1.4.17}$$

In the first mode, the two masses move as if rigidly connected (Fig. 1.51). Their motions are then in phase. In the second mode, however, they move in antiphase, and the central point of the coupling spring is a node (Fig. 1.52); that is, it suffers no displacement. It may be readily shown that the system is then equivalent to one in which the coupling spring is replaced by two identical springs of stiffness 2S anchored at their connection.

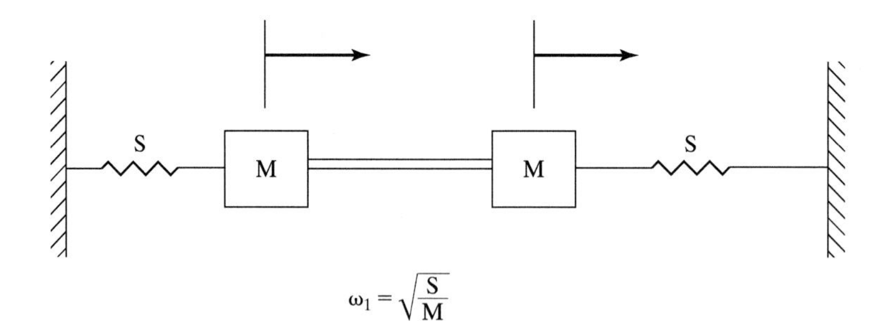

Fig. 1.51 In the lowest natural mode of a two degree of freedom system, the masses move in unison.

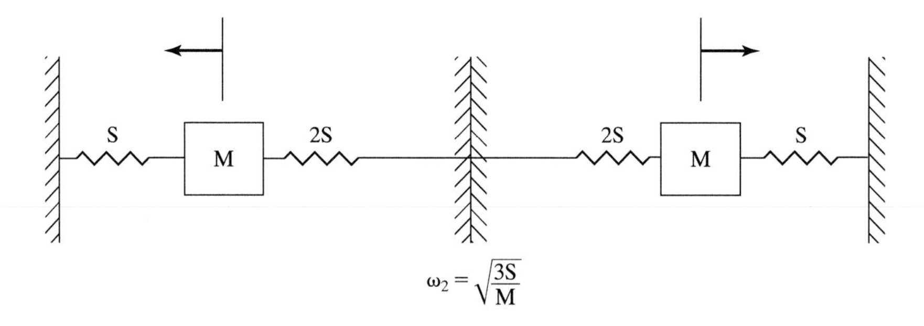

Fig. 1.52 In the second mode of a two degree of freedom system the masses move in opposition.

1.4.2 Forced Oscillation

We will now consider the situation in which a force is applied to one of the two masses, as shown in Fig. 1.53. The equations of motion are then

$$M_1\ddot{x}_1 + S_1x_1 - S_c(x_2 - x_1) = F_0e^{j\omega t} \tag{1.4.18}$$

and

$$M_2\ddot{x}_2 + S_2x_2 + S_c(x_2 - x_1) = 0 \tag{1.4.19}$$

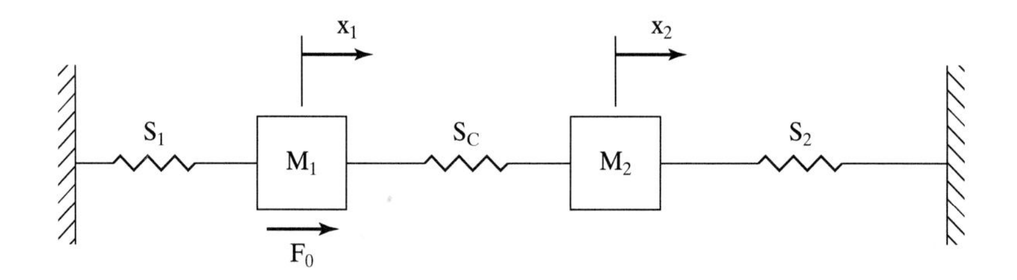

Fig. 1.53 Forced oscillation of a two degree of freedom system.

where we also assume that the applied force is harmonic. Again, for a steady-state solution, we may assume

$$x_1 = X_1 e^{j\omega t} \quad \text{and} \quad x_2 = X_2 e^{j\omega t} \tag{1.4.20}$$

and substitute into the equations of motion, Eqs. (1.4.18) and (1.4.19). Thus,

$$X_1\left\{-M_1\omega^2 + S_1 + S_c\right\} - X_2 S_c = F_0 \tag{1.4.21}$$

and

$$X_2\left\{-M_2\omega^2 + S_2 + S_c\right\} - X_1 S_c = 0 \tag{1.4.22}$$

Solving for X_1 and X_2 in terms of F_0 yields

$$X_1 = \frac{F_0\left\{S_2 + S_c - M_2\omega^2\right\}}{\left\{S_1 + S_c - M_1\omega^2\right\}\left\{S_2 + S_c - M_2\omega^2\right\} - S_c^2} \tag{1.4.23}$$

and

$$X_2 = \frac{S_c F_0}{\left\{S_1 + S_c - M_1\omega^2\right\}\left\{S_2 + S_c - M_2\omega^2\right\} - S_c^2} \tag{1.4.24}$$

Notice that the denominator in these expressions is the same as the left side of Eq. (1.4.11). Hence, when $\omega = \omega_1$ or ω_2, X_1 and X_2 become infinite; that is, there is a resonance at the natural frequencies. In particular, we again examine the case in which $M_1 = M_2 = M$ and $S_1 = S_2 = S_c = S$. Then

$$\frac{X_1}{F_0} = \frac{2S - M\omega^2}{M^2\omega^4 - \omega^2(4SM) + 3S^2} = \frac{\frac{1}{M}\left(2\frac{S}{M} - \omega^2\right)}{\left(\omega^2 - \omega_1^2\right)\left(\omega^2 - \omega_2^2\right)} \tag{1.4.25}$$

where

$$\omega_1^2 = \frac{S}{M} \quad \text{and} \quad \omega_2^2 = \frac{3S}{M} \tag{1.4.26}$$

Note that $X_1 = 0$, when $\omega^2 = 2S/M$.
Thus for values of $\omega^2 < S/M$, $X_1 > 0$,

$$S/M < \omega^2 < 2S/M, X_1 < 0$$

$$2S/M < \omega^2 < 3S/M, X_2 > 0$$

and

$$\omega^2 > 3S/M, X_2 < 0 \tag{1.4.27}$$

The variation of X_1 with ω/ω_1 is thus as shown in Fig. 1.54. Similarly,

$$\frac{X_2}{F_0} = \frac{S/M^2}{(\omega^2 - \omega_1^2)(\omega^2 - \omega_2^2)} \tag{1.4.28}$$

X_2 never becomes zero. For

when $\omega^2 < \omega_1^2$, then $X_2 > 0$,

when $\omega_1^2 < \omega^2 < \omega_2^2$, then $X_2 < 0$,

and when $\omega^2 > \omega_2^2$, then $X_2 > 0$. (1.4.29)

The resonant absorber is another example of a forced two degree of freedom system in which the stiffness S_2 of Fig. 1.53 is zero (see Fig. 1.55). The equations for the displacement amplitudes then become

$$X_1 = \frac{F_0\{S_c - M_2\omega^2\}}{\{S_1 + S_c - M_1\omega^2\}\{S_c - M_2\omega^2\} - S_c^2} \tag{1.4.30}$$

and

$$X_2 = \frac{S_cF_0}{\{S_1 + S_c - M_1\omega^2\}\{S_c - M_2\omega^2\} - S_c^2} \tag{1.4.31}$$

In this case, X_1 is zero when $\omega^2 = S_c/M_2$. We can conclude that the amplitude of the driven mass remains zero despite the applied force. This is the principle of the resonant absorber used in synchronous machines and devices that run on constant frequency alternating current (e.g., hair clippers). This type of device could not be used on gasoline engines or on other variable speed machines, since it is restricted to the fixed frequency $(S_c/M_2)^{1/2}$. The energy supplied to the system is all stored in the resonant mass, which dissipates it gradually through the damping that exists in practice.

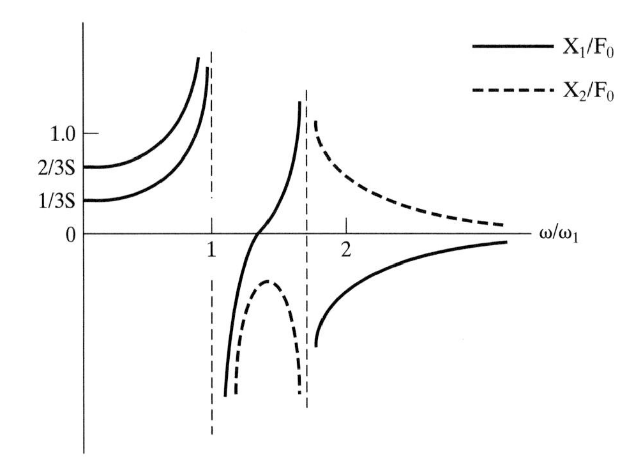

Fig. 1.54 Amplitude of motion versus frequency.

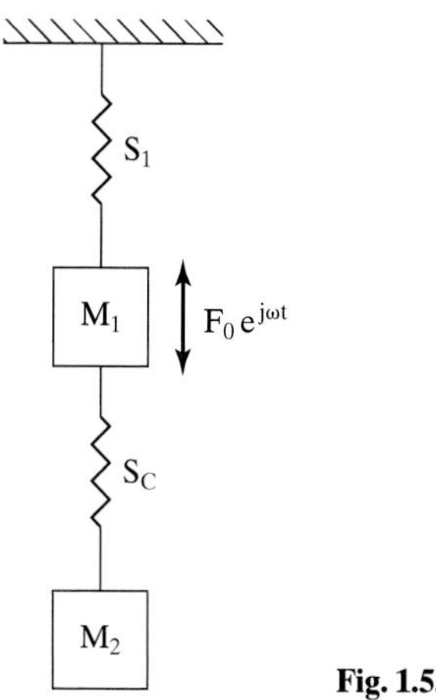

Fig. 1.55 Resonant absorber.

1.5 Multidegree of Freedom Systems

1.5.1 Natural Modes

When it comes to analyzing a system with many degrees of freedom, we have to write down as many equations of motion as there are degrees of freedom. There are n natural frequencies, each associated with its own mode shape, for a system having n degrees of freedom. This method of analysis becomes very cumbersome, so we resort to matrix techniques as a form of shorthand to simplify the manipulation. Solving large arrays of equations using matrix techniques is done conveniently using a computer program, so these procedures have become very important in analyzing complicated vibrating systems.

Consider a simple n degree of freedom system such as the one shown in Fig. 1.56. The equation of motion of the ith mass is

$$M_i\ddot{x}_i - S_i x_{i-1} + (S_i + S_{i+1})x_i - S_{i+1}x_{i+1} = F_i, \quad i = 2, 3, \ldots, n - 1 \qquad (1.5.1)$$

The equations of motion of masses M_1 and M_n can be derived from this equation by setting $i = 1$, along with $x_0 = 0$ and $i = n$ with $x_{n+1} = 0$; that is,

$$M_1\ddot{x}_1 + (S_1 + S_2)x_1 - S_2x_2 = F_1 \qquad (1.5.2)$$

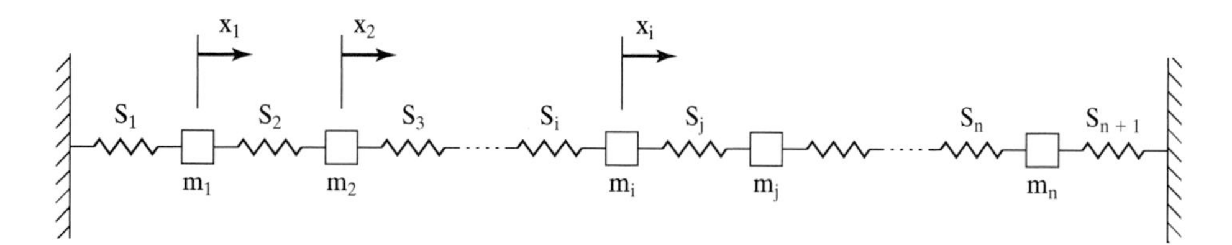

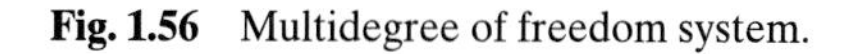
Fig. 1.56 Multidegree of freedom system.

and

$$M_n\ddot{x}_n - S_n x_{n-1} + (S_n + S_{n+1})x_n - S_{n+1}x_{n+1} = F_n \tag{1.5.3}$$

Eqs. (1.5.1) to (1.5.3) can be expressed in matrix form (see Appendix G) as

$$[M]\{\ddot{x}\} + [S]\{x\} = \{F\} \tag{1.5.4}$$

where [M] and [S] are called the mass and stiffness matrix, respectively, and are given by

$$[M] = \begin{bmatrix} M_1 & 0 & 0 & \cdots & 0 \\ 0 & M_2 & 0 & \cdots & 0 \\ \\ 0 & 0 & 0 & \cdots & M_n \end{bmatrix} \tag{1.5.5}$$

and

$$[S] = \begin{bmatrix} (S_1 + S_2) & -S_2 & 0 & \cdots & 0 \\ -S_2 & (S_1 + S_3) & -S_3 & \cdots & 0 \\ \\ 0 & \cdots & \cdots & \cdots & (S_n + S_{n+1}) \end{bmatrix} \tag{1.5.6}$$

where $\{\ddot{x}\}$, $\{x\}$, and $\{F\}$ are the displacement, acceleration, and force vectors given by

$$\{\ddot{x}\} = \begin{bmatrix} \ddot{x}_1 \\ \ddot{x}_2 \\ \\ \ddot{x}_n \end{bmatrix} \qquad \{x\} = \begin{bmatrix} x_1 \\ x_2 \\ \\ x_n \end{bmatrix} \qquad \{F\} = \begin{bmatrix} F_1 \\ F_2 \\ \\ F_n \end{bmatrix} \tag{1.5.7}$$

We can simply use uppercase letters and write the matrix equation as

$$M\ddot{x} + Sx = F \tag{1.5.8}$$

For free vibrations, F is zero and thus

$$M\ddot{x} + Sx = 0 \tag{1.5.9}$$

If we multiply Eq. (1.5.9) by M^{-1}, which is the inverse matrix of M, then

$$MM^{-1}\ddot{x} + SM^{-1}x = 0 \tag{1.5.10}$$

But

$$MM^{-1} = I \text{ (known as the unit matrix)} \tag{1.5.11}$$

then

$$I\ddot{x} + M^{-1}Sx = 0 \tag{1.5.12}$$

Assuming a harmonic solution:

$$\ddot{x} = -\omega^2 x \tag{1.5.13}$$

then

$$[M^{-1}S - I\omega^2]x = 0 \tag{1.5.14}$$

The characteristic equation of the system is the determinant equated to zero; that is,

$$M^{-1} S - I\omega^2 = 0 \tag{1.5.15}$$

The roots ω_i^2 of the characteristic equation are called the eigenvalues, and they are the natural frequencies of the system. By substituting ω_i^2 into the matrix equation Eq. (1.5.14), we obtain the corresponding mode shape X_i, which is called an eigenvector.

1.5.2 Examples of Multidegree of Freedom Systems

We will start by considering again a two degree of freedom system as shown in Fig. 1.57 to illustrate the process of obtaining eigenvalues. The equation of motion can be expressed in matrix notation as follows:

$$\begin{bmatrix} 2M & 0 \\ 0 & M \end{bmatrix} \begin{Bmatrix} \ddot{x}_1 \\ \ddot{x}_2 \end{Bmatrix} + \begin{bmatrix} 3S & -S \\ -S & S \end{bmatrix} \begin{Bmatrix} x_1 \\ x_2 \end{Bmatrix} = \begin{Bmatrix} 0 \\ 0 \end{Bmatrix} \tag{1.5.16}$$

Now

$$M^{-1} = \begin{bmatrix} 1/2M & 0 \\ 0 & 1/M \end{bmatrix} \tag{1.5.17}$$

and multiplying Eq. (1.5.16) by M^{-1} and letting $\ddot{x} = -\omega^2 x$, we obtain

$$\begin{bmatrix} 1 & 0 \\ 0 & 1 \end{bmatrix} \begin{Bmatrix} -\omega^2 & x_1 \\ -\omega^2 & x_2 \end{Bmatrix} + \begin{bmatrix} 3S/2M & -S/2M \\ -S/M & S/M \end{bmatrix} \begin{Bmatrix} x_1 \\ x_2 \end{Bmatrix} = \begin{Bmatrix} 0 \\ 0 \end{Bmatrix} \tag{1.5.18}$$

Hence

$$\begin{bmatrix} (3S/2M - \omega^2) & -S/2M \\ -S/M & (S/M - \omega^2) \end{bmatrix} \begin{Bmatrix} x_1 \\ x_2 \end{Bmatrix} = \begin{Bmatrix} 0 \\ 0 \end{Bmatrix} \tag{1.5.19}$$

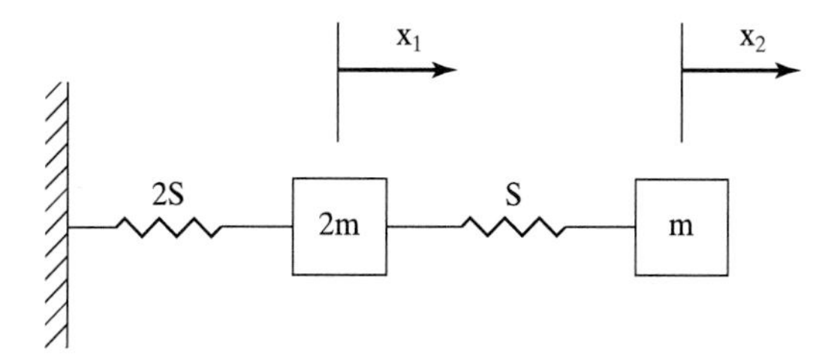

Fig. 1.57 A two degree of freedom system used to illustrate the matrix approach.

The characteristic equation is

$$\omega^4 - \{5S/2M\}\omega^2 + \{S/M\}^2 = 0 \tag{1.5.20}$$

From this equation, the eigenvalues can be found to be

$$\omega_1^2 = S/2M \quad \text{and} \quad \omega_2^2 = 2S/M \tag{1.5.21}$$

The eigenvectors are found by substituting the above values of ω_1^2 and ω_2^2 into the matrix equation. The first eigenvector is

$$x_1 = \begin{Bmatrix} 0.5 \\ 1.0 \end{Bmatrix} \tag{1.5.22}$$

and the second eigenvector is

$$x_2 = \begin{Bmatrix} -1.0 \\ 1.0 \end{Bmatrix} \tag{1.5.23}$$

The two normal modes are shown in Fig. 1.58.

The same technique can be used to solve multidegree of freedom problems, and the vibration of large, complicated objects can be reduced to a multidegree of freedom vibration problem by supposing the object to be broken up into small parts, each one of which can be treated as a point mass. This is known as the finite element technique, and with the aid of a finite element computer program, we can predict how a complicated object will vibrate at different frequencies. In the examples thus far, the masses were arranged along a line, while extended objects are three dimensional, but this is a complication that is easily taken care of in principle. We will not discuss the details of the programming of such problems (which are covered in books on the finite element technique and are often contained in proprietory commercial software packages), but we will show the results of a particular problem to illustrate what may be accomplished in this way.

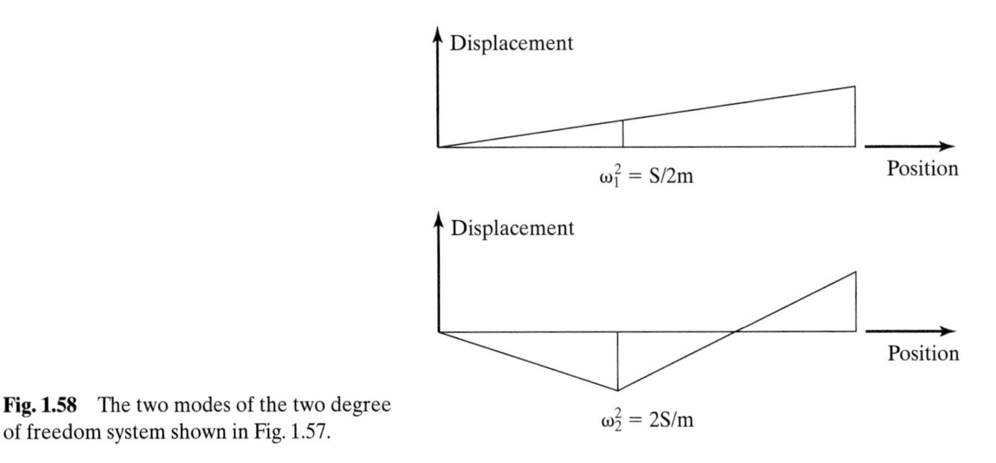

Fig. 1.58 The two modes of the two degree of freedom system shown in Fig. 1.57.

A railroad wheel is like a large flat bell and rings with a pleasing sound when struck by a hammer. If it is cracked, however, the sound is dissonant, and the difference in the sound can be easily recognized by the ear. Railroad workers used to inspect wheels in this way. As part of a research project to design an automatic wheel inspection system, Dousis et al. (1978) wanted to find the resonance frequencies of the wheel. Fig. 1.59 shows the wheel modeled using finite elements. The modes of vibration obtained by finding the eigenvalues of this system are shown in Fig. 1.60. It is important to realize that the accuracy of the results of the finite element technique depend on the accuracy of the assumption that the elements can be treated as point masses. As a rule of thumb, the element should be less than one tenth of a wavelength of sound in the material of the object being modeled. This means that the results are unreliable at high frequencies. Greater accuracy can be attained with a

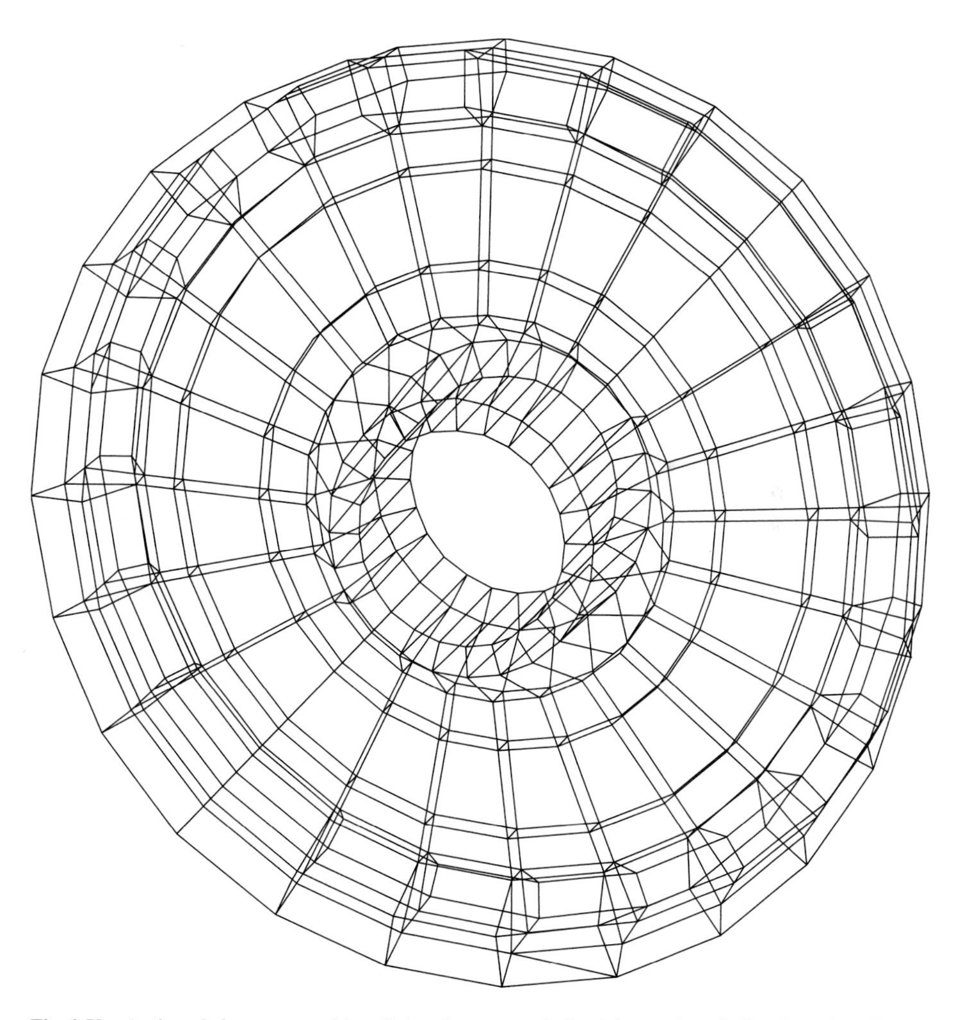

Fig. 1.59 A plot of elements used in a finite element analysis of the modes of vibration of a railroad wheel. The hub is assumed to be fixed. (After Nagy, 1974.)

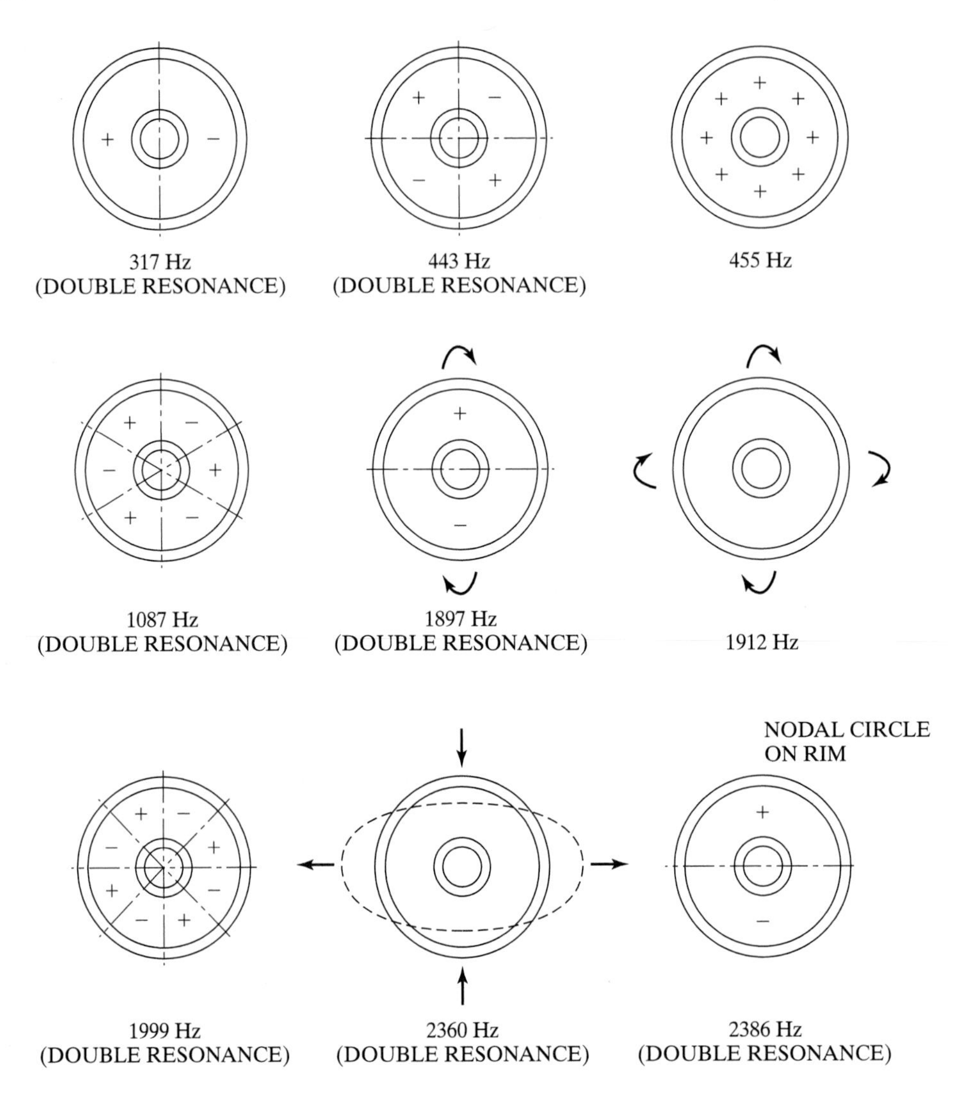

Fig. 1.60 Mode shapes obtained with finite element analysis of the wheel shown in Fig. 1.59. The hub is fixed. (After Nagy, 1974.)

smaller element size, but this accuracy is offset by an increase in the number of elements and, consequently, of computing time.

The finite element technique is of comparatively recent origin, relying as it does on the power of the computer to produce results. Before its invention, such insight as we might have obtained into the vibrations of a railroad wheel came from comparing it with a circular plate or a ring, both of which had been analyzed using knowledge of wave motion in continuous media. In the succeeding chapters, we will study wave motion, starting with the simpler cases and gradually introducing more complex situations as the book progresses. The foregoing remarks do not mean that wave analysis approaches have been invalidated by the development of the finite element technique, since they are the starting point for even more powerful computational techniques. At the outset, however, our emphasis is on the understanding that we can achieve, rather than on computation. In the next section, we will start the study of wave motion with the case of a continuous distribution of mass along a line.

1.6 Vibration of a One-Dimensional Continuum: Waves

1.6.1 Dynamics of a Transmission Line

So far we have considered lumped constant systems in which masses may be considered to reside at points in space and where the strain is constant along the lengths of springs. Many important systems however are such that mass and stiffness are distributed along lines, over areas, or throughout volumes. The simplest case is that of a line such as a fluid-filled pipe, a solid rod, or a stretched string in which there is a continuous distribution of mass and stiffness. Such a system is called a transmission line, because as we shall see, it may be used to transmit a mechanical disturbance. Let the mass per unit length be ρ_L and the stiffness of unit length be σ. Then the mass of an element δx will be $\rho_L \delta x$ and its stiffness, S, will be given by

$$\sigma = S/n = S\delta x \tag{1.6.1}$$

since the stiffness of n identical springs, each of stiffness k, in series is k/n. Thus, S becomes $\sigma/\delta x$ and the element may be represented as shown in Fig. 1.61, where the mass of each element is divided into two portions for symmetry.

Now if the rest of the line on the left of the element exerts a force F upon it, for mathematical consistency we write the force exerted by the element on the rest of the line to the right as $F + (\partial F/\partial x)\,dx$. Consequently, by Newton's third law an equal opposite force $-(F + (\partial F/\partial x)dx)$ is exerted by the line to the right on the element. Thus,

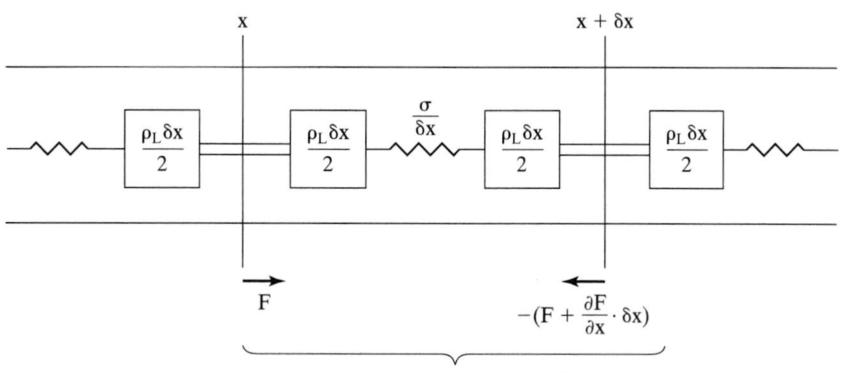

Fig. 1.61 Element of a transmission line.

assuming the displacements at x and x + dx are η and $(\eta + \partial\eta/\partial x.dx)$, the equations of motion of the two component masses within the element may be written

$$F + \frac{\sigma}{\delta x}\frac{\partial \eta}{\partial x}\delta x = \frac{\rho_L \delta x}{2}\frac{\partial^2 \eta}{\partial t^2} \tag{1.6.2}$$

and

$$-\left[F + \frac{\partial \eta}{\partial x}\delta x\right] - \frac{\sigma}{\delta x}\frac{\partial \eta}{\partial x}\delta x = \frac{\rho_L \delta x}{2}\frac{\partial^2}{\partial t^2}\left[\eta + \frac{\partial \eta}{\partial x}\delta x\right] \tag{1.6.3}$$

Neglecting terms of order $(dx)^2$ and adding:

$$-\frac{\partial F}{\partial x}\delta x = \rho_L \delta x \frac{\partial^2 \eta}{\partial t^2} \tag{1.6.4}$$

Now by substracting Eq. (1.6.3) from Eq. (1.6.2)

$$2F + \frac{\partial F}{\partial x}\delta x + 2\sigma\frac{\partial \eta}{\partial x} = 0 \tag{1.6.5}$$

and letting $\delta x \rightarrow 0$

$$F = -\sigma\frac{\partial \eta}{\partial x} \tag{1.6.6}$$

Differentiating Eq. (1.6.6) with respect to x

$$\frac{\partial F}{\partial x} = -\sigma\frac{\partial^2 \eta}{\partial x^2} \tag{1.6.7}$$

and substituting into Eq. (1.6.4) yields

$$\frac{\partial^2 \eta}{\partial x^2} = \frac{\rho_L}{\sigma}\frac{\partial^2 \eta}{\partial t^2} \tag{1.6.8}$$

It follows from Eqs. (1.6.6) and (1.6.4) that

$$\frac{\partial^2 F}{\partial t^2} = -\sigma\frac{\partial}{\partial x}\frac{\partial^2 \eta}{\partial t^2} = \frac{\sigma}{\rho_L}\frac{\partial^2 F}{\partial x^2} \tag{1.6.9}$$

or rearranging,

$$\frac{\partial^2 F}{\partial x^2} = \frac{\rho_L}{\sigma}\frac{\partial^2 F}{\partial t^2} \tag{1.6.10}$$

Thus, both the force and the displacement obey the same differential equation, which is in fact called the wave equation, for reasons to be apparent in the next section. The constant σ/ρ_L is usually called c^2, where

$$c = \sqrt{\sigma/\rho_L} \tag{1.6.11}$$

and is the velocity of wave propagation, as we will see in the next section. Before proceeding to the solution of the wave equation, we will note the value of c for some specific cases.

For a fluid-filled pipe of cross-sectional area A, the stiffness per unit length

$$\sigma_f = BA \tag{1.6.12}$$

Now $\rho_L = \rho A$, where ρ = mass density, and thus for a fluid

$$c_f = \sqrt{\frac{BA}{\rho A}} = \sqrt{\frac{B}{\rho}} \tag{1.6.13}$$

For a solid rod of unit length, $\sigma_s = YA$ (see Eq. (1.2.16)) and thus $c_s = \sqrt{Y/\rho}$. For a gas, $B = kP$ and $c_g = \sqrt{kP/\rho}$. The fluctuations during the passage of a sound wave are extremely rapid so that k should be taken to be its adiabatic value γ, the ratio of specific heats. When Newton first derived this expression for the velocity of sound, he incorrectly used the isothermal modulus, for which k is unity, resulting in a value for the velocity that was too low. It follows from the ideal gas law that the ratio of the pressure to the density is RT, where R is the gas constant and T, the absolute temperature. Consequently, we obtain an alternative expression $\sqrt{\gamma RT}$ for the velocity in a gas, from which it may be concluded that the velocity depends on the square root of the absolute temperature, but is independent of the pressure.

In the case of a fluid-filled pipe, it is usual to express the wave equation in terms of the excess pressure, p (as distinct from the equilibrium or ambient pressure), and the particle velocity, v (as distinct from the wave or propagation velocity). Thus, from Eq. (1.6.8):

$$\frac{\partial^2 v}{\partial x^2} = \frac{1}{c^2}\frac{\partial^2 v}{\partial t^2} \tag{1.6.14}$$

and from Eq. (1.6.10)

$$\frac{\partial^2 p}{\partial x^2} = \frac{1}{c^2}\frac{\partial^2 p}{\partial t^2} \tag{1.6.15}$$

It also follows from Eq. (1.6.4) that the equation of motion may be written

$$\frac{\partial p}{\partial x} = -\rho\frac{\partial^2 \eta}{\partial t^2} = -\rho\frac{\partial v}{\partial t} \tag{1.6.16}$$

and that from Eq. (1.6.6)

$$p = -\rho c^2 \frac{\partial}{\partial x} \int v dt \tag{1.6.17}$$

1.6.2 One-Dimensional Waves

A wave is a disturbance that travels while retaining a recognizable shape. Thus in the case we have been discussing, the particle displacement, velocity, and acceleration as well as the force are all "disturbances" to the equilibrium situation. Such waves are said to be longitudinal because the motion is in the same direction as the wave propagates. The surface of the ocean or the position of a point on a string are other well-known examples in which a disturbance can give rise to wave motion, although in those cases the particle motion is transverse to the propagation direction. We have chosen to study the transmission line first because the application of the principles of mechanics to that case is particularly simple, although from a historical standpoint wave motion on the water surface and on strings invited attention before it was realized that a similar phenomenon occurs in the transmission of sound in pipes. D'Alembert pointed out that if there is a disturbance of some kind on a line in the x direction, then at a certain instant of time, t, this disturbance can be written as $f(x-ct)$, where c is a constant. It is easy to see that this function will travel with a constant velocity equal to c. For example, in Fig. 1.62a we see some hypothetical disturbance at time $t = 0$. Consider the value of the function at the origin. Now consider the value at the same point at a time t later. This value will then be $f(-ct)$; that is, it will be the same as the value of the function at a distance ct to the left of the origin in Fig. 1.62a. It follows that the entire curve will be uniformly shifted by a distance ct to the right to obtain the function at the later time, as in Fig. 1.62b. A similar argument can be used to show that a disturbance represented by the function $g(x + ct)$ will be a wave that will travel to the left.

Now it follows that

$$\frac{\partial^2 f(x - ct)}{\partial t^2} = c^2 f''(x - ct) \tag{1.6.18}$$

and that

$$\frac{\partial^2 f(x - ct)}{\partial x^2} = f''(x - ct) \tag{1.6.19}$$

so that $y = f(x - ct)$ must satisfy the wave equation. Similarly, $y = g(x + ct)$ is also a solution, as is the summation

$$y = f(x - ct) + g(x + ct) \tag{1.6.20}$$

There is another method of obtaining a solution that we can illustrate by considering the wave equation for velocity:

$$\frac{\partial^2 v}{\partial x^2} = \frac{1}{c^2}\frac{\partial^2 v}{\partial t^2} \tag{1.6.21}$$

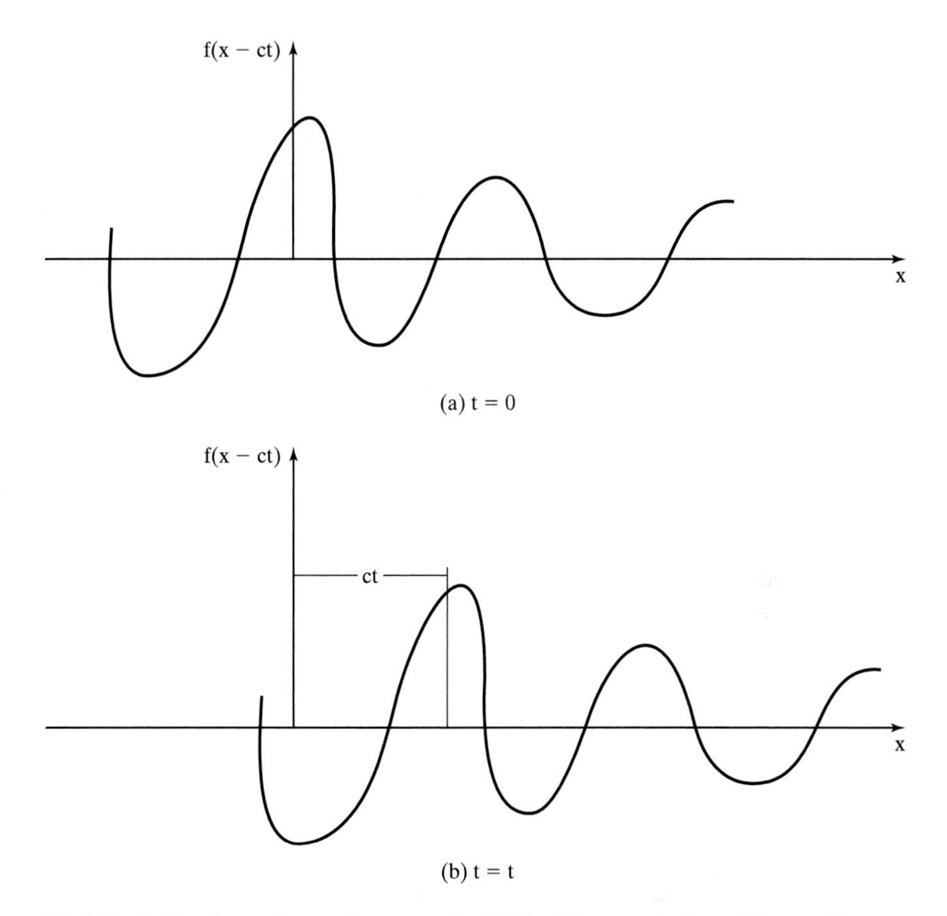

Fig. 1.62 A disturbance f(x − ct) moves to the right by distance ct in elapsed time t, without change of shape. Any such function is called a wave, and will satisfy the wave equation.

In this case, we assume that a separable solution exists such that

$$v = X(x)T(t) \tag{1.6.22}$$

then substituting Eq. (1.6.22) into Eq. (1.6.21) yields

$$T\frac{\partial^2 X}{\partial x^2} = \frac{X}{c^2}\frac{\partial^2 T}{\partial t^2} \tag{1.6.23}$$

and dividing by XT

$$\frac{1}{X}\frac{\partial^2 X}{\partial x^2} = \frac{1}{c^2 T}\frac{\partial^2 T}{\partial t^2} \tag{1.6.24}$$

Now the right side of this equation is a function of t only, while the left side is a function of x only. But x and t are independent variables, so both sides of the equation must be equal to a constant, $-k^2$, say.

Then

$$\frac{\partial^2 X}{\partial x^2} = -k^2 X \qquad (1.6.25)$$

and

$$\frac{\partial^2 T}{\partial t^2} = -c^2 k^2 T \qquad (1.6.26)$$

These equations have the same form as the harmonic motion equation, and letting

$$ck = \omega \text{ or } c = \omega/k \qquad (1.6.27)$$

we obtain the solution of Eq. (1.6.27):

$$T = Ae^{j\omega t} + Be^{-j\omega t} \qquad (1.6.28)$$

For a physically meaningful solution, $B = 0$. Similarly, from Eq. (1.6.25)

$$T = Ae^{j\omega t} + Be^{-j\omega t} \qquad (1.6.29)$$

and finally

$$v = A_1 e^{j(\omega t + kx)} + B_1 e^{j(\omega t - kx)} \qquad (1.6.30)$$

This expression then is a general solution of the wave equation, which is both consistent with the D'Alembert solution given in Eq. (1.6.20) and is separable. It is helpful at this point to interpret the two terms in Eq. (1.6.30). Now,

$$e^{j(\omega t + kx)} = \cos(\omega t + kx) + j\sin(\omega t + kx) \qquad (1.6.31)$$

At time $t = 0$, the first term is a wave as shown in Fig. 1.63a. There is a complete cycle over a distance λ, the wavelength, where

$$k\lambda = 2\pi \text{ or } k = \frac{2\pi}{\lambda} \qquad (1.6.32)$$

Now at a time dt later, there will be a slight phase change and the wave will be given by the dashed curve. We may describe the situation by saying that the wave has travelled in the negative x direction. The maximum has been shifted in the interval by δx, so the propagation velocity $= \delta x/\delta t$.
But

$$k(\lambda - \delta x) + \omega\delta t = 2\pi \qquad (1.6.33)$$

and thus from Eq. (1.6.27)

$$\delta x/\delta t = \omega/k = c \qquad (1.6.34)$$

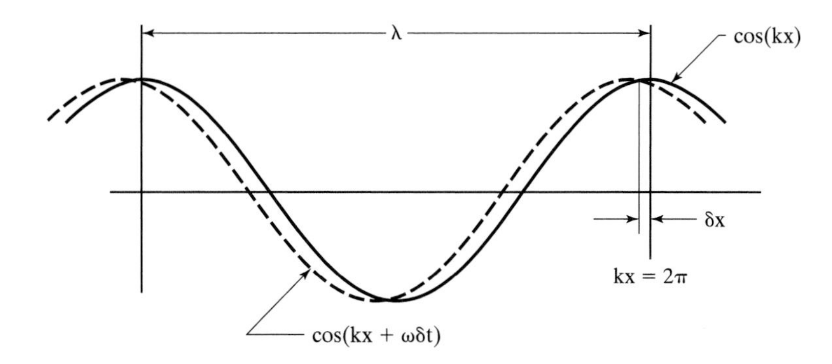

Fig. 1.63a Backward travelling cosine wave.

Eq. (1.6.34) demonstrates again that c is the propagation velocity. Similar behavior (travel in the backward x direction) is also shown by the function $\sin(\omega t + kx)$. Thus, we can conclude that $e^{j(\omega t + kx)}$ represents harmonic waves travelling in the backward or negative x direction. On the other hand,

$$e^{j(\omega t - kx)} = \cos(\omega t - kx) + j\sin(\omega t - kx) \tag{1.6.35}$$

and the functions $\cos(kx)$ and $\cos(kx - \omega\delta t)$, which are equivalent to $\cos(-kx)$ and $\cos(\omega\delta t - kx)$, respectively, are shown in Fig. 1.63b. In this case, then, the wave travels in the forward or positive x direction. A similar result is obtained for $\sin(\omega t - kx)$, and we conclude that $e^{j(\omega t - kx)}$ represents waves travelling in the forward x direction.

Returning to the general solution, Eq. (1.6.30), we see that it contains waves travelling in both the forward and backward directions. These are known as progressive waves. The values of the arbitrary constants A_1 and B_1 depend on the boundary conditions imposed on the line. Consider for instance a semi-infinite line driven by a piston with velocity $u = e^{j\omega t}$ (Fig. 1.63c). In this case, there is no way in which a wave travelling in the backward x direction could arise, so it is immediately seen that $A_1 = 0$ and

$$v = B_1 e^{j(\omega t - kx)} \tag{1.6.36}$$

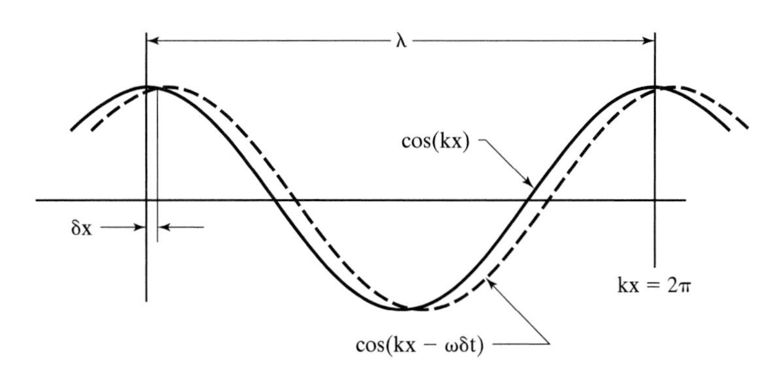

Fig. 1.63b Forward travelling cosine wave.

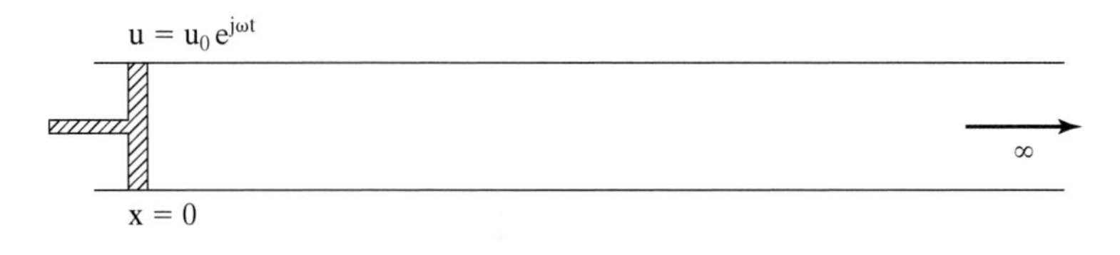

Fig. 1.63c A piston oscillating at one end of a fluid-filled tube produces a progressive wave which travels to the right.

When $x = 0$, $v = u_o$, so $B_1 = u_o$ and

$$v = u_0 e^{j(\omega t - kx)} \tag{1.6.37}$$

at any point along the pipe. From Eq. (1.6.17)

$$p = -\rho c^2 \frac{\partial}{\partial x} \int v\, dt = -\rho c^2 \frac{\partial}{\partial x}\left(\frac{v}{j\omega}\right) = \rho c^2 \frac{k}{\omega} v \tag{1.6.38}$$

That is,

$$p/v = \rho c \tag{1.6.39}$$

Note that the pressure and particle velocity are in phase. The quantity ρc is known as the specific acoustic impedance, an important property in determining the acoustical behavior of a medium.

1.6.3 Energy Transport in a Progressive Wave

Considering again the element of the transmission line shown in Fig. 1.61, we realize that it possesses both kinetic and potential energy. The kinetic energy is given by

$$T = \frac{1}{2} m v^2 \delta x \tag{1.6.40}$$

Consequently, the kinetic energy per unit volume of the element is

$$\tau = \frac{T}{A\delta x} = \frac{\rho v^2}{2} \tag{1.6.41}$$

Similarly, the potential energy due to extension of the spring is

$$U = \frac{1}{2}\frac{\sigma}{\delta x}\left(\frac{\partial \eta}{\partial x}\delta x\right)^2 = \frac{1}{2}\sigma \delta x\left(\frac{\partial \eta}{\partial x}\right)^2 \tag{1.6.42}$$

For a fluid-filled pipe, $\sigma = BA$, and from Eq. (1.6.6),

$$\left(\frac{\partial \eta}{\partial x}\right)^2 = \left(\frac{\rho A}{\sigma}\right)^2 \tag{1.6.43}$$

and thus

$$U = \frac{1}{2}A\delta x\frac{p^2}{B} \tag{1.6.44}$$

and the potential energy per unit volume

$$u = \frac{1}{2}\frac{p^2}{B} \tag{1.6.45}$$

For a solid rod, we would obtain the result

$$u = \frac{1}{2}\frac{p^2}{Y} \tag{1.6.46}$$

Note that so far these results are general; that is, they have not been restricted to the case of a plane wave. When this limitation is made, we have

$$p = \rho c v = \rho\left(\frac{B}{\rho}\right)^{1/2} v = (\rho B)^{1/2}\, v \tag{1.6.47}$$

and thus

$$u = \frac{1}{2}\rho v^2 = \tau \tag{1.6.48}$$

and kinetic and potential energies per unit volume are equal. The total energy per unit volume (the energy density) is

$$\varepsilon = \tau + u = \rho v^2 \tag{1.6.49}$$

Plotting the energy density at a given point as a function of time, remembering to take only the real part of v—namely, $u_o \cos(\omega t - kx)$—the average value of the energy density is easily shown to be

$$\bar{\varepsilon} = \rho {u_0}^2/2 = \rho u_{rms}^2 \tag{1.6.50}$$

A similar plot of the energy may be made as function of distance (Fig. 1.64). Since ε depends on v^2, and v, the particle velocity, propagates as a wave with velocity c, it follows that the energy density also propagates with velocity c.

The intensity (I) of a sound wave is defined as the average energy flowing across unit cross-sectional area in unit time. In the case of our plane wave travelling along the transmission line, the intensity is easily calculated by realizing that all the energy contained in a column of length c will flow past a given point in unit time (see Fig. 1.65). The average energy in this column is $\bar{\varepsilon}cA$, thus the intensity is $\bar{\varepsilon}c$.
Thus, from Eq. (1.6.38)

$$I = \bar{\varepsilon}c = \rho c\frac{u_0^2}{2} = \frac{p_0^2}{2\rho c} \tag{1.6.51}$$

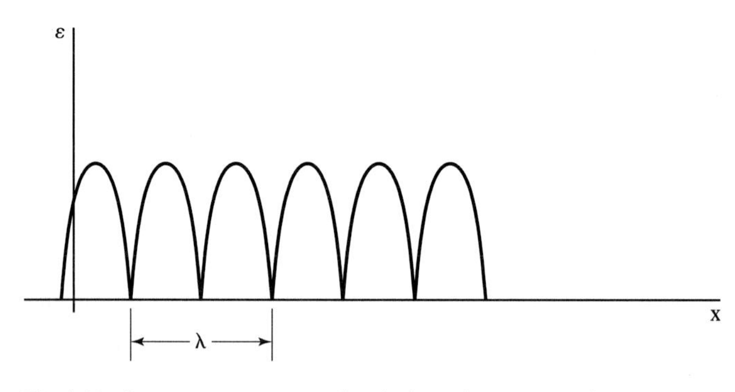

Fig. 1.64 Instantaneous energy density in a plane progressive wave.

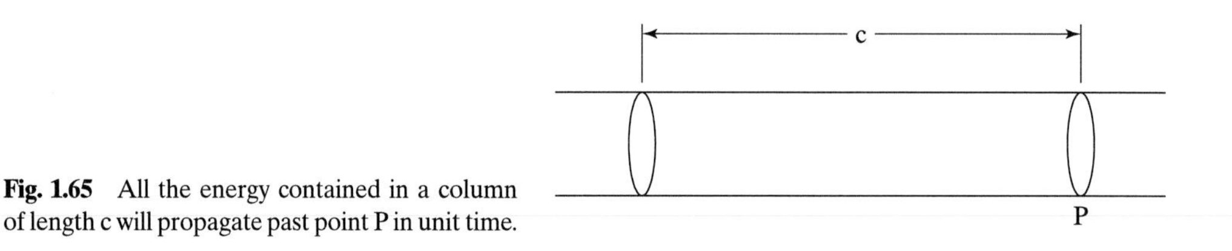

Fig. 1.65 All the energy contained in a column of length c will propagate past point P in unit time.

This result is valid for an unbounded plane wave. In fact (we state without proof), it is true for a wavefront of any shape (e.g., spherical, cylindrical).

1.6.4 Sound Pressure Level

At the frequencies at which the ear is most sensitive, the minimum pressure amplitude that it can detect is about 0.0002 dyne/cm^2 or microbars. This level is used as a reference, termed p_r. Above the threshold value, the sensation of loudness is proportional to the logarithm of the pressure amplitude (or the intensity) and consequently the sound pressure level (L) of a sound whose pressure amplitude is p_o is defined by

$$L = 20 \log_{10}(p_o/p_r) = 20 \log_{10}p_o - 20 \log_{10}p_r \quad (1.6.52)$$

The units of L are decibels (abbreviation dB), which were chosen as a unit because 1 dB is about the minimum increment in sound pressure level detectable by the human ear. Some typical sound pressure level readings are given in Table 1.2. Note that L can also be written in terms of the intensity:

$$L = 10 \log_{10}(I/I_r) \quad (1.6.53)$$

The reference intensity level used most commonly is 10^{-12} W/m^2. There are other interesting points that can be made about the measurement of sound pressure level. For instance, the intensity surrounding a "point" source Q (watts) radiating spherical waves is given by

$$I = Q/4\pi r^2 \quad (1.6.54)$$

Table 1.2 Typical Sound Pressure Levels

Typical Q (watts)	Pressure p_o in bar	Typical SPL	Source
10^{7}		200	Saturn Rocket
10^{5}		180	Military turbo jet with afterburner
			4-propellor airliner
10^{3}		160	
10^{1}		140	
	1 mbar → Threshold of pain		
			Pneumatic chipper
10^{-1}		120	
	100 μbar →		Automobile horn (1m)
10^{-3}		100	
			Inside subway train (New York)
	10 μbar →		
			Inside bus
10^{-5}		80	Appliances: washing machine, dryer, dishwasher, vacuum cleaner
			Traffic
	1 μbar →		
			Conversation
10^{-7}		60	
			Office
	0.1 μbar →		
			Living room
10^{-9}		40	Very soft whisper
	0.01 μbar →		Library
			Bedroom at night
10^{-11}		20	
			Broadcasting studio
	0.001 μbar →		
10^{-13}	Threshold of hearing (0.0002 μbar)	0	

It follows that if the sound pressure level due to a source, Q, is measured, it will depend on the inverse square law in Eq. 1.6.54; thus

$$L = 10 \log_{10}\frac{Q}{2\pi r^2 I_r} = \text{constant} - 10 \log_{10} r^2 = \text{constant} - 20 \log_{10} r \quad (1.6.55)$$

It follows that if r is doubled, the sound pressure level will decline by 6 db. It is to be emphasized that this law applies only where the source is effectively a point, which may be decided by a rule of thumb according to whether its dimensions are small compared with a sound wavelength in air.

Another interesting point concerns the addition of sound pressure levels (see Fig. 1.66). Suppose a source Q_1 produces L_1 by itself, and another source Q_2 produces L_2 by itself. What level will be produced by the two sources operating simultaneously? Now the total energy flow through the point O is

$$I = I_1 + I_2 \quad (1.6.56)$$

thus

$$10 \log_{10}(I/I_r) = 10 \log_{10}(\{I_1 + I_2\}/I_r) \quad (1.6.57)$$

that is,

$$L = 10 \log_{10}\left[\text{antilog}\left(\frac{L_1}{10}\right) + \text{antilog}\left(\frac{L_2}{10}\right)\right] \quad (1.6.58)$$

Note that

$$L \neq L_1 + L_2 \quad (1.6.59)$$

Let us take as an example $L_1 = 90$ dB and $L_2 = 80$ dB. Then

$$I_1/I_r = \text{antilog}(90/10) = 10^9 \quad (1.6.60)$$

and

$$I_2/I_r = \text{antilog}(80/10) = 10^8 \quad (1.6.61)$$

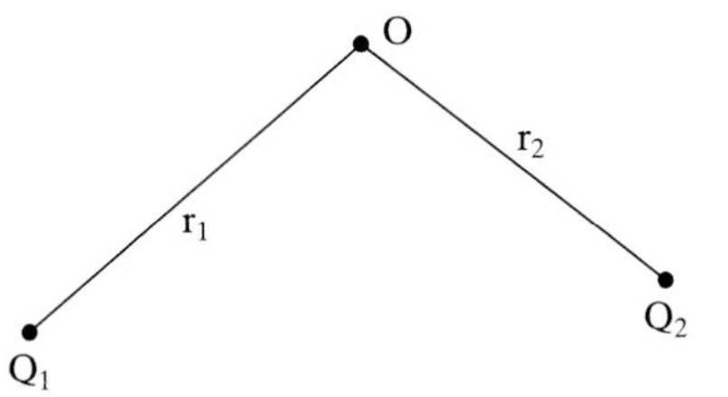

Fig. 1.66 Source Q_1 produces sound level L_1 at O, and source Q_2 produces L_2. What level will they produce when operating together?

and thus

$$\frac{I}{I_r} = \frac{I_1 + I_2}{I_r} = 11.0 \times 10^8 \tag{1.6.62}$$

Then

$$L = 10 \log I/I_r = 10 \times 9.041 = 90.41 \text{ dB} \tag{1.6.63}$$

which represents only a very small increment over the value of L_1.

1.6.5 Transmission and Reflection at a Boundary

Suppose a plane wave travelling in the positive x direction encounters a boundary between two media of different specific acoustic impedances, $\rho_1 c_1$ and $\rho_2 c_2$. It is of interest to calculate the strengths of the reflected and transmitted waves (Fig. 1.67). In the first medium, the disturbance obeys the wave equation

$$\frac{\partial^2 v}{\partial x^2} = \frac{1}{c_1^2}\frac{\partial^2 v}{\partial t^2} \tag{1.6.64}$$

having a general solution

$$v_1 = A_1 e^{j(\omega t - k_1 x)} + B_1 e^{j(\omega t + k_1 x)} \tag{1.6.65}$$

In the second medium, the disturbance obeys the wave equation

$$\frac{\partial^2 v}{\partial x^2} = \frac{1}{c_2^2}\frac{\partial^2 v}{\partial t^2} \tag{1.6.66}$$

having a general solution

$$v_2 = A_2 e^{j(\omega t - k_2 x)} + B_2 e^{j(\omega t + k_2 x)} \tag{1.6.67}$$

But the second medium is semi-infinite, with no boundary to cause a reflection so that B_2 is zero. The relations between the arbitrary constants A_1, B_1, and A_2 determine the

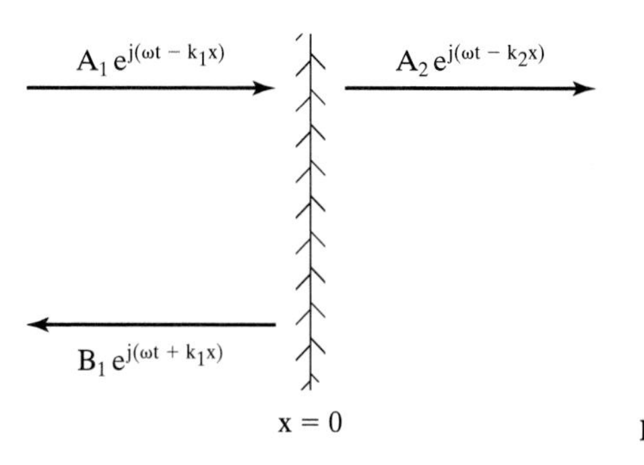

Fig. 1.67 Reflection at a boundary.

reflection and transmission strengths and are decided by the boundary conditions. These are

1. continuity of displacement and velocity, and
2. continuity of pressure.

The first arises because there can be no separation or interpenetration of the two media at the boundary. The second can be understood by considering the free-body diagram of forces acting on a thin layer of material, where p_1 and p_2 are the pressures on either side of the layer (Fig. 1.68). The net force on the layer is $A(p_1 - p_2) = M\ddot{\eta}$. If the layer is vanishingly small, $M = 0$ and $p_1 = p_2$. The boundary between two media may be considered as such a vanishing layer. Before we can apply the boundary conditions, it is necessary to express pressure in terms of the particle velocity. This we can do using the basic equation of motion, (Eq. 1.6.16), which we repeat for convenience:

$$\frac{\partial p}{\partial x} = -\rho \frac{\partial v}{\partial t} \tag{1.6.68}$$

Hence, from Eqs. (1.6.65) and (1.6.67)

$$p_1 = \rho_1 c_1 A_1 e^{j(\omega t - k_1 x)} - \rho_1 c_1 B_1 e^{j(\omega t + k_1 x)} \tag{1.6.69}$$

and

$$p_2 = \rho_2 c_2 A_2 e^{j(\omega t - k_2 x)} \tag{1.6.70}$$

Alternatively, we might use the result that for a plane progressive wave travelling in the positive x direction, $p/v = \rho c$, as given by Eq. (1.6.39). However, we must then remember and verify that for a plane progressive wave travelling in the negative x direction, $p/v = -\rho c$.

Applying the boundary condition of continuity of velocity at $x = 0$ yields

$$A_1 + B_1 = A_2 \tag{1.6.71}$$

and continuity of pressure yields

$$\rho_1 c_1 (A_1 - B_1) = \rho_2 c_2 A_2 \tag{1.6.72}$$

Eliminating A_2 from Eqs. (1.6.71) and (1.6.72) yields

$$A_1(\rho_1 c_1 - \rho_2 c_2) = B_1(\rho_1 c_1 + \rho_2 c_2) \tag{1.6.73}$$

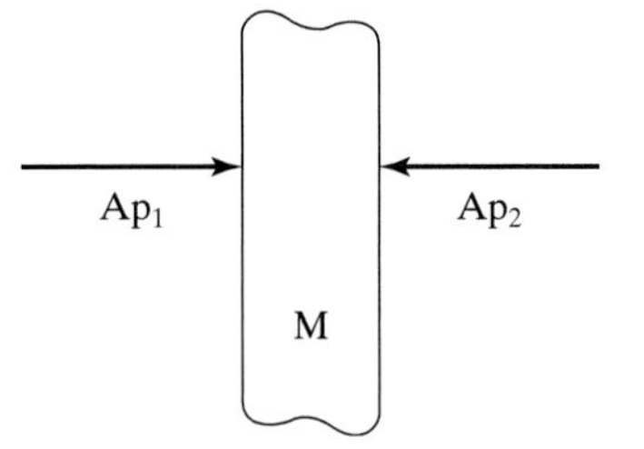

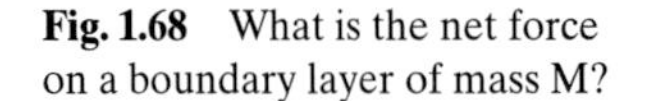

Fig. 1.68 What is the net force on a boundary layer of mass M?

Now the sound power reflection coefficient, R, is defined as the ratio of reflected to incident intensity. Thus

$$R = \frac{I_R}{I_i} = \frac{\rho_1 c_1 B_1^2/2}{\rho_1 c_1 A_1^2/2} = \left(\frac{\rho_1 c_1 - \rho_2 c_2}{\rho_1 c_1 + \rho_2 c_2}\right)^2 \quad (1.6.74)$$

Eliminating B_1 from Eq. (1.6.71) yields

$$A_2 = A_1\left(1 + \frac{\rho_1 c_1 - \rho_2 c_2}{\rho_1 c_1 + \rho_2 c_2}\right) = A_1 \frac{2\rho_1 c_1}{\rho_1 c_1 + \rho_2 c_2} \quad (1.6.75)$$

The sound power transmission coefficient, T, is given by

$$T = \frac{I_t}{I_i} = \frac{\rho_2 c_2 A_2^2/2}{\rho_1 c_1 A_1^2/2} = \frac{4\rho_1 c_1 \rho_2 c_2}{(\rho_1 c_1 + \rho_2 c_2)^2} \quad (1.6.76)$$

Now from Eqs. (1.6.74) and (1.6.76), it follows that

$$R + T = \frac{I_R + I_t}{I_i} = 1 \quad (1.6.77)$$

and hence

$$I_i = I_R + I_t \quad (1.6.78)$$

implying that all the energy incident is either reflected or transmitted, which is to be expected considering the principle of conservation of energy. We will note some further interesting implications. Let

$$\rho_2 c_2/\rho_1 c_1 = r_{12} \quad (1.6.79)$$

then

$$R = \left(\frac{1 - r_{12}}{1 + r_{12}}\right)^2 \quad (1.6.80)$$

Thus

$$R = 0 \quad \text{if} \quad \rho_1 c_1 = \rho_2 c_2 \quad (1.6.81)$$

that is, a boundary between two different media (with different densities and sound velocities) may be "invisible" to sound, provided the two media have the same specific acoustic impedance. Further, $R = 1$ if either $r_{12} = 0$ or $r_{12} = \infty$; that is, either

$$\rho_1 c_1 >> \rho_2 c_2 \quad \text{or} \quad \rho_2 c_2 >> \rho_1 c_1 \quad (1.6.82)$$

The first case might be approximated by a solid-to-gas boundary and the second by a gas-to-solid boundary. In this first case, $(r_{12} = 0)$, the boundary is said to be pressure-release or pressure-relief or free. In the opposite case, the boundary is rigid.

REFERENCES AND FURTHER READING

Blevins, R.D. 1979. *Formulas for Natural Frequency and Mode Shape*. Van Nostrand.

Dimorogonas, A. 1996. *Vibration for Engineers*. 2nd ed. Prentice Hall.

Dousis, D.A., K. Nagy, and R.D. Finch. 1978. "Detection of Flaws in Railroad Wheels Using Acoustic Signatures." 100 *ASME J. Eng. Ind.* 459–467.

Nagy, K. 1974. "Feasibility Study of Flaw Detection in Railway Wheels Using Acoustic Signatures." PhD Thesis, University of Houston.

Roark, R.J. 1989. *Formulas for Stress and Strain*. 5th ed. McGraw Hill.

Thomson, W.T. 1966. *Vibration Theory and Applications*. Allen and Unwin.

Timoshenko, S., D.H. Young, and W. Weaver. 1974. *Vibration Problems in Engineering*. Wiley.

PROBLEMS

1.1 Prove that for the set of rotating phasors, the following identity holds:

$$ae^{j\omega t} + ae^{j(\omega t+\delta)} + ae^{j(\omega t+2\delta)} + \cdots + ae^{j(\omega t+n\delta)} = Re^{j(\omega t+e)}$$

where

$$R = \frac{a \sin(n+1)\,\delta/2}{\sin \delta/2}$$

and

$$\varepsilon = n\delta/2$$

1.2 A siren, situated 170 m from a vertical cliff, emits a tone that rises from zero to 340 Hz in 5 s and then drops uniformly to zero in the same length of time. Make a quantitative statement as to what will be observed by a person standing by the siren. How would the situation change as this person walks toward or away from the cliff? Assume the velocity of sound in the air is 340 m/s.

1.3 Show that the natural angular frequency of a mass m on a short spring of mass M and stiffness S is given by

$$\omega = \left[S/(m + M/3)\right]^{1/2}$$

1.4 Consider a single harmonic oscillator with M = 0.5 kg, stiffness S = 10 N/m, and amplitude A = 3 cm.

a. What is the total energy of this oscillator?
b. What is its maximum speed?
c. What is the speed when x = 1.8 cm?
d. What are the kinetic and potential energies when x = 1.8 cm?

1.5 The natural frequency of a Helmholtz resonator is given by the equation

$$f = c\frac{\sqrt{A/Vl}}{2\pi}$$

where c (speed of sound) = 344 m/s, A = area of the opening, V is the volume of air enclosed, and *l* is the length of the neck:

a. Determine the natural frequency of an acoustic guitar if the enclosed volume of air = 0.05 m^3, the area of the opening is 2.0×10^{-3} m^2, and the length of the neck is 5.1×10^{-3} m.
b. What is the best way to increase the frequency of the Helmholtz resonance for the guitar?
c. If electronics are installed inside the guitar and reduce the volume of air enclosed by 0.03 m^3, will this increase or decrease the Helmholtz resonance frequency?

1.6 An anechoic room is supported on 36 air springs, which are pressurized to 115 psi. Assume the active area of each spring is 300 in^2 and that the effective volume of each spring is 3000 in^3. If the room weighs 320 tn., calculate its natural frequency in Hz. The measured value is 1.37 Hz; if your value differs from this, suggest an explanation.

1.7 A helical steel spring having a coil diameter of 1 in. is made of 50 turns of 1/8 in. diameter wire. What will be its lowest frequency if it is fixed at one end? Will it have any other natural frequencies? Explain. (Assume for the properties of steel a density of 7700 kg/m^3 and a shear modulus of 8.3×10^{10} N/m^2.)

1.8 A machine weighing 1000 lb. is supported on springs with a static deflection of 0.2 in. If the machine has a rotating imbalance of 20 lb per in., neglect damping and determine

a. the force transmitted to the floor at 1200 rpm.
b. the amplitude at this speed.

1.9 A damped harmonic oscillator of mass M, damping constant R, and spring stiffness S is at rest in its equilibrium position when a driving force, $F_o \sin \omega t$, is applied at t = 0. With this set of initial conditions, will the steady state be achieved immediately or will some time elapse before the steady state is reached?

1.10 A loudspeaker is modeled as a driven harmonic oscillator with values of damping R = 2.4 kg/s, stiffness S = 3500 N/m, and mass M = 0.04 kg.

a. Find the resonant frequency of the loudspeaker.
b. At a frequency of 55 Hz, what is the mechanical impedance Z?
c. Above what frequency can this system be considered a mass-controlled oscillator?
d. If the speaker is driven by a harmonic driving force at a frequency of 55 Hz, what will be the phase difference between the driving force and the velocity? Which one leads?

1.11 A 2 kg mass is connected to a linear spring having a stiffness S = 3600 N/m. There is a weak damping force R = 6.2×10^{-3} kg/s. The mass is driven by a force = 0.03 N. What is the resonant frequency of the mass? Also show that a sufficient resonant amplitude can result even though the driving force is weak. What is the quality factor Q for this oscillator?

1.12 An instrument weighing 0.6 kg is mounted on springs to isolate it from vibration of the table on which it is placed. The table vibrates with an amplitude of 3.6 mm at 50 Hz. The instrument can tolerate a maximum amplitude of 0.6 mm. What should be the maximum stiffness of the springs used in the mounting? Assume there is negligible damping and the system is modeled as a single degree of freedom system.

1.13 For the multidegree of freedom system shown in Fig. P 1.13

a. Find the mass and stiffness matrix.
b. Is the system of equations dynamically coupled, statically coupled, or both?

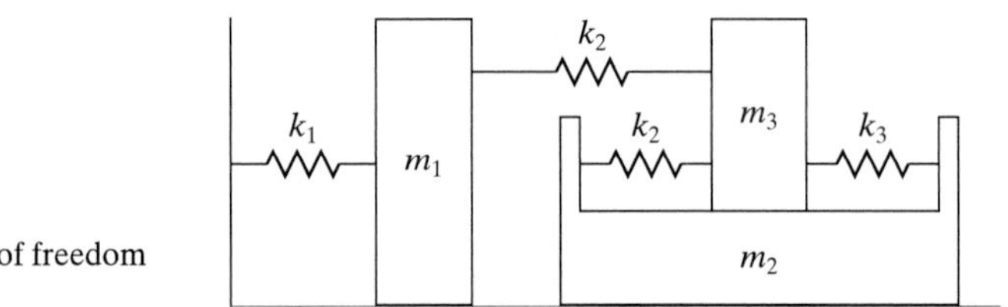

Fig. P1.13 Multidegree of freedom system for Problem 1.13.

1.14 The point A of the system shown in Fig. P1.14 moves according to the formula

$$y = A_1 \sin(f_1 t) + A_2 \sin(f_2 t)$$

$A_1(=0.015\text{ m})$ and $A_2(=0.01\text{ m})$ are the amplitudes of this motion and $f_1(=35\text{ rad/s})$ and $f_2(=30\text{ rad/s})$ are the corresponding frequencies. The mass $M_1(=12\text{ kg})$ and $M_2(=24\text{ kg})$. The length is $l_1(=1.2\text{ m})$ and $l_2(=2.4\text{ m})$, the stiffness for the spring $S(=11000\text{ N/m})$, and the resistance of the damper $R(=120\text{ Ns/m})$. The other input assumptions are $E_1 = E_2 = 2 \times 10^{11}\text{ N/m}^2$, $J_1 = J_2 = 1.0 \times 10^{-8}\text{ m}^4$. Determine

a. The differential equations of motion.
b. The natural frequencies of the free motion.
c. The steady-state motion of the system due to kinematic excitation y.
d. The exciting force at point A required to maintain steady-state motion.

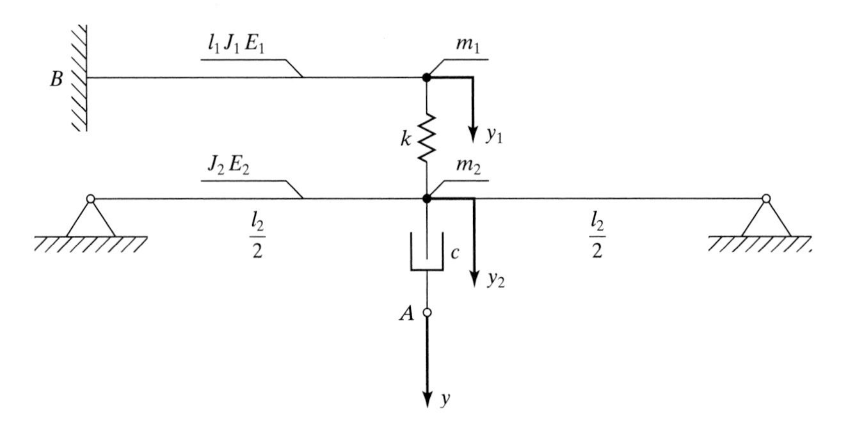

Fig. P1.14 System for Problem 1.14.

Chapter 2

Linear Systems

2.1 Fourier Analysis

In Chapter 1, we considered vibrations and waves that are harmonic, or sinusoidal, in character. Fourier realized that many other types of disturbances can be considered summations of harmonic forms. To see how this is so, consider the superposition of a number of harmonic motions whose periods are related by multiples (T, T/2, T/3, . . .). During one complete period of the first motion, the second will have completed two periods; the third, three; and so on. Clearly, the resultant of the superposition will be periodic with period T. The form of the motion within one period will depend on the amplitudes of the various components. Obviously, we could synthesize any periodic motion in this way or, working in reverse, it is possible to analyze such well-known functions as square waves, sawtooths, and triangular waves. The procedure for doing this was first worked out by Fourier. Suppose that some motion is compounded of various harmonic components:

$$F(t) + a_0 = a_1 \cos \omega t + a_2 \cos 2\omega t + \cdots + b_1 \sin \omega t + b_2 \sin 2\omega t + \cdots \tag{2.1.1}$$

It simplifies matters to let $\omega t = \theta$ so that $F(t) = f(\theta)$, then

$$f(\theta) = a_0 + \sum_{n-1}^{\infty} a_n \cos n\theta + \sum_{n-1}^{\infty} b_n \sin n\theta \tag{2.1.2}$$

This change of variable redirects our attention from the periods (T, T/2, T/3, etc.) to the angular frequencies that must follow the sequence (ω, 2ω, 3ω, etc.). Using θ as a variable is also advantageous in that the results are not restricted to functions of time. Periodic spatial variations (waves) could also be covered by the analysis.

Any periodic function may be represented by the above expansion, provided the function

1. remains finite, and
2. is single valued.

These two provisions are known as the Dirichlet conditions. Mechanical oscillations normally meet these conditions, since material objects cannot have infinite displacements or more than one value of displacement at a time.

We will now find the values of the coefficients in Eq. (2.1.2) for a function $f(\theta)$ which has periodicity 2π. To find a_0, integrate from $-\pi$ to $+\pi$, thus

$$\int_{-\pi}^{\pi} f(\theta)\, d\theta = a_0 \int_{-\pi}^{\pi} d\theta + \sum_{n=1}^{\infty} a_n \int_{-\pi}^{\pi} \cos n\theta\, d\theta + \sum_{n=1}^{\infty} a_n \int_{-\pi}^{\pi} \sin n\theta\, d\theta \tag{2.1.3}$$

Hence

$$a_0 = \frac{1}{2\pi} \int_{-\pi}^{\pi} f(\theta)\, d\theta \tag{2.1.4}$$

To find a_n, multiply throughout by $\cos m\theta$ and integrate from $-\pi$ to $+\pi$, then

$$\int_{-\pi}^{\pi} f(\theta) \cos m\theta\, d\theta = a_0 \int_{-\pi}^{\pi} \cos m\theta\, d\theta + \sum a_n \int_{-\pi}^{\pi} \cos n\theta \cos m\theta\, d\theta$$
$$+ \sum b_n \int_{-\pi}^{\pi} \sin n\theta \cos m\theta\, d\theta \tag{2.1.5}$$

Now it is readily shown that

$$\int_{-\pi}^{\pi} \cos n\theta \cos m\theta\, d\theta = \begin{cases} 0 & \text{if } n \neq m \\ \pi & \text{if } n = m \end{cases} \tag{2.1.6}$$

and that

$$\int_{-\pi}^{\pi} \sin n\theta \cos m\theta\, d\theta = \begin{cases} 0 & \text{if } n \neq m \\ \pi & \text{if } n = m \end{cases} \tag{2.1.7}$$

Consequently

$$\int_{-\pi}^{\pi} f(\theta) \cos m\theta\, d\theta = a_m \pi \tag{2.1.8}$$

Thus

$$a_n = \frac{1}{\pi} \int_{-\pi}^{\pi} f(\theta) \cos n\theta\, d\theta \tag{2.1.9}$$

and by a similar process

$$b_n = \frac{1}{n}\int_{-\pi}^{\pi} f(\theta)\sin n\theta\, d\theta \tag{2.1.10}$$

We can illustrate the theorem by its application to the following examples:

a. Sawtooth (see Fig. 2.1), in which

$$f(\theta) = \theta,\ -\pi < \theta < \pi \tag{2.1.11}$$

Using Eqs. (2.1.4), (2.1.9), and (2.1.10), it is easy to show that

$$a_0 = 0 \tag{2.1.12}$$

$$a_n = 0 \tag{2.1.13}$$

and

$$b_n = -\frac{2}{n}\cos n\pi \tag{2.1.14}$$

But

$$\cos n\pi = (-1)^n \tag{2.1.15}$$

so that

$$b_n = \frac{2}{n}(-1)^n \tag{2.1.16}$$

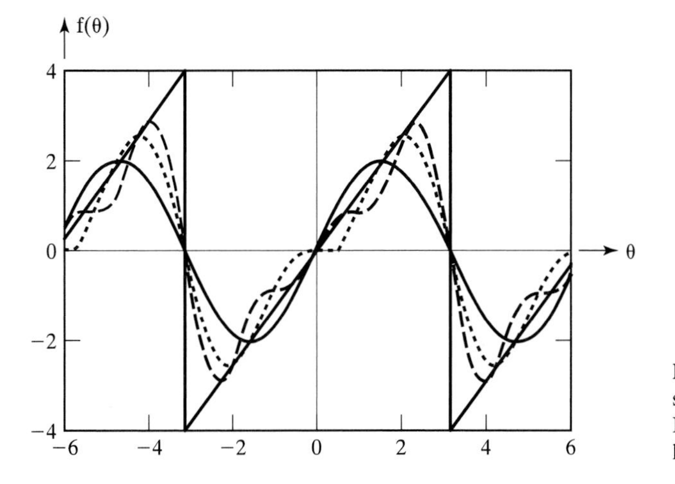

Fig. 2.1 Fourier analysis of a sawtooth. Solid curve: first term. Dotted curve: two terms. Broken line: three terms.

and the expansion becomes

$$f(\theta) = \sum_{n=1}^{\infty}(-1)^{n+1}\frac{2}{n}\sin n\theta$$
$$= 2\left[\sin\theta - \frac{\sin 2\theta}{2} + \frac{\sin 3\theta}{3} - \cdots\right] \quad (2.1.17)$$

Fig. 2.1 shows how even the summation of the first three terms begins to resemble the sawtooth.

b. Square wave (see Fig. 2.2), in which

$$f(\theta) = \begin{cases} 0, -\pi < \theta < 0 \\ 1, 0 < \theta < \pi \end{cases} \quad (2.1.18)$$

Following the same procedures as for the sawtooth, it can be readily shown that $a_0 = 1/2$ and

$$f(\theta) = \frac{1}{2} + \frac{2}{\pi}\left\{\sin\theta + \frac{\sin 3\theta}{3} + \frac{\sin 5\theta}{5} + \cdots\right\} \quad (2.1.19)$$

Hence, we have only the odd harmonics, but their relative values are the same as in the expansion of the sawtooth. Again, we see the superposition of the first three terms in Fig. 2.2.

There are interesting points to notice in connection with these examples. First, the value of a_0 is the "average" value of the expanded function. Second, in the two cases examined, the cosine terms dropped out of the expansion for the following reason.

A function is said to be odd if

$$f(\theta) = -f(-\theta) \quad (2.1.20)$$

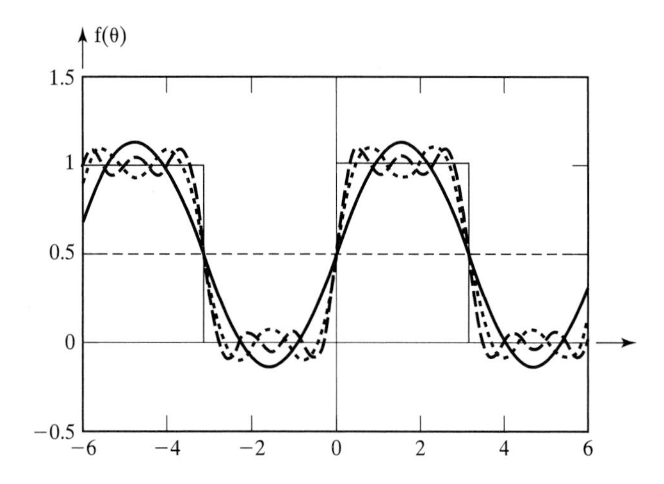

Fig. 2.2 Fourier analysis of a square wave. Dashed line at 0.5 is the first term. (This term is the "DC" value of the square wave.) Solid curve is the sum of two terms. Dotted curve: three terms. Broken curve: four terms.

and even if

$$f(\theta) = +f(-\theta) \tag{2.1.21}$$

Examples of odd and even functions are shown in Figs. 2.3 and 2.4. Odd functions may be represented by sine terms only and even functions by cosine terms only—which may be proved easily in the following way:

$$\begin{aligned} a_0 &= \frac{1}{2\pi}\int_{-\pi}^{\pi} f(\theta)\,d\theta = \frac{1}{2\pi}\int_{-x}^{0} f(\theta)\,d\theta + \frac{1}{2\pi}\int_{0}^{\pi} f(\theta)\,d\theta \\ &= -\frac{1}{2\pi}\int_{0}^{-\pi} f(\theta)\,d\theta + \frac{1}{2\pi}\int_{0}^{\pi} f(\theta)\,d\theta \\ &= -\frac{1}{2\pi}\int_{0}^{\pi} f(-\theta)\,d(-\theta) + \frac{1}{2\pi}\int_{0}^{\pi} f(\theta)\,d\theta \\ &= \frac{1}{2\pi}\int_{0}^{+\pi} f(-\theta)\,d\theta + \frac{1}{2\pi}\int_{0}^{\pi} f(\theta)\,d\theta \end{aligned} \tag{2.1.22}$$

If $f(\theta)$ is odd, $a_0 = 0$. If $f(\theta)$ is even,

$$a_0 = \frac{1}{\pi}\int_{0}^{\pi} f(\theta)\,d\theta \tag{2.1.23}$$

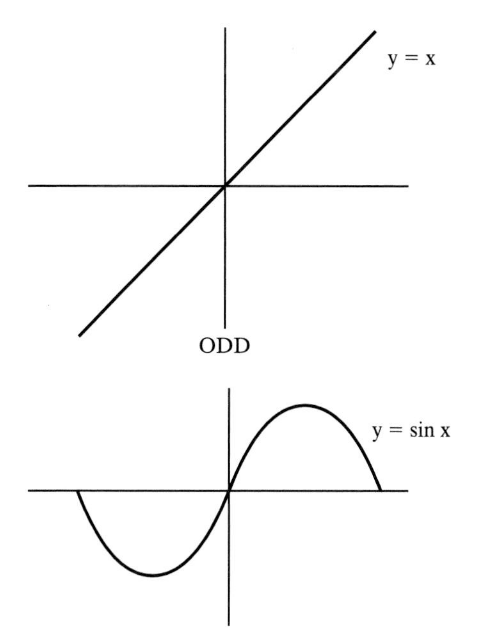

Fig. 2.3 For odd functions $f(x) = -f(-x)$.

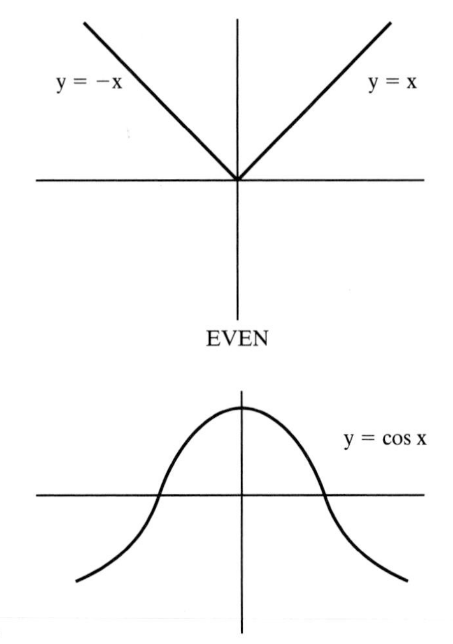

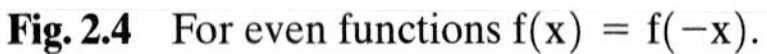
Fig. 2.4 For even functions $f(x) = f(-x)$.

We readily interpret this result geometrically if we realize that the areas under the curve in the two half ranges are equal and of the same sign if the function is even, but of opposite sign if it is odd, as shown in Fig. 2.5. Since

$$a_0 = \frac{1}{\pi} \int_0^{\pi} f(\theta) \cos n\theta \, d\theta \tag{2.1.24}$$

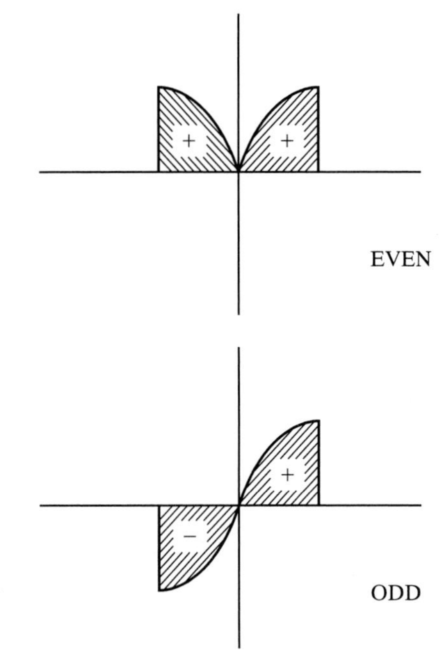

Fig. 2.5 Over the whole range $-\pi$ to $+\pi$, the area under the curve in the two half ranges is equal. Thus the full range integral for an odd function is zero, and twice the value for the half range for an even function.

if $f(\theta)$ is odd, then $f(\theta)\cos n\theta$ is also odd and $a_n = 0$, but if $f(\theta)$ is even, then $f(\theta)\cos n\theta$ is even and

$$a_n = \frac{2}{\pi}\int_{-\pi}^{\pi} f(\theta)\cos n\theta\, d\theta \tag{2.1.25}$$

Similarly,

$$b_n = \frac{1}{\pi}\int_{-\pi}^{\pi} f(\theta)\sin n\theta\, d\theta \tag{2.1.26}$$

and if $f(\theta)$ is odd, then $f(\theta)\sin n\theta$ is even and

$$b_n = \frac{2}{\pi}\int_{0}^{\pi} f(\theta)\sin n\theta\, d\theta \tag{2.1.27}$$

but if $f(\theta)$ is even, then $f(\theta)\sin n\theta$ is odd, so $b_n = 0$.

Over a half range 0 to π or $-\pi$ to 0, it is possible to represent a function by either a sine series or a cosine series by pretending that the function is either odd or even. In the sine series case, for example, we would have

$$f(\theta) = \sum_{n=1}^{\infty} b_n \sin n\theta \; (0 < \theta < \pi) \tag{2.1.28}$$

and b_n is given by Eq. (2.1.27). If we assume $f(\theta)$ to be even, then

$$f(\theta) = a_0 + \sum a_n \cos n\theta \tag{2.1.29}$$

where a_0 and a_n are given by Eqs. (2.1.23) and (2.1.25).

The difference shows up outside the chosen half range, where one expansion results in an odd function and the other in an even one. Expansion in a sine series yields an odd function; expansion in a cosine series yields an even function. Note that a_0 is not zero.

In general, we might want to find a Fourier series to represent $f(\theta)$ in any range $(0 < \theta < \alpha)$ or $(-\alpha < \theta < \alpha)$. There is no problem if $\alpha < \pi$, since we can then change the variable, thus

$$\theta = \frac{2\phi}{\pi}, \text{ so that } \phi = \frac{\pi\theta}{2} \tag{2.1.30}$$

Then

$$f(\theta) = f\left(\frac{2\phi}{\pi}\right) = g(\phi) \tag{2.1.31}$$

Taking the full range $-\alpha < \theta < \alpha$ as an example, we have $-\pi < \theta < \pi$, and using the previous results gives

$$g(\phi) = f(\theta) = a_0 + \sum a_n \cos n\phi + \sum b_n \sin n\phi \tag{2.1.32}$$

Then

$$\begin{aligned} a_0 &= \frac{1}{2\pi}\int_{-\pi}^{\pi} g(\varphi)\, d\varphi = \frac{1}{2\pi}\int_{-\pi}^{\pi} f\left(\frac{\alpha\varphi}{\pi}\right) d\varphi \\ &= \frac{1}{2\alpha}\int_{-\alpha}^{\alpha} f(\theta)\, d\theta \end{aligned} \tag{2.1.33}$$

Further,

$$a_n = \frac{1}{\pi}\int_{-\pi}^{\pi} f\left(\frac{2\varphi}{\pi}\right) \cos(n\varphi)\, d\varphi = \frac{1}{\alpha}\int_{-\alpha}^{\alpha} f(\theta) \cos\left(\frac{n\pi\theta}{\alpha}\right) d\theta \tag{2.1.34}$$

and

$$b_n = \frac{1}{\alpha}\int_{-\alpha}^{\alpha} f(\theta) \sin\left(\frac{n\pi\theta}{\alpha}\right) d\theta \tag{2.1.35}$$

Finally, the full expansion is

$$f(\theta) = a_0 + \sum_n a_n \cos\frac{n\pi\theta}{\alpha} + \sum_n b_n \sin\frac{n\pi\theta}{\alpha} \tag{2.1.36}$$

2.2 Complex Form of Fourier Series

Any finite, periodic function may be expanded as a series of sinusoids as in Eq. (2.1.2) If we now let

$$a_n = p_n + q_n \tag{2.2.1}$$

and

$$b_n = jp_n + q_n \tag{2.2.2}$$

then

$$\begin{aligned} f(\theta) &= a_n + \sum_{n=1}^{\infty} (p_n e^{jn\theta} + q_n e^{jn\theta}) \\ &= \sum_{n=1}^{\infty} c_n e^{jn\theta} \end{aligned} \tag{2.2.3}$$

If θ is a phase angle, then Eq. (2.2.3) states that any finite, periodic function can be expanded as a series of phasors. To evaluate the coefficient c_n, we multiply both sides of Eq. (2.2.3) by $e^{-jm\theta}$, and integrate from $-\pi$ to $+\pi$. Then if $n = m$

$$\int_{-\pi}^{\pi} f(\theta)e^{-jm\theta}\, d\theta = \sum_{n=-\infty}^{\infty} c_n \int_{-\pi}^{\pi} e^{j(n-m)\theta}\, d\theta = 2\pi c_n \tag{2.2.4}$$

Thus

$$c_n = \frac{1}{2\pi}\int_{-\pi}^{\pi} f(\theta)e^{-jn\theta}\, d\theta \tag{2.2.5}$$

The complex Fourier series may also be used in a general range, $-\alpha < \theta < \alpha$, by a change of variable. Let

$$\phi = \frac{\pi\theta}{\alpha} \tag{2.2.6}$$

Then θ lies in the range $-\pi < \theta < \pi$ and

$$f(\theta) = f\left(\frac{\alpha\phi}{\pi}\right) = \sum_{n=-\infty}^{\infty} c_n e^{jn\phi} \tag{2.2.7}$$

where

$$c_n = \frac{1}{2\pi}\int_{-\pi}^{\pi} f\left(\frac{\alpha\phi}{\pi}\right)e^{-jn\varphi}\, d\phi = \frac{1}{2\alpha}\int_{-\alpha}^{\alpha} f(\theta)e^{-jn\pi\theta/\alpha}\, d\theta \tag{2.2.8}$$

The complex Fourier series in the general range can be rewritten as follows:

$$f(\theta) = \sum_{n=-\infty}^{\infty} e^{jn\pi\theta/\alpha} \tag{2.2.9}$$

$$= \frac{1}{2\alpha}\sum_{n=-\infty}^{\infty}\int_{-\alpha}^{\alpha} f(\varphi)e^{jn\pi(\theta-\varphi)/\alpha}\, d\varphi \tag{2.2.10}$$

2.3 Fourier Integral Theorem

Physically realizable variables generally do not repeat in a periodic manner. In this case, the range of the Fourier integration becomes infinite. Then

$$f(\theta) = \sum_{\alpha\to\infty}\sum_{n=-\infty}^{\infty}\frac{1}{2\alpha}\int_{-\alpha}^{\alpha} f(\varphi)e^{jn\pi(\theta-\varphi)/\alpha}\, d\varphi \tag{2.3.1}$$

Let

$$\pi/\alpha = h \tag{2.3.2}$$

then

$$f(\theta) = \sum_{n=1}^{\infty} hg(nh) = \int_0^{\infty} g(k)\, dk \tag{2.3.3}$$

But

$$\lim_{h\to\theta} \sum_{n=1}^{\infty} hg(nh) = \int_0^{\infty} g(k)\, dk \tag{2.3.4}$$

from the definition of integration and as shown in Fig. 2.6. Let

$$g(nh) = g(k) = \int_{-\pi/h}^{\pi/h} f(\varphi) e^{jk(\varphi-\theta)}\, d\varphi \tag{2.3.5}$$

then from Eqs. (2.3.3) and (2.3.4)

$$\begin{aligned} f(\theta) &= \frac{1}{2\pi}\int_{-\infty}^{\infty} g(k)\, dk = \frac{1}{2\pi}\int_{-\infty}^{\infty} g(k)\, dk = \frac{1}{2\pi}\int_{-\infty}^{\infty} dk \int_{-\infty}^{\infty} f(\varphi) e^{jk(\varphi-\theta)}\, d\varphi \\ &= \frac{1}{2\pi}\int_{-\infty}^{\infty} e^{-jk\theta}\, dk \int_{-\infty}^{\infty} f(\varphi) e^{jk(\varphi-\theta)}\, d\varphi \end{aligned} \tag{2.3.6}$$

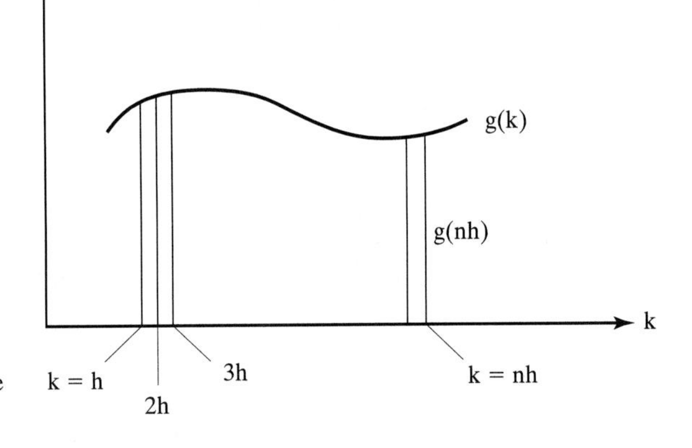

Fig. 2.6 The sum of the areas of the strips is equal to the integral.

From this result we can derive what is known as the Fourier integral theorem, namely, that any finite, single-valued function may be written as

$$f(\theta) = \frac{1}{\sqrt{2\pi}} \int_{-\infty}^{\infty} F(k)\, e^{-jk\theta}\, dk \tag{2.3.7}$$

where F(k) is called the complex Fourier transform of $f(\theta)$ and is given by

$$F(k) = \frac{1}{\sqrt{2\pi}} \int_{-\infty}^{\infty} f(\varphi)\, e^{jk\varphi}\, d\varphi \tag{2.3.8}$$

Note that the factor $1/2\pi$ can be associated with either of the integrals in the transform pair or partitioned between them as $1/\sqrt{2\pi}$, as in Eqs. (2.3.7) and (2.3.8).

Consider a certain function of time, f(t). From the Fourier integral theorem we may write

$$f(t) = \frac{1}{\sqrt{2\pi}} \int_{-\infty}^{\infty} F(\omega)\, e^{-j\omega t}\, d\omega \tag{2.3.9}$$

We have chosen to replace the dummy variable k with ω, the angular frequency. Eq. (2.3.9) can be interpreted to mean that any finite, single-valued function of time can be considered to be composed of a distribution of phasors with continuously varying frequency. $F(\omega)$ is then called the complex frequency spectrum of f(t). The squared amplitude of $F(\omega)$ is called the power spectral density.

Consider the Fourier transform pair

$$f(\theta) = \frac{1}{\sqrt{2\pi}} \int_{-\infty}^{\infty} F(k) e^{-jk\theta}\, dk \tag{2.3.10}$$

and

$$F(k) = \frac{1}{\sqrt{2\pi}} \int_{-\infty}^{\infty} f(\varphi) e^{jk\varphi}\, d\varphi \tag{2.3.11}$$

F(k) is the Fourier transform of $f(\theta)$ in the k domain, while $f(\theta)$ is the inverse Fourier transform of F(k) in the θ or φ domain.

Now we may also write

$$\begin{aligned} f(\theta) &= \frac{1}{2\pi} \int_{-\infty}^{\infty} \int_{-\infty}^{\infty} f(\varphi) e^{jk\varphi}\, d\varphi e^{-jk\theta}\, dk \\ &= \frac{1}{2\pi} \int_{-\infty}^{\infty} \int_{-\infty}^{\infty} f(\varphi) \cos k(\varphi - \theta)\, d\varphi dk \\ &\quad + \frac{j}{2\pi} \int_{-\infty}^{\infty} \int_{-\infty}^{\infty} f(\varphi) \sin k(\varphi - \theta)\, d\varphi dk \end{aligned} \tag{2.3.12}$$

Then reversing the order of integration,

$$f(\theta) = \frac{1}{2\pi}\int_{-\infty}^{\infty} f(\varphi)\int_{-\infty}^{\infty} \cos k(\varphi - \theta)\, dk d\varphi + \frac{j}{2\pi}\int_{-\infty}^{\infty} f(\varphi)\int_{-\infty}^{\infty} \sin k(\varphi - \theta)\, dk d\varphi \tag{2.3.13}$$

But $\sin k(\varphi - \theta)$ is an odd function, so the second integral vanishes. Then

$$\begin{aligned} f(\theta) &= \frac{1}{2\pi}\int_{-\infty}^{\infty}\int_{-\infty}^{\infty} f(\varphi) \cos k(\varphi - \theta)\, dk d\varphi \\ &= \frac{1}{\pi}\int_{0}^{\infty}\int_{-\infty}^{\infty} f(\varphi) \cos k(\varphi - \theta)\, dk d\varphi \\ &= \frac{1}{\pi}\int_{0}^{\infty}\int_{-\infty}^{\infty} f(\varphi) \cos k\varphi \cos k\theta\, dk d\varphi \\ &\quad + \frac{1}{\pi}\int_{0}^{\infty}\int_{-\infty}^{\infty} f(\varphi) \sin k\varphi \sin k\theta\, dk d\varphi \end{aligned} \tag{2.3.14}$$

Then if $f(\varphi)$ is odd:

$$f(\theta) = \frac{2}{\pi}\int_{0}^{\infty}\int_{-\infty}^{\infty} f(\varphi) \sin k\varphi \sin k\theta\, dk d\varphi \tag{2.3.15}$$

and if $f(\varphi)$ is even:

$$f(\theta) = \frac{2}{\pi}\int_{0}^{\infty}\int_{0}^{\infty} f(\varphi) \cos k\varphi \cos k\theta\, dk d\varphi \tag{2.3.16}$$

Eqs. (2.3.15) and (2.3.16) are known as Fourier sine and cosine integrals, respectively.

2.4 Pulses and Wavetrains

The Fourier integral can be used to analyze certain functions of considerable importance in acoustics. We will consider first the rectangular pulse (see Fig. 2.7):

$$\begin{aligned} f(\theta) &= 0,\ |\theta| > \frac{L}{2} \\ &= a,\ |\theta| < \frac{L}{2} \end{aligned} \tag{2.4.1}$$

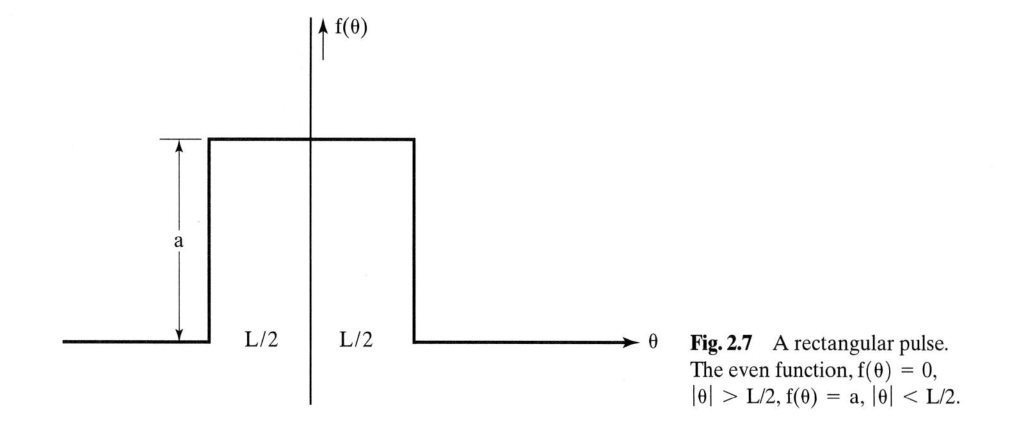

Fig. 2.7 A rectangular pulse. The even function, $f(\theta) = 0$, $|\theta| > L/2$, $f(\theta) = a$, $|\theta| < L/2$.

Since this is an even function, we will use the Fourier cosine integral given by Eq. (2.3.16) and obtain

$$f(\theta) = \frac{2}{\pi}\int_0^\infty F(k)\cos k\theta\, dk \tag{2.4.2}$$

where

$$\begin{aligned} F(k) &= \int_0^\infty f(\varphi)\cos k\varphi d\varphi \\ &= \int_0^{L/2} a\cos k\varphi d\varphi = a\frac{\sin kL/2}{2} \end{aligned} \tag{2.4.3}$$

The rectangular pulse is thus built up of an infinite set of cosine waves with amplitudes given by $2\varphi k/\pi$ having values of k ranging from 0 to ∞, as shown in Fig. 2.8.

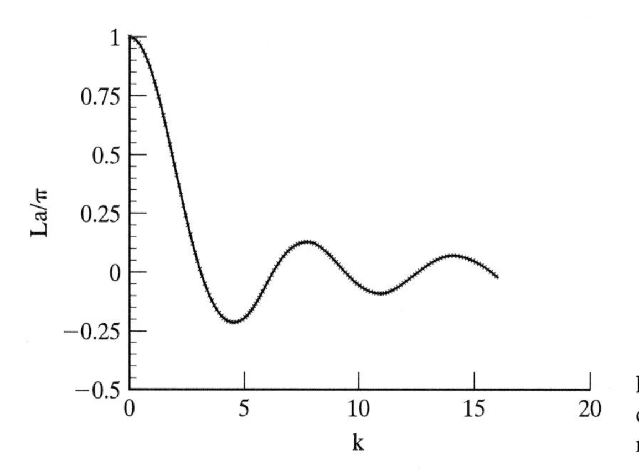

Fig. 2.8 Amplitude of the cosine waves of constant k required to produce the rectangular pulse of Fig. 2.7.

As a second example, we will consider the rectangular wave pulse, sometimes called a wave train or wavelet, as illustrated in Fig. 2.9:

$$\begin{aligned} f(\theta) &= 0, |\theta| > \frac{L}{2} \\ &= \cos k_0\theta, |\theta| < \frac{L}{2} \end{aligned} \tag{2.4.4}$$

This result is also an even function, and again we use the Fourier cosine integral from Eq. (2.3.16) so that

$$f(\theta) = \frac{2}{\pi}\int_0^\infty F(k)\cos k\theta\, dk \tag{2.4.5}$$

where

$$\begin{aligned} F(k) &= \int_0^\infty f(\varphi)\cos k\varphi\, d\varphi \\ &= \int_0^{L/2} \cos k_0\varphi \cos k\varphi\, d\varphi \end{aligned} \tag{2.4.6}$$

$$\begin{aligned} &= \frac{1}{2}\int_0^{L/2} [\cos(k_0 + k)\varphi + \cos(k_0 - k)\varphi]\, d\varphi \\ &= \frac{1}{2}\left[\frac{\sin(k_0 + k)L/2}{k_0 + k} + \frac{\sin(k_0 - k)L/2}{k_0 - k}\right] \end{aligned} \tag{2.4.7}$$

For a pulse of high-frequency sound,

$$\frac{\sin(k_0 + k)L/2}{k_0 + k} \to 0 \tag{2.4.8}$$

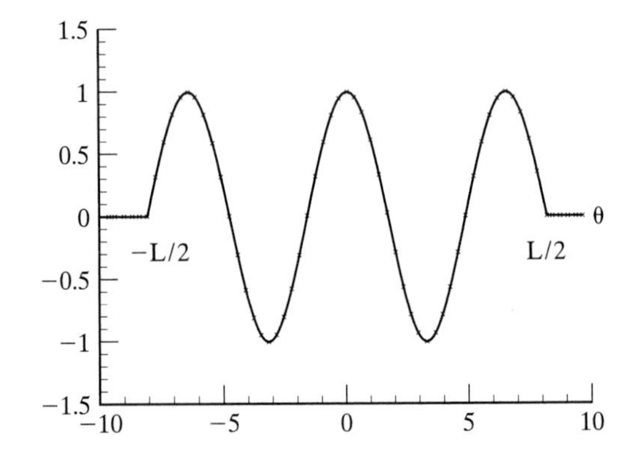

Fig. 2.9 Rectangular wave pulse. Even function.

$$\begin{aligned} f(\theta) &= 0, & |\theta| &> L/2 \\ &= \cos k_0\theta, & |\theta| &< L/2 \end{aligned}$$

Then,

$$F(k) = \frac{L}{4} \frac{\sin(k_0 - k)L/2}{(k_0 - k)L/2} \tag{2.4.9}$$

This result is shown in Fig. 2.10. Notice that the bulk of the Fourier components lies in the range

$$-\pi < \frac{(k_0 - k)}{2} < \pi \tag{2.4.10}$$

and to represent the wave pulse satisfactorily, we need a range $(k_0 - k)$ about k_0 such that

$$\Delta k \geq 2\pi/L \tag{2.4.11}$$

where $\Delta k = k_0 - k$. But the length of the pulse $L = c\tau$, where c is the sound velocity and τ is the pulse duration, thus

$$\Delta k = \frac{2\pi}{c\tau} \tag{2.4.12}$$

and

$$\frac{\Delta k}{k_0} = \frac{1}{f_0\tau} \tag{2.4.13}$$

where f_0 is the center frequency of the pulse. If the pulse is n cycles long,

$$\tau = n/f_0 \tag{2.4.14}$$

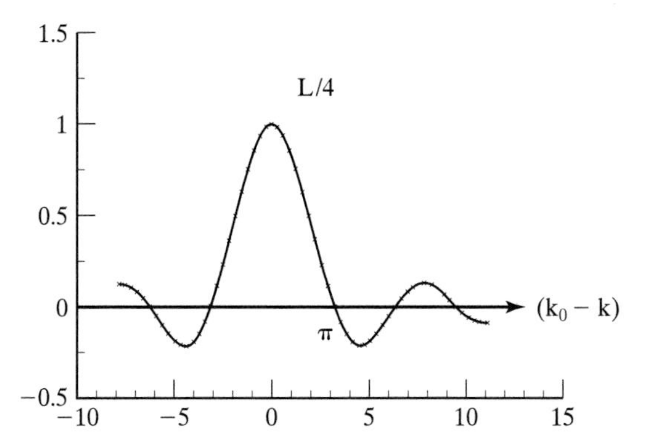

Fig. 2.10 Amplitude of the cosine waves of constant k required to produce the rectangular wave pulse of Fig. 2.9. Note that the abscissa is $(k_0 - k)$.

and

$$\frac{\Delta k}{k_0} = \frac{1}{n} = \frac{\Delta f}{f_0} \tag{2.4.15}$$

For example, consider a 100 kHz pulse, then

$$\Delta f = \frac{10^5}{n} \tag{2.4.16}$$

For $n = 5$, $\Delta f = 2.10^4$, and the frequency components must lie within 20 kHz of the pulse center frequency. However, for $n = 1$, $\Delta f = 10^5$, and the spread is from 0 to 200 kHz. In other words, the longer the pulse, the sharper its resolution in the frequency domain.

2.5 Phase and Group Velocity

The terms "wave package" or "group" are sometimes used to describe a pulse of the type we just discussed to convey the concept that the pulse in fact is composed of an assembly of harmonic waves. If the velocity of propagation depends on frequency, the velocity of the group will be different from the velocity of any one of its components. The velocity of a component is then called the "phase" velocity. This situation has been compared to the way a caterpiller moves by tucking in its tail and making a ripple move along its body to its head. The velocity of the ripple (the phase velocity) is greater than that of the whole caterpiller (the group). The problem can be analyzed using the Fourier integral theorem as follows.

Suppose that the phase velocity depends on frequency, so that

$$c = c(\omega) \tag{2.5.1}$$

or we might say that the frequency depends on wavelength or that

$$\omega = \omega(k) \tag{2.5.2}$$

A pulse will be given by

$$f(\theta) = \frac{1}{\sqrt{2\pi}} \int_{-\infty}^{\infty} F(k) e^{jk\theta}\, dk \tag{2.5.3}$$

Letting

$$\theta = x - ct \tag{2.5.4}$$

$$f(\theta) = \frac{1}{\sqrt{2\pi}} \int_{-\infty}^{\infty} F(k) e^{j(kx-\omega t)}\, dk \tag{2.5.5}$$

Assume the pulse is centered about a certain angular frequency ω_0 with an associated propagation constant k_0 and that only a finite range Δk on either side of k_0 is involved. Then we can make the following expansion:

$$\begin{aligned} kx - \omega t &= k_0 x - \omega_0 t + (\delta k)x - (\delta\omega)t \\ &= k_0 x - \omega_0 t + (\delta k)\left[x - \left(\frac{\delta\omega}{\delta k}\right)_0 t + \cdots\right] \end{aligned} \tag{2.5.6}$$

Now the integral in Eq. (2.5.5) will be constant if

$$x - \left(\frac{\delta\omega}{\delta K}\right)_0 t \tag{2.5.7}$$

Thus at points where

$$\frac{x}{t} = \left(\frac{\delta\omega}{\delta k}\right)_0 t \tag{2.5.8}$$

the wave package will appear to be the same; that is, the group will have a velocity

$$\begin{aligned} u &= \left(\frac{\partial\omega}{\partial k}\right)_0 = \left(\frac{\partial(ck)}{\partial k}\right)_0 = c_0 + k_0\left(\frac{\partial c}{\partial k}\right)_0 \\ &= c_0 + \frac{2\pi}{\lambda_0}\left(\frac{\partial c}{\partial\lambda}\right)_0\left(\frac{\partial\lambda}{\partial k}\right)_0 = c_0 + \frac{2\pi}{\lambda_0}\left(\frac{-2\pi}{\lambda_0^2}\right)^{-1}\left(\frac{\partial c}{\partial\lambda}\right)_0 \\ &= c_0 - \lambda_0\left(\frac{\partial c}{\partial\lambda}\right)_0 \end{aligned} \tag{2.5.9}$$

This result is important for acoustics, because it implies that waves of different frequency (or wavelength) will spread out as the package travels.

2.6 The Laplace Transform

The Laplace transform may be regarded as an extension of the Fourier integral theorem. Recollect that any periodic function can be written as a sum of phasors with frequencies that are multiples of the fundamental frequency of the function in question. Thus from Eq. (2.2.3):

$$f(t) = \sum_{n=-\infty}^{\infty} c_n e^{jn\omega t} \tag{2.6.1}$$

c_n is then the amplitude of the nth harmonic. If the function is not periodic, it may be shown that it is still possible to represent it as a sum of phasors, in this case as an integral of an infinite continuum of phasors, and we have from Eq. (2.3.9):

$$f(t) = \frac{1}{\sqrt{2\pi}} \int_{-\infty}^{\infty} F(\omega) e^{-j\omega t}\, d\omega \tag{2.6.2}$$

where $F(\omega)$ is the complex Fourier transform of $f(t)$ and is given by

$$F(\omega) = \frac{1}{\sqrt{2\pi}} \int_{-\infty}^{\infty} f(t) e^{j\omega t}\, dt \tag{2.6.3}$$

We have shown how various functions, such as rectanglar pulses and wave pulses, can be "Fourier analyzed" in this way. But certain functions present a difficulty when we attempt to do this; one is the unit step function (Fig. 2.11), which we will consider next.

The unit step function is defined by

$$u(t) = \begin{cases} 0, & t < 0 \\ 1, & t > 0 \end{cases} \tag{2.6.4}$$

The unit step function is said to be a one-sided function of time. For this case, the Fourier transform pair becomes "unilateral":

$$u(t) = \frac{1}{2\pi} \int_{-\infty}^{\infty} F(\omega) e^{-j\omega t}\, d\omega \tag{2.6.5}$$

and

$$F(\omega) = \int_{0}^{\infty} e^{j\omega t}\, dt \tag{2.6.6}$$

Note that for convenience we incorporate the factor $\frac{1}{2\pi}$ in Eq. (2.6.5).

Thus, for the unit step function,

$$F(\omega) = \left[\frac{e^{j\omega t}}{j\omega}\right]_0^{\infty} = \left[\frac{\cos \omega t + j \sin \omega t}{j\omega}\right]_0^{\infty} \tag{2.6.7}$$

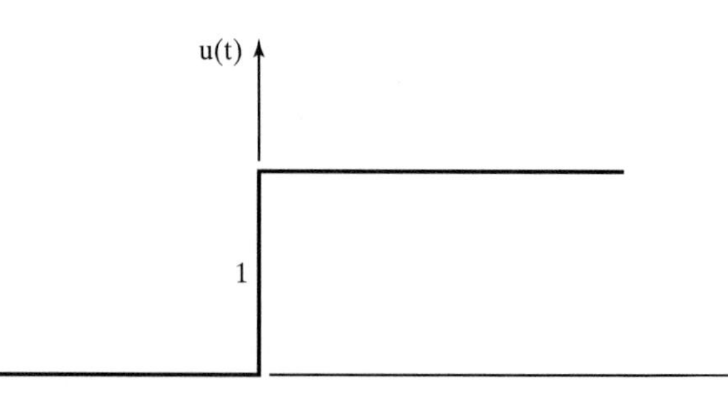

Fig. 2.11 The unit step function.

This result is meaningless because sine and cosine oscillate as t approaches infinity. The problem arises because our function has a value at infinity. To force it to approach zero at infinity, we try the device of multiplying the step function by an exponential decay. Hence

$$f(t) = e^{-\sigma t}u(t) \tag{2.6.8}$$

and

$$F(\omega) = \int_0^{\infty} e^{-\sigma t}e^{j\omega t}\,dt = \left[\frac{e^{-(\sigma - j\omega)t}}{-(\sigma - j\omega)}\right]_0^{\infty} = \frac{1}{\sigma - j\omega} \tag{2.6.9}$$

Now it follows that

$$f(t) = \frac{1}{2\pi}\int_{-\infty}^{\infty} \frac{e^{-j\omega t}}{\sigma - j\omega}d\omega \tag{2.6.10}$$

Rationalizing and separating odd and even functions, we obtain

$$f(t) = \frac{1}{\pi}\int_0^{\infty} \frac{\sigma \cos \omega t}{\sigma^2 + \omega^2}d\omega + \frac{1}{\pi}\int_0^{\infty} \frac{\omega \sin \omega t}{\sigma^2 + \omega^2}d\omega \tag{2.6.11}$$

Letting $\omega = \sigma z$ in the first integral gives

$$f(t) = \frac{1}{\pi}\int_0^{\infty} \frac{\cos(\sigma z t)}{1 + z^2}dz + \frac{1}{\pi}\int_0^{\infty} \frac{\omega \sin \omega t}{\sigma^2 + \omega^2}d\omega \tag{2.6.12}$$

Now let $\sigma \rightarrow 0$, then $f(t) \rightarrow u(t)$, and we obtain

$$u(t) = \frac{1}{2} + \frac{1}{\pi}\int_0^{\infty} \frac{\sin \omega t}{\omega}d\omega \tag{2.6.13}$$

Note that 1/2 is the "DC" term and $1/\omega$ is the Fourier transform.

We can generalize this procedure for any one-sided function of time f(t) by multiplying both sides of the Fourier transform pair by $e^{-\sigma t}$ so that

$$f(t)e^{-\sigma t} = \frac{1}{2\pi}\int_{-\infty}^{\infty} F(\omega)e^{-\sigma t}e^{-j\omega t}\,d\omega \tag{2.6.14}$$

and

$$F(\omega)e^{-\sigma t} = \int_{-\infty}^{\infty} f(t)e^{-\sigma t}e^{j\omega t}\,dt \tag{2.6.15}$$

The function defined by Eq. (2.6.15) is called the Laplace transform of f(t). It is usually written in terms of the so-called complex frequency or Laplace variable $s = \sigma - j\omega$.

Rewriting Eqs. (2.6.14) and (2.6.15):

$$f(t) = \frac{1}{2\pi}\int_{-\infty}^{\infty} L[f(t)]e^{st}\,ds \tag{2.6.16}$$

and

$$L[f(t)] = \int_{0}^{\infty} f(t)e^{-st}\,dt \tag{2.6.17}$$

The Laplace transform can be interpreted as prescribing the amplitude of a continuum of phasors with damping used in the representation of a one-sided function. It is named for Pierre Simon de Laplace (1749–1827), who first used it in developing the theory of probability. The Laplace transform is the basis of the operational calculus invented by Oliver Heaviside at the end of the nineteenth century. A rigorous discussion of the subject is beyond the scope of this book, but texts such as Wylie (1960) can be referred to for further details.

2.7 Simple Results with Laplace Transforms

1. The *exponential*, $e^{-\alpha t}$, has a Laplace transform given by

$$L[e^{-\alpha t}] = \int_{0}^{\infty} e^{-\alpha t}e^{-st}\,dt = \left[\frac{e^{-(s+\alpha)t}}{-(s+\alpha)}\right]_{0}^{\infty} = \frac{1}{s+\alpha} \tag{2.7.1}$$

2. For *cos bt*, say we have

$$L[\cos bt] = \int_{0}^{\infty} \cos bt\, e^{-st}\,dt = I \tag{2.7.2}$$

If we designate the value of the integral in Eq (2.7.2) as I, then the expression may be integrated by parts: By letting $u = \cos bt$ and $dv = e^{-st}$, we obtain

$$I = \left[\frac{-e^{-st}\cos bt}{s}\right]_{0}^{\infty} + \frac{b}{s}\left[\frac{e^{-st}\sin bt}{s}\right]_{0}^{\infty} - \frac{b^2}{s^2}\int_{0}^{\infty} e^{-st}\cos bt\,dt \tag{2.7.3}$$

But now we see that the integral on the right side is itself I, and thus

$$I\left(\frac{s^2+b^2}{s^2}\right) = \left[\frac{e^{-st}}{s^2}[b\sin bt - s\cos bt]\right]_{0}^{\infty} \tag{2.7.4}$$

and we finally obtain

$$I = L[\cos bt] = \frac{s}{s^2+b^2} \tag{2.7.5}$$

3. The *sine* function is treated in a way similar to the cosine, giving

$$L[\sin bt] = \frac{b}{s^2 + b^2} \tag{2.7.6}$$

4. The so-called *unit impulse* is the rectangular pulse shown in Fig. 2.12, which always has unit area. The Laplace transform is

$$L[f(t)] = \int_0^a \frac{1}{a} e^{-st}\,dt = \frac{1 - e^{-sa}}{as} \tag{2.7.7}$$

In the limit where the duration approaches zero (i.e., $a \to 0$), the unit impulse is then called a delta function, which may be defined by its value at a general time t_0,

$$\delta(t - t_0) = \begin{cases} 0, & t \neq t_0 \\ 1, & t = t_0 \end{cases} \tag{2.7.8}$$

The result for the Laplace transform is now indeterminate, and we have to use L'Hopital's rule:

$$L[f(t)] = \lim_{a\to 0} \frac{\frac{d}{da}[1 - e^{-sa}]}{\frac{d}{da}(as)} = 1 \tag{2.7.9}$$

5. The Laplace transform of a *derivative* is given by

$$L[f'(t)] = \int_0^\infty f'(t)e^{-st}\,dt \tag{2.7.10}$$

This result may be integrated by parts, letting $u = e^{-st}$ and $dv = f'(t)\,dt$, so that

$$L[f'(t)] = [e^{-st}f(t)]_0^\infty + \int_0^\infty f(t)e^{-st}\,dt = -f(0^+) + s\,L[f(t)] \tag{2.7.11}$$

This result is sometimes referred to as the initial value theorem, and can be shown to hold even if f (t) has discontinuities or if there are discontinuities in its derivative.

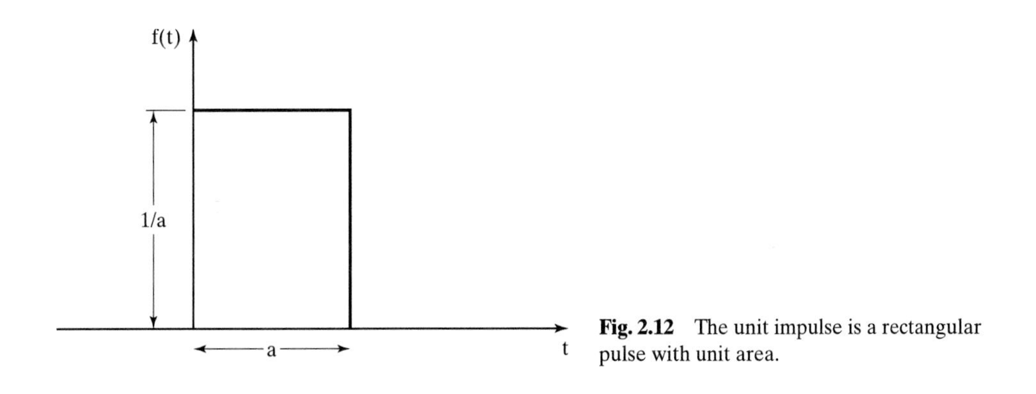

Fig. 2.12 The unit impulse is a rectangular pulse with unit area.

2.8 Transients: Impulse Response and Convolution

When a force is suddenly applied to an object, it will undergo a transient motion. Such transients are very important in a number of engineering applications. When we ride an elevator from one floor to the next, we would like it to accelerate and decelerate smoothly and come to rest without noticeable overshooting and oscillating. We would put similar requirements on the movement of a robot arm or an aileron on an aircraft. The Laplace transform is a powerful tool in the analysis of transient motion. We will begin with a simple example, the response of a mass-spring system to an impulse I (see Fig. 2.13). We will start out in a familiar way and then show how to introduce the Laplace transform.

An impulsive force occurring at time $t = 0$ can be described using a delta function as follows:

$$F(t) = f(t)\delta(t) \tag{2.8.1}$$

and hence the impulse given to the system,

$$I = \int_0^\infty F(t)\, dt = f(0) \tag{2.8.2}$$

But from Newton's second law, during the impact,

$$M\ddot{x} + R\dot{x} + Sx = F(t) \tag{2.8.3}$$

and after the impact,

$$M\ddot{x} + R\dot{x} + Sx = 0 \tag{2.8.4}$$

But this is the equation for a simple harmonic motion. The situation is similar to the one we discussed in Chapter 1 except that we then assumed the mass had been given an initial displacement. In this case we are assuming that the system starts at rest, and since the duration of the force is very short there will not be enough time to achieve

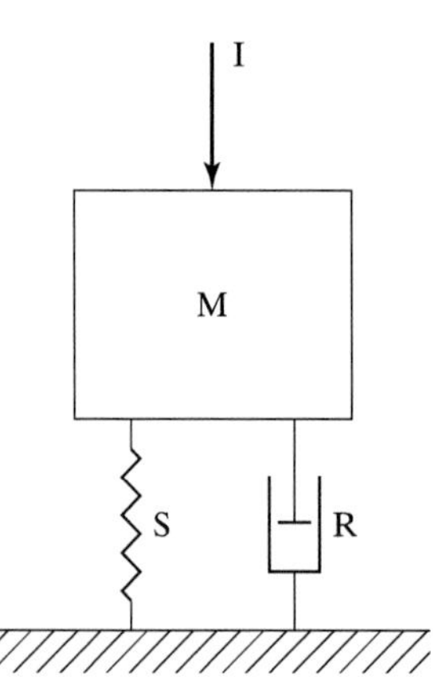

Fig. 2.13 A one degree of freedom system receives an impulse I. How will it respond?

any significant displacement. The effect of the impact will be to produce a change in momentum, so that in a very short time the mass will achieve a velocity given by

$$v_0 = \frac{1}{M} \tag{2.8.5}$$

If we now follow the same procedure as we did in Section 1.2.5, namely, to assume a solution for the equation of motion Eq. (2.8.4) such as

$$v = v_0 e^{-\alpha t} e^{j\omega_D t} \tag{2.8.6}$$

on substitution of Eq. (2.8.6) into Eq. (2.8.4), we find as before that

$$\alpha = \frac{R}{2M} \tag{2.8.7}$$

and

$$\omega_D^2 = \frac{S}{M} - \frac{R^2}{4M^2} \tag{2.8.8}$$

Eq. (2.8.6) could be called the impulse response of the one degree of freedom system.

We can arrive at the same result by another way. Consider the full equation of motion Eq. (2.8.3). Both sides of this equation are functions of time, and thus we may take the Laplace transform of both sides. Before the impulse begins, both displacement and velocity will be zero, so that using the results of Section 2.7 we have

$$[s^2 M + Rs + S]\, L[x] = I \tag{2.8.9}$$

Hence

$$L[x] = \frac{I}{[s^2 M + Rs + S]} = \frac{I/M}{[s^2 + rs + S/M]} \tag{2.8.10}$$

$$= \frac{I/M}{\left(s + \frac{r}{2} + \lambda\right)\left(s + \frac{r}{2} - \lambda\right)} \tag{2.8.11}$$

where $r = 2\alpha$ and $\lambda = \sqrt{\frac{r^2}{4} - \frac{S}{M}}$

By using partial fractions, we can now show that

$$L[x] = \frac{jI/(2M\omega_D)}{[s + \alpha + j\omega_D]} - \frac{jI/(2M\omega_D)}{[s + \alpha - j\omega_D]} \tag{2.8.12}$$

Hence, from Eq. (2.7.1),

$$x(0) = \frac{Ie^{-\alpha t} \sin \omega_D t}{M\omega_0} \tag{2.8.13}$$

Finally, differentiating to obtain the velocity:

$$\dot{x} = \frac{I}{M} e^{-\alpha t} \cos(\omega_D t) - \alpha x \tag{2.8.14}$$

The impulse response is important because any arbitrary excitation f(t) can be considered a series of impulses. We can again use the delta function to describe this. If we multiply the arbitrary excitation by $\delta(t - t_0)$ and integrate, we will obtain the impulse at t_0, thus

$$f(t_0) = \int_0^{\infty} f(t)\delta(t - t_0)\, dt \tag{2.8.15}$$

Hence, if we know the impulse response of the system, $g(t - t_0)$ at any time after an impulse at t_0, then we can calculate the response at time t due to the entire arbitrary excitation, up to that time, by integrating

$$x(t) = \int_0^t f(\xi) g\left(t - \xi\right) d\xi \tag{2.8.16}$$

This is called a convolution integral. There is a useful theorem stating that the Laplace transform of the convolution of two functions is equal to the product of the Laplace transforms of the two functions. The proof proceeds as follows. Let the two functions be f(t) and g(t). The Laplace transform of their convolution is then

$$Q = L\left[\int_0^t f(\xi)g(t - \xi)\, d\xi\right] = \int_0^{\infty}\left[\int_0^t f(\xi)g(t - \xi)\, d\xi\right]e^{-st}\, dt \tag{2.8.17}$$

Now

$$u(t - \xi) = \begin{cases} 1, t > \xi \\ 0, t < \xi \end{cases} \tag{2.8.18}$$

hence

$$f(\xi)g(t - \xi)u(t - \xi) = \begin{cases} f(\xi)g(t - \xi), t > \xi \\ 0, t < \xi \end{cases} \tag{2.8.19}$$

Thus, the inner integral in Eq. (2.8.17) can be extended to infinity by inserting the factor $u(t - \xi)$ in the integrand; that is,

$$Q = \int_0^\infty \left[\int_0^t f(\xi) g(t - \xi) u(t - \xi)\, d\xi \right] e^{-st}\, dt \tag{2.8.20}$$

Reversing the order of integration, noting that $f(\xi)$ does not depend on t,

$$Q = \int_0^\infty f(\xi) \left[\int_0^\infty g(t - \xi) u(t - \xi) e^{-st}\, dt \right] d\xi = \int_0^\infty f(\xi) \left[\int_\lambda^\infty g(t - \xi) e^{-st}\, dt \right] d\xi \tag{2.8.21}$$

Now let $t - \xi = \tau$, so that $dt = d\tau$, and then

$$\begin{aligned} Q &= \int_0^\infty f(\xi) \left[\int_\lambda^\infty g(\tau) e^{-s(\tau+\xi)}\, d\tau \right] d\xi = \int_0^\infty f(\xi) e^{-s\xi} \left[\int_\lambda^\infty g(\tau) e^{-s\tau}\, d\tau \right] d\xi \\ &= \left[\int_0^\infty f(\xi) e^{-s\xi}\, d\xi \right] \left[\int_\lambda^\infty g(\tau) e^{-s\tau}\, d\tau \right] = l[f(t)] l[g(t)] \end{aligned} \tag{2.8.22}$$

The convolution integral is often written

$$x(t) = \int_\lambda^\infty f(\xi) g(t - \xi)\, d\xi = f(t) * g(t) \tag{2.8.23}$$

and thus the convolution theorem may be expressed as follows:

$$L[f(t)] L[g(t)] = f(s) g(s) = L[f(t) * g(t)] = L[g(t) * f(t)] \tag{2.8.24}$$

2.9 Use of Laplace Transforms in Solving Equations

Suppose that we have the equation

$$ay'' + by' + cy = f(t) \tag{2.9.1}$$

Take the Laplace transform of both sides:

$$a\, L(y'') + b\, L(y') + c\, L(y) = L[f(t)] \tag{2.9.2}$$

Hence, from Eq. (2.7.11),

$$a\{s^2 L(y) - sy_0 - y'_0\} + b\{sL(y) - y_0\} + cL(y) = L[f(t)] \tag{2.9.3}$$

where y'_0 and y_0 are the initial values of y' and y. Solving for L (y):

$$L(y) = \frac{L[f(t)] + (as + b)y_0 + ay'_0}{as^2 + bs + c} \tag{2.9.4}$$

Now f(t) is a given function of t, hence we can find its Laplace transform from tables, and we can find L (y) from Eq. (2.9.4). We can then finally obtain y, the solution of Eq. (2.9.1), from Laplace transform tables.

But any physical system may be thought of as a device whereby an input is transformed to an output. Setting up the differential equations for the system, and assuming zero initial conditions when f(t) begins to act, solving for the Laplace transform of the output y (t):

$$L(y(t)) = \frac{L[f(t)]}{z(s)} \tag{2.9.5}$$

Here $1/z(s) = w(s)$ and is called a transfer function.

Most engineering systems can be represented by differential equations, so this procedure has wide applicability. It is often convenient to think of what transpires with the system as a process, which can be represented as a "black box," with an input and an output as shown in Fig. 2.14. The input is f(t), or what is equivalent, its Laplace transform, L[f(t)], or f (s), the function in the "Laplace domain." The output is y (s), in the Laplace domain. Some of this nomenclature arose among electrical engineers, but it has been adopted in many other disciplines, from computer science to political science. In a way, it is a modern version of the old idea of cause and effect in physics, or of stimulus and response from biology. Frequently, it is not clear which physical variable is the cause and which is the effect (as shown in Fig. 2.15).

The transfer function may also be regarded as a generalization of the concept of admittance, the ratio of current (output) and voltage (input) in an electrical circuit. A circuit with a high admittance will experience a high current value when subjected to a given voltage. The reciprocal of the transfer function is equivalent to the impedance of the circuit. The voltage and current are measured at a given port of the circuit. The concept of

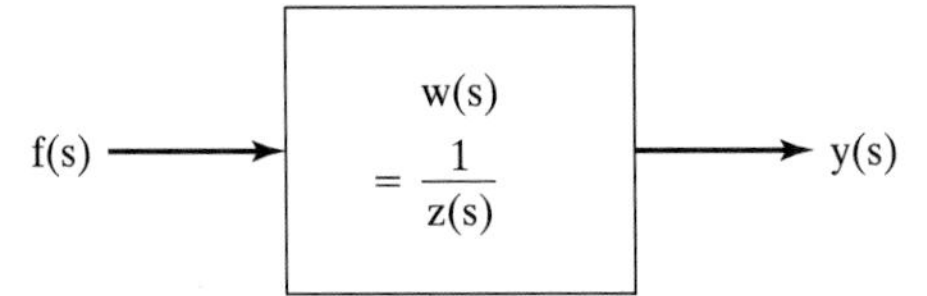

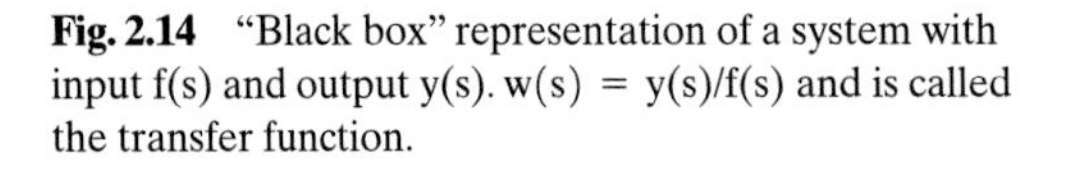
Fig. 2.14 "Black box" representation of a system with input f(s) and output y(s). $w(s) = y(s)/f(s)$ and is called the transfer function.

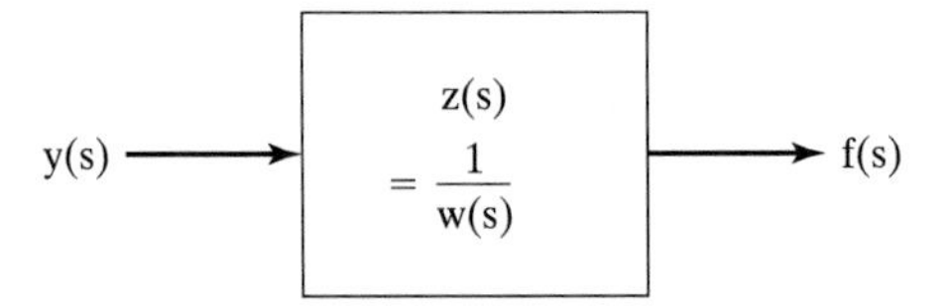

Fig. 2.15 If y(s) is regarded as the input, then f(s) is the output.

mechanical impedance was first introduced by Webster, who defined it as

$$z_M = \frac{F}{v} \tag{2.9.6}$$

where F is the force acting on a mechanical system, and v is the velocity, measured at the point of application of the force.

In acoustical systems we define an acoustic impedance as given by

$$z_A = \frac{p}{vA} \tag{2.9.7}$$

Another quantity that is sometimes found useful is the specific acoustic impedance

$$z_{AS} = \frac{P}{v} \tag{2.9.8}$$

In the case of a piston oscillating in a pipe, the impedance may be described using any of the above definitions. Thus, mechanical impedance:

$$z_M = \frac{F}{v} \tag{2.9.9}$$

acoustic impedance:

$$z_A = \frac{p}{vA} = \frac{F}{A^2 v} = z_M A^2 \tag{2.9.10}$$

and the specific acoustic impedance:

$$z_{As} = \frac{p}{v} = A z_A = \frac{z_M}{A} \tag{2.9.11}$$

We can also use the idea of a transfer function to link the response at a remote point in a system to a stimulus at a proximate point.

2.10 Stability

One of the most important applications of the transfer function technique is in the analysis of the stability of systems. A stable system has a transient response that dies away with time. However, systems with transient responses that increase with time can exist and they are said to be unstable. Obviously, an unstable response in a system such as a control in an aircraft could be disastrous. The subject of controls and stability analysis, which is closely associated with the names of Nyquist and Bode, grew in importance and extent in the past century. Here, we can discuss only briefly how this topic is tackled.

Assume that the transfer function of the system can be expressed as a ratio of polynomials in s, as in Eq. (2.9.5). We will make the argument for the general case of an nth order system. This implies that the characteristic equation is formed by setting the denominator equal to zero, so that by writing it in factored form we have

$$D(s) = \prod_{i=1}^{n} (s - r_i) \tag{2.10.1}$$

The roots r_i may be real, complex, or zero. If the roots are complex, they will always occur in conjugate pairs. It follows then that the output of the system can be written in the Laplace domain as follows:

$$y(s) = \frac{C_1}{(s - r_1)} + \frac{C_2}{(s - r_2)} + \cdots + \frac{C_n}{(s - r_n)} \tag{2.10.2}$$

Reverting to the time domain, we obtain the various transient modes of the system:

$$y(t) = C_1 e^{r_1 t} + C_2 e^{r_2 t} + \cdots + C_n e^{r_n t} \tag{2.10.3}$$

The shape of the mode is determined by the location of the root in the complex plane, while the value of the coefficients C_i depends on the input and the numerator polynomial. For stable systems, the transients die away with time; for unstable ones, they grow to infinity. For stability, every mode must contain a negative exponential. The various possibilities are illustrated in Figs. 2.16 and 2.17.

2.11 Transfer Functions of Simple Systems

We will now apply the concepts of the previous sections to the modeling of physical systems, starting with the familiar simple elements.

1. Mass

If the mass represented in Fig. 2.18a is acted on by a force f(t), then

$$f(t) = M\ddot{x}(t) \tag{2.11.1}$$

Taking the Laplace transform of both sides, assuming a zero initial value, we have

$$f(s) = Ms^2 x(s) \tag{2.11.2}$$

from which it follows that the transfer function is

$$w(s) = \frac{x(s)}{f(s)} = \frac{1}{Ms^2} \tag{2.11.3}$$

which may be represented as a block diagram as in Fig. 2.18b. Alternatively, we may regard the displacement as the input and the force as the output, with a block diagram as in Fig. 2.18c.

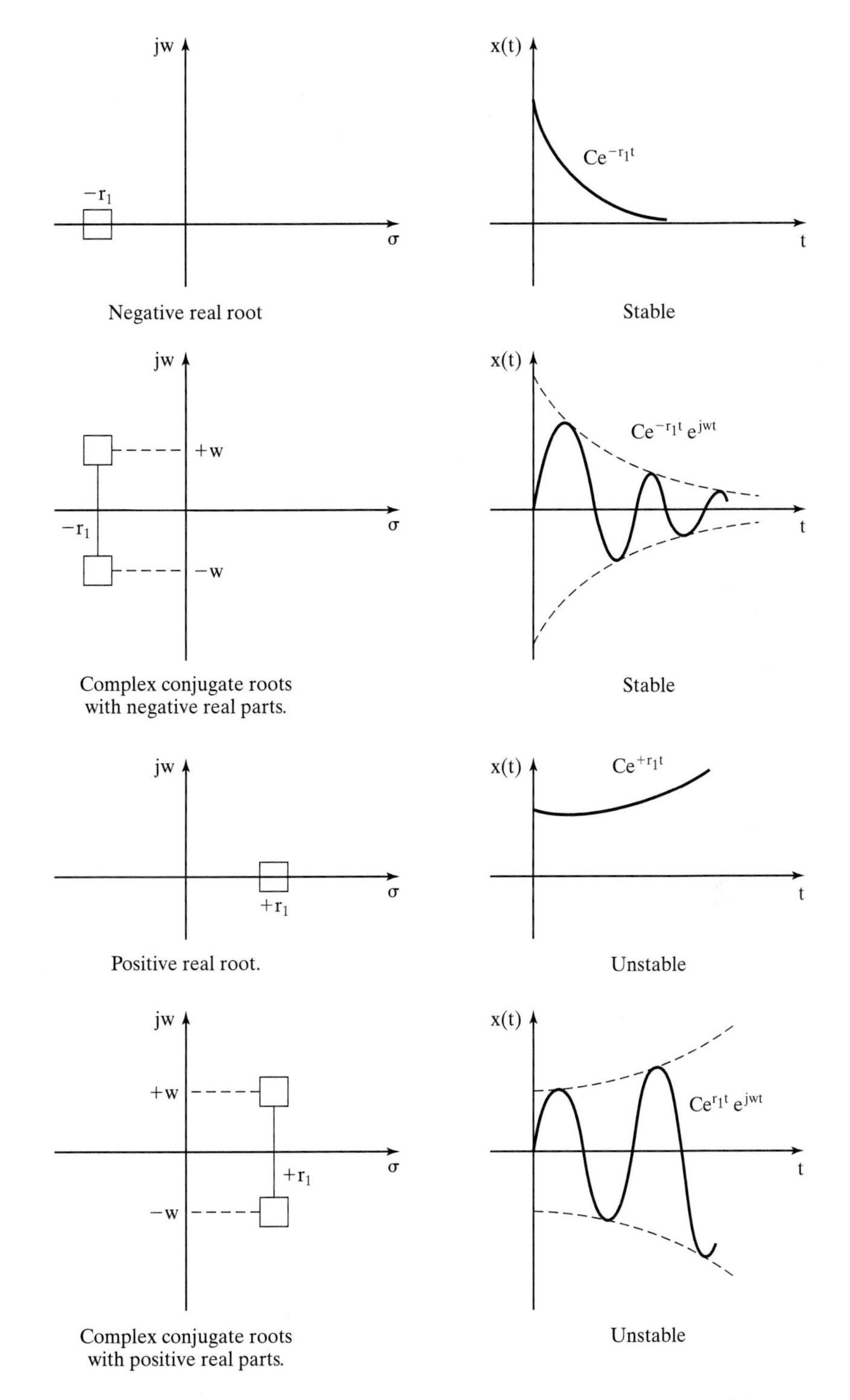

Fig. 2.16 Location of roots on the complex plane and associated mode shapes. (After Hale, 1988.)

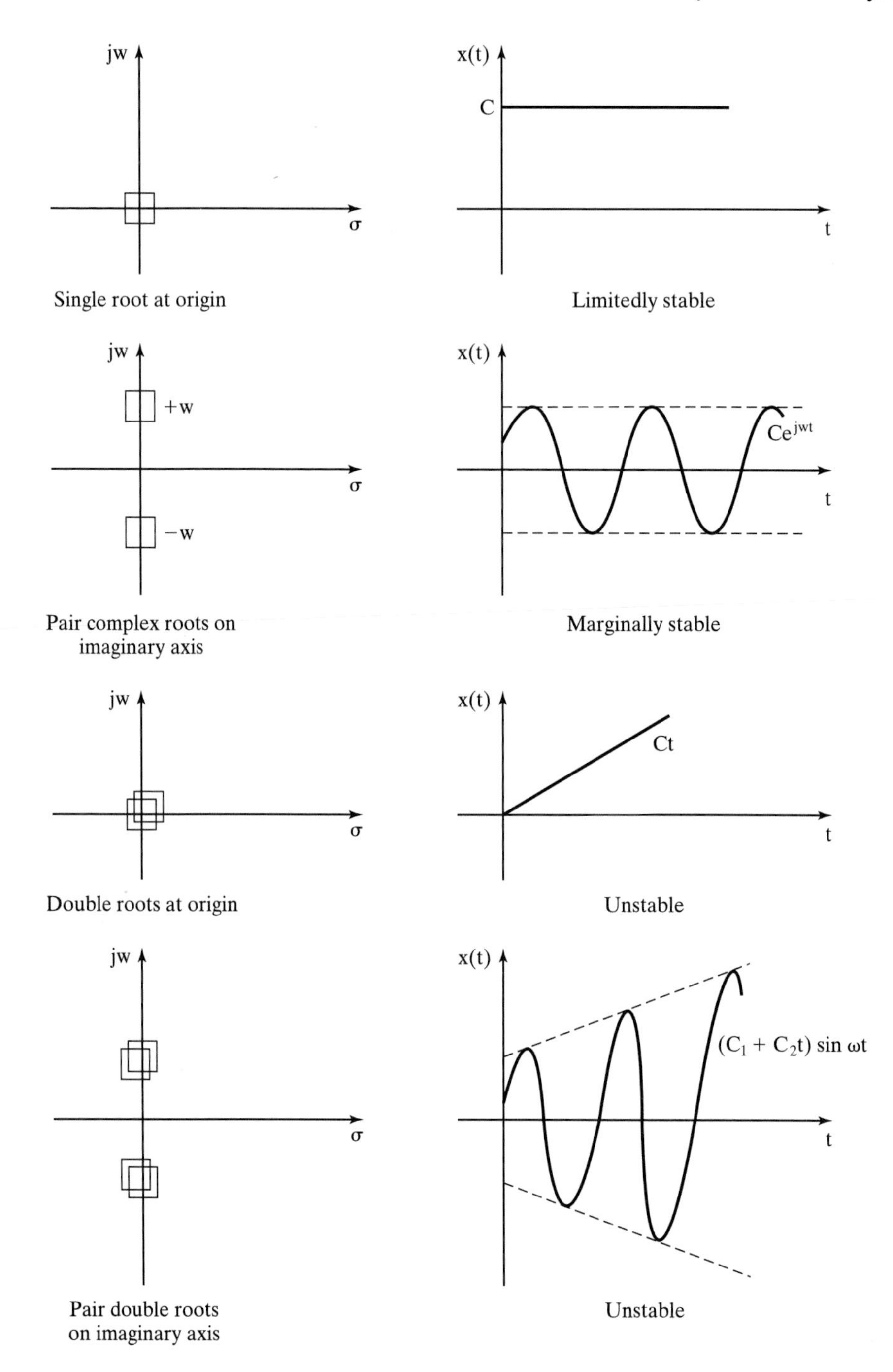

Fig. 2.17 Location of roots on the complex plane and associated mode shapes. (After Hale, 1988.)

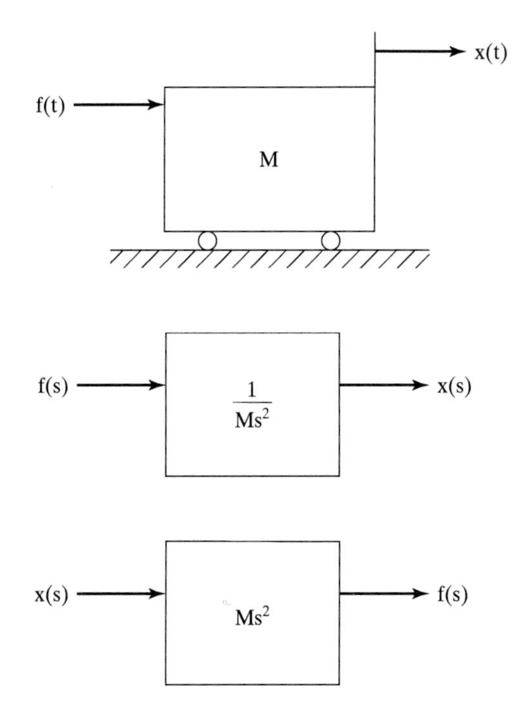

Fig. 2.18a Mass M is acted on by force f(t).

Fig. 2.18b Transfer function with force regarded as input.

Fig. 2.18c Transfer function with displacement regarded as input.

2. Spring

For the spring shown in Fig. 2.19a, if acted on by a force f(t), then

$$f(t) = S\,x(t) \tag{2.11.4}$$

Taking the Laplace transform, assuming a zero initial value, we have

$$f(s) = S\,x(s) \tag{2.11.5}$$

from which it follows that the transfer function is

$$w(s) = \frac{x(s)}{f(s)} = \frac{1}{S} \tag{2.11.6}$$

which may be represented as a block diagram as in Fig. 2.19b, or, regarding the displacement as the input and the force as the output, as in Fig. 2.19c.

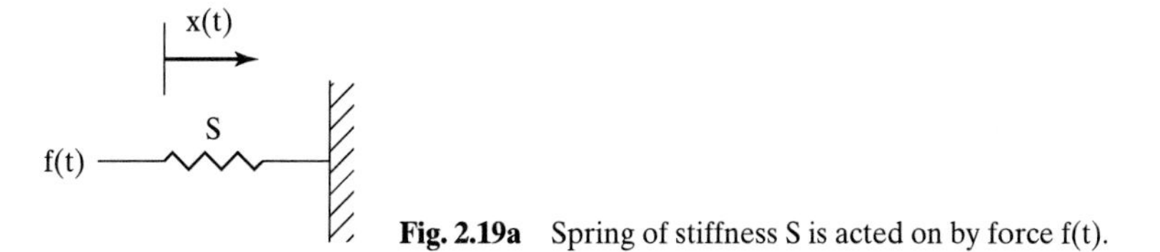

Fig. 2.19a Spring of stiffness S is acted on by force f(t).

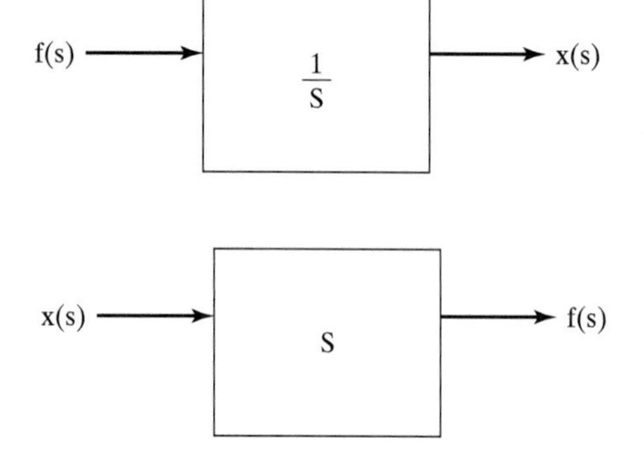

Fig. 2.19b Transfer function with force regarded as input.

Fig. 2.19c Transfer function with displacement regarded as input.

3. Damper
   With a very similar procedure, we have for the damper

$$f(t) = R\dot{x}(t) \tag{2.11.7}$$

$$f(s) = Rs\, x(s) \tag{2.11.8}$$

and transfer function

$$w(s) = \frac{x(s)}{f(s)} = \frac{1}{Rs} \tag{2.11.9}$$

with block diagram representation as in Fig. 2.20.

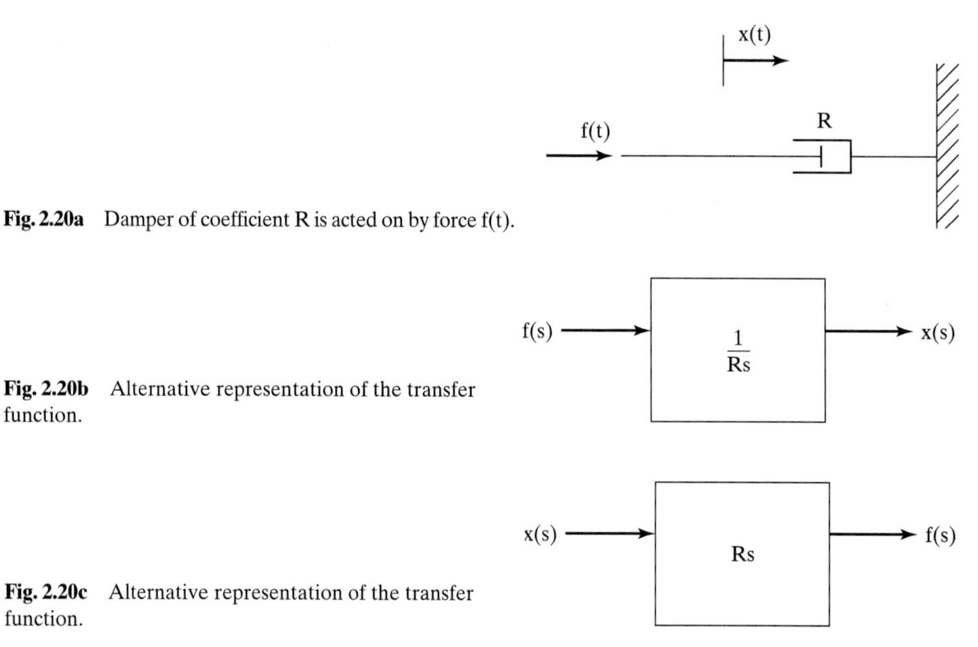

Fig. 2.20a Damper of coefficient R is acted on by force f(t).

Fig. 2.20b Alternative representation of the transfer function.

Fig. 2.20c Alternative representation of the transfer function.

4. One Degree of Freedom System
Fig. 2.21a shows a one degree of freedom system combining the three elements discussed above, and for which the equation of motion is

$$f(t) = M\ddot{x}(t) + R\dot{x}(t) + S\,x(t) \tag{2.11.10}$$

Taking the Laplace transform of both sides of this equation gives

$$f(s) = [Ms^2 + Rs + S]\,x(s) \tag{2.11.11}$$

and

$$w(s) = \frac{x(s)}{f(s)} = \frac{1}{[Ms^2 + Rs + S]} \tag{2.11.12}$$

Hence the block diagram must be as shown in Fig. 2.21b. If we regard the transmitted force as the output, the block diagram becomes that shown in Fig. 2.22. It is interesting to note that the transfer function can also be constructed from the block diagram of the elements. Thus, putting the individual elements together, we have the diagram in Fig. 2.23a.

The presence of the spring and damper produce feedback. But a system with a feedback loop is equivalent to a simple feed forward system, as shown in Fig. 2.23b, so that Fig. 2.23a reduces in two stages to Figs. 2.23c and 2.23d, which is identical to Fig. 2.21b.

The treatment of simple electrical elements with zero initial conditions proceeds along lines precisely similar to those used for their mechanical counterparts. Thus, for a

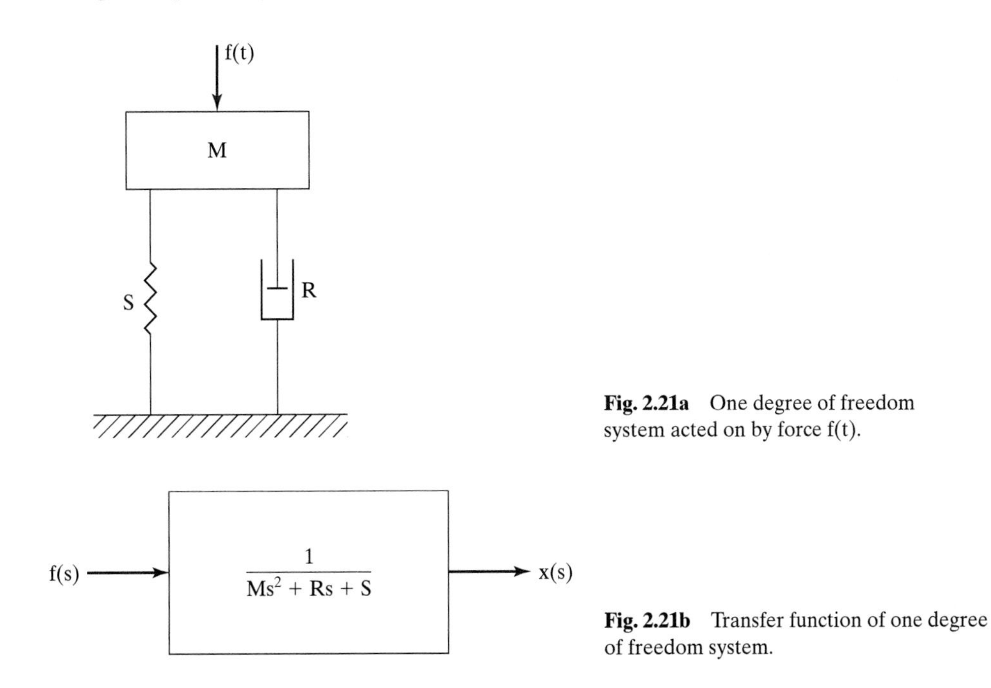

Fig. 2.21a One degree of freedom system acted on by force f(t).

Fig. 2.21b Transfer function of one degree of freedom system.

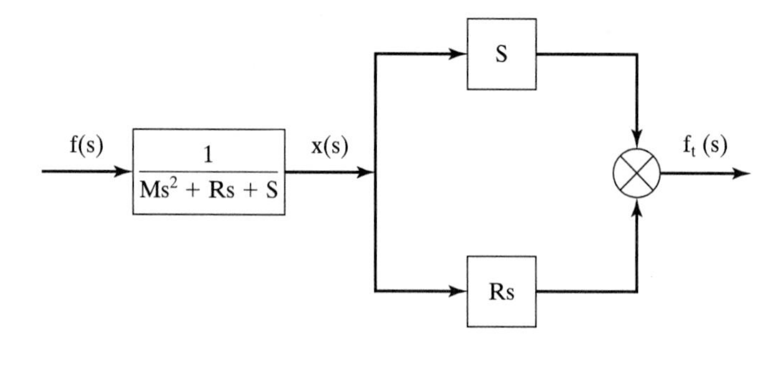

Fig. 2.22 Block diagram to obtain transmitted force $f_t(s)$.

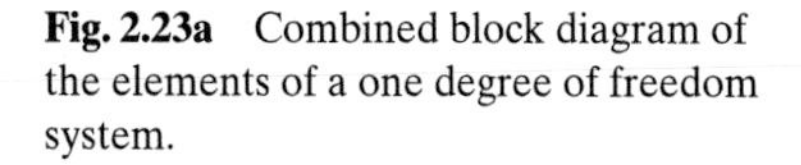

Fig. 2.23a Combined block diagram of the elements of a one degree of freedom system.

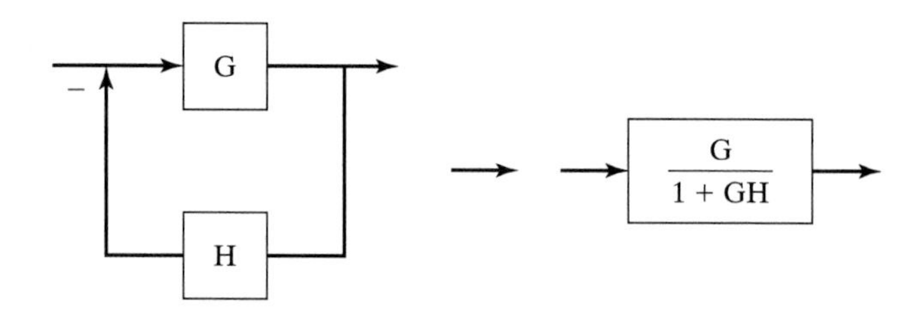

Fig. 2.23b A system with a feedback loop is equivalent to a feed forward system.

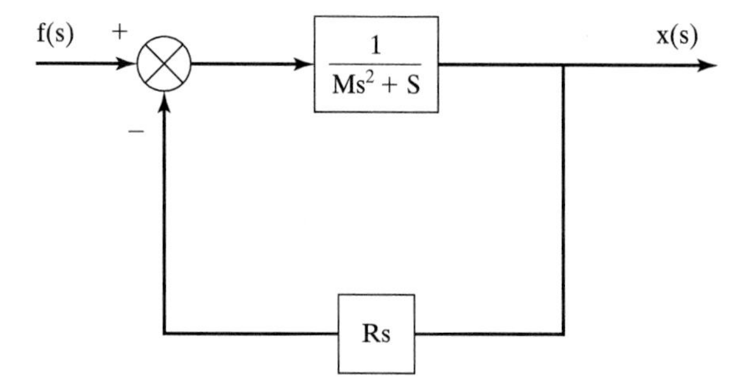

Fig. 2.23c Reduction of 2.23a using 2.23b.

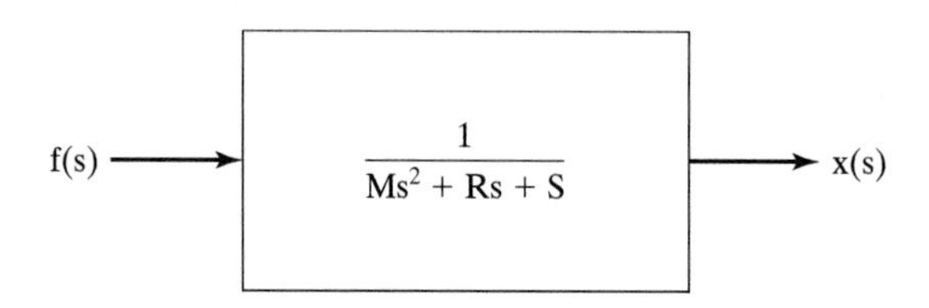

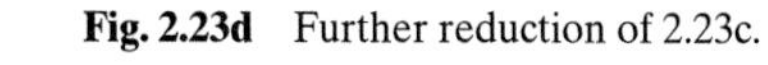

Fig. 2.23d Further reduction of 2.23c.

resistor we have

$$E(t) = RT(t) \tag{2.11.13}$$

or

$$E(s) = RT(s) \tag{2.11.14}$$

For the capacitor,

$$I(t) = C\dot{E}(t) \tag{2.11.15}$$

and

$$I(s) = CsE(s) \tag{2.11.16}$$

While for the inductor,

$$E(t) = L\dot{I}(t) \tag{2.11.17}$$

and

$$E(s) = LsI(s) \tag{2.11.18}$$

The transfer functions of these elements are shown in Fig. 2.24. For the series combination of the three elements:

$$E(t) = RI(t) + L\dot{I}(t) + \frac{q(t)}{C} \tag{2.11.19}$$

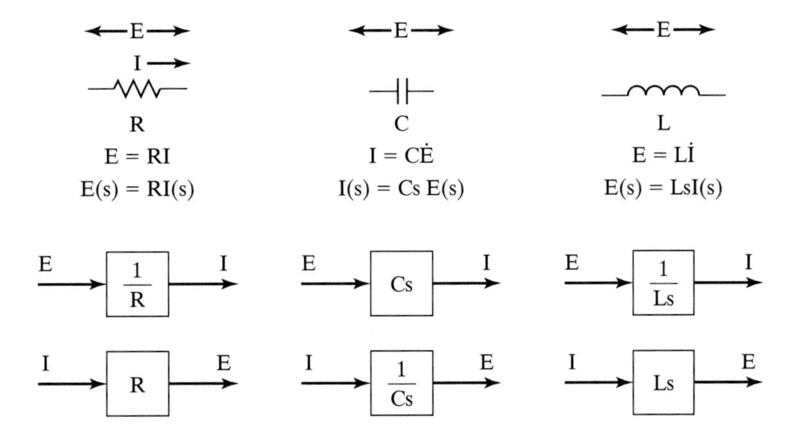

Fig. 2.24 Electrical elements and their transfer functions.

and

$$E(s) = RI(s) + LsI(s) + \frac{I(s)}{sC} \tag{2.11.20}$$

The transfer function is thus

$$\frac{I(s)}{E(s)} = \frac{s}{Ls^2 + Rs + \frac{1}{C}} \tag{2.11.21}$$

The block diagram representations are shown in Fig. 2.25. We can make the analogy with the mechanical system exact if we treat q, the charge, as the output, in which case the transfer function becomes

$$\frac{q(s)}{E(s)} = \frac{1}{Ls^2 + Rs + \frac{1}{C}} \tag{2.11.22}$$

which is illustrated by the block diagram of Fig. 2.26.

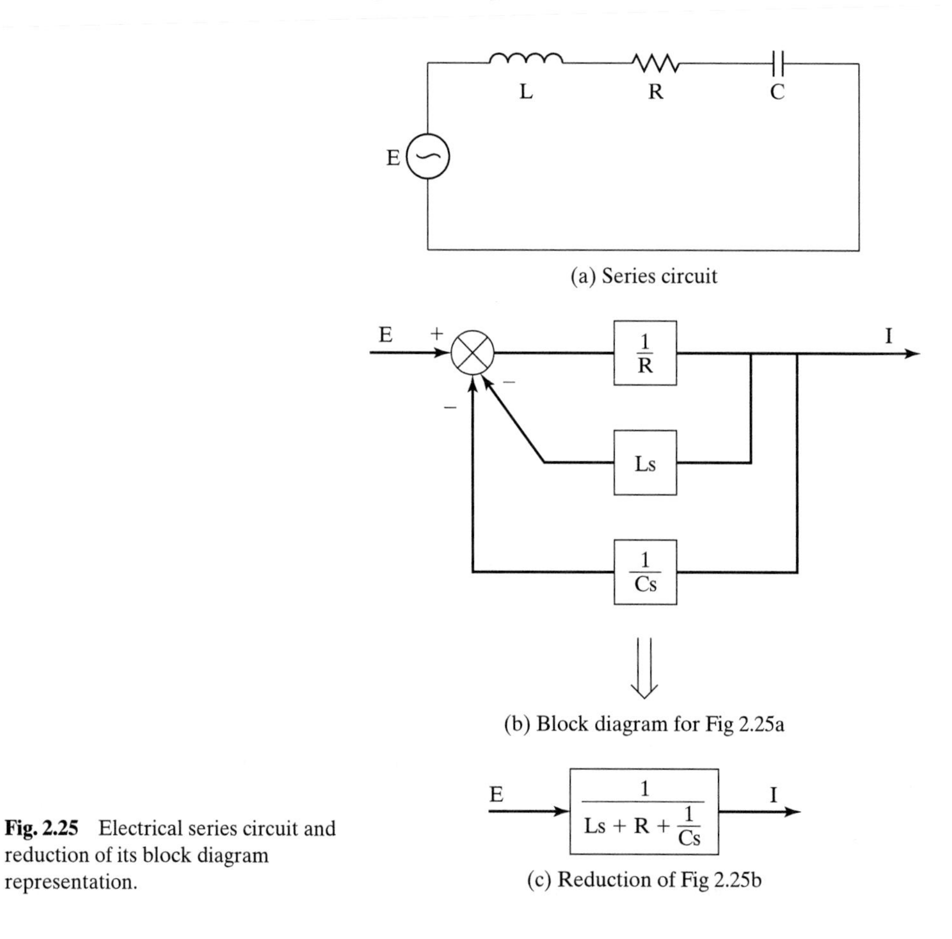

Fig. 2.25 Electrical series circuit and reduction of its block diagram representation.

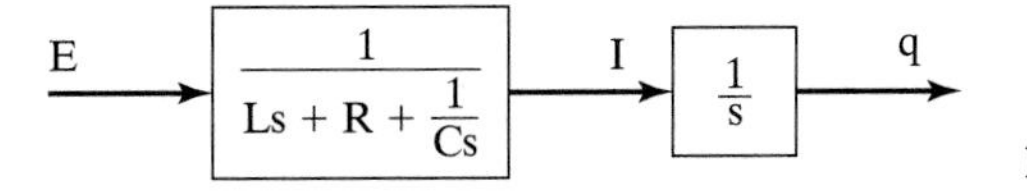

Fig. 2.26 Block diagram regarding charge q as output.

2.12 Linear Electromechanical Transducer

Block diagram analysis is particularly helpful in analyzing systems that combine electrical and mechanical elements, such as certain transducers. A transducer is a device that converts energy from one form to another. Of the devices of interest, by far the most important class is the electromechanical transducers, although some hydropneumatic sound sources, such as whistles and sirens, are in use. There are also devices involving heat energy in one stage of the transduction, such as hot wire microphones or the Rijke tube, but these are not often encountered. The major subclasses of transducers in vibrations and acoustics are shown in Fig. 2.27. Most electroacoustic transducers are reversible (i.e., they can be used to convert electrical to acoustic energy and vice versa), but there are some irreversible devices, most notably the carbon microphones used for many years in telephone handsets. The reversible electroacoustic transducers are subdivided according to whether they are linked by magnetic or electric flux. In the former category we have the electrodynamic devices (moving coil, ribbon, and wire) and the magnetostrictive class. The former depend on the force felt by a current carrying conductor in a magnetic field and the back electromotive force generated by magnetic flux cutting a conductor. The magnetostrictive effect is the slight change in length experienced by a magnetic material during magnetization, a phenomenon found both

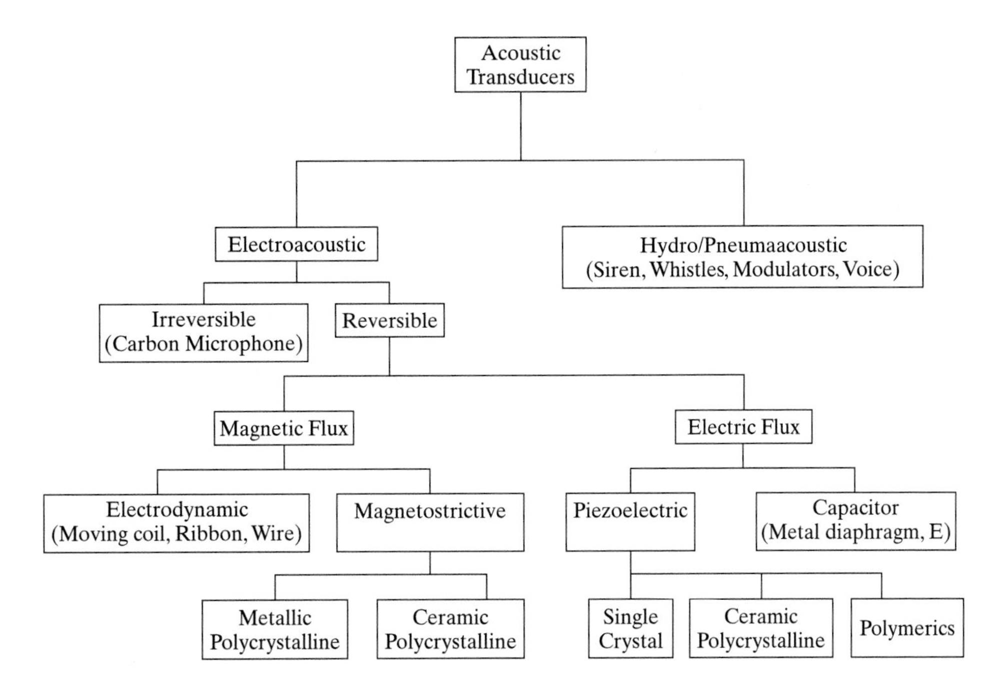

Fig. 2.27 Classification of Acoustic Transducers.

in metallic ferromagnets and in the ceramic materials known as ferrites. The electric-flux-linked family is divided into capacitor types and piezoelectrics. The piezoelectric effect—or change in length due to imposition of a voltage and its inverse—is found in single crystals, polycrystalline ceramics, and polymers. These mechanisms will be discussed in more detail in Chapter 7.

Since our greatest concern is with electroacoustic transducers, we will start by reviewing the basic concepts of electrical and mechanical systems, which are fundamental to all linear theories of electroacoustic transduction. Such a transducer is in fact the simplest form of electromagnetic system. Motors and generators are forms of transducers, although we do not usually think of them as connected to acoustics. However, when motors are used to produce limited movements—as in the movement of a robot arm—the main design problem is avoiding undesirable transient vibrations. Similarly, we often use an electromechanical device to produce a linear motion, in which case the device is called an actuator, and again a primary consideration is the control of the motion to avoid transients. Control theory offers an excellent entry into the modern theory of transducers.

Fig. 2.28 is a schematic representation of an electromechanical transducer. The right side is the series analogous equivalent circuit for the mechanical system. The left side is the electrical circuit that would describe the transducer if it were "blocked" (i.e., prevented from moving). The effect of the coupling between the two sides is to induce a voltage proportional to velocity in the electrical circuit and a force proportional to current in the mechanical circuit. If the ratios of voltage/velocity and force/current are constant, then the device is linear, as we will assume hereafter. The constants of proportionality for these two cases are known as the transduction or coupling coefficients, T_{em} and T_{me} (the subscripts should be read as "effect due to cause"). For example, in the case of an electrodynamic transducer employing a coil of total winding length l (m) free to move in a magnetic field B (Webers), the force on the coil when carrying a current i (amperes) is Bli. Thus, $T_{me} = Bl$. When the coil has velocity v, there is a back emf in the circuit so that $T_{em} = -Bl$; that is,

$$T_{me} = -T_{em} \tag{2.12.1}$$

The basic equations governing the behavior of the equivalent circuit are thus

$$E = Z_e I + T_{em} v \tag{2.12.2}$$

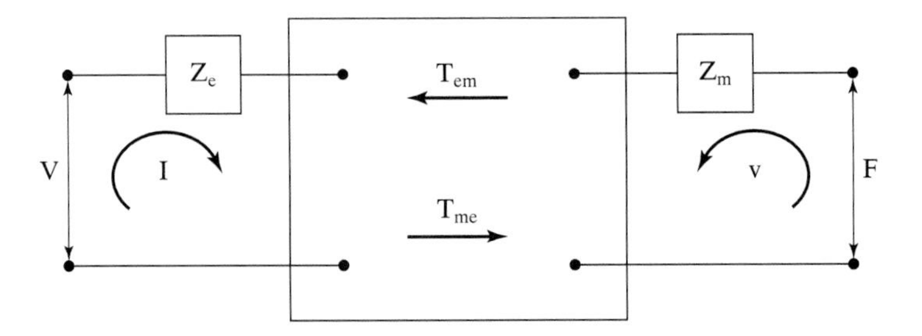

Fig. 2.28 Schematic representation of an electromechanical transducer.

and

$$F = T_{me}I + Z_m v \tag{2.12.3}$$

where E and F, I and v are phasors of the same frequency. Suppose the transducer is used as a driver (in other words, an electrical voltage produces mechanical motion), then F = 0 in Eq. (2.12.3), and by eliminating v we obtain:

$$\frac{E}{I} = Z_e - \frac{T_{em}T_{me}}{Z_m} \tag{2.12.4}$$

But E/I is the electrical driving point impedance, so we can see from Eq. (2.12.4) that the effect of the transducer is to introduce an extra electrical impedance known as the motional impedance:

$$Z_{mot} = -\frac{T_{em}T_{me}}{Z_m} \tag{2.12.5}$$

Thus

$$\frac{E}{I} = Z_{ee} = Z_e + Z_{mot} \tag{2.12.6}$$

z_e is sometimes referred to as the blocked impedance (i.e., the electrical impedance that would be measured if the transducer were prevented from moving). The equivalent circuit can now be redrawn as shown in Fig. 2.29. Now the mechanical impedance (see Section 1.3.2) may be separated into resitive and reactive parts:

$$Z_m = R_m + jX_m \tag{2.12.7}$$

Consider the case of a lumped constant single degree of freedom mechanical system, then

$$X_m = M\omega - \frac{S}{\omega} \tag{2.12.8}$$

where M is the mass and S is its stiffness, as in Fig. 2.21a. The situation can also be analyzed in terms of system elements, as shown in Fig. 2.30. When these elements are combined, we

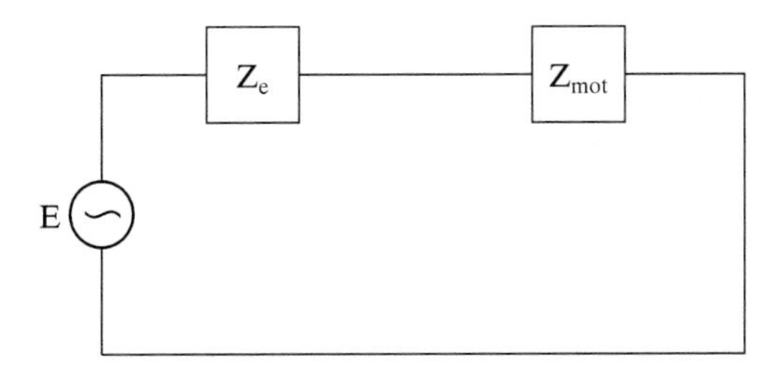

Fig. 2.29 Equivalent circuit of Fig. 2.28 used as a driver (i.e., F = 0). Z_{mot} = motional impedance.

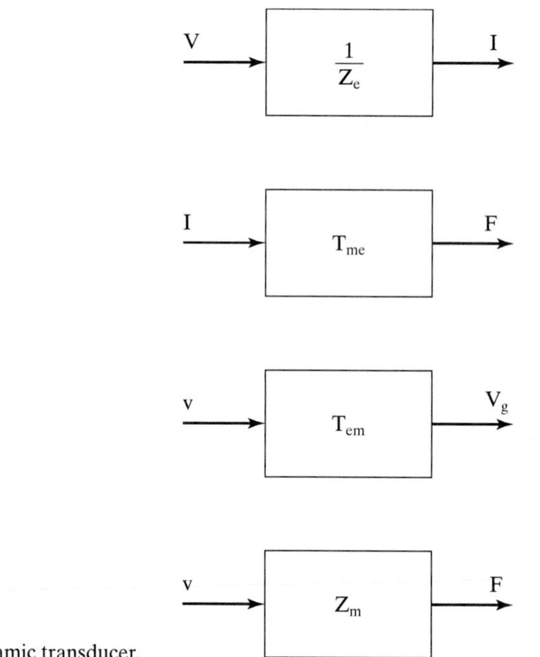

Fig. 2.30 Elements of an electrodynamic transducer.

obtain the block diagram shown in Fig. 2.31a, which reduces in steps as shown in Figs. 2.31b and c to a single transfer function:

$$w(s) = \frac{1}{Z_e + \dfrac{T_{me}T_{em}}{Z_m + Z_R}} \tag{2.12.9}$$

This result is consistent with the interpretation of a motional impedance as in Eq. (2.12.5), for the case of a driver with no applied force that experiences a radiation load Z_R. Z_m is the mechanical impedance of the driver alone. When the device is used to produce a voltage from an applied force, it is said to be a sensor, accelerometer, or microphone, depending on the exact configuration. We will analyze some of these cases in Chapter 6.

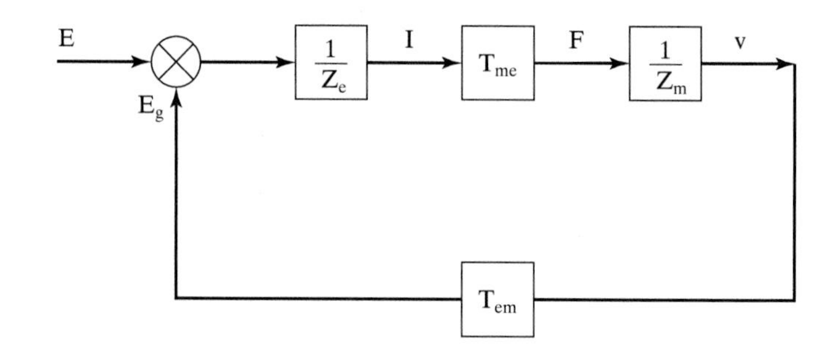

Fig. 2.31a Block diagram of electrodynamic transducer.

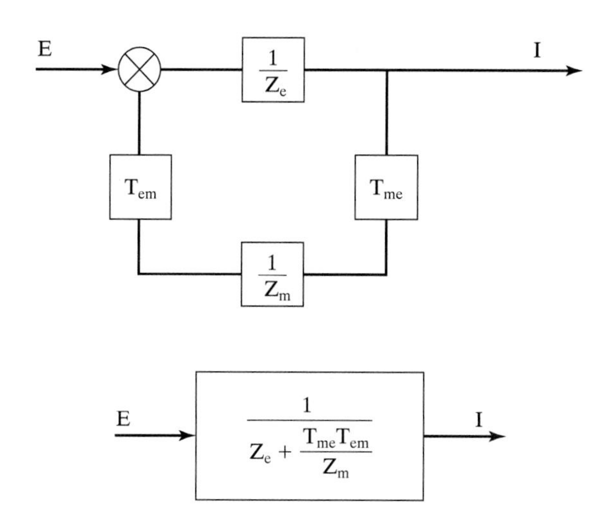

Fig. 2.31b Reduction of Fig. 2.31a.

Fig. 2.31c Reduction of Fig. 2.31b.

REFERENCES AND FURTHER READING

Hale, F.J. 1988. *Introduction to Control System Analysis and Design*. 2nd ed. Prentice Hall.
Padulo, L., and M.A. Arbib. 1974. *System Theory*. Saunders.
Raven, F.H. 1987. *Automatic Control Engineering*. 4th ed. McGraw Hill.
Wylie, C.R. 1960. *Advanced Engineering Mathematics*. 2nd ed. McGraw Hill.

PROBLEMS

2.1 A string is stretched between two rigid supports, and its midpoint is pulled to one side in a direction perpendicular to the length of the string. What harmonics will be excited when the string is released? What will be the relative intensities of these harmonics?

2.2 Let $\phi(t)$ be a function that is zero outside the range $(-1, 1)$ and let the Fourier transform of $\phi(t)$ be

$$\Phi(\phi) = \int_{-1}^{?} \phi(t)e^{j\omega t}\,dt$$

By repeated differentiation of $\Phi(1)$ show that

$$\Phi(t^n) = (j)^n 2 \frac{d^n\sigma(\omega)}{d\omega^n}$$

where $\sigma(\omega) = (\sin\omega)/\omega$. Also show that

$$\Phi(e^{nj\pi t}) = 2\sigma(\omega + n\pi)$$

2.3 The velocity of propagation of waves over a liquid surface is c, given by

$$c^2 = \frac{g\lambda}{2\pi} + \frac{2\pi\sigma}{\rho\lambda}$$

where λ is the wavelength, g; ρ and σ being constants. Show that the limiting values of the group velocities of pulses of very short and very long wavelengths are 3c/2 and c/2, respectively.

2.4 Sketch the signal that results from convolving the following pair of functions:

$$x(0) = 2, x(-1) = 1 = x(1), x(n) = 0, \text{ for all other n}$$

and

$$y(0) = 2, y(-1) = 1 = y(1), y(n) = 0, \text{ for all other n}$$

2.5 Prove that the Laplace transform of an integral is given by

$$L\left[\int_a^t f(t)\,dt\right] = \frac{1}{s}L[f(t)] + \int_a^0 f(t)\,dt$$

2.6 Find the Laplace transform of the following functions:

a. $\frac{1}{t^3}$

b. $\frac{1}{t^2 + 5}$

c. $\frac{2t + 5}{t^2 + 9}$

2.7 The sawtooth waveform shown in Fig. P2.7a is applied to a circuit as shown in Fig. P2.7b. The sawtooth waveform has an amplitude of 1 v and a period of 1 s.

a. Find the exponential Fourier coefficients of the sawtooth wave.
b. Find the transfer function $H(\omega) = Y(\omega)/X(\omega)$.

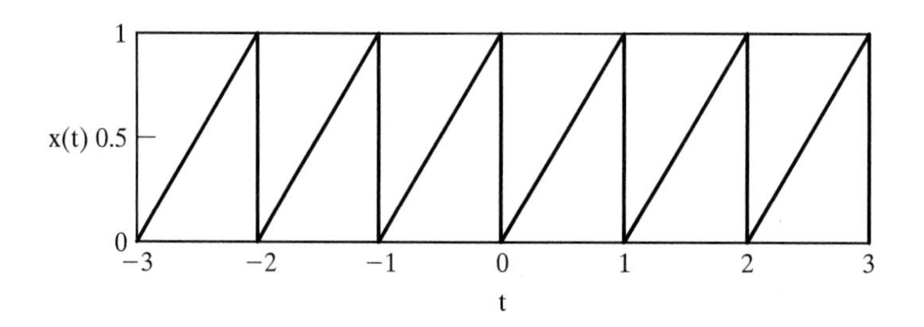

Fig. P2.7a Sawtooth waveform for Problem 2.7.

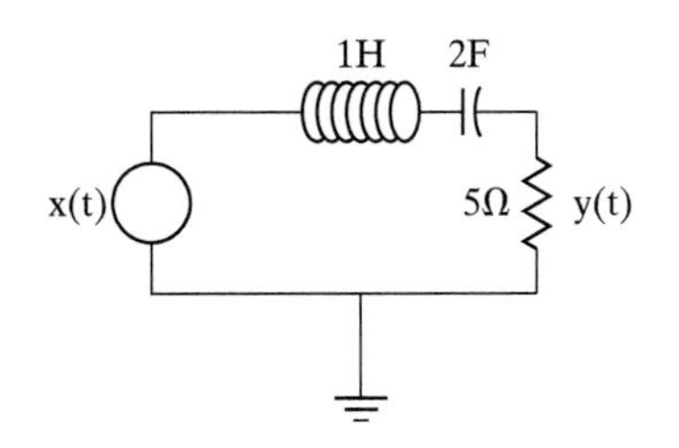

Fig. P2.7b Circuit for Problem 2.7.

c. Represent x(t) by its exponential Fourier series.
d. Represent y(t) by its exponential Fourier series.

2.8 Consider the differential equation

$$y''(x) + 3y(x) = [1 + e^{jkx/2}]^{-1}$$

where x and k are real.

a. Determine k so that y(x) is periodic with period equal to 2x.
b. For the values of k determined above, find the periodic y(x) by expanding it in a complex Fourier series in the range $-\pi$ to π.

2.9 A damped harmonic oscillator has a damping term proportional to the velocity of the oscillator mass (5 kg) with a damping coefficient equal to 4.0 Ns/m. The spring constant is 20 N/m.

a. Determine the resultant force on the oscillator mass as a function of time.
b. If the oscillator is at rest at t = 0, and displaced by the amount x = 0.1 m from equilibrium, find the algebraic equation satisfied by the Laplace transform of x(t).
c. Find the inverse Laplace transform to the equation in part b.

2.10 Find the transfer function between the output voltage across the capacitor y(t) and the input voltage u(t) for the circuit shown in Fig. P2.10. Determine the response y(t) for R = 1 ohm, L = 0.1 H, and C = 0.5 F if u(t) is a unit step function.

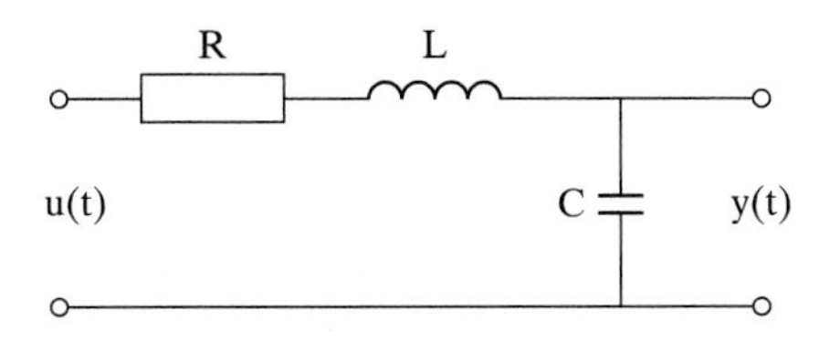

Fig. P2.10 Circuit for Problem 2.10.

2.11 Find the transfer function for the mechanical system shown in Fig. P2.11. The mass m = 1 kg, the spring constant K = 4 kg/s^2, the friction constant b = 2.4 kg/s, and the external force (input function) u(t) = 2 N. Calculate the response y(t).

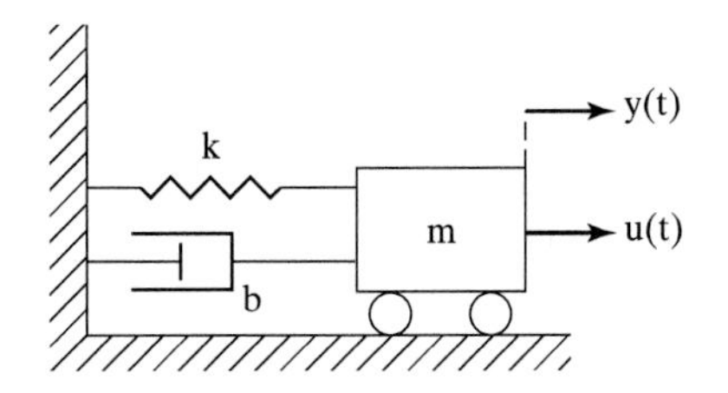

Fig. P2.11 Mechanical system for Problem 2.11.

2.12 Determine the output Y(s) of the block diagram shown in Fig. P2.12.

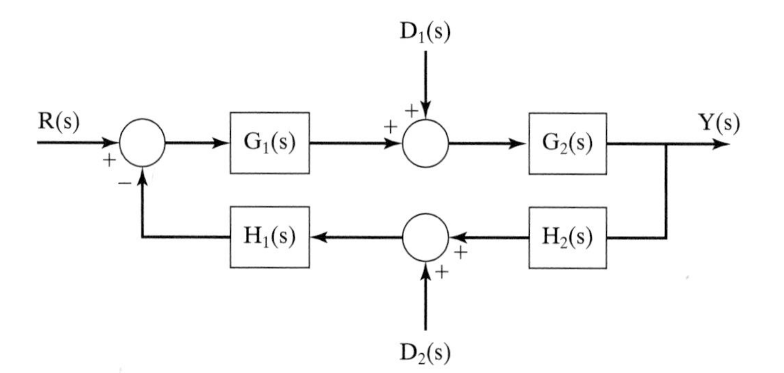

Fig. P2.12 Block diagram for Problem 2.12.

2.13 The applied force F is the input to the system shown in Fig. P2.13, and the output is the displacement x.

a. Determine the transfer function.
b. What is the steady-state response for an applied force $F(t) = 10\cos(t + 1)$ N?

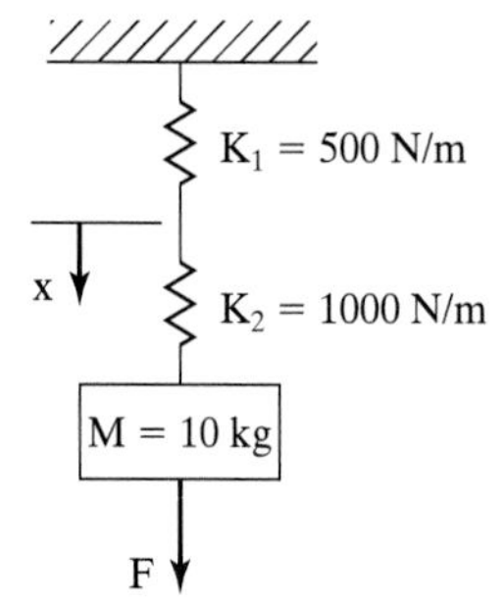

Fig. P2.13 System for Problem 2.13.

2.14 Convert the Laplace function Y(s) to the time domain Y(t) where

$$Y(s) = \frac{5}{s} + \frac{12}{s^2 + 4} + \frac{3}{2 + 2 - 3j} + \frac{3}{s + 2 + 3j}$$

2.15 For the following input and transfer function, find the response in time.

Transfer function (m/N):

$$\frac{x(s)}{F(s)} = \frac{s + 2}{(s + 3)(s + 4)}$$

Input function (N):

$$F(t) = 5$$

Chapter 3 Waves in Fluids

3.1 Radiation in Three Dimensions

We discussed wave motion along a line in Chapter 1, and now we must generalize that treatment to three dimensions. The motions of a fluid medium are governed by the principle of continuity of mass, Newton's laws, and the equation of state of the fluid. First we will consider continuity of mass. The components of velocity at point Q in Fig. 3.1 are u, v, and w and the density is ρ. The net gain of mass in the elementary cube due to flow in the x direction, per unit time, is

$$\left\{\left[\rho u - \frac{\partial(\rho u)}{\partial x}\frac{\delta x}{2}\right] - \left[\rho u + \frac{\partial(\rho u)}{\partial x}\frac{\delta x}{2}\right]\right\}\delta y \delta z = -\frac{\partial(\rho u)}{\partial x}\delta x \delta y \delta z \qquad (3.1.1)$$

So the total gain in mass due to flow in all three directions is

$$-\left[\mathbf{i}\frac{\partial}{\partial x} + \mathbf{j}\frac{\partial}{\partial y} + \mathbf{k}\frac{\partial}{\partial z}\right]\cdot(\rho \mathbf{v})\,\delta x \delta y \delta z = \frac{\partial \rho}{\partial t}\delta x \delta y \delta z \qquad (3.1.2)$$

from continuity of mass. Finally,

$$\nabla\cdot(\rho \mathbf{v}) + \frac{\partial \rho}{\partial t} = 0 \qquad (3.1.3)$$

This is the equation of continuity of mass.

Acoustics being the study of mechanical oscillations, we restrict ourselves to the case of oscillatory velocities, allowing Eq. (3.1.3) to be simplified. Consider flow in the

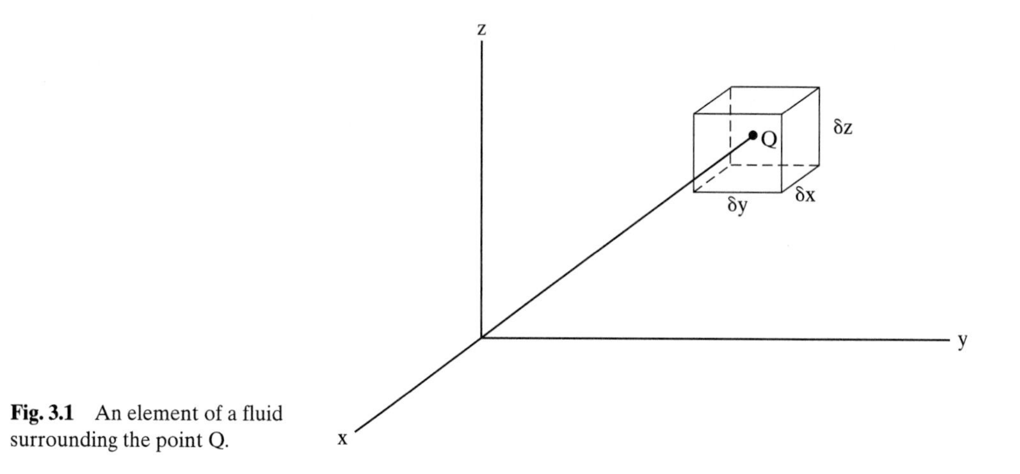

Fig. 3.1 An element of a fluid surrounding the point Q.

x direction:

$$\frac{\partial(\rho u)}{\partial x} = \rho\frac{\partial u}{\partial x} + u\frac{\partial \rho}{\partial x} = \rho u\left[\frac{1}{u}\frac{\partial u}{\partial x} + \frac{1}{\rho}\frac{\partial \rho}{\partial x}\right] \tag{3.1.4}$$

Now, while u is oscillatory about zero, ρ varies about a mean value, ρ_o, such that $(\rho - \rho_o)/\rho$ typically does not much exceed 10^{-3}. However, $(u - u_o)/u$ may be much larger than 10^{-3}. Thus, for sound,

$$\frac{1}{\rho}\frac{\partial \rho}{\partial x} << \frac{1}{u}\frac{\partial u}{\partial x} \tag{3.1.5}$$

and

$$\frac{\partial(\rho u)}{\partial x} = \rho_0\frac{\partial u}{\partial x} \tag{3.1.6}$$

since $\rho = \rho_0$. Finally, we may write the three-dimensional equation of continuity as

$$\frac{\partial \rho}{\partial t} + \rho_0 \nabla \cdot \bar{v} = 0 \tag{3.1.7}$$

Now, by inspecting Fig. 3.2 we see that

$$P\delta y\delta z - \left[P + \frac{\partial P}{\partial x}\cdot \delta x\right]\delta y\delta z = -\frac{\partial P}{\partial x}\delta x\delta y\delta z \tag{3.1.8}$$

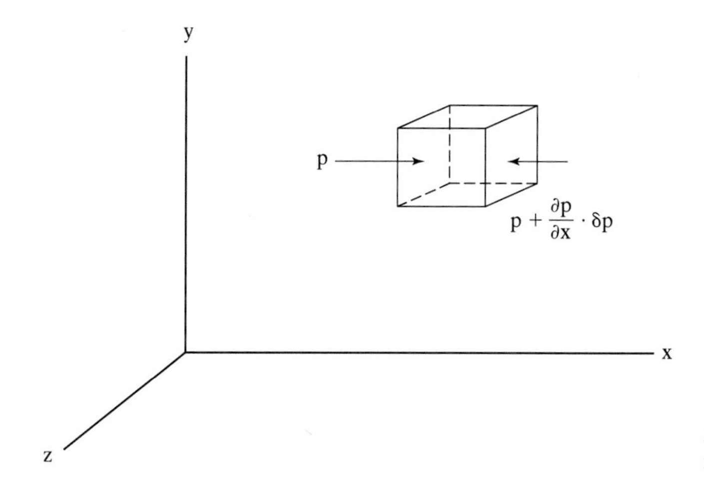

Fig. 3.2 The balance of pressure in the x direction.

So, in three dimensions the net force on the element is

$$\mathbf{F} = [-\text{grad } P]\, \delta x \delta y \delta z \tag{3.1.9}$$

Consequently, applying Newton's second law of motion:

$$[-\text{grad } P]\, \delta x \delta y \delta z = \rho \delta x \delta y \delta z \frac{D\mathbf{v}}{Dt} \tag{3.1.10}$$

or

$$D\mathbf{v}/Dt = -1/\rho \nabla P \tag{3.1.11}$$

Here, the operator D/Dt is the hydrodynamic derivative and allows for the difference in velocity at different points in space; that is,

$$\frac{D\mathbf{v}}{Dt} = \frac{\partial \mathbf{v}}{\partial t} + u\frac{\partial \mathbf{v}}{\partial x} + \frac{\partial \mathbf{v}}{\partial y} + w\frac{\partial \mathbf{v}}{\partial z} \tag{3.1.12}$$

Now the spatial variation terms are negligible in the acoustic case, and since the variation in ρ is small, we may write

$$\frac{D\mathbf{v}}{Dt} = \frac{\partial \mathbf{v}}{\partial t} = -\frac{1}{\rho}\nabla P \tag{3.1.13}$$

This is the equation of motion.

Finally, we must discuss the equation of state. Suppose, for instance, that the motions we are considering occur adiabatically in an ideal gas, then

$$PV^{\gamma} = \text{constant} \tag{3.1.14}$$

and

$$(P + \delta P)(V + \delta V)^{\gamma} = PV^{\gamma} \tag{3.1.15}$$

and if δV and δp are small, using the binomial expansion,

$$PV^{\gamma}\left(1 + \frac{\delta P}{P}\right)\left(1 + \gamma\frac{\delta V}{V}\right) = PV^{\gamma} \tag{3.1.16}$$

so that

$$PV^{\gamma} + V^{\gamma}\,\delta P + \gamma PV^{\gamma-1}\,\delta V = PV^{\gamma} \tag{3.1.17}$$

or

$$\delta P = -\gamma P\frac{\delta V}{V} \tag{3.1.18}$$

Now $\delta V/V$ is the volume strain due to stress δP, so for adiabatic changes in a perfect gas, the bulk modulus of elasticity is δP. Even if the equation of state is not known, we may assume for small strains that there is a proportionality between strain and stress; that is,

$$\delta P = -B\,\delta V/V \tag{3.1.19}$$

Furthermore, for a fixed mass of fluid,

$$m = \rho V = (\rho + \delta\rho)(V + \delta V) = \rho V + \rho\delta V + V\delta\rho \tag{3.1.20}$$

Thus

$$\delta\rho/\rho = -\delta V/V \tag{3.1.21}$$

and

$$\delta P = \frac{B\delta\rho}{\rho_0} \tag{3.1.22}$$

In the acoustic approximation, ρ is almost constant (ρ_o, say) and the excess pressure is proportional to the excess density. Letting the excess pressure $= \delta P = p_e = P - P_o$ and excess density $= \delta\rho = \rho_\varepsilon = \rho - \rho_o$, we have

$$p = (B/\rho_o)\rho_e \tag{3.1.23}$$

We can now use Eqs. (3.1.7), (3.1.13), and (3.1.23) to obtain the wave equation. First we differentiate Eq. (3.1.23) with respect to time:

$$\frac{\partial p_e}{\partial t} = \frac{B}{\rho_0}\frac{\partial \rho_e}{\partial t} \tag{3.1.24}$$

Now

$$\frac{\partial \rho}{\partial t} = \frac{\partial(\rho_0 + \rho_e)}{\partial t} = \frac{\partial \rho_e}{\partial t} \tag{3.1.25}$$

Thus, from Eq. (3.1.7),

$$\frac{\rho_0}{B}\frac{\partial p_e}{\partial t} = -\rho_0 \nabla \cdot \mathbf{v} \tag{3.1.26}$$

and again differentiating with respect to time:

$$\frac{\rho_0}{B}\frac{\partial^2 p_e}{\partial t^2} = -\rho_0 \nabla \cdot \frac{\partial \mathbf{v}}{\partial t} \tag{3.1.27}$$

Thus, from Eq. (3.1.13),

$$\frac{\rho_0}{B}\frac{\partial^2 p_e}{\partial t^2} = \nabla \cdot \nabla P \tag{3.1.28}$$

and since

$$\nabla P = \nabla(P_0 + p) \tag{3.1.29}$$

it follows that

$$\frac{\rho_0}{B}\frac{\partial^2 p}{\partial t^2} = \nabla^2 p \tag{3.1.30}$$

which is a wave equation with propagation velocity

$$c = \sqrt{\frac{B}{\rho_0}} \tag{3.1.31}$$

Thus

$$\nabla^2 p = \frac{1}{c^2}\frac{\partial^2 p}{\partial t^2} \tag{3.1.32}$$

Now if we define velocity potential φ by

$$\mathbf{v} = -\nabla\varphi \tag{3.1.33}$$

it follows that, by using Eq. (3.1.13),

$$\frac{\partial \mathbf{v}}{\partial t} = -\frac{\partial}{\partial t}\nabla\varphi = -\nabla\frac{\partial\varphi}{\partial t} = -\frac{1}{\rho_0}\nabla p \tag{3.1.34}$$

Thus

$$p = \rho_0\frac{\partial\varphi}{\partial t} \tag{3.1.35}$$

and from Eqs. (3.1.32) and (3.1.35)

$$\nabla^2\frac{\partial\varphi}{\partial t} = \frac{1}{c^2}\frac{\partial^3\varphi}{\partial t^3} \tag{3.1.36}$$

or

$$\nabla^2\varphi = \frac{1}{c^2}\frac{\partial^2\varphi}{\partial t^2} \tag{3.1.37}$$

Finally, we can quote two vector identities

$$\nabla \times \nabla\Psi = 0 \tag{3.1.38}$$

where ψ is a scalar and

$$\nabla \times \nabla \times \mathbf{A} = \nabla\nabla\cdot\mathbf{A} - \nabla\cdot\nabla\mathbf{A} \tag{3.1.39}$$

where **A** is a vector. It follows from Eqs. (3.1.33) and (3.1.38) that

$$-\nabla \times \nabla\varphi = \nabla \times \mathbf{v} = 0 \tag{3.1.40}$$

Thus

$$\nabla\nabla\cdot\mathbf{v} = \nabla\cdot\nabla\mathbf{v} = \nabla^2\mathbf{v} \tag{3.1.41}$$

But from Eq. (3.1.37)

$$\nabla\cdot\nabla^2\varphi = \frac{1}{c^2}\frac{\partial^2\nabla\varphi}{\partial t^2} \tag{3.1.42}$$

That is,

$$\nabla \cdot \nabla \cdot \nabla \varphi = \frac{1}{c^2}\frac{\partial^2 \mathbf{v}}{\partial t^2} \tag{3.1.43}$$

Thus

$$\nabla^2 \mathbf{v} = \frac{1}{c^2}\frac{\partial^2 \mathbf{v}}{\partial t^2} \tag{3.1.44}$$

In other words, the particle velocity also obeys a wave equation—in this case a vector wave equation.

3.2 Solutions of the Wave Equation

There are as many solutions of the wave equation as there are boundary conditions. Because it is useful to be able to write the wave equation in coordinates appropriate to the geometry of a particular problem, we will now review the expression of the Laplacian operator in the major coordinate systems.

In Cartesian coordinates,

$$\nabla^2 p = \frac{\partial^2 p}{\partial x^2} + \frac{\partial^2 p}{\partial y^2} + \frac{\partial^2 p}{\partial z^2} \tag{3.2.1}$$

In cylindrical coordinates (u, z, φ), as in Fig. 3.3,

$$\nabla^2 p = \frac{\partial^2 p}{\partial u^2} + \frac{1}{u}\frac{\partial p}{\partial u} + \frac{1}{u^2}\frac{\partial^2 p}{\partial \varphi^2}\frac{\partial^2 p}{\partial z^2} \tag{3.2.2}$$

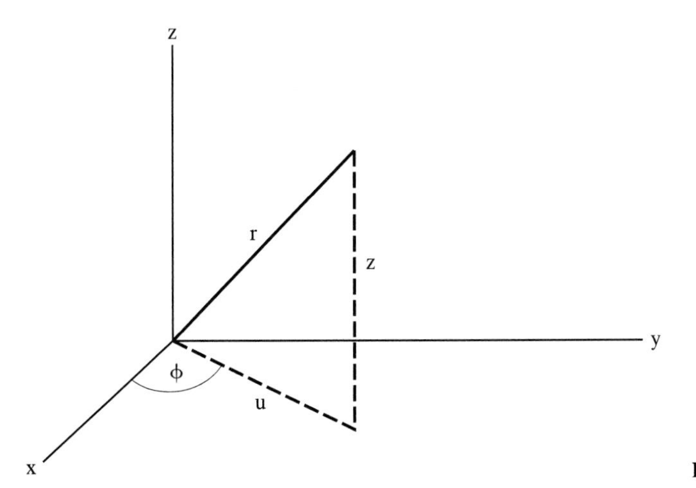

Fig. 3.3 Cylindrical coordinates.

In spherical polar coordinates (r, θ, φ), shown in Fig. 3.4,

$$\nabla^2 p = \frac{1}{r^2}\frac{\partial}{\partial r}\left(r^2 + \frac{\partial p}{\partial r}\right) + \frac{1}{r^2 \sin\theta}\frac{\partial}{\partial \theta}\left(\sin\theta \frac{\partial p}{\partial \theta}\right) + \frac{1}{r^2 \sin^2\theta}\frac{\partial^2 p}{\partial \varphi^2} \tag{3.2.3}$$

One of the simplest and most important solutions of the wave equation arises in the case of complete spherical symmetry when the wave equation reduces to

$$\frac{1}{r^2}\frac{\partial}{\partial r}\left(r^2 \frac{\partial p}{\partial r}\right) = \frac{1}{c^2}\frac{\partial^2 p}{\partial t^2} \tag{3.2.4}$$

or

$$2\frac{\partial p}{\partial r} + r\frac{\partial^2 p}{\partial r^2} = \frac{r}{c^2}\frac{\partial^2 p}{\partial t^2} \tag{3.2.5}$$

Thus

$$\frac{\partial^2}{\partial r^2}(rp) = \frac{1}{c^2}\frac{\partial^2 (rp)}{\partial t^2} \tag{3.2.6}$$

This equation is very similar to the plane wave equation

$$\frac{\partial^2 p}{\partial x^2} = \frac{1}{c^2}\frac{\partial^2 p}{\partial t^2} \tag{3.2.7}$$

except that x is replaced by r and p by rp, so a general solution will be

$$p = \frac{Ae^{j(\omega t - kr)}}{r} + \frac{Be^{j(\omega t + kr)}}{r} \tag{3.2.8}$$

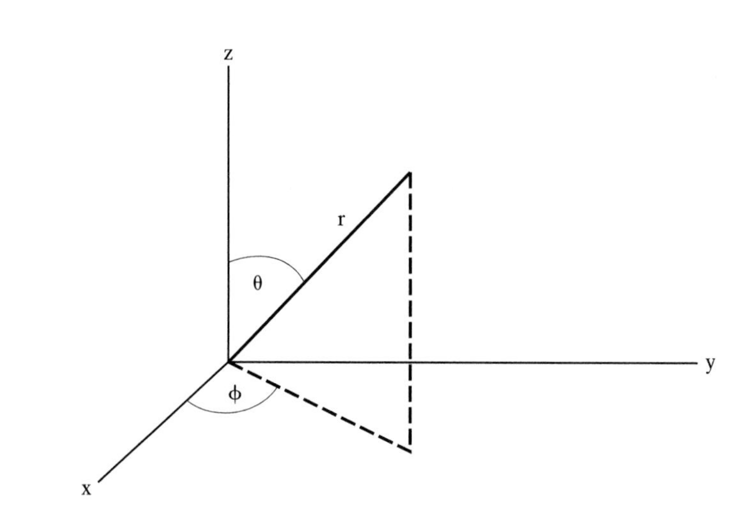

Fig. 3.4 Spherical polar coordinates.

where

$$\omega/k = c \tag{3.2.9}$$

The first term is a radially diverging wave, and the second is radially converging. Radially converging waves are not frequently encountered in reality, and we will restrict our discussion to the diverging case:

$$p = (A/r)e^{j(\omega t - kr)} \tag{3.2.10}$$

Now, from Eq. (3.1.13),

$$\frac{\partial \mathbf{v}}{\partial t} = -\frac{1}{\rho_0}\frac{\partial p}{\partial r} \tag{3.2.11}$$

Thus

$$v = \frac{p}{j\omega\rho_0}(1/r + jk) \tag{3.2.12}$$

It follows that

$$\frac{p}{v} = \frac{j\omega\rho_0 r}{1 + jkr} = z \tag{3.2.13}$$

which is the specific acoustic impedance by definition.

We see that

$$z = \frac{\rho_0 ckre^{j\theta}}{\sqrt{1 + k^2r^2}} \tag{3.2.14}$$

where

$$\tan\theta = 1/kr \tag{3.2.15}$$

So that p and v are not in phase except as $\theta \rightarrow 0$ (which we see from Eq. (3.2.15) occurs at distances from the origin much greater than a wavelength). Note also that, as contrasted with plane waves, the specific acoustic impedance for spherical waves is complex:

$$z = \rho_0 c\frac{k^2r^2}{1 + k^2r^2} + j\rho_0 c\frac{kr}{1 + k^2r^2} \tag{3.2.16}$$

where the first term is the specific acoustic resistance and the second term, the specific acoustic reactance. An appreciation of the physical significance of these quantities may be obtained as follows.

A thin spherical shell is subjected to a radially symmetrical driving pressure p, so that $\int p dA = F_o e^{jwt}$. If the mass of the shell is m, its resistance, R (due to internal losses in the material of the shell), and its stiffness, S (due to the elastic properties of the shell), then the equation of motion in vacuo will be

$$m\ddot{\eta} + R\dot{\eta} + S\eta = F_0 e^{j\omega t} \tag{3.2.17}$$

Now suppose the sphere is hollow but enclosed in a fluid medium. A train of radial waves will emanate from its spherical surface, of radius a, and the specific acoustic impedance of these waves will be

$$z_a = \rho_0 c \frac{k^2a^2}{1 + k^2a^2} + j\rho_0 c \frac{ka}{1 + k^2a^2} \tag{3.2.18}$$

Thus, the pressure exerted by the sphere on the medium is

$$p = vz_a = \dot{\eta} z_a \tag{3.2.19}$$

and the force exerted by the sphere on the medium is $4\pi a^2 \dot{\eta} z_a$. By Newton's third law, the medium must react on the sphere with force

$$F_R = 4\pi a^2 \dot{\eta} z_a \tag{3.2.20}$$

Now we have for the equation of motion:

$$m\ddot{\eta} + R\dot{\eta} + S\eta = F_0 e^{j\omega t} + F_R \tag{3.2.21}$$

and since $\ddot{\eta} = j\omega\eta$,

$$\left(m + \frac{4\pi a^3 \rho_0}{1 + k^2a^2}\right)\ddot{\eta} + \left(R + \frac{4\pi a^4 \rho_0 c k^2}{1 + k^2a^2}\right)\dot{\eta} + S\eta = F_0 e^{j\omega r} \tag{3.2.22}$$

The effect of the radiation is to add to the mass and resistance of the shell amounts:

$$m_r = \frac{4\pi a^3 \rho_0}{1 + k^2a^2} \tag{3.2.23}$$

and

$$R_r = \frac{4\pi a^4 \rho_0 c k^2}{1 + k^2a^2} \tag{3.2.24}$$

At low frequencies (i.e., ka >> 1),

$$m_r \rightarrow 4\pi a^3 \rho_0 = \text{3 times the mass of fluid displaced} \tag{3.2.25}$$

and $R_r \rightarrow 0$.

At high frequencies (i.e., ka >> 1), $m_r \to 0$ and

$$R_r \to 4\pi\, a^2 \rho_0 c = \rho_0 c \text{ times the surface area of the sphere} \tag{3.2.26}$$

The last result could be anticipated from a knowledge of plane waves, whose specific acoustic impedance is ρc. At high frequencies, the curvature of the surface is negligible in comparison with the wave length.

3.3 Intensity of Spherical Waves

The intensity of a spherical wave is the average rate at which work is done on the fluid medium over unit area of the wavefront; in other words, if in Eq. (3.2.10) we let

$$A/r = p_0 \tag{3.3.1}$$

we have from Eqs. (3.2.13) and (3.2.14):

$$I = \frac{1}{T}\int_0^T pv\,dt = \frac{1}{T}\int_0^T p_0 \cos(\omega t - kr)\, u_0 \cos(\omega t - kr - \theta)\, dt = \frac{p_0 u_0}{2}\cos\theta \tag{3.3.2}$$

where $\cos\theta$ is analogous to the power factor of a simple oscillator and

$$u_0 = \frac{A\sqrt{1 + k^2r^2}}{r\rho c k r} = \frac{p_0}{\rho c}\frac{\sqrt{1 + k^2r^2}}{kr} \tag{3.3.3}$$

But $\tan\theta = 1/kr$ (see Eq. (3.2.15)), so

$$\cos\theta = \frac{kr}{\sqrt{1 + k^2r^2}} \tag{3.3.4}$$

and

$$u_0 = p_0/(\rho c \cos\theta) \tag{3.3.5}$$

Thus

$$I = \frac{p_0^2}{2\rho c} \tag{3.3.6}$$

which is exactly the result for plane waves. However, in this case, p_0 depends inversely on r so the intensity obeys an inverse square law. It follows that the energy from a spherical source crossing a surface of radius r in unit time is

$$W = 4\pi r^2 I = 2\pi A^2/\rho c \tag{3.3.7}$$

which does not depend on r, as we would expect.

3.4 Simple Source or Monopole

Let the radius of a spherical radiator be a and the velocity at any point on its surface be

$$u_s = u_0 \cos \omega t = R_e(u_0 e^{j\omega t}) \tag{3.4.1}$$

We will assume this motion gives rise to spherical waves in the surrounding medium, whose acoustic pressure will be

$$p = \frac{A}{r} e^{j\omega t - kr} \tag{3.4.2}$$

Then the particle velocity

$$v = p/z \tag{3.4.3}$$

But from Eq. (3.2.16),

$$z = \rho ckr(kr + j)/(1 + k^2r^2) \tag{3.4.4}$$

Now at the radius of the sphere

$$v = u_s \tag{3.4.5}$$

which is a boundary condition ensuring that the sphere's surface and the medium stay together. Thus

$$v = u_0 e^{j\omega t} = \frac{A(1 + k^2a^2)e^{j(\omega t - ka)}}{a\rho cka(ka + j)} \tag{3.4.6}$$

and we determine the arbitrary constant A that was introduced in Eq. (3.2.8):

$$A = \frac{a^2 u_0 \rho ck(ka + j)\, e^{jka}}{1 + k^2a^2} \tag{3.4.7}$$

Most sources of spherical waves of interest to us are small compared with a wavelength; that is,

$$ka << 1 \tag{3.4.8}$$

then

$$e^{jka} = \cos ka + j\sin ka \rightarrow 1 + jka \tag{3.4.9}$$

and

$$A = j\rho cka^2 u_0 \tag{3.4.10}$$

Then

$$p = \frac{j\rho cka^2u_0e^{j(\omega t-kr)}}{r} \tag{3.4.11}$$

So we have found the magnitude of the acoustic pressure in terms of the source properties. In practice it makes little difference whether or not the shape of the source is spherical, provided its dimensions are small compared with a wavelength and provided all points on the surface of the source move in phase. Consequently, it is convenient to define the strength of a source Q as the maximum rate of volume displacement; that is,

$$Q = \int u_0 \cdot dS \tag{3.4.12}$$

Then for our case

$$Q = 4\pi a^2 u_0 \tag{3.4.13}$$

and from Eq. (3.4.11),

$$p = \frac{j\rho ckQ}{4\pi r} \cdot e^{j(\omega t-kr)} \tag{3.4.14}$$

thus

$$p_0 = \frac{\rho ckQ}{4\pi r} \tag{3.4.15}$$

and the intensity

$$I = \frac{p_0^2}{2\rho c} = \frac{\rho ck^2Q^2}{32\pi^2r^2} \tag{3.4.16}$$

The total radiated power is

$$W = 4\pi r^2 I = \rho ck^2Q^2/8\pi \tag{3.4.17}$$

In terms of source strength, the velocity potential gives a simple result. Since

$$\partial\varphi/\partial t = +p/\rho \tag{3.4.18}$$

$$\varphi = \frac{Qe^{j(\omega t-kr)}}{4\pi r} \tag{3.4.19}$$

The monopole is one of the most important sound sources and describes any situation in which there is a fluctuation of density or pressure at a point. Examples of such a behavior are provided by sirens, organ pipes, pulsed jets, and the human voice.

3.5 The Dipole or Doublet Source

Another very common type of sound source, whose characteristics may be developed readily as an extension of our analysis of the monopole, is the dipole or doublet. In this situation, two monopoles of opposite strength—in other words, out of phase—are situated close to one another. Immediately, we may write for the velocity potential at a point P, as shown in Fig. 3.5,

$$\varphi = \frac{Qe^{j(\omega t - kr_2)}}{4\pi r_2} - \frac{Qe^{j(\omega t - kr_1)}}{4\pi r_1} \tag{3.5.1}$$

It is assumed the poles are close (i.e., $r >> l$), so

$$\left.\begin{aligned} r_2 &= r + (l/2)\cos\theta \\ r_1 &= r - (l/2)\cos\theta \end{aligned}\right\} \tag{3.5.2}$$

Since we may further assume that kl is small (i.e., $kl << 1$), we have

$$\varphi = \frac{-Ql}{4\pi r} e^{j(\omega t - kr)} \cos\theta \left[\frac{1}{r} + jk\right] \tag{3.5.3}$$

It is customary to define $-Ql/4\pi$ as μ—the so-called dipole moment—so

$$\varphi = \frac{\mu}{r}\left[\frac{1}{r} + jk\right] \cos\theta \, e^{j(\omega t - kr)} \tag{3.5.4}$$

At zero frequency, $\omega = 0$ and $k = 0$, giving

$$\varphi = \frac{\mu}{r^2}\cos\theta \tag{3.5.5}$$

which is the case of a static dipole potential familiar in electricity and magnetism and which, in fluid mechanics, corresponds to a source and a sink at a short separation. The

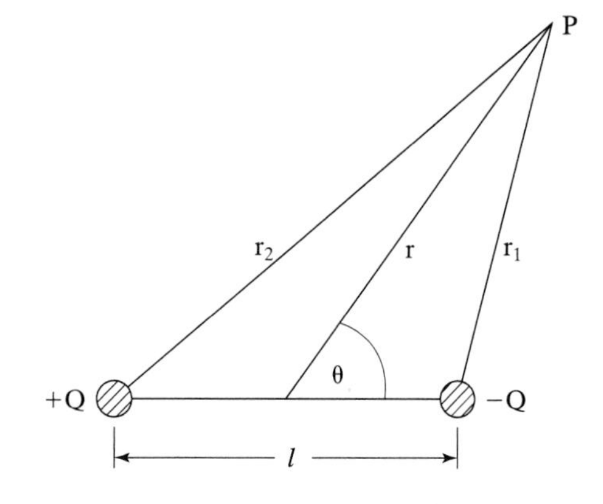

Fig. 3.5 A dipole source. l is much less than a wavelength and much less than distance r to a far field observation point P.

particle velocity at any time may be divided into radial and transverse components v_r and v_t given by

$$v_r = -\partial\varphi/\partial r = \frac{2\mu\cos\theta}{r^3} \tag{3.5.6}$$

and

$$v_t = -(1/r)\,\partial\varphi/\partial\theta = \frac{\mu\sin\theta}{r^3} \tag{3.5.7}$$

Now on the dipole axis ($\theta = 0$) the flow is entirely radial, and on its equator ($\theta = \pi/2$) the flow is entirely transverse. The streamlines are as shown marked by arrows in Fig. 3.6 and where the surfaces are equipotentials.

We now return to the oscillatory case. The acoustic pressure

$$p = \rho\frac{\partial\varphi}{\partial t} = j\frac{\omega\rho\mu}{r}\left[\frac{1}{r} + jk\right]\cos\theta\, e^{j(\omega t - kr)} \tag{3.5.8}$$

In the far field when $r >> \lambda$ or $kr >> 1$,

$$p = \frac{-\mu\omega k\rho}{r}\cos\theta\, e^{j(\omega t - kr)} \tag{3.5.9}$$

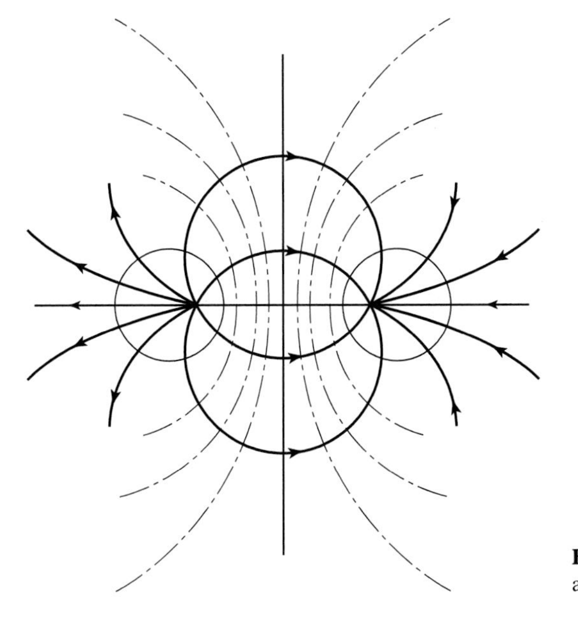

Fig. 3.6 Equipotentials and streamlines around a dipole.

This result is similar to that for a monopole, in respect to the radial dependence, but with angular variation. It follows that the intensity

$$I_D = \frac{\rho c k^4 \mu^2 \cos^2 \theta}{2r^2} \tag{3.5.10}$$

If for some fixed value of r, in the far field, a polar plot is made for I_0, the result is the figure eight shown in Fig. 3.7. The sound is most intense along the axis of the dipole and minimal at the equator. The total average power radiated may be found by integrating the intensity over an area enclosing the dipole; that is,

$$W = \int I_D \, ds \tag{3.5.11}$$

Let the dipole axis coincide with the z axis and integrate in spherical polar coordinates, then, from Fig. 3.8, we see that

$$ds = r^2 \sin \theta \, d\theta d\varphi \tag{3.5.12}$$

so that

$$W_D = \int_0^{2x} \int_0^{x} \frac{\rho c \mu^2 k^4 \cos^2 \theta \sin \theta}{2} d\theta d\varphi = \frac{2}{3} \pi \mu^2 \rho c k^4 \tag{3.5.13}$$

It is instructive to compare this result with the power W_M radiated from a simple source (monopole) of strength Q, which is given by Eq. (3.4.17). Thus

$$W_D/W_M = \frac{k^2 l^2}{3} \tag{3.5.14}$$

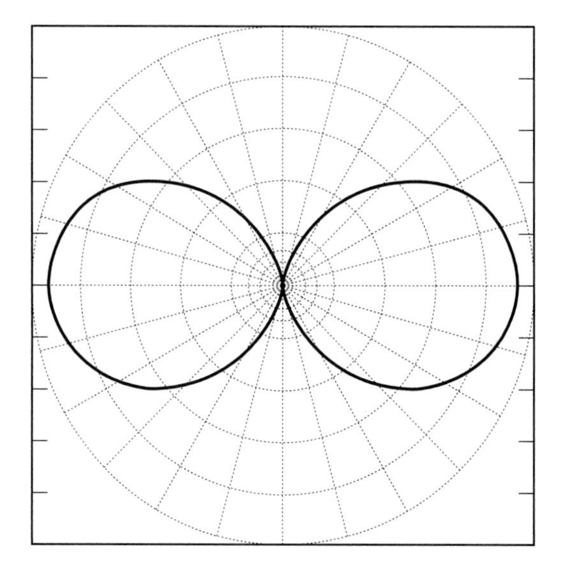

Fig. 3.7 Polar plot of far field intensity of radiation from a dipole.

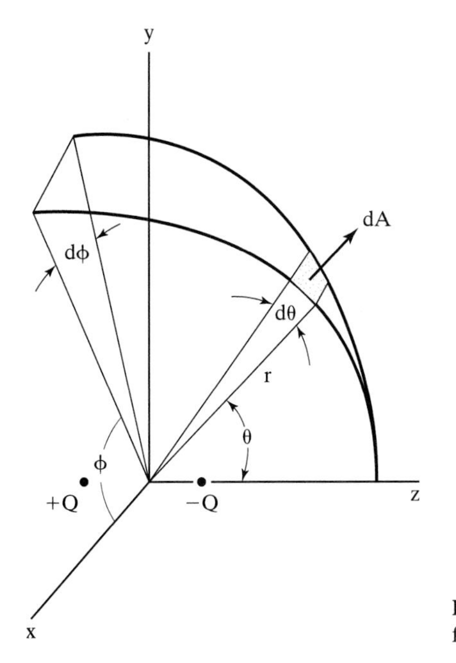

Fig. 3.8 Coordinate system for integrating power radiated from a dipole.

so that when the wavelength is greater than about $3l$, the dipole is a less efficient radiator than the monopole. Now, any vibrating object (e.g., a wire or an unbaffled loudspeaker cone) will have a dipole moment, and the dipole length, l, will be of the order of the amplitude of displacement. Since such displacement amplitudes do not generally exceed a few millimeters, it follows that dipole radiators in air are inefficient in the audible range. This is the principal reason for mounting loudspeakers in baffles. Propellor lift noise is an example of dipole radiation.

3.6 The Linear Array

In progressing from the monopole to the dipole, we have noticed increasing directionality or directivity of the radiation pattern. High directivity is obviously an advantage in any form of echo ranging, and we might guess that increasing the number of sources will assist in achieving this aim. Consequently, we will now consider the case of sources of the same strength but of varying phase, distributed evenly along a line, as in Fig. 3.9. We assume the observation point P is at a great distance from the array so radii from all sources may be regarded as parallel. The velocity potential at P is obtained by summing the velocity potentials due to all individual sources; that is,

$$\varphi_p = \sum_{m=1}^{n} \frac{Q}{4\pi r_m} \cdot e^{j(\omega t - kr_m - \delta_m)} \tag{3.6.1}$$

But since P is distant,

$$r_m = r_1 - (m - 1)\, d \cos\theta \tag{3.6.2}$$

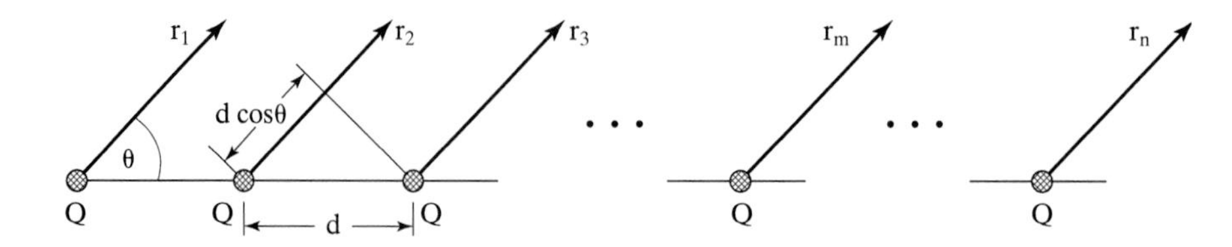

Fig. 3.9 A linear array of point sources separated by distance d. The mth source has phase δ_m relative to that of the first source. Radii $r_1, r_2, \ldots r_n$ are approximately parallel, since the observation point P is far away.

Then

$$\varphi_p = \sum_{m=1}^{n} \frac{Qe^{j(\omega t - kr_1)}}{4\pi r_1\left(1 - \dfrac{(m-1)\,d\cos\theta}{r_1}\right)} \cdot e^{j(k\overline{m-1}\,d\cos\theta - \delta_m)} \tag{3.6.3}$$

Now nd/r_1 is small (i.e., the observation point is at a distance much greater than the length of the array). Consequently, we may safely neglect the quantity

$$\frac{\overline{m-1}\,d\cos\theta}{r_1}$$

in the denominator, so

$$\varphi_p = \frac{Qe^{j(\omega t - kr_1)}}{4\pi r_1} \cdot \sum_{m=1}^{n} e^{j(k\,\overline{m-1}\,d\cos\theta - \delta_m)} \tag{3.6.4}$$

This result resembles the spherical wave with an angular dependence contained in the summation. It is much simplified if the phase difference is progressive; that is,

$$\delta_m = (m-1)\delta \tag{3.6.5}$$

In this case, the summation is easily accomplished by addition of vectors on the complex plane or by realizing that the summation is a geometric progression, so that

$$\varphi = \frac{Q}{4\pi r_1} e^{j(\omega t - kr_1)} \frac{\sin n\beta/2}{\sin \beta/2} e^{j(n-1)\beta/2} \tag{3.6.6}$$

where

$$\beta = kd\cos\theta - \delta \tag{3.6.7}$$

while

$$p = \rho\frac{\partial\varphi}{\partial t} = j\frac{k\rho c Q}{4\pi r_1} e^{j(\omega t - kr_1)} \frac{\sin n\beta/2}{\sin \beta/2} e^{j(n-1)\beta/2} \tag{3.6.8}$$

The intensity is given by

$$I = \frac{|p|^2}{2\rho c} = \frac{\rho c k^2 Q^2}{32\pi^2 r_1^2} \frac{\sin^2(n\beta/2)}{\sin^2(\beta/2)} \tag{3.6.9}$$

From this general expression we can rather easily reproduce the results for the monopole and dipole. In the case of the monopole for instance, $n = 1$, and we get

$$I_M = \frac{\rho c k^2 Q^2}{32\pi^2 r_1^2} \tag{3.6.10}$$

We have a dipole if $n = 2$ and $\delta = \pi$, then

$$I_D = \frac{\rho c k^2 Q^2}{32\pi^2 r_1^2} \cdot 4\cos^2\beta/2 = \frac{\rho c k^2 Q^2}{32\pi^2 r_1^2} \cdot 4\sin^2\left(\frac{kd\cos\theta}{2}\right) \tag{3.6.11}$$

If $kd << 1$,

$$I_D = \frac{\rho c k^2 Q^2}{2(4\pi)^2 r_1^2} \cdot k^2 d^2 \cos^2\theta = \frac{\rho c k^4 \mu^2 \cos^2\theta}{2r_1^2} \tag{3.6.12}$$

Using Eq. (3.6.10), we may write for the array in general

$$I = I_M \cdot \frac{\sin^2(n\beta/2)}{\sin^2(\beta/2)} \tag{3.6.13}$$

or

$$I = I_M \cdot \frac{\sin^2 n\gamma}{\sin^2\gamma} \tag{3.6.14}$$

where $\gamma = \beta/2$.

A polar plot of I versus γ (Fig. 3.10) has maxima that lie on an ellipse-like curve with major axes I_M and $n^2 I_M$. γ does not correspond to the true space angle θ, since

$$\gamma = \frac{\beta}{2} = \frac{kd\cos\theta}{2} - \frac{\delta}{2} \tag{3.6.15}$$

Fig. 3.11 illustrates polar plots of intensity from various linear arrays in air at 1000 Hz, with a source separation of 10 cm and a distance to the observation point of 10 m. There is no phase difference between the sources, which are of equal strength and very small radius. We see that the directivity of the beam increases with the number of sources. Note that in these plots the levels are normalized to the maximum level, which will also increase with the square of the number of sources from Eq. (3.6.14).

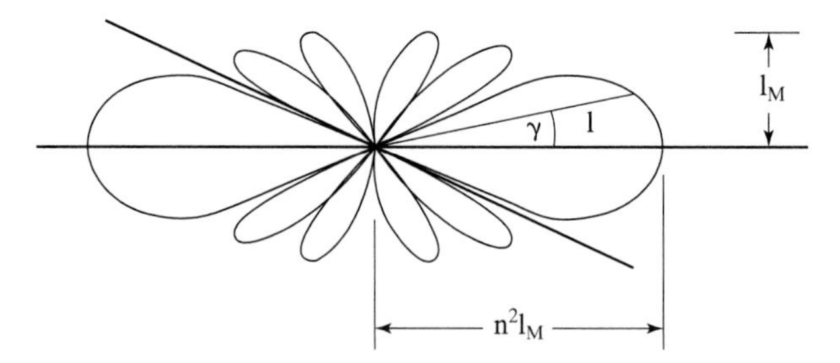

Fig. 3.10 A polar plot of intensity versus phase angle γ, which does not directly correspond to the angle in space.

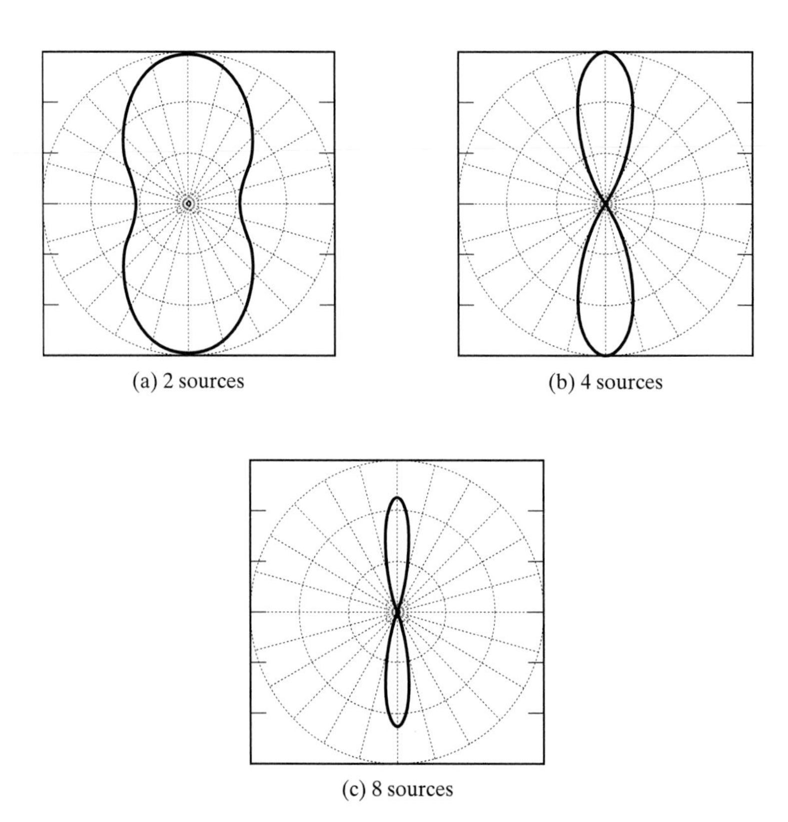

Fig. 3.11 Polar plots of intensity from linear arrays. Shows the effect of increasing the number of sources. The beams are symmetrical about the horizontal axes.

The beams are actually symmetrical about the array axis, which corresponds to the horizontal axes in Fig. 3.11.

Fig. 3.12 is obtained for the same conditions as in Fig. 3.11, except that now the number of sources remains constant at eight, and there is a constant progressive phase difference between the sources. Up to 90 degrees, we can sweep the beam by varying this phase angle, which is similar to the action of weather radars. At greater angles, we obtain an

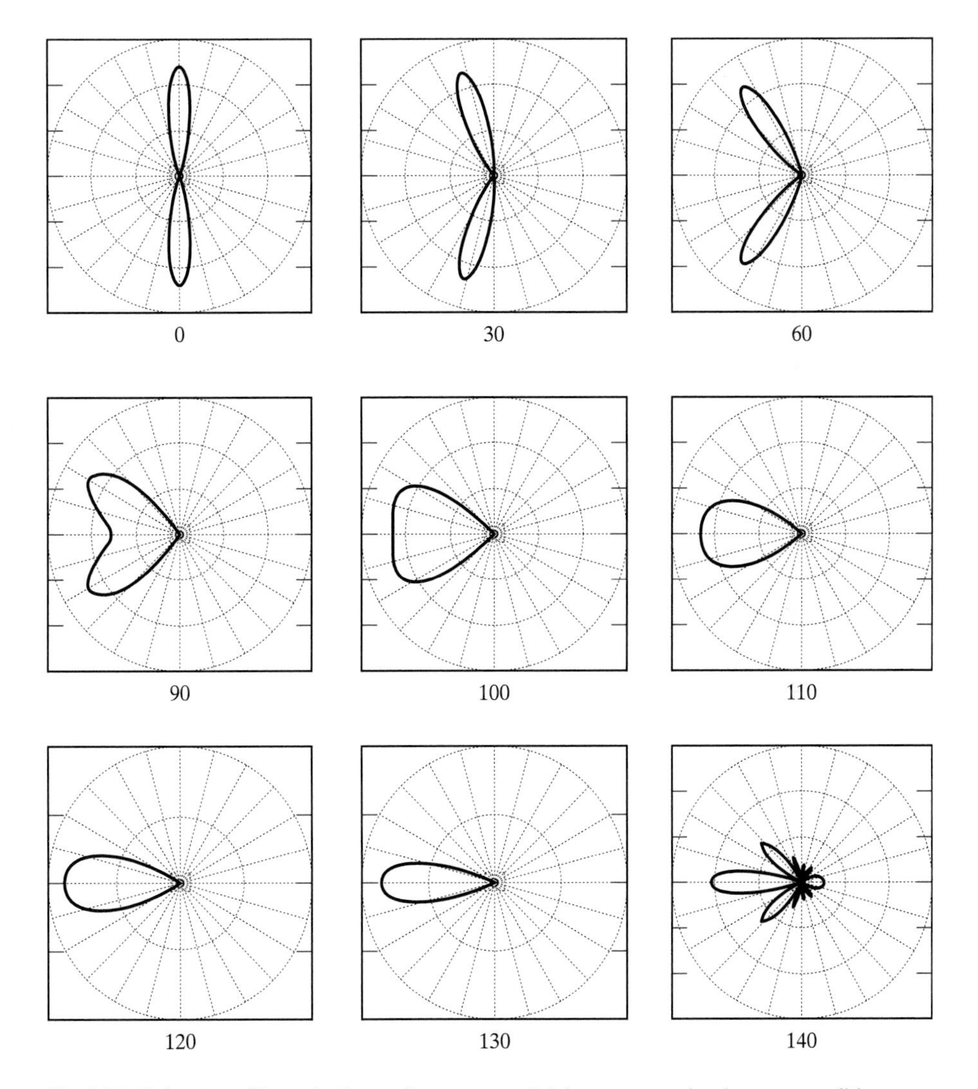

Fig. 3.12 Polar plot of intensity from a linear array of eight sources under the same conditions as for Fig. 3.11. There is a constant, progressive phase difference between the sources, as indicated. At the smaller phase angles, the beam is swept. With greater phase angles, we produce an "end-fired" array. Phase angle in degrees.

"end-fired" array. The side lobes can sometimes be a problem, which can be overcome by "shading" the array (i.e., varying the source strength according to certain distributions). This analysis may be extended to arrays on curves, in two dimensions, and continuous distribution of source strength. Arrays of great sophistication and very high directivities (beam widths of fractions of a degree) have been built. It is beyond the scope of this book to study them in detail, but Urick (1975) can be consulted for further details.

3.7 Huygens' Principle

We have seen that due to an array of point sources the acoustic field may be treated by taking the superposition of spherical waves emanating from each source. This result leads us to accept the plausibility of a useful and historically important development known as Huygens' principle, which may be stated as follows:

> Each point on a wavefront (called the primary) may be regarded as a source of secondary hemispherical waves which propagate in the forward direction and whose envelope at any time constitutes a new primary wavefront.

Huygens formulated the principle in 1690 on the basis of physical intuition. Subsequently, he and others used it to provide a framework to explain a wide variety of propagation phenomena. The quantitative formulation of the principle that we will use was made by Fresnel more than one hundred years later. Rigorous justification for its use—in terms of the fundamental wave equation—invokes Green's theorem, which was a still later development.

In the original semiquantitative form, the principle was propounded as a basis for a wave theory of light. The principle clearly accounts for the diffraction or bending or spreading of waves (see Fig. 3.13), which can be demonstrated with a ripple tank, and was recognized to occur with light. The new theory, however, had to be able to account for everything covered by the old rectilinear propagation theory of optics. The demonstration

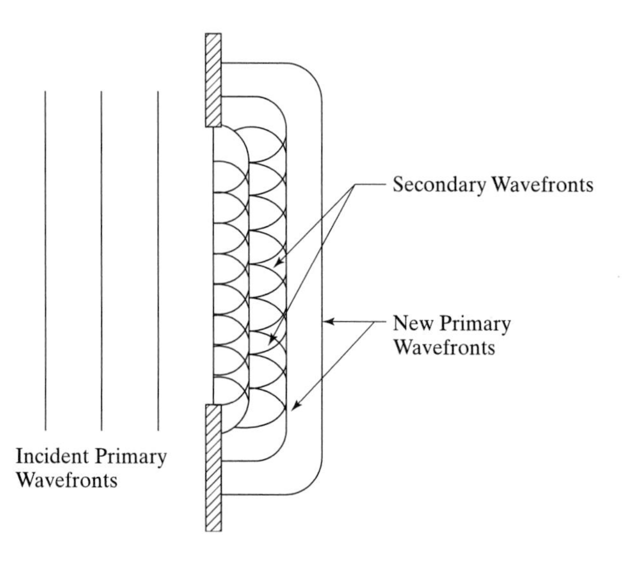

Fig. 3.13 Diffraction of waves interpreted using Huygens' principle.

that the simple laws of reflection and refraction hold for wave motion was not only a prerequisite for the establishment of the wave theory of light but is of interest to us because it carries the not immediately obvious implication that the same laws must be true for acoustical propagation. We will therefore examine certain simple cases in detail.

3.7.1 Reflection of a Plane Wave at a Plane Boundary

As shown in Fig. 3.14, let ABC be the incident wavefront. AD, BG, and CF are normals to the wavefront and correspond to *rays* in the older geometrical theory. Let DJ be a tangent to a secondary wave from C, which travels out from C for the same length of time it takes the point A on the incident wavefront to reach the point D on the reflecting surface. Therefore, both D and J lie on the reflected wavefront. But triangles CJD and CFD are congruent (right angles CJD and CFD; common hypotenuse CD and equal sides CJ and CF), so

$$\angle \mathrm{CDJ} = \angle \mathrm{CDF}$$

Therefore, triangles GHD and GED are congruent ($\angle$GHD = $\angle$GED = 90°; $\angle$GDH = $\angle$GDE, (proved) and common side GD), so that

$$\mathrm{GH} = \mathrm{GE}$$

Since the secondary wave from G, which may be regarded as a general point, also has JD as a tangent, JD is therefore the envelope of all secondary waves arising from points of intersection of the incident wave and the reflecting surface. Thus, a plane wave incident on a plane surface reflects as a plane wave, and since triangles CAD and CJD are congruent, the angle of incidence $\angle$ACD equals the angle of reflection $\angle$JDC. In ray optics or acoustics, these angles are usually defined as the angles between incident and reflected rays and a normal to the surface.

3.7.2 Refraction of Plane Waves at a Plane Boundary

Referring to Fig. 3.15, we see that plane wave AC is incident on the boundary CD between media of sound velocities c_1 and c_2. In time t, ray a reaches D, while the secondary wave

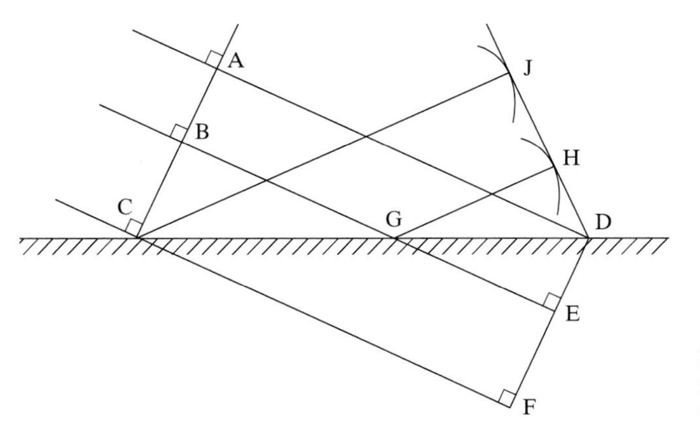

Fig. 3.14 Construction to demonstrate reflection of a plane wave at a rigid plane surface.

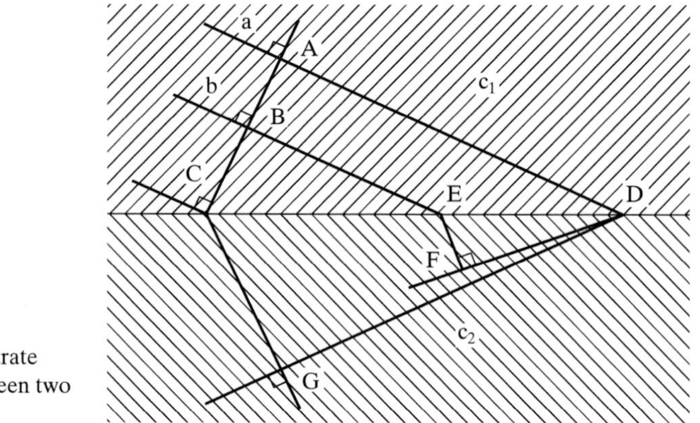

Fig. 3.15 Construction to demonstrate refraction at a plane interface between two media.

from C reaches G. DG is the tangent to this secondary wave. In time t′, ray b reaches E, therefore

$$\mathrm{BE} = c_1 t'$$

and EF = the radius of the secondary wave from E, so

$$\mathrm{EF} = c_2(t - t')$$

Hence

$$\frac{\mathrm{EF}}{\mathrm{CG}} = 1 - \frac{t'}{t} = 1 - \frac{\mathrm{BE}}{\mathrm{AD}}$$

$$= \frac{\mathrm{AD} - \mathrm{BE}}{\mathrm{AD}} = \frac{\mathrm{CD} - \mathrm{CE}}{\mathrm{CD}} = \frac{\mathrm{ED}}{\mathrm{CD}}$$

Therefore

$$\frac{\mathrm{EF}}{\mathrm{ED}} = \frac{\mathrm{CG}}{\mathrm{CD}}$$

and

$$\sin\angle\mathrm{EDF} = \sin\angle\mathrm{CDG}$$

Thus, F lies on GD.

GD is the refracted wavefront, which is also a plane. It follows as well that

$$\frac{\sin\angle\mathrm{ACD}}{\sin\angle\mathrm{CDG}} = \frac{\mathrm{AD/CD}}{\mathrm{CG/CD}} = \frac{c_1 t}{c_2 t} = \frac{c_1}{c_2}$$

This ratio is a constant irrespective of the angle of incidence ∠ACD, so we find that wave refraction is subject to Snell's law, which states:

> The ratio of the sine of the angle of incidence to the sine of the angle of refraction is a constant.

These angles are usually understood to be measured between incident and refracted rays and a normal to the surface.

The phenomenon of refraction is of considerable importance in underwater acoustics. The velocity of sound in seawater depends on temperature, salinity, and depth. A typical variation of sound velocity with depth is shown in Fig. 3.16a. We may use Snell's law to trace the paths of rays through successive horizontal layers. Rays passing into layers of increasing sound velocity are "bent" away from the normal. If this process continues, there will come a point when the ray is travelling horizontally and a slight further change will cause the ray to turn back. Figs. 3.16b and 3.16c show typical ray tracings from sources located in regions I and II of the velocity profile. Note the shadow zones and the formation of a sound channel around the minimum at 1000 m.

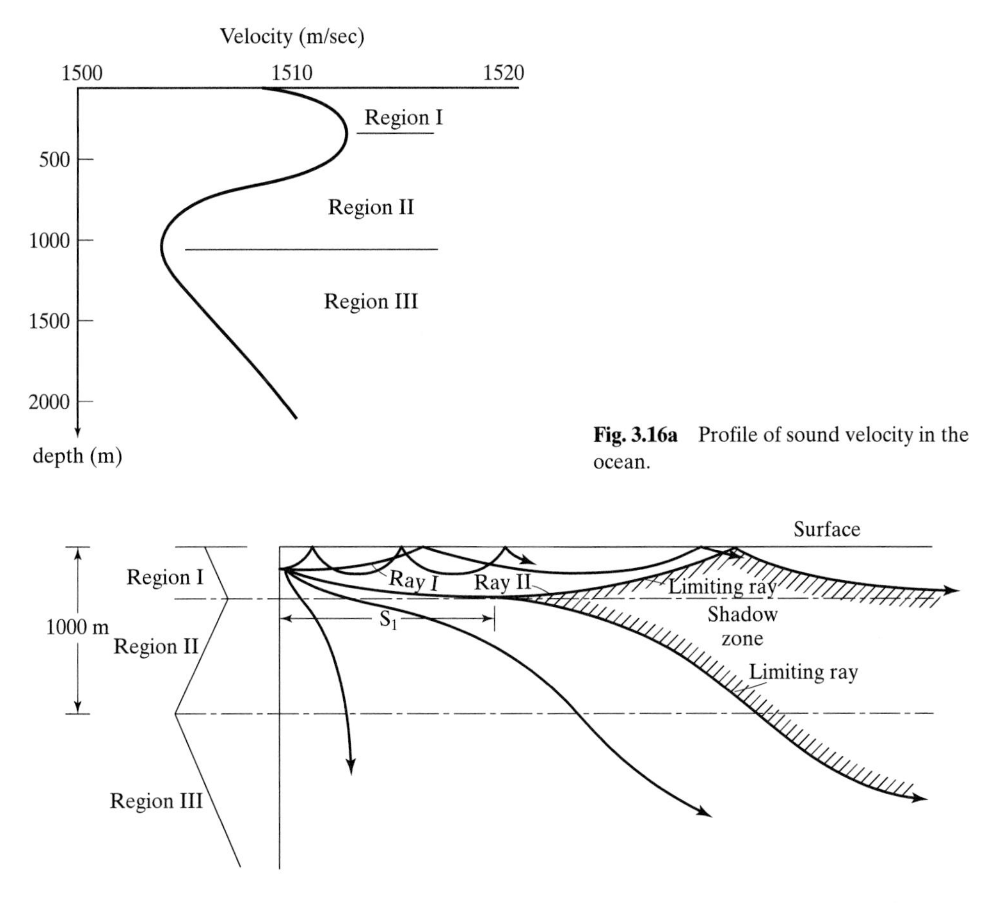

Fig. 3.16a Profile of sound velocity in the ocean.

Fig. 3.16b The ray diagram corresponding to the velocity profile of Fig. 3.16a. (After Tucker and Gazey, 1966.)

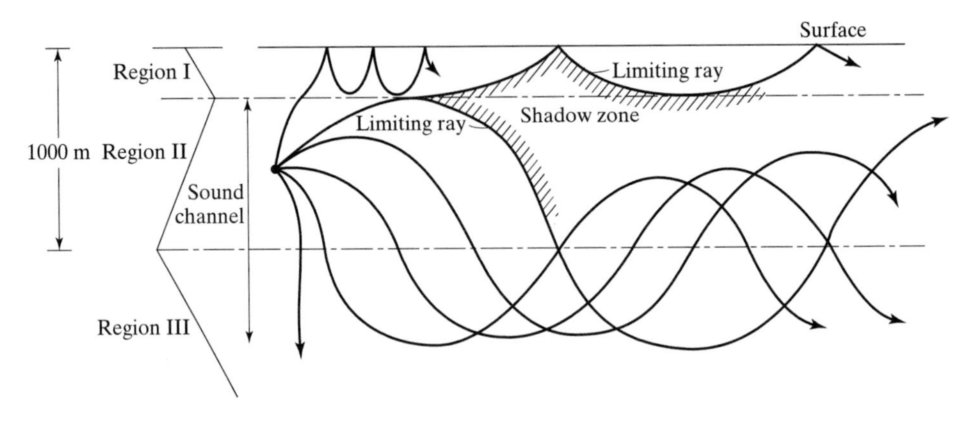

Fig. 3.16c A ray diagram showing the formation of a sound channel. (After Tucker and Gazey, 1966.)

3.7.3 Reflection of Spherical Waves on a Plane Surface

We need only look at Fig. 3.17 to realize that the reflected wavefront will be the exact replica of the portion of the incident wave that would lie beyond the boundary were it not present. The reflected wave will therefore appear to emanate from the point I, which is said to be the image of the source S. An extended source may be regarded as a number of point sources giving rise to spherical waves: So the extended source will also be imaged behind the boundary. Images below the seabed under shallow water are a nuisance in sonar operation.

3.7.4 Reflection of Spherical Waves at a Concave Spherical Surface

The full solution for the reflection of spherical waves at a concave spherical surface is quite complicated. We will be satisfied with a solution valid when all spherical sectors involved have large radii of curvature (i.e., the mirrors are more like saucers than cups). We will make the so-called sagittal approximation, which states that if AB is a

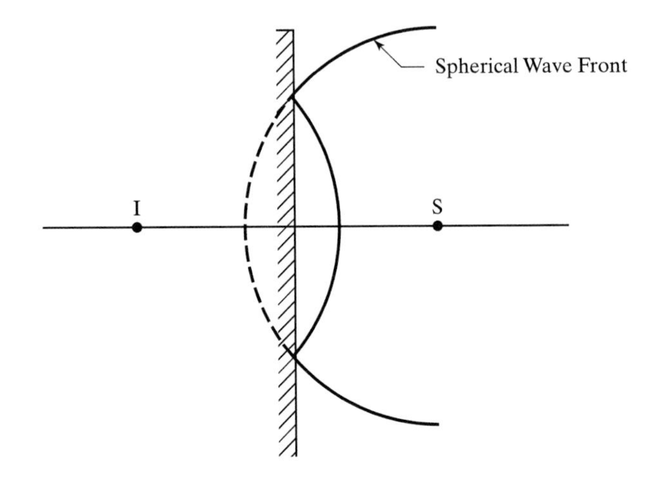

Fig. 3.17 Reflection of a spherical wave from a plane surface.

short chord of a circle (see Fig. 3.18) of radius r, then

$$OM = \frac{AM^2}{2r}$$

Now, consider the situation shown in Fig. 3.19, where

$$\begin{aligned} PN &= PM + MN \\ &= PM + AK \end{aligned}$$

For small apertures, $AK = PQ$, so $PN = PM + PQ$. Further, $PQ = PM - QM$, so $PN = 2PM - QM$. Using the sagittal approximation,

$$\frac{AM^2}{2v} = \frac{2AM^2}{2r} - \frac{AM^2}{2u}$$

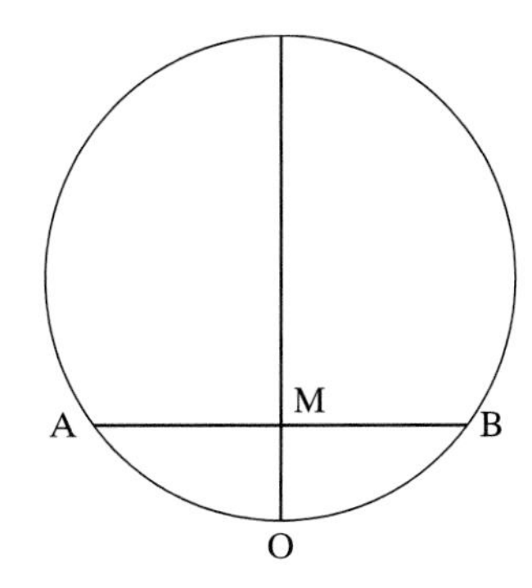

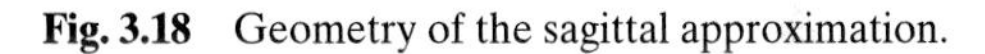
Fig. 3.18 Geometry of the sagittal approximation.

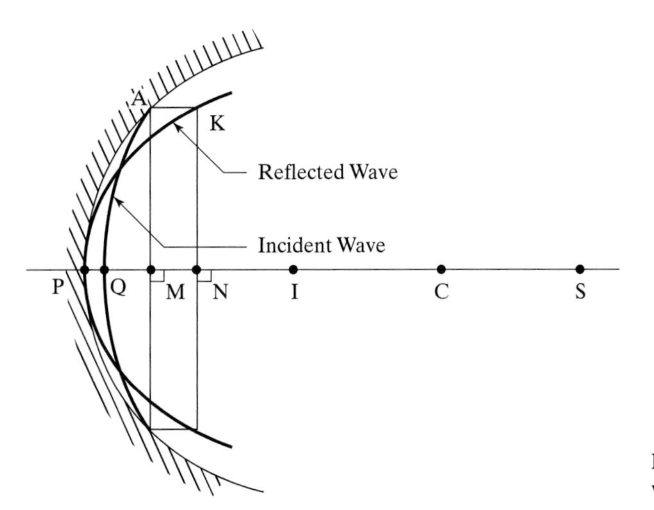

Fig. 3.19 Reflection of a spherical wave at a concave surface.

where r = CP = radius of mirror, u = source distance = PS, and PI = v = distance to the center of curvature of the reflected wave, the so-called image distance. Hence we have

$$\frac{1}{u} + \frac{1}{v} = \frac{1}{r} \tag{3.7.1}$$

If the source is at infinity v = r/2, and this distance is defined as the focal length of the mirror, f. Then

$$\frac{1}{u} + \frac{1}{v} = \frac{1}{f} \tag{3.7.2}$$

This formula applies to mirrors of small aperture and thin lenses, as we will show next. The focal point of the mirror is a distance f or r/2 from P. If a source is located at this point, F, then u = f. Thus, 1/v is zero and parallel rays will emerge from the reflector (see Fig. 3.20). If a source is located at C, then u = 1/r, and v = u (i.e., the source and its image are coincident). In architectural acoustics, it is found that any concave curved wall will approximate a small aperture mirror and there will be a point at which a speaker will hear his own echo very distinctly. The engineering lecture theater at the University of Houston was built with curved side walls designed so that the center of curvature of one wall is located on the opposite wall (see Fig. 3.21). A hand clap at this point produces an excellent flutter echo. A few feet to either side of this point the echo completely vanishes. Concave surfaces of this sort, and as particularly exemplified by domes, should be avoided where there is concern for the acoustics of a room.

3.7.5 The Thin Lens

The procedure for explaining the behavior of a thin lens again employs the sagittal approximation. The ability of the lens to converge or diverge rays depends on the difference in propagation velocities of the lens material and the surrounding medium. In Fig. 3.22, a point source is located at O and rays from O pass through the lens and converge at image

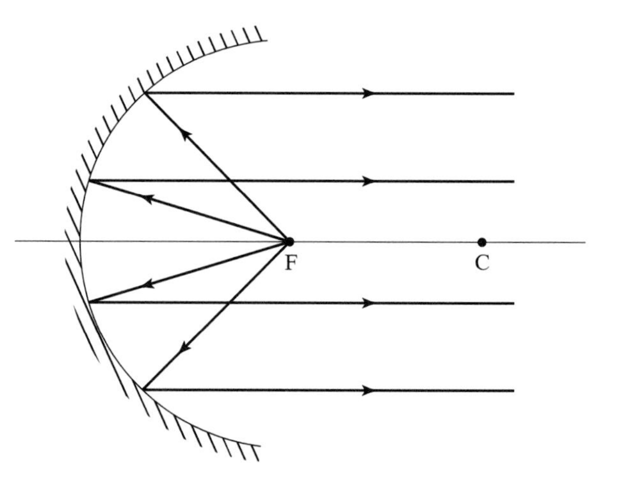

Fig. 3.20 A source situated at the focal point of a concave mirror produces parallel rays.

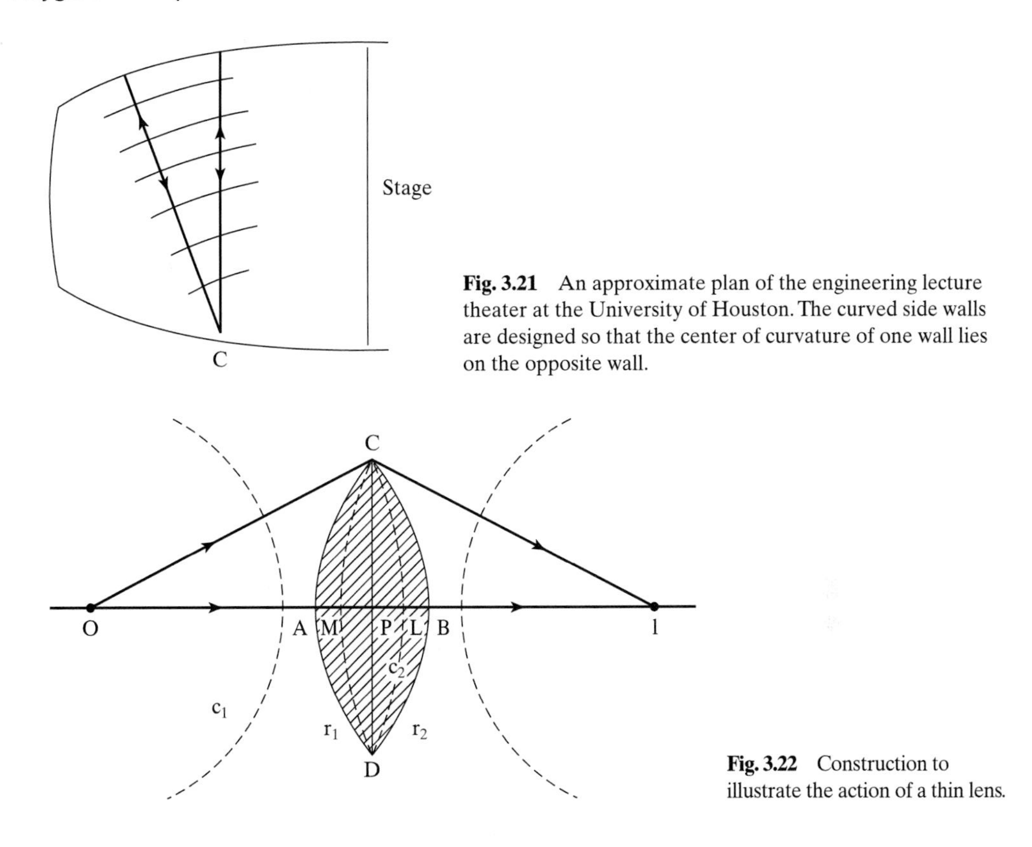

Fig. 3.21 An approximate plan of the engineering lecture theater at the University of Houston. The curved side walls are designed so that the center of curvature of one wall lies on the opposite wall.

Fig. 3.22 Construction to illustrate the action of a thin lens.

point I. The time taken to reach I from O is the same whether the ray passes via C or via AB. CLD is the position that would be occupied by a diverging wave from O, and CMD is the position that would be occupied by a wavefront converging on I. Now

$$\frac{\mathrm{OC}}{c_1} + \frac{\mathrm{C1}}{c_1} = \frac{\mathrm{OA}}{c_1} + \frac{\mathrm{AB}}{c_2} + \frac{\mathrm{B1}}{c_1}$$

Let $c_1/c_2 = \mu$, then

$$\begin{aligned}
&\mathrm{OC} + \mathrm{CI} = \mathrm{OA} + \mu\mathrm{AB} + \mathrm{BI} \\
&\mathrm{OL} + \mathrm{MI} = \mathrm{OA} + \mu\mathrm{AB} + \mathrm{BI} \\
&(\mathrm{OL} - \mathrm{OA}) + (\mathrm{MI} - \mathrm{BI}) = \mu\mathrm{AB} \\
&\mathrm{AL} + \mathrm{MB} = \mu\mathrm{AB} \\
&\mathrm{AP} + \mathrm{PL} + \mathrm{MP} + \mathrm{PB} = \mu(\mathrm{AP} + \mathrm{PB})
\end{aligned}$$

We now use the sagittal approximation and find that

$$\frac{\mathrm{PC}^2}{2r_1} + \frac{\mathrm{PC}^2}{2u} + \frac{\mathrm{PC}^2}{2v} + \frac{\mathrm{PC}^2}{2r_2} = \mu\left(\frac{\mathrm{PC}^2}{2r_1} + \frac{\mathrm{PC}^2}{2r_2}\right)$$

where r_1 and r_2 are the radii of curvature of the lens surfaces, and u and v are again object and image distances. Thus

$$\frac{1}{u} + \frac{1}{v} = (\mu - 1)\left(\frac{1}{r_1} + \frac{1}{r_2}\right) \tag{3.7.3}$$

Again we define focal length, f, as the image distance for an object at infinity, and then

$$\frac{1}{f} = (\mu - 1)\left(\frac{1}{r_1} + \frac{1}{r_2}\right) \tag{3.7.4}$$

and, as in the case of the curved mirror,

$$\frac{1}{u} + \frac{1}{v} = \frac{1}{f} \tag{3.7.5}$$

Acoustic lenses have been constructed to operate at both audio and ultrasonic frequencies. In the first case, the sound velocity difference is supplied by causing the sound to travel through enclosed paths of various lengths, and such lenses resemble microwave lenses. In the ultrasonic case, the lens may be cut entirely from a uniform block of material (e.g., lucite) and would operate in a medium of different sound velocity (e.g., water). Such lenses closely resemble the familiar optical lenses.

3.8 Rectilinear Propagation, the Zone Plate, and Diffraction

We have now seen that Huygens' principle will account for the simple laws of reflection and refraction familiar in the geometrical theories of optics and acoustics. The question that remains is: Why does visible light, if it is a form of wave motion, appear to travel in straight lines, whereas audible sound, clearly a form of wave motion, is able to bend or diffract around obstacles with no difficulty? Acoustic shadow zones are not a very familiar phenomenon. Fresnel developed the following argument in the late eighteenth century to answer this question.

Referring to Fig. 3.23a, ABCD is an advancing plane wavefront. OP is perpendicular to ABCD. We now construct circles with radii OR, OS, OT, etc. such that

$$\begin{aligned} RP - OP &= \lambda/2 \\ SP - RP &= \lambda/2 \\ TP - SP &= \lambda/2, \text{ etc.} \end{aligned}$$

In this way, the area of the wavefront is divided into a number of concentric zones, the first of which is the circle of radius OR whose area is

$$\pi(RP^2 - OP^2) = \pi(RP - OP)(RP + OP)$$

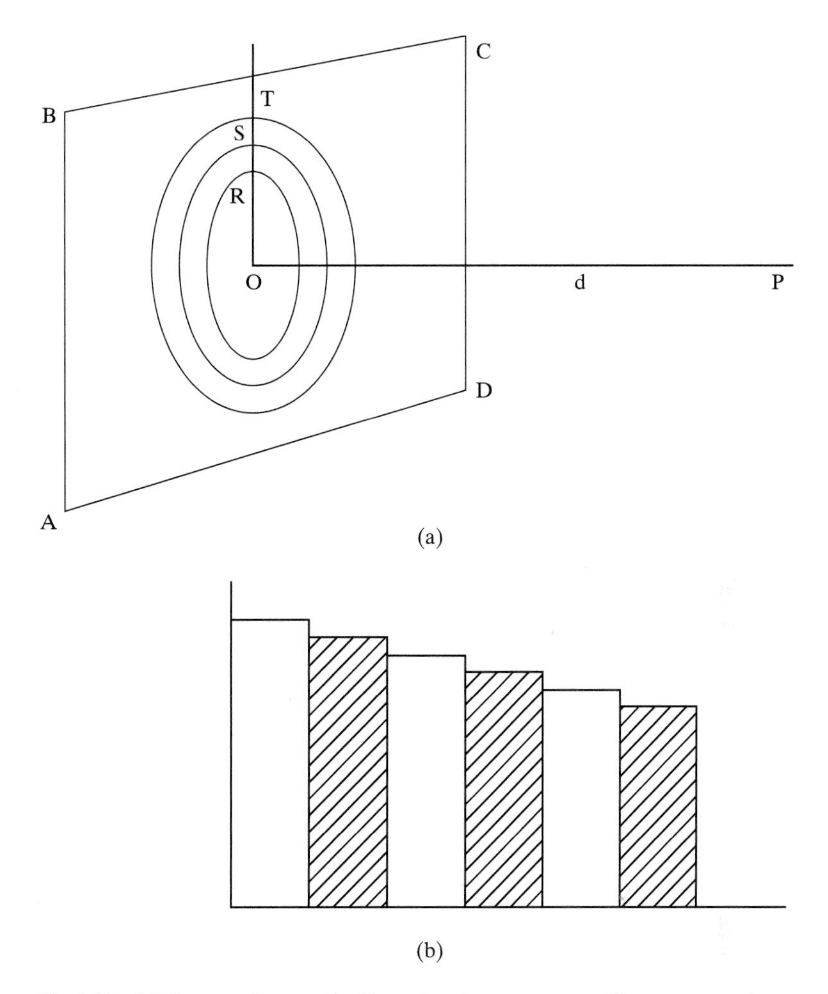

Fig. 3.23 (a) Construction used by Fresnel to demonstrate rectilinear propagation. Imaginary zones are drawn on advancing wavefront ABCD. (b) The amplitudes of radiation arrive at P from successive zones and alternate in and out of phase.

But if OP (d) is large compared with a wavelength (λ), the area of this first zone $\cong \pi(\lambda/2)(2d) \cong \pi\lambda d$. Similarly, the area of the second zone is

$$\begin{aligned}\pi OS^2 - \pi OR^2 &= \pi(SP^2 - OP^2) - \pi\lambda d \\ &= \pi(SP + OP)(SP - OP) - \pi\lambda d \\ &= \pi 2\lambda - \pi\lambda d = \pi\lambda d\end{aligned}$$

In turn, we see that the area of each successive zone is $\pi\lambda d$, provided its radius is much less than d.

The amplitudes of the disturbances arriving at P from successive rings are represented by the areas of the strips in Fig. 3.23b. Although the zonal areas are the same, the resulting amplitudes will not be because of inverse square law falloff due to the increasing distance of each zone due to P. To a first approximation, there is a phase change of π from zone to zone, resulting in near, but not complete, cancellation of the disturbances from successive zones. It may be seen, in fact, that the resultant effective at P is due to half the area of the first zone; that is, due to a circle of radius $(d\lambda/2)^{1/2}$.

Now consider the effect of blocking off the advancing wavefront everywhere except for a small circular aperture, as shown in Fig. 3.24a. The resultant amplitude at P is due only to small fractions of several zones whose relative contributions are as indicated by the strips in Fig. 3.24b. The resultant amplitude at P will be much smaller than from the unobstructed wavefront, and almost zero in fact, *unless* the aperture lies *inside* the *first zone*. In this case, the amplitude at P will actually be greater than from the unobstructed wave.

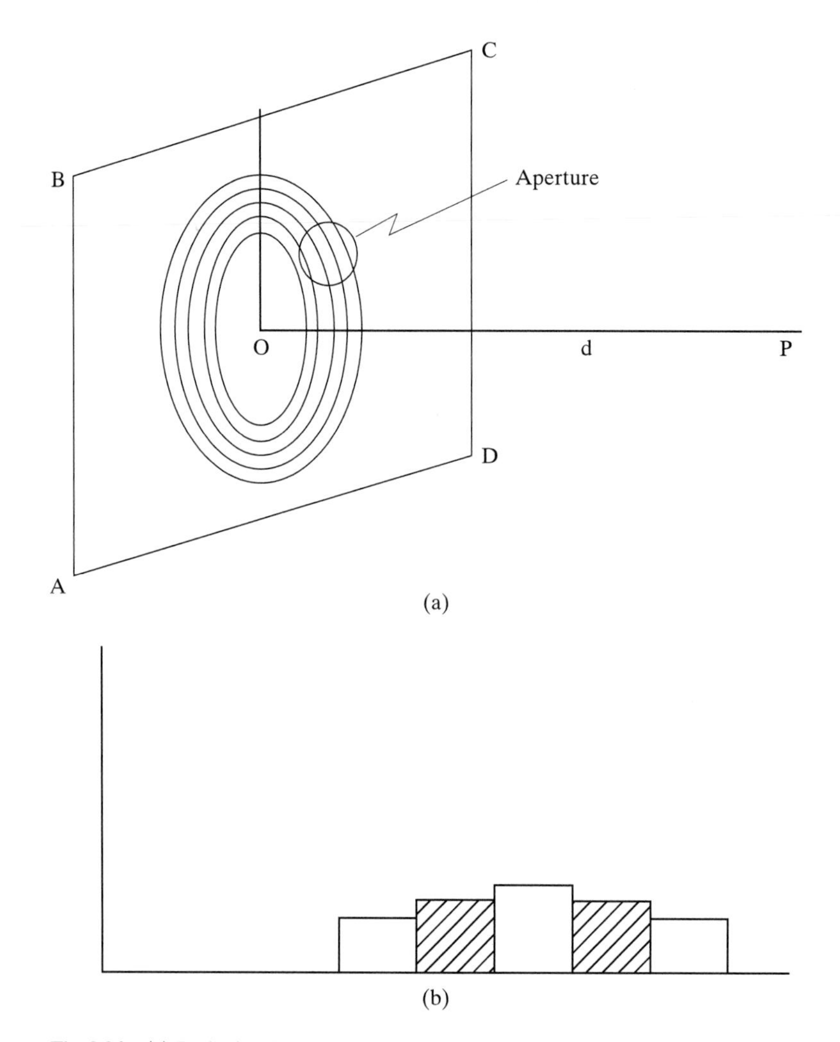

Fig. 3.24 (a) Radiation from Fresnel zones is blocked everywhere except from the aperture shown. (b) Amplitudes of radiation arriving at P from successive zones in the aperture.

It may thus be seen that if P lies in the geometrical projection of the aperture, there will be a resultant amplitude at P but this amplitude will be less the farther P lies outside this geometrical projection. It also follows that the larger the aperture is relative to the first zone, the sharper will be the delineation of the radiation beam. As a criterion of how large an aperture has to be to give good rectilinear propagation, we can calculate the radius of the first zone, namely $(d\lambda)^{1/2} = r_z$. For example, if $d = 100$ cm: for yellow light, $\lambda = 4 \times 10^{-5}$ cm and $r_z - 0.06$ cm; while for sound in air, a typical wavelength $\lambda = 10^2$ cm and $r_z = 10^2$ cm.

Thus, yellow light will propagate in a rectilinear manner from an aperture of radius about a millimeter, while to obtain rectilinear propagation with audible sound requires apertures of the order of a meter in radius. It only makes sense to speak of acoustic shadow zones at very high frequencies (in the ultrasonic range) or where the propagation distances are very large, as in underwater work. This is not to say that ray tracing will not give some sensible results even when the dimensions involved are comparable to a wavelength, which is often the situation in architectural acoustics, for instance. Here it would often be quite impossible to obtain full solutions of the wave equation, while ray acoustics at least gives some idea of the behavior of the sound field.

The zone plate was originally a geometrical construction in a mathematical argument. Now, however, it is the basis of a practical design of lens. Again referring to Fig. 3.23, we can argue that if the radiation from alternate zones could be blocked off, there would be only constructive interference at P and an enhancement of the amplitude over that from the complete wavefront. Suppose XY in Fig. 3.25 is such a lens with the even-numbered zones blocked off. Rays from O reach I through the central zone and

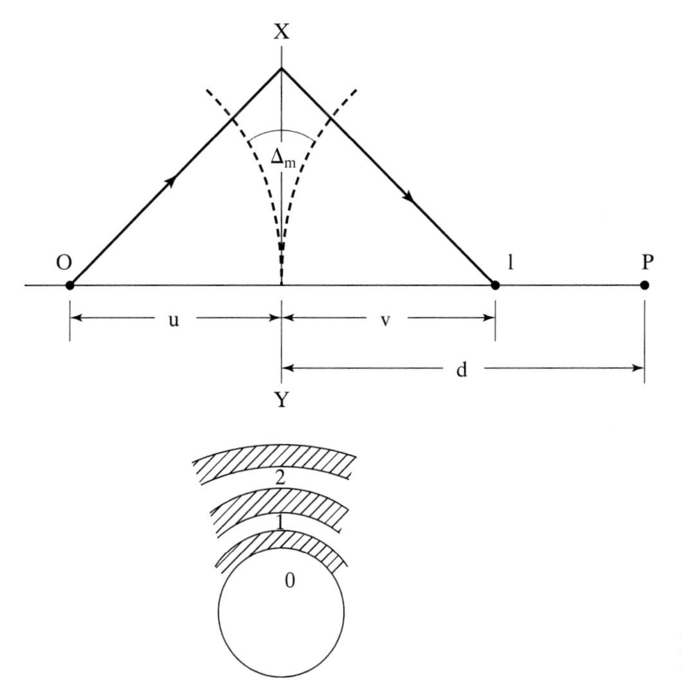

Fig. 3.25 Construction to explain the action of a Fresnel lens.

through zone m. For reinforcement,

$$\Delta_m = m\lambda \approx \frac{r_m^2}{2u} + \frac{r_m^2}{2v} \tag{3.8.1}$$

using the sagittal approximation. Thus

$$\frac{1}{u} + \frac{1}{v} = \frac{2m\lambda}{r_m^2} \tag{3.8.2}$$

But

$$\pi r_m{}^2 = 2m\lambda\pi d \tag{3.8.3}$$

so

$$\frac{1}{u} + \frac{1}{v} = \frac{1}{d} \tag{3.8.4}$$

Thus, the plate with alternate blocked zones behaves like a lens of focal length d.

We have so far discussed how Huygens' principle can be used to derive the important results of geometrical optics and to explain the rectilinear propagation of light. It also provided the understanding for other phenomena clearly beyond the realm of geometrical optics and which constituted around 1800, a new discipline known as physical optics. An example of these developments, the diffraction grating, is important in both optics and acoustics. The grating illustrated in Fig. 3.26 consists of a large number of parallel linear obstructions or reflectors, which in the optical case, for example, may consist of rulings in the silvering of a mirror. If the radiation passes through the grating, it is said to be of the transmission type; if it is reflected, of the reflection type. In neither case are the simple laws of refraction and reflection obeyed. We will first consider a transmission grating.

In Fig. 3.26a, the path difference between rays (1) and (2) = BC + CD; that is,

$$d[\sin(i) + \sin(\theta)]$$

If these rays are eventually brought together—either through slight initial convergence or, if they were initially parallel, by the action of a lens—they will reinforce if

$$d[\sin(i) + \sin(\theta)] = m\lambda, \ m = \text{integer} \tag{3.8.5}$$

Here we see that for a given angle of incidence there are several transmission angles for reinforcement. In a transmission grating, the angle of incidence is usually zero; that is,

$$d \sin\theta = m\lambda \tag{3.8.6}$$

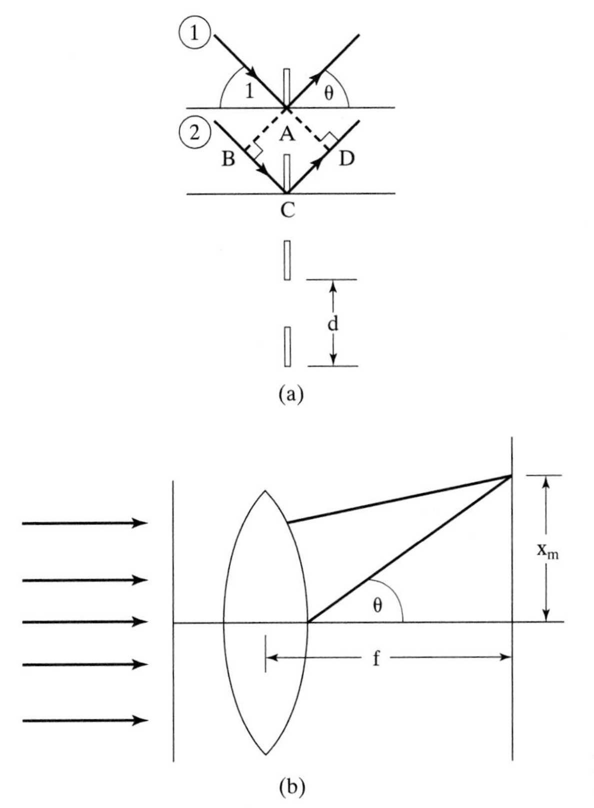

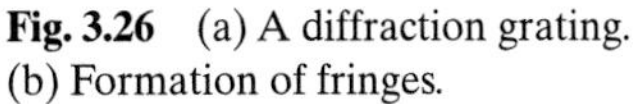
Fig. 3.26 (a) A diffraction grating. (b) Formation of fringes.

If a long focal length lens is used to focus the transmitted radiation as in Fig. 3.26b, the position of the mth intensity maximum, x_m, is given by

$$\frac{x_m}{f} \approx \sin \theta_m \tag{3.8.7}$$

Thus

$$x_m = \frac{mf\lambda}{d} \tag{3.8.8}$$

and

$$\Delta x = x_m - x_{m-1} = \frac{f\lambda}{d} \tag{3.8.9}$$

These intensity maxima are called fringes, so it is seen that by measuring the fringe separation it is possible to determine λ.

The theory of the diffraction grating may be applied to the Schlieren optical system (Fig. 3.27) in which there is an interaction between optical and acoustical waves. In

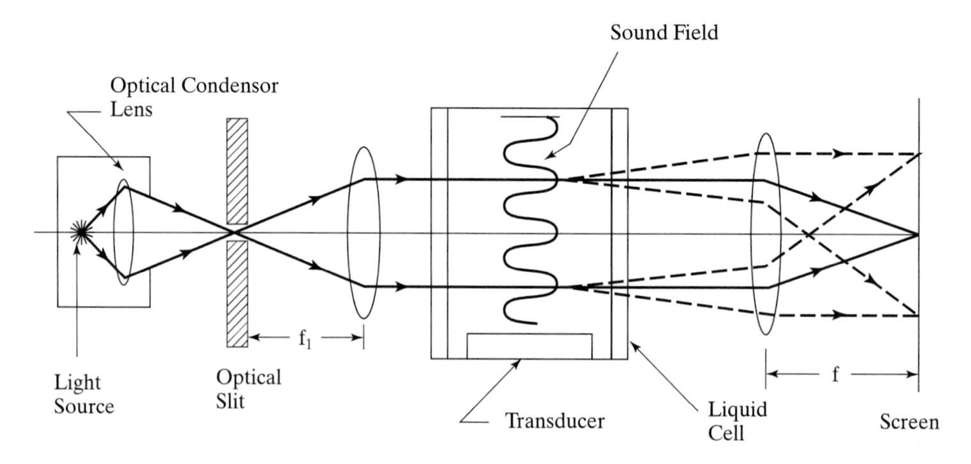

Fig. 3.27 Diffraction of light by an acoustic standing wave.

the system shown we arrange for a plane standing wave in the liquid cell. The maxima of excess density of the sound wave are λ_0 apart, where λ_0 is the sound wavelength, and they act as a diffraction grating for the incident light, which has to be monochromatic. Multiple optical images of the slit are formed on the screen, and if their spacing is Δx, it follows from Eq. (3.8.9) that

$$\Delta x = \frac{f\lambda}{\lambda_0} \tag{3.8.10}$$

Hence, λ_0 and the velocity of sound may be determined to great accuracy.

In a reflection grating (see Fig. 3.28), the path difference is

$$d[\sin(i) - \sin(\theta)] = m\lambda \tag{3.8.11}$$

for reinforcement.

The cable grid floor sometimes used in an anechoic room presents an interesting example of a reflection diffraction grating (see Fig. 3.29). The effect is most noticeable at grazing incidence ($i = 90°$) when the wires present a large cross section to the radiation. In that case,

$$d(1 - \sin\theta) = m\lambda \tag{3.8.12}$$

and

$$1 - \frac{m\lambda}{d} = \sin\theta \tag{3.8.13}$$

for reinforcement. Now if $d >> \lambda$,

$$\sin\theta \cong 1, \theta \cong 90° \tag{3.8.14}$$

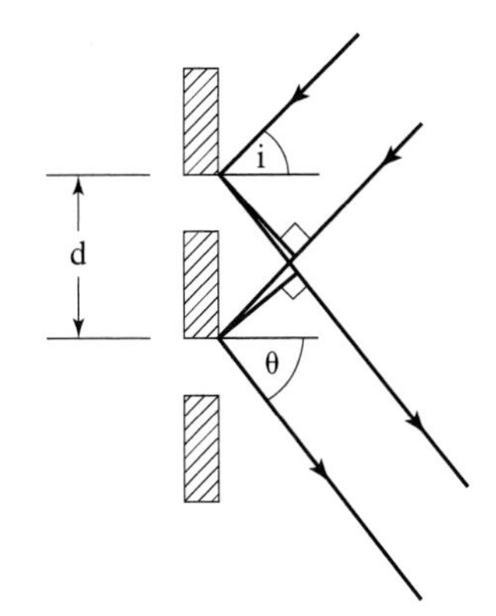

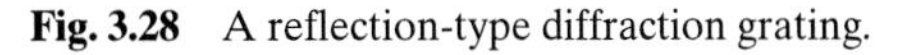
Fig. 3.28 A reflection-type diffraction grating.

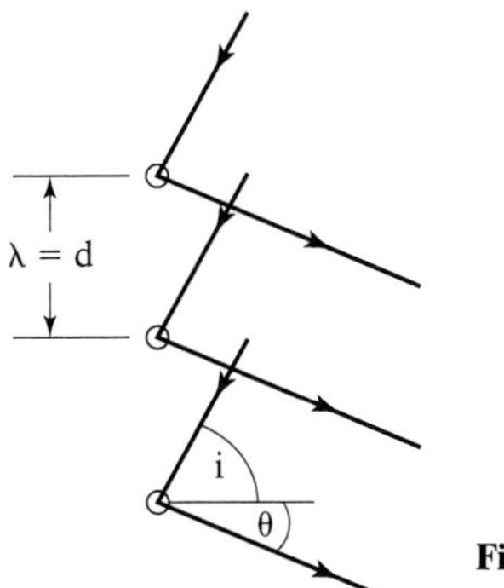

Fig. 3.29 The wires of a cable grid floor in an anechoic chamber act as a reflection grating.

and the radiation is not deflected. But as $d \rightarrow \lambda$, we see that when $m = 1$,

$$\sin\theta = 0$$

and

$$\theta = 0$$

so the floor is a strong reflector. For a 2 inch square grid floor, this condition occurs at about 6.5 KHz. At this frequency, strong deviations from inverse square law falloff are observed using the arrangement shown in Fig. 3.30.

3.9 Fresnel's Formulation of Huygens' Principle

As formulated in the preceding sections, Huygens' principle is semiquantitative in that it does not permit prediction of the amplitude of the radiation. To overcome this deficiency, Fresnel assumed that the complex amplitude of radiation at P due to an element of a wavefront dS (see Fig. 3.31) could be written

$$d\psi = A\psi_o \frac{e^{-jkr}}{r} dS \tag{3.9.1}$$

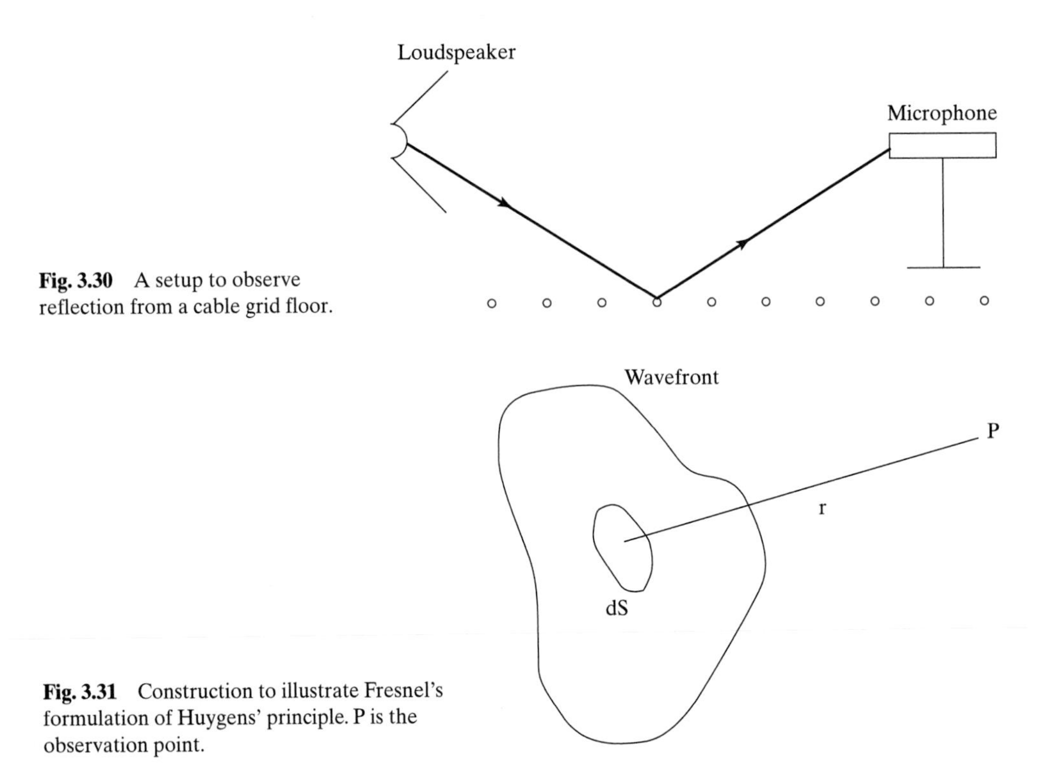

Fig. 3.30 A setup to observe reflection from a cable grid floor.

Fig. 3.31 Construction to illustrate Fresnel's formulation of Huygens' principle. P is the observation point.

where ψ_o is the amplitude at the wave surface and A is a constant of proportionality. Consequently, the total complex amplitude at P is given by

$$\psi = \int\int_s A\psi_o \frac{e^{-jkr}}{r} dS \tag{3.9.2}$$

Consider now a plane wave of amplitude ψ_o incident on an aperture in a rigid plane baffle, as in Fig. 3.32. Then with the situation shown, the amplitude at P will be

$$\psi = \int\int_s A\psi_o \frac{e^{-jkr}}{r} h\, dh\, d\psi \tag{3.9.3}$$

But

$$r^2 = h^2 + z^2 \tag{3.9.4}$$

$$rdr = hdh \tag{3.9.5}$$

and

$$\frac{hdh}{r} = dr \tag{3.9.6}$$

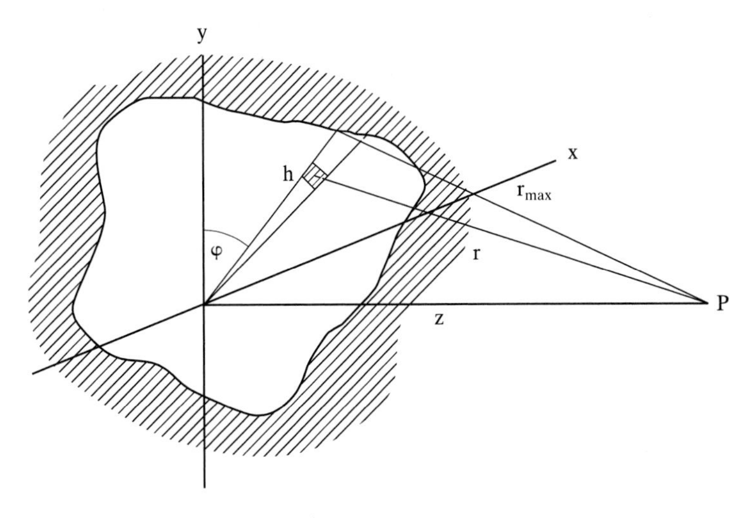

Fig. 3.32 A plane wave is incident on an aperture in a rigid baffle situated in the x-y plane.

thus

$$\psi = \int_{\varphi}\int_{r=z}^{r=r_{max}} A\psi_o e^{-jkr}\, dr\, d\varphi$$
$$= A\psi_o \int_{\varphi}\left\{\frac{e^{-jkz}}{jk} - \frac{e^{-jkz_{max}}}{jk}\right\} d\varphi \tag{3.9.7}$$

The evaluation with respect to φ depends on whether P lies inside or outside the geometrical beam.

1. If P lies inside the geometrical beam or the projection of the wavefront, then

$$\psi = A\psi_o \frac{e^{-jkz}}{jk}\int_o^{2\pi} d\varphi - \frac{A\psi_o}{jk}\int_o^{2\pi} e^{-jkr_{max}}\, d\varphi \tag{3.9.8}$$

Now if r_{max} were to vary randomly by a large number of wavelengths, the second term would have to vanish and the complex amplitude would be

$$\psi = \frac{A}{jk} a e^{-jkz} 2\pi = \frac{A\lambda}{j} a e^{-jkz} \tag{3.9.9}$$

Under these circumstances, the radiation propagates as a plane wave. But when $z = 0$, we have

$$\psi = \psi_o \tag{3.9.10}$$

and we determine the constant of proportionality

$$A = \frac{j}{\lambda} \tag{3.9.11}$$

Restating the Huygens-Fresnel principle, we have

$$\psi = \int\int_s \frac{j\psi_o}{\lambda}\frac{e^{-jkr}}{r}dS \tag{3.9.12}$$

Note that since $j = e^{j\pi/2}$, the secondary spherical waves are $\pi/2$ advanced in phase over the primary wave. Returning to the case of the bounded plane, the complex amplitude at a point in the geometrical beam is

$$\psi = \psi_o e^{-jkz} - \frac{\psi_o}{2\pi}\int_o^{2\pi} e^{-jkr_{max}}\, d\varphi \tag{3.9.13}$$

The first term is the amplitude that would result with purely rectilinear or geometrical propagation, while the second term is variously called the diffraction, boundary, or edge wave.

2. If P lies outside the geometrical beam, as in Fig. 3.33, the limits of integration for φ are different. We now have

$$\psi = \frac{\psi_o e^{-jkz}}{2\pi}\int_\varphi d\varphi - \frac{\psi_o}{2\pi}\int_o^{2\pi} e^{-jkr_{max}}\, d\varphi \tag{3.9.14}$$

The limits are in fact that $\varphi_{final} = \varphi_{initial}$, and thus

$$\int_\varphi d\varphi = 0 \tag{3.9.15}$$

and

$$\psi = -\frac{\psi_o}{2\pi}\int_\varphi e^{-jkr_{max}}\, d\varphi \tag{3.9.16}$$

that is, only the diffraction wave is present.

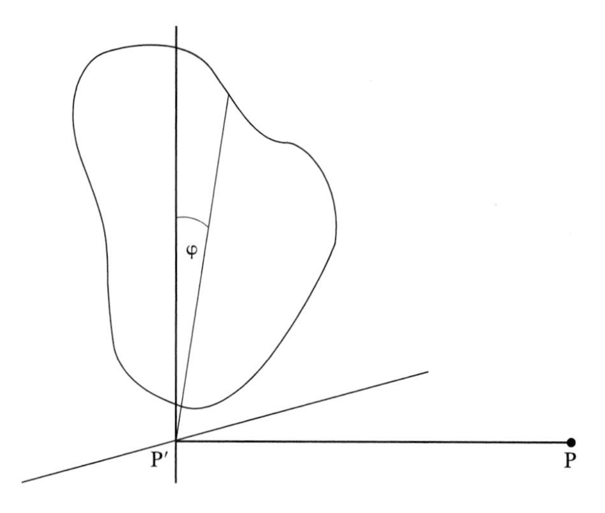

Fig. 3.33 The observation point P lies outside the geometrical projection of the aperture.

3.10 Kirchoff Radiation Theory

One problem with the Huygens-Fresnel theory is that a point on the primary wavefront has to be assumed to give rise to hemispherical secondaries. If completely spherical waves were assumed, radiation from a given primary would proceed in both forward *and backward* directions. It is not clear if the secondary waves are pressure or velocity waves. In 1882, Kirchoff was able to show that a form of Huygens' principle could be established on an exact basis, starting from the wave equation.

Consider the vectors **E** and **F** such that

$$\mathbf{E} = \varphi\nabla\psi \tag{3.10.1}$$

and

$$\mathbf{F} = \psi\nabla\varphi \tag{3.10.2}$$

where φ and ψ are two spatially dependent scalar quantities.

Now, Gauss' theorem states that

$$\int\int_s \mathbf{E}\cdot d\mathbf{S} = \int\int\int_v \nabla\cdot\mathbf{E}\, dV \tag{3.10.3}$$

and

$$\int\int_s \mathbf{F}\cdot d\mathbf{S} = \int\int\int_v \nabla\cdot\mathbf{F}\, dV \tag{3.10.4}$$

where the integration is applied to a closed volume. Hence

$$\begin{aligned}\int\int_s \varphi\nabla\psi\cdot d\mathbf{S} &= \int\int\int_v \nabla\cdot(\varphi\nabla\psi)\, dV \\ &= \int\int\int_v (\nabla\varphi\cdot\nabla\psi + \varphi\nabla^2\psi)\, dV\end{aligned} \tag{3.10.5}$$

and

$$\int\int_s \psi\nabla\varphi\cdot d\mathbf{S} = \int\int\int_v (\nabla\psi\cdot\nabla\varphi + \psi\nabla^2\varphi)\, dV \tag{3.10.6}$$

Subtracting Eq. (3.10.6) from Eq. (3.10.5), we obtain Green's theorem:

$$\int\int_s [\phi\nabla\psi - \psi\nabla\varphi]\cdot d\mathbf{S} = \int\int\int_v (\varphi\nabla^2\psi - \psi\nabla^2\varphi)\, dV \tag{3.10.7}$$

If we now stipulate that φ and ψ are solutions of the Helmholtz equation, that is,

$$(\nabla^2 + k^2)\varphi = 0 \quad \text{and} \quad (\nabla^2 + k^2)\psi = 0 \tag{3.10.8}$$

we obtain

$$\int\int_s [\phi\nabla\psi - \psi\nabla\varphi]\cdot d\mathbf{S} = 0 \tag{3.10.9}$$

Eq. (3.10.9) is sometimes referred to as Helmholtz's extension of Green's theorem. Kirchoff's contribution was the realization that the theorem could be used to derive a form of Huygens' principle.

Consider the surface of integration shown in Fig. 3.34 where S_2 is a spherical surface, radius ε, centered on the point P and connected to S_1 by a vanishingly small tube. The outward normals on these two surfaces are as shown. Then

$$\int\int_{S_1}[\psi\nabla\varphi - \varphi\nabla\psi]\cdot d\mathbf{S}_1 + \int\int_{S_2}[\psi\nabla\varphi - \varphi\nabla\psi]\cdot d\mathbf{S}_2 = 0 \qquad (3.10.10)$$

Let φ be a solution of the wave equation, due to a point source at P, then over surface S_2

$$\varphi = \frac{Ae^{j(\omega t - k\varepsilon)}}{\varepsilon} \qquad (3.10.11)$$

and

$$\nabla\varphi = -\mathbf{e}_r\frac{A}{\varepsilon^2}(1 + jk\varepsilon)e^{j(\omega t - k\varepsilon)} \qquad (3.10.12)$$

Over surface S_2, $\mathbf{n} = -\mathbf{e}_r$; thus, on substituting Eqs. (3.10.11) and (3.10.12) into the second term of Eq. (3.10.10), we obtain:

$$\int_\Omega\left[-\mathbf{e}_r\psi\frac{A}{\varepsilon^2}(1 + jk\varepsilon)e^{j(\omega t - k\varepsilon)} - \mathbf{e}_r\frac{A}{\varepsilon}e^{j(\omega t - k\varepsilon)}\frac{\partial\psi}{\partial n}\right]\cdot(-\mathbf{e}_r\varepsilon^2\,d\Omega) \qquad (3.10.13)$$

Evaluating this integral and letting $\varepsilon \to 0$ yields

$$\lim_{\varepsilon\to 0}\left[-\mathbf{e}_r\psi_p\frac{A}{\varepsilon^2}(1 + jk\varepsilon)e^{j(\omega t - k\varepsilon)} - \mathbf{e}_r\frac{A}{\varepsilon}e^{j(\omega t - k\varepsilon)}\left(\frac{\partial\psi}{\partial n}\right)_p\right]\cdot(-\mathbf{e}_r 4\pi\varepsilon^2) = 4\pi A\psi_p e^{j\omega t} \qquad (3.10.14)$$

where ψ_p is the value of ψ at P. Hence, returning to Eq. (3.10.10), we have

$$\psi_p e^{j\omega t} = \frac{1}{4\pi A}\int\int_{S_1}[\phi\nabla\psi - \psi\nabla\varphi]\cdot d\mathbf{S}_1 \qquad (3.10.15)$$

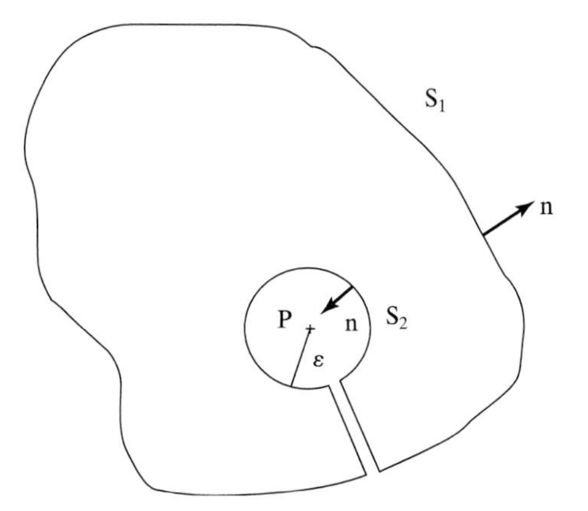

Fig. 3.34 Surface of integration for Kirchoff Radiation Theory.

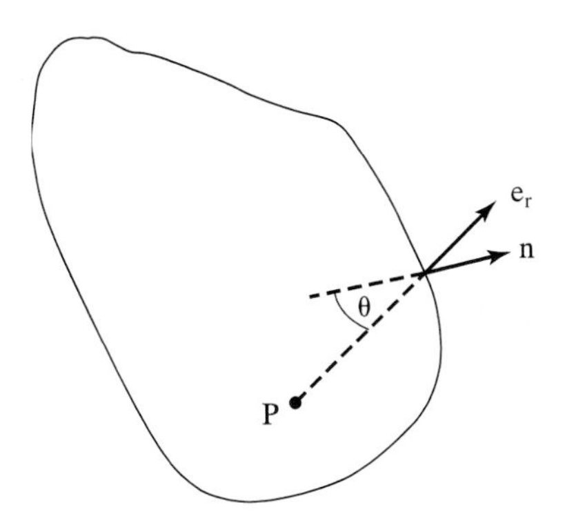

Fig. 3.35 Unit vectors e_r and n on surface S_1.

If **r** is the radius vector from P to any point on S_1,

$$\varphi = \frac{A}{\mathbf{r}} e^{j(\omega t - kr)} \tag{3.10.16}$$

Then

$$\psi_p e^{j\omega t} = \frac{1}{4\pi} \int\int_{S_1} \left[\frac{e^{j(\omega t - kr)}}{r} \frac{\partial \psi}{\partial n} - \psi \frac{\partial}{\partial r} \left(\frac{e^{j(\omega t - kr)}}{r} \right) \cos\theta \right] dS \tag{3.10.17}$$

where θ is the angle between $\mathbf{e}_r$ and **n** as shown in Fig. 3.35. Eq. (3.10.17) is often referred to as the Kirchoff integral formula. The first term on the right side may be interpreted as arising from the summation of secondary spherical wavelets due to elementary monopole sources with strengths proportional to $\partial\varphi/\partial n$, which in turn is proportional to the particle velocity. The second term may be interpreted as arising from the summation of secondary wavelets due to dipole sources with moments proportional to ψ, which in turn is proportional to pressure. Thus, we can see a problem with Kirchoff's formulation of Huygens' principle; namely, that it requires specification of both pressure and velocity over the radiating surface. It is usually not convenient to do this, and a more convenient formulation was produced by Rayleigh and Sommerfeld.

3.11 Rayleigh-Sommerfeld Formulation of Radiation

The Rayleigh-Sommerfeld formulation is applied to the special case of vibration over a plane. The only condition for φ in Eq. (3.10.9) is that it satisfy the Helmholtz equation, so we may choose it to be the solution for a dipole with one source located at P (see Fig. 3.36) and a second source at the same distance behind the plane. If the sources are of the same strength, but out of phrase

$$\varphi = \psi \frac{A}{r} e^{j(\omega t - kr)} - \frac{A}{r'} e^{j(\omega t - kr')} \tag{3.11.1}$$

A solution to Eq. (3.10.9) may now be obtained taking the integration surface to be the plane S_1, a hemisphere S_3, closed by the plane and S_2, a small sphere centered

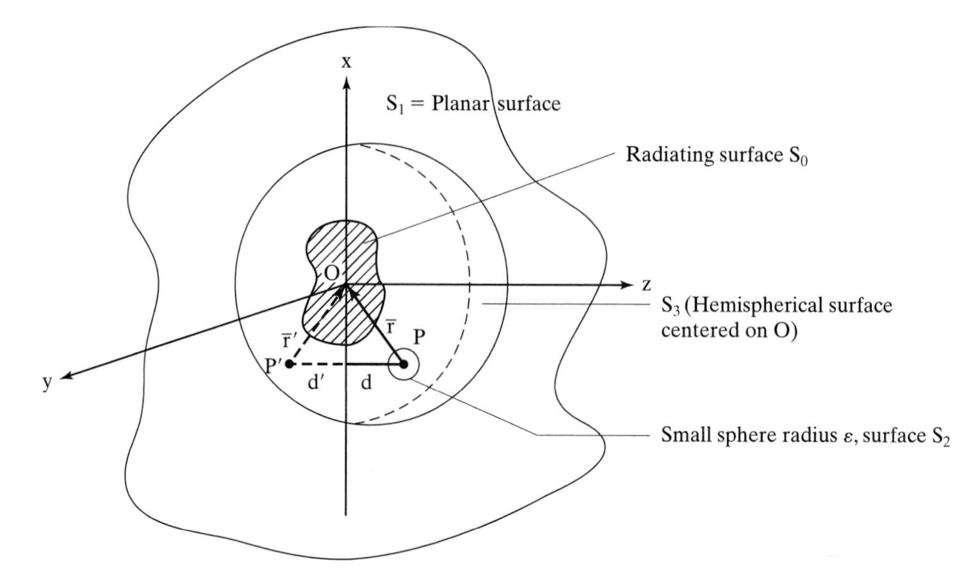

Fig. 3.36 Construction for the Rayleigh-Sommerfeld radiation theory.

on P. This construction is as shown in Fig. 3.36. Now assume that φ and ψ and their derivatives decrease at least as fast as spherical waves over S_3 (this is sometimes called the Sommerfeld radiation condition). The integral over that surface is thus zero as the hemisphere approaches infinity. But

$$\left(\frac{\partial\varphi}{\partial n}\right)_{S_2} = -\left(\frac{1}{\varepsilon} + jk\right)\frac{A}{\varepsilon}e^{j(\omega t - k\varepsilon)} - \frac{\partial}{\partial n}\left(\frac{e^{j(\omega t - kr')}}{r'}\right)_{S_2} \tag{3.11.2}$$

In the limit as $\varepsilon \rightarrow 0$,

$$(\varphi)_{S_2} = \frac{A}{\varepsilon}e^{j(\omega t - k\varepsilon)}$$

and

$$\left(\frac{\partial\varphi}{\partial n}\right)_{S_2} = -\left(\frac{1}{\varepsilon} + jk\right)\varphi_{S_2}e_r \tag{3.11.3}$$

Thus

$$\int\int_{S_2}\left(\varphi\frac{\partial\psi}{\partial n} - \psi\frac{\partial\varphi}{\partial n}\right)dS = I_2$$

$$= \left[\frac{A}{\varepsilon}\left(\frac{\partial\psi}{\partial n}\right)_p - \psi_p\left(\frac{1}{\varepsilon} + jk\right)\frac{A}{\varepsilon}(-e_r)\right]e^{j(\omega t - k\varepsilon)}.4\pi\varepsilon^2(-e_r) \tag{3.11.4}$$

Now

$$\lim_{\varepsilon \to 0} I_2 = -4\pi\psi_p A e^{j\omega t} \tag{3.11.5}$$

Thus, finally we obtain from Eq. (3.10.9)

$$\psi_p e^{j\omega t} = \frac{1}{4\pi A} \int\int_{S_1} \left(\varphi \frac{\partial \psi}{\partial n} - \psi \frac{\partial \varphi}{\partial n} \right) dS \tag{3.11.6}$$

This result is identical to Eq. (3.10.15) derived with a monopole Green's function, but φ and $\partial\varphi/\partial n$ are now different. The boundary conditions over the plane surface S_1 are now much easier to apply.

$$\frac{\partial \varphi}{\partial n} = -\left[jk + \frac{1}{r}\right] A \frac{e^{j(\omega t - kr)}}{r} \cos(\mathbf{n}, \mathbf{r}) + \left[jk + \frac{1}{r'}\right] A \frac{e^{j(\omega t - kr')}}{r'} \cos(\mathbf{n}, \mathbf{r}') \tag{3.11.7}$$

where $\cos(\mathbf{n}, \mathbf{r})$ is the cosine of the angle between $\mathbf{n}$ and $\mathbf{r}$. The second term is positive, since $\mathbf{r}'$ is directed into the surface of integration. Now over S_1, $\mathbf{r} = \mathbf{r}'$ and

$$\cos(\mathbf{n}, \mathbf{r}') = -\cos(\mathbf{n}, \mathbf{r}) \tag{3.11.8}$$

Thus $\varphi = 0$ and

$$\frac{\partial \varphi}{\partial n} = -2\left[jk - \frac{1}{r}\right] A \frac{e^{j(\omega t - kr)}}{r} \cos(\mathbf{n}, \mathbf{r}) \tag{3.11.9}$$

Substitution into Eq. (3.11.6) yields

$$\psi_p = -\frac{2}{4\pi} \int\int_{S_1} \psi \frac{\partial}{\partial r} \left[jk + \frac{1}{r}\right] \left(\frac{e^{j(\omega t - kr)}}{r} \right) \cos(\mathbf{n}, \mathbf{r})\, dS \tag{3.11.10}$$

Thus, the velocity potential at P is given by the summation of radiation from a distribution of dipole radiators with moments proportional to the pressure over the radiating surface.

An alternative, and more useful, choice of Green's function has the two point sources vibrating in phrase, in which case

$$\varphi = A \frac{e^{j(\omega t - kr)}}{r} + A \frac{e^{j(\omega t - kr)}}{r'} \tag{3.11.11}$$

and

$$\frac{\partial \varphi}{\partial n} = -\left[jk + \frac{1}{r}\right] A \frac{e^{j(\omega t - kr)}}{r} \cos(\mathbf{n}, \mathbf{r}) - \left[jk + \frac{1}{r'}\right] A \frac{e^{j(\omega t - kr)}}{r'} \cos(\mathbf{n}, \mathbf{r}') \tag{3.11.12}$$

Now when $r = r'$, $\partial\varphi/\partial n$ is zero and $\varphi = (2A/r)e^{j(\omega t - kr)}$.

Eq. (3.11.6) will still hold true so that if $\partial\varphi/\partial n$ is zero over all of S_1 except over S_0, we obtain

$$\psi_p = -\frac{1}{2\pi}\iint_{S_0} \frac{e^{j(\omega t - kr)}}{r}\frac{\partial\psi}{\partial n}\,dS = \frac{1}{2\pi}\iint_{S_0} \frac{e^{j(\omega t - kr)}}{r} V_n\,dS \qquad (3.11.13)$$

Assuming harmonic time dependence, Eq. (3.11.13) yields for the acoustic pressure at P,

$$p(P) = j\frac{\rho c}{\lambda} e^{j\omega t}\iint_{S_0} \frac{e^{-jkr}}{r} v_n\,dS \qquad (3.11.14)$$

where v_n is the normal particle velocity over S_0. We will now use this relationship to find the pressure field due to a plane piston radiator in an infinite rigid baffle.

3.12 Piston Radiator: Near Field

The piston radiator (see Fig. 3.37), as we have already indicated, is a topic of great importance in acoustics. We start by calculating the pressure field along its axis. From Eq. (3.11.14), following the same reasoning as in Section 3.9, the acoustic pressure at a point in the geometrical projection of the radiating area will be

$$\begin{aligned} p &= v_n \rho c e^{-jkz} - \frac{v_n \rho c}{2\pi}\int_0^{2\pi} e^{jkr_m}\,d\varphi \\ &= v_n \rho c\left[e^{-jkz} - e^{-jkr_m}\right] \end{aligned} \qquad (3.12.1)$$

We see then that the pressure is given by the difference of two phasors, and from Eq. (3.12.1) and Fig. 3.38,

$$\frac{p_0^2}{v_n^2\rho^2 c^2} = 2\left(1 - \cos k\overline{r_m - z}\right) \qquad (3.12.2)$$

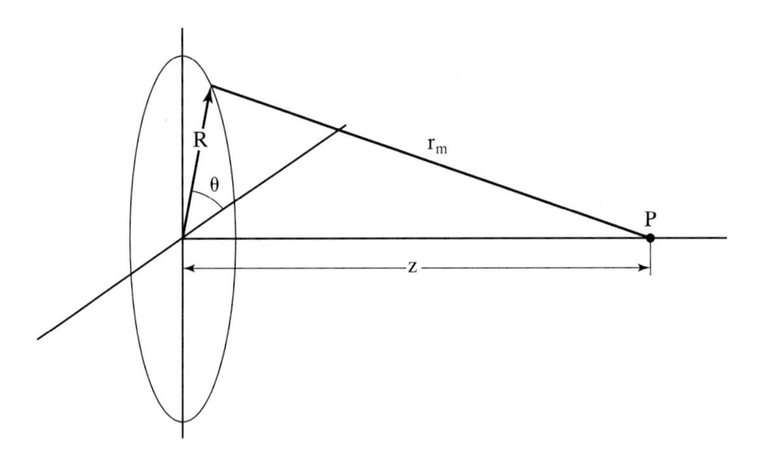

Fig. 3.37 A piston radiator, radius R, in an infinite baffle.

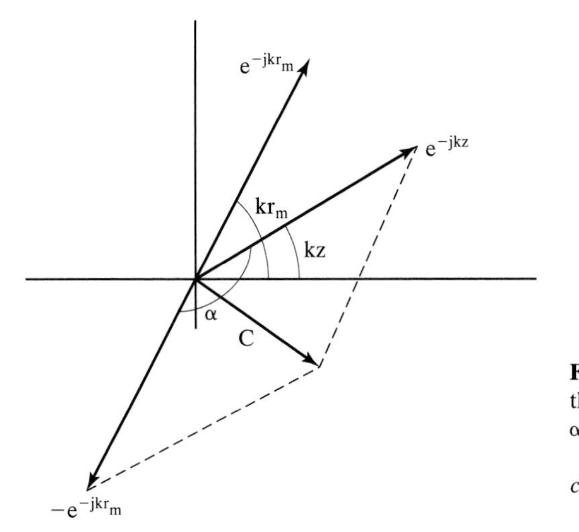

Fig. 3.38 Construction to find the intensity on the axis of the piston radiator.
$\alpha = kt + \pi - kr_m$
$= \pi - (kr_m - kt)$
$c^2 = 1^2 + 1^2 + 2\cos\alpha$
$= 2 - 2\cos k(r_m - z)$

Thus, the intensity

$$I = \frac{p_0^2}{2\rho c} = 2v_n^2\rho c \sin^2\frac{k}{2}(r_m - z) \tag{3.12.3}$$

The intensity distribution is shown in Fig. 3.39. We see that the intensity has maxima when

$$\frac{k}{2}(r_m - z) = (2n + 1)\frac{\pi}{2}$$

or

$$r_m - z = \left(n + \frac{1}{2}\right)\lambda \tag{3.12.4}$$

There are minima when

$$\frac{k}{2}(r_m - z) = m\pi$$

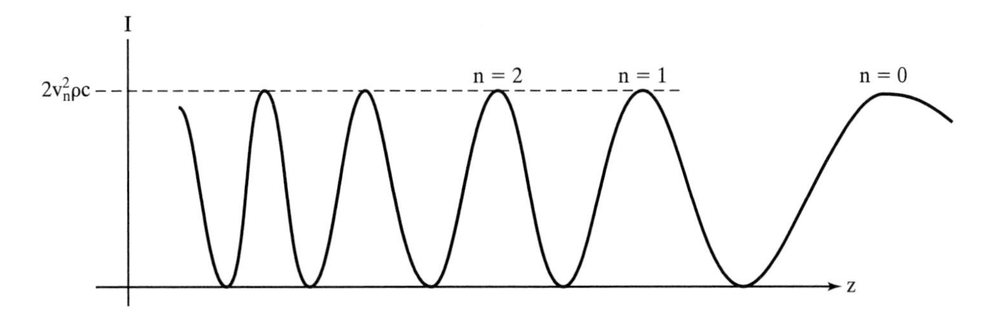

Fig. 3.39 The intensity along the axis of a piston radiator.

or

$$r_m - z = m\lambda \tag{3.12.5}$$

It follows that at the maxima the intensity is $2v_n^2\rho c$, which is four times the intensity in a plane wave with particle velocity v_n (i.e., the intensity to be expected from geometrical acoustics). The farthest maximum occurs when n is zero, so that

$$r_m - z = \frac{\lambda}{2} \tag{3.12.6}$$

But since $R^2 + Z^2 = r_m^2$,

$$R^2 \simeq (r_m - z)2z$$

and thus,

$$z \simeq \frac{R^2}{\lambda} \tag{3.12.7}$$

This value of z can be used as a criterion for entry into the far field (i.e., the region in which the intensity falls with an inverse square law dependence). We can see that this is the case, since beyond the farthest maximum, $(r_m - z) \simeq 0$, and

$$I_{ff} \to 2v_n^2\rho c\frac{k^2}{4}(r_m - z)^2 \to v_n^2\rho c k^2\left(\frac{R^2}{2z}\right)^2$$

or

$$I_{ff}z^2 \to v_n^2\rho c k^2 R^4$$

a constant. Thus

$$I_{ff} = \frac{\text{constant}}{z^2} \tag{3.12.8}$$

Eq. (3.12.8) is a statement of the inverse square law. The far field distance is given for the following examples:

- **a.** loudspeaker of diameter 25 cm at 1000 Hz (5 cm)
- **b.** an ultrasonic transducer of diam 2.5 cm at 100 kHz (5 cm)
- **c.** light of wavelength 5000Å (1Å = 10^{-8} cm) passing through an aperture 2.5 mm in diameter (3.1 mm)

3.13 Piston Radiator: Far Field

We will consider an off-axis point, in the x-z plane, for convenience, as shown in Fig. 3.40. We have

$$p = \int\int j\frac{\rho c v_n}{\lambda}\frac{e^{-jkr}}{r}dA \tag{3.13.1}$$

But dA = hdhd$d$, thus

$$p = j\frac{\rho c v_n}{\lambda}\int\int\frac{e^{-jkr}}{r}hdhd\varphi \tag{3.13.2}$$

But in the far field

$$r \simeq r'$$

and

$$r' \simeq r_o - h\cos\varphi\sin\alpha \tag{3.13.3}$$

Note that this is the same approximation that we made for the far field of an array in Section 3.6. Hence we obtain

$$p \simeq j\frac{\rho c v_n}{\lambda}\frac{e^{-jkr_o}}{\mathbf{r}_o}\int_{r=0}^{R}\int_{\varphi=0}^{2\pi} e^{jkh\sin\alpha\cos\varphi}d\varphi hdh \tag{3.13.4}$$

But

$$\int_{o}^{2\pi} e^{jz\cos\omega}\cos(m\omega)d\omega = 2\pi j^m J_m(z) \tag{3.13.5}$$

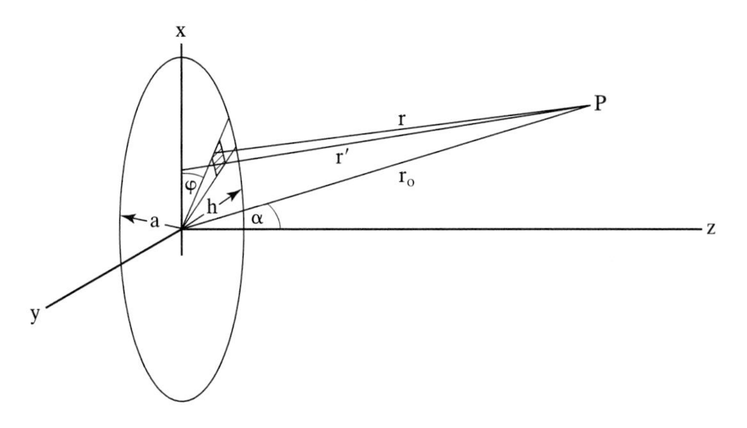

Fig. 3.40 Geometry of the piston radiator.

so that

$$p = j\frac{\rho c v_n}{\lambda}\frac{e^{-jkr_o}}{r_o}\int_o^R 2\pi J_o(kh \sin \alpha)hdh \tag{3.13.6}$$

and since

$$\int zJ_o(z)dz = zJ_1(z) \tag{3.13.7}$$

it follows that

$$p = j\frac{\rho c v_n}{\lambda}\frac{e^{-jkr_o}}{r_o}2\pi R^2\left[\frac{J_1(kR\sin \alpha)}{kR\sin \alpha}\right] \tag{3.13.8}$$

Here the dependence on e^{-jkr_o}/r_o determines an inverse square law dependence, and the factor $J_1(kR\sin \alpha)/kR\sin \alpha$ determines the angular dependence. This variation of the latter function is shown in Fig. 3.41. Note that the first minimum occurs when $kR\sin \alpha = 3.83$; that is,

$$\sin \alpha = 0.61\frac{\lambda}{R} \tag{3.13.9}$$

Two beam patterns are shown in Fig. 3.42. Note that if $R >> \lambda$, the beam is highly directional. This case also corresponds to the situation in which the Fresnel region extends a great distance, since $z \simeq R^2/\lambda$. These two cases correspond to typical circumstances for optical and acoustical propagation. Since Rayleigh's time, much effort has been devoted to finding velocity distributions over the face of the piston that

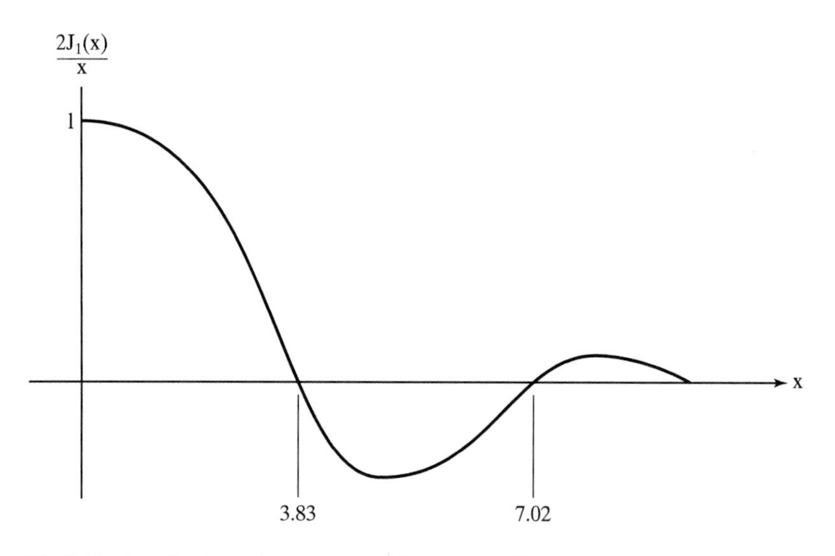

Fig. 3.41 Angular dependence factor of the piston radiator.

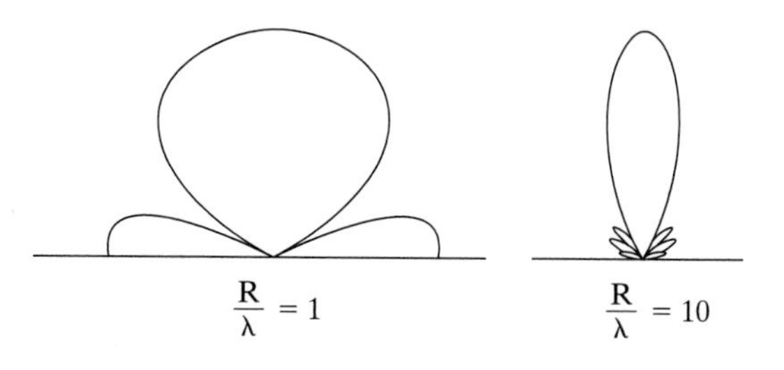

Fig. 3.42 Radiation patterns from a circular piston at two different ratios of radius to wavelength.

result in elimination of the first side lobe and drastic reductions in the other side lobes. It has proven impossible, however, to reduce the angular width of the beam by such "juggling" with the velocity distribution.

3.14 Piston Radiator: Radiation Impedance

If a piston is driven with a sinusoidal force $F_o e^{j\omega t}$, then the actual motion will depend on the radiation reactance of the medium, as we have seen in the case of a spherical oscillator. The calculation of the radiation reactance is not so easy in this case, but Rayleigh (1896) showed that it may be carried out as follows. Referring to Fig. 3.43, let dp be the acoustic pressure at dA′ due to motion at dA. Then the total acoustic pressure at dA is

$$p = \int\int \frac{j}{\lambda} \rho c v_n \frac{e^{-jkr}}{r} dA \tag{3.14.1}$$

The reaction force on the piston is then

$$\begin{aligned} F_R &= -\int\int p dA' \\ &= -j\frac{\rho c v_n}{\lambda} \int\int dA' \int\int \frac{e^{-jkr}}{r} dA \end{aligned} \tag{3.14.2}$$

But the reaction on dA′ due to dA is equal to the reaction on dA due to dA′. Hence the integration that follows is twice as great as if it were performed by summing the force

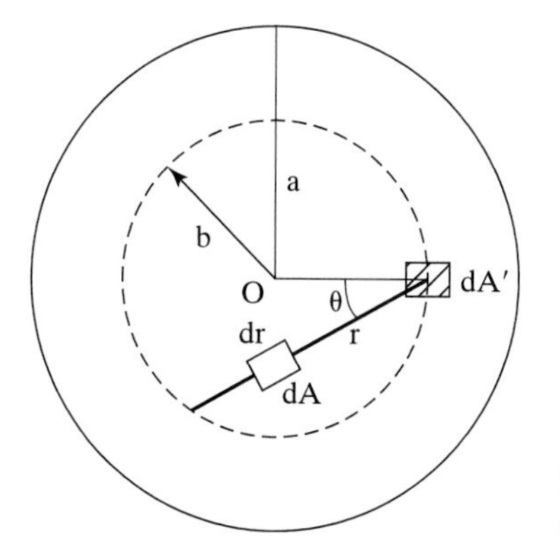

Fig. 3.43 Geometry for calculating the radiation impedance of a piston.

due to each pair of elements only once. It is easier, however, to carry out the integration this way.

The element of area

$$dA' = rd\theta dr \tag{3.14.3}$$

It may seem from Fig. 3.43 that the maximum distance in the direction of θ in a circle of radius b is 2bcos θ, and thus we may cover the whole area of this circle by integrating with respect to r with limits 0 to 2bcos θ and with respect to θ from $-\pi/2$ to $\pi/2$. The integration of dA is extended to cover the whole piston by putting

$$dA = bdbd\varphi \tag{3.14.4}$$

and taking the limits of b from 0 to a, and of φ from 0 to 2π. Thus, we have from Eq. (3.14.2),

$$F_R = -j\frac{2\rho c v_n}{\lambda}\int_0^a bdb\int_0^{2\pi} dd\int_{-\pi/2}^{\pi/2} d\theta\int_0^{2b\cos\theta} e^{-jkr}dr \tag{3.14.5}$$

The result of the integration is

$$F_R = -\rho c\pi a^2 v_n[R_1(2ka) + jX_1(2ka)] \tag{3.14.6}$$

where $R_1(x)$ and $X_1(x)$ are the functions given by Eqs. (3.14.7) and (3.14.8) and illustrated in Fig. 3.44. Thus

$$R_1(x) = \frac{x}{2\cdot 4} - \frac{x^4}{2\cdot 4^2\cdot 6} + \frac{x^6}{2\cdot 4^2\cdot 6^2\cdot 8} - \cdots \tag{3.14.7}$$

and

$$X_1(x) = \frac{4}{\pi}\left(\frac{x}{3} - \frac{x^3}{3^2 5} + \frac{x^5}{3^2\cdot 5^2\cdot 7} - \cdots\right) \tag{3.14.8}$$

At low frequencies ($a << \lambda$), the impedance is given by the approximate relation

$$Z_r = \frac{\rho c\pi a^2 k^2 a^2}{2} + j\frac{\rho c\pi a^2 8ka}{3\pi} \tag{3.14.9}$$

If the piston, instead of being set in a baffle, is mounted at the mouth of a pipe, we have a situation first solved analytically by Levine and Schwinger (1948). The low frequency mechanical radiation impedance then becomes

$$Z_{rp} = \rho c\pi a^2(0.23)\, k^2a^2 + j\rho c\pi a^2(0.6)\, ka \tag{3.14.10}$$

This result explains the long-known necessity to introduce an end correction in designing organ pipes.

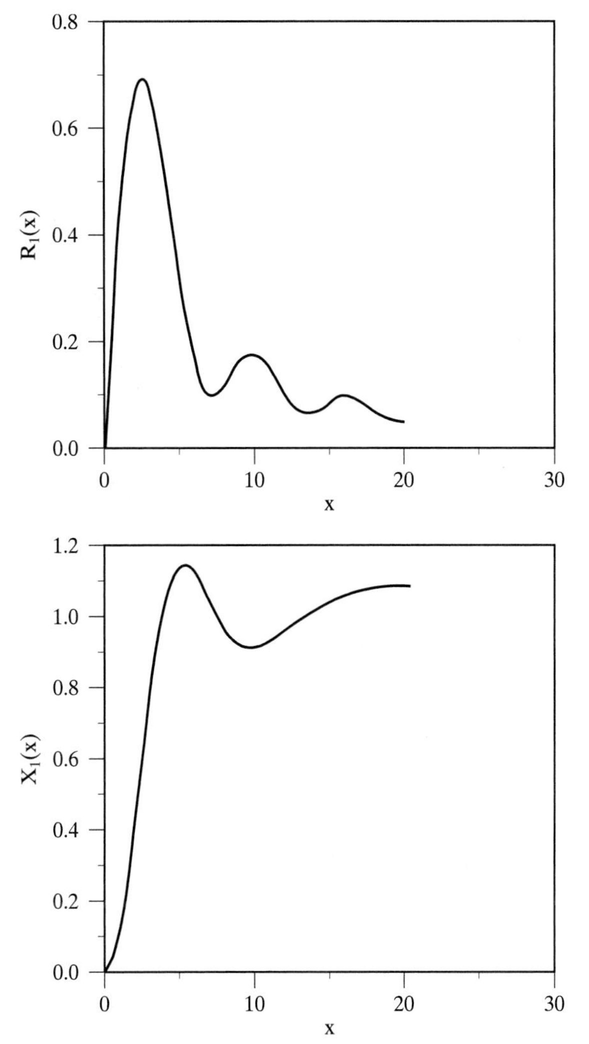

Fig. 3.44 Impedance functions for the piston radiator.

3.15 Diffraction Around a Circular Obstacle

Fig. 3.45 shows a plane wave incident on a circular disk-shaped obstacle in the x-y plane. Clearly, in the absence of the obstacle, the pressure at P would be that due to a plane wave and the situation could be predicted by a "geometrical" theory of acoustics. Let us call this pressure p_g. It follows that p_g could be obtained by integration of the effects of secondary waves from areas D and S in Fig. 3.45, or

$$p_g = \int\int_D + \int\int_S \tag{3.15.1}$$

Consequently, we deduce that the pressure due to the area of the wavefront S is

$$p_s = p_g - \int\int_D \tag{3.15.2}$$

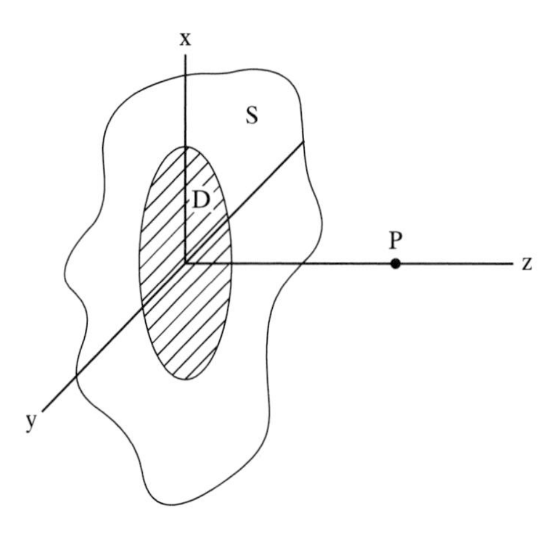

Fig. 3.45 Construction for discussing diffraction around a circular obstacle.

But the pressure due to D is given by Eq. (3.12.1), which may be expressed as

$$p_o = p_g - v_n \rho c e^{-jkrm} \tag{3.15.3}$$

Hence it follows from these two equations that the scattered wave is given by

$$p_s = v_n \rho c e^{-jkrm} \tag{3.15.4}$$

Note this result presumes that the observation point P is on the axis of the disk, and r_m is the distance from P to the rim of the disk. The scattered wave is said to be an edge wave.

It follows that there will be a relatively intense insonified line down the axis of the shadow region. Off-axis r_m varies significantly. During World War I, large disks were used to find the direction of incoming aircraft by listening at an axial observation point behind the disk.

REFERENCES AND FURTHER READING

Levine, H., and J. Schwinger. 1948. "On the Radiation of Sound From an Unflanged Circular Pipe." 73 *Phys. Rev.* 383–406.

Strutt, J. W. (Baron Rayleigh). [1896] 1945. *The Theory of Sound.* Vol. 2. 2nd ed. Reprinted: Dover Publications. [See section 278.]

Tucker, D. G., and B. K. Gazey. 1966. *Applied Underwater Acoustics.* Pergamon Press.

Urick, R. J. 1975. *Principles of Underwater Sound.* 2nd ed. McGraw Hill.

PROBLEMS

3.1 A car is travelling at a speed of 65 mph toward a tunnel through a hill. If the car blows its horn at a frequency of 100 Hz and the sound reflects from the hill, what will be the frequency of the reflected sound wave that the driver of the car hears?

3.2 Fig. P3.2 shows the pressure oscillations at a fixed position for a triangular sound wave. Answer the following questions referring to this waveform.

a. What is the period of the triangle sound wave?

b. Starting with the fundamental, list the frequencies of the first four elements of the harmonic series of pure tones that can be used to represent the triangle wave.

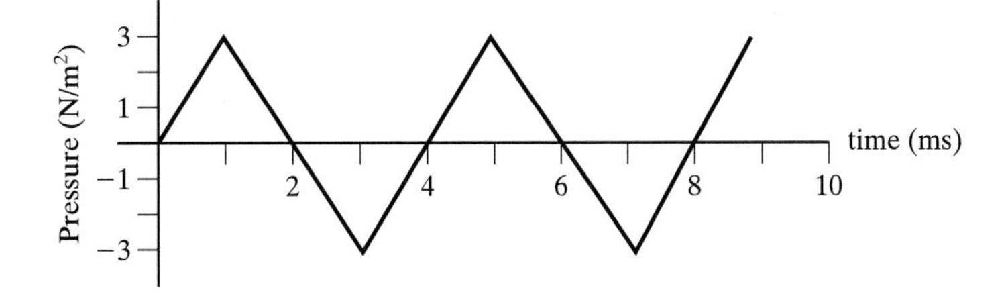

Fig. P3.2 Pressure oscillations for a triangular sound wave.

c. What is the amplitude of the triangle wave?
d. What is the sound level associated with the triangle wave?

3.3 Derive an expression for the speed of sound waves in a gas-filled pipe. How is the speed of sound in a gas dependent on a) temperature, b) static pressure? What is the speed of sound inside a gasoline engine cylinder just after combustion, when the pressure is 200 times atmospheric and the temperature is 1000°C, if the gas mixture has a value of $\gamma = 1.35$ and would have a density of 0.0014 g/cm^3 at 0°C and atmospheric pressure?

3.4 Find the particle velocity, the velocity potential, the energy density, and the intensity for a pressure wave given by

$$\mathbf{p} = p_0 e^{j(\omega t - kx)}$$

3.5 Derive expressions for the average energy density and intensity of a plane sound wave in a fluid.

The threshold of human hearing is said to be 10^{-12} W/m^2. Referred to this standard, two sound pressure levels are +40 and +130 dB. Calculate the pressure amplitudes of the two sounds, and assuming they are plane progressive waves, calculate their velocity and displacement amplitudes, assuming the density of air to be 10^{-3} g/cm^3 and that $\gamma = 1.4$, the ambient pressure is atmospheric, and the frequency is 1000 Hz. What would be the sound pressure level if the two waves were superimposed?

3.6 A flute is designed to play middle C (264 Hz) with all finger holes closed at 20°C. The velocity of sound at 20°C is 343 m/s.

a. How far should the open end of the tube be from the mouthpiece?
b. What frequency will this closed fingering produce at 10°C. The velocity of sound at 10°C is 337 m/sec.

3.7 A wave on a string is represented by the transverse displacement function $y(x,t) = 0.42 \sin(7.6x + 94t)$ meters. Determine

a. The direction the wave is travelling.
b. Its amplitude, frequency, wavelength, and phase velocity.
c. The maximum speed of the string particles.

3.8 A 600 Hz tuning fork sets up a standing wave in a string clamped at both ends. The wave speed on the string is 400 m/sec. The standing wave has four loops and an amplitude of 2 mm.

a. What is the length of the string?
b. Write an expression for the displacement of the string as a function of x and t.

3.9 Students in a wrecking yard decide to make some sound instruments from the wrecked pieces.

a. One student makes a steel guitar by stretching a strong wire over a length of 0.3 m. What is the speed of the sound in the wire required to produce a fundamental note of 440 Hz in air?
b. Another student cuts a length of open pipe to produce the same note. What is the shortest length of the pipe?
c. What is the shortest length of a similar pipe, closed at one end, that can also produce this musical note?
d. When the open and semiclosed pipes are played by blowing across them, the notes produced have the same pitch, but sound very different. Explain why.

3.10 Find the far-field angular radiation patterns for intensity radiated in air by a dipole source with a moment of 1 mm at frequencies of 1, 100, 10,000, and 1,000,000 Hz. Compare the intensity maxima at these frequencies.

3.11 Light of wavelength 750 nm passes through a slit 1.0×10^{-3} mm wide. How wide is the central maximum in

a. Degrees.
b. In centimeters on a screen 20 cm away.

3.12 A Young's double slit experiment is illuminated with coherent blue light at a wavelength of 0.5 μ and coherent red light at 0.6 μ wavelength. The slits are 100 μ apart and 2 μ wide. The pattern is viewed on a screen 10 meters away. (Assume small angle approximation.)

a. For the blue light, what is the distance, y, of the first maximum above the central maximum?
b. What is the separation at the screen between the first blue and first red maximum?
c. If a very thin film of soap (index = 1.5, thickness = 0.03 μ) is placed over the upper slit in part a illuminated with blue light, will the central maximum move up or down?
d. By how much?

3.13 A fog signal is used to warn of hazards at sea. The best frequency to use is determined to be 400 Hz and for energy efficiency it is necessary to beam the sound in a horizontal plane. Explain how this might be done with a linear array. Determine the spacing between sources for the case of eight elements.

3.14 What do you understand by Huygens' principle? Show that the action of a thin lens may be explained by Huygens' principle and prove that the focal length, f, of a thin lens is given by

$$\frac{1}{f} = (\mu - 1)\left(\frac{1}{r_1} - \frac{1}{r_2}\right)$$

Where r_1 and r_2 are the radii of curvature of the lens faces, and

$$\mu = \frac{c_1}{c_2}$$

where c_1 and c_2 are the sound velocities in the surrounding medium and the lens respectively.

3.15 Two glass plates in air are inclined slightly to each other, forming a wedge-shaped volume between them. The top plate is 10 cm long and touches the bottom plate at the left, as shown in Fig. P3.15. The separation of the plates at the right edge is 0.1 mm. The plates are illuminated

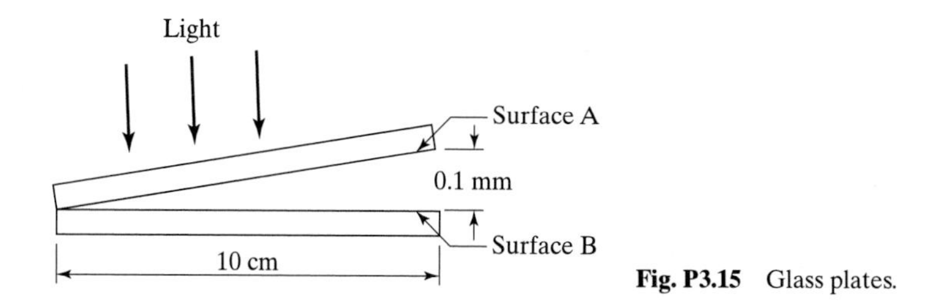

Fig. P3.15 Glass plates.

from above with light at a wavelength $\lambda = 5.0 \times 10^{-7}$ m. The resulting interference pattern is viewed from above. Consider only the interference of the reflected light from the bottom of the top plate and the top of the bottom plate (i.e., surfaces A and B in the diagram).

a. What is the spatial separation between the dark fringes?

b. If the air is removed from the wedge region, by how much will the total number of fringes change?

3.16 Assume that a loudspeaker of diameter 20 cm is mounted in a rigid baffle and moves so as to approximate a piston radiator. Sketch the radiation patterns of intensity in air at frequencies of 1, 100, 10,000, and 1,000,000 Hz. Compare the half angles at which the first intensity minima occur at these frequencies.

3.17 A piston mounted on one side of an infinite baffle radiates into air. The radius of this piston ($a = 0.15$m) is driven at a frequency such that $\lambda = \pi a$.

a. If the maximum displacement amplitude of the piston is 0.00025 m, how much acoustic power is radiated?

b. What is the axial intensity at a distance of 3.2 m?

c. What is the radiation mass?

Chapter 4

Pipes and Horns

4.1 Webster's Equation

A horn is assumed to be a column of varying cross sections enclosed in loss-free, rigid walls and having a diameter considered small compared with the wavelength of the sound propagated (see Fig. 4.1). The propagation medium is assumed to be homogeneous, isotropic, and dissipationless. It is also assumed that the surfaces over which pressure amplitude and phase are constant (i.e., the wavefronts within the horn) are approximately plane.

The equation of motion for an arbitrary element of thickness dx, at a distance x measured from the throat is

$$-\partial p/\partial x = \rho_o \, \partial v/\partial t \tag{4.1.1}$$

where p is the acoustic pressure, ρ_o is the equilibrium density, and v is the particle velocity. The equation of continuity is in this case

$$A(x)\, \partial\rho/\partial t + \partial/\partial x \,(A(x)\rho v) = 0 \tag{4.1.2}$$

where A(x) is the cross-sectional area at x, assumed to have a negligible variation along dx, and ρ is the instantaneous density, given by

$$\rho = \rho_o(1 + s)$$

where s is the condensation. Since $s << 1$, the small disturbances of linear acoustics $\partial s/\partial x$ can be neglected, so that Eq. (4.1.2) becomes

$$A\frac{\partial s}{\partial t} + A\frac{\partial v}{\partial x} + v\frac{\partial A}{\partial x} = 0 \tag{4.1.3}$$

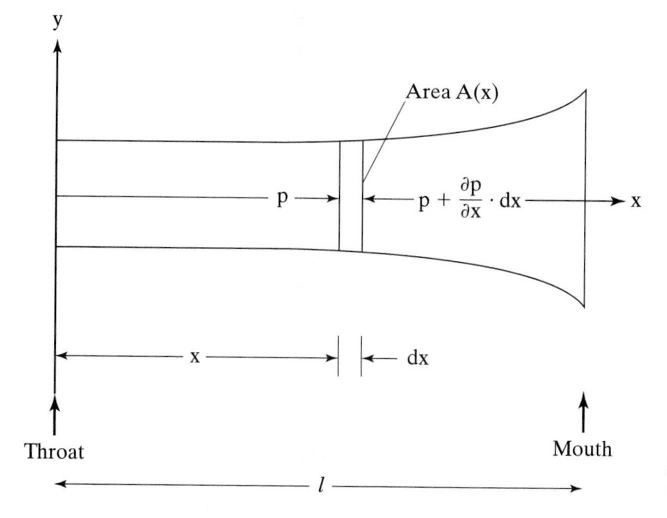

Fig. 4.1 Volume element for plane wave horn theory.

The elastic behavior of a fluid has the form

$$p = Bs = \rho_o c^2 s \tag{4.1.4}$$

where B is the bulk modulus and c is the sound velocity. It follows from Eqs. (4.1.1, 4.1.3), and (4.1.4) that

$$\frac{\partial}{\partial x}(\ln A)\frac{\partial p}{\partial x} + \frac{\partial^2 p}{\partial x^2} = \frac{1}{c^2}\frac{\partial^2 p}{\partial t^2} \tag{4.1.5}$$

This relation, which is usually called Webster's equation, may be alternatively expressed in terms of the condensation (s), particle displacement (η), or velocity potential (ϕ) as follows:

$$\frac{\partial}{\partial x}(\ln A)\frac{\partial s}{\partial x} + \frac{\partial^2 s}{\partial x^2} = \frac{1}{c^2}\frac{\partial^2 s}{\partial t^2} \tag{4.1.6}$$

$$\frac{\partial}{\partial x}\left[\frac{1}{A}\cdot\frac{\partial}{\partial x}(A\eta)\right] = \frac{1}{c^2}\frac{\partial^2 \eta}{\partial t^2} \tag{4.1.7}$$

or

$$\frac{\partial}{\partial x}(\ln A)\frac{\partial \phi}{\partial x} + \frac{\partial^2 \phi}{\partial x^2} = \frac{1}{c^2}\frac{\partial^2 \phi}{\partial t^2} \tag{4.1.8}$$

4.2 Propagation in Pipes

For a pipe or rod, the cross section is uniform, and Webster's equation reduces to a simple plane wave equation:

$$\frac{\partial^2 p}{\partial x^2} = \frac{1}{c^2}\frac{\partial^2 p}{\partial t^2} \tag{4.2.1}$$

If a harmonic time dependence is assumed (at angular frequency ω, say), then letting $\omega/c = k$, we obtain the Helmholtz equation

$$\frac{\partial^2 p}{\partial x^2} = k^2 p = 0 \tag{4.2.2}$$

The general solution of Eq. (4.2.2) is

$$Ae^{jkx} + Be^{-jkx}$$

so that the general solution of Eq. (4.2.1) is

$$p = Ae^{j(\omega t + kx)} + Be^{j(\omega t - kx)} \tag{4.2.3}$$

The first term represents waves travelling in the backward x direction, and the second term, waves travelling in the forward x direction.

Now for the uniform cross section, the particle velocity is obtained from Eq. (4.1.1), thus

$$\partial v/\partial t = -1/\rho \, \partial p/\partial x \tag{4.2.4}$$

so

$$v = \frac{1}{\rho c}\left[-Ae^{j(\omega t + kx)} + Be^{j(\omega t - kx)}\right] \tag{4.2.5}$$

Now let

$$A/B = -e^{-2\varphi} \tag{4.2.6}$$

and rewrite Eqs. (4.2.3) and (4.2.5), thus

$$p = 2B \sinh(\varphi - jkx)e^{j\omega t - \varphi} \tag{4.2.7}$$

and

$$v = \frac{2B}{\rho c}\cosh(\varphi - jkx)\, e^{j\omega t - \varphi} \tag{4.2.8}$$

Hence, the specific acoustic impedance at any point along the line is given by

$$z = \frac{p}{v} = \rho c \tanh(\varphi - jkx) \tag{4.2.9}$$

Now the definition of φ given by Eq. (4.2.6) allows for change of both amplitude and phase on reflection (i.e., φ in general is a complex quantity). Bearing this in mind, we may let

$$\varphi - jkx = \pi(\alpha + j\beta) \tag{4.2.10}$$

so that Eq. (4.2.9) may be rewritten as:

$$\frac{Z}{\rho c} = \tanh\left[\pi(\alpha + j\beta)\right] \tag{4.2.11}$$

To fully determine the pressure, particle velocity, or impedance, the arbitrary constants A and B, or φ, their ratio, must be found from the boundary conditions at mouths and junctions. Consider a vanishingly thin layer at any point along the line. From Newton's second law, since the layer has no mass, the pressure must be the same on either side of the layer. Further, since the material on either side of the layer neither separates nor interpenetrates, the density must be continuous. For a uniform line, this means that the particle velocity must be continuous. At a junction (Fig. 4.2), the volume velocity must be continuous, as illustrated. For other types of termination, the guiding principle is that the same amount of material flows out of the tube as approaches the mouth. For instance, for a baffled or unbaffled tube (Figs. 4.3 and 4.4) radiating into space, continuity of velocity may be assumed. Since most problems involve one or the other of these two cases, we will derive the radiation impedance for these terminations. The problem of a plane piston in an infinite rigid baffle was first tackled by Rayleigh, and the results were given in Sections 3.12 and 3.13 in Chapter 3.

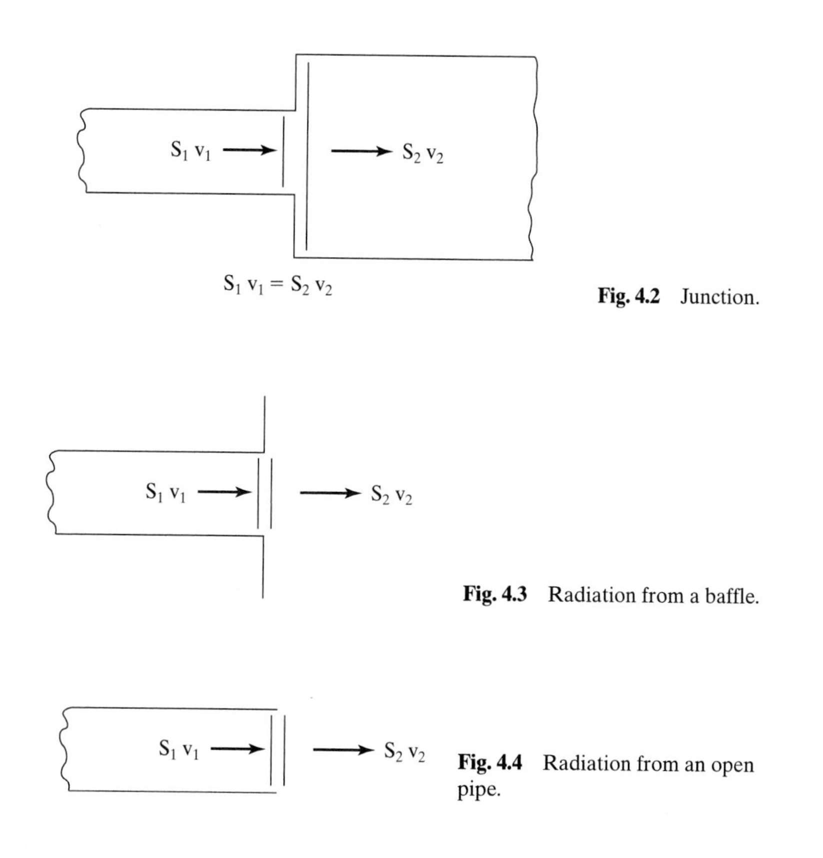

Fig. 4.2 Junction.

Fig. 4.3 Radiation from a baffle.

Fig. 4.4 Radiation from an open pipe.

4.3 Impedance Transformation by a Tube

Section 4.2 shows how the specific acoustic impedance at the mouth of a tube may be determined. Applying the boundary condition of continuity of specific acoustic impedance at the mouth gives, from Eq. (4.2.9),

$$z_m = \rho c \tanh(\varphi - jkl) \tag{4.3.1}$$

Eq. (4.3.1) in effect determines φ in terms of z_m, the specific acoustic impedance at the mouth. Thus

$$\frac{z_m}{\rho c} = \frac{\tanh \varphi - \tanh jkl}{1 - \tanh \varphi \tanh jkl} \tag{4.3.2}$$

so

$$\tanh \varphi = \frac{z_m/\rho c + \tanh jkl}{1 + (z_m/\rho c) \tanh jkl} \tag{4.3.3}$$

But from Eq. (4.2.9), the specific acoustic impedance at the throat, z_t, is given by

$$z_t/\rho c = \tanh \varphi \tag{4.3.4}$$

Hence, from Eqs. (4.3.3) and (4.3.4)

$$\frac{z_t}{\rho c} = \frac{z_m + j\rho c \tan kl}{\rho c + jz_m \tan kl} \tag{4.3.5}$$

The following cases of Eq. (4.3.5) are of special interest.

1. If the line is matched (i.e., $z_m = \rho c$), then $z_t = \rho c$.
2. $kl = (2n - 1)\pi/2$ (i.e., $l = (2n - 1)\lambda/4$)

   where

   $$n = 1, 2, 3, \ldots$$

   then

   $$\tan kl \rightarrow \pm\infty$$

   and

   $$\left(\frac{z_t}{\rho c}\right)_{\frac{\lambda}{4}} = \frac{\rho c}{z_m} \tag{4.3.6}$$

   that is, the throat impedance is the inverse of the mouth impedance.

3. $kl = n\pi$ (i.e., $l = n\lambda/2$)

where

$$n = 1, 2, 3, \ldots$$

then

$$\tan kl = 0$$

and

$$\left(\frac{z_t}{\rho c}\right)_{\frac{\lambda}{2}} = \frac{z_m}{\rho c} \tag{4.3.7}$$

that is, the throat impedance equals the mouth impedance.

When kl is not an exact multiple of $\pi/2$, the calculation of z_t is more difficult. This calculation is greatly facilitated by the use of a Smith chart, a tool invented precisely for this type of calculation (see Fig. 4.5). The normal rectangular grid of the reduced impedance plane is transformed into two sets of orthogonal circles, with the point (1,0) of the rectangular grid placed at the center of the chart. Recalling Eq. (4.2.11)

$$z/\rho c = \tan\left[\pi(\alpha + j\beta)\right]$$

for such an impedance function, points with constant α lie on concentric circles centered on (1,0) and points with constant β lie on straight lines radiating from (1,0) as indicated. Since, from Eq. (4.2.10),

$$\beta = \frac{\mathrm{Im}(\varphi)}{\pi} - \frac{2x}{\lambda}$$

it follows that as x increases β diminishes. Hence, to find $z_t/\rho c$ for a lossless line as assumed here, the point corresponding to $z_m/\rho c$ is plotted on the chart. A circle centered on (1,0) is drawn through the point and a line corresponding to an increment of $2l/\lambda$ in β then intersects the circle at $z_t/\rho c$. For convenience, the chart is marked with an equivalent wavelength scale around its outer edge. Since, as shown by Eq. (4.3.7), one-half wavelength of travel brings z_t back to z_m, one complete circuit of the Smith chart is equivalent to one-half wavelength. One-quarter wavelength is equivalent to one half circuit. The Smith chart has fallen into disuse recently in favor of computer programs, but it is still useful as a way of visualizing the form of the transformation.

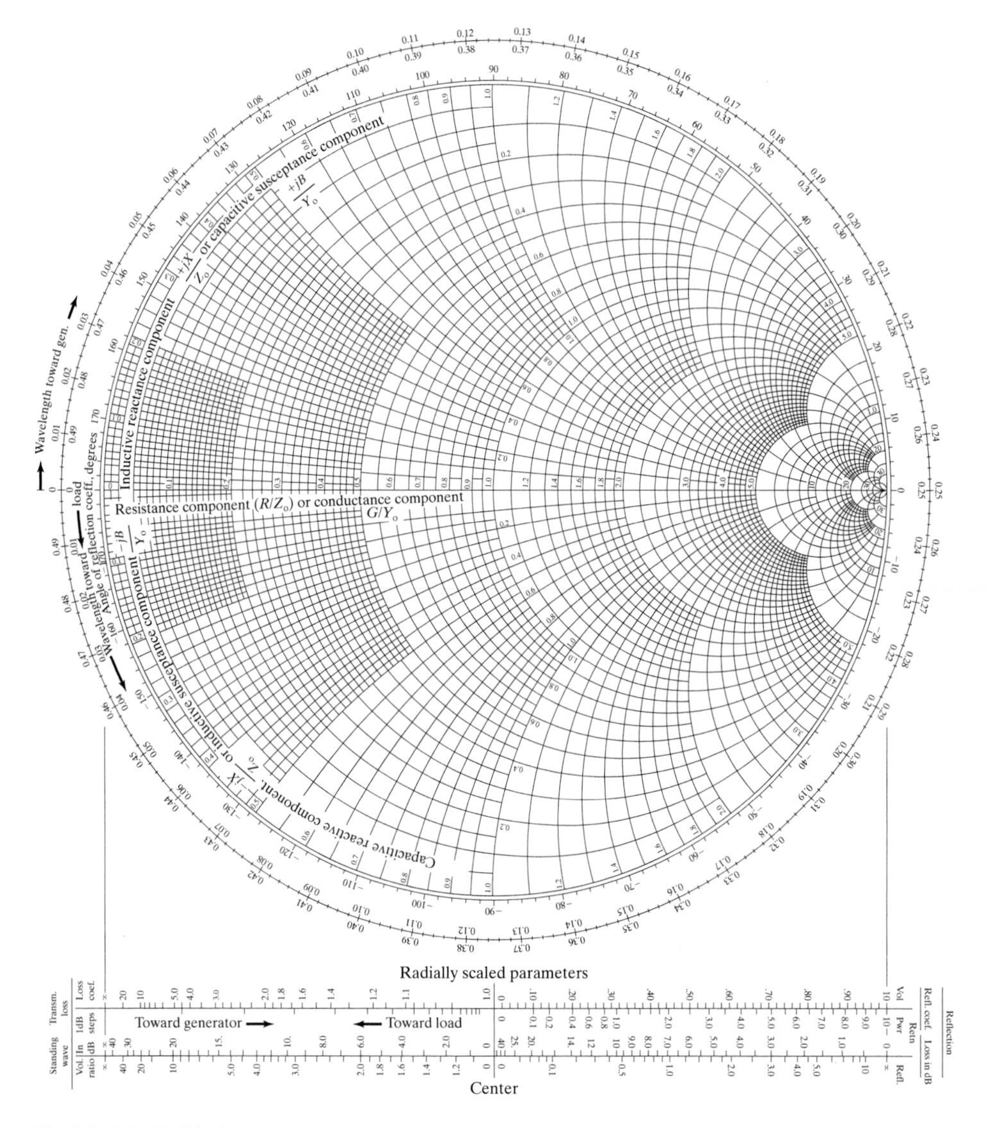

Fig. 4.5 The Smith chart.

Fig. 4.6 shows the transformation of resistance by a tube, and Fig. 4.7 the transformation of reactance. For a constant input force, the radiated power is proportional to r_t/z_t^2, a quantity plotted in Fig. 4.8. Note the large number of resonance lines.

4.4 Standing Wave Ratio

The pressure inside the tube is given by Eq. (4.2.3):

$$p = Ae^{j(\omega t+kx)} + Be^{j(\omega t-kx)}$$

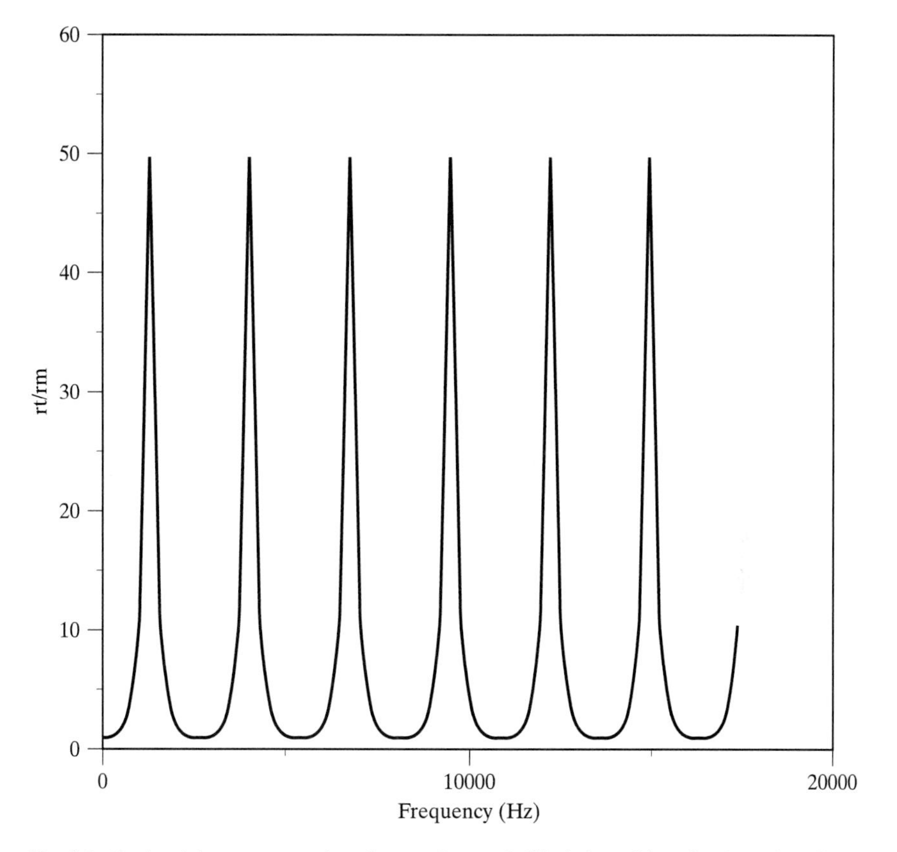

Fig. 4.6 Ratio of throat-to-mouth resistance for an air-filled pipe with radius 1 cm; length $2\pi \times$ diameter.

Defining the reflection coefficient, r, as the ratio of the amplitudes of reflected (negatively travelling) to incident (positively travelling) waves at the tube mouth, we have

$$\left(\frac{p_-}{p_+}\right)_{x=l} = \frac{Ae^{jkl}}{Be^{-jkl}} = \frac{A}{B} e^{j2kl} = re^{j2\delta} \tag{4.4.1}$$

where 2δ is the phase change on reflection. Hence

$$\frac{A}{B} = re^{j(2\delta - 2kl)} = re^{j2\Delta} = -e^{-2\varphi} \tag{4.4.2}$$

Rewriting Eq. (4.2.3) and using Eq. (4.4.2):

$$p = B\left[e^{-jkx} + re^{j(kx+2\Delta)}\right]e^{j\omega t} \tag{4.4.3}$$

Thus, p consists of two phasor components that may be added on the complex plane.

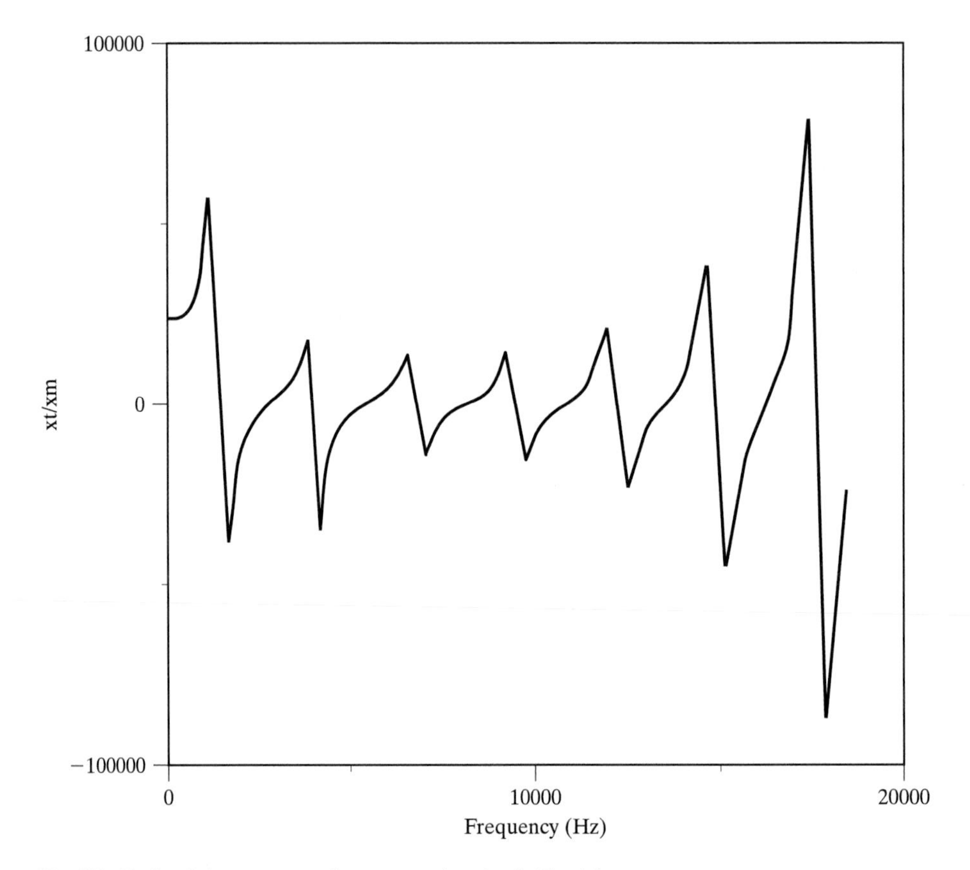

Fig. 4.7 Ratio of throat-to-mouth reactance for pipe in Fig. 4.6.

The resultant phasor (Fig. 4.9) is

$$p = BR\, e^{j\phi} e^{j\omega t} \tag{4.4.4}$$

where

$$R = [1 + r^2 + 2r \cos(2\Delta + 2kx)]^{1/2} \tag{4.4.5}$$

Now ϕ simply represents a phase lag, so the maximum and minimum of pressure amplitude are given by

$$\begin{aligned} p_{max} &= B(1 + r) \\ p_{min} &= B(1 - r) \end{aligned} \tag{4.4.6}$$

Hence, the standing wave ratio (SWR) is

$$SWR = \frac{p_{max}}{p_{min}} = \frac{1 + r}{1 - r} \tag{4.4.7}$$

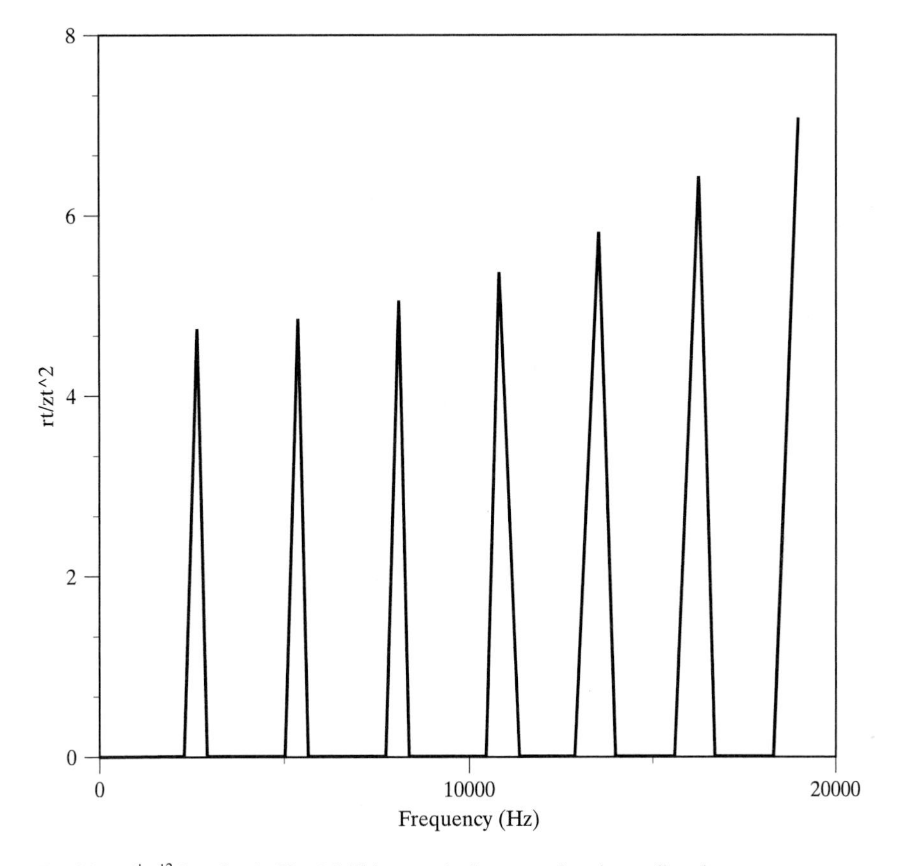

Fig. 4.8 $r_t/|z_t|^2$ for pipe in Fig. 4.6. This quantity is proportional to radiated power.

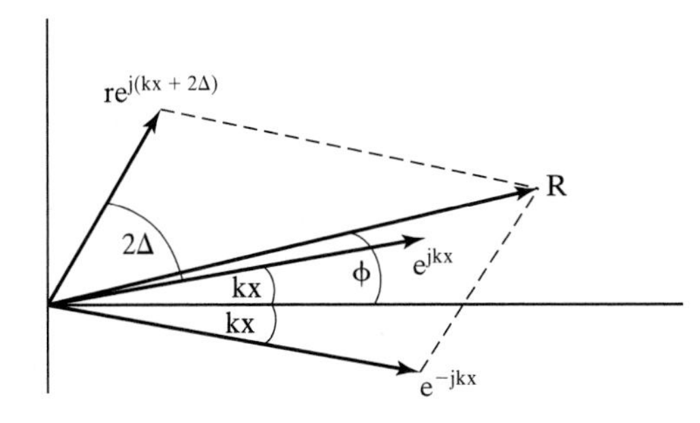

Fig. 4.9 Phasor components of pressure in a standing wave tube.

We can obtain further insight into the events within the tube from an alternative treatment of Eq. (4.4.3) by adding and subtracting re^{-jkx} within the bracket, thus

$$P = B(1 - r)e^{j(\omega t - kx)} + 2Br\cos(kx + \Delta)e^{j(\omega t + \Delta)} \tag{4.4.8}$$

Hence, the pressure inside the tube may be considered to be a travelling wave whose amplitude is $B(1 - r) = p_{min}$, and a stationary wave whose maximum amplitude is 2Br (see Fig. 4.10). The maximum pressure of the combined travelling and stationary waves

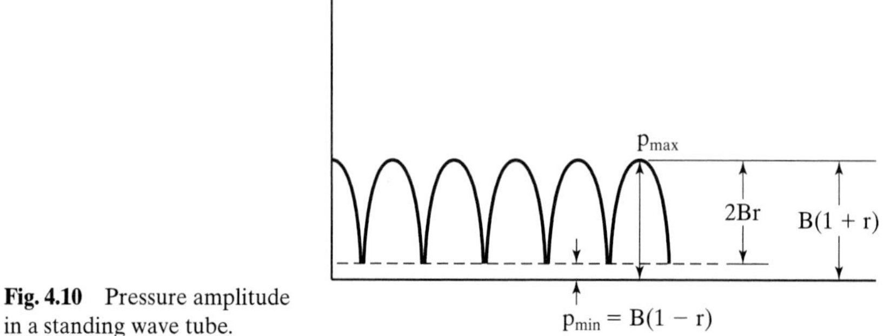

Fig. 4.10 Pressure amplitude in a standing wave tube.

is B(1 + r), as before. The combination of stationary and travelling waves is called a standing wave. Note that the pressure maxima occur when, from Eq. (4.4.5),

$$2\Delta + 2kx = n2\pi, \quad n = 0, 1, 2, \ldots$$
$$n = -1, -2, \ldots$$

that is, when

$$x = n\pi/k - \Delta/k$$
$$= n\lambda/2 - \partial/k + 1$$

Now an open tube is approximately pressure release so 2δ is about π. Then the positions of the pressure maxima are

$$x_m = n\,\lambda/2 - \lambda/4 + 1$$
$$l = \lambda/4, \quad x_m = nl/2 = 0, \lambda/2, \text{etc.}$$

These results are used in the impedance tube technique, as described in Section 9.8.

4.5 Salmon Horns

When the cross-sectional area of an acoustic conduit varies, we have a horn rather than a pipe. The function of a horn is to transform impedance. It was not until 1946 that Vincent Salmon tackled the problem of deciding the optimum shape for a horn, a process he named "horn synthesis." Salmon proposed that a good horn would have a minimum back reflection from the walls so that with an infinite horn, to avoid the complication of reflection from the mouth, a purely progressive wave would be obtained within the horn. Consequently, constancy of energy flow for a plane progressive wave requires the acoustic pressure to take the form

$$p = \frac{B}{y} e^{j(\omega t - \phi)} \tag{4.5.1}$$

where B is a constant

$$y^2 = S/\pi$$

and $\phi(x)$ is a phase angle. Substituting Eq. (4.5.1) into Webster's equation—Eq. (4.1.5)—leads to

$$\left(\frac{\partial \phi}{\partial x}\right)^2 + \frac{1}{y}\cdot\frac{\partial^2 y}{\partial x^2} - \frac{\omega^2}{c^2} + j\frac{\partial^2 \phi}{\partial x^2} = 0 \tag{4.5.2}$$

Now, since both real and imaginary parts of this equation must be zero,

$$\partial^2\phi/\partial x^2 = 0$$

so that it is possible to write

$$\partial\phi/\partial x = \tau\,\omega/c \tag{4.5.3}$$

where τ is a constant. Substituting this result into the real part of Eq. (4.5.2) yields

$$\partial^2 y/\partial x^2 - (\omega/c)^2(1 - \tau^2)\,y = 0 \tag{4.5.4}$$

A solution of this equation is

$$y = y_o\left[\cosh\left(\frac{x}{h}\right) + T\sinh\left(\frac{x}{h}\right)\right] \tag{4.5.5}$$

where $y = y_o$ when $x = 0$, and

$$h^2 = \frac{c^2}{\omega^2(1 - \tau^2)} \tag{4.5.6}$$

and T is an arbitrary constant:

1. When $T = 1$, then

$$y = y_o e^{x/h} \tag{4.5.7}$$

   so that the shape is *exponential*.
2. When $T = h/x_o$, then

$$y = y_o[\cosh(x/h) + (h/x_o)\sinh(x/h)]$$

   and if h becomes infinite,

$$y = y_o[1 + (x/x_o)] = mx + y_o \tag{4.5.8}$$

   where $m = y_o/x_o$, so that the horn is *conical*. In addition, note that $\tau = 1$.

3. When $T = 0$, then

$$y = y_o \cosh(x/h) \tag{4.5.9}$$

so that the horn is *catenoidal.*

We will use the exponential, conical, and catenoidal horns to illustrate the manner in which Webster's theory explains the function of real horns. Note that these three horns are frequently encountered in practice.

4.6 Properties of Infinite Horns

It follows from Eqs. (4.5.1) and (4.5.3) that

$$p = \frac{B}{y} e^{j\omega(t - \tau x/c)} \tag{4.6.1}$$

and substituting for τ, using Eq. (4.5.6)

$$p = \frac{B}{y} e^{j\left[\omega t - \frac{x}{ch}\sqrt{\omega^2 h^2 - c^2}\right]} \tag{4.6.2}$$

so that the propagation velocity

$$v = \frac{c}{\left(1 - \frac{1}{k^2 h^2}\right)^{1/2}} \tag{4.6.3}$$

where $k = \omega/c$. Consequently, there is a cutoff frequency given by

$$\omega_c = \frac{c}{h}$$

Thus, horns act as high pass filters, except if h is infinite, as in the case of a conical horn, when the cutoff frequency is actually zero. Below the cutoff frequency, there is no true wave propagation, merely an evanescent motion at the horn throat. Above the cutoff frequency, there is propagation at a phase velocity greater than that in free space, except if h is infinite, in which case $v = c$.

The velocity potential Φ is given by

$$\Phi = \frac{1}{\rho}\int p\,dt = -j\frac{p}{\omega\rho}$$

$$= -j\frac{B}{\rho\omega y}\exp j\omega\left(t - \frac{x}{c}\right)$$

so that the particle velocity is given by

$$v = (\rho/p)(\omega\tau/c - (j/y)\, dy/dx)$$

Thus, the impedance is

$$z = \frac{p}{u} = \frac{\rho c}{\tau - \frac{j}{yk}\frac{dy}{dx}}$$

Hence, the throat impedances for some infinite horns are

$$\rho c/(1 - j/kx_o) \qquad \text{(conical)}$$

$$\rho c/(\tau - j/kh) \qquad \text{(exponential)}$$

$$\rho c/\tau = \rho c/(1 - \omega_c^2/\omega^2)^{0.5} \qquad \text{(catenoidal)} \qquad (4.6.4)$$

4.7 Finite Horns: Impedance Transformation

In practice, horns have to be of finite length so that in general it is not permissible to ignore reflections from the mouth. One procedure for obtaining the impedance of infinite horns has been given by Morse, who showed that Salmon's formula for the cross-sectional radius of a good horn, as given by Eq. (4.5.5), may be rewritten as follows:

$$y = (y_o/\cosh\varepsilon)\cosh\,(x/h + \varepsilon)$$

where $T = \tanh\,\varepsilon$. Thus, the cross-sectional area will be given by

$$A = (A_o/\cosh^2\varepsilon)\cosh^2\,(x/h + \varepsilon) \qquad (4.7.1)$$

Substituting this expression into Webster's equation for pressure—Eq. (4.1.5)—and assuming harmonic time dependence gives

$$\partial^2 p/\partial x^2 + 2/h\, \tanh\,(x/h + \varepsilon)\, \partial p/\partial x + k^2 p = 0$$

Making the substitution

$$y = p\cosh\,(x/h + \varepsilon)$$

then leads to

$$\frac{\partial^2 y}{\partial x^2} + k^2\tau^2 y = 0$$

The general solution is then

$$y = Ae^{-jk\tau x} + Be^{jk\tau x}$$

where A and B are arbitrary constants so that waves travelling in both the forward and backward directions are included. An alternative method of writing this general solution is

$$y = A \sinh(\varphi + jk\tau x)$$

where φ is given by the relation

$$-B/A = e^{-2\varphi}$$

so that finally

$$p = \frac{-Pe^{j\omega t} \sinh(\varphi + jk\tau x)}{\cosh\left(\frac{x}{h} + \varepsilon\right)}$$

where P is an arbitrary constant. The particle velocity is then given by

$$v = -(j/\rho\omega)(\partial p/\partial x)$$

$$v = \frac{p}{\rho c}\left[\tau \coth(\varphi + jk\tau x) + \frac{j}{kh}\tanh\left(\frac{x}{h} + \varepsilon\right)\right]$$

and thus the specific acoustic impedance

$$z = \frac{\rho}{v} = \rho c\left[\tau \coth(\varphi + jk\tau x) + \frac{j}{kh}\tanh\left(\frac{x}{h} + \varepsilon\right)\right]^{-1} \quad (4.7.2)$$

For a horn of a given shape containing a known material, all quantities in Eq. (4.7.2) are known except for ϕ, which depends on the impedance at the mouth. Thus, to evaluate z some assumption must be made regarding the impedance at the mouth, which Morse, for example, took to be that of a plane piston radiator in an infinite baffle.

We can see that when $h \rightarrow \infty$, z in Eq. (4.7.2) gives the impedance of a tube. This result suggests an approximate method for solving Eq. (4.7.2) when h is not infinite. If the flare is not great, the term including $\tanh(x/h + \varepsilon)$ may be neglected at the mouth. Then

$$z_m = \frac{\rho c}{\tau}\tanh(\varphi + jk\tau l)$$

But this expression is analogous to the impedance of a tube of length l and with phase velocity c/τ. Consequently, z may be compared with tabulated impedances for a plane piston and ϕ determined. The impedance at the throat may then be determined by inserting the value for ϕ into Eq. (4.7.2) with $x = 0$. This procedure may be carried out using a Smith chart.

Olson (1957) has obtained expressions for throat impedances of various horn types as functions of the impedance at the mouth by an alternative procedure. For an exponential horn, for example, $\varepsilon \rightarrow \infty$ then Eq. (4.7.2) becomes

$$\frac{z}{\rho c} = \frac{1}{\tau \coth(\varphi + jk\tau x) + \dfrac{j}{kh}} \tag{4.7.3}$$

and at the mouth the reduced radiation impedance is

$$\frac{z_m}{\rho c} = \left[\tau \coth(\varphi + j\tau kl) + \frac{j}{kh}\right]^{-1}$$

Introducing φ obtained from this boundary condition into expression Eq. (4.7.3) and setting $x = 0$ gives the throat impedance

$$\frac{z_t}{\rho c} = \frac{-\dfrac{1}{\tau} + j\dfrac{z_m}{\rho c}\left[\dfrac{1}{\tau kh} - \cot(\tau kl)\right]}{-\tau \cdot \dfrac{z_m}{\rho c}\left[1 + \dfrac{1}{\tau^2 k^2 h^2}\right] - j\left[\cot(\tau kl) + \dfrac{1}{\tau kh}\right]} \tag{4.7.4}$$

Introducing Olson's notation

$$1/\tau kh = \tan\theta$$

$$\tau k = b$$

Eq. (4.7.4) becomes

$$\frac{z_t}{\rho c} = \frac{\sin(bl) + j\left(\dfrac{z_m}{\rho c}\right)\cos(\theta + bl)}{\dfrac{z_m}{\rho c}\sin(bl) + j\cos(\theta - bl)} \tag{4.7.5}$$

Values for resistance, reactance, and radiated power calculated according to Eq. (4.7.5) for an actual exponential horn are shown in Figs. 4.11, 4.12, and 4.13.

The price paid for the improvement in low frequency response when a finite horn is used as an impedance matching device for a transducer is the appearance of sharp peaks in the frequency response of the system. These peaks are characteristic of finite horns and the result of internal resonances due to reflections from impedance mismatch at the horn mouth. Olson showed that for a family of exponential horns having constant throat dimensions and flare parameter, but with different lengths, these resonance peaks become more pronounced with a decrease in the length of the horn to the extent that the impedance-versus-frequency plots at the resonant frequencies depart as much as one order of magnitude from the theoretical infinite horn plot. The finite horn is, however, superior to a pipe, which has many more resonances, as shown in Fig. 4.8.

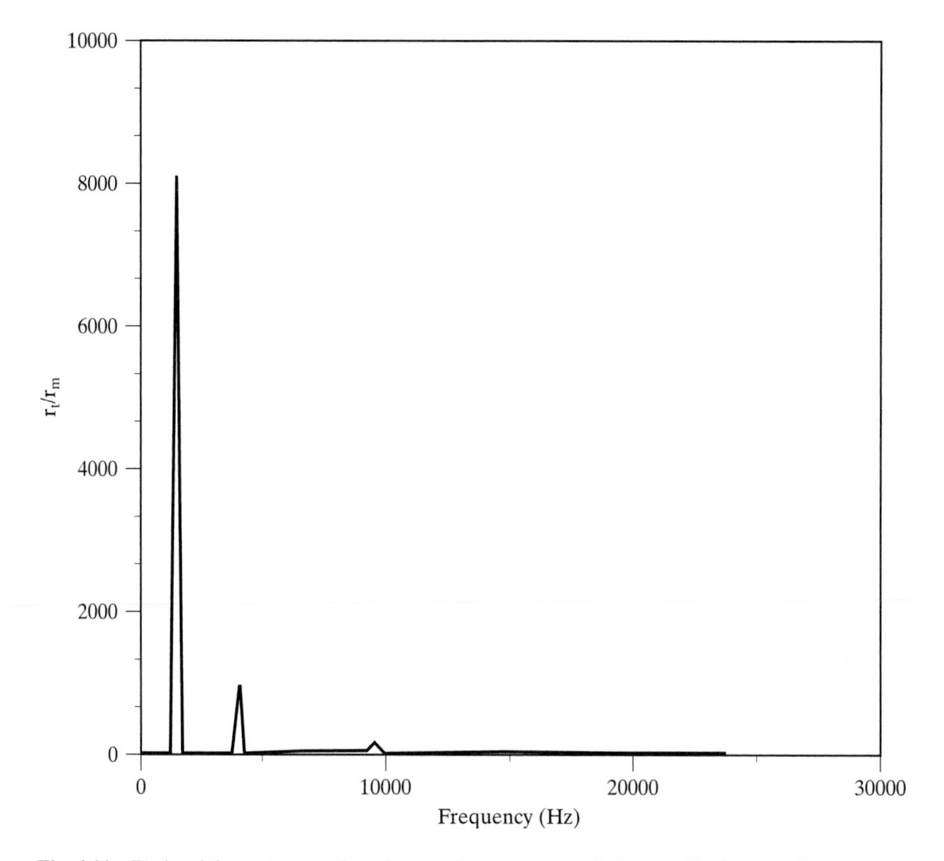

Fig. 4.11 Ratio of throat-to-mouth resistance for an exponential horn with throat radius 1 cm; length $2\pi \times$ radius, h = 0.5.

4.8 Solid Horns

Solid horns are used in a variety of applications. In industry, for example, horns are used in "drills" to machine brittle components; in medicine, horns are also used as "drills" to remove plaque and decay in dentistry and also in opthalmology for cataract removal. The principles of the operation of solid horns have been described by Neppiras (1958) and Eisner and Scager (1965).

The action of a solid horn is similar to that of a fluid in a flaring pipe. The motion of an element of the solid is described by the same equations of motion and continuity as for describing the fluid, Eqs. (4.1.1) and (4.1.2). The appropriate equation of elasticity is different, however. For a rod, the excess force on the element is given by

$$F = -YA\,\partial\eta/\partial x \tag{4.8.1}$$

The condensation, s, is approximately $-\partial\eta/\partial x$, and hence the excess pressure

$$p = Ys \tag{4.8.2}$$

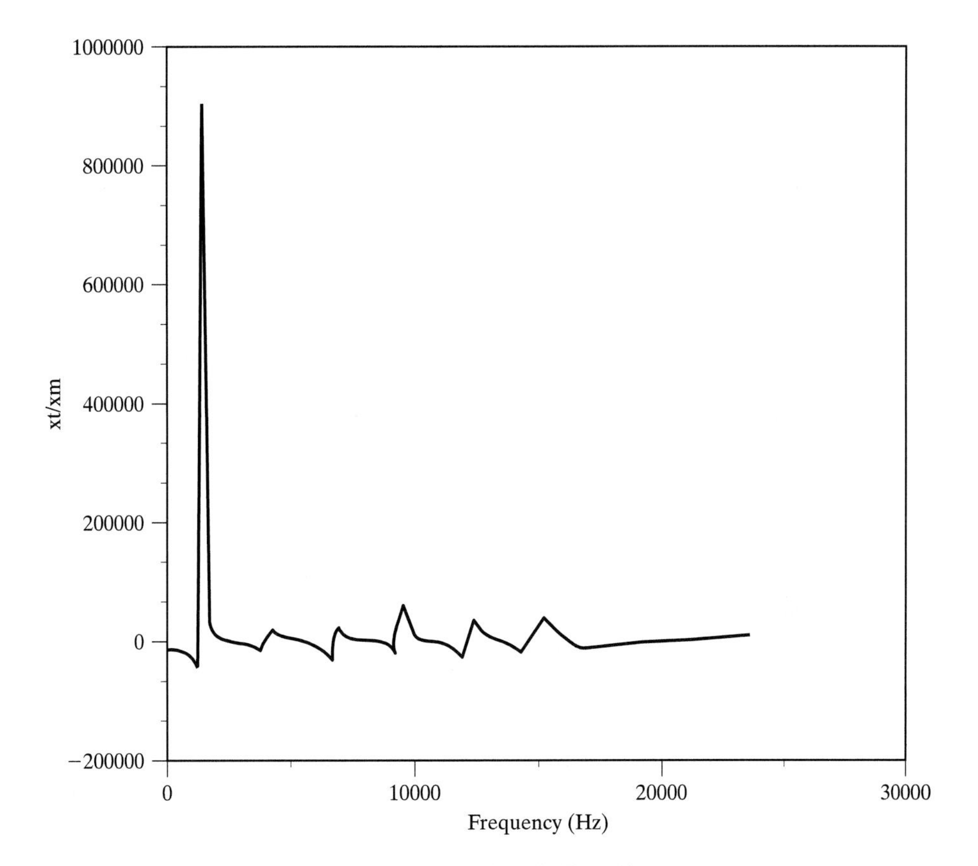

Fig. 4.12 Ratio of throat-to-mouth reactance for horn in Fig. 4.11.

This equation now replaces Eq. (4.1.4) and hence propagation in a solid horn can be treated in a manner exactly analogous to the way fluid-filled horns were handled by replacing the bulk modulus by Young's modulus.

There is one important difference regarding the boundary conditions at junctions, however. At the boundary between two solid sections A and B, the force must be continuous, as well as the displacement (and consequently the velocity). Thus

$$F_A = F_B \tag{4.8.3}$$

and

$$v_A = v_B \tag{4.8.4}$$

Letting the areas be A_A and A_B,

$$p_A = F_A/A_A \tag{4.8.5}$$

so that

$$F_A = A_A p_A = A_B p_B = F_B \tag{4.8.6}$$

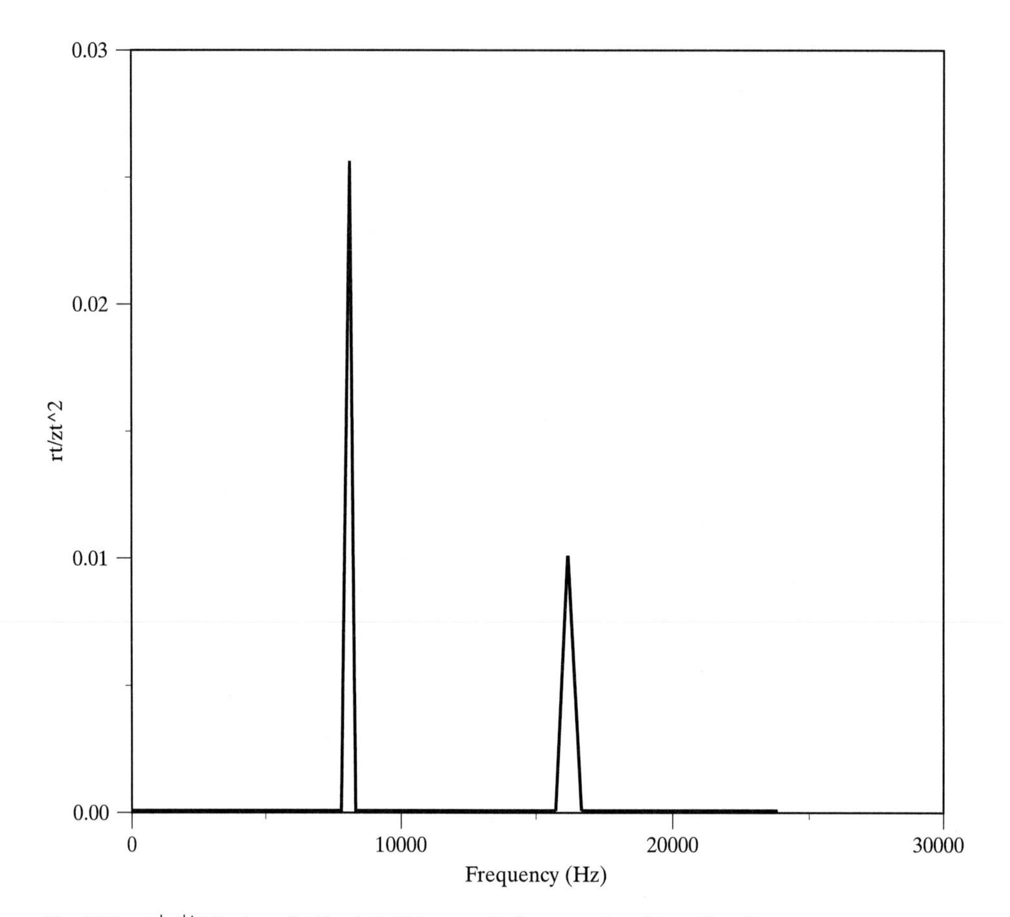

Fig. 4.13 rt/|zt|^2 for horn in Fig. 4.11. This quantity is proportional to radiated power.

and

$$\frac{A_A p_A}{v_A} = \frac{A_B p_B}{v_B} \tag{4.8.7}$$

or

$$\frac{z_A}{z_B} = \frac{A_B}{A_A} \tag{4.8.8}$$

Thus, the mechanical impedance is continuous, but there is a discontinuity in specific acoustic impedance according to the ratio of the areas.

4.9 Sinusoidal Horns

Consider again Eq. (4.5.4). We assumed that $\tau > 1$ in order to obtain the solution in Eq. (4.5.5), giving the well-known horn shapes. Suppose, however, that $\tau > 1$. In this event, the solution of Eq. (4.5.4) is

$$y = y_o\left[\cos\frac{x}{h_1} + T\sin\frac{x}{h_1}\right] \tag{4.9.1}$$

where T is an arbitrary constant and

$$h_1^2 = c^2/\omega^2\ (\tau^2 - 1) \tag{4.9.2}$$

In particular, if

$$y_o = G \sin\left(\frac{x_o}{h_1}\right) \tag{4.9.3}$$

and

$$y_o T = G \cos\left(\frac{x_o}{h_1}\right) \tag{4.9.4}$$

then Eq. (4.9.1) can be written

$$y = G \sin\ (x + x_o)/h_1 \tag{4.9.5}$$

The horn is formed by rotation of this profile around the x axis.

When the question arose as to whether sinusoidal horns had been discovered empirically, as in the case of Salmon horns, attention was directed toward certain musical instruments featuring globular bells, such as the English horn and the Been or Pungi (the instrument used by snake charmers in India). The English horn in its modern form was first used by Rossini in 1829 in the opera *William Tell*. The snake charmer's horn has been used since ancient times. Because of the relative precision of its manufacture, a study was made of the English horn. The internal profile of the bell of the English horn was obtained by a gamma ray photograph (Fig. 4.14). The best-fitting sine curve is $y = 0.029 \times \sin\ (0.85x)$ (see Fig. 4.15).

Fig. 4.14 Gamma ray photograph of the bell of the English horn. (After Nagarkar and Finch, 1971.)

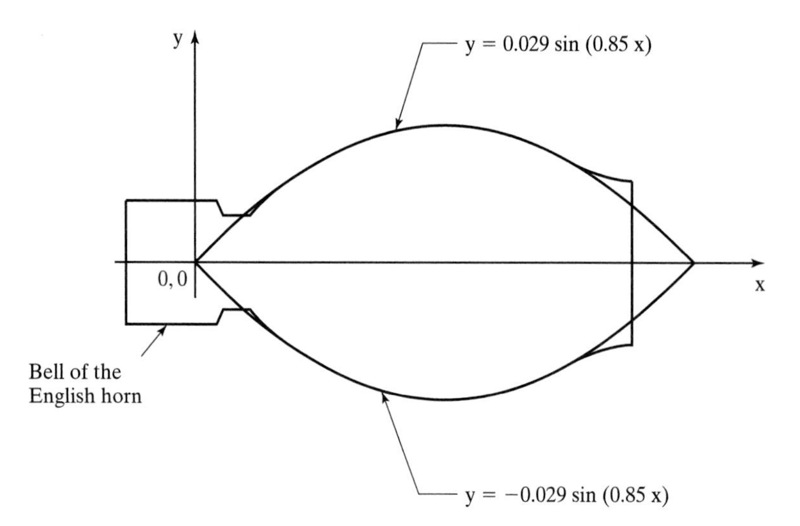

Fig. 4.15 Profile of the bell of an English horn, taken from the photograph in Fig. 4.14 and fitted by the curve y (inches) = 0.029 sin (0.85x). (After Nagarkar and Finch, 1971.)

REFERENCES AND FURTHER READING

Eisner, E. 1967. 41 *J. Acoust. Soc. Am.* 1126–1146. [Provides further historical information and an excellent bibliography on horns.]

Eisner, E., and J.S. Seager. 1965. "A Longitudinally Resonant Stub for Vibrations of Large Amplitude." 3 *Ultrasonics* 88–98.

Nagarkar, B.N., and R.D. Finch. 1971. "Sinusoidal Horns." 50 *J. Acoust. Soc. Am.* 23–31.

Neppiras, E.A. 1958. "Design of Ultrasonic Machine Tools." The Institution of Mechanical Engineers. Section 4, Paper 6.

Olson, H.F. 1957. *Acoustical Engineering.* Van Nostrand.

PROBLEMS

4.1 An air-filled pipe is 1 m in length and 10 cm in diameter. Assuming that it radiates sound into the open air at one end according to the Levine and Schwinger theory, estimate the acoustic impedance at the other end at 300, 1000, 3000, and 10,000 Hz.

4.2 Estimate the standing wave ratios in the pipe described in Problem 4.1 at the frequencies listed.

4.3 A wooden pipe is to be used as a musical instrument. The pipe is 25 cm in length and 1 cm in diameter. What is its fundamental resonance frequency when radiating into air? How would this frequency change if a 0.5 cm diameter hole were to be drilled in the pipe wall at 20 cm from the throat?

4.4 A megaphone comprises a cone 2 cm in diameter at its throat and 20 cm in diameter at the mouth. Calculate the length for a resonance frequency of 500 Hz.

4.5 An air-filled trumpet is 1 m in length and 10 cm in diameter at its mouth. The profile of the interior is exponential and the throat has a 1 cm diameter. Assuming that it radiates sound into the open air at the mouth according to the Levine and Schwinger theory, estimate the acoustic impedance at the throat at 300, 1000, 3000, and 10,000 Hz.

4.6 An exponential brass horn is to be used to vibrate an object that can be approximated to a point mass and whose weight is 1 kg. Assuming the vibration frequency is to be 3 Hz and that the horn is to be turned out of 2 in. diameter bar stock, calculate the length of the horn for operation at resonance if the mass is to be situated at the throat of the horn whose diameter is to be 0.25 in.

4.7 The bell of an English horn is an air-filled cavity fitting an approximately sinusoidal profile whose inner dimensions are 2 cm in diameter at the throat, 3 cm in diameter at the mouth, and 8 cm in length. If the maximum internal diameter is 6 cm, find the equation of the profile and sketch the resistance and reactance of the horn as a function of frequency. Estimate the frequency of resonance.

4.8 A horn comprises two solid cylindrical sections joined axially. Each section is 5 cm in length and the diameters are 1 cm and 1 mm. Estimate the amplitude gain for longitudinal vibrations if the horn material is

a. brass
b. steel
c. titanium

4.9 A fog signal is to operate at a resonance frequency of 440 Hz. The device is to be fabricated from two sections of pipe having diameters of 1 in. and 2 in. Given that the two sections will be sealed by a flat annulus, estimate the two pipe lengths.

4.10 The one-dimensional Schrodinger equation of quantum mechanics for a free particle may be written

$$-\frac{\partial^2\psi}{\partial x^2} = \frac{2m}{\hbar^2}(E - V)\psi = \left(\frac{2\pi}{\lambda}\right)^2\psi$$

where ψ is the wave function of the particle, m its mass, E its energy, V is the potential function in which it moves, and

$$\hbar = h/2\pi$$

where h is Planck's constant. Compare this equation with Webster's equation for a harmonic wave and find expressions for the wavelength in both cases.

Chapter 5

Audio Frequency Generators

5.1 Introduction

Much of the early development of audio frequency generators was carried out for situations in which the reception was by a human listener. In reproducing speech or music from an electrical signal (from a recording or broadcast radio wave), the ideal aim is to have a transducer with a flat frequency response over the audio range. The designer of such devices also aims to maximize the electroacoustic efficiency and to minimize the directivity of the radiated sound. Most loudspeakers employ the moving coil mechanism, and the same principle is carried over to the design of shakers. For certain warning devices (such as fog signals, sirens, and train signals), the general objective is to produce a sound recognizable to a listener at a distance. In these devices, a flat frequency response is not required, and a very high power output at a single frequency can meet the design objectives. In these cases, the electroacoustic efficiency can be well above 50%. In air-modulated generators, a moving coil mechanism is used to modify an air flow, in which case considerable acoustic energy derives from decompression of air. We will start by considering a simple direct radiator loudspeaker. Our treatment follows the presentation developed by Kinsler et al. (1982).

5.2 Direct Radiator Loudspeaker

A simple direct radiator loudspeaker (Fig. 5.1) consists of a coil of length l (m) wound on a former attached to a light-weight cone or diaphragm of total mass M. The coil moves in a magnetic field of B Wb/m^2 in strength. The cone is supported by a ring of corrugations of stiffness S. The cone is thus idealized to a one degree of freedom system.

Let the radius of the speaker cone be a and assume it is mounted in a baffle. The total mechanical impedance is then

$$Z_m = Z_r + Z_c \tag{5.2.1}$$

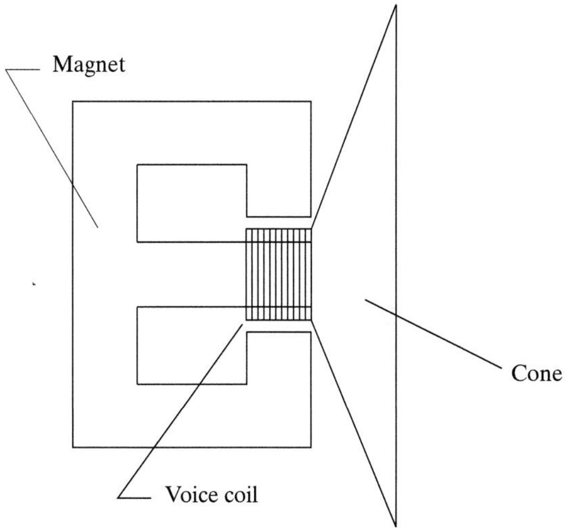

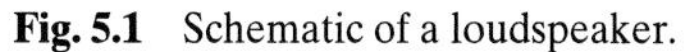
Fig. 5.1 Schematic of a loudspeaker.

where Z_r is the mechanical impedance due to acoustic radiation, and Z_c is the lumped mechanical impedance of the cone; that is,

$$Z_c = R_m + j\,(\omega M - S/\omega) \tag{5.2.2}$$

The mechanical damping, R_m, principally represents losses due to flexing of the diaphragm supports. At higher driving frequencies, higher order vibrational modes of the diaphragm occur and these could be accounted for in Z_c. The total radiation impedance is

$$Z_r = R_r + jX_r \tag{5.2.3}$$

Ignoring loading on the rear surface of the cone, for a baffled piston,

$$R_r = \rho c \pi a^2 R_1(2ka) \tag{5.2.4}$$

and

$$X_r = \rho c \pi a^2 X_1(2ka) \tag{5.2.5}$$

as given in Section 3.14. In MKS units, the force driving the cone is

$$F = Bl\,I \tag{5.2.6}$$

where B is the magnetic flux in Wb/m^2 and l is the total length of voice coil winding. Applying the general principles of electromechanical transduction (see Section 2.12),

$$T_{me} = -T_{em} = Bl = \varphi \tag{5.2.7}$$

Here, Bl or φ is called the flux linkage. Then, the driving point impedance,

$$Z_{ee} = Z_e + Z_M \tag{5.2.8}$$

where

$$Z_e = R_e + jL\omega \tag{5.2.9}$$

R_e being the electrical resistance and L, the inductance of the coil. Z_M is the motional impedance. Thus

$$\begin{aligned} Z_M &= \varphi^2/Z_m \\ &= \frac{\varphi^2}{R_r + R_m + j\left(M\omega - \frac{S}{\omega} + X_r\right)} \\ &= \frac{\varphi^2}{Z_m^2}\left[\left(R_r + R_m - j\left(M\omega - \frac{S}{\omega} + X_r\right)\right)\right] \\ &= R_M + j\,X_M \end{aligned} \tag{5.2.10}$$

The motional resistance consists of two parts:

$$R_{Mr} = \frac{\varphi^2 R_r}{Z_m^2} \tag{5.2.11}$$

which is associated with the dissipation of energy through radiation, and

$$R_{Mm} = \frac{\varphi^2 R_m}{Z_m^2} \tag{5.2.12}$$

which is associated with mechanical damping of the cone. We can now draw the equivalent electrical circuit as shown in Fig. 5.2. Hence, the electroacoustic efficiency (acoustic power radiated / electric power consumed) is

$$\eta = \frac{R_{Mr}}{R_{Mr} + R_{Mm} + R_e} = \frac{\varphi^2 R_r}{\varphi^2(R_r + R_m) + R_e Z_m^2} \tag{5.2.13}$$

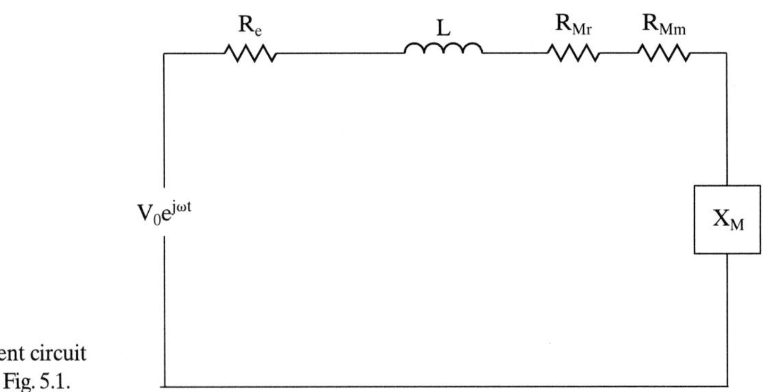

Fig. 5.2 Equivalent circuit of loudspeaker in Fig. 5.1.

The acoustic power radiated in watts is

$$W = \frac{I_0^2 R_{Mr}}{2} = \frac{\varphi^2 R_r I_0^2}{2 Z_m^2} \quad (5.2.14)$$

With applied voltage V,

$$I = \frac{V}{Z_{ee}} = \frac{V}{[(R_e + R_M)^2 + (L\omega + X_M)^2]^{1/2}} \quad (5.2.15)$$

and

$$W = \frac{\varphi^2 R_r V^2}{2 Z_m^2 Z_{ee}^2} \quad (5.2.16)$$

Increasing the flux linkage φ or Bl increases R_r and consequently the efficiency.

5.3 Typical Direct Radiator Parameters

Typical values of the parameters for a direct radiator loudspeaker are as follows:

voice coil mass $m_c = 0.0015$ kg
piston mass = 0.0085 kg
piston radius = 0.1 m (8 in speaker)
$S = 2000$ N/m
$R_m = 1$ mechanical ohm
$L = 0.2$ mH
$R_e = 5$ ohm
$l = 5$ m
$B = 0.9$ Wb/m2
$\varphi = Bl = 4.5$ Wb/m

The frequency dependence of the mechanical impedance is shown in Fig. 5.3, the electroacoustic efficiency in Fig. 5.4, and the motional and electrical input impedances in Figs. 5.5 and 5.6, respectively.

At low frequencies, $2ka \ll 1$, and we have from Section 3.14

$$R_1(2ka) \approx \frac{(2ka)^2}{8} \quad (5.3.1)$$

$$X_1(2ka) \approx \frac{4\,(2ka)}{3\pi} \quad (5.3.2)$$

that is, when

$$\lambda > 4\pi a$$

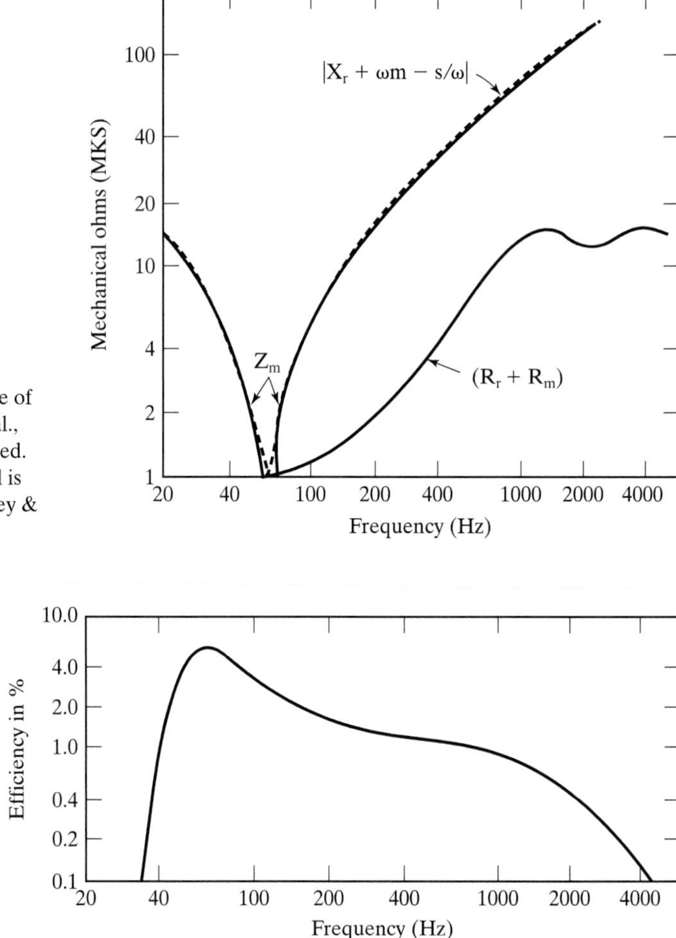

Fig. 5.3 Mechanical impedance of loudspeaker. (From Kinsler et al., *Fundamentals of Acoustics*, 3rd ed. Copyright © 1982. This material is used by permission of John Wiley & Sons, Inc.)

Fig. 5.4 Electroacoustic efficiency of loudspeaker. (From Kinsler et al., *Fundamentals of Acoustics*, 3rd ed. Copyright © 1982. This material is used by permission of John Wiley & Sons, Inc.)

in other words, when

$$\lambda > 1.2\text{m}$$

then

$$f < \frac{330}{1.2} \quad \text{or} \quad 275 \text{ Hz}$$

and

$$R_r \approx 2.2 \times 10^{-5} f^2 \text{ and } X_r \approx 0.02 f$$

When $2ka > 4$, $R_1(2ka) = 1$ and $R_r = 13$, $f > 110$ Hz.

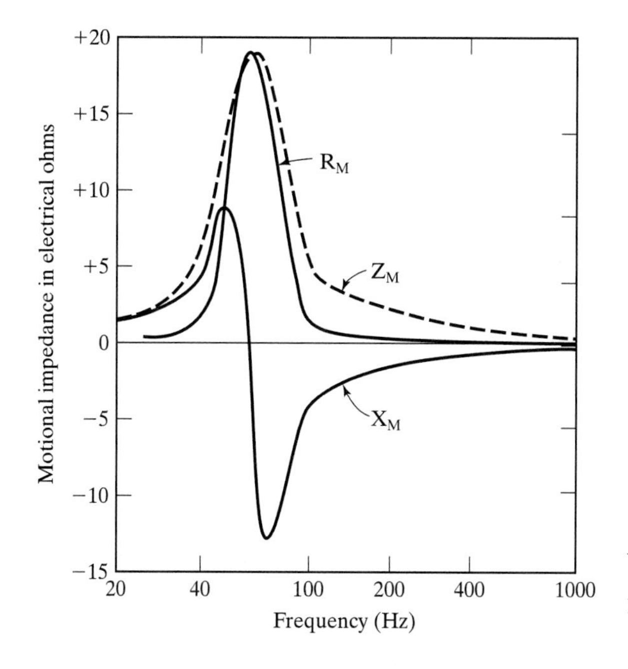

Fig. 5.5 Motional impedance of loudspeaker. (From Kinsler et al., *Fundamentals of Acoustics*, 3rd ed. Copyright © 1982. This material is used by permission of John Wiley & Sons, Inc.)

The frequency of mechanical resonance is given by

$$\left(X_r + M\omega_0 - \frac{S}{\omega_0}\right) = 0 \tag{5.3.3}$$

that is,

$$f_0 \approx 62 \text{ Hz}$$

Note that between 200 Hz and 2000 Hz the impedance is approximately constant and equal to R_e (5Ω). Many loudspeaker manufacturers specify this impedance as the rated value. Curves A, B, and C in Fig. 5.7 show the acoustic output in watts, computed from

A. $W = B^2l^2 R_r I^2/Z_m{}^2$ (5.3.4)

keeping I constant.

B. $\eta = R_{Mr}/(R_M + R_e)$ and $W_0 = \eta \times W_e$ (5.3.5)

where W_e is the electric power which is kept constant.

C. $W = \dfrac{B^2l^2 R_r V^2}{Z_m^2 Z_{ee}^2}$ (5.3.6)

keeping voltage constant.

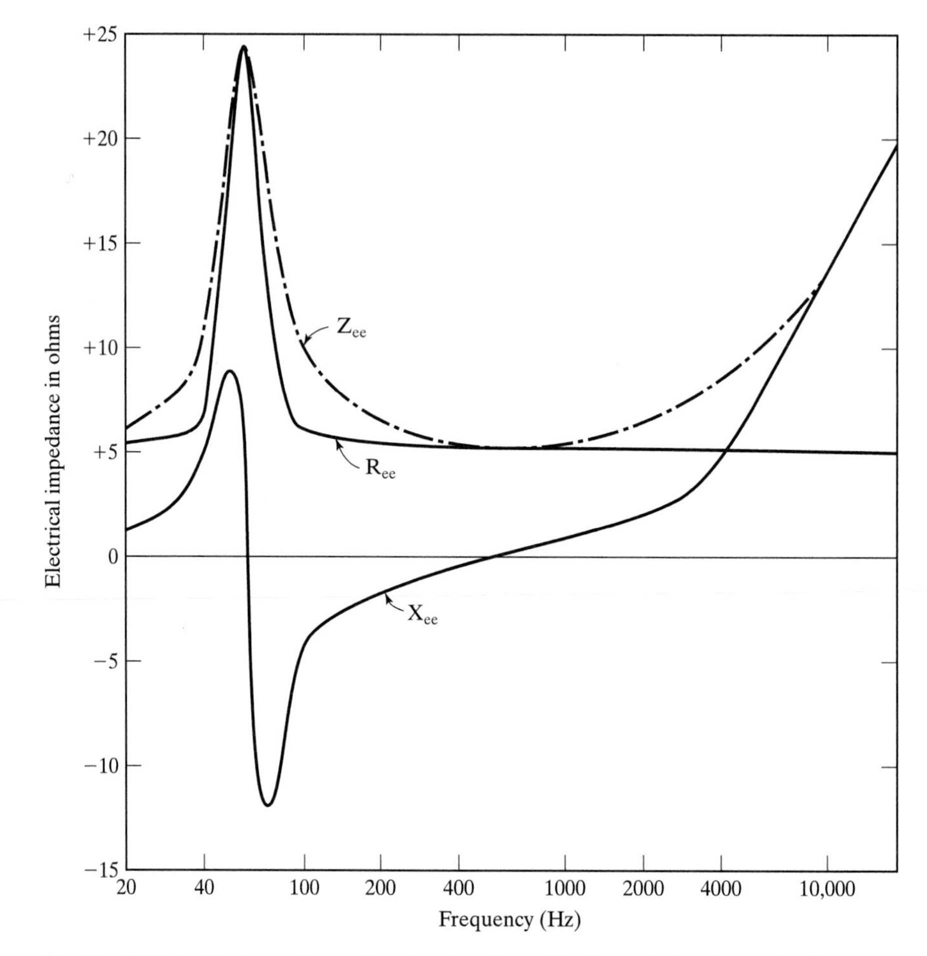

Fig. 5.6 Total electrical impedance of loudspeaker. (From Kinsler et al., *Fundamentals of Acoustics*, 3rd ed. Copyright © 1982. This material is used by permission of John Wiley & Sons, Inc.)

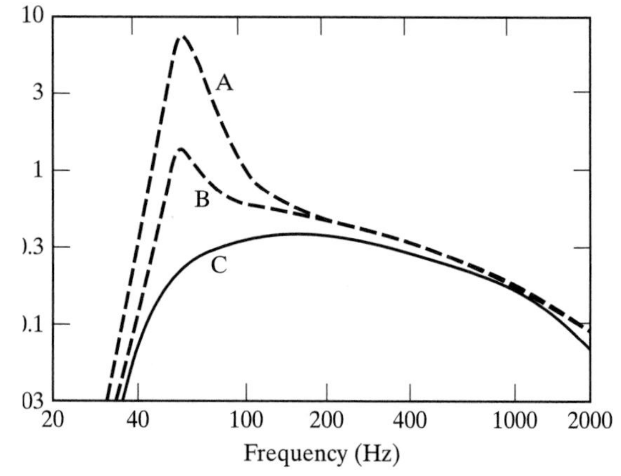

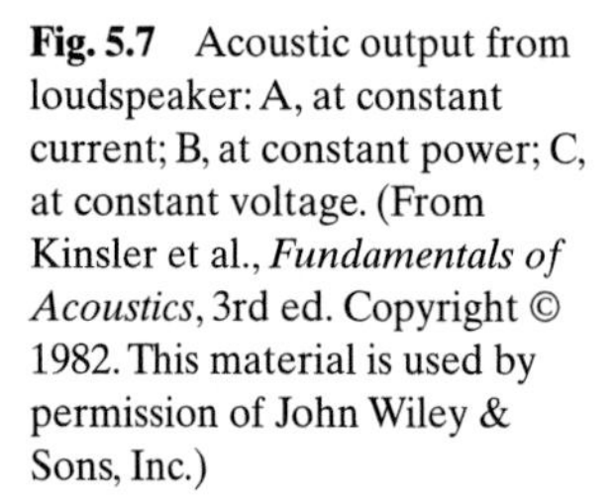
Fig. 5.7 Acoustic output from loudspeaker: A, at constant current; B, at constant power; C, at constant voltage. (From Kinsler et al., *Fundamentals of Acoustics*, 3rd ed. Copyright © 1982. This material is used by permission of John Wiley & Sons, Inc.)

The peaks in A and B are undesirable for high-fidelity reproduction. The smoothest acoustic output is obtained by an input intermediate between B and C. A big change in impedance at resonance causes a drop in power because of mismatch, and therefore in practice the actual output is usually as required.

5.4 Impedance Loops

The variation of Z_m with frequency can be exhibited by plotting the locus of the head of the impedance vector on the complex plane. If R_m is frequency independent, this locus will be a straight line as shown in Fig. 5.8. The mechanical admittance, Y_m, can be plotted similarly, as in Fig. 5.9. It follows that we may write

$$Y_m = \frac{1}{Z_m} = \frac{R_m - jX_m}{R_m^2 + X_m^2} \tag{5.4.1}$$

that is,

$$Y_m = R_m|Y_m|^2 - jX_m|Y_m|^2 \tag{5.4.2}$$

$$= G_m - jB_m \tag{5.4.3}$$

Now the phase angle of the admittance vector is given by

$$\cos\theta = \frac{G_m}{|Y_m|} \tag{5.4.4}$$

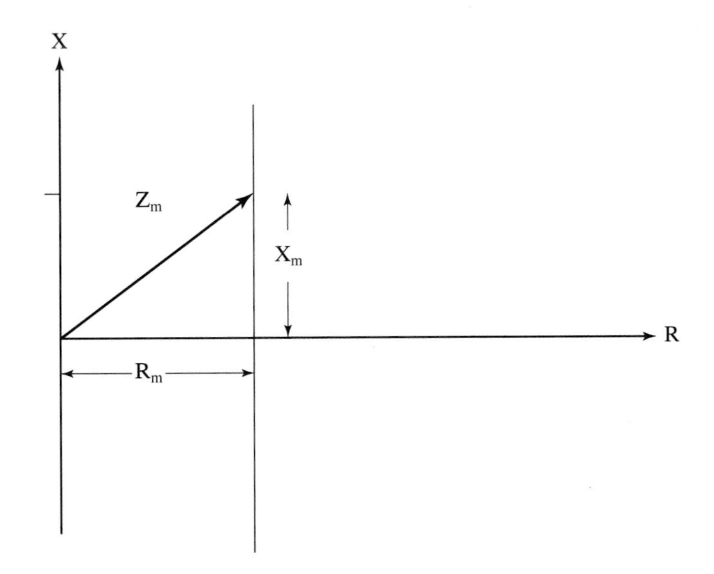

Fig. 5.8 Impedance vector on the complex plane.

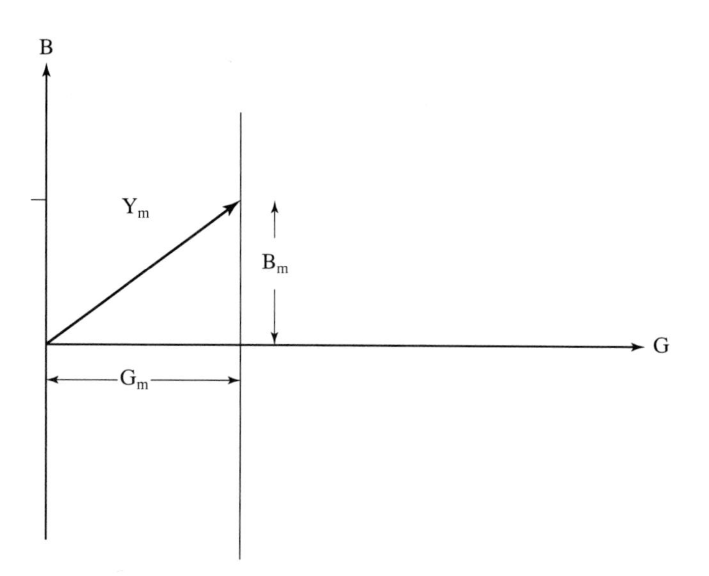

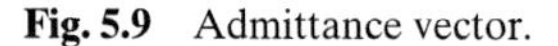

Fig. 5.9 Admittance vector.

so that from Eqs. (5.4.2), (5.4.3), and (5.4.4):

$$\cos\theta = R_m|Y_m| \tag{5.4.5}$$

or

$$|Y_m| = \frac{\cos\theta}{R_m} \tag{5.4.6}$$

But Eq. (5.4.6) is the equation of a circle in polar coordinates having diameter $1/R_m$. Thus, the admittance vector head locus is a circle on a complex plane as shown in Fig. 5.10. The circle intersects the real axis when $B_m = X_m = 0$ (i.e., when $\omega^2 = S/M = \omega_o^2$. At the quadrantal frequencies ω_1 and ω_2, the real and imaginary parts of Y_m are equal; that is,

$$R_m = X_m(\omega_1, \omega_2)$$

or

$$R_m = -M\omega_1 + S/\omega_1 \tag{5.4.7}$$

and

$$R_m = M\omega_2 - S/\omega_2 \tag{5.4.8}$$

Subtracting Eq. (5.4.7) from Eq. (5.4.8),

$$\omega_1 \omega_2 = S/M = \omega_0^2 \tag{5.4.9}$$

Multiplying Eq. (5.4.7) by ω_1 and Eq. (5.4.8) by ω_2 and adding,

$$\omega_2 - \omega_1 = R_m/M \tag{5.4.10}$$

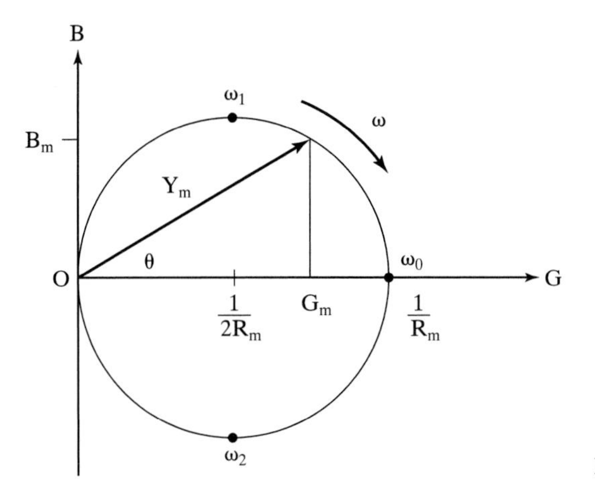

Fig. 5.10 Admittance loop.

The motional impedance locus is obtained by multiplying the mechanical admittance by $-T_{em}T_{me}$ as in Eq. (5.4.7). In general, the Z_{mot} locus is then a circle, with a diameter $-T_{em}T_{me}$ times as great as that of the Y_m circle, and also with a possible rotation, since $T_{em}T_{me}$ may contain a phase change. Finally, the driving point electrical impedance is assembled by adding Z_e and Z_{mot}. The locus of Z_e will be a straight line, similar to that of Z_m. This superposition is shown in Fig. 5.11.

In practice, systems can rarely be reduced to one degree of freedom equivalents, and other resonances may show up in impedance loci as subsidiary loops. An impedance loop for the loudspeaker considered in Section 5.3 is plotted in Fig. 5.12. The diameter of the loop is

$$\frac{-T_{me}T_{em}}{R_m} = \frac{\varphi^2}{R_m} = (4.5)^2 = 20.25 \text{ ohm}$$

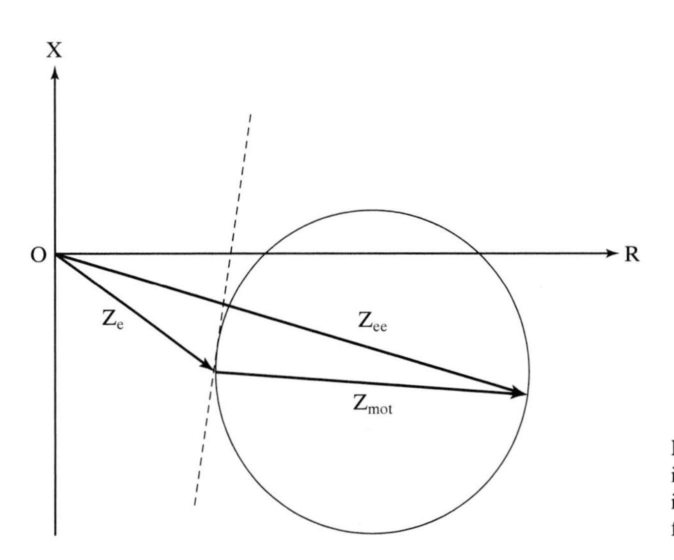

Fig. 5.11 Impedance loop including blocked electrical impedance and allowing for frequency variation of damping.

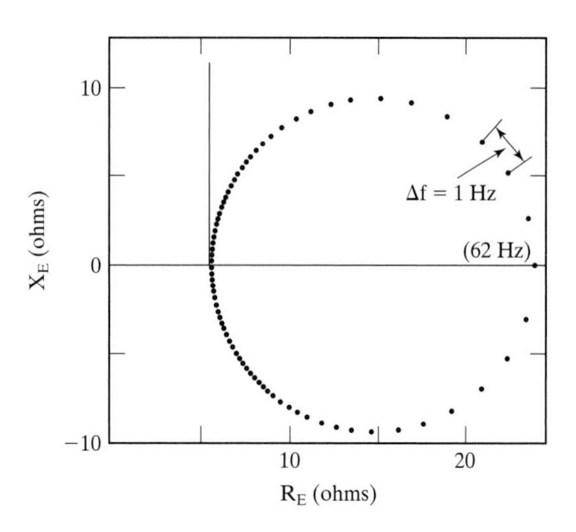

Fig. 5.12 Impedance loop for loudspeaker.

5.5 Improving on the Typical Speaker

For reproducing speech and music, a flat frequency response is required over a wider frequency range than is obtained with the typical loudspeaker we have just examined. However, the measures necessary to improve the device at high and low frequencies are different.

If we assume a constant voltage drive, the power is given by

$$W = \frac{B^2 l^2 R_r V^2}{Z_m^2\, Z_{ee}^2} \tag{5.5.1}$$

At high frequencies, Z_m depends principally on $M\omega$, and Z_{ee} on $L\omega$. The mass of the voice coil can be reduced by using aluminum wire and a thermo-setting epoxy to bind the coil together, instead of winding it on a former. However, reducing the radius of the cone is offset to some extent because the radiation resistance is proportional to the fourth power of the radius.

A reduction of the effective mass of the cone happens naturally to some extent because at higher frequencies the cone vibrates in higher order modes, and outer zones have relatively low amplitude. The effect can be enhanced by making corrugations in the cone as in Fig. 5.13. At low frequencies, Z_m depends primarily on S/ω, and Z_{ee} on R_e, so the following measures may be taken:

1. Increase a to increase R_r and a^2. This measure is not very effective because M increases and thus Z_m increases.
2. Decrease S to give a lower f_0. This measure is also not effective because the amplitude increases and brings in harmonic distortion, which cannot be balanced by an opposite nonuniformity in electrical performance, as can nonuniformity in frequency response.
3. Use a baffle or a cabinet mount to increase radiation resistance and to permit use of speakers, producing large amplitudes without distortion.

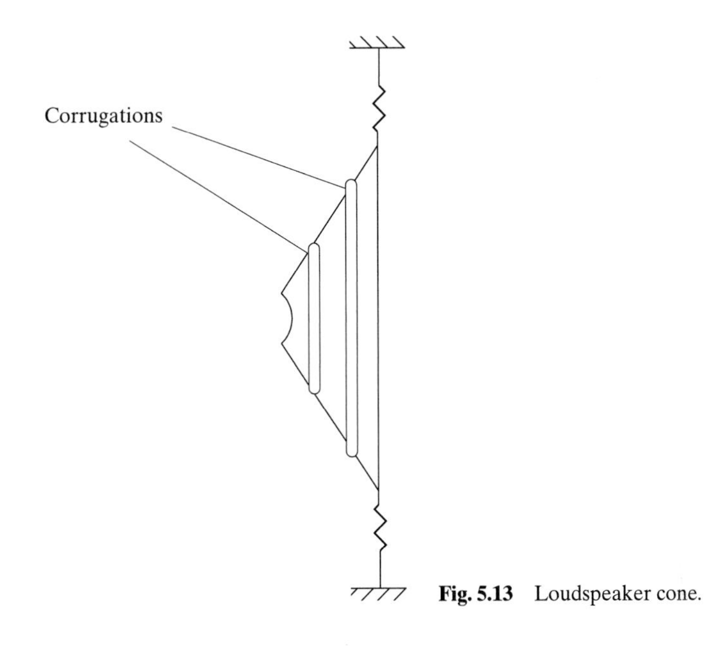

Fig. 5.13 Loudspeaker cone.

Generally, good high and low frequency responses are incompatible and it is common to use two speakers, a tweeter and woofer with an electrical crossover network.

5.6 Effect of Voice Coil Parameters

From Eq. (5.2.13), we have the efficiency:

$$\eta = \frac{B^2 l^2 R_r}{B^2 l^2 (R_r + R_m) + R_e Z_m^2}$$

but

$$B^2 l^2 (R_r + R_m) << Z_m^2 R_e$$

so that

$$\eta = \frac{B^2 l^2 R_r}{R_e Z_m^2} \tag{5.6.1}$$

Thus, to increase η we must either increase B (i.e., use a powerful magnet with a small air gap) or increase l (i.e., increase the air gap). There is a trade-off between these two effects, which means there is an optimum air gap. If we need to reduce M, we use aluminum; if we need to reduce R_e, we use copper. No greater gain is realized by reducing the wire diameter d, because

$$R_e \, \alpha \frac{l}{d^2} \text{ and } d^2 l \, \alpha \, M \tag{5.6.2}$$

If M is kept constant,

$$R_e \,\alpha\, \frac{l^2}{M} \tag{5.6.3}$$

and does not, in other words, depend on d, so there is no gain in efficiency.

5.7 Use of a Baffle

In practice baffles cannot be of infinite size. However, some sort of baffle (Fig. 5.14) must be used or the piston will behave like a doublet or a dipole. The power radiated from a doublet W_D, compared with the power radiated from a monopole of the same strength, W_M, is given by

$$\frac{W_D}{W_M} = \frac{l^2k^2}{3} = \frac{l^2 4\pi^2}{3\lambda^2} \tag{5.7.1}$$

If $l << \lambda$, then

$$W_D << W_M \tag{5.7.2}$$

Without a baffle we would also suffer a loss of directionality. As kl approaches unity, the theory of the dipole no longer applies and the intensity along the axis will vary with frequency. The question then arises: What is a minimum effective baffle size? Consider a circular baffle of radius b. At low frequencies, when $kb << l$, $ka << l$ so the speaker acts as a simple source of strength:

$$Q = v_0 \pi a^2 \tag{5.7.3}$$

where v_0 is the velocity amplitude. After travelling distance b, radiation from either side of the speaker rounds the baffle. Thus, such a speaker may be considered a dipole of separation b.

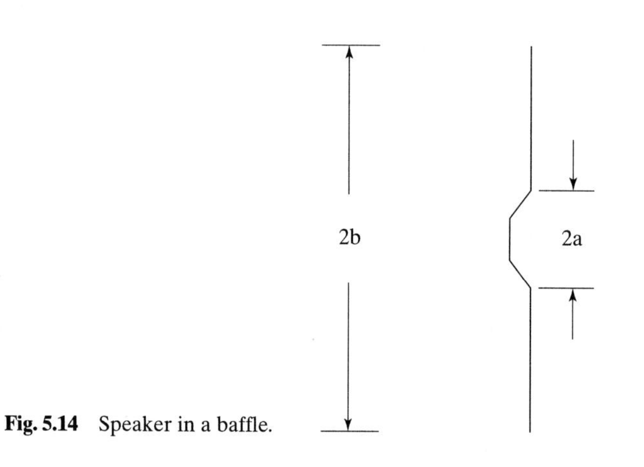

Fig. 5.14 Speaker in a baffle.

But

$$\frac{W_D}{W_M} = \frac{k^2b^2}{3} \tag{5.7.4}$$

hence

$$R_r = \frac{k^2l^2}{3} R_r \tag{5.7.5}$$

where, at low frequencies,

$$R_r = \rho_0 c\pi a^2 R_1(2ka) = \rho c\pi k^2 a^4/2 \tag{5.7.6}$$

A baffle works quite well for k^2b^2 up to 3, in other words, the output will be reduced if

$$kb < \sqrt{3}$$

that is

$$f < \frac{\sqrt{3}\,c}{2\pi b} = f_c \tag{5.7.7}$$

(i.e., as we increase b, f_c is lowered). We therefore make b sufficiently large that $f_c = f_0$, because below this resonant frequency the output fails off rapidly anyway.

For the loudspeaker considered previously

$$b = \frac{\sqrt{3}\,c}{2\pi \times 62} = 1.46 \text{ m} \approx 5 \text{ ft}$$

This result is still too large to be practical, and values of about 2 feet are used without excessive loss in power output.

5.8 Loudspeaker Cabinets

An alternative to the use of a baffle is a cabinet, as shown in Fig. 5.15.

1. *Closed back.* A closed back cabinet is similar to a Helmholtz resonator. With rigid walls, there is no radiation from the back of the speaker, so the enclosure is sometimes called an infinite baffle cabinet. The stiffness is increased by the enclosed air. The additional stiffness is

$$S_c = \frac{\gamma P A^2}{V} \tag{5.8.1}$$

If the loudspeaker in Section 5.3 is mounted in a cabinet of dimensions $0.5 \times 0.5 \times 0.2$ m, then $S_c = 2850$ n/m, which would raise the resonant frequency from 62 Hz to 96 Hz, and consequently the low frequency output would be reduced. To offset this result, we could reduce the mounting stiffness and increase the mass of the cone. In this way, speakers have been produced with an acceptable acoustic output down to 40 Hz in a 50 cm^3 cabinet. But doing this harms the high frequency response, so such a speaker should be used only as a woofer.

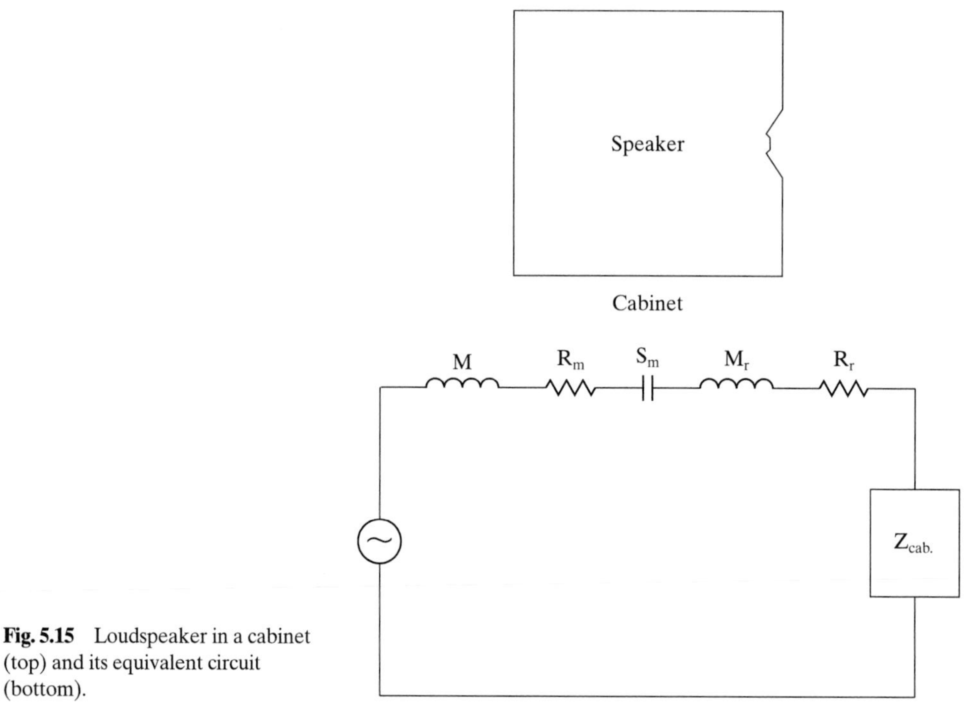

Fig. 5.15 Loudspeaker in a cabinet (top) and its equivalent circuit (bottom).

To eliminate resonances in the cabinet, it should be lined with a sound absorber, which would also lower γ (the ratio of specific heats) and hence the stiffness and the resonance frequency, thus further improving the low frequency characteristics.

2. *Open Back.* An open back cabinet is characteristic of most radio and television sets. At low frequencies, the path length back-front is too short and the radiated power drops. The cabinet acts as a short pipe or transmission line whose impedance is zero at certain frequencies. When this occurs, the output is enhanced, resulting in an unnatural booming sound.
3. *Vented box.* One type of cabinet is designed to radiate the sound from the back of the speaker in the forward direction by "inverting" the phase and increasing the radiation resistance. Such a cabinet is called a vented box, or bass-reflex cabinet, as illustrated in Fig. 5.16. The acoustic impedance is then that of an air spring given by

$$S_b = \frac{\rho c^2 A_H^2}{V} \tag{5.8.2}$$

in parallel with the impedance of the vent, and at low frequencies

$$M_H = \rho(l_H + 1.7a_H)\,\pi a_H^2 \tag{5.8.3}$$

and

$$R_H = \pi a_H^2 \rho c \frac{(2ka_H)^2}{8} = \frac{(\pi a_H)^2 \rho c k^2}{2\pi} \tag{5.8.4}$$

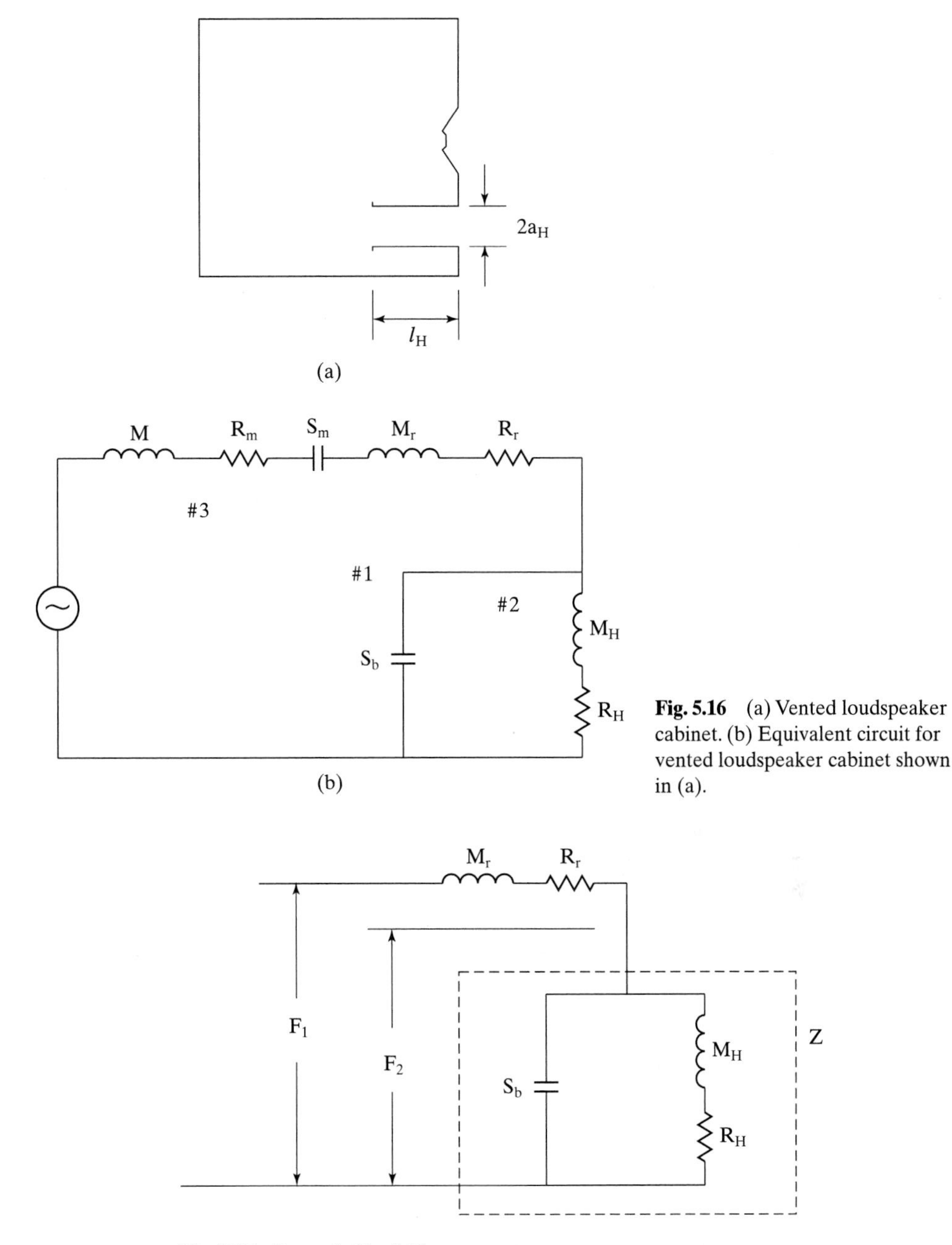

Fig. 5.16 (a) Vented loudspeaker cabinet. (b) Equivalent circuit for vented loudspeaker cabinet shown in (a).

Fig. 5.17 Forces in Fig. 5.16.

For phase inversion, the phase in branch 2 must be 180 degrees different from that in branch 3 (see Fig. 5.17). At high frequencies, that is, when

$$\frac{S_b}{\omega} < M_H\omega, M_r \rightarrow 0$$

consequently

$$Z \approx \frac{S_b}{j\omega} \tag{5.8.5}$$

so

$$\frac{F_2}{F_1} = \frac{S_b/j\omega}{R_r + S_b/j\omega} = \frac{[-jS_b/\omega][R_r + jS_b/\omega]}{R_r^2 + S_b^2/\omega^2} \tag{5.8.6}$$

Hence, $\tan\alpha = -R_r\omega/S_b$, where α is the phase difference between the two branches and $\alpha \approx -180°$.

At low frequencies,

$$Z \approx jM_H\omega + R_H$$

and

$$\tan\alpha = \frac{M_H\omega(R_r + R_H) - R_H(M_r + M_H)\,\omega}{M_H\omega^2(M_r + M_H) + R_H(R_r + R_H)}$$
$$\approx \frac{\omega(M_H R_r - R_H M_r)}{M_H\omega^2(M_r + M_H)} \approx 0 \tag{5.8.7}$$

in other words, the radiation from the vent is in phase with the radiation from the speaker. Thus, if the cabinet is designed to have a resonant frequency somewhat lower than that of the speaker, it is possible to achieve better low frequency response than with a speaker mounted in a baffle. At high frequencies, the impedance of branch 2 is high and the impedance of branch 1 is low. Thus, the speaker then radiates as if in a closed cabinet and must be lined with a sound absorber. The walls of a speaker cabinet should be constructed of wood that is at least one-half inch thick to ensure mechanical rigidity; for a bass-reflex cabinet, the walls should be even thicker, as well as braced.

5.9 Horn Loudspeakers

As we have seen, the efficiency and power radiation can be improved by increasing the radiation resistance. We can do so by using a horn (see Fig. 5.18), as in public address systems, or by using a tubular coupler, as in certain fog signals. Again, the essence of the device is the force exerted on a current-carrying coil. The transduction coefficient is $\varphi = Bl$, where B is the magnetic flux and l is the total length of the winding. The value of φ can be increased by using a high-field strength magnet, decreasing the gap size, and using rectangular cross-section wire on a large diameter former in order to increase l. The diaphragm is usually dome shaped and is held in place by a toroidal suspension, which may have one or more corrugations. It is desirable to use a light-weight material for the diaphragm, and both phenolic and aluminum diaphragms are available. The phenolic diaphragms are made by stamping cloth impregnated with resin. The connecting tube permits the horn to be bolted easily onto the front of the transducer. In some models,

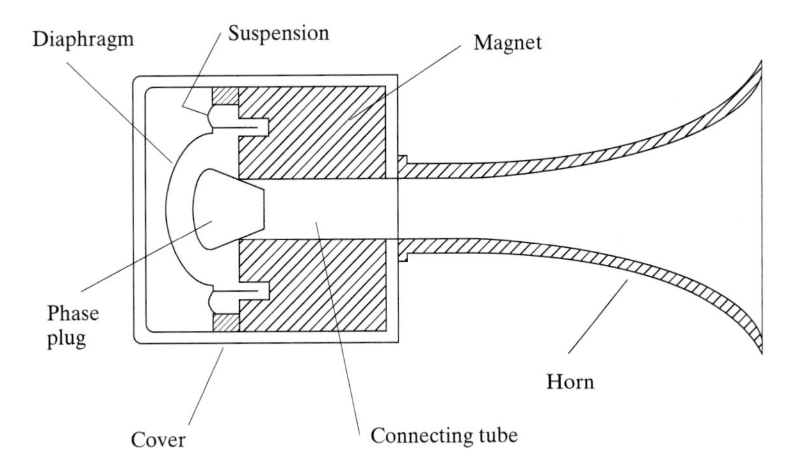

Fig. 5.18 A horn loudspeaker. (After Finch and Higgins, 1976.)

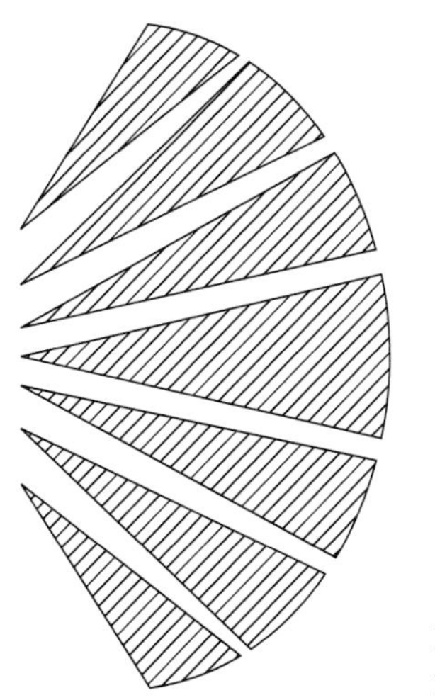

Fig. 5.19 Cross section of a phase plug with constant areal variation.

there is a phase plug (Fig. 5.19) to ensure that at high frequencies the phase of sound from different points on the diaphragm arrives in phase at the mouth of the connecting tube. The phase plug acts like a series of concentric annular horns of various flares.

5.10 Maximizing Efficiency: General Remarks

By bolting horns or pipes of various dimensions onto the connecting tube of the transducer, we alter the values of R_{Mm} and R_{Mr}. Further details of the procedure described in this section can be found in Finch and Higgins (1976). Suppose the design requirement is to do this in a way that will maximize the efficiency, as in the case of a fog signal.

$$\eta = \frac{\varphi^2 R_{Mr}}{\varphi^2 R_{Mr} + \varphi^2 R_{Mm} + R_e |Z_m|^2}$$

where

$$|Z_m|^2 = (R_{Mm} + R_{Mr})^2\left(M\omega - \frac{S}{\omega} + X_{Mr}\right)^2 \qquad (5.10.1)$$

To maximize η with respect to X_{Mr} requires X_{Mr} to be zero. To maximize with respect to R_{Mr} requires that

$$\frac{\partial \eta}{\partial R_{Mr}} = 0 \qquad (5.10.2)$$

which yields

$$R_{Mr\eta} = \left[\frac{\varphi^2 R_{Mm}}{R_e} + R_{Mm}^2 + \left(M\omega - \frac{S}{\omega}\right)^2\right]^{1/2} \qquad (5.10.3)$$

The efficiency can be further maximized by operating at the natural frequency of the driver when

$$\omega_0 = \left(\frac{S}{M}\right)^{1/2} \qquad (5.10.4)$$

This result, however, may introduce phasing problems if the transmitter is a member of an array.

5.11 Maximizing Efficiency: Tubular Couplers

One method for maximizing efficiency using a tubular coupler is illustrated in Fig. 5.20. A tube of length $(\lambda/4 - \epsilon)$ will have zero reactance at its throat, assuming the mouth impedance is given by the theory of Levine and Schwinger (who showed that at low

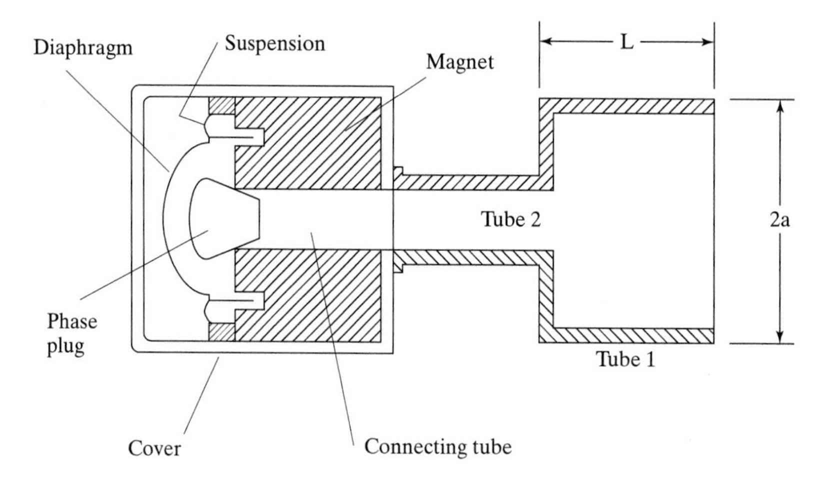

Fig. 5.20 Fog signal tuned for matched output. (After Finch and Higgins, 1976.)

frequencies $\varepsilon = 0.61a$). According to Eqs. (5.3.1) and (5.3.2), the specific acoustic impedance at the mouth of the first tube Z_{m1} will be given by

$$Z_{m1}/\rho c = (0.61\ ka)^2 + j(0.611\ ka) \tag{5.11.1}$$

Hence, from Eq. (4.3.5), the impedance at the throat will be

$$Z_{t1}/\rho c = \rho c/Z_{m1} = 1/(0.61\ ka)^2 \tag{5.11.2}$$

Note that this assumption cannot be precisely correct, since Tube 1 is not exactly $\lambda/4$ in length. Although a more accurate calculation should take this into account, we will use the approximation here to illustrate the procedure. Transforming across the junction of Tubes 1 and 2, the specific acoustic impedance at the mouth of the connecting Tube 2 is given by

$$Z_{m2} = S_2 Z_{t1}/S_1 \tag{5.11.3}$$

where S_2 and S_1 are the areas of Tubes 1 and 2. Transforming to the throat of Tube 2 using Eq. (4.3.6) yields

$$Z_{t2} = Z_{m2} \tag{5.11.4}$$

The mechanical impedance at the diaphragm due to radiation is

$$R_{Mr} = S_D Z_{t2}/S_2 \tag{5.11.5}$$

where S_D is the diaphragm area. Finally, combining Eqs. (5.11.1) through (5.11.5) yields

$$R_{Mr} = S_D \rho c/[\pi(0.372)k^2a^4] \tag{5.11.6}$$

Equating this derived radiation load with the value required for maximum efficiency from Eq. (5.10.3) gives

$$a^4 = \frac{S_D \rho c}{\pi(0.372)\ k^2}\left[\frac{\varphi^2 R_{Mm}}{R_e} + R_{Mm}^2 + \left(M\omega\frac{S}{\omega}\right)^2\right]^{1/2} \tag{5.11.7}$$

This result will apply only at a single frequency, and the device would be used in applications where this is no problem (e.g., in fog signals).

This coupler configuration is not the only solution to the problem of maximizing efficiency, but it is one of the simplest. It is possible to use other combinations of $\lambda/4$ multiple tubes. Several couplers can be attached to a single transducer, or several transducers can be used to drive a single coupler via a manifold. An advantage of tubular couplers is relative ease of fabrication. Tubular couplers can also be designed for power matching or ρc coupling.

Horns of various flares can also be used as couplers. In doing so, the design procedure is similar to that described in this section, but the impedance transformation is given by Eq. (4.7.3) rather than Eq. (4.3.5). A horn will be used where a relatively wide bandwidth is necessary, as in the reproduction of speech or music.

REFERENCES AND FURTHER READING

Finch, R.D., and P.W. Higgins. 1976. "Optimizing the Monotone Performance of Electrodynamic Drives Using Tubular Couplers." 60 *J. Acoust. Soc. Am.* 937–943.

Kinsler, L.E., A.R. Frey, A.B. Coppens, and J.V. Sanders. 1982. *Fundamentals of Acoustics*. 3rd ed. Wiley.

PROBLEMS

5.1 A circular piston, 10 cm in diameter, is mounted in a rigid baffle and radiates sound into air at high frequency. The mechanical radiation load is then

$$Z_r = \rho c \pi a^2$$

where a is the radius and ρc is the characteristic impedance of air. The pressure amplitude in the far field is given by

$$p_0 = \frac{\rho c k \pi a^2 U_0}{2\pi r}\left[\frac{2J_1(x)}{x}\right]$$

where

$k = 2\pi/\lambda$, λ = wavelength

U_0 = piston velocity amplitude

$x = ka\sin\theta$, θ = angle from the perpendicular axis

passing through the piston's center

a. Calculate the velocity amplitude of the piston if it is to radiate 0.5 mw of acoustic power at standard temperature and pressure at 10 kHz.

b. Calculate the pressure amplitude at a perpendicular distance of 3 m. What would be the sound pressure level referenced to 20 μ Pa?

5.2 If the piston in Problem 5.1 has a mass of 15 gm and is attached to the baffle with a suspension of stiffness 2000 N/m with negligible damping, what force amplitudes will be necessary to produce the velocity amplitudes calculated in Problem 5.1?

5.3 The piston in Problems 5.1 and 5.2 is now to be considered the cone of a loudspeaker. The voice coil operates in a magnetic field of strength 1 Wb/m^2. The length of the wire is 7.5 m, its inductance is 0.5 mH, and its resistance is 10 ohms. Draw an equivalent circuit, assuming the speaker is driven with an electrical generator of emf E and output impedance Z_g. Indicate the value of Z_g for maximum electromagnetic power transformation. With this value of Z_g, find the value of E necessary to produce 0.5 mw of radiated power under the conditions put forth in Problem 5.1.

5.4 Considering the equivalent circuit in Problem 5.3, if the electrical resistance of the voice coil and the mechanical damping of the suspension cannot be altered, to what value should the load be transformed to obtain the maximum output power at a given value of E? Describe how the radiation resistance on the piston could be transformed using a horn or a single step tubular coupler.

5.5 A direct radiator dynamic loudspeaker has a total mass of 0.01 kg (the cone and voice coil) and operates in a magnetic field of flux density 1 Wb/m^2. The radius of the speaker is 0.1 m, its mechanical resistance is 1 kg/s, its radiation resistance is 2 kg/s, and the stiffness of the cone system is 2000 N/m. The length of the voice coil is 7.5 m, its inductance is 0.5 mH, and its resistance is 10 ohm. Compute the electroacoustic efficiency at a frequency of 200 Hz.

5.6 When the voice coil of a certain loudspeaker is blocked, the resistive component of its input impedance is 5 ohm. When flush mounted in a large wall and driven at its frequency of mechanical resonance, the resistive component is found to be 10 ohm. When removing the speaker from the wall in order to eliminate its acoustic radiation loading, the input resistance is 12 ohm at the frequency of mechanical resonance. What is the electroacoustic efficiency of the speaker at resonance when mounted in the wall?

5.7 Design a maximum efficiency system to operate at 1000 Hz.

5.8 What would determine the upper frequency limit for a maximum efficiency radiator of the type described in Section 5.11? Explain your answer.

5.9 Design an electrodynamic shaker to vibrate a 1 kg lumped mass load at 50 Hz.

5.10 What design changes could be made to optimize the efficiency of the device you proposed in Problem 5.9?

Chapter 6

Sensors: Microphones and Accelerometers

6.1 Introduction

The selection of an acoustic sensor depends partly on the nature of the signal to be received and partly on the processing to be performed on the transduced electrical signal. The sensor may have to operate in a gas (usually air), in a liquid (e.g., in water, when the detector is called a hydrophone), or on the surface of a solid (e.g., an accelerometer on a structure or a seismometer on the Earth's surface). If we want to detect speech or music, the sensor (which will then be a microphone) should have a flat frequency response in the audio range and a low noise level. For probing the sea, the atmosphere, or the Earth, we need a voltage signal output for processing. The received signal is frequently weak, consequently the major problem is that of improving the signal-to-noise ratio. Since acoustical background noise is often localized, it can be lessened by designing a receiving beam having high directivity or by shielding the sensors and protecting them from wind or currents that could cause turbulence. Assuming that adequate beam shape can be achieved by setting microphones in arrays or by using reflectors and shields, the signal-to-noise ratio can be improved either by increasing the signal strength or by selecting a microphone with minimal internal noise. The most convenient way to analyze such factors is with a noise figure equation, such as the sonar equation.

The best way to classify sensors is in terms of the transduction mechanism. A common type of microphone uses the change in resistance of carbon granules under pressure, as used in telephone receivers. Although they are cheap and sensitive compared with other types, they generate relatively high broadband noise levels and harmonic distortion. Furthermore, to be kept in optimum condition, a carbon microphone needs to be physically rotated regularly to prevent the granules from aggregating. Consequently, the carbon microphone is being phased out, leaving the selection of microphones to be made from among the electrodynamic, condenser (or electret), or piezoelectric types, which are all reversible transducers. We will discuss the equivalent circuit of such microphones in general terms, but the piezoelectric sensor is so important that Chapter 7 will examine it in greater depth.

6.2 Equivalent Circuit of Reversible Microphones

Recall that the basic Eqs. (2.12.2) and (2.12.3) (discussed in Chapter 2) governing the behavior of a transducer are

$$E = Z_e I + T_{em} v \tag{6.2.1}$$

and

$$F = T_{me} I + Z_m v \tag{6.2.2}$$

When the transducer is used as a microphone, there is no applied voltage, E. Consequently, the equivalent circuit can be drawn as in Fig. 6.1a and Fig. 6.1b. Thus, from Eq. (6.2.1), the electromotive force generated on the electrical side, E_g, is given by

$$E_g = T_{em} v = -Z_e I \tag{6.2.3}$$

where Z_e is the electrical impedence, $R_e + jX_e$.

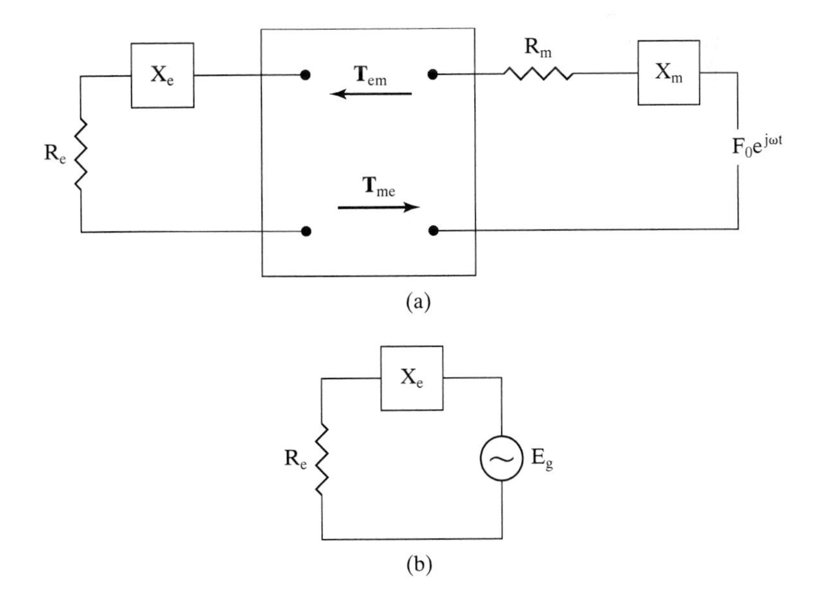

Fig. 6.1 (a) Representation of the equivalent circuit of a transducer; (b) reduction of Fig. 6.1b.

But the accessible voltage, E_a, is the voltage across the electrical resistance R_e, thus

$$E_a = IR_e = -(E_g/Z_e)\,R_e \tag{6.2.4}$$

Since we generally need to maximize the accessible voltage, R_e is usually arranged to be much greater than X_e, so that $R_e \sim R_e$. The same result would be obtained if R_e were infinite (i.e., if the circuit were open). Consequently, the accessible voltage

$$E_a \rightarrow E_g \text{ and } I \rightarrow 0 \tag{6.2.5}$$

Hence, from Eq. (6.2.2) the velocity is

$$v = \frac{F_0 e^{j\omega t}}{Z_m} \tag{6.2.6}$$

From Eqs. (6.2.3) and (6.2.6)

$$E_g = \frac{T_{em} F_o e^{j\omega t}}{Z_m} \tag{6.2.7}$$

If we now let $F_0 = p_0 A$, where A is the active area of the transducer,

$$E_{go} = \frac{T_{em} p_0 A}{Z_m} \tag{6.2.8}$$

and finally the "open circuit voltage sensitivity"

$$M = \frac{E_{go}}{p_0} = \frac{T_{em} A}{Z_m} \tag{6.2.9}$$

The equivalent circuit can now be represented as shown in Fig. 6.1b.
It is often convenient to use a logarithmic measure called the sensitivity level, which is defined by

$$SL = 20 \log_{10} \frac{M}{M_{ref}} \tag{6.2.10}$$

where M_{ref} is a reference sensitivity, usually 1 v/μ bar.

We will now discuss the transduction coefficients T_{me} and mechanical impedances Z_m of various types of microphones.

6.3 Electrodynamic (Moving Coil) Microphone

The electrodynamic microphone, frequently used in intercom systems, is in principle of the same construction as the electrodynamic driver. Consequently, the sensitivity of an uncoupled microphone will, using Eq. (6.2.9), be

$$M_E = \frac{BlA}{R_m + j\left(M\omega - S/_{\omega}\right)} = \frac{BlA}{|Z_m|e^{j\alpha}} \tag{6.3.1}$$

and neglecting phase changes,

$$M_E = \frac{BlA}{\left[R_m^2 + \left(M\omega - S/_{\omega}\right)^2\right]^{1/2}} \tag{6.3.2}$$

For broadband flat response, the damping term R_m should be large; but for high sensitivity, $|Z_m|$ needs to be small. A solution to this problem was suggested by Wente and Thuras (1931) with a construction of the two degree of freedom type shown in Fig. 6.2a. The series-analogous equivalent circuit is shown in Fig. 6.2b. By choosing the values of the parameters,

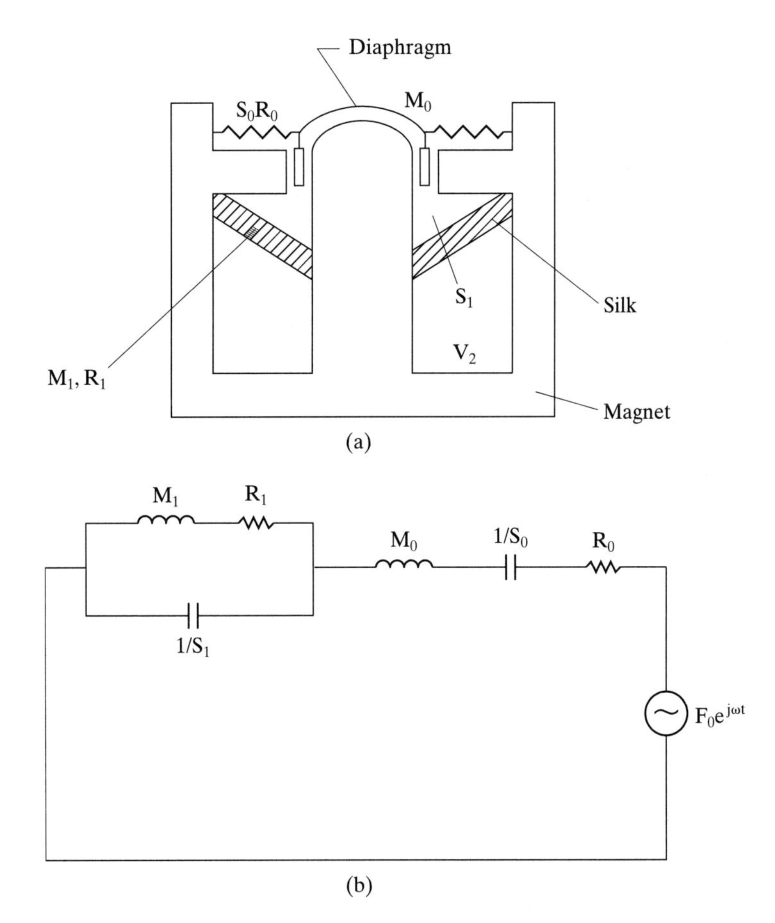

Fig. 6.2 (a) The Wente-Thuras moving coil microphone. (b) The series-analogous equivalent circuit for the Wente-Thuras moving coil microphone.

a fairly broadband frequency response can be obtained. In commercial electrodynamic transducers, Bl is approximately 15 (10 m $\times$ 1.5 Wb/m^2) and A is about 4 $\times$ 10^{-4} m^2, giving a sensitivity level of about -92 dB re 1 v/μ bar (a bar is 1 atmosphere).

6.4 Condenser (Capacitor) Microphone

The condenser or capacitor microphone was invented by Wente (1917). The basic principle of the device is that the capacitance of a parallel plate capacitor can be altered by changing the spacing (see Fig. 6.3a-b). If there is a polarizing voltage on the capacitor, the altered capacitance will result in an alteration of the stored charge. Probe microphones with diaphragms as small as one-eighth inch are commercially available, as are

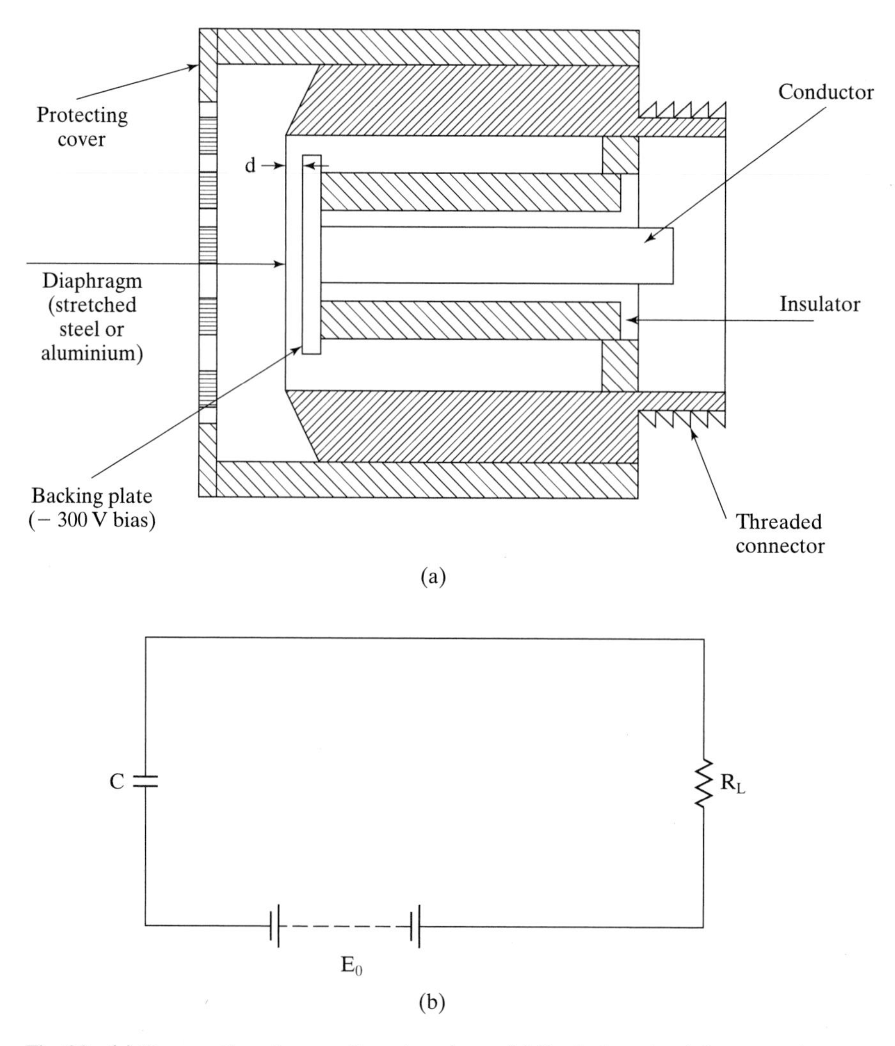

Fig. 6.3 (a) Cross section of a capacitor microphone. (b) Equivalent circuit for a capacitor microphone.

units as large as three-feet square. We will analyze a circular diaphragm type of radius a. Let the plate separation be d and the polarizing voltage be E_0. The capacitance is then given in SI units by

$$C = \epsilon_0 A/d \tag{6.4.1}$$

Hence, if

$$d = d_0 + \eta_0 e^{j\omega t} \tag{6.4.2}$$

for small displacements

$$\begin{aligned} C &= \frac{\epsilon_0 A}{d_0}\left(1 - \frac{\eta_0}{d_0} e^{j\omega t}\right) \\ &= C_0 - C_1 e^{j\omega t} \end{aligned} \tag{6.4.3}$$

But

$$E_0 - IR_L = \frac{1}{C}\int I dt \tag{6.4.4}$$

Assuming

$$I = I_0 e^{j(\omega t - \alpha)} \tag{6.4.5}$$

it follows that

$$E_0 C_0 - E_0 C_1 e^{j\omega t} - I_0 R_L e^{j(\omega t-\alpha)}(C_0 - C_1 e^{j\omega t}) = \int I dt \tag{6.4.6}$$

Differentiating with respect to time and retaining only first-order terms,

$$-j\omega E_0 C_1 - j\omega I_0 R_L C_0 e^{-j\alpha} = I_0 e^{-j\alpha} \tag{6.4.7}$$

hence

$$I_0 = \frac{-E_0 C_1 e^{j\alpha}}{C_0\left(R_L + \dfrac{1}{jC_0\omega}\right)} \tag{6.4.8}$$

Thus, the capacitor microphone acts like a generator of electromotive force $E_0 C_1/C_0$ and internal capacitance C_0. From Eq. (6.4.3), we can see that this transduced voltage is

$$E_g = \frac{E_0 C_1}{C_0} = \frac{E_0 \eta_0}{d_0} \tag{6.4.9}$$

and for harmonic response,

$$T_{em} = E_g/v = E_0/\omega d_0 \tag{6.4.10}$$

At low frequencies, the mechanical impedance of the diaphragm is stiffness dominated; that is,

$$Z_m = -jS/\omega, \text{ where } S = 8\pi T \tag{6.4.11}$$

T being the tension. Hence, the open circuit voltage sensitivity from Eq. (6.2.9) is given by

$$M_c = \frac{T_{em}A}{Z_m} = \frac{E_0 a^2}{8 d_0 T} \tag{6.4.12}$$

In a typical example of a capacitor microphone, we have

aluminum diaphragm $= 4 \times 10^{-3}$ cm thick
radius a $= 1$ cm
tension T $= 20{,}000$ n/m
d $= 4 \times 10^{-3}$ cm
$E_0 = 300$ V

so the sensitivity $= 4.7 \times 10^{-4}$ V/μ bar or -66.6 dB re 1 V/μ bar. If the mass of the diaphragm is σ per unit area, then (as shown in Section 6.5) the upper limit for frequency independent response is

$$f = \frac{1}{2\pi a}\sqrt{\frac{T}{\sigma}} = 6840 \text{ Hz} \tag{6.4.13}$$

A typical response is shown in Fig. 6.4 for a classical microphone, the Western Electric 640AA, which is arranged so that viscous damping of air in the slots in the backing plate becomes effective at the limiting frequency. Thus, the rise in response at the peak is minimized. Capacitor microphones range in diameter from one-eighth inch to one inch. The

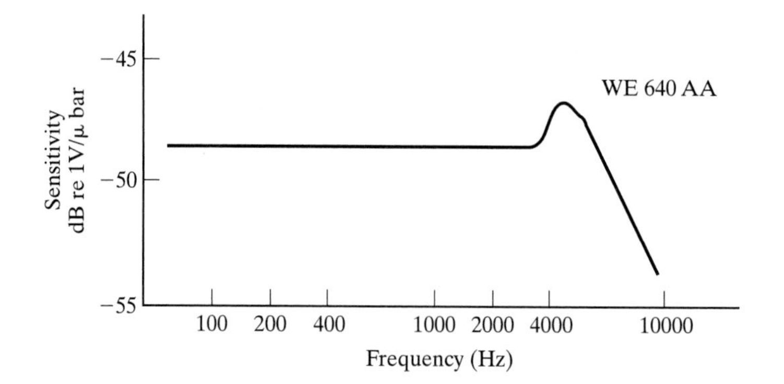

Fig. 6.4 The sensitivity of the Western Electric 640AA capacitor microphone.

smaller ones have a higher natural frequency and thus a wider dynamic range at the expense of lower sensitivity. Reverting to the microphone we just considered in the above numerical example, $C_0 = 70\ \mu\mu F$ and the impedance $= 1/C_0\omega$, at say 100 Hz, $= 23$ ohms. Now the capacitance of cable $= 40$ to $20\ \mu\mu F/ft$. Hence, only a few feet of cable could effectively alter C_0 and affect the sensitivity. To avoid this outcome, it is necessary to use a preamplifier with a capacitor microphone.

6.5 Vibration Sensors

Microphones are pressure sensors. Closely related are transducers designed to measure either linear or angular displacement, velocity, or acceleration. A vibrometer is a device sensing either displacement or velocity, while an accelerometer produces an output proportional to acceleration.

If the amplitude of a vibration is large enough and the frequency low enough, the vibration can be measured directly. As the frequency increases, provided the amplitude is still large, the vibration can be measured by the wedge method: A wedge of paper is stuck to the vibrating object, and at the point where the double image crosses, the width of the wedge is twice the amplitude (see Fig. 6.5). The frequency of a vibration can be found by adjusting the position of a load on a cantilever until a resonance is found (Fig. 6.6a). A wire coiled in a housing that can be extended to various lengths can be used as an alternative to this method (Fig. 6.6b). An elementary accelerometer that can determine if a predetermined acceleration level has been reached is illustrated in Fig. 6.7.

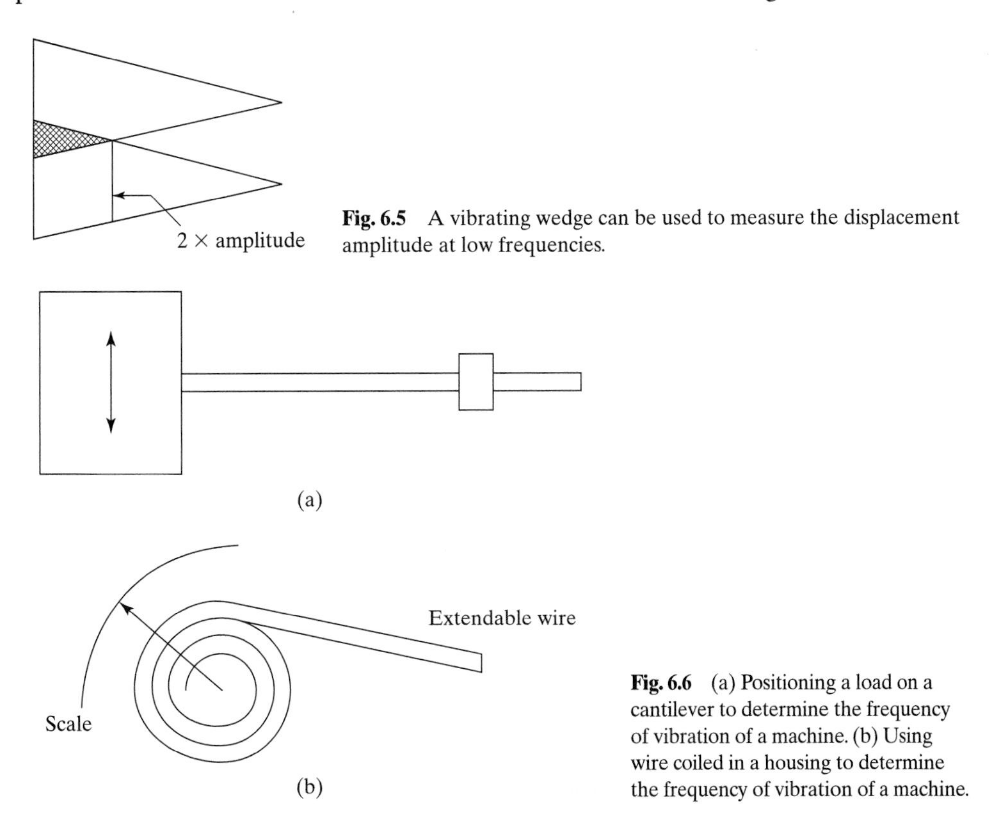

Fig. 6.5 A vibrating wedge can be used to measure the displacement amplitude at low frequencies.

Fig. 6.6 (a) Positioning a load on a cantilever to determine the frequency of vibration of a machine. (b) Using wire coiled in a housing to determine the frequency of vibration of a machine.

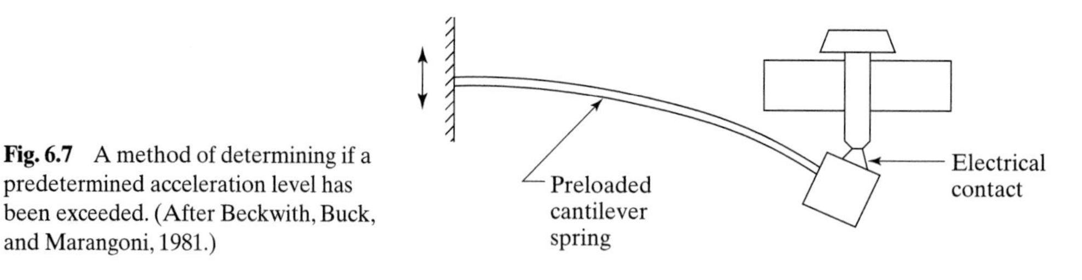

Fig. 6.7 A method of determining if a predetermined acceleration level has been exceeded. (After Beckwith, Buck, and Marangoni, 1981.)

6.6 The Seismometer

The seismometer is a variation of a moving coil microphone in which the case is excited by vibration of the support, as shown in Fig. 6.8. In some designs, the coil remains stationary while the magnet moves, which presents a problem similar to that of isolation from base vibrations, as we discussed in Section 1.3.5 in Chapter 1. In this case, however, the output of the instrument depends on the relative velocity of the mass and case. It is simplest to solve for this directly by rewriting the equation of motion in terms of the relative displacement:

$$\eta = x - y \tag{6.6.1}$$

Thus, the force on the mass may be written

$$-S\eta - R\dot{\eta} = M\ddot{x} = M\ddot{\eta} + M\ddot{y} \tag{6.6.2}$$

or

$$\begin{aligned} M\ddot{\eta} + R\dot{\eta} + S\eta &= -M\ddot{y} \\ &= M\omega^2 y \\ &= M\omega^2 y_0 e^{j\omega t} \end{aligned} \tag{6.6.3}$$

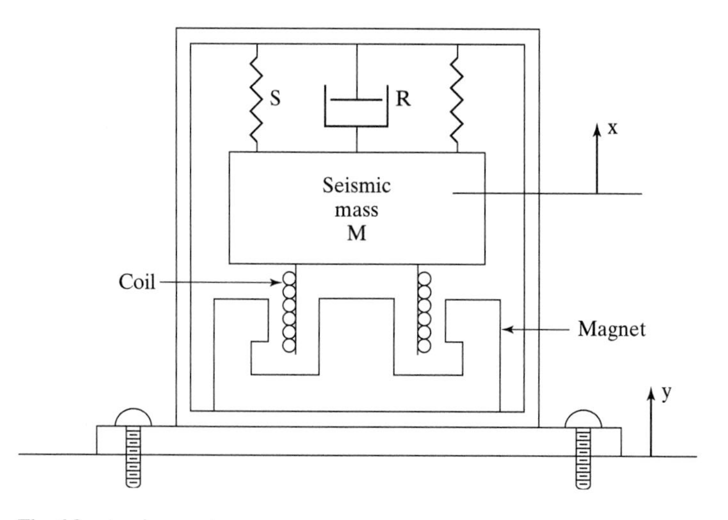

Fig. 6.8 A seismometer.

Eq. (6.6.3) is now the equation of a forced oscillation with a forcing function

$$F_0 e^{j\omega t} = M\omega^2 y_0 e^{j\omega t} \tag{6.6.4}$$

hence, from Eq. (1.3.13), the steady-state solution is

$$\dot{\eta} = \frac{M\omega^2 y_0 e^{j(\omega t - \alpha)}}{\sqrt{\left(M\omega - \dfrac{S}{\omega}\right)^2 + R^2}} \tag{6.6.5}$$

where

$$\alpha = \tan^{-1}\frac{\left(M\omega - \dfrac{S}{\omega}\right)}{R} \tag{6.6.6}$$

Thus

$$\eta_0 = \frac{M\omega y_0}{\sqrt{\left(M\omega - \dfrac{S}{\omega}\right)^2 + R^2}} \tag{6.6.7}$$

and following Section 1.2.3 by letting

$$\xi = R \Big/ R_{cr} = \frac{R}{2\sqrt{SM}} \tag{6.6.8}$$

we can show that

$$\eta_0 = \frac{y_0\left(\dfrac{\omega}{\omega_0}\right)^2}{\sqrt{\left[\left(\dfrac{\omega}{\omega_0}\right)^2 - 1\right]^2 + \left[2\xi\left(\dfrac{\omega}{\omega_0}\right)\right]^2}} \tag{6.6.9}$$

Now η_0/y_0 varies with ω/ω_0, as shown in Fig. 6.9. Damping about 65% to 70% of the critical value gives the best response, and the instrument should be operated when used as a vibrometer well above its natural frequency. The phase shift between the displacements of the mass and case is

$$\beta = \alpha + \pi/2 \tag{6.6.10}$$

so that

$$\tan\beta = \frac{-1}{\tan\alpha} \tag{6.6.11}$$

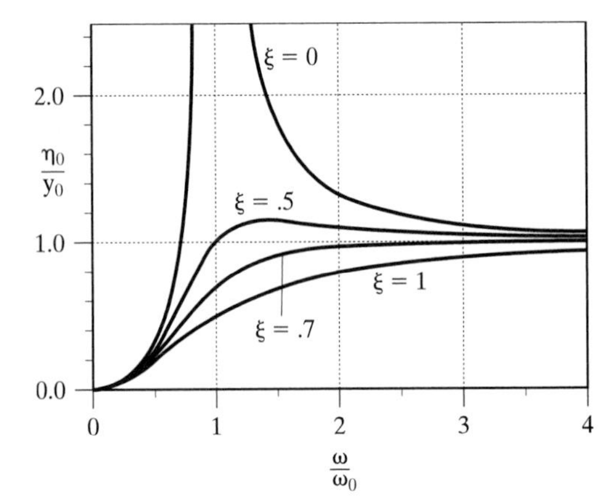

Fig. 6.9 Seismometer response at different damping levels.

and

$$\beta = \tan^{-1}\frac{R}{\left(\dfrac{S}{\omega} - M\omega\right)} \tag{6.6.12}$$

The phase shift may also be written in terms of dimensionless ratios

$$\beta = \tan^{-1}\left[\frac{2\xi\left(\dfrac{\omega}{\omega_0}\right)}{1 - \left(\dfrac{\omega}{\omega_0}\right)^2}\right] \tag{6.6.13}$$

This function behaves as shown in Fig. 6.10. It would be desirable to have zero phase at all frequencies, but since this is not possible, the relation for $\xi = 0.7$ is a fair compromise.

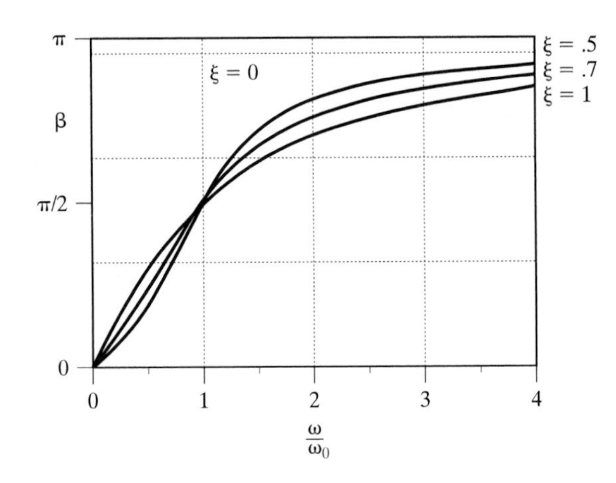

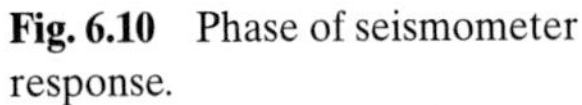

Fig. 6.10 Phase of seismometer response.

6.7 Accelerometers

The same construction used for the seismometer can be used for an accelerometer. Thus, from Eq. (6.6.9), the relative displacement may be written

$$\eta_0 = \frac{\ddot{y}_0}{\omega_0^2\sqrt{\left[\left(\frac{\omega}{\omega_0}\right) - 1\right]^2 + \left[2\xi\left(\frac{\omega}{\omega_0}\right)\right]^2}} \tag{6.7.1}$$

where $\ddot{y}_0$ is the acceleration amplitude of the case. To obtain an output independent of frequency, the quantity

$$q = \frac{1}{\sqrt{\left[\left(\frac{\omega}{\omega_0}\right)^2 - 1\right]^2 + \left[2\xi\left(\frac{\omega}{\omega_0}\right)\right]^2}} \tag{6.7.2}$$

has to be constant. From Fig. 6.11, we see that this occurs only if $\omega/\omega_0 < 0.4$ and $\xi \cong 0.7$. Thus, when used as an accelerometer, the device is operated below its natural frequency.

A very compact, rugged, and reliable accelerometer with a very high resonant frequency, and thus a broad operating range, can be achieved using piezoelectric material to make the supporting spring. Two piezoelectric accelerometers are shown in Fig. 6.12. The output impedance of a piezoelectric accelerometer is high, and it needs to be connected to an impedance transforming amplifier or a charge amplifier.

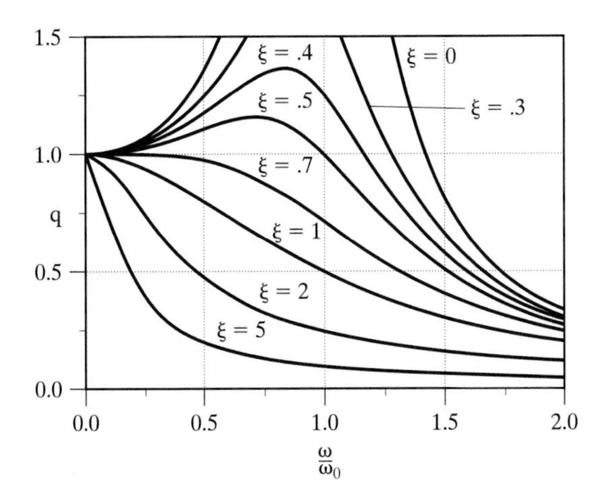

Fig. 6.11 The quantity q has to be constant for frequency independent accelerometer output.

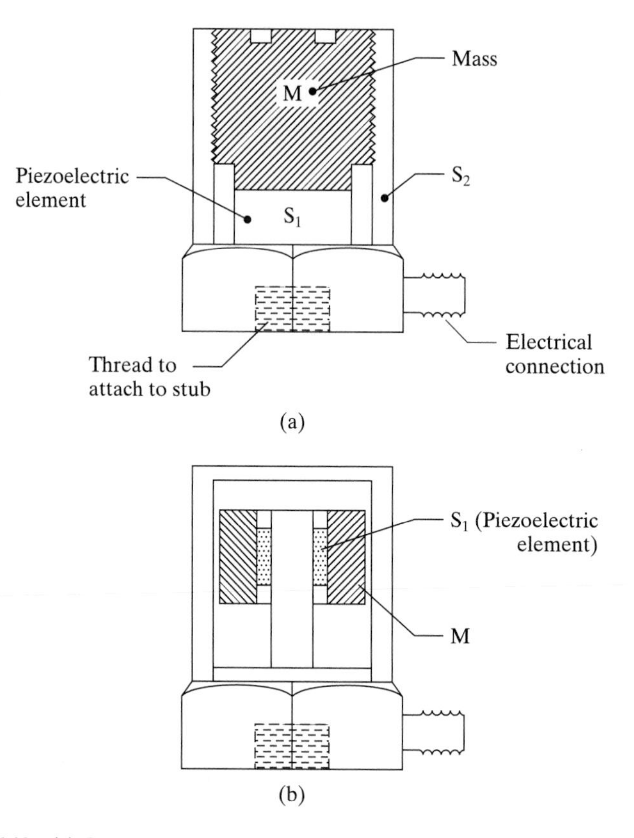

Fig. 6.12 (a) Compression-type piezoelectric accelerometer. (b) Shear-type piezoelectric accelerometer.

6.8 New Transducers

Several new classes of piezoelectric polymers have recently been developed, as well as some new magnetostrictive materials, including Terfenol, which is several times as effective as older materials. Over the last decade, research and development in the area of transduction has become increasingly important for three reasons. First, there is an increasing emphasis on the use of electronic techniques in acoustics, as in active noise reduction and in robotics, and these procedures require reliable and accurate transducers. Second, although electronic parts in the last decade have become much faster, more reliable, and cheaper, transducers have not changed greatly during the same period. Thus, transducers are now the limiting hardware in most setups. Third, recent research has introduced applications of technology that strain the limits of currently available transducers. State-of-the-art transducers may be roughly classed into three types: fiber optic, polymeric, and solid state. Fiber optic devices are very useful in many applications because of their versatility and sensitivity. In their simplest application as optic levers, as shown in Fig. 6.13, one or more glass fibers emit light and one or more receive reflected or transmitted light. As for polymeric devices, during the past decade, there has been an exponential growth in the number of new polymeric materials available. Many of these polymers, polymer composites, and copolymers are suitable transduction materials because they are

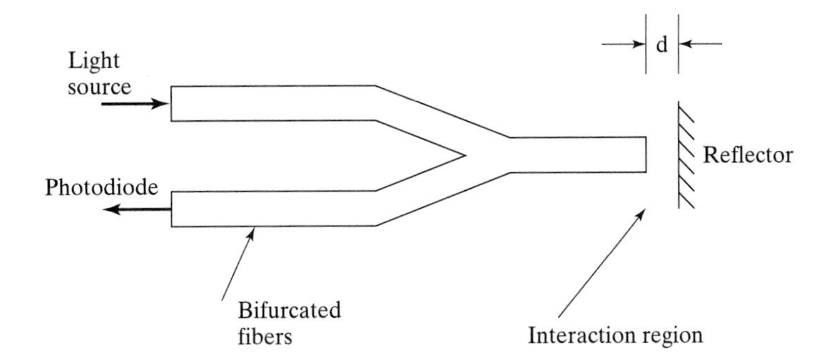

Fig. 6.13 Fiber optic lever transducer.

magnetic, piezoelectric, electrostrictive, magnetostrictive, or piezoresistive. In fact, polymer science has progressed to the point where it is possible, within limits, to choose the mechanical and electrical properties of a material independently. Solid-state transducers are covered in the next chapter.

There has also been a growing trend in industry toward miniaturizing mechanical and electrical parts. To monitor operations that use such parts it is also necessary to miniaturize transducers. The technique most likely to produce miniature sensors and actuators has already been established for VLSI (very large scale integration) chips of silicon. These developments have the potential to make available inexpensive, miniature transducers with high sensitivity and low noise.

REFERENCES AND FURTHER READING

Beckwith, T.G., N.L. Buck, and R.D. Marangoni. 1981. *Mechanical Measurements*. 3rd ed. Addison-Wesley.

Thomson, W.T. 1988. *Theory of Vibration with Applications*. 3rd ed. Prentice Hall.

Wente, E.C. 1917. "A Condenser Transmitter as a Uniformly Sensitive Instrument for the Absolute Measurement of Sound Intensity." 10 *Phys. Rev.* 39–63.

Wente, E.C., and A.L. Thuras. 1931. "Moving Coil Telephone Receivers and Microphones." 3 *J. Acoust. Soc. Am.* 44–55.

PROBLEMS

6.1 Consider a transducer described by

$$E = 10I + 4V$$

and

$$F = 2V - 4I$$

where E, I, V, and F are measured in the MKS system.

a. What is the blocked electrical impedance?

b. What is the electrical impedance if the transducer has no external load?

c. What is the electrical impedance if the transducer is working into a load of $5 + 2j$ mechanical ohms?

d. If the transducer is open circuited on the electrical connections, what would be the sensitivity defined as output voltage per input velocity?

6.2 Describe the action of the following and derive expressions for their sensitivities:

a. a capacitor microphone
b. an accelerometer
c. a seismometer

A microphone reads 1 mv for an incident pressure level of 120 dB re 1 μbar. Find the sensitivity level of the microphone re 1 v/μbar.

6.3 A capacitor microphone diaphragm of 0.02 m radius and 0.00002 m spacing between the diaphragm and the backing plate is stretched to a tension of 10,000 N/m.

a. If the polarizing voltage is 200 v, what is the low frequency open-circuit voltage response of the microphone in v/N/m^2?
b. What is the corresponding response level in dB relative to 1 volt per microbar?
c. When acted upon by a sound pressure of 1 N/m^2 amplitude, what is the amplitude of the average displacement of the diaphragm?
d. What voltage will be generated in a load resistance of 5 megohm by this capacitor microphone at a frequency of 100 Hz when acted upon by a sound wave of 10 μbar pressure amplitude?

6.4 The density of air in a duct varies at constant temperature. Short duration sound pulses are produced by a loudspeaker flush mounted in one wall. The speaker has characteristics as given in Problems 5.1 through 5.4. The sound is sensed by a microphone of sensitivity M mounted in the opposite wall at a distance of 3 m from the loudspeaker. At a constant driving frequency, the output voltage is V_M. What is the dependence of V_M on the density?

6.5 The background noise at the microphone of Problem 6.4 is wide band. It is decided to digitize the microphone signal and perform a fast fourier transform to filter out noise. What sampling rate should be used to obtain a 5 Hz resolution at 10 kHz? Explain your answer. What other methods could be used to reduce noise?

6.6 The sound received by a microphone is coming from a turbine.

a. Could this sound be used to monitor blade condition?
b. How would you process the signal?
c. What other signals could be used to monitor the turbine's performance?

6.7 A seismometer with a natural frequency of 2 Hz and damping ratio 70% of critical is used to measure displacement amplitude and phase at 200 Hz, 20 Hz, and 2 Hz. Comment on the accuracy of the measurements at each frequency and estimate the correction factor that should be applied to each measured value.

6.8 An accelerometer with a natural frequency of 20 kHz has a damping ratio 7% of critical. Determine the correction factor to be applied in measurements of acceleration amplitude and phase at 2 kHz and 22 kHz.

6.9 Displacement amplitude is measured with an accelerometer and found to be 80 microns at a phase angle of 80° at 3000 rpm. The accelerometer has a natural frequency of 5 kHz and 10% damping ratio. Find the correct value for the measurement.

6.10 Amplitude measurements of an earthquake are made in the frequency range between 0.1 Hz and 5 Hz. Assuming the measurements are made with a seismometer with natural frequency 0.05 Hz and negligible damping, determine the maximum percentage error in the measurements.

Chapter 7

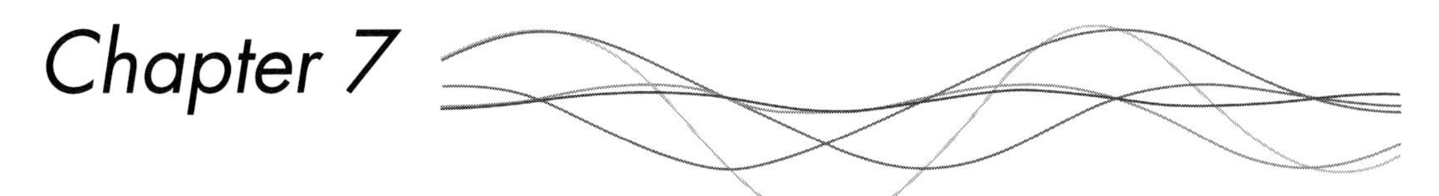

Piezoelectric Transducers

7.1 Postulates of the Electromagnetic Theory

A piezoelectric transducer employs a solid-state pressure-sensitive element. For a thorough understanding of the subject, it is best to start with the basic postulates of the electromagnetic theory. Feynman, Leighton, and Sands (1965) and Jackson (1962) each provide a review of these fundamentals, which are contained in Maxwell's equations:

$$\nabla \cdot \mathbf{E} = \frac{\rho}{\epsilon_0} \tag{7.1.1}$$

$$\nabla \times \mathbf{E} = -\frac{\partial \mathbf{B}}{\partial t} \tag{7.1.2}$$

$$\nabla \cdot \mathbf{B} = 0 \tag{7.1.3}$$

$$c^2\, \nabla \times \mathbf{B} = \frac{\partial \mathbf{E}}{\partial t} + \frac{\mathbf{j}}{\epsilon_0} \tag{7.1.4}$$

where

- $\mathbf{E}$ = electric field vector (V/m), defined as the force experienced by a unit positive test charge
- ρ = charge density (C/m^3)
- ϵ_0 = a constant known as the permittivity of free space = 8.85×10^{-12} F/m
- $\mathbf{B}$ = magnetic field vector (N/A-m)
- t = time (s)
- c^2 = constant = square of the velocity of light
- $c = 3 \times 10^8$ m/s
- $\mathbf{j}$ = current density (A/m^2)

Applying Gauss' theorem to Eq. (7.1.1)

$$\int_V \nabla \cdot \mathbf{E} dv = \int_S \mathbf{E} \cdot ds = \int_V \frac{\rho}{\epsilon_0} dv \tag{7.1.5}$$

Thus, the net outward electric flux from a closed volume is equal to the net positive charge contained in the volume. This result can be used to derive Coulomb's law, as shown in Fig. 7.1.

Similarly, from Eq. (7.1.3), we might say that the density of magnetic charge is zero, or that there is no such thing as a free magnetic pole. We now believe that all magnetic fields arise from electric currents or changing electric fields even in permanent magnets, where circulating electrons, which are in effect currents, cause the magnetic fields. The zero divergence of the magnetic field was discovered by Ampere. When $\partial \mathbf{E}/\partial t = 0$, from Stoke's theorem, Eq. (7.1.4) yields

$$\int_S \nabla \times \mathbf{B} \cdot d\mathbf{S} = \int_\ell \mathbf{B} \cdot d\ell = \int_S \frac{\mathbf{j}}{\varepsilon_0 c^2} \cdot d\mathbf{S} \tag{7.1.6}$$

or

$$\varepsilon_0 c^2 \int \mathbf{B} \cdot dl = \mathbf{I} \tag{7.1.7}$$

where **I** is the total current. This gives Ampere's law for a long current-carrying conductor (see Fig. 7.2):

$$\mathbf{B} = \frac{\mathbf{I}}{2\pi\epsilon_0 c^2 r} \tag{7.1.8}$$

The term $\partial \mathbf{E}/\partial t$ in Eq. (7.1.4) was added by Maxwell, who realized that "displacement" currents would flow upon changing the electric field. Experimental confirmation was obtained by Helmholtz.

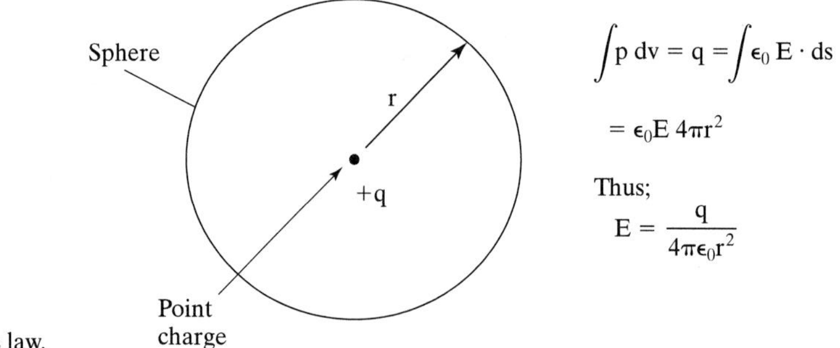

Fig. 7.1 Coulomb's law.

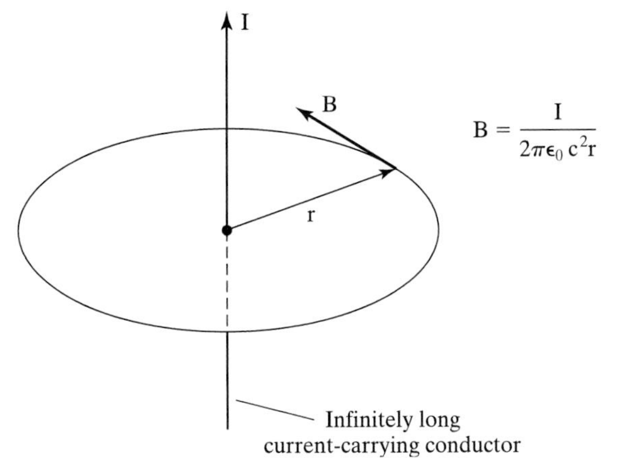

Fig. 7.2 Ampere's law.

Applying Stoke's theorem to Eq. (7.1.2)

$$\int \mathbf{E} \cdot d\mathbf{l} = -\frac{\partial}{\partial t} \int \mathbf{B} \cdot d\mathbf{S} \tag{7.1.9}$$

or

$$\epsilon = -\frac{\partial \Phi}{\partial t} \tag{7.1.10}$$

where Φ is the magnetic flux through a closed loop and ϵ is the electromagnetic force around the loop. In other words, Eq. (7.1.2) is an expression of Faraday's law of electromagnetic induction.

Maxwell's achievement was more than to summarize the known facts in electricity and magnetism and add the concept that a magnetic field could be produced by a changing electric field. For he further realized that a magnetic field so produced would itself be changing in time and in turn produce an electric field (as per Eq. (7.1.2)). It seemed to him that a disturbance of either the electric field or the magnetic field could result in a self-sustained oscillation. We can illustrate this possibility for empty space where $\rho = \mathbf{j} = 0$, so that

$$\nabla \cdot \mathbf{E} = 0 \tag{7.1.11}$$

$$\nabla \times \mathbf{E} = -\frac{\partial \mathbf{B}}{\partial t} \tag{7.1.12}$$

$$\nabla \cdot \mathbf{B} = 0 \tag{7.1.13}$$

$$c^2 \nabla \times \mathbf{B} = \frac{\partial \mathbf{E}}{\partial t} \tag{7.1.14}$$

Taking the curl of both sides of Eq. (7.1.12) yields

$$\nabla \times (\nabla \times \mathbf{E}) = -\frac{\partial}{\partial t}(\nabla \times \mathbf{B}) \tag{7.1.15}$$

Using the expansion

$$\nabla \times \nabla \times \mathbf{A} = \nabla\nabla \cdot \mathbf{A} - \nabla \cdot \nabla\mathbf{A} \tag{7.1.16}$$

and Eqs. (7.1.11) and (7.1.14) gives

$$-\nabla \cdot \nabla\mathbf{E} = -\frac{\partial}{\partial t}\left(\frac{1}{c^2}\frac{\partial \mathbf{E}}{\partial t}\right) \tag{7.1.17}$$

or

$$\frac{\partial^2 \mathbf{E}}{\partial t^2} = c^2\nabla^2\mathbf{E} \tag{7.1.18}$$

which is a wave equation with propagation velocity c. A similar procedure gives

$$\frac{\partial^2 \mathbf{B}}{\partial t^2} = c^2\nabla^2\mathbf{B} \tag{7.1.19}$$

Consider a plane wave of **E** or **B** travelling in the x direction, so that neither **E** nor **B** are functions of y or z; then, from Eq. (7.1.11),

$$\frac{\partial \mathbf{E}_x}{\partial x} = 0 \tag{7.1.20}$$

or from Eq. (7.1.13),

$$\frac{\partial \mathbf{B}_x}{\partial x} = 0 \tag{7.1.21}$$

Thus, E_x and B_x do not vary with x, so that electromagnetic waves are always transverse.

Many analogies may be drawn between electromagnetic and mechanical radiation of different types, but one of the most important historically was between electromagnetic waves and mechanical waves travelling in a hypothetical ether. Since the waves had to be transverse, the ether had to be incompressible, so that (as we will see in Chapter 13)

$$\nabla \cdot \overline{\Delta} = \left(\frac{dV}{v}\right) = 0 \tag{7.1.22}$$

However, the reflection of shear waves at a boundary results in the production of a dilatation wave, and mechanical waves suffer entrainment in a moving medium. Electromagnetic waves show neither of these effects, and eventually the attempt to understand electromagnetic propagation as a form of mechanical radiation had to be abandoned. It is interesting to note that much of the early theory of mechanical shear wave propagation was motivated by the effort to understand electromagnetic propagation.

The establishment of the basic mechanics of deformable bodies and the beginnings of the ether theory date from the 1820s and are associated with such names as Navier, Poisson, Cauchy, and Green. The final collapse of the ether theory with the Michelson-Morley experiment heralded the profound theoretical departures introduced by Lorentz and Einstein, among others.

From the practical point of view, electromagnetic waves are frequently a problem for the experimenter in acoustics because a piece of equipment that carries an alternating voltage acts as an "antenna." Any conductor in the vicinity can then act as a receiver of these waves, which will generate voltages in the receiving conductor at the same frequency as the radiation. To prevent this problem with electromagnetic "pickup," it is necessary to ground potential transmitters and receivers. Pickup at the 60 Hz power line frequency is a very common problem, for example.

7.2 Dipoles

Piezoelectric transducers are solid-state materials that produce electric fields due to mechanical stress and vice versa. These effects occur because the material contains electric dipoles. It is therefore appropriate to review some aspects of electrostatic dipole theory, in which it is assumed that the fields do not vary with time. By applying Eq. (7.1.1) to a spherical surface about a point charge q (see Fig. 7.1), we can readily see that the force F between two point charges q_1 and q_2 in free space is given by Coulomb's law, which states that

$$\mathbf{F} = \frac{q_1 q_2 \mathbf{e}_r}{4\pi\epsilon_0 r^2} \tag{7.2.1}$$

where $\mathbf{e}_r$ is a unit vector directed along the line joining the charges, which have a separation r. The electric field **E** is defined as the force felt by a charge of unit magnitude and thus the electrostatic potential at a point is defined as the work done in bringing a unit charge to that point:

$$\varphi = -\int_{\infty}^{r} \mathbf{E} \cdot dl \tag{7.2.2}$$

The negative sign comes about because in order to move the charge without accelerating it we have to impose a force slightly greater than **E** and opposed to it in direction. In the special case that the field is due to a point charge q, then the potential at position r_1, as in Fig. 7.3, is

$$\begin{aligned} \varphi &= -\int_{\infty}^{r_1} \frac{q\mathbf{e}_r}{4\pi\epsilon_0 r^2} \cdot d\ell = -\int_{\infty}^{r_1} \frac{q}{4\pi\epsilon_0 r^2} dr \\ &= -\frac{q}{4\pi\epsilon_0}\left[-\frac{1}{r}\right]_{\infty}^{r_1} = \frac{q}{4\pi\epsilon_0 r_1} \end{aligned} \tag{7.2.3}$$

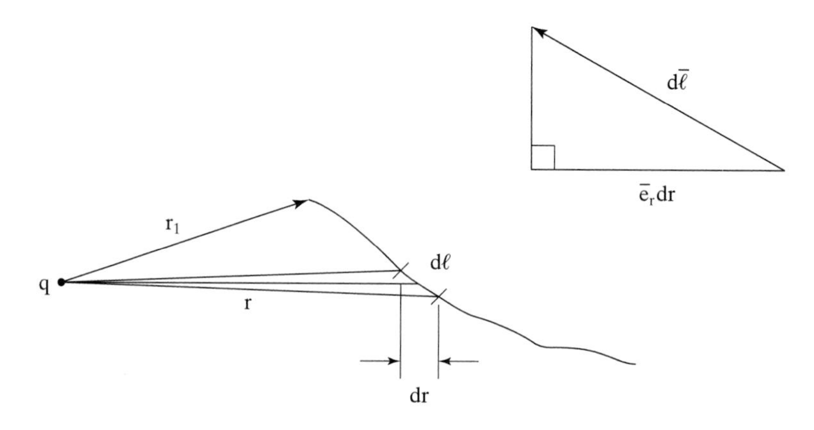

Fig. 7.3 Derivation of the potential at r_1 due to a point charge q. Note that $\bar{e}_r.d\bar{\ell}$ is negative, as is dr, so that $\bar{e}_r.d\bar{\ell} = dr$.

Note that $\mathbf{e_r} \cdot d\ell$ is negative; but $\mathbf{dr}$ is also negative, so that $\mathbf{e_r} \cdot d\ell = dr$. Consequently, the potential difference between two points A and B at position vectors $\mathbf{r}_A$ and $\mathbf{r}_B$ is

$$\varphi_A - \varphi_B = -\int_{r_B}^{r_A} \mathbf{E} \cdot \mathbf{d}\boldsymbol{l} \tag{7.2.4}$$

Since the potential must be single valued at a point, it follows that the integral of $\mathbf{E}$ around a closed loop is zero; that is,

$$\oint \mathbf{E} \cdot \mathbf{d}\boldsymbol{l} = 0 \tag{7.2.5}$$

From Eq. (7.2.4), it follows that the electric field may be written as the gradient of the potential field; in other words,

$$\mathbf{E} = -\nabla\varphi \tag{7.2.6}$$

From Eqs. (7.1.1) and (7.2.6), we obtain Poisson's equation:

$$\nabla^2\varphi = -\rho/\epsilon_0 \tag{7.2.7}$$

There are free electric charges (electrons) within conducting materials (metals) that permit the flow of electricity. On the other hand, insulators, or dielectrics, have their positive and negative charges closely bound, so that currents cannot flow. In insulators, however, since the charged particles occur in pairs, they constitute electric dipoles. Consider such a dipole, shown in Fig. 7.4. The potential due to both charges is

$$\varphi_P = \frac{q}{4\pi\epsilon_0}\frac{(r_2 - r_1)}{r_1 r_2} \tag{7.2.8}$$

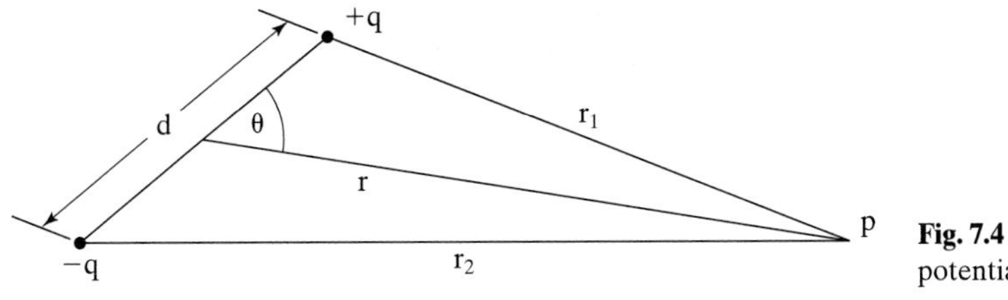

Fig. 7.4 Derivation of the potential at P due to a dipole.

Using a far field approximation,

$$\varphi_P = \frac{qd\cos\theta}{4\pi\epsilon_0 r^2} \tag{7.2.9}$$

We define the dipole moment, **p**, as a vector directed from the negative to the positive charge, of magnitude qd (i.e., $\mathbf{p} = \mathbf{n}\,q\,d$), where **n** is a unit vector. Now $\mathbf{e_r}\cdot\mathbf{n} = \cos\theta$, and thus

$$\varphi_P = \frac{\mathbf{p}\cdot\mathbf{e}_r}{4\pi\epsilon_0\mathbf{r}^2} \tag{7.2.10}$$

7.3 Polarization

We will now consider a medium containing a large number, N, of identical dipoles per unit volume as shown in Fig. 7.5. The potential at some point P outside the medium will be

$$\varphi_\mathbf{P} = \frac{1}{4\pi\epsilon_0}\int_V \frac{N\mathbf{p}\cdot\mathbf{e}_r}{r^2}dV \tag{7.3.1}$$

Let

$$N\mathbf{p} = \mathbf{P} \tag{7.3.2}$$

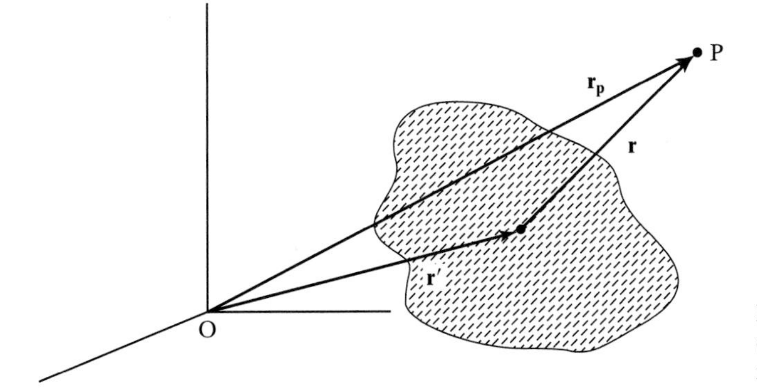

Fig. 7.5 Potential at point P outside a medium containing many dipoles.

where $\mathbf{P}$ is the dipole moment per unit volume, and thus

$$\varphi_{\mathbf{P}} = \frac{1}{4\pi\epsilon_0}\int_V \frac{\mathbf{P}\cdot\mathbf{e}_r}{\mathbf{r}^2}dV \tag{7.3.3}$$

Now taking the origin at the center of a dipole

$$\nabla\left(\frac{1}{r}\right) = -\frac{\mathbf{e}_r}{\mathbf{r}^2} \tag{7.3.4}$$

but since

$$\mathbf{r}' + \mathbf{r} = \mathbf{r}_{\mathbf{P}} \text{ (a constant)} \tag{7.3.5}$$

$$\delta\mathbf{r}' = -\delta\mathbf{r} \tag{7.3.6}$$

so it follows, taking the origin at O as in Fig. 7.5, that

$$\nabla'\left(\frac{1}{r}\right) = -\nabla\left(\frac{1}{r}\right) = \frac{\mathbf{e}_r}{\mathbf{r}^2} \tag{7.3.7}$$

and thus

$$\varphi_{\mathbf{P}} = \frac{1}{4\pi\epsilon_0}\int \mathrm{P}\cdot\nabla'\left(\frac{1}{r}\right)dV \tag{7.3.8}$$

But

$$\nabla\cdot\left(\frac{\mathbf{P}}{r}\right) = \frac{1}{r}\nabla\cdot\mathbf{P} + \mathbf{P}\cdot\nabla\left(\frac{1}{r}\right) \tag{7.3.9}$$

hence

$$\varphi_{\mathbf{P}} = \frac{1}{4\pi\epsilon_0}\left[\int_V \nabla'\cdot\left(\frac{\mathbf{P}}{r}\right)dV - \int_V \frac{\nabla'\cdot\mathbf{P}}{r}dV\right] \tag{7.3.10}$$

Using the divergence theorem, the first integral is converted to a surface integral

$$\varphi_{\mathbf{P}} = \frac{1}{4\pi\epsilon_0}\left[\int_S \frac{\mathbf{P}\cdot d\mathbf{S}}{r} - \int_V \frac{\nabla'\cdot\mathbf{P}}{r}dV\right] \tag{7.3.11}$$

By comparison with Eq. (7.2.3), we see that the total field is the sum of two components:

a. a volume charge distribution of density $-\nabla\cdot\mathbf{P}$, and

b. a surface charge density equal to the normal component of $\mathbf{P}$ at the surface of the dielectric.

We can illustrate this situation by referring to a parallel plate capacitor (Fig. 7.6a). When a slab of dielectric material is inserted between the plates, the dipoles tend to align with the imposed field. The effect is as though the positive and negative charges were separated by a small distance, δ. Consequently, there will be a +ve charge on one surface and a negative charge on the opposite surface (see Fig. 7.6b). The surface charge density is then P, since $P = Nq\delta$, q being the positive charge in the dipole. Thus, the charge/unit area must be $Nq\delta = \sigma$. For the capacitor, the electric field is uniform so that there is no volume charge in the slab, and thus

$$\nabla \cdot \mathbf{P} = 0 \tag{7.3.12}$$

If, however, the field was not uniform, and consequently **P** was not uniform, more charge might flow out over some portions of the surface than flows in elsewhere, which would result in a net volume charge.

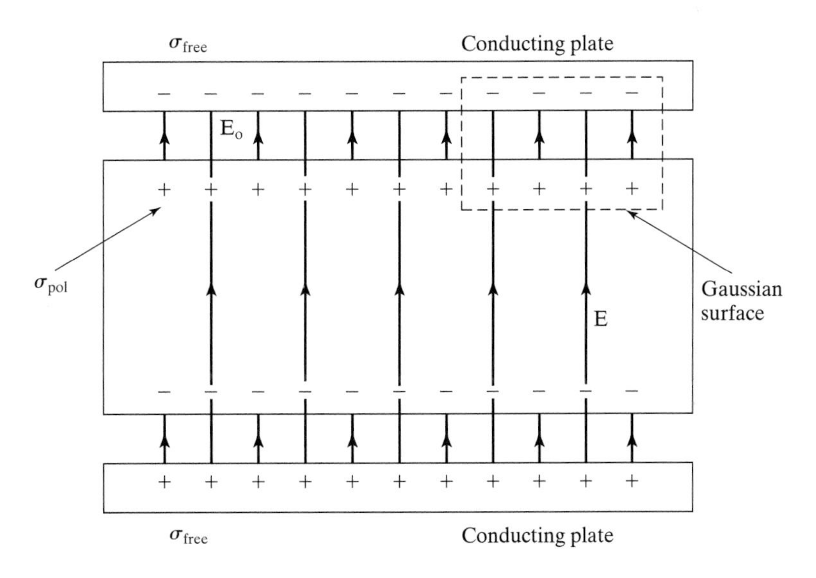

Fig. 7.6a Polarization of an insulator in a parallel plate capacitor.

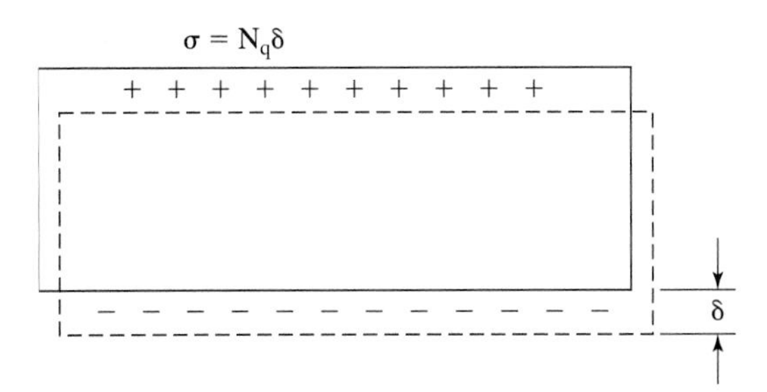

Fig. 7.6b Polarization may be regarded as the separation of positive and negative charges by a small amount, δ.

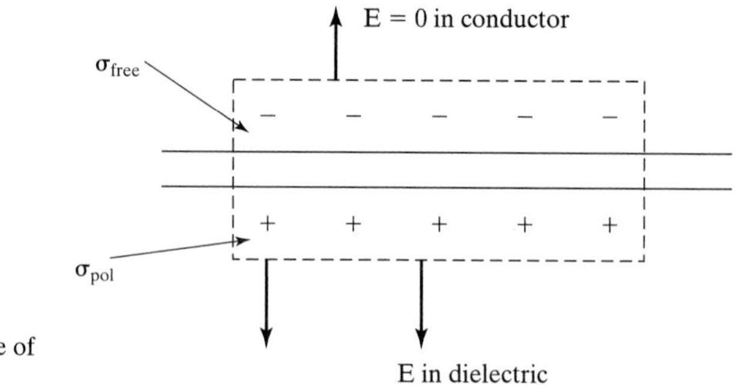

Fig. 7.6c The Gaussian surface of integration.

Since the charge that flows out of the dielectric is $\int_S \mathbf{P} \cdot d\mathbf{S}$, it follows that the accumulated charge within a volume must be

$$\int \rho dV = -\int_S \mathbf{P} \cdot d\mathbf{S} = -\int_V \nabla \cdot \mathbf{P} dV \tag{7.3.13}$$

or in the case of the nonuniform field

$$-\nabla . \mathbf{P} = \rho \tag{7.3.14}$$

It is also interesting to relate this result to the capacitance of the parallel plates shown in Fig. 7.6a. Applying Gauss' law to the surface shown in Fig. 7.6c, since

$$\int_S \mathbf{E} \cdot d\mathbf{S} = \int_V \frac{\rho}{\epsilon_0} dV = \int_S \frac{\sigma}{\epsilon_0} d\mathbf{S} \tag{7.3.15}$$

it follows that

$$E = \frac{\sigma_{free} - \sigma_{pol}}{\epsilon_0} \tag{7.3.16}$$

hence

$$E = \frac{\sigma_{free} - P}{\epsilon_0} \tag{7.3.17}$$

If an electric field is imposed on a dielectric material, the dipole moment may be induced by the action of the field. Assuming a linear relationship, we may write

$$P = \epsilon_0 \chi E \tag{7.3.18}$$

where χ is termed the susceptibility.

Hence, from Eq. (7.3.17)

$$E = \frac{\sigma_{free} - \chi\epsilon_0 E}{\epsilon_0} \tag{7.3.19}$$

so

$$E = \frac{\sigma_{free}}{\epsilon_0(1 + \chi)} \tag{7.3.20}$$

But the potential difference between the plates is $V = Ed$, and thus

$$C = \frac{Q}{V} = \frac{A\sigma_{free}}{V} = \frac{A\epsilon_0(1 + \chi)}{d} = kC_0 \tag{7.3.21}$$

where C_o is the capacitance in free space and

$$k = 1 + \chi = \text{dielectric constant} \tag{7.3.22}$$

Inside the medium, the volume charge distribution gives rise to a field E_p, where from Eq. (7.1.1)

$$\nabla \cdot \mathbf{E_P} = -\frac{\nabla \cdot \mathbf{P}}{\epsilon_0} \tag{7.3.23}$$

If there was also a free charge density ρ within the material, the total field E would be given by

$$\nabla \cdot \mathbf{E} = \frac{\rho - \nabla \cdot \mathbf{P}}{\epsilon_0} \tag{7.3.24}$$

Let

$$\mathbf{D} = \epsilon_0 \mathbf{E} + \mathbf{P} \tag{7.3.25}$$

where **D** is called the displacement field vector. Then, from Eq. (7.3.24)

$$\nabla \cdot \mathbf{D} = \epsilon_0 \nabla \cdot \mathbf{E} + \nabla \cdot \mathbf{P} = \rho \tag{7.3.26}$$

Note that in free space, $\mathbf{D} = \epsilon_0 \mathbf{E}$ and Eq. (7.3.26) reduces to Eq. (7.1.11). If P is proportional to E,

$$\mathbf{D} = \epsilon_0 \mathbf{E} + \epsilon_0 \chi \mathbf{E} = \epsilon_0 k \mathbf{E} \tag{7.3.27}$$

k is called the relative dielectric constant and $\epsilon_0 k$ is the dielectric constant. In ferroelectric materials (e.g., barium titanate), the relationships between **D** and **E** exhibit hysteresis (Fig. 7.7). We must therefore take care in defining at which point on the curve the various dielectric constants are measured.

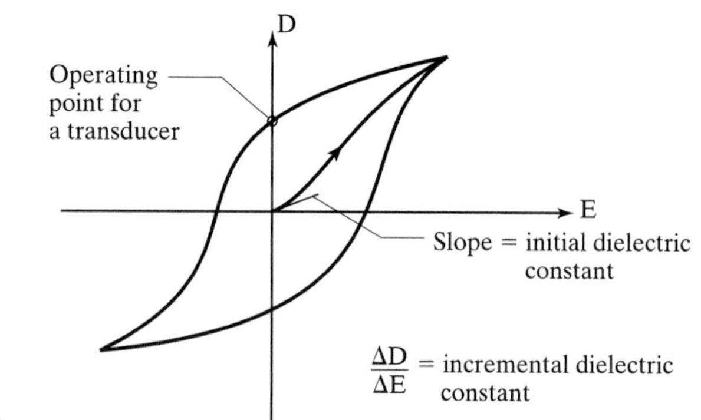

Fig. 7.7 Hysteresis curve for the polarization of a ferroelectric material.

7.4 Energy of Polarization

The incremental work done in increasing the charge density by $\delta\rho(r)$ in a medium may be written as

$$\delta W = \int_V \delta\rho(r)\,\varphi(r)\,dv \tag{7.4.1}$$

But the charge density and displacement fields are related by Eq. (7.3.22). Thus

$$\nabla\cdot\delta D = \delta\rho \tag{7.4.2}$$

hence

$$\delta W = \int_V \{\nabla\cdot\delta D(r)\}\,\varphi(r)\,dv \tag{7.4.3}$$

But

$$\nabla\cdot\{\varphi(r)\delta\mathbf{D}(r)\} = (\nabla\cdot\delta\mathbf{D}(r))\varphi(r) + \delta\mathbf{D}(r)\cdot\nabla\varphi(r) \tag{7.4.4}$$

So

$$\begin{aligned}\delta W &= \int_V [\nabla\cdot\{\varphi(r)\delta\mathbf{D}(r)\} - \delta\mathbf{D}(r)\cdot\nabla\varphi(r)]dV \\ &= \int_S \varphi(r)\delta\mathbf{D}(r)\,dS + \int_V \delta\mathbf{D}(r)z\cdot\mathbf{E}dV\end{aligned} \tag{7.4.5}$$

using the divergence theorem and the fact that the electric field is the negative gradient of the potential. But if we take the surface integration far enough outside the localized charge distribution, $\varphi(r)$ will be zero and hence

$$\delta W = \int_V \mathbf{E}\cdot\delta\mathbf{D}\,dv \tag{7.4.6}$$

7.5 Mechanisms of Polarization

Atoms that are electrically neutral can have an electric dipole moment if there is a separation of the positively charged nucleus and the negatively charged electron cloud. An atom with an excess or a deficiency of charge is called an ion, and it may combine with other ions to form a molecule, which may then exhibit a dipole moment due to the separation of the ions of different charges. Molecular dipole moments then tend to be larger than atomic dipole moments. In an unordered assembly of atoms or molecules, the orientations of the dipoles will be completely random. If we apply an external electric field to the assembly, the dipoles will tend to line up with the field, as discussed in the previous section, but usually this alignment will disappear again once the field has been removed. But there are several ways in which a solid can become permanently polarized. For instance, a material such as a wax, which contains long molecules with dipole moments, can be polarized with an electric field while melted and then allowed to cool while still under the influence of the field. Such polarized material is called an electret (by analogy with the better known "magnet"). The polarization is not usually observable, however, because charges from the air settle on the surfaces of the wax and cancel out the polarization charge layers.

Another way in which a permanent dipole moment can be built up in a solid is through crystallization. For example, ionic crystals consist of positive and negative ions and the electrostatic forces between the ions contribute to the configuration of the crystal. Such a crystal has a permanent electric dipole moment. This is the case with quartz, for example. The importance of this permanent dipole moment to acoustics is that the application of pressure to the crystal will result in charge concentrations on the crystal's surface (i.e., in the production of an electric field). This is called the piezoelectric effect, and the production of mechanical stress by electric fields is called the inverse piezoelectric effect. Fig. 7.8a shows a quartz crystal indicating its crystallographic axes. A bar cut from the crystal as shown is said to be x-cut and the electronic field is applied in the x direction (Fig. 7.8b). Table 7.1 shows that twenty of the possible thirty-two crystal classes have no center of symmetry and thus exhibit piezoelectric effects. Pyroelectric materials are a subclass of the piezoelectrics, which contain a dipole moment in their unstrained unit cell. They are said to be "pyroelectric"—"pyro" is Greek for

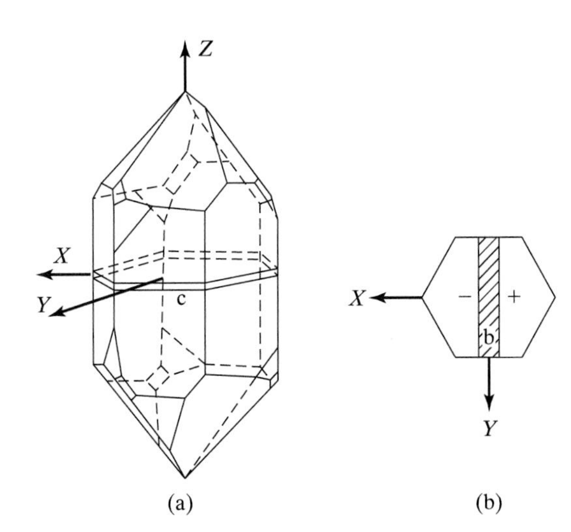

Fig. 7.8 (a) A quartz crystal showing crystallographic axes. (b) An x-cut bar is exposed to an electric field in the x direction.

Table 7.1 Classification of crystals by properties of polarization. Numbers of crystal classes in parentheses.

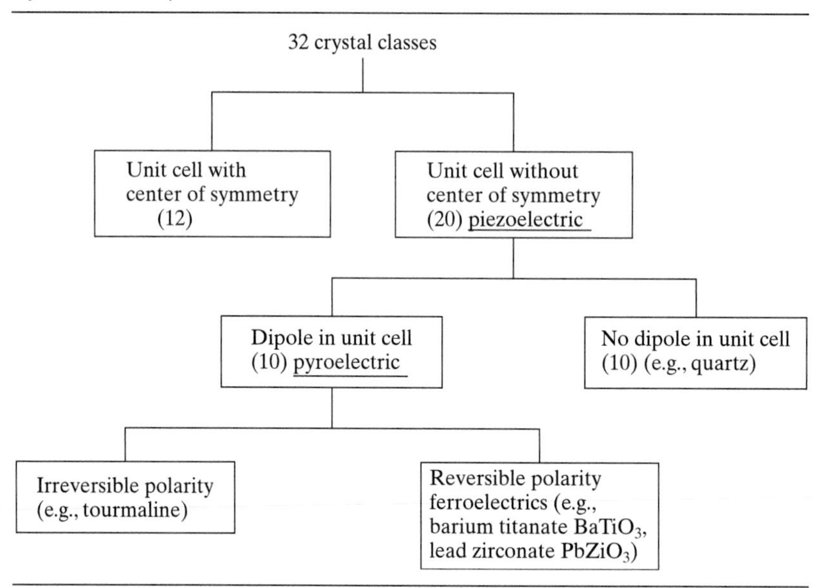

fire—because they produce charges under the action of heat. As with the electrets, these charges are soon neutralized when exposed to the atmosphere.

There is also a class of materials called ferroelectrics where molecular dipole moments can be reoriented to lock in with the dipole moments of neighboring atoms. For example, in barium titanate ($BaTiO_3$), which has a perovskite structure, this phenomenon is believed to be largely due to the fact that the Ti ion is only loosely bound to the center of the unit cell (see Fig. 7.9). It can be readily moved in one direction or another. When it takes up an off-center position, there is a net separation of charges within the unit cell. Increasing thermal agitation eventually prevents it from remaining in any off-center position. The temperature at which this happens is called the Curie temperature and is 118° C for barium titanate. Above this temperature, barium titanate is an ordinary dielectric with a large dielectric constant.

The regions in which the dipole moments are aligned are known as domains. One possible arrangement of domains that is energetically preferred is shown in Fig. 7.10. Under an applied electric field, the domains whose polarization direction is energetically preferred tend to grow at the expense of the nonpreferred domains. Saturation occurs when the entire material comprises one single domain. Ferroelectric materials are extremely useful, since a pure crystal is not required to obtain a macroscopic polarized sample. In fact, the ferroelectrics may be made into customized shapes by working them as ceramics. Lead zirconate titanate (PZT) is another example of a ferroelectric material.

Magnetic polarization arises from the circulation of the electron cloud in the atom. A current loop produces a magnetic field similar to that of a bar magnet, with a strong dipole moment. In ferromagnetic materials, the dipole moments tend to align with an applied magnetic field and then remain in a permanent state of magnetization by locking in with the polarization in neighboring atoms over extensive domains. Once

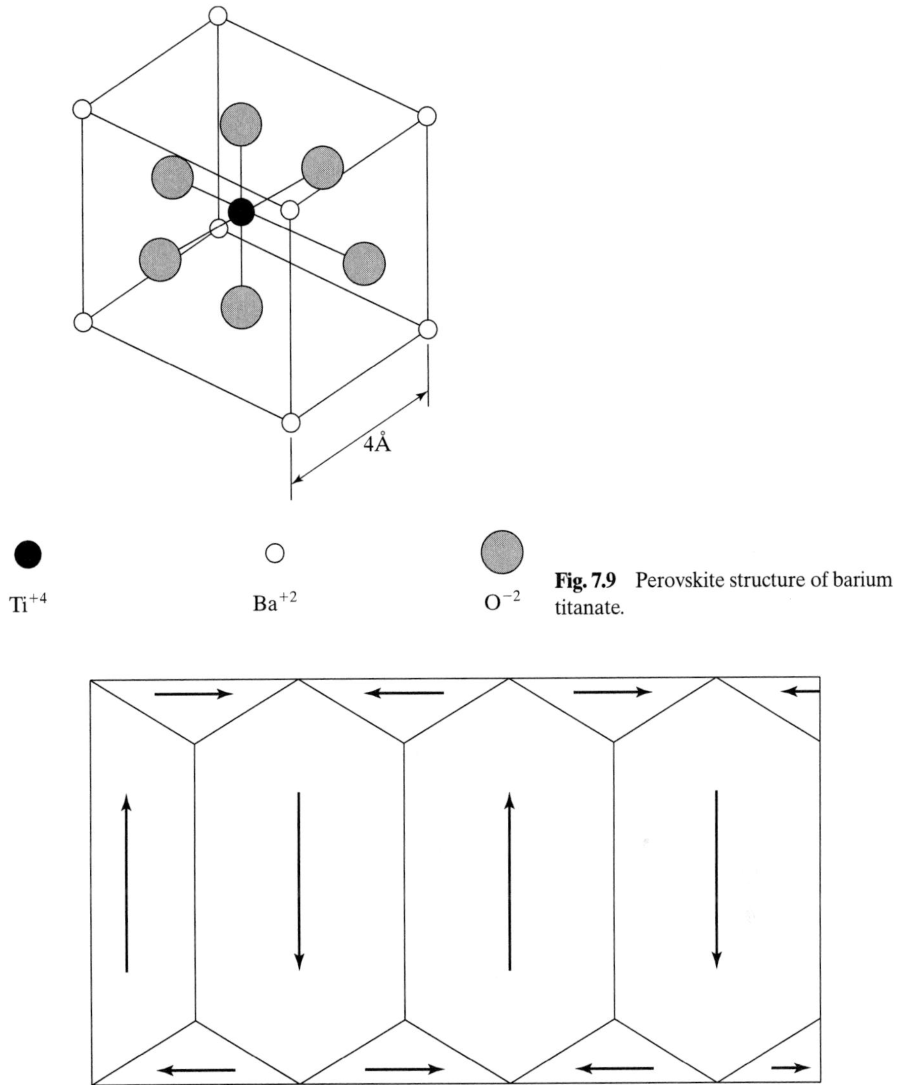

Fig. 7.9 Perovskite structure of barium titanate.

Fig. 7.10 Domains of polarization.

again, as with ferroelectricity, there is a saturation effect. The magnetostrictive effect is the production of a magnetic field by compression of a magnetically polarized material.

7.6 Maxwell's Equations in a Dielectric

To rewrite Maxwell's equations inside a dielectric material, remember that imposition of an electric field causes polarization, and there will be a polarization charge as a result:

$$\rho_{\text{polarization}} = -\nabla \cdot \mathbf{P} \tag{7.6.1}$$

Hence, Maxwell's first equation gives

$$\nabla \cdot \mathbf{E} = \frac{\rho_{\text{free}} + \rho_{\text{pol}}}{\epsilon_0} \tag{7.6.2}$$

or

$$\nabla \cdot (\epsilon_0 \mathbf{E} + \mathbf{P}) = \rho_{\text{free}} \tag{7.6.3}$$

Now, we define the electric displacement by

$$\mathbf{D} = \epsilon_0 \mathbf{E} + \mathbf{P} \tag{7.6.4}$$

and

$$\mathbf{P} = \epsilon_0 \chi \mathbf{E} \tag{7.6.5}$$

Thus

$$\mathbf{D} = \epsilon_0 (1 + \chi) \mathbf{E} = \epsilon_0 \mathbf{k}\, \mathbf{E} \tag{7.6.6}$$

where χ = susceptibility and k = the relative dielectric constant. Thus, from Eq. (7.6.3)

$$\nabla \cdot \mathbf{D} = \rho_{\text{free}} \tag{7.6.7}$$

Now, while **P** is changing there will be currents due to movement of charges. This affects Maxwell's fourth equation:

$$\begin{aligned} c^2 \nabla \times \mathbf{B} &= \frac{\partial \mathbf{E}}{\partial t} + \frac{\mathbf{j}_{\text{free}} + \mathbf{j}_{\text{pol}}}{\epsilon_0} \\ &= \frac{\partial \mathbf{E}}{\partial t} + \frac{\dot{\mathbf{P}}}{\epsilon_0} + \frac{\mathbf{j}_{\text{free}}}{\epsilon_0} \end{aligned} \tag{7.6.8}$$

hence

$$\epsilon_0 c^2 \nabla \times \mathbf{B} = \frac{\partial}{\partial t}(\epsilon_0 \mathbf{E} + \mathbf{P}) + \mathbf{j}_{\text{free}}$$

and finally

$$\epsilon_0 c^2 \nabla \times \mathbf{B} = \frac{\partial \mathbf{D}}{\partial t} + \mathbf{j}_{\text{free}} \tag{7.6.9}$$

Eqs. (7.6.7) and (7.6.9) replace Eqs. (7.1.1) and (7.1.4) for dielectric materials.

7.7 Kelvin's Theory of Piezoelectricity

There are certain propositions that can be developed about piezoelectricity on thermodynamic grounds, regardless of the exact mechanism of polarization. Such a theory was produced by Lord Kelvin. The theory is developed in terms of the variables mechanical stress σ, strain ε, electric field E, and displacement D. We have to relate these variables in a way that will exhibit the coupling between electric and elastic fields. We choose independent variables, one mechanical and one electrical, and then express the remaining variables as functions of these. This selection can be made in four different ways. The following is one way:

$$\sigma = \sigma(\varepsilon, D) \qquad E = E(\varepsilon, D) \tag{7.7.1}$$

and expanding by Taylor's theorem to linear terms gives

$$d\sigma = \left(\frac{\partial \sigma}{\partial \varepsilon}\right)_D d\varepsilon + \left(\frac{\partial \sigma}{\partial D}\right)_\varepsilon dD$$

$$dE = \left(\frac{\partial E}{\partial \varepsilon}\right)_D d\varepsilon + \left(\frac{\partial E}{\partial D}\right)_\varepsilon dD \tag{7.7.2}$$

$(\partial\sigma/\partial\varepsilon)_D$ is an elastic stiffness constant at constant displacement, c_D. $(\partial E/\partial D)_\varepsilon$ or β_ε is the susceptance or the reciprocal of the dielectric constant. The internal energy of a piece of piezoelectric material is a function of strain (mechanical) and displacement (electrical); that is,

$$U = U(\varepsilon, D)$$

and

$$dU = \left(\frac{\partial U}{\partial \varepsilon}\right)_D d\varepsilon + \left(\frac{\partial U}{\partial D}\right)_\varepsilon dD \tag{7.7.3}$$

Under adiabatic conditions, $dU = dW$ (first law of thermodynamics). But for a unit cube of material, if the strain is increased by a small amount $d\varepsilon$, at constant displacement, the work done per unit volume is Stress $\times$ Area $\times$ Extension; that is,

$$dW_D = \sigma d\varepsilon \tag{7.7.4}$$

hence

$$\sigma = \left(\frac{\partial U}{\partial \varepsilon}\right)_D \tag{7.7.5}$$

For a distribution of electric dipoles, it is shown in Section 7.4 that the energy required to create an electric field and to polarize the dipoles, when the material is fixed in position and the field is varied, is given by

$$dW_\varepsilon = \mathbf{E} \cdot d\mathbf{D} \tag{7.7.6}$$

hence

$$E = \left(\frac{\partial U}{\partial D}\right)_{\varepsilon} \tag{7.7.7}$$

so

$$\left(\frac{\partial \sigma}{\partial D}\right)_{\varepsilon} = \frac{\partial^2 U}{\partial D \partial \varepsilon} = \frac{\partial^2 U}{\partial \varepsilon \partial D} = \left(\frac{\partial E}{\partial \varepsilon}\right)_{D} = -h \tag{7.7.8}$$

a so-called electromechanical constant. Thus, the basic equations can be written

$$\begin{aligned} d\sigma &= c_D d\varepsilon - h dD \\ dE &= -h d\varepsilon + \beta_{\varepsilon} dD \end{aligned} \tag{7.7.9}$$

which is only one of the four ways of relating the field variables. When other selections are made, we obtain the other electromechanical constants:

$$\begin{aligned} d &= \left(\frac{\partial \varepsilon}{\partial E}\right)_{\sigma} = \left(\frac{\partial D}{\partial \sigma}\right)_{E} \\ e &= \left(\frac{\partial \sigma}{\partial E}\right)_{\varepsilon} = \left(\frac{\partial D}{\partial \varepsilon}\right)_{E} \\ g &= \left(\frac{\partial E}{\partial \sigma}\right)_{D} = \left(\frac{\partial \varepsilon}{\partial D}\right)_{\sigma} \end{aligned} \tag{7.7.10}$$

Which set of equations we choose to work with depends on the nature of the problem to be solved.

A complication exists in that the field variables are tensors. The elastic, electric, and piezoelectric constants are matrix elements. The usual matrix notation with two suffixes is used to relate the field variables. The common modes of transducer operation, however, involve only a few of these matrix elements.

Based on the postulates of this section, we can easily see that a simple thickness mode slab of area A and thickness T, used as a receiver, has open circuit sensitivity:

$$M_p = \frac{hT}{c_D} \tag{7.7.11}$$

Thus, the low frequency sensitivity of a piezoelectric disc is not frequency dependent and increases with increasing thickness. However, the sensitivity of such a piezoelectric disc would be relatively low, and in typical applications arrangements are made to obtain a mechanical resonance at frequencies in or close to the operating range. In this circumstance, the dimensions of the device are comparable to a wavelength and we have a distributed system.

7.8 Distributed Piezoelectric Systems

A distributed system may also be thought of as one in which the elastic stresses and strains, if not the electric field and flux, are variable along the length or thickness of the transducer. The electromechanical equations must then be taken as referring only to an elemental slice parallel to the major faces of the crystal. Most often we are interested only in pure elastic modes, so we will assume that only one mode is excited in the transducer and that the stresses and strains occur in one direction only.

As we intend propagating an elastic wave in the crystal, we must first establish a wave equation for a piezoelectric solid. Newton's law gives

$$\rho \frac{\partial^2 \xi}{\partial t^2} = \frac{\partial \sigma}{\partial x} \tag{7.8.1}$$

where x is the oscillatory displacement. Substituting from Eq. (7.7.9.) gives, for a longitudinal wave, for example:

$$\rho \frac{\partial^2 \xi}{\partial t^2} = c_{11}^D \frac{\partial^2 \xi}{\partial x^2} - h \frac{\partial D}{\partial x} \tag{7.8.2}$$

All piezoelectric crystals are electrical insulators and cannot conduct free charge, so that $\nabla \cdot D = 0$. Thus, for a plane wave $dD/dx = 0$, and the equation reduces to

$$\rho \frac{\partial^2 \xi}{\partial t^2} = c_{11}^D \frac{\partial^2 \xi}{\partial x^2} \tag{7.8.3}$$

This is identical to the one-dimensional wave equation derived previously in Section 1.6.1 with propagation velocity

$$c_t^D = \left(\frac{c_{11}^D}{\rho} \right)^{1/2} \tag{7.8.4}$$

7.9 Thickness Vibrations of Piezoelectric Plates

Mason (1935–1965) applied the theory of the previous two sections to the thickness vibrations of piezoelectric plates. Suppose the plate shown in Fig. 7.11 is polarized in the x_3 direction. At some instant, the stresses are such as to cause a positive linear strain in the x_3 direction, increasing with time. The forces and velocities will then be as shown in Fig. 7.11. When a voltage is applied as shown, then

$$D_1 = D_2 = 0 \tag{7.9.1}$$

and

$$\frac{\partial D_3}{\partial x_3} = 0 \tag{7.9.2}$$

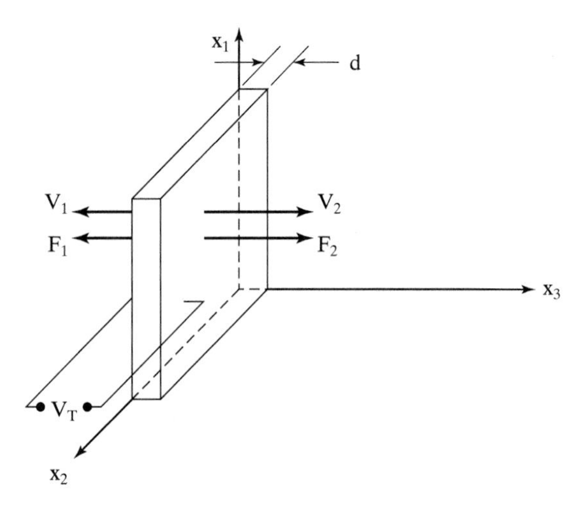

Fig. 7.11 Piezoelectric plate with applied forces and voltage.

For a thickness expander plate, the changes in the field variables are all in the linear region, so we may drop the incremental notation throughout the equations; thus

$$\sigma_3 = c_{33}^D \varepsilon_3 - h_{33} D_3 \tag{7.9.3}$$

and

$$E_3 = -h_{33}\varepsilon_3 + \beta_{33}^{\varepsilon} D_3 \tag{7.9.4}$$

If the voltage is oscillatory, there will be wave motion in the plate and the displacement, ξ, will have a general solution:

$$\xi_3 = \left[A \sin\frac{\omega x_3}{c_t^D} + B \cos\frac{\omega x_3}{c_t^D}\right] e^{j\omega t} \tag{7.9.5}$$

The constants A and B have to be found from the boundary conditions. We will assume that all the field and circuit parameters are phasors with frequency ω. Now,

$$\dot{\xi}_3 = -v_1 \quad \text{at} \quad x_3 = 0 \tag{7.9.6}$$

and

$$\dot{\xi}_3 = v_2 \quad \text{at} \quad x_3 = d \tag{7.9.7}$$

When $x_3 = 0$,

$$\dot{\xi} = -v_1 = j\omega B \tag{7.9.8}$$

thus

$$B = \frac{-v_1}{(j\omega)} \tag{7.9.9}$$

When $x_3 = d$,

$$\dot{\xi}_3 = v_2 = \left[A \sin\frac{\omega d}{c_t^D} - \frac{v_1}{j\omega}\cos\frac{\omega d}{c_t^D}\right] j\omega \tag{7.9.10}$$

and

$$A = \frac{1}{j\omega}\left[\frac{v_2}{\sin\dfrac{\omega d}{c_t^D}} + \frac{v_1}{\tan\dfrac{\omega d}{c_t^D}}\right] \tag{7.9.11}$$

Hence, from Eq. (7.9.3)

$$\begin{aligned} F &= \sigma_3 A_0 = A_0[c_{33}^D \varepsilon_3 - h_{33} D_3] \\ &= A_0\left[c_{33}^D \frac{\partial \xi_3}{\partial x_3} - h_{33} D_3\right] \\ &= A_0 c_{33}^D \left[\frac{\omega A}{c_t^D}\cos\frac{\omega x_3}{c_t^D} - \frac{\omega B}{c_t^D}\sin\frac{\omega x_3}{c_t^D}\right] - A_0 h_{33} D_3 \end{aligned} \tag{7.9.12}$$

When $x_3 = 0$,

$$F = F_1 = A_0 c_{33}^D \frac{\omega A}{c_t^D} - A_0 h_{33} D_3 \tag{7.9.13}$$

and from Eqs. (7.9.11) and (7.9.13),

$$F_1 = \frac{A_0 c_{33}^D}{j c_t^D}\left[\frac{v_2}{\sin\dfrac{\omega d}{c_t^D}} + \frac{v_1}{\tan\dfrac{\omega d}{c_t^D}}\right] - A_0 h_{33} D_3 \tag{7.9.14}$$

Let

$$z_0 = A_0(\rho c_{33}^D)^{1/2} \tag{7.9.15}$$

then

$$\frac{A_0 c_{33}^D}{c_t^D} = \frac{A_0 c_{33}^D}{(c_{33}^D/\rho)^{1/2}} = z_0 \tag{7.9.16}$$

and hence

$$F_1 = \left[\frac{z_0}{j\tan\frac{\omega d}{c_t^D}} \cdot v_1 + \frac{z_0}{j\sin\frac{\omega d}{c_t^D}} \cdot v_2 \right] - A_0 h_{33} D_3 \tag{7.9.17}$$

Now from Maxwell's equation, Eq. (7.6.9),

$$\frac{\partial D}{\partial t} = -\frac{I}{A_0} \tag{7.9.18}$$

and

$$-A_0 D_3 = \frac{I}{j\omega} \tag{7.9.19}$$

Thus

$$F_1 = \left[\frac{z_0}{j\tan\frac{\omega d}{c_t^D}} \cdot v_1 + \frac{z_0}{j\sin\frac{\omega d}{c_t^D}} \cdot v_2 \right] + \frac{h_{33} I}{j\omega} \tag{7.9.20}$$

When $x_3 = d$, with some manipulation we obtain

$$F = F_2 = \left[\frac{z_0}{j\sin\frac{\omega d}{c_t^D}} \cdot v_1 + \frac{z_0}{j\tan\frac{\omega d}{c_t^D}} \cdot v_2 \right] + \frac{h_{33} I}{j\omega} \tag{7.9.21}$$

The potential difference across the plate is obtained by substituting into Eq. (7.9.4)

$$\begin{aligned} V_T &= -\int_0^d E dx = \int_0^d h_{33} \varepsilon_3 dx - \int_0^d \beta_{33}^{\varepsilon} D_3 dx \\ &= \int_0^d h_{33} \frac{\partial \xi_3}{\partial x} dx - \beta_{33} D_3 d \\ &= h_{33} \left[\xi_3 \right]_0^d + \frac{\beta_{33} d I}{j\omega A_0} \end{aligned} \tag{7.9.22}$$

Now using

$$\dot{\xi}_3 = -v_1 \quad \text{at} \quad x_3 = 0$$

and

$$\dot{\xi}_3 = v_2 \quad \text{at} \quad x_3 = d$$

$$\xi_3 = \frac{-v_1}{j\omega} \quad \text{at} \quad x_3 = 0$$

and

$$\xi_3 = \frac{v_2}{j\omega} \quad \text{at} \quad x_3 = d$$

we obtain

$$V_T = \frac{h_{33}}{j\omega}[v_1 + v_2] + \frac{I}{j\omega C_0} \tag{7.9.23}$$

where

$$C_0 = \frac{A_0}{\beta_{33}^{\varepsilon}\, d} \tag{7.9.24}$$

Now consider the circuit shown in Fig. 7.12. It follows that

$$F_2 = v_2 Z_2 + (v_1 + v_2)Z_3 + V_2 \tag{7.9.25}$$

$$F_1 = v_1 Z_1 + (v_1 + v_2)Z_3 + V_2 \tag{7.9.26}$$

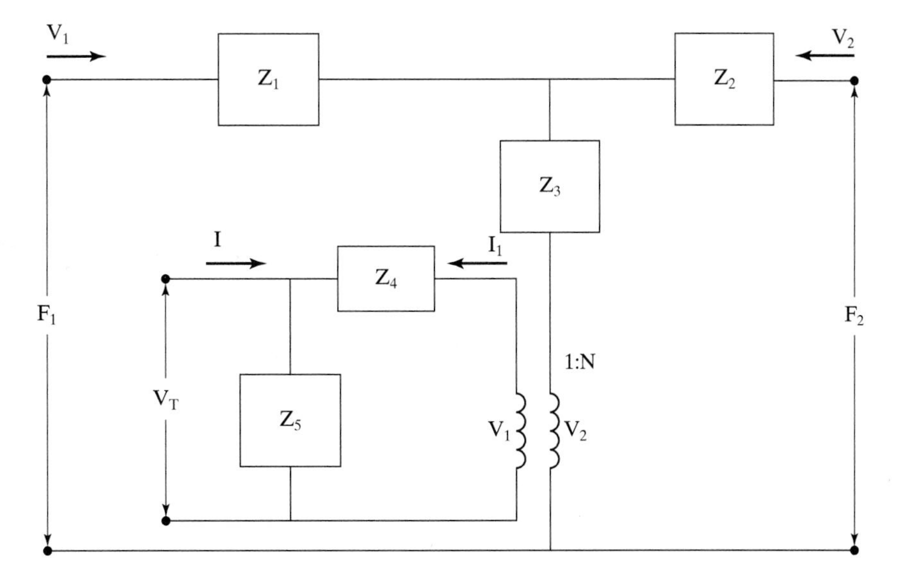

Fig. 7.12 Equivalent circuit of a piezoelectric slab.

$$V_T = (I + I_1)Z_5 \tag{7.9.27}$$

$$V_1 = I_1Z_4 + (I + I_1)Z_5 \tag{7.9.28}$$

$$V_2 = NV_1 \tag{7.9.29}$$

and

$$N(v_1 + v_2) = I_1 \tag{7.9.30}$$

Eliminating V_2, I_1, and V_1 gives three equations:

$$F_2 = v_2Z_2 + (v_1 + v_2)Z_3 + NIZ_5 + N^2(v_1 + v_2)(Z_4 + Z_5) \tag{7.9.31}$$

$$F_1 = v_1Z_1 + (v_1 + v_2)Z_3 + NIZ_5 + N^2(v_1 + v_2)(Z_4 + Z_5) \tag{7.9.32}$$

and

$$V_T = [1 + N(v_1 + v_2)]Z_5 \tag{7.9.33}$$

Comparing Eqs. (7.9.20), (7.9.21), and (7.9.23) with Eqs. (7.9.31), (7.9.32), and (7.9.33) gives

$$Z_5 = \frac{1}{jC_0\omega} \tag{7.9.34}$$

hence

$$\frac{N(v_1 + v_2)}{jC_0\omega} = \frac{h_{33}(v_1 + v_2)}{j\omega} \tag{7.9.35}$$

and

$$N = C_0h_{33} \tag{7.9.36}$$

Rewriting Eq. (7.9.31):

$$F_2 = v_1\left[Z_3 + C_0^2h_{33}^2\left(Z_4 + \frac{1}{j\omega C_0}\right)\right] + v_2\left[Z_2 + Z_3 + C_0^2h_{33}^2\left(Z_4 + \frac{1}{j\omega C_0}\right)\right] + \frac{h_{33}I}{j\omega} \tag{7.9.37}$$

and comparing with Eq. (7.9.21) gives

$$\frac{Z_0}{j\sin\dfrac{\omega d}{c_t^D}} = Z_3 + C_0^2h_{33}^2\left[Z_4 + \frac{1}{j\omega C_0}\right] \tag{7.9.38}$$

and

$$\frac{Z_0}{j\tan\frac{\omega d}{c_t^D}} = Z_2 + Z_3 + C_0^2 h_{33}^2\left[Z_4 + \frac{1}{j\omega C_0}\right] \tag{7.9.39}$$

Rewriting Eq. (7.9.32) and comparing with Eq. (7.9.20) gives

$$\frac{Z_0}{j\tan\frac{\omega d}{c_t^D}} = Z_1 + Z_3 + C_0^2 h_{33}^2\left[Z_4 + \frac{1}{j\omega C_0}\right] \tag{7.9.40}$$

From Eqs. (7.9.39) and (7.9.40)

$$Z_1 = Z_2 \tag{7.9.41}$$

A solution results if

$$Z_4 = -\frac{1}{j\omega C_0} \tag{7.9.42}$$

Then, from Eq. (7.9.38)

$$Z_3 = \frac{Z_0}{j\sin\frac{\omega d}{c_t^D}} \tag{7.9.43}$$

and from Eq. (7.9.40)

$$\begin{aligned} Z_1 = Z_2 &= \frac{Z_0}{j\tan\frac{\omega d}{c_t^D}} - \frac{Z_0}{j\sin\frac{\omega d}{c_t^D}} \\ &= \frac{Z_0}{j}\left[\frac{1}{\tan\alpha} - \frac{1}{\sin\alpha}\right] \\ &= \frac{Z_0}{j}\left[\frac{\cos\alpha - 1}{\sin\alpha}\right] \end{aligned} \tag{7.9.44}$$

But

$$\frac{\cos\alpha - 1}{\sin\alpha} = \frac{\cos^2(\alpha/2) - \sin^2(\alpha/2) - 1}{2\sin(\alpha/2)\cos(\alpha/2)} = \frac{-2\sin^2(\alpha/2)}{2\sin(\alpha/2)\cos(\alpha/2)} = -\tan(\alpha/2) \tag{7.9.45}$$

and thus, from Eq. (7.9.44)

$$Z_1 = Z_2 = jZ_0 \tan\frac{\omega d}{2C_t^D} \tag{7.9.46}$$

The results of this section are summarized in Fig. 7.13.

We see that this circuit contains an electromechanical transformer with a "turns ratio," which is the product of the piezoelectric h-constant and the static capacitance of the crystal ($N = h_{33}C_0$). This turns ratio is therefore a dimensioned quantity, the dimensions being such as to transform mechanical into electrical quantities and vice versa. On one side of the transformer are purely mechanical quantities; on the other side, electrical. The negative capacitance arises from the distributed-constant nature of the problem. It can be ignored for low-coupling materials like quartz, but must be included for high-coupling ceramics.

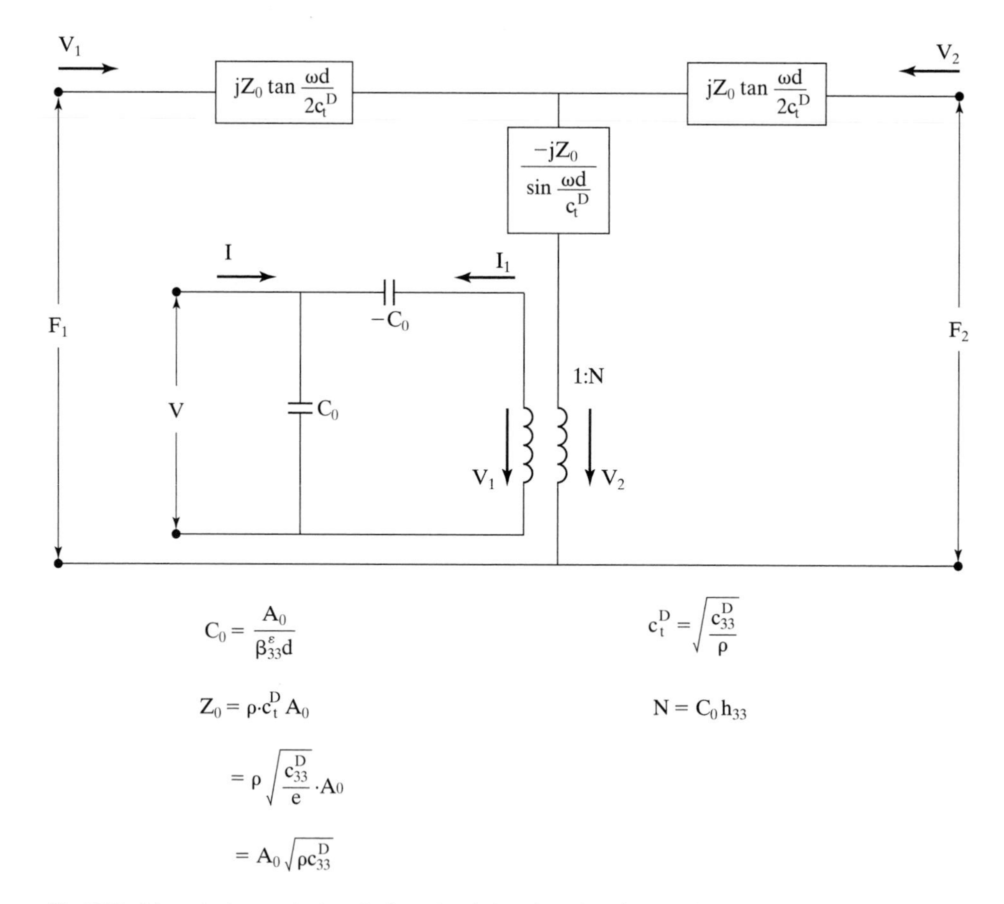

$$C_0 = \frac{A_0}{\beta_{33}^{\varepsilon} d}$$

$$c_t^D = \sqrt{\frac{c_{33}^D}{\rho}}$$

$$Z_0 = \rho \cdot c_t^D A_0$$

$$N = C_0 h_{33}$$

$$= \rho\sqrt{\frac{c_{33}^D}{e}} \cdot A_0$$

$$= A_0\sqrt{\rho c_{33}^D}$$

Fig. 7.13 Mason's six-terminal equivalent circuit for piezoelectric transducer operated as a thickness expander. The same circuit also represents the element operated in the thickness shear mode if we imagine the axis rotated so that x_1 becomes the thickness direction with x_3 in the plane of the plate. c_t is thus replaced by c_s.

Note that there are three ports (six terminals): two are mechanical, which refer to the two crystal faces (either or both of which may carry a load), and one is electrical, where either an active electromotive force and current are applied (when the transducer is operated as a transmitter) or the terminals are closed by an electrical impedance (when used as a receiver).

Note also that because of the symmetrical nature of the electromechanical constants and the reciprocity implied by them, the same equivalent circuit can be used to represent the transducer when it is being used as either as a transmitter or a receiver. In some applications, the same transducer is often required to function as both transmitter and receiver.

The mechanical part of the circuit is simply the representation of a mechanical transmission line, as in Section 4.3, the elements being frequency-dependent. It is necessary to retain this general form of circuit, since in echo ranging work, short pulses are used, which implies that the transducer must pass a wide band of frequencies.

In operation, we can terminate two of the three outlet ports, thereby converting the circuit into a two-terminal network. We will now consider how to use an equivalent circuit of this sort.

In most equipment only one of the mechanical ports is used; that is, we send out and receive ultrasonic signals from one face of the element only, the other face being either left free to vibrate in air or supplied with a permanent lossy backing. In either case, the back face is subjected to a permanent load, which may be zero or very large, but is constant at all times.

7.10 Effects of Damping

The effects of damping can be included in our analysis quite easily by adding terms to Eqs. (7.9.3) and (7.9.4), thus

$$\sigma_3 = c_{33}^D \epsilon_3 + s_{33}\frac{\partial \epsilon_3}{\partial t} - h_{33} D_3 \tag{7.10.1}$$

and

$$E_3 = -h_{33}\epsilon_3 + \beta_{33}^{\epsilon} D_3 + \frac{1}{r_{33}}\frac{\partial D_3}{\partial t} \tag{7.10.2}$$

where β_{33} is a mechanical damping term and r_{33} is the electrical conductivity. For harmonic oscillations,

$$\frac{\partial \epsilon_3}{\partial t} = j\omega\epsilon_3 \quad \text{and} \quad \frac{\partial D_3}{\partial t} = j\omega D_3 \tag{7.10.3}$$

Now if we let

$$\tilde{c}_{33}^D = c_{33}^D + j\omega s_{33} \tag{7.10.4}$$

and

$$\tilde{\beta}_{33}^{\epsilon} = \beta_{33}^{\epsilon} + \frac{j\omega}{r_{33}} \tag{7.10.5}$$

The tilde now denotes a complex quantity. Then Eqs. (7.10.1) and (7.10.2) can be written in the same form as Eqs. (7.9.3) and (7.9.4):

$$\sigma_3 = \tilde{c}_{33}^{D}\epsilon_3 - h_{33}D_3 \tag{7.10.6}$$

and

$$E_3 = -h_{33}\epsilon_3 + \tilde{\beta}_{33}^{\epsilon}D_3 \tag{7.10.7}$$

We now have a complex sound velocity to replace the expression in Eq. (7.8.4), thus

$$\tilde{c}_t^D = \left[\frac{c_{33}^D}{\rho}\left\{1 + \frac{j\omega s_{33}}{c_{33}^D}\right\}\right]^{1/2} \tag{7.10.8}$$

The former analysis leading to the equivalent circuit is still valid except that the susceptance, elastic constant, and sound velocity are replaced by their complex counterparts.

7.11 Low Frequency Limit: Coupling Factor

Fig. 7.14 is the equivalent circuit of a piezoelectric slab, having a backing with impedance Z_B and radiating into a load with impedance Z_L. As before,

$$Z_a = jZ_0 \tan\frac{\omega d}{2\tilde{c}_t^D}$$

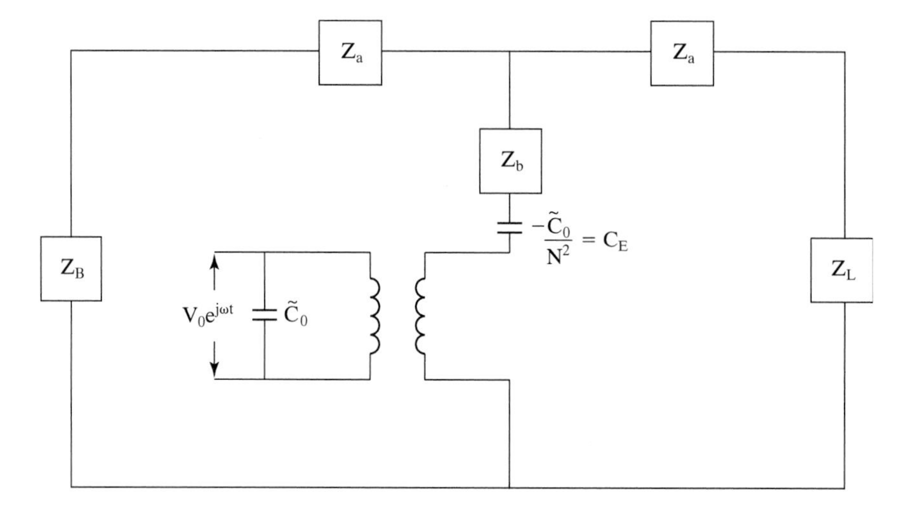

Fig. 7.14 Equivalent circuit of a piezoelectric slab with backing Z_B and load Z_L.

and

$$Z_b = \frac{-jZ_0}{\sin \frac{\omega d}{\tilde{c}_t^D}} \tag{7.11.1}$$

Fig. 7.14 can be reduced as shown in Fig. 7.15. First consider the element $\tilde{C}_0$, which has impedance

$$Z_1 = \frac{1}{j\tilde{C}_0\omega} = \frac{\tilde{\beta}_{33}^{\epsilon} d}{jA_\omega} = -j\frac{\beta_{33}^{\epsilon} d}{A\omega}\left[1 + \frac{j\omega}{\beta_{33}^{\epsilon} r_{33}}\right]$$
$$= \frac{-j}{C_0\omega} + \frac{d}{Ar_{33}} = \frac{1}{jC_0\omega} + R_e \tag{7.11.12}$$

where

$$R_e = \frac{d}{Ar_{33}} = \text{electrical resistance} \tag{7.11.3}$$

Fig. 7.15 can be further reduced using the identity circuits of Fig. 7.16. Hence, we obtain Fig. 7.17, since in this case M = 2 and

$$Z_a + Z_b = jZ_0 \tan\left(\frac{\omega d}{2c_t^D}\right) - \frac{jZ_0}{\sin\left(\frac{\omega d}{c_t^D}\right)} = -jZ_0 \cot\left(\frac{\omega d}{c_t^D}\right) = Z_c \tag{7.11.4}$$

For a freely suspended slab, Z_B and Z_L are approximately equal so that the impedance of the branch containing $Z_L - Z_B$ is much less than the branch containing 2(Za + Zb),

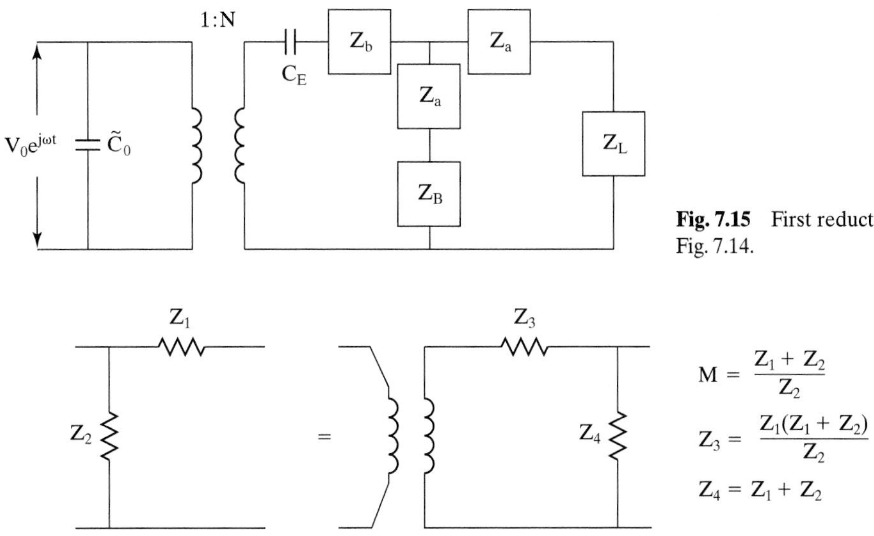

Fig. 7.15 First reduction of Fig. 7.14.

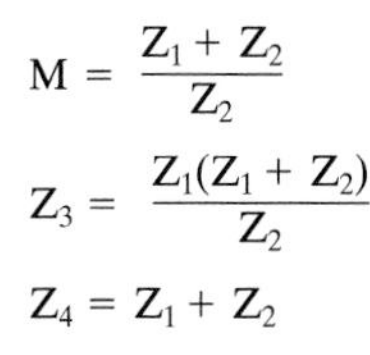

Fig. 7.16 Identity circuits permitting reduction of Fig. 7.15.

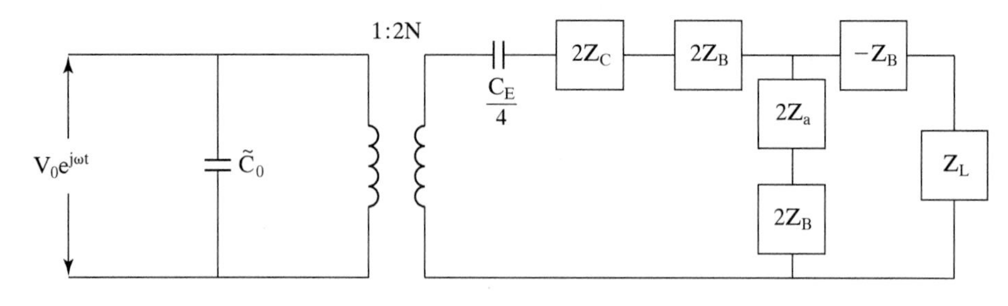

Fig. 7.17 Reduction of Fig. 7.15 using identity circuits of Fig. 7.16. $Z_c = Z_a + Z_b$.

which can thus be neglected, as in Fig. 7.18. At very low frequencies,

$$Z_c \approx -j\frac{Z_0 c_t^D}{\omega d} = -j\frac{S}{\omega} \tag{7.11.5}$$

where S is the stiffness of the slab; that is,

$$S = \frac{Ac_{33}^D}{d} \tag{7.11.6}$$

Finally, by letting

$$\frac{S}{2N^2} = \frac{1}{C_1} \tag{7.11.7}$$

and

$$\frac{1}{4N^2}\left[\frac{As_{33}}{d} + R_B + R_L\right] = R_1 \tag{7.11.8}$$

where R_B and R_L are the resistances of the backing and load, the equivalent circuit is reduced to that shown in Fig. 7.19. Now R_e is small and $-C_0$ and C_1, being a series combination, are equivalent to

$$C_T = \frac{-C_0C_1}{C_1 - C_0} \tag{7.11.9}$$

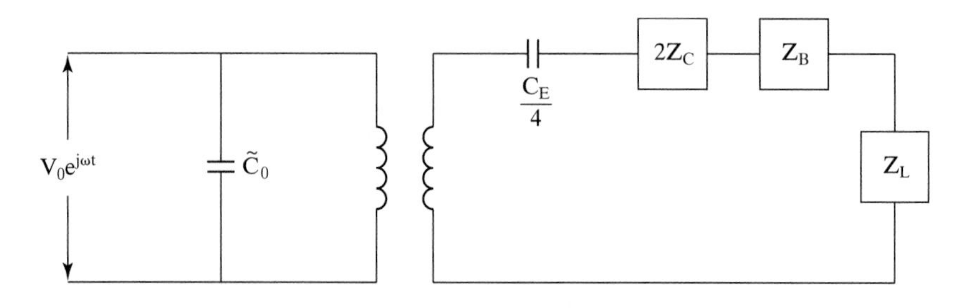

Fig. 7.18 If $Z_B \triangleq Z_L$ in Fig. 7.17, it reduces to this form.

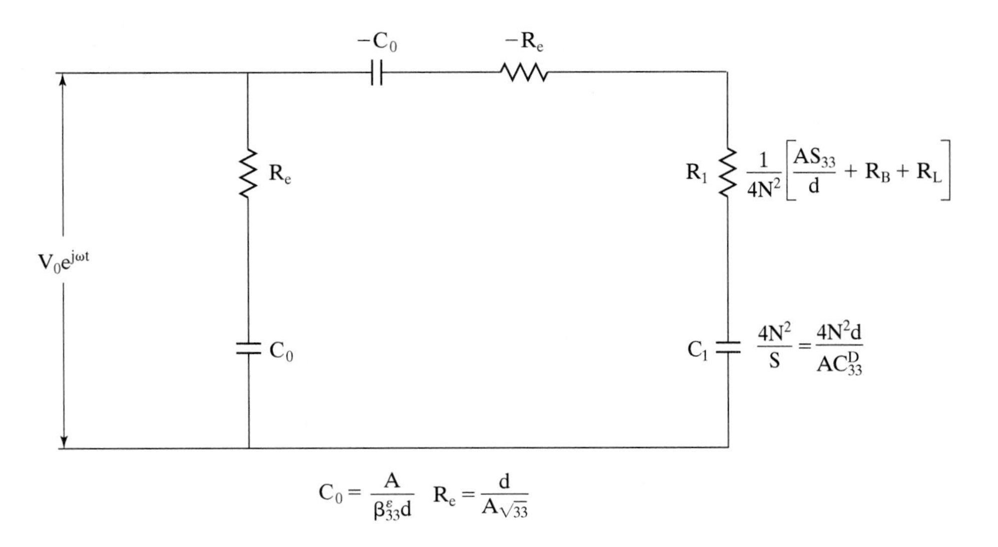

Fig. 7.19 Equivalent circuit at low frequency.

If C_0 is much greater than C_1, then

$$C_T = C_1 \tag{7.11.10}$$

The capacitance dominates this circuit, which effectively reduces to Fig. 7.20. Hence

$$\frac{\text{Energy stored mechanically}}{\text{Total energy stored}} = \frac{C_1}{C_1 + C_0} = k^2 \tag{7.11.11}$$

where k is called the coupling factor. It follows that

$$\frac{1}{k^2} = \frac{C_1 + C_0}{C_1} = 1 + \frac{C_0}{C_1} \tag{7.11.12}$$

and thus

$$\frac{C_1}{C_0} = \frac{k^2}{1 - k^2} \tag{7.11.13}$$

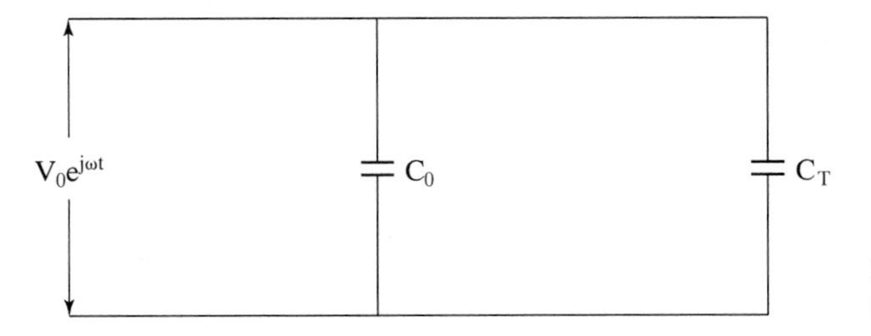

Fig. 7.20 Approximate equivalent circuit at low frequency.

Table 7.2

Material	Coupling Factor(%)
Quartz (x cut)	10
Rochelle salt (45° x cut)	54
ADP (45° z cut)	29
Lithium sulphate (y cut)	35
Barium titanate	40
Tourmaline	10
PZT5	70
PVDF film	19

k can be determined experimentally by a method described in Section 7.12 and values for various materials are given in Table 7.2.

A useful result, which can be proved from Eqs. (7.8.4), (7.9.36), (7.11.6), (7.11.7), (7.11.10), and (7.11.13), relates the coupling factor and the turns ratio, thus

$$\frac{N^2}{C_0\omega} = \frac{\rho c_t^D A k_{33}^2}{kd(1 - k_{33}^2)} \tag{7.11.14}$$

This equation can be used to determine the turns ratio and hence $h_{33}(= N^2/C_0)$. k is the propagation constant c_t^D/ω.

7.12 Equivalent Circuits Valid Near Resonance

A resonance of the element occurs when it is about a half wavelength in thickness; that is,

$$d = \frac{\lambda_R}{2} = \frac{c_t^D}{2f_R} = \frac{\pi c_t^D}{\omega_R} \tag{7.12.1}$$

or

$$\frac{\omega_R d}{2c_t^D} = \frac{\pi}{2} \tag{7.12.2}$$

Hence, approximate values of Z_C are derived as follows:

$$Z_c = -jZ_0 \cot\frac{(\omega_R + \Delta\omega)d}{2c_t^D} \tag{7.12.3}$$

or

$$Z_c = -jZ_0 \cot\left(\frac{\pi}{2} + \frac{\Delta\omega}{\omega_R}\frac{\pi}{2}\right) = jZ_0\frac{\pi}{2}\frac{\Delta\omega}{\omega_R} \tag{7.12.4}$$

In the vicinity of resonance, the impedance can be approximated by a series combination of capacitance C_1 and inductance L_1 having impedance

$$Z = j\frac{2\Delta\omega}{C_1\omega_R^2} \tag{7.12.5}$$

hence, comparing Eqs. (7.12.4) and (7.12.5) gives

$$C_1 = \frac{4}{\pi Z_0 \omega_R} \tag{7.12.6}$$

and

$$L_1 = \frac{1}{C_1\omega_R^2} \tag{7.12.7}$$

In the same frequency range, Z_a approximates an antiresonant parallel combination, which is as follows:

$$\begin{aligned} Z_a &= jZ_0 \tan\left(\frac{\omega d}{2c_t^D}\right) = jZ_0 \tan\frac{(\omega_R + \Delta\omega)\, d}{2c_t^D} \\ &= jZ_0 \tan\left(\frac{\pi}{2} + \frac{\Delta\omega}{\omega_R}\frac{\pi}{2}\right) \approx -jZ_0\frac{2\omega_R}{\pi\Delta\omega} \end{aligned} \tag{7.12.8}$$

Near resonance, the impedance of an antiresonant circuit can be shown to be

$$Z = -j\frac{\omega_A^2 L_2}{2\Delta\omega} \tag{7.12.9}$$

Thus, comparing Eqs. (7.12.8) and (7.12.9), we have

$$L_2 = \frac{8Z_0}{\pi\omega_A}$$

and

$$C_2 = \frac{1}{L_2\omega_A^2} = \frac{\pi}{8Z_0\omega_A} \tag{7.12.10}$$

The equivalent circuit is thus reduced to the form shown in Fig. 7.21. Near resonance again, the impedance of the shunt branch is relatively large, and the circuit reduces to the form shown in Fig. 7.22. The internal damping of the element is now represented by R. Since for an unloaded resonator, $Z_B \approx Z_L \approx 0$, the circuit can be represented as shown in Fig. 7.23. This situation is sometimes referred to as an inertia drive and is important both in measuring the transducer's properties and in frequency selective and control circuits. The mechanical elements have been transferred across the transformer. This circuit was first obtained by K.S. Van Dyke.

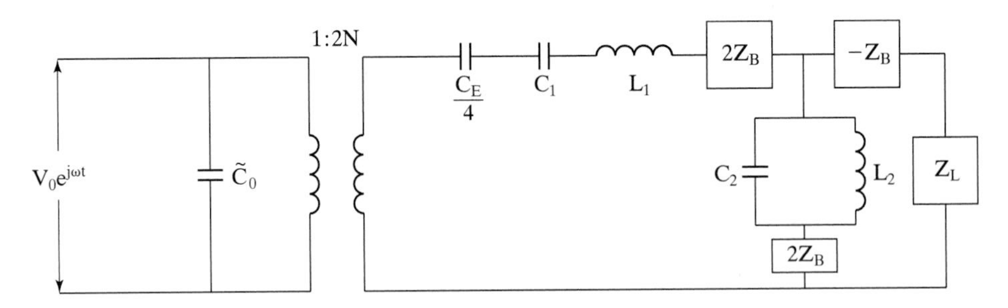

Fig. 7.21 Reduction of Fig. 7.20, valid near the resonance of the element.

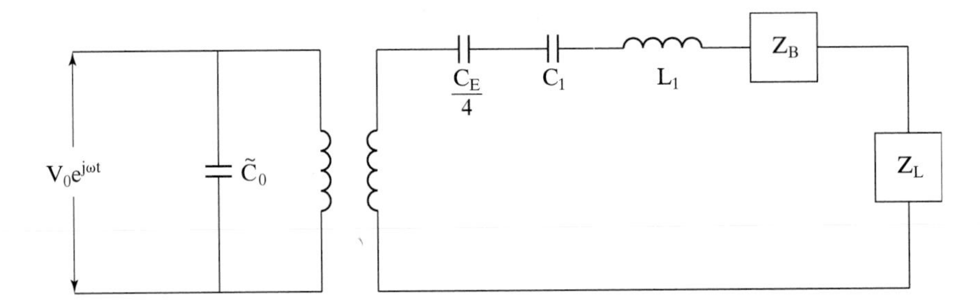

Fig. 7.22 Reduction of Fig. 7.21, valid near resonance.

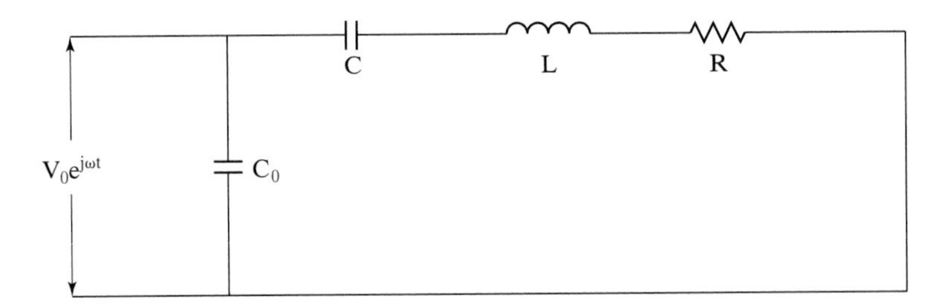

Fig. 7.23 Final reduction of Fig. 7.22 (Van Dyke circuit).

7.13 Resonance: Admittance Loops

As mentioned in Sections 7.11 and 7.12, the thickness mode resonator (and the Van Dyke circuit that describes it) is important because of its many applications and because it is the means of measuring the piezoelectric constants. The same circuit can be used to describe higher order modes, although the parameters are different for each mode. The circuit can be drawn to include several modes as parallel branches, since the impedance of a given branch is effectively infinite at frequencies differing from the resonant value. The circuit can be modified to include the effect of a gap in the event the electrodes are separated from the slab itself.

There are several ways in which the vibrator may be excited. It could be set into a vibration, which is then allowed to decay freely. This is equivalent to putting an initial charge on C, there being no connection between the terminals. The charge will then oscillate in a series circuit $RLCC_1$ with a frequency given by

$$\omega_2 = \frac{1}{LC_f} - \frac{R^2}{4L^2} \text{ where } C_f = \frac{CC_0}{C + C_0} \tag{7.13.1}$$

This situation is called the parallel resonance. Another way to excite the vibrator is with the terminals shorted; in this case C_0 will not participate in the oscillation, which will have the slightly different frequency

$$\omega^2 = \frac{1}{LC} - \frac{R^2}{4L^2} \tag{7.13.2}$$

This is known as the series resonance. The normal way of using the vibrator, however, is with a variable frequency generator connected across its terminals. The network then has two characteristic modes of oscillation: the series resonances and the parallel resonances. It is best to explain these and other matters in terms of impedance and admittance loops. We introduced these concepts in Section 5.4 in Chapter 5 in our discussion of the loudspeaker. In that case, the impedance loop was the natural one to use, since the equivalent circuit was a series combination. Because the piezoelectric vibrator is best represented by the parallel combination of the Van Dyke circuit, the natural form to use in this case is the admittance vector. The total admittance of the vibrator is seen from Fig. 7.23 to be

$$Y' = Y + Y_0 = g - jb - jb_0 \tag{7.13.3}$$

where

$$Y_0 = -jb_0 = j\omega C_0 \tag{7.13.4}$$

and where Y is the admittance of the RLC branch. Now as we know from our discussion in Section 5.4, with variation of frequency the locus of the head of the admittance vector for a series RLC combination is a circular loop with diameter 1/R. We also saw that

$$g = \frac{R}{Z^2} \text{ and } b = \frac{X}{Z^2} \tag{7.13.5}$$

Consequently, the admittance vector in this case is as shown in Fig. 7.24, where A is the origin for Y (the vector representing the RLC branch) while O is the origin for the vector Y' (representing the whole network, including the C_0 in the shunt to the RLC branch). For the slight variation of frequency around the resonance, A is essentially a fixed point. It follows that

$$Y' = \frac{R}{Z^2} - j\left(\frac{X}{Z^2} - \omega C_0\right) \tag{7.13.6}$$

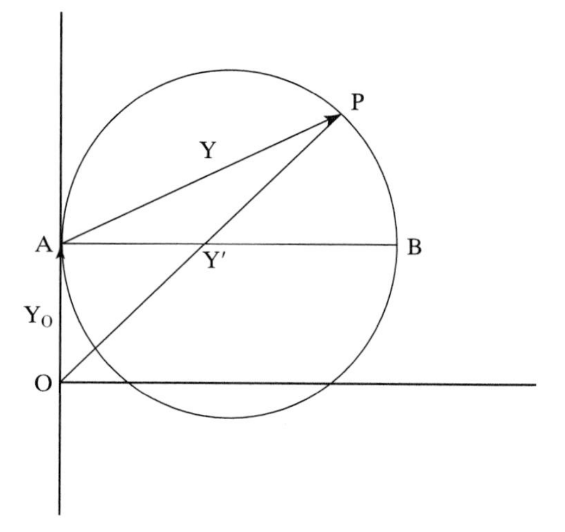

Fig. 7.24 Admittance diagram for thickness vibrator.

Now the network may be regarded as a series combination with impedance given by

$$Z' = \frac{1}{Y'} = R_s + jX_s \tag{7.13.7}$$

where it may be shown that

$$R_s = \frac{R}{(1 - \omega C_0 X)^2 + \omega^2 C_0^2 R^2} \tag{7.13.8}$$

and

$$X_s = \frac{X - \omega C_0(X^2 + R^2)}{(1 - \omega C_0 X)^2 + \omega^2 C_0^2 R^2} \tag{7.13.9}$$

We are now ready to identify the critical frequencies of the vibrator, as shown in Fig. 7.25. The point B corresponds to the resonance of the RLC branch so that

$$\omega = \omega_0 = \frac{1}{\sqrt{LC}} \tag{7.13.10}$$

Points P_1 and P_2 in Fig. 7.25 are the quadrantal frequencies at which the resistance and reactance of the RLC branch are equal. The resonances of the whole network occur when the reactance is zero (i.e., at points P_3 and P_4 in Fig. 7.25). It follows from Eq. (7.13.9) that these are given by the solutions of the expression:

$$X - \omega C_0(X^2 + R^2)$$

$$= L\omega - \frac{1}{C\omega} - \omega C_0 R^2 - \omega C_0\left(L\omega - \frac{1}{C\omega}\right)^2 = 0 \tag{7.13.11}$$

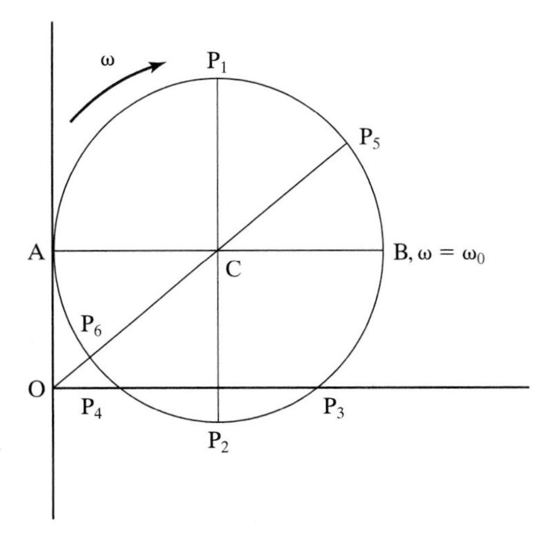

Fig. 7.25 Critical frequency points for thickness vibrator.

After some manipulation in which higher order terms in R^2 are neglected, the two resonance frequencies are found to be given by

$$\omega_s = 2\pi f_s \approx \frac{1}{\sqrt{LC}}\left(1 + \frac{R^2 C_0}{2L}\right) \tag{7.13.12}$$

and

$$\omega_p = 2\pi f_p \approx \frac{1}{\sqrt{LC}}\left(1 + \frac{C}{2C_1} - \frac{R^2(C_0 + C)}{2L}\right) \tag{7.13.13}$$

It follows that

$$\omega_p - \omega_s \approx \frac{1}{\sqrt{LC}}\left(\frac{C}{2C_1} - \frac{R^2(2C_0 + C)}{2L}\right) \tag{7.13.14}$$

When the damping is very small

$$\omega_s \approx \frac{1}{\sqrt{LC}} \text{ and } \omega_p \approx \frac{1}{\sqrt{LC}}\left(1 + \frac{C}{2C_0}\right) \tag{7.13.15}$$

and

$$\frac{\omega_p - \omega_s}{\omega_s} \approx \frac{C}{2C_0} \tag{7.13.16}$$

These results agree with the values for zero damping from Eqs. (7.13.1) and (7.13.2). Eq. (7.13.14) provides us with a way of measuring the coupling factor defined in Eq. (7.11.12). An alternative method is to measure the difference between the frequencies at points P_5

and P_6 in Fig. 7.25, where the admittance has its maximum and minimum values, respectively. In this case, the result is shown by Cady to be similar to that in Eq. (7.13.16):

$$\frac{\omega_n - \omega_m}{\omega_m} \approx \frac{C}{2C_0} \tag{7.13.17}$$

7.14 Transducers for Echo Sounding

Most ultrasonic equipment for nondestructive evaluation (NDE) or medical diagnostics operates with pulses of ultrasound, often very short pulses. These pulses may be rectangular or with a sharp rise and exponential decay. The reason for using short pulses is to achieve high resolution, to enable echoes from objects situated close together to be separated easily, and to shorten the "dead" zone. But the shorter the pulse, the wider the band of frequencies needed to pass it without distortion; very short pulses theoretically require an infinite bandwidth. Without entering into all the reasons, we can say that, if we want to avoid adding additional damping in the form of constraints, the maximum bandwidth obtainable from a resonant transducer is determined by the electromechanical coupling factor k. With a matched electrical termination (i.e., with the static capacitance of the crystal tuned out), the fractional bandwidth, $\Delta\omega/\omega$, cannot exceed $k/(1 - k)^{1/2}$. This is approximately equal to k for the sort of coupling values that are achieved in practice. The bandwidth can be increased beyond this value only by adding mechanical damping or, to some extent, by reducing the output impedance of the drive generator (i.e., driving nearer to constant voltage). This last resort usually increases the electrical losses severely.

Ceramic transducers, because of their high coupling, can be operated without any backing and still have sufficient bandwidth to pass short pulses. But for transducers with low coupling, like quartz, a lossy backing is very desirable if short pulses are to be generated. A typical transducer is shown in Fig. 7.26a.

An efficient mechanical backing should achieve two things. First, it should be an effective resistive mechanical load on the transducer. Second, its geometry should be such that the pulse propagated into it does not return to the crystal to interfere with

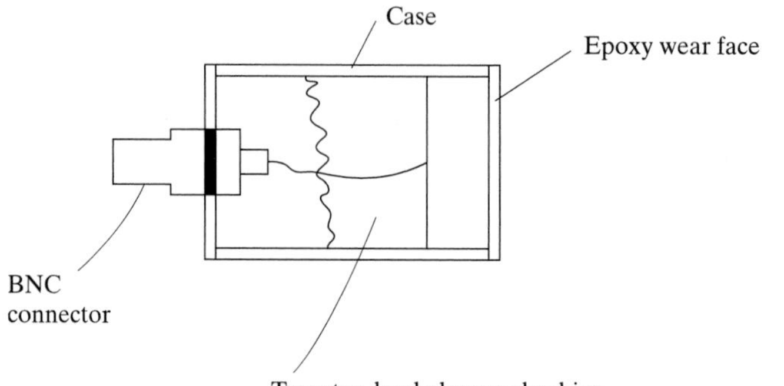

Fig. 7.26a Typical ultrasonic transducer for NDE or medical diagnostics.

the transmitted pulse. This means that the backing should be of a sufficient length (measured in terms of wavelengths) to absorb the lower frequency components of the pulse. As much as possible, the backing material should be matched acoustically to the material of the transducer. Lossy backings of this sort are represented in the equivalent circuit as a high impedance, approximately resistive at all frequencies, placed across one of the mechanical outlets, as indicated in Fig. 7.26b.

No backing is required with high-coupling transducers, and the appropriate terminals of the equivalent circuit are just shorted out. The situation also arises sometimes in ultrasonic applications where energy transfer into the test medium is important. In this case, electroacoustic efficiency becomes an important consideration, and transducers would be operated as highly resonant devices, using very long pulses. In these cases also, of course, the transducer would be left unbacked.

An important use of the equivalent circuit is to permit prediction of the effect of any type of backing or loading of any sort in the response of the transducer, in order to show immediately what types of pulse can be passed with tolerable distortion.

For propagating compressional waves, the work face of the transducer is coupled to the solid load through a liquid film, usually thin oil. In the rarer cases where shear waves are needed, the contact surfaces are either left dry and slightly roughened or a viscoelastic oil or grease is used for coupling. More rarely, an easily detachable rigid bond is used. When shear or surface waves are required, a transducer operated in the thickness-expander mode is commonly used, generating compressional waves that are mode converted at the interface by directing the beam at an appropriate angle.

The load imposed by the test material itself may be very variable. Except where the workpiece is very thin or can resonate at the transducer frequency, the load is often heavy and approximately resistive ($Z_L = (\rho c A)_L$).

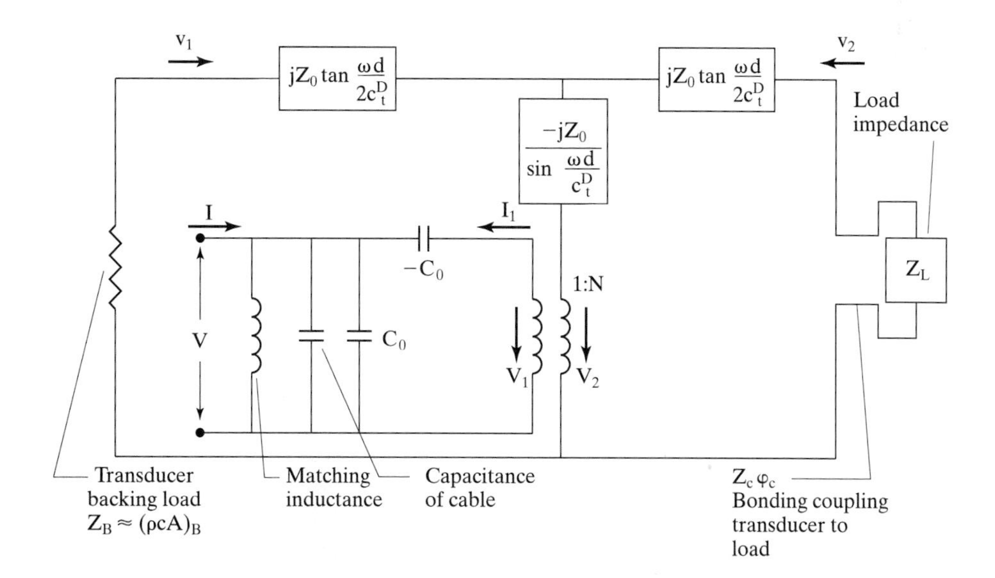

Fig. 7.26b Equivalent circuit for piezoelectric transducer operated as thickness expander, used as a transmitter with both faces loaded, and electrical terminal matched.

The thickness of the oil bond is of great significance, especially at higher carrier frequencies, and has an important effect on the frequency response. The thickness of the bond can be kept down to a small fraction of the wavelength at the frequencies normally used in nondestructive testing. Nevertheless, because of the large difference in acoustic impedance between the liquid bond and the solids on either side of it, it will still have a very significant effect on the performance of the loaded transducer. With the aid of the equivalent circuit, and a knowledge of the various impedances and dimensions, the effect of bonds of any known average thickness and characteristic impedance can be predicted. For example, Fig. 7.27 shows the calculated effect of the thickness of oil bond on the sensitivity of an x-cut quartz crystal coupled to a large steel test piece, as in a typical nondestructive evaluation operation. The abscissa is the phase thickness of the quartz crystal and is therefore proportional to frequency. The curves show that the response changes from a symmetrical form with wide effective bandwidth for zero thickness of the bond to a highly asymmetric form as the bond thickness increases, with narrower bandwidths. The effect is very great even when thickness is small compared with the wavelength. The frequency for maximum response changes with bond thickness, due to the fact that the oil is essentially a reactive load on the transducer. The performance shown by these theoretical curves is fully confirmed from measurement. In the equivalent circuit, the oil bond is represented as a short transmission line. Here, Z_c represents $jZ_c \tan(\omega\ell/v_c)$, where v_c is the velocity of the compressional waves in the bond material and ℓ is the thickness of the bond.

The coaxial cable connection between the transducer and generator is also a transmission line, but very short compared with the wavelength, and therefore represented as a lumped capacitance. This added capacitance will reduce (but usually only slightly) the sensitivity of the transducer.

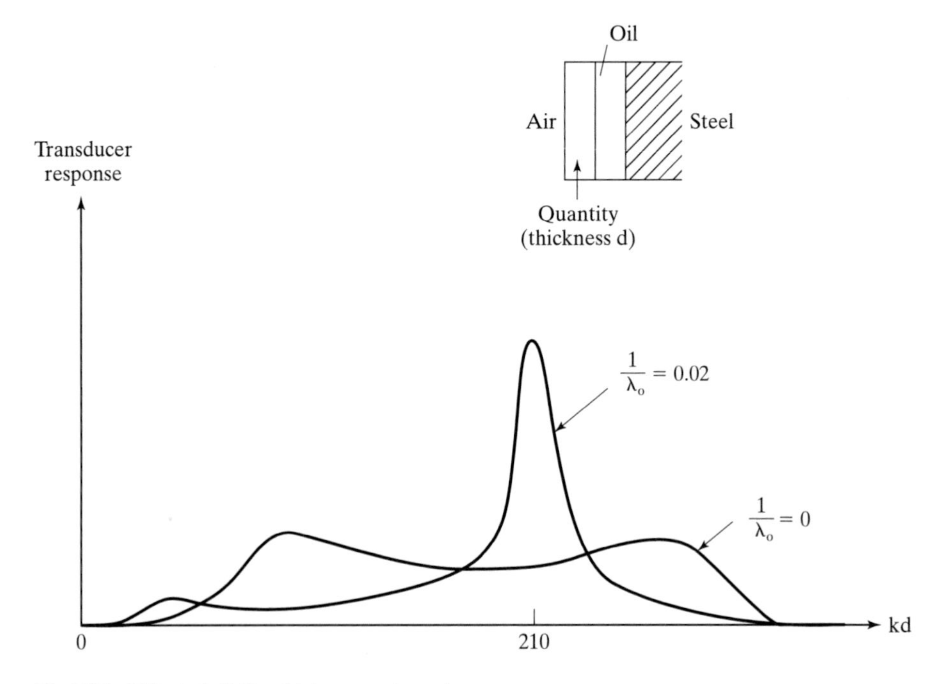

Fig. 7.27 Effect of oil film thickness on transducer response.

Although we are not concerned here with the design of the electronic generator, we should note that proper design of the amplifiers is as important as that of the transducer from the point of view of achieving bandwidth. In other words, it is clearly no use giving the transducer a wide-band response if the amplifiers cannot pass the band.

In a typical NDE operation, the transducer emits a pulse (operating as a transmitter), which is reflected by a flaw and picked up by the same transducer (now operating as a receiver). We are usually interested not in electroacoustic efficiency, but in the overall sensitivity; that is, the ratio of received/transmitted signal amplitude, a voltage ratio. The transmitter and receiver sensitivities can be expressed separately as the ratio of a mechanical to an electrical variable at the transducer terminals, and both can be expressed in terms of any one of the electromechanical constants (d, e, g, h), along with one or another of the elastic or dielectric constants. The overall sensitivity will be proportional to the product of the transmitter and receiver sensitivities. However, the amplitude of the received pulse will depend largely on what happens to the pulse during propagation, the effects of attenuation, and finite reflectivity from the flaw. These are usually unknowable and incalculable, and it is therefore impossible, in the general case, to predict the overall sensitivity and therefore the size of the received pulse.

In most ultrasonic work, detection is required over short distances and short pulses are used. But sometimes long transmission paths are involved in lossy media (e.g., in geological work and oil-well logging). In these cases, low frequencies must be used (in the range 20 kHz to 200 kHz) and electroacoustic efficiency becomes an important consideration. Highly resonant transducers are necessary as no transducer backing is used, the second pair of mechanical terminals being shorted out in the equivalent circuit. The load, often approximately resistive, is placed across the other pair of terminals. In a case of single-frequency operation such as this, a great simplification for the equivalent circuit representation is possible. The circuit elements can be lumped; that is, they can be evaluated at a single frequency. Transforming the mechanical elements over the electromechanical transformer gives a simple two-terminal electrical equivalent similar to that shown in Fig. 7.26. In this case, the motional elements contain contributions from the internal transducer impedance and the load resistance.

7.15 Conclusion

It is very common, especially in transmitters, to operate with two piezoelectric elements glued back to back and poled in opposite directions—a so-called bimorph. The reason is that the electrode at their common face can then carry the driving voltage while the external electrodes that are connected to the other equipment can be grounded. Electrically, the two elements are connected in parallel. A worthwhile exercise would be to prove that the equivalent circuit in this case may be represented as in Fig. 7.28. A discussion of the sandwich construction and double quarter wave horn used in many applications has been given by Bangviwat et al. (1991).

The equivalent circuits of many other configurations have been worked out, including multi-element thickness expander combinations and various types of flexural transducers. Berlincourt, Curran, and Jaffe (1965) discuss these in more detail. Impedance matching is necessary for maximum power transfer to a radiation load and there are various methods for doing this. Magnetostrictive transducers are used in the low ultrasonic

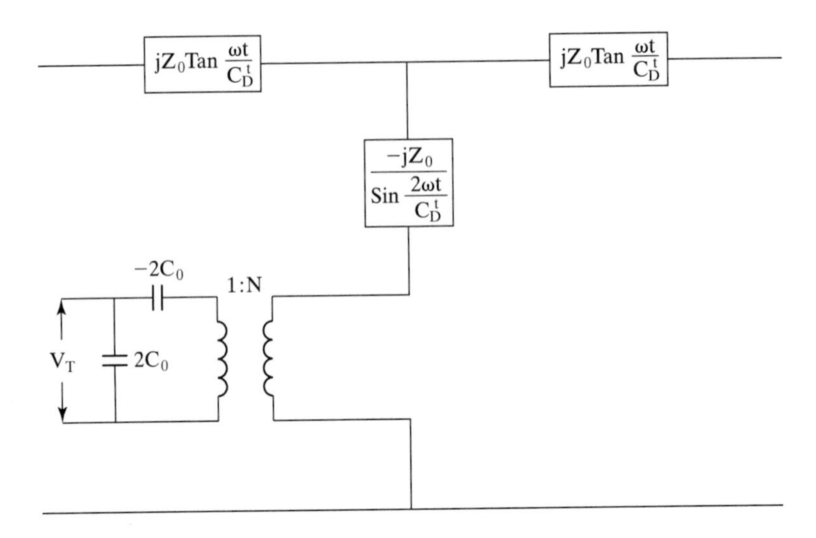

Fig. 7.28 Equivalent circuit of a bimorph.

range for power transmission. These devices have many similarities to their piezoelectric counterparts. Power transducers used in underwater work have been described by Wilson (1988) and Woollett (1989).

REFERENCES AND FURTHER READING

Bangviwat, A., H.K. Ponnekanti, and R.D. Finch. 1991. "Optimizing the Performance of Piezoelectric Drivers That Use Stepped Horns." 90 *J. Acoust. Soc. Am.* 1223–1229.

Berlincort, D.A., D.R. Curran, and H. Jaffe. 1965. "Piezoelectric and Piezomagnetic Materials and Their Function in Transducers." Vol. 1, Part A of *Physical Acoustics*, edited by W.P. Mason. Academic Press.

Cady, W.G. 1964. *Piezoelectricity*. Vols. 1 & 2. Dover.

Feynman, R.P., R.B. Leighton, and M. Sands. 1965. *The Feynman Lectures on Physics*. Vol. 2. Addison Wesley.

Hazen, R.M. June 1988. "Perovskites." *Sci. Am.* 74–81.

Hutson, A.R., and D.L. White. 1962. "Elastic Wave Propagation in Piezoelectric Semiconductors. 33 *J. Appl. Phys.* 40–47.

Jackson, J.D. 1962. *Classical Electrodynamics*. Wiley.

Jaffe, B., W.R. Cook, Jr., and H. Jaffe. 1971. *Piezoelectric Ceramics*. Academic Press.

Katz, H.W. ed. 1959. *Solid State Magnetic and Dielectric Devices*. Wiley.

Kino, G.S. 1987. *Acoustic Waves*. Prentice Hall.

Kyame, J.J. 1954. "Conductivity and Viscosity Effects on the Wave Propagation in Piezoelectric Material." 26 *J. Acoust. Soc. Am.* 990–993.

Martin, G.E. 1964. "On the Theory of Segmented Electromechanical Systems." 36 *J. Acoust. Soc. Am.* 1366–1370.

Mason, W.P. 1935. "An Electrochemical Representation of a Piezoelectric Crystal Used as a Transducer." 23 *Proc. IRE* 1252–1263.

Mason, W.P. 1948. *Electromechanical Transducers and Wave Filters*. 2nd ed. Van Nostrand.

Mason, W.P. 1950. *Piezoelectric Crystals and Their Application to Ultrasonics*. Van Nostrand.

Mason, W.P. 1965. "Use of Piezoelectric Crystals and Mechanical Resonators in Filters and Oscillators." Vol. 1, Part A of *Physical Acoustics*, edited by W.P. Mason. Academic Press.

Merkulov, L.G., and A.V. Kharitonov. 1959. "Theory and Analysis of Sectional Concentrators." 5 *Sov. Phys. Acoust.* 183-190.
Wilson, O.B. 1988. *Introduction to Theory and Design of Sonar Transducers.* Peninsular Publishing.
Woollett, R.S. 1989. *The Flexural Bar Transducer.* Naval Underwater Systems Center, Newport, R.I.

PROBLEMS

7.1 A piezoelectric transducer has mechanical impedance $Z_m = 1000 + j\,(0.4\omega - 2.10^{10}/\omega)$ and clamped electrical impedance $Z_0 = 8 - j.10^8/\omega$. What is its mechanical resonance frequency and Q-value offload? From measurements of the electrical impedance through resonance, the diameter of the motional circle is found to be 4000 ohm. Write the linear electromechanical equations and the four terminal equivalent circuit. What effect would loading the transducer with an impedance $Z_L = 500 + j0 \cdot 1\omega$ have on the resonance frequency and Q-value? What impedance would be measured for the loaded vibrator at the electrical terminals at resonance?

7.2 **a.** A circular PZT 5 transducer is 2 cm^2 in area and 1 mm thick. It is silver coated on each face. Calculate the frequencies f_1 and f_2 for maximum and minimum impedance when the transducer is operated in air.

b. The transducer is now glued to one end of a very long aluminum bar of the same cross-sectional area and used to produce longitudinal waves in the bar. Calculate the impedance, including the series capacity at the resonance frequency f_1, and the efficiency of power conversion from a generator with a 100 ohm output impedance. Estimate the transducer's acoustic bandwidth.

7.3 Now suppose that the bar in Problem 7.2 is 2.5 cm in length. What would be the input impedance at f_1 with a continuous wave drive? Assume this impedance is reactive.

7.4 The transducer described in Problem 7.1 is now loaded on its front surface by contact with water.

a. Calculate the impedance under these circumstances, including the series capacitance at f_1, and the efficiency of power conversion from a generator with a 100 ohm output impedance.

b. Calculate the bandwidth and Q factor of the transducer under these conditions.

7.5 **a.** Give a brief explanation of the piezoelectric effect.

b. Produce an equivalent circuit for a thin slab of piezoelectric material. How could such a transducer be used as an accelerometer?

c. Explain why a piezoelectric accelerometer needs to be used in conjunction with a charge amplifier.

7.6 An accelerometer comprises a circular disc of PZT5, 2 mm in thickness, mounted beneath a circular steel backing 1 cm in thickness. Estimate the resonant frequency of the device. What frequency range would you recommend for its use?

7.7 **a.** Draw the Mason equivalent circuit for a slab of piezoelectric material of thickness d. Define any quantities you use in the circuit.

b. Show how this equivalent circuit can be reduced to a branch with an RLC combination in parallel with a branch containing C_0, the capacitance of the slab (Van Dyke circuit).

c. Show that the coupling factor of the material of the slab can be measured using an admittance loop, by finding either the difference between the resonance frequencies or the difference between the frequencies at the maximum and the minimum of admittance.

d. Describe an application of the thickness mode transducer.

7.8 Prove that the Π and T networks are equivalent if

$$Z_1 = \frac{Z_A Z_B}{Z_B + 2Z_A} \text{ and } Z_2 = \frac{Z_A^2}{Z_B + 2Z_A}$$

Use this theorem to derive the equivalent circuit of a bimorph (i.e., two piezoelectric slabs cemented back to back so that they are polarized in opposite directions), as shown in Fig. P7.8.

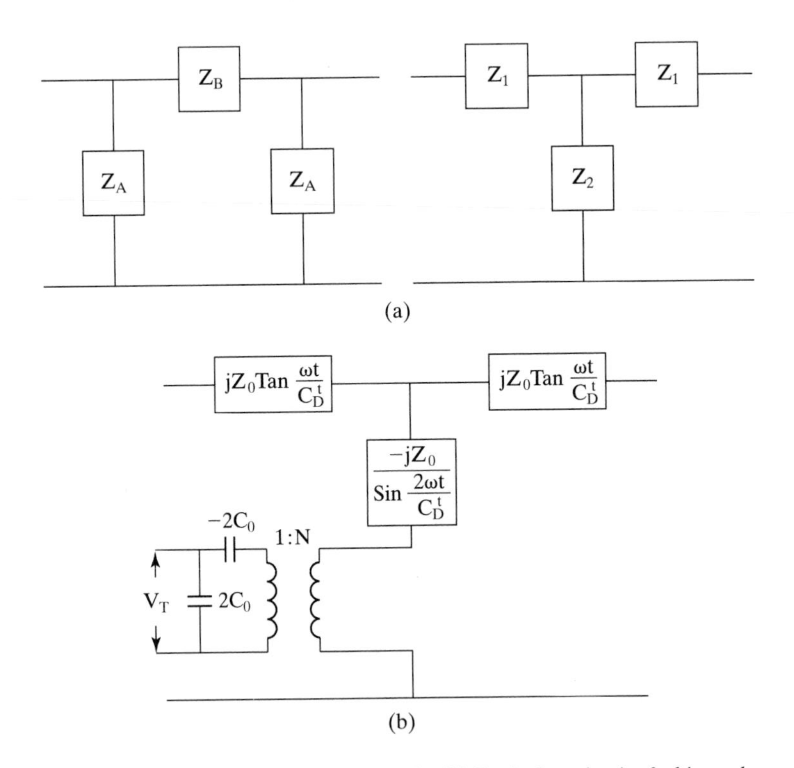

Fig. P7.8 (a) Equivalent Π and T networks. (b) Equivalent circuit of a bimorph.

7.9 Show how to generalize the results of Problem 7.8 to a case of n elements cemented back to back.

7.10 a. Derive a general expression for the low frequency receiving response of a thickness vibrator of area S and thickness t, used as a microphone, when the sound pressure acts on both faces of the element.

b. What is the numerical value of this type of receiving response for a barium titanate wafer of 0.0004 m^2 cross section and 0.002 m thickness?

Chapter 8

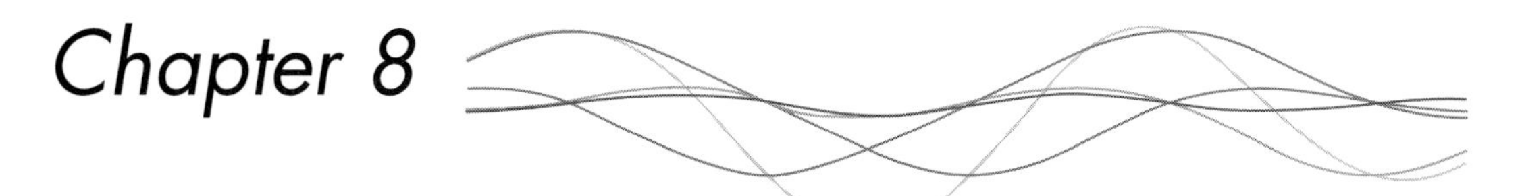

Instrumentation and Signal Processing

8.1 Acoustic Signals

The preceding chapters discussed the nature of elementary acoustic systems and the transducers that allow them to interface with electrical circuits. We will now discuss the nature of the instrumentation and signal processing involved in these circuits. This chapter presents an overview of the more salient features of these subjects; Beckwith, Buck, and Marangoni (1982) provide deeper insight on instrumentation, and Oppenheim and Schafer (1975) on signal processing.

Acoustic disturbances vary continuously with time, and thus transducers produce continuously varying electrical signals. These are examples of "analog" signals, as opposed to "digital" signals, which we will discuss in Section 8.2. Until recently, any processing was carried out directly on the analog signal. For example, we often have to deal with the problem of electrical noise; that is, unwanted signals. If these signals have a frequency content different from the wanted signal, we can eliminate them by using a filter. An electrical filter is a device with a definite frequency response; in other words, almost every electrical device can be regarded as a filter. Filters may be low pass, high pass, or bandpass (Figs. 8.1a–d), as well as active or passive. Active filters contain powered components. For the low pass circuit shown in Fig. 8.2a,

$$\frac{V_0}{V_i} = \frac{-j/(C\omega)}{jL\omega - j/(C\omega)} = \frac{1}{1 - LC\omega^2} = \frac{1}{1 - \left(\dfrac{\omega}{\omega_0}\right)^2} \tag{8.1.1}$$

where $\omega_0^2 = 1/(LC)$. This relationship is shown in Fig. 8.3. The inversion of V_0/V_i above frequency ω_0 simply means that the output is out of phase with the input above resonance. If resistance is also considered, then the response curve will be given by Fig. 8.4. A pass-band in which the upper cutoff frequency is twice the lower cutoff frequency is called an octave band. One-third and one-tenth octave bands are also frequently used, and the

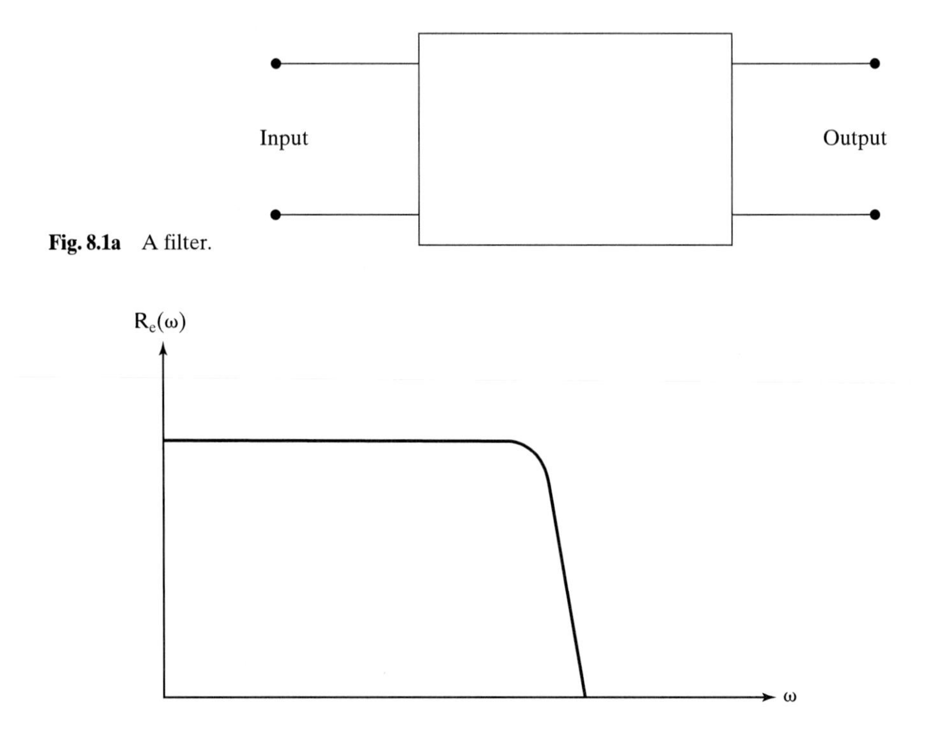

Fig. 8.1a A filter.

Fig. 8.1b Low pass filter characteristic.

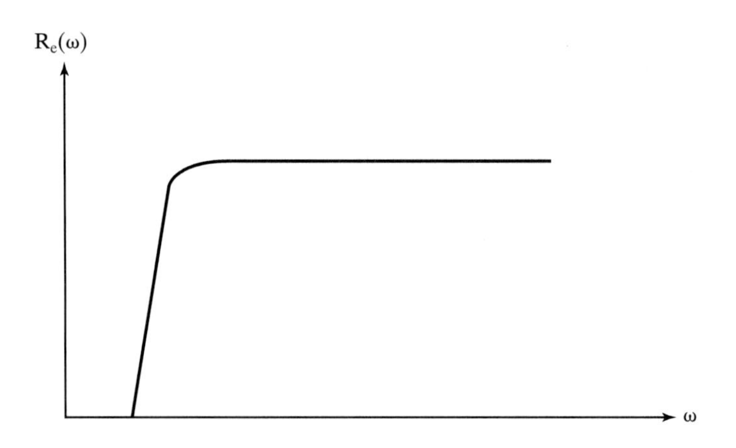

Fig. 8.1c High pass filter characteristic.

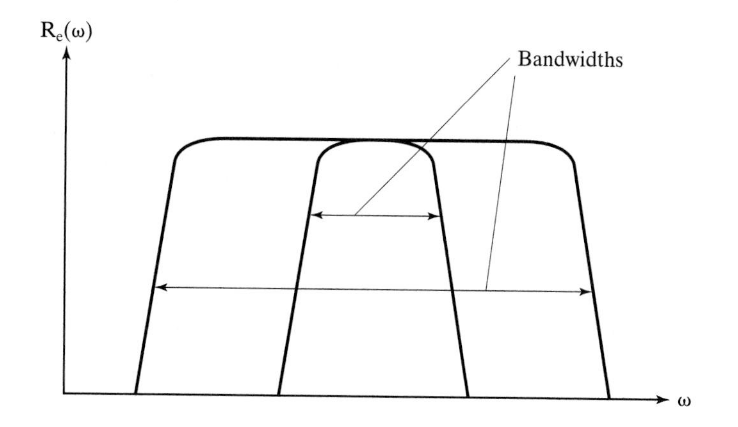

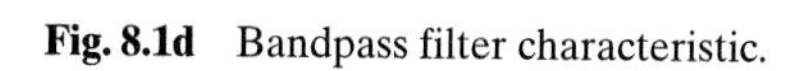
Fig. 8.1d Bandpass filter characteristic.

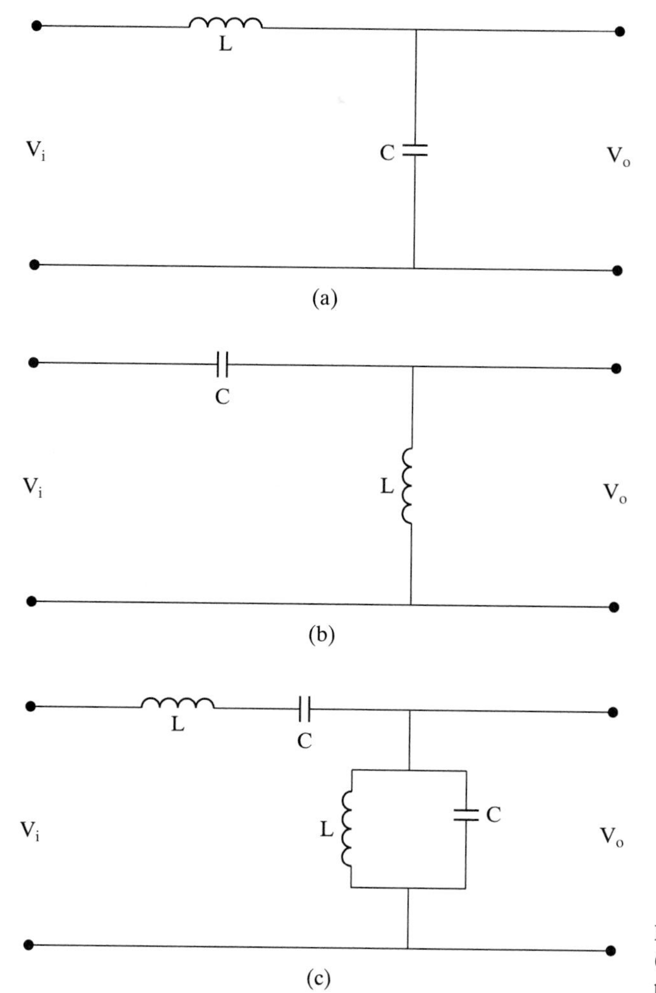

Fig. 8.2 Filter circuits: (a) Low pass; (b) High pass; (c) Bandpass.

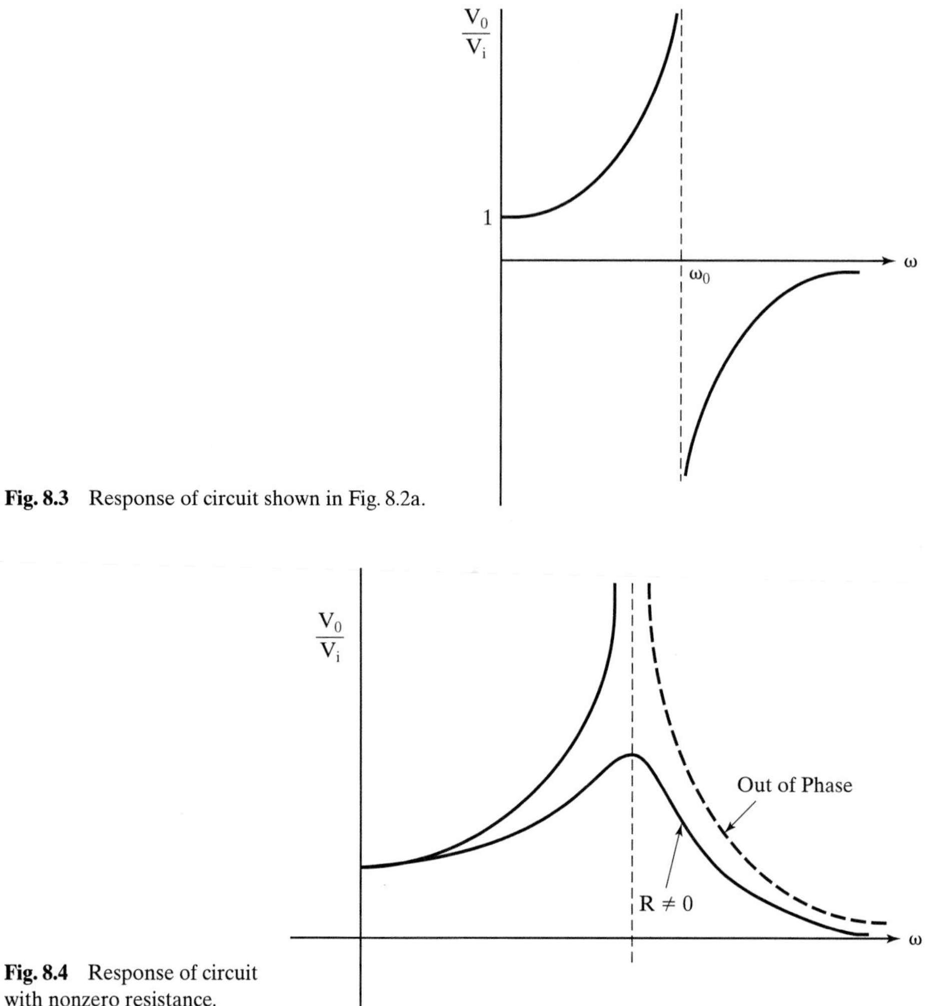

Fig. 8.3 Response of circuit shown in Fig. 8.2a.

Fig. 8.4 Response of circuit with nonzero resistance.

cutoffs are $2^{1/3}$ and $2^{1/10}$ times the lower. Some standard center frequencies for octave and third octave band filters are listed in Table 8.1.

Simple circuits can be used as differentiators and integrators. For example, for the differentiator shown in Fig. 8.5a,

$$I = \frac{V_i}{R + \frac{1}{jC\omega}} = \frac{V_0}{R} \tag{8.1.2}$$

hence

$$\frac{V_0}{V_i} = \frac{R}{R + \frac{1}{jC\omega}} = \frac{jRC\omega}{1 + jRC\omega} \tag{8.1.3}$$

Table 8.1 Center and Approximate Cutoff Frequencies for Standard Set of Contiguous Octave and One-third Octave Bands Covering the Audio Frequency Range

	Frequency, Hz					
	Octave			One-third Octave		
Band	Lower Band Limit	Center	Upper Band Limit	Lower Band Limit	Center	Upper Band Limit
12	11	16	22	14.1	16	17.8
13				17.8	20	22.4
14				22.4	25	28.2
15	22	31.5	44	28.2	31.5	35.5
16				35.5	40	44.7
17				44.7	50	56.2
18	44	63	88	56.2	63	70.8
19				70.8	60	89.1
20				89.1	100	112
21	88	125	177	112	125	141
22				141	160	178
23				178	200	224
24	177	250	355	224	250	282
25				282	315	355
26				355	400	447
27	355	300	710	447	500	562
28				562	630	708
29				708	800	891
30	710	1,000	1,420	891	1,000	1,122
31				1,122	1,250	1,413
32				1,413	1,600	1,778
33	1,420	2,000	2,840	1,778	2,000	2,239
34				2,239	2,500	2,818
35				2,818	3,150	3,548
36	2,840	4,000	3,680	3,548	4,000	4,467
37				4,467	5,000	5,623
38				5,623	6,300	7,079
39	5,680	8,000	11,360	7,079	8,000	8,913
40				8,913	10,000	11,220
41				11,220	12,500	14,130
42	11,360	16,000	22,720	14,130	16,000	17,780
43				17,780	20,000	22,390

Source: After ANSI, 1984.

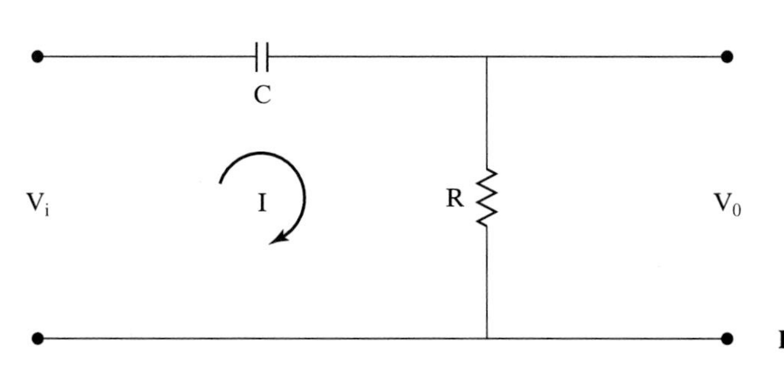

Fig. 8.5a A differentiator.

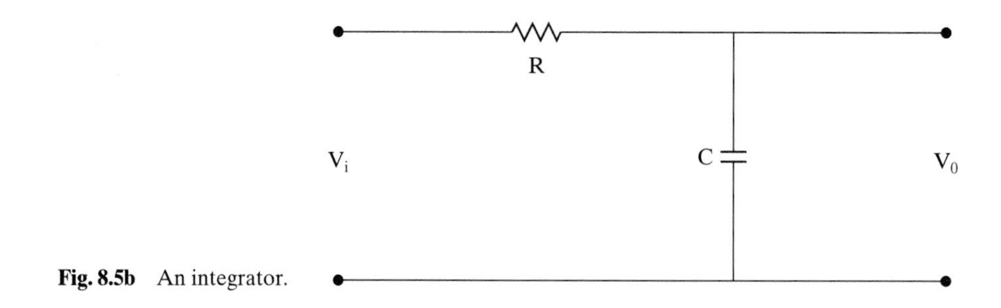

Fig. 8.5b An integrator.

Let RC = τ = time constant of the circuit. If $RC\omega \ll 1$ (i.e., the time constant is short), then $V_o = jRC\omega V_i$; and if V_i is harmonic (or a summation of harmonics), then the output is proportional to the differential of the input. For the integrator shown in Fig. 8.5b,

$$I = \frac{V_i}{R + \frac{1}{jC\omega}} = V_0 jC\omega \tag{8.1.4}$$

thus

$$\frac{V_0}{V_i} = \frac{\frac{1}{jC\omega}}{R + \frac{1}{jC\omega}} = \frac{1}{jRC\omega + 1} \tag{8.1.5}$$

Now if the time constant is long (i.e., $RC\omega \gg 1$), then

$$V_0 = \frac{V_i}{jRC\omega} \tag{8.1.6}$$

and if V_i is harmonic, then V_0 is proportional to the integral of the input.

8.2 Digital Signals

The processing of acoustical signals historically used analog techniques. Analog signals are continuous functions. During the 1970s it became widely realized that there are considerable advantages to processing signals in digital form. Such signals contain information as a series of "bits," or two logic levels that are commonly zero and 5 volts. A 12 V level is sometimes used in place of 5 V. An example of a digital signal is shown in Fig. 8.6. The "bit rate" is determined by a crystal-controlled oscillator or clock. A sequence of bits is called a "word," and an 8-bit word in particular is termed a "byte." Since most acoustical signals originate in analog form, some type of analog-to-digital converter (A/D or ADC) is required. Digital methods are used increasingly because digital instruments can be easily interfaced with and controlled by digital computers, which helps in data reduction. The

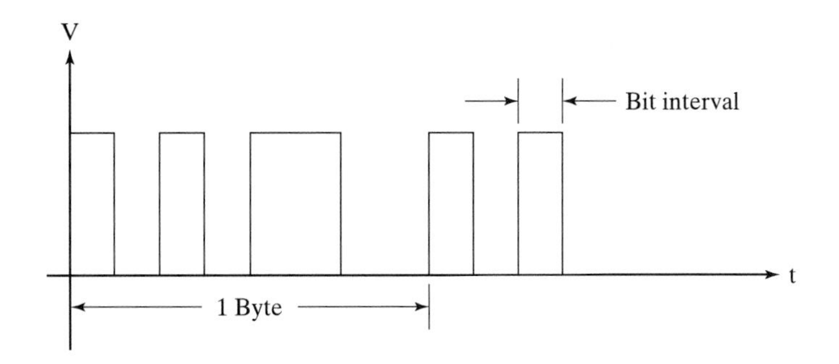

Fig. 8.6 A digital signal.

other very powerful reason for using digital techniques in acoustics is that digital signals are very insensitive to noise because the information is not contained in the signal's amplitude and phase, as is the case with an analog signal, but rather in the sequencing of the bits. It is relatively difficult to obscure this sequencing with noise. Voice signals are now converted to digital form before being sent over telephone wires, and the best sound recordings are digital. Several codes are used to represent alphanumeric data, of which the binary codes are the most common, as shown in Table 8.2.

The Morse code was historically important, but its elements were of uneven length (dot-dash). The most important code today is the ASCII code (American Standard Code for Information Interchange), which has a 7-bit primary form (see Table 8.3).

We will now discuss some of the elementary propositions involving digital signals. The sequence $x(-2), x(-1),\ x(0),\ x(1),\ x(2),\ \ldots$ is represented algebraically as $\{x(n)\}$, formally, where x(n) is the nth number in the sequence, and graphically as in

Table 8.2 Binary Codes

Decimal Digit	Four-Bit Binary	Binary Coded Decimal (BCD)
0	0000	0000
1	0001	0001
2	0010	0010
3	0011	0011
4	0100	0100
5	0101	0101
6	0110	0110
7	0111	0111
8	1000	1000
9	1001	1001
10	1010	1010
11	1011	1011
12	1100	1100
13	1101	1101
14	1110	1110
15	1111	1111

Table 8.3 The American Standard Code for Information Interchange (ASCII)

Binary	Hexadecimal	Character	Binary	Hexadecimal	Character
000 0000	00	NUL			
Nonprinting control characters					
010 0000	20	Space	011 1000	38	8
010 0001	21	!	011 1001	39	9
010 0010	22	“	011 1010	3A	:
010 0011	23	#	011 1011	3B	;
010 0100	24	$	011 1100	3C	<
010 0101	25	%	011 1101	3D	=
010 0110	26	&	011 1110	3E	>
010 0111	27	'	011 1111	3F	?
010 1000	28	(	100 0000	40	@
010 1001	29	)	100 0001	41	A
010 1010	2A	*	100 0010	42	B
010 1011	2B	+	100 0011	43	C
010 1100	2C	,	100 0100	44	D
010 1101	2D	-	100 0101	45	E
010 1110	2E	.	100 0110	46	F
010 1111	2F	/	100 0111	47	G
011 0000	30	0	100 1000	48	H
011 0001	31	1	100 1001	49	I
011 0010	32	2	100 1010	4A	J
011 0011	33	3	100 1011	4B	K
011 0100	34	4	100 1100	4C	L
011 0101	35	5	100 1101	4D	M
011 0110	36	6	100 1110	4E	N
011 0111	37	7	100 1111	4F	O
101 0000	50	P	110 1000	68	h
101 0001	51	Q	110 1001	69	i
101 0010	52	R	110 1010	6A	j
101 0011	53	S	110 1011	6B	k
101 0100	54	T	110 1100	6C	l
101 0101	55	U	110 1101	6D	m
101 0110	56	V	110 1110	6E	n
101 0111	57	W	110 1111	6F	o
101 1000	58	X	111 0000	70	p
101 1001	59	Y	111 0001	71	q
101 1010	5A	Z	111 0010	72	r
101 1011	5B	[	111 0011	73	s
101 1100	5C	\	111 0100	74	t
101 1101	5D	]	111 0101	75	u
101 1110	5E	^	111 0110	76	v
101 1111	5F	-	111 0111	77	w
110 0000	60	‘	111 1000	78	x
110 0001	61	a	111 1001	79	y
110 0010	62	b	111 1010	7A	z
110 0011	63	c	111 1011	7B	{
110 0100	64	d	111 1100	7C	
110 0101	65	e	111 1101	7D	}
110 0110	66	f	111 1110	7E	~
110 0111	67	g	111 1111	7F	Rub Out

Fig. 8.7. For brevity, it is normal to represent the sequence by x(n). A simple digital signal is the delta function or unit sample impulse (Fig. 8.8), defined by

$$\begin{aligned} \delta(n) &= 1, n = 0 \\ &= 0, n \neq 0 \end{aligned} \tag{8.2.1}$$

A delayed unit impulse $\delta(n - k)$ is delayed by k samples (Fig. 8.9). Similarly, $\delta(n + k)$ is a unit impulse advanced by k samples (Fig. 8.10). A more general definition is

$$\begin{aligned} \delta(n - k) &= 1, n = k \\ &= 0, n \neq k \end{aligned} \tag{8.2.2}$$

Then

$$\begin{aligned} \delta(n - 0) = \delta(n) &= 1, n = 0 \\ &= 0, n \neq 0 \end{aligned}$$

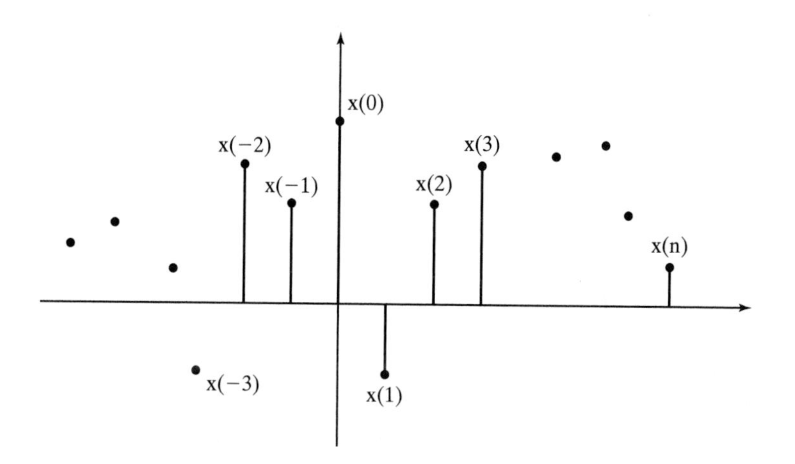

Fig. 8.7 The digital signal or sequence x(n).

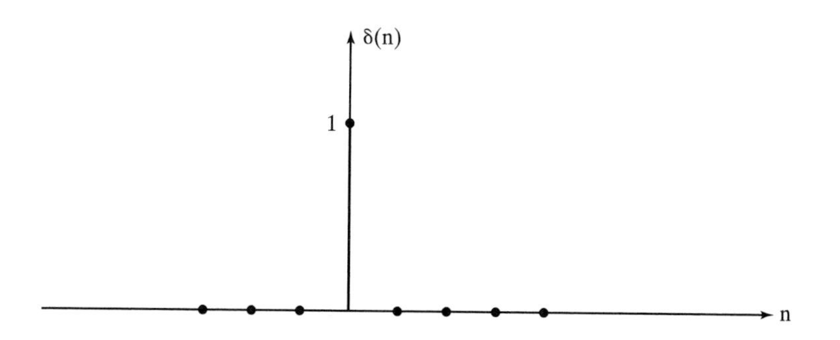

Fig. 8.8 The unit impulse or delta function δ(n).

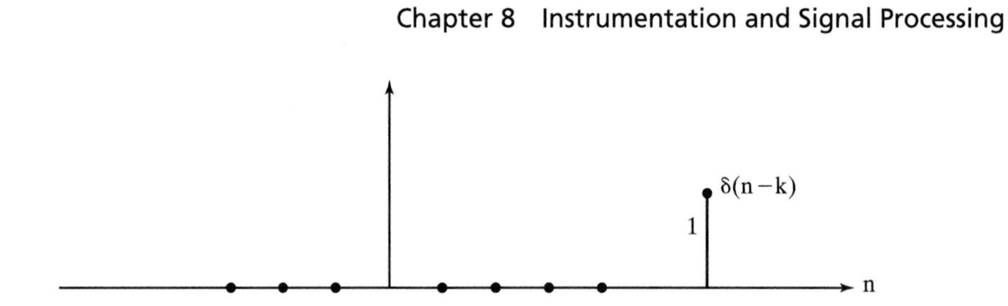

Fig. 8.9 The unit impulse delayed by k samples $\delta(n - k)$.

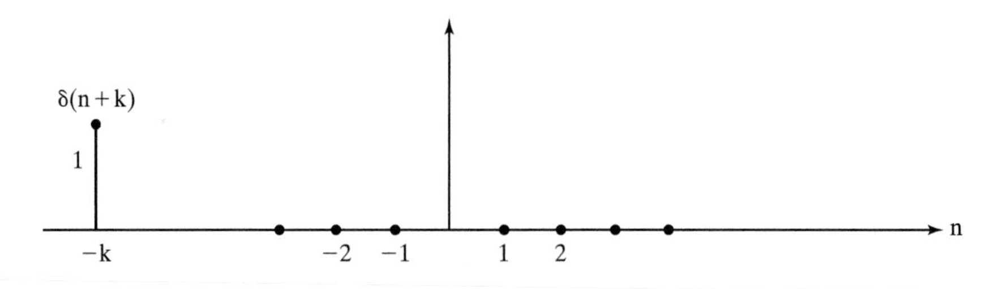

Fig. 8.10 The unit impulse delayed by k samples $\delta(n + k)$.

A unit step sequence (Fig. 8.11) is defined by

$$\begin{aligned} u(n) &= 0, n < 0 \\ &= 1, n \geq 0 \end{aligned} \tag{8.2.3}$$

Any arbitrary sequence $\{x(n)\}$ can be represented by a summation of weighted unit impulses:

$$\sum_{k=-\infty}^{\infty} x(k)\, \delta(n - k) \tag{8.2.4}$$

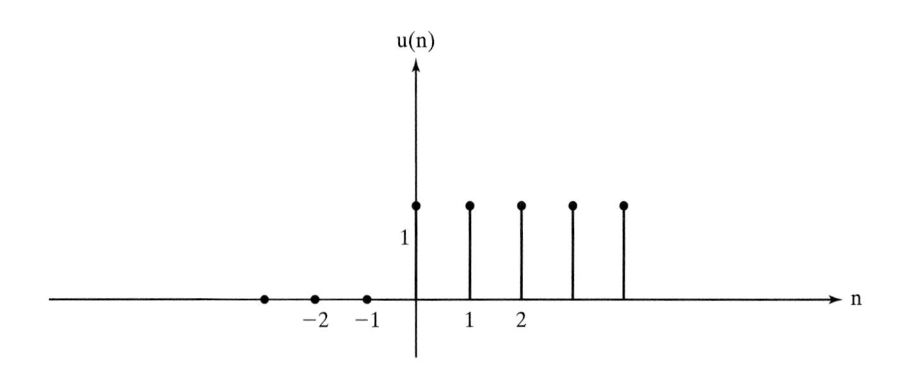

Fig. 8.11 The unit step sequence u(n).

Consider the rotating phasor, $e^{jn\omega}$, then

$$\cos n\omega = R_e\,(e^{jn\omega})$$

and

$$\sin n\omega = I_m\,(e^{jn\omega}) \tag{8.2.5}$$

where ω is the number of radians swept out by the phasor per sampling interval. Cosnω and sinnω are said to be *sinusoidal sequences*. Signals are sometimes classified as

1. One-sided to the future (i.e., a signal with a definite start but no cessation);
2. One-sided to the past (i.e., a signal with no start but a definite cessation); and
3. Two-sided or finite length (Fig. 8.12).

Signals are also classified as causal or noncausal. For causal signals, $x(n) = 0,\ n < 0$; for noncausal signals, $x(n) \neq 0,\ n < 0$.

The "energy" of a sequence is defined as

$$E = \sum_{n=-\infty}^{\infty} |x(n)|^2 \tag{8.2.6}$$

A digital signal is said to be *stable* if it is absolutely summable; that is,

$$E = \sum_{n=-\infty}^{\infty} |x(n)| < \infty \tag{8.2.7}$$

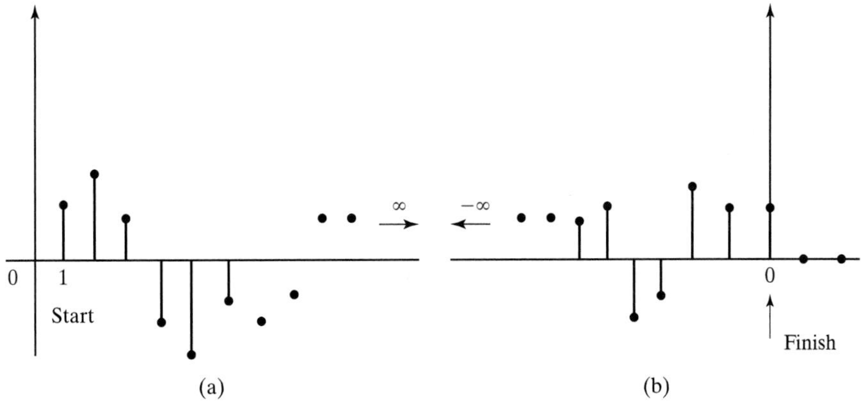

Fig. 8.12 One-sided signals: (a) Signal to the future; (b) Signal to the past.

An alternative, but not equivalent, criterion is that

$$E = \sum_{n=-\infty}^{\infty} |x(n)|^2 < \infty \tag{8.2.8}$$

Finite energy sequences are sometimes referred to as energy signals. Examples of energy signals are the unit impulse and the causal exponential sequence

$$\begin{aligned} x(n) &= 1, \quad n < 0 \\ &= a^n, n \geq 0 \end{aligned} \tag{8.2.9}$$

where a is real and $|a| < 1$. Another useful classification is a division into periodic and aperiodic signals. The sequence x(n) is periodic if there is an integer p so that

$$x(n) = x(n + p) \tag{8.2.10}$$

Signals may also be divided into *deterministic versus random*. A deterministic signal is determined in advance, whereas a random signal is one for which there is uncertainty before it occurs.

A *wavelet* is causal and of finite energy (i.e., it starts at time $t = 0$ and dies away). The causal exponential sequential a_n for $-1 < a < 1$ and $n \geq 0$ is an example of a wavelet. This wavelet is of infinite length, but some wavelets are finite; that is, beyond some index they are of zero value.

A system (or filter or operator) can be represented as a prescribed relationship between two quantities called the system input $\{x(n)\}$ and the system output $\{y(n)\}$, which can be represented diagramatically as follows:

$$\{x(n)\} \rightarrow \{y(n)\} \tag{8.2.11}$$

A system may be a physical entity consisting, for example, of a few simple mechanical or electrical components (either circuit or logic elements) or of something much more complex, such as a computer. Although human beings and their social arrangements may be thought of as systems, we will confine our discussion to the simpler natural and artificial systems. These simpler systems can be represented in a mathematical formalism, and we usually divide the formal representations into linear and nonlinear systems. The system is said to be linear if superposition holds; that is,

i. If $x_1(n) \rightarrow y_1(n)$

and

$$x_2(n) \rightarrow y_2(n)$$

then

$$x_1(n) + x_2(n) \rightarrow y_1(n) + y_2(n) \tag{8.2.12}$$

and

ii. If $x(n) \rightarrow y(n)$

then

$$c\,x(n) \rightarrow c\,y(n) \qquad (8.2.13)$$

where c is in general a complex constant. Conditions (i) and (ii) are said to be the additive and multiplicative properties of a system. The system is said to be shift invariant if the relationship between x(n) and y(n) is independent of the index n. If n is associated with time, the system is time invariant.

A causal system is one whose output does not depend on future values of the input.

8.3 Impulse Response and Convolution

The central problem of digital analysis is determining the response of a system to a given input or excitation, which can be done in various ways:

1. By formal solution of the difference equations characterizing the system, or
2. By the use of various transform techniques, or
3. By analysis based on the response to elementary signals and use of the principle of superposition.

The third approach is limited to linear systems. In the method of convolution, which is based on this approach, any arbitrary input signal is resolved into a set of input impulses. The system's response to each member of the set of input impulses is computed, and the system's total response is obtained by the superposition of the separate responses. The expression that gives the output is called the convolution sum.

Let the response of a linear shift (time) invariant system (Fig. 8.13) to a unit impulse $\delta(n)$ be h(n). Symbolically, we represent this as

$$\delta(n) \rightarrow h(n) \qquad (8.3.1)$$

Since the system is shift invariant,

$$\delta(n-k) \rightarrow h(n-k) \qquad (8.3.2)$$

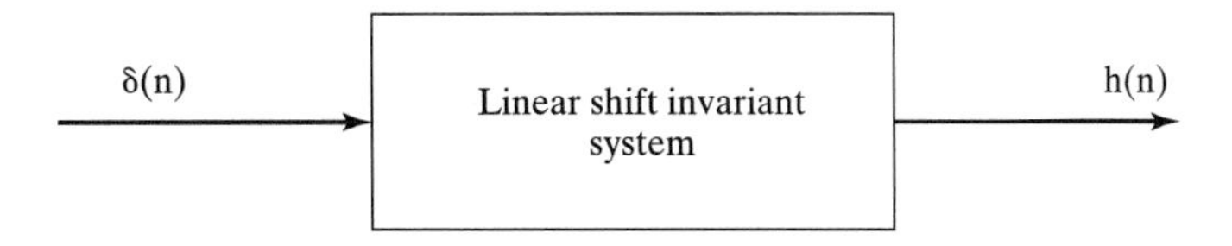

Fig. 8.13 h(n) is the response of a linear shift invariant system to a unit impulse.

Now any arbitrary signal can be resolved into a set of unit impulses; that is,

$$x(n) = \sum_{k=-\infty}^{\infty} x(k)\delta(n - k) \tag{8.3.3}$$

Using the multiplicative property of linear systems,

$$x(k)\,\delta(n - k) \rightarrow x(k)\,h(n - k) \tag{8.3.4}$$

and from the additive property:

$$\sum_{k=-\infty}^{\infty} x(k)\delta(n - k) \rightarrow \sum_{k=-\infty}^{\infty} x(k)h(n - k) \tag{8.3.5}$$

But the left side is x(n); hence

$$y(n) = \sum_{k=-\infty}^{\infty} x(k)h(n - k) \tag{8.3.6}$$

and is the convolution sum. We use a shorthand notation for this summation:

$$y(n) = x(n) * h(n) = \sum_{k=-\infty}^{\infty} x(k)h(n - k) \tag{8.3.7}$$

y(n) is said to be the convolution of x(n) and h(n). Letting $m = n - k$, it follows that

$$y(n) = \sum_{m=-\infty}^{\infty} h(m)x(n - m) \tag{8.3.8}$$

so that

$$y(n) = x(n) * h(n) = h(n) * x(n) \tag{8.3.9}$$

in other words, convolution is commutative.

It is helpful to study an example, such as the convolution for the system with a rectangular wave impulse:

$$h(n) = u(n) - u(n - 4) \tag{8.3.10}$$

and with the input

$$\begin{aligned} x(n) &= n, 0 \leq n \leq 3 \\ &= 0, n > 3 \text{ or } n < 0 \end{aligned} \tag{8.3.11}$$

First, we will perform the convolution $x(n) * h(n)$ or

$$y(n) = \sum_{k=-\infty}^{\infty} x(k)h(n - k) \tag{8.3.12}$$

Treat k as the independent variable and n as a parameter. Convolution then consists of four operations:

1. FOLD the signal h(k) to obtain h(−k) (Fig. 8.14b).
2. SHIFT the signal −k for each value of n to obtain h(n − k).
3. MULTIPLY x(k) and h(n − k) term by term for each value of n (Figs. 8.15a–8.15i).
4. SUM the products over all k to obtain y(n) (Fig. 8.15j).

It is easy to show that the distributive law holds for convolution:

$$[x_1(n) + x_2(n)] * h(n) = x_1(n) * h(n) + x_2(n) * h(n) \tag{8.3.13}$$

The associative law is also followed:

$$[x_1(n) * x_2(n)] * h(n) = x_1(n) * [x_2(n) * h(n)] \tag{8.3.14}$$

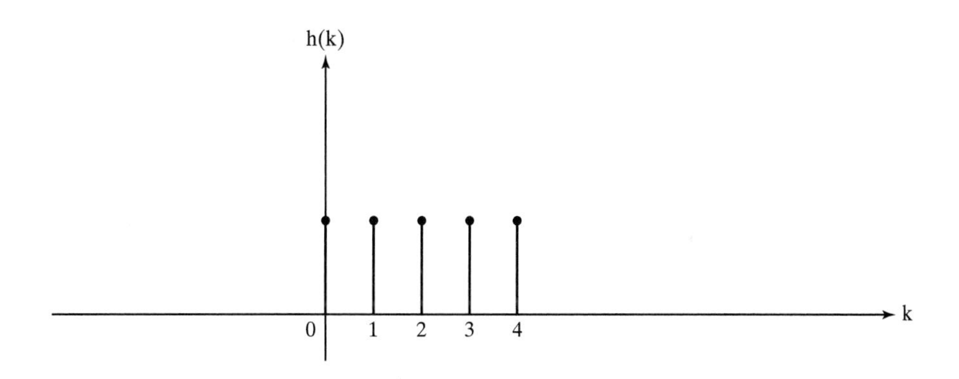

Fig. 8.14a Folding a signal: Impulse response h(k) of a certain system is a rectangular wave.

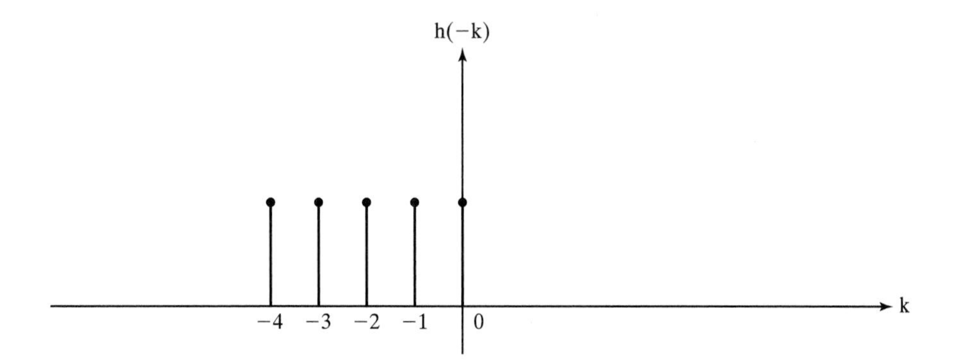

Fig. 8.14b Folding a signal: Folded version of h(k) yields h(−k).

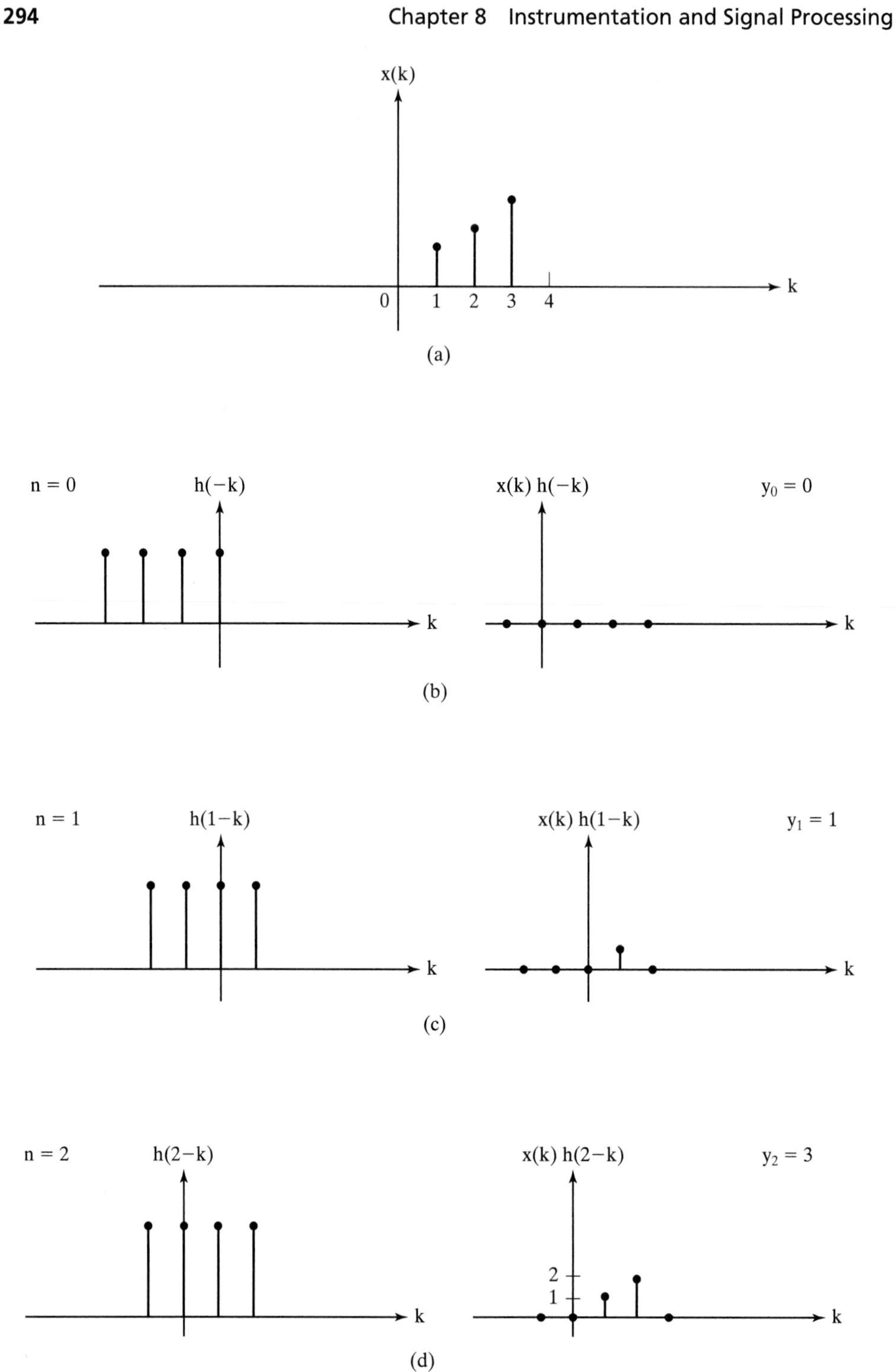

Fig. 8.15 Forming a convolution summation. (After Oppenheim and Schafer, 1975.)

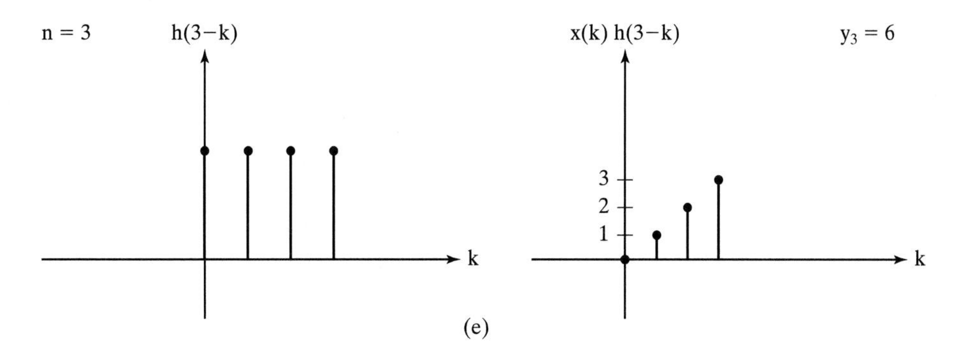

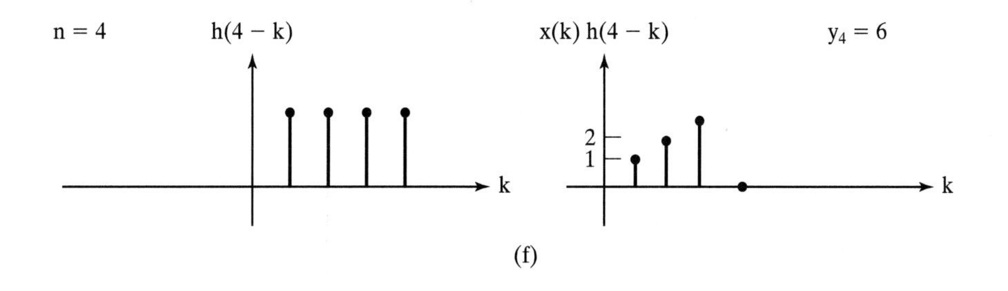

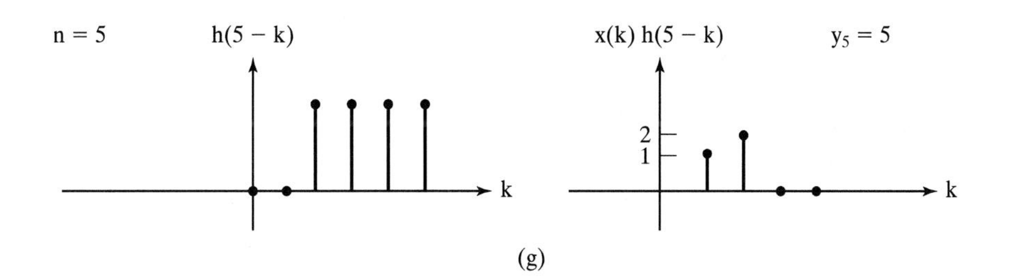

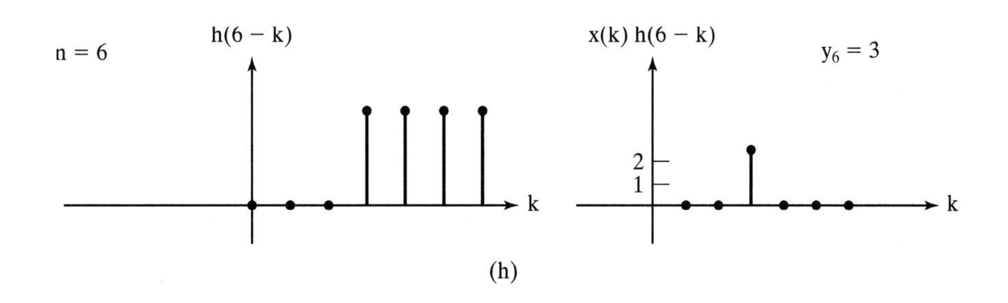

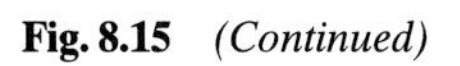
Fig. 8.15 *(Continued)*

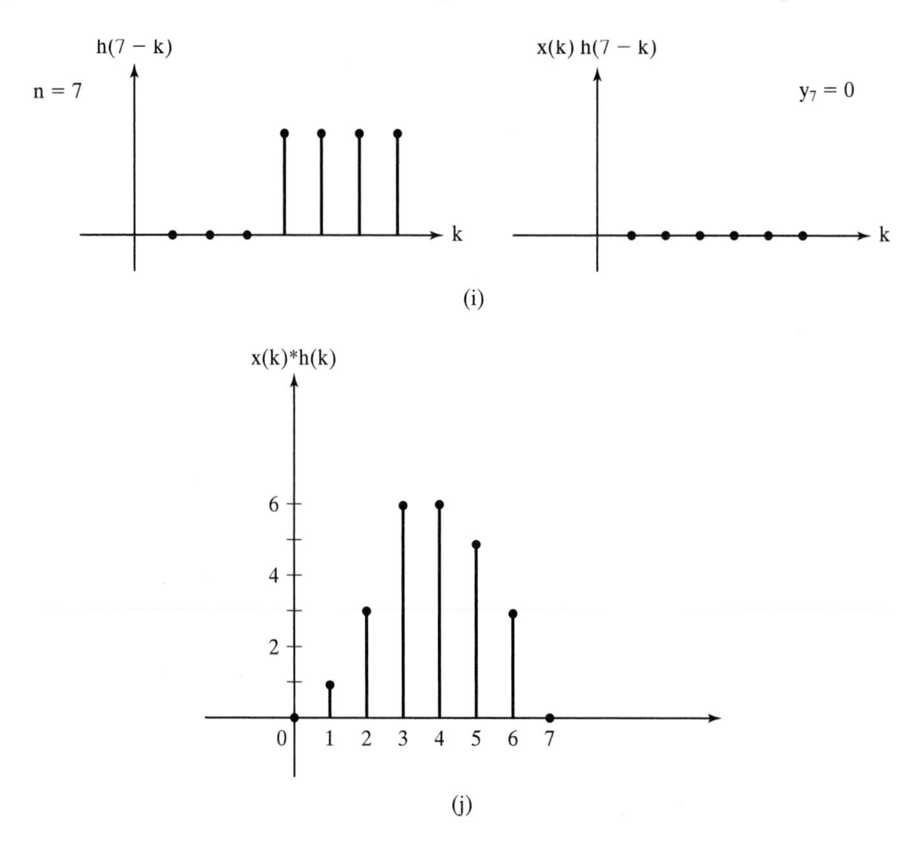

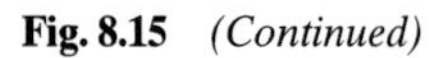
Fig. 8.15 *(Continued)*

Some other properties of convolution sums can be obtained by convolution with unit impulses and unit step sequences. Thus, we see that

$$x(n) = \sum_{k=-\infty}^{\infty} x(k)\delta(n - k) = x(n) * \delta(n) \tag{8.3.15}$$

Consider now

$$u(n) * x(n) = \sum_{k=-\infty}^{\infty} u(k)x(n - k) \tag{8.3.16}$$

Since

$$u(n) = \sum_{m=0}^{\infty} \delta(n - m) \tag{8.3.17}$$

it follows that

$$u(n) * x(n) = \sum_{k=-\infty}^{\infty} \sum_{m=0}^{\infty} \delta(k - m)x(n - k) = \sum_{m=0}^{\infty} \sum_{k=-\infty}^{\infty} \delta(k - m)x(n - k) \tag{8.3.18}$$

Letting $j = k - m$,

$$u(n) * x(n) = \sum_{m=0}^{\infty} \sum_{j=-\infty}^{\infty} \delta(j)x(n - m - j)$$

$$= \sum_{m=0}^{\infty} \sum_{j=-\infty}^{\infty} x(j)\delta(n - m - j) = \sum_{m=0}^{\infty} x(n - m) \qquad (8.3.19)$$

The convolution of two analog functions can be expressed in terms of an integral. For example, if h(t) is the impulse response of a mechanical system, then the response to a generalized forcing function x(t) will be given by

$$y(t) = \int x(\tau)h(t - \tau)d\tau = x(t) * h(t) \qquad (8.3.20)$$

8.4 Amplification and Electronics

The gain of an amplifier is defined as the ratio of output to input. We usually speak of gain for voltages, but we can also use the term in connection with currents or power. Gain is sometimes expressed in decibels:

$$\text{Gain (dB)} = 20 \log_{10}\left[\frac{V_0}{V_i}\right] \qquad (8.4.1)$$

Amplifiers are used in acoustical measurements to provide voltage or power gain or impedance transformation. The first electronic amplifiers to be engineered successfully were vacuum tubes, which are now used very infrequently. Most instrumentation amplification elements are solid state. The basic transistor is a three-element device similar to the triode vacuum tube.

Transistors can be miniaturized and incorporated into integrated circuits. Over time, the complexity of integrated circuits increased until a whole computer could be incorporated into one element, or "chip." Such a chip forms the heart of every "micro" computer or "micro" processor.

The two semiconductors of the greatest importance in electronics are silicon and germanium. Each has four valence electrons. The crystal structure has a tetrahedral pattern in which each atom shares one valence electron with each of four neighbors. At absolute zero, these electrons are tightly bound, there are no free carriers, and silicon is an insulator. At room temperature, a few electrons will have the energy required to become free. There will also be some broken covalent bonds. The region of a broken bond is termed a "hole." We may consider conduction in semiconductors to be due to two separate and independent types of "particles": the negatively charged electrons (n/m^3) and the positively charged holes (p/m^3).

Pure silicon is a poor conductor. We can increase its conductivity in two ways. One way is to raise its temperature as in devices called thermistors. The second way is by "doping" it with impurities, yielding devices that can be "n doped" with a pentavalent element (antimony, phosphorus, or arsenic) or "p doped" with a trivalent element (aluminum, boron, gallium, or indium). Semiconductors, pure or doped, are bilateral (i.e., current will flow in them in either direction). However, at a p-n junction, there is a unilateral carrier density gradient and current will flow only one way. Such a device is called a junction diode. In the junction gate form of field effect transistor (FET), a thin conducting channel exists between two p-n junctions (Fig. 8.16a). The arrow in the symbol in Fig. 8.16b always points to the n-type material. The current from the source to the collector or drain for a given voltage V_{DS} depends on the dimensions of the channel. If the p-n junctions are reverse biased by applying a voltage to the gate, the conductance is dramatically affected (see Fig. 8.17). Thus the FET can be used as an amplifier.

An operational amplifier (op. amp.) is a particular type of integrated circuit (IC) which amplifies a DC differential voltage. By DC we mean that the operational amplifier will operate over a frequency range down to zero. Doing so usually involves at least four stages of amplification and it typically has a gain without feedback of 200,000. It is represented in its ideal form in Fig. 8.18a.

Both V_n and V_p are voltages measured relative to ground. The function of an operational amplifier is to produce an output voltage defined by

$$V_0 = A\,(V_p - V_n) \tag{8.4.2}$$

V_p is termed the noninverted voltage and is associated with the positive symbol. V_n is termed the inverted voltage and is associated with the negative symbol. Operational amplifiers have essentially an infinite input impedance.

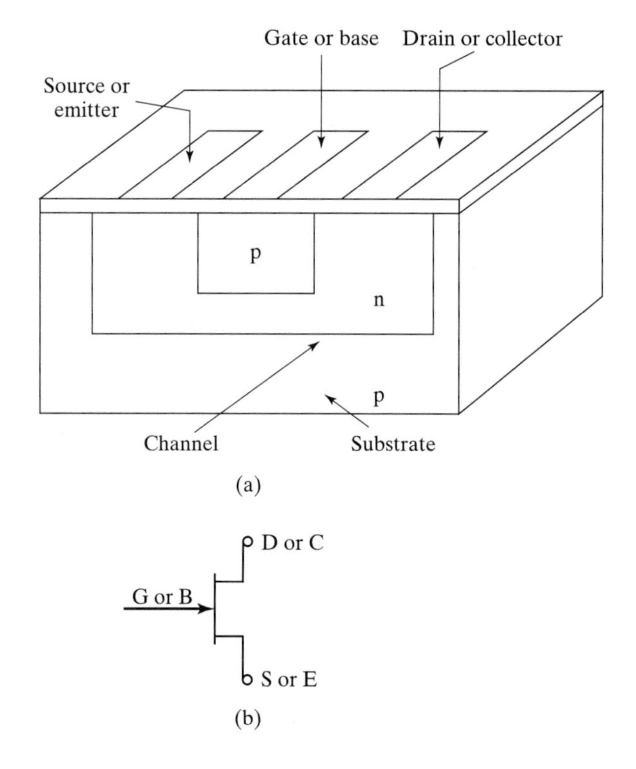

Fig. 8.16 Symbolic representation of a transistor. (a) A field effect transistor. (b) A transistor.

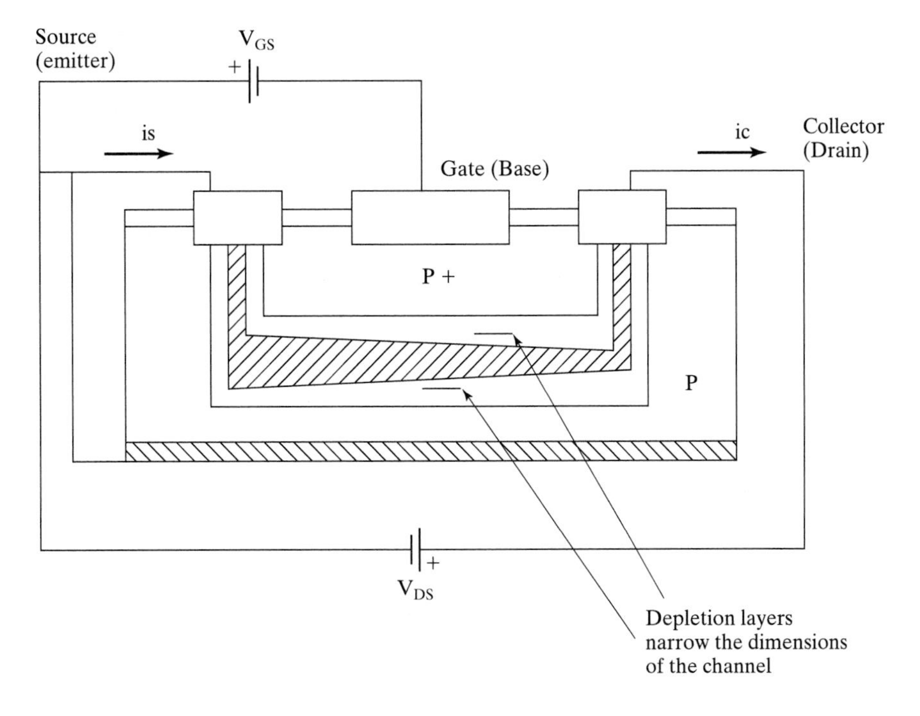

Fig. 8.17 Connections for a field effect transistor.

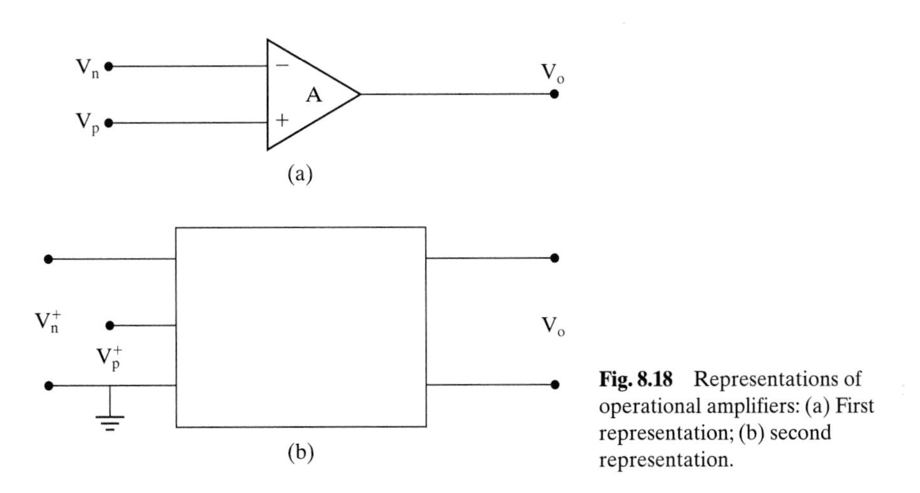

Fig. 8.18 Representations of operational amplifiers: (a) First representation; (b) second representation.

One of the basic uses of an operational amplifier is in an inverting circuit (as in Fig. 8.19). The input signal is applied to the negative or inverting terminal, and the non-inverting terminal is grounded. The output voltage is fed back through resistance R_F. $V_i \ll V_o$ and approaches 0; $i_i = V_i/R_i$ and approaches 0. Thus, the sum of current into node a is

$$i_1 + i_F = \frac{V_1}{R_1} + \frac{V_0}{R_F} = 0 \tag{8.4.3}$$

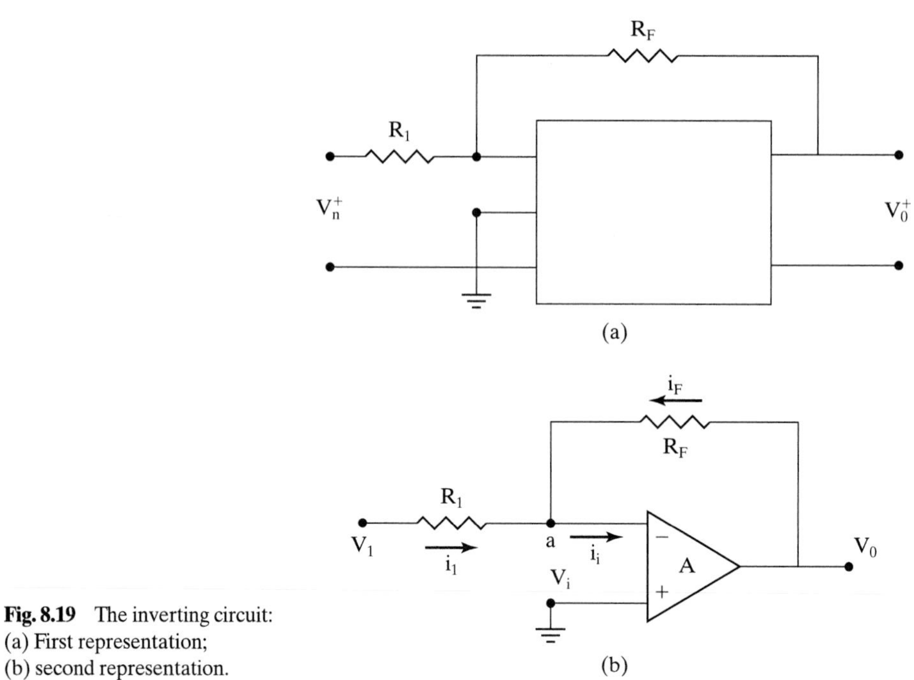

Fig. 8.19 The inverting circuit: (a) First representation; (b) second representation.

thus

$$\frac{V_0}{V_1} = A = \frac{-R_F}{R_1} \tag{8.4.4}$$

Therefore, the output has a sign opposite to the input.

Another important use of operational amplifiers is in voltage amplification. The voltage output of most acoustic transducers is quite low, so we need a means of voltage amplification. Although the operational amplifier was originally designed to amplify DC signals, it was realized that it could be used to amplify pure AC voltages also. One way to do this is with a simple modification of the basic inverting circuit in which a capacitor is added to the input resistance (as in Fig. 8.20). Now

$$\frac{V_0}{V_i} = A = \frac{R_F}{R_1 + \dfrac{1}{jC\omega}}$$

$$= -\frac{R_F C\omega}{R_1 C\omega - j} = \frac{R_F}{R_1}\left(\frac{R_1 C\omega}{R_1 C\omega - j}\right)$$

$$= -\frac{R_F}{R_1}\left(\frac{j\omega}{j\omega - \dfrac{1}{R_1 C_1}}\right) \tag{8.4.5}$$

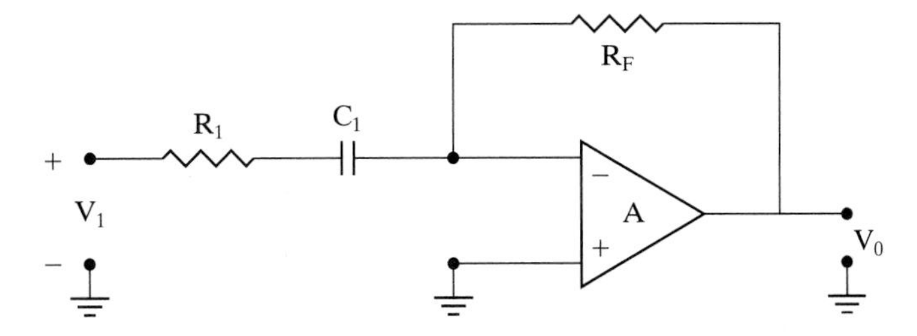

Fig. 8.20 AC amplifier using an operational amplifier. (After Beckwith, Buck, and Marangoni, 1982.)

So the DC gain, as ω approaches 0, will be zero, but the high frequency gain will approach $-R_F/R_1$. The cutoff frequency is given by

$$\omega_c = \frac{1}{R_1 C_1} \tag{8.4.6}$$

The last application of operational amplifiers that we will consider is the charge amplifier. Some transducers, including capacitor microphones and piezoelectric devices, produce a charge that varies in proportion to the transduced quantity. They may be connected to the high-input impedance operational amplifier to produce a usable output voltage. The equivalent circuit of such a transducer may be represented as shown in the dashed box in Fig. 8.21. The varying charge is

$$\Delta q = \Delta C_1 E \tag{8.4.7}$$

connected to the varying input of an operational amplifier, and flowing onto the feedback capacitor C_F. Hence

$$V_0 = -\frac{\Delta C_1 E}{C_F} \tag{8.4.8}$$

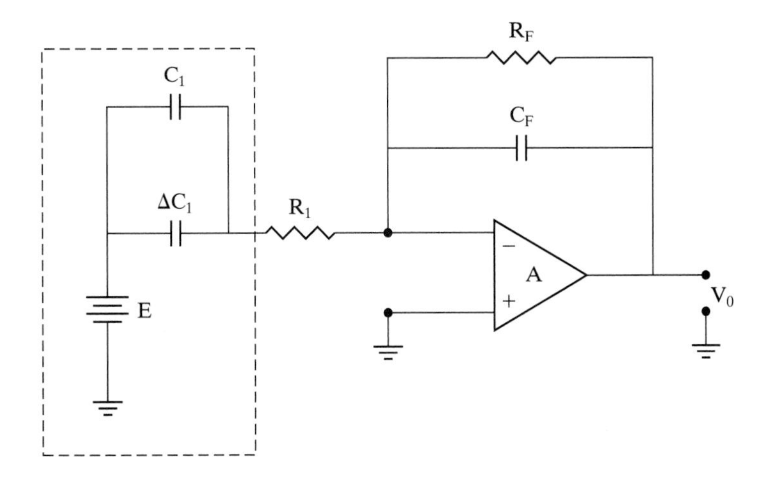

Fig. 8.21 A charge amplifier. (After Beckwith, Buck, and Marangoni, 1982.)

The gain or sensitivity is then

$$\frac{V_0}{\Delta C_1} = -\frac{E}{C_F} \tag{8.4.9}$$

8.5 Basic Digital Circuit Elements

Basic logic elements consist of switches or transistors (see Fig. 8.22). Many special purpose chips have been developed, each consisting of a combination of many circuit elements. Shorthand symbols have been devised for some of the more common ones, as shown in Fig. 8.23. Medium-scale integrated (MSI) chips may contain 50 to 100 elements, and large-scale integrated (LSI) chips may have more than 100 elements. For example, a NAND gate actually contains four transistors, three resistors, and a diode, as shown in Fig. 8.24. NAND gates with as many as eight inputs are available, permitting 256 input combinations. The AND gate is on when both inputs are on. The small circle in the symbol indicates negation. The NAND gate is the negation of an AND gate. The OR gate is on when either input or both are on, and the NOR is its negation. The exclusive OR gate is on when either one or the other is on, but not both. The exclusive NOR gate is the negation of the OR gate. Finally, note that the **inverter** can be used as a NOT gate. It can be shown that all operations in Boolean logic can be built up from AND, OR, and NOT operations.

There are several groups of integrated chips. Common groups are

resistor-transistor-logic (RTL)
diode-transistor-logic (DTL)
transistor-transistor-logic (TTL)
complementary-metal-oxide-semiconductor (CMOS)

For example, the TTL family consists of more than 150 different types of chips. They all look outwardly similar but each has a different function.

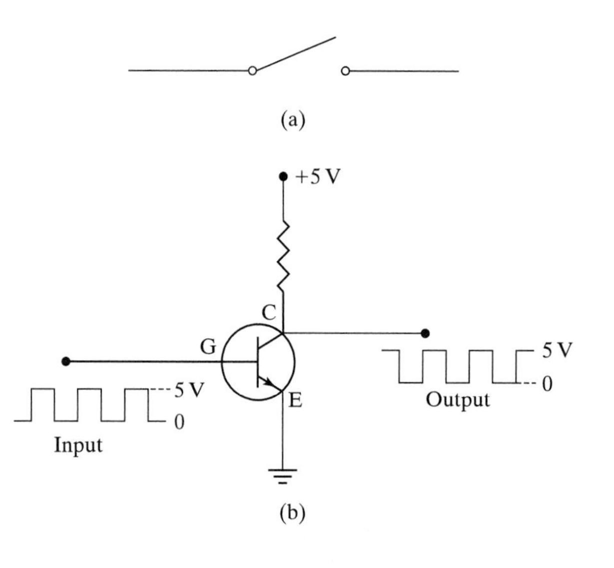

Fig. 8.22 Basic logic circuits. (a) Switch. (b) Transistor used as a diode. (After Beckwith, Buck, and Marangoni, 1982.)

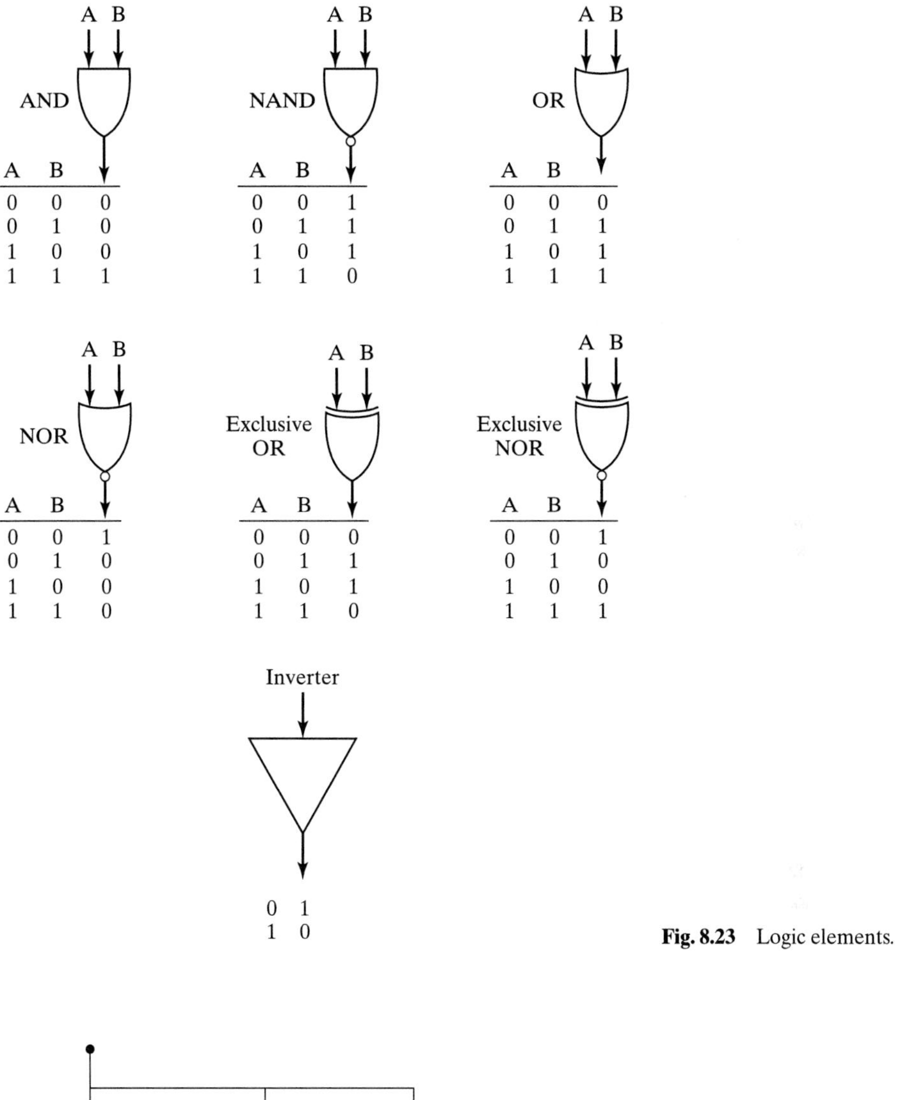

Fig. 8.23 Logic elements.

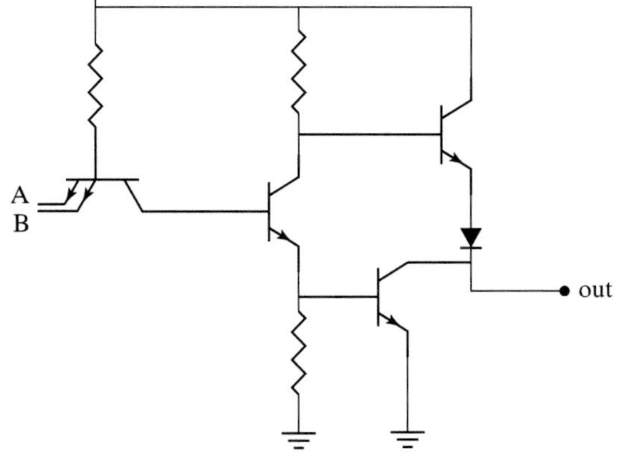

Fig. 8.24 Circuit of a NAND gate.

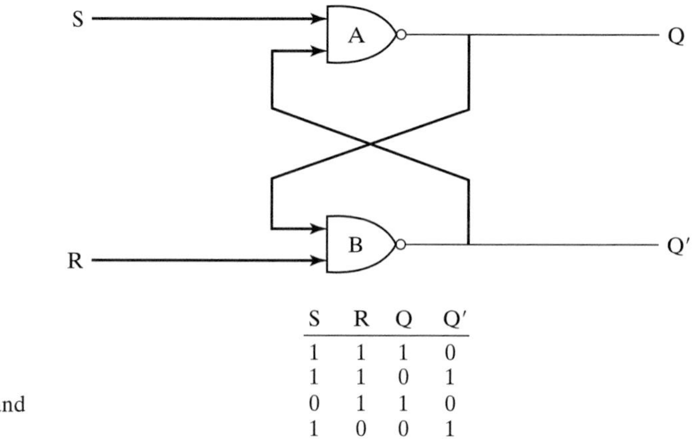

S	R	Q	Q′
1	1	1	0
1	1	0	1
0	1	1	0
1	0	0	1

Fig. 8.25 A flip-flop circuit and associated Truth Table.

A flip-flop circuit, which is sometimes called a latch unit and is used as a switch debouncer, is shown in Fig. 8.25. Its operation can be understood by verifying the consistency of the status represented in the associated truth table.

Alphanumeric readouts require one of the following:

a. A cold-cathode nixie tube,

b. A neon tube,

c. A liquid crystal diode (LCD), or

d. A light-emitting diode (LED).

The cold-cathode nixie tube and the neon tube require relatively high voltages, so they have been superseded. The LCD requires very low power and so it is good for watches, but it needs external illumination. Hence, the LED is used most commonly.

8.6 Signal Processing Electronics

An events counter is shown in Fig. 8.26. Input pulses in serial form arrive at pin 14 on the TTL 7490 chip, a decade counter. Output goes to a decoder-driver, TTL 7447, which controls an LED display, and to pin 11, which also provides an output pulse after 9 counts, which goes to pin 14 of the next decade counter. Pins 3, 4, and 5 of the 7447 are used to reset blank zeros and to provide decimal points. Fig. 8.27 shows a frequency counter. The AND gate conducts for 1/2 sec every second, so that the events counter gives a frequency reading. Wave shaping can be done by converting a sinusoidal signal to square wave pulses using a Schmidt trigger. Another type of IC chip (e.g., TTL 74121) is the oscillator or multivibrator; it can produce a square wave output whose frequency can be varied from 0.1 Hz to 100 kHz, depending on the configuration of an outboard control circuit. Additional outboard circuit elements can be used to convert the output to triangular or linear ramp wave forms. Fig. 8.28 shows an IC multiplexer (e.g., 74150), which selects from among a number of inputs (often sixteen) according to control signals. It is similar to a single-pole sixteen-position switch. A demultiplexer (Fig. 8.29) works in reverse (i.e., it decides to which of sixteen outputs an input shall be routed). Multiplexing permits periodic sampling of many variables that may require monitoring. Data processing at a distance can be conducted

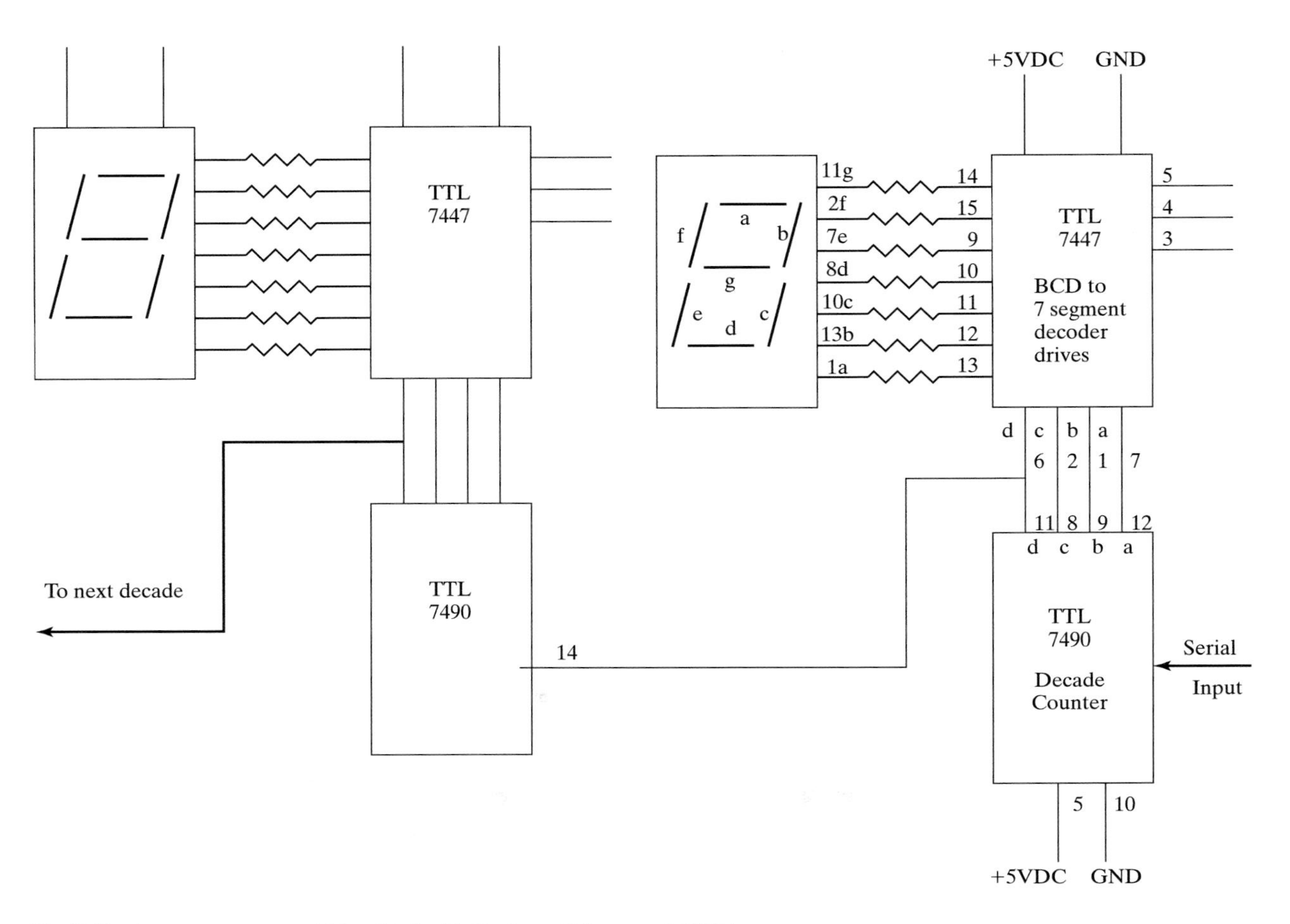

Fig. 8.26 An events counter. (After Beckwith, Buck, and Marangoni, 1982.)

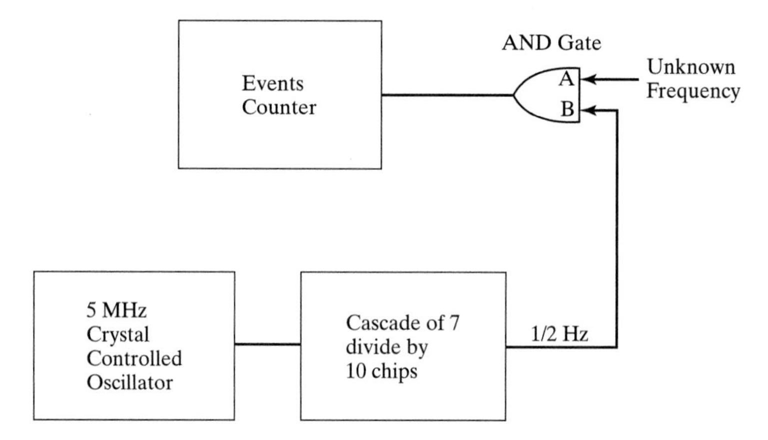

Fig. 8.27 A frequency counter.

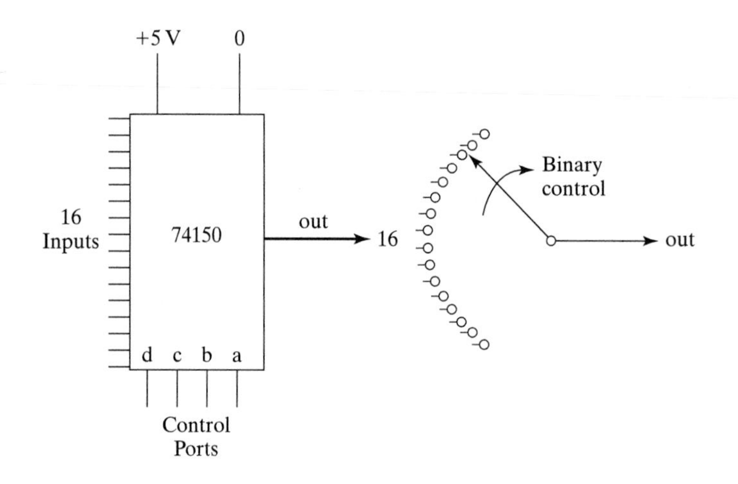

Fig. 8.28 A multiplexer.

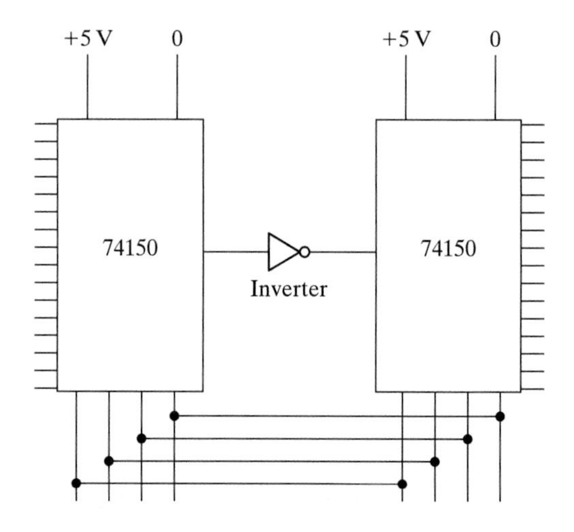

Fig. 8.29 A multiplexer-demultiplexer combination.

economically by using a multiplexer-demultiplexer combination with synchronized controls, as in Fig. 8.29, but there are then only five instead of sixteen circuits. Another way of achieving synchronization is to use a universal asynchronous receiver/transmitter (UART), which converts serial to parallel data and vice versa. It is then possible to use separate clocks, which do not have to be exactly synchronized.

Modern acoustical processing devolves around obtaining digital signals from analog signals, a process in which the electronic computer is essential. A computer is any device that performs logical arithmetic operations, under the control of a program, that can be stored in a memory unit. The essential architecture of computers has been the same since the time of Charles Babbage's analytical engine of the 1860s, a totally mechanical device. In Babbage's machine, the arithmetic operations were performed in a part of the machine he called the "mill," and the results of the calculation were transferred to the "store." Today, we would call the mill the central processing unit (CPU) and the store would be the memory.

It seems that the first true digital electronic computer was constructed by Atanasoff at Iowa State in 1942. This machine carried out special purpose calculations; the first general purpose machine was the ENIAC, built at the end of World War II. Borrough's Company and IBM were also early entrants in the computer business with the UNIVAC and Mark I. The mainframe computers of today are the descendants of these "first-generation" machines. The first-generation machines were constructed using vacuum tubes, which were notoriously unreliable, producing heat (which attracted moths who would perish inside the machine—hence the term "debugging"). The second generation of machines used transistors and the third generation, integrated circuits. The fourth-generation machines were built around "microprocessors," in which the CPU was itself embodied on a chip.

We will now discuss the digital-to-analog (D/A) converter, which is easier to understand than the analog-to-digital converter. The series of switches (Fig. 8.30) are in reality solid-state gates (e.g., AND ICs) and their function can be controlled by the successive bits in a data byte. When the switch is closed, a current flows whose magnitude

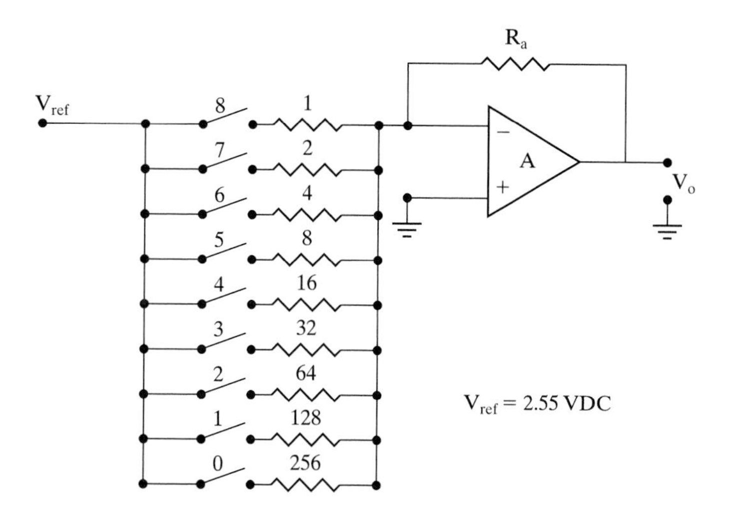

Fig. 8.30 Digital-to-analog converter. (After Beckwith, Buck, and Marangoni, 1982.)

is proportional to its positional number. The summation of these currents is an analog rendition of the data byte. The gain of the operational amplifier is set to control the output (e.g., gain gives a maximum of 2.55 V).

The circuit for the A/D converter is the same as for the D/A converter with the addition of a voltage comparator (see Fig. 8.31). The switches are closed in sequence, producing a steadily rising digital voltage at the operational amplifier. Suitably amplified, this voltage then appears at the voltage comparator, which conducts when the analog input voltage matches the rising digital reference and then flags the computer that the analog signal matches the digital one, and this value can be recorded in the memory. The cycle is repeated and another sample of the analog voltage is obtained.

One of the most important operations in acoustics is the analysis of the frequency components of a signal. Some of the earliest work of this kind was done by direct acoustic analysis, using arrays of Helmholtz resonators. With the advent of electroacoustics, it became more convenient to analyze the signal electrically, and arrays of narrow-band electronic filters were used instead of Helmholtz resonators. Later, tunable filters were constructed that could be mechanically swept through a frequency range. This was a slow process and required that the signal not vary during the sampling period. More rapidly changing signals had to be tape recorded and replayed several times to give a long enough duration of quasi stationary signal.

The so-called real time analyzers were developed to overcome this difficulty. A typical design for a real time analyzer is shown in Fig. 8.32. The sawtooth signal generator produces a voltage that rises on a ramp, thus causing the voltage-controlled oscillator to produce a pure tone output that rises in frequency. The mixer combines this tone with the input signal to yield sum and difference frequencies. The intermediate frequency bandpass filter passes only one of the difference frequencies. The detector rectifies the signal from the bandpass amplifier and thus there will be a spike produced on the cathode ray oscilloscope screen at a point on the trace corresponding to the frequency of the input signal. The modern method of spectral analysis employs digital processing, which we will discuss in Section 8.7.

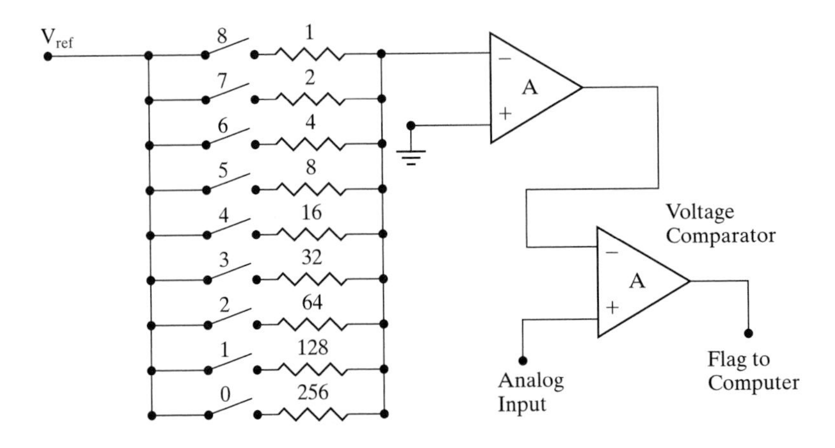

Fig. 8.31 Analog-to-digital converter. (After Beckwith, Buck, and Marangoni, 1982.)

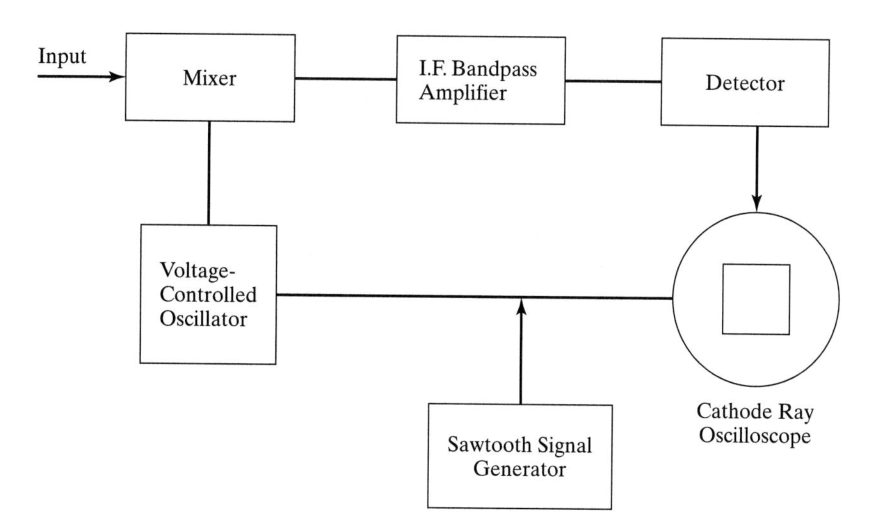

Fig. 8.32 Sweep frequency real time spectrum analyzer.

8.7 The Discrete Time Fourier Transform

Sinusoidal or complex exponential sequences are particularly important, because the steady-state response of a linear shift invariant (LSI) system to a sinusoid is another sinusoid of the same frequency. This may be proved as follows. Let

$$x(n) = e^{j\omega n}, -\infty < n < \infty \tag{8.7.1}$$

Then the output will be

$$\begin{aligned} y(n) &= \sum_{k=-\infty}^{\infty} h(k)e^{j\omega(n-k)} \\ &= e^{j\omega n} \sum_{k=-\infty}^{\infty} h(k)e^{-j\omega k} \end{aligned} \tag{8.7.2}$$

Let

$$H(e^{j\omega}) = \sum_{k=-\infty}^{\infty} h(k)e^{-j\omega k} \tag{8.7.3}$$

then

$$y(n) = H(e^{j\omega})e^{j\omega n} \tag{8.7.4}$$

that is, $y(n)$ is a complex exponential with a complex amplitude $H(e^{j\omega})$. The function $H(e^{j\omega})$ that describes the change in complex amplitude as a function of ω is called the frequency response of the system whose unit sample response is $h(n)$. It may be divided into real and imaginary parts:

$$H(e^{j\omega}) = H_R(e^{j\omega}) + jH_I(e^{j\omega}) \tag{8.7.5}$$

or

$$H(e^{j\omega}) = |H(e^{j\omega})|e^{j\arg[H(e^{j\omega})]} \tag{8.7.6}$$

Since a sinusoid can be expressed as a linear combination of complex exponentials, the frequency response also expresses the response to a sinusoidal input. Consider

$$x(n) = A\cos(\omega_0 n + \phi) = \frac{A}{2}e^{j\phi}e^{j\omega_0 n} + \frac{A}{2}e^{-j\phi}e^{-j\omega_0 n} \tag{8.7.7}$$

The response to the first term is

$$y_1(n) = H(e^{j\omega_0})\frac{A}{2}e^{j\phi}e^{j\omega_0 n} \tag{8.7.8}$$

If h(n) is real, since

$$y(n) = \sum_{k=-\infty}^{\infty} x(k)h(n-k) \tag{8.7.9}$$

the response to the second term of Eq. (8.7.7) is the complex conjugate of the response to the first term. So the complete response is

$$\begin{aligned} y(n) &= \frac{A}{2}[H(e^{j\omega_0})e^{j\phi}e^{j\omega_0 n} + H(e^{-j\omega_0})e^{-j\phi}e^{-j\omega_0 n}] \\ &= A|H(e^{j\omega_0})|\cos(\omega_0 n + \phi + \theta) \end{aligned} \tag{8.7.10}$$

where

$$\theta = \arg[H(e^{j\omega_0})] \tag{8.7.11}$$

Thus, we see that $H(e^{j\omega_0})$ may be regarded as the response to a sinusoid. $H(e^{j\omega_0})$ is a continuous function of frequency and must be a periodic function of ω, since $e^{j(\omega+2\pi)} = e^{j\omega}$. This period is 2π (i.e., the frequency response is the same at ω and $\omega + 2\pi$). Let us consider as an example:

$$\begin{aligned} h(n) &= 1, 0 \le n \le N-1 \\ &= 0, \text{elsewhere} \end{aligned} \tag{8.7.12}$$

(see Fig. 8.33). Thus, the frequency response is

$$H(e^{j\omega_0}) = \sum_{n=0}^{N-1} e^{-j\omega n} = \frac{1-e^{j\omega N}}{1-e^{j\omega}} = \frac{\sin(\omega N/2)}{\sin \omega/2}e^{-j(N-1)\omega/2} \tag{8.7.13}$$

This is the same summation we encountered in the theory of radiation from arrays (see Section 3.6 in Chapter 3). The magnitude and phase of $H(e^{j\omega_0})$ is periodic in ω as shown

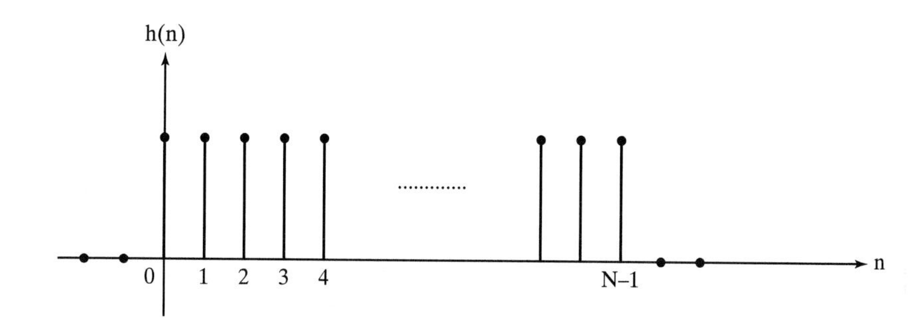

Fig. 8.33 A rectangular impulse response.

in Fig. 8.34 for the case N = 5. $H(e^{j\omega_0})$ can be represented by a Fourier series. In fact, it follows from the definition in Eq. (8.7.3)

$$\sum_{k=-\infty}^{\infty} h(k)e^{-j\omega k} = H(e^{j\omega})$$

that the frequency response is a Fourier series whose coefficients correspond to the unit impulse response h(n). Hence, from Eq. (2.2.5):

$$h(n) = \frac{1}{2\pi}\int_{-\pi}^{\pi} H(e^{j\omega})e^{j\omega n}\,d\omega \tag{8.7.14}$$

where

$$H(e^{j\omega}) = \sum_{n=-\infty}^{\infty} h(n)\,e^{-j\omega n} \tag{8.7.15}$$

The Fourier transform pair can be written only if the series converges. The representation of a sequence by a Fourier transform, as in Eq. (8.7.14), is not restricted to the unit sample response. Thus, for a general sequence x(n), the Fourier transform is defined as

$$X(e^{j\omega}) = \sum_{n=-\infty}^{\infty} x(n)\,e^{-j\omega n} \tag{8.7.16}$$

and the inverse Fourier transform is

$$x(n) = \frac{1}{2\pi}\int_{-\pi}^{\pi} X(e^{j\omega})\,e^{j\omega n}\,d\omega \tag{8.7.17}$$

$X(e^{j\omega})$ is called the discrete time Fourier transform. It is a continuous function of frequency. From Eq. (8.7.17) we may regard x(n) as an integral of complex exponentials.

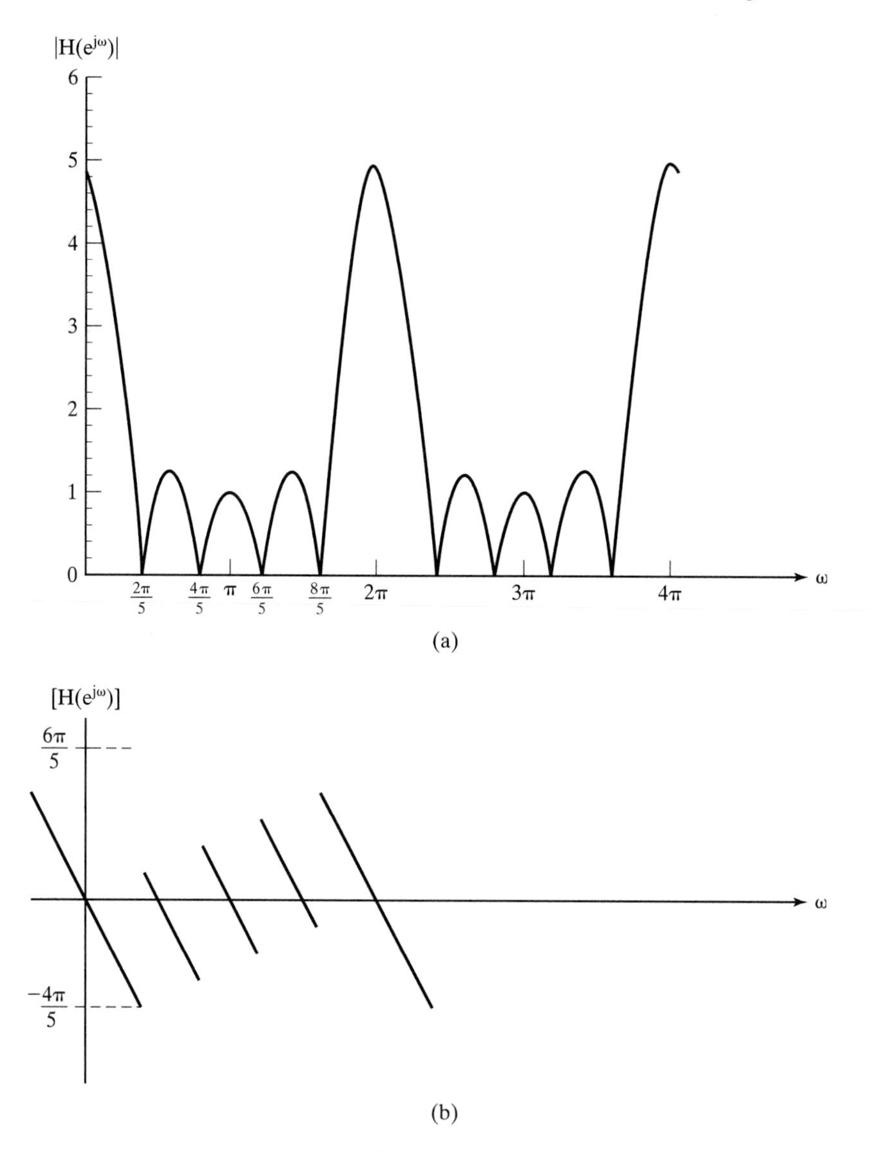

Fig. 8.34 Magnitude and phase of $H(e^{j\omega})$: (a) magnitude; (b) phase. (After Oppenheim and Schafer, 1975.)

Thus, for each value of n, the input will be

$$X\,(e^{j\omega})\,e^{j\omega n}\,d\omega$$

in the frequency interval ω to $\omega + d\omega$. The response to this part of the input will therefore be

$$H\,(e^{j\omega})\,X\,(e^{j\omega})\,e^{j\omega n}\,d\omega$$

Thus, for the entire range of frequencies, the output will be given by

$$y(n) = \frac{1}{2\pi}\int_{-\pi}^{\pi} H\,(e^{j\omega})\,X\,(e^{j\omega})\,e^{j\omega n}\,d\omega \tag{8.7.18}$$

and the Fourier transform of the output becomes

$$Y(e^{j\omega}) = H(e^{j\omega})\,X(e^{j\omega}) \tag{8.7.19}$$

We now see that the convolution of two signals in the time domain is paralleled by the multiplication of the Fourier transforms in the frequency domain, or

$$x(t) * h(t) \leftrightarrow X(\omega)H(\omega) \tag{8.7.20}$$

It follows that, because of the symmetry of the time and frequency domains in the Fourier transform, multiplication of two time domain functions is equivalent to convolution of their Fourier transforms. For example, the ideal low pass (box-car) filter has a frequency response

$$\begin{aligned} H(e^{j\omega}) &= 1, \quad |\omega| \le \omega_{co} \\ &= 0, \quad \omega_{co} < |\omega| < \pi \end{aligned} \tag{8.7.21}$$

as illustrated in Fig. 8.35. It is assumed that $H(e^{j\omega})$ is periodic outside the range $-\pi$ to $+\pi$. Hence, the impulse response is found:

$$h(n) = \frac{1}{2\pi}\int_{-\omega_{co}}^{\omega_{co}} e^{j\omega n}\,d\omega = \frac{\sin \omega_{co} n}{\pi n} \tag{8.7.22}$$

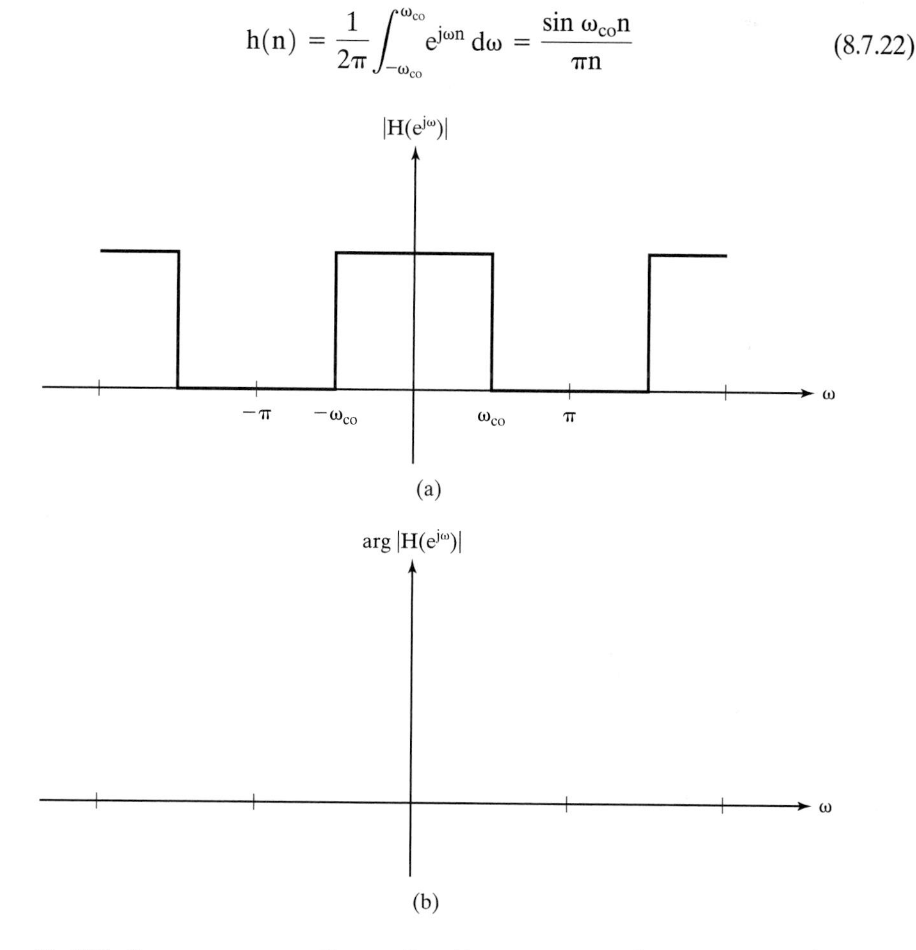

Fig. 8.35 Frequency response of box-car filter: (a) magnitude of $H(e^{j\omega})$; (b) phase of $H(e^{j\omega})$. (After Oppenheim and Schafer, 1975.)

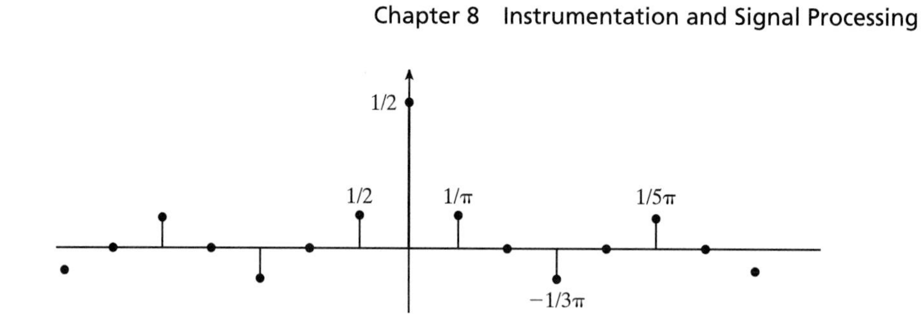

Fig. 8.36 Impulse response of a box-car filter. (After Oppenheim and Schafer, 1975.)

and is shown in Fig. 8.36. The ideal low pass filter is not causal, since its impulse response depends on values of n less than zero. It is also unstable. But it is conceptually important.

8.8 Sampling of Continuous Time Signals

We now know that it is advantageous in acoustics to use digital signals, but how frequently should an analog signal be sampled to get a useful digital signal? The answer depends on what is to be done with the digital signal. One very important application is the computation of a Fourier transform. We will see in this section that if we do not sample sufficiently rapidly, then the calculated Fourier transform will be defective. The theory that established the correct sampling rate was proposed by Shannon.

Consider an analog signal with Fourier representation

$$\begin{aligned} x_a(t) &= \frac{1}{2\pi}\int_{-\infty}^{\infty} X_a(j\Omega)\, e^{j\Omega t}\, d\Omega \\ X_a(j\Omega) &= \int_{-\infty}^{\infty} x_a(t) e^{-j\Omega t}\, dt \end{aligned} \tag{8.8.1}$$

If the sampling period is T, we may derive a sequence

$$x(n) = x_a(nT)$$

The sampling frequency or rate $= 1/T$. We want to relate $X_a(j\Omega)$, the continuous-time Fourier transform of $x_a(t)$ to $X(e^{j\omega})$, to the discrete time Fourier transform of x(n). Now from Eq. (8.8.1)

$$x(n) = x_a(nT) = \frac{1}{2\pi}\int_{-\infty}^{\infty} X_a(j\Omega)\, e^{j\Omega nT}\, d\Omega \tag{8.8.2}$$

and from the discrete time Fourier transform from Eq. (8.7.17)

$$x(n) = \frac{1}{2\pi}\int_{-\pi}^{\pi} X(e^{j\omega})\, e^{j\omega n}\, d\omega \tag{8.8.3}$$

Expand Eq. (8.8.2) as a sum of integrals over intervals $2\pi/T$. Thus

$$x(n) = \frac{1}{2\pi}\sum_{r=-\infty}^{\infty}\int_{(2r-1)\pi/T}^{(2r+1)\pi/T} X_a(j\Omega)e^{j\Omega nT}\,d\Omega \tag{8.8.4}$$

Each term in the sum can be reduced to an integral over the range $-\pi/T$ to $+\pi/T$ by change of variables; in other words, let

$$\Omega = \phi + \frac{2\pi r}{T}, \text{ thus when } \Omega = (2r+1)\frac{\pi}{T}, \text{ then } \phi = \frac{\pi}{T}, \text{ etc.}$$

Therefore

$$\int_{(2r-1)\pi/T}^{(2r+1)\pi/T} X_a(j\Omega)e^{j\Omega nT}\,d\Omega = \int_{-\pi/T}^{\pi/T} X_a\left(j\phi + j\frac{2\pi r}{T}\right)e^{j\phi\, nT}\,e^{j2\pi\, rn}\,d\phi \tag{8.8.5}$$

Now, let $\phi = \Omega$, then

$$x(n) = \frac{1}{2\pi}\sum_{r=-\infty}^{\infty}\int_{-\pi/T}^{\pi/T} X_a\left(j\Omega + j\frac{2\pi r}{T}\right)e^{j\Omega nT}\,e^{j2\pi\, rn}\,d\Omega \tag{8.8.6}$$

Interchange the order of integration and summation and note that $e^{j2\pi\, rn} = 1$, since r and n are integers, and substitute $\Omega = \omega/T$:

$$x(n) = \frac{1}{2\pi}\int_{-\pi}^{\pi}\left[\frac{1}{T}\sum_{r=-\infty}^{\infty} X_a\left(j\frac{\omega}{T} + j\frac{2\pi r}{T}\right)\right]e^{j\omega n}\,d\omega \tag{8.8.7}$$

Eq. (8.8.7) is identical in form to Eq. (8.8.3). Hence

$$X(e^{j\omega}) = \frac{1}{T}\sum_{r=-\infty}^{\infty} X_a\left(\frac{j\omega}{T} + j\frac{2\pi r}{T}\right) \tag{8.8.8}$$

or

$$X(e^{j\Omega T}) = \frac{1}{T}\sum_{r=-\infty}^{\infty} X_a\left(j\Omega + j\frac{2\pi r}{T}\right) \tag{8.8.9}$$

These equations establish a relation between the continuous time Fourier transform and the discrete time Fourier transform of a sequence derived by sampling. For example, the Fourier transform of a continuous function is shown in Fig. 8.37a. In Fig. 8.37b, we have the discrete time Fourier transform but the sampling period is too large. The

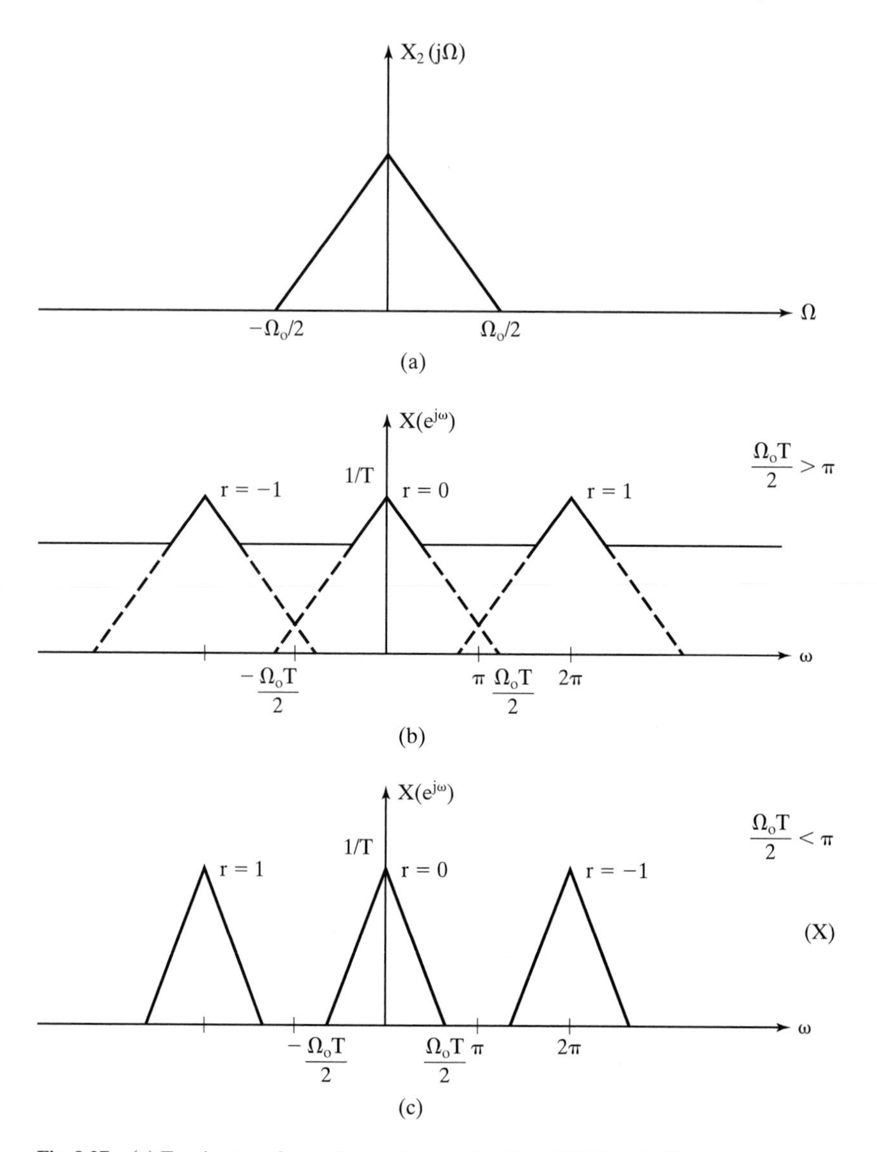

Fig. 8.37 (a) Fourier transform of a continuous function. (b) Discrete Fourier transform where sampling period is too large. Aliasing results from sampling at a rate less than the Nyquist criterion ($\omega_s = \Omega_o$). (c) Results from sampling at a rate greater than the Nyquist criterion. (After Oppenheim and Schafer, 1975)

shifted versions of $X_a(j\Omega)$ overlap. This phenomenon, where in effect a high frequency in $X_a(j\Omega)$ takes on the identity of a lower frequency, is called *aliasing*. From Fig. 8.37c, it is clear that if $\Omega_0/2 < \pi/T$, the highest angular frequency in the analog specimen is $\Omega_H = \Omega_0/2$, so that the highest frequency is $\Omega_H/2\pi$, and $f_H < 1/2T$. In other words, if we sample at a rate at least twice the highest frequency in $X_a(j\Omega)$, then $X(e^{j\omega})$ is identical to $X_a(\omega/T)$ in the interval $-\pi \leq \omega \leq \pi$. Then we can recover $x_a(t)$ from the samples $x_a(nT)$. This sampling rate is called the *Nyquist rate*. The point is illustrated for the case of a sinusoid in Fig. 8.38.

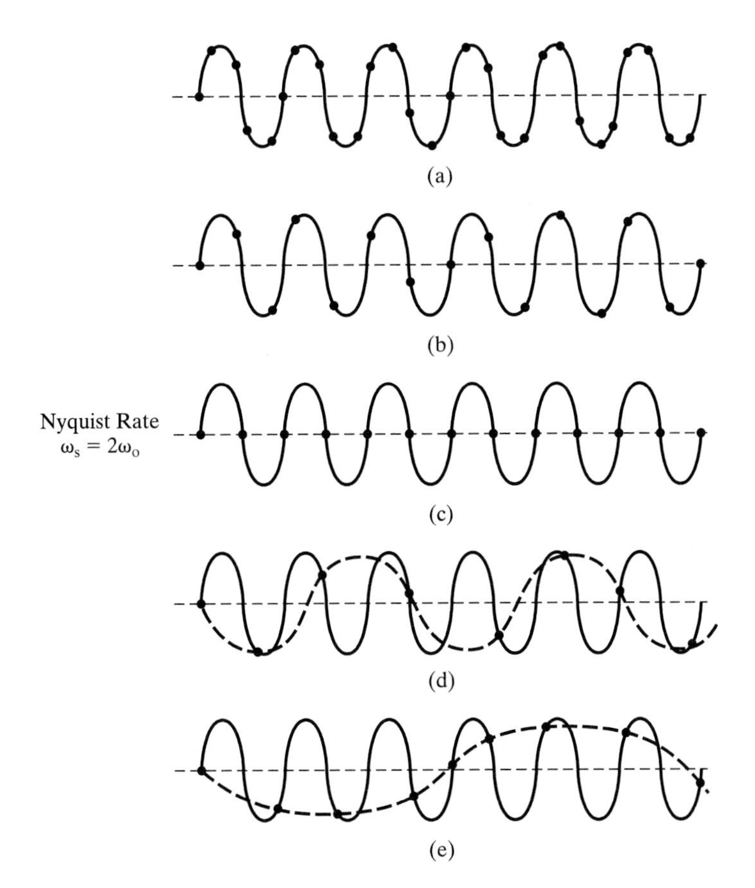

Fig. 8.38 Sampling of a sinusoid. In a and b, the sampling rate is faster than the Nyquist criterion ($\omega_s = 2\omega_o$). In c, the sampling is at the Nyquist rate. In d and e, the rate is less than the Nyquist rate and aliasing occurs.

8.9 The Fast Fourier Transform

A periodic function can be represented by a series of sinusoids, called a Fourier series. The frequencies of the sinusoids are separated by discrete intervals. A nonperiodic, or finite duration, function is represented by a continuum of sinusoids. Thus, if x(n) is a nonperiodic sequence, then its Fourier transform, the discrete time Fourier transform, is given by Eq. (8.7.16) and will be a continuous function of frequency. However, as a practical matter, most sequences have a finite duration, and it is then possible to produce a representation of that function in the form of a discrete Fourier series. This is done by "pretending" that the function is periodic beyond its actual duration. Such a series is called a discrete Fourier transform. The DFT will be periodic, but we simply ignore the part of it outside the basic range.

The DFT is given by Eqs. (8.7.16) and (8.7.17). To evaluate the integral in Eq. (8.7.17) using a computer, we convert it to a summation. Suppose the duration of the signal is N samples; if we let

$$\omega = \frac{2\pi k}{N}, \; k = 0, 1, \ldots (N-1) \tag{8.9.1}$$

then

$$d\omega = \frac{2\pi}{N}dk \tag{8.9.2}$$

The integration is equivalent to a summation if $dk = 1$. Notice also that the integration limits of Eq. (8.7.17) could be written from 0 to 2π, since the integrand is periodic outside the range $-\pi$ to π. Now when $\omega = 2\pi$ and $k = N$, Eq. (8.7.17) may be rewritten as follows:

$$x(n) = \frac{1}{N}\sum_{k=0}^{N-1} X(k)e^{j\frac{2\pi nk}{N}}, n = 0, 1, \ldots, (N-1) \tag{8.9.3}$$

Since x(n) is a causal signal, the summation in Eq. (8.7.16) has values only in the range $n = 0, 1, \ldots, (N-1)$, so it may also be rewritten

$$X(k) = \frac{1}{N}\sum_{n=0}^{N-1} x(n)e^{-j\frac{2\pi nk}{N}}, \; k = 0, 1, \ldots, (n-1) \tag{8.9.4}$$

Eq. (8.9.4) is the DFT, and Eq. (8.9.3) is the inverse DFT or IDFT.

Note that the expressions for the DFT and the IDFT differ only regarding the sign of the exponent of the phasor and a scale factor 1/N. Hence, the computational procedures for one can be readily adapted to compute the other. The most straightforward method of calculating the DFT is to apply the definition directly. Thus, since x(n) contains N values, the computation of X(k), for a particular value of k, requires N complex multiplications. Since k also has N values, a total of N^2 complex multiplications are required all told. The computation time is thus proportional to N^2. For large values of N, the computation time can become very long. The situation was eased when Cooley and Tukey (1965) published an algorithm that has become known as the fast Fourier transform (FFT). The FFT is an algorithm that reduces the computing time of Eqs. (8.9.3) and (8.9.4) to a time proportional to Nlog(N). A historical review of the discovery of the FFT can be found in Cooley, Lewis, and Welch (1967).

One type of FFT, known as a "decimation-in time" algorithm, is based on decomposing the sequence x(n) into successively smaller subsequences. This process can be understood by first considering the DFT of a two-point sequence; that is, $x(n) = x(0)$, $x(1)$. From Eq. (8.9.4), the DFT is

$$\begin{aligned} X(k) &= \sum_{n=0}^{1} x(n)e^{-j\frac{2\pi nk}{2}} \\ &= x(0) + x(1)\, e^{-j\pi k}, \text{ for } k = 0, 1 \end{aligned} \tag{8.9.5}$$

Hence

$$\begin{aligned} X(0) &= x(0) + x(1) \\ X(1) &= x(0) - x(1) \end{aligned} \tag{8.9.6}$$

where we have used the result

$$e^{-jk\pi} = (-1)^k$$

Notice that no multiplications are necessary to perform the two-point DFT. A four-point DFT can be reduced to two-point DFTs as follows:

$$x(n) = x(0), x(1), x(2), x(3)$$

Hence

$$\begin{aligned} X(k) &= \sum_{n=0}^{3} x(n)\, e^{-j\frac{2\pi nk}{4}} \\ &= x(0) + x(1)e^{-j\frac{2\pi 1k}{4}} + x(2)e^{-j\frac{2\pi 2k}{4}} + x(3)e^{-j\frac{2\pi 3k}{4}} \end{aligned} \tag{8.9.7}$$

for $k = 0, 1, 2, 3$. We can rewrite Eq. (8.9.7) as

$$X(k) = [x(0) + x(2)e^{-j\pi k}] + e^{-j\frac{2\pi k}{4}}[x(1) + x(3)e^{-j\pi k}]$$

So we can see that the four-point DFT has been reduced to the weighted sum of two two-point DFTs. The frequency sequence is given by

$$\begin{aligned} X(0) &= x(0) + x(2) + (x(1) + x(3)) \\ X(1) &= (x(0) - x(2)) + e^{-j\frac{2\pi}{4}}(x(1) - x(3)) \\ X(2) &= (x(0) + x(2)) + e^{-j\frac{4\pi}{4}}(x(1) + x(3)) \\ X(3) &= (x(0) - x(2)) + e^{-j\frac{6\pi}{4}}(x(1) - x(3)) \end{aligned} \tag{8.9.8}$$

Similarly, the DFT of an eight-point sequence can be written as

$$\begin{aligned} X(k) = {} & (x(0) + x(4)e^{-j\pi k}) + e^{-j\frac{2\pi k}{8}}(x(1) + x(5)e^{-j\pi k}) \\ & + e^{-j\frac{4\pi k}{8}}(x(2) + x(6)e^{-j\pi k}) + e^{-j\frac{6\pi k}{8}}(x(3) + x(7)e^{-j\pi k}) \end{aligned} \tag{8.9.9}$$

for $k = 0, 1, 2, \ldots, 7$. We may represent a two-point DFT as shown in Fig. 8.39. Hence, we are able to draw a flow chart for the computation of a four-point DFT as shown in Fig. 8.40. To perform an eight-point DFT, additional shuffling is required, as shown in Fig. 8.41.

From these examples we can see that any N-point sequence can be shifted to generate N/2 two-point FFTs provided N can be expressed as a power of 2. By carrying out the superposition and weighting efficiently, we have a fast computational procedure for computing the DFT. Efficient algorithms to compute the DFT for a general N-point sequence have now been written.

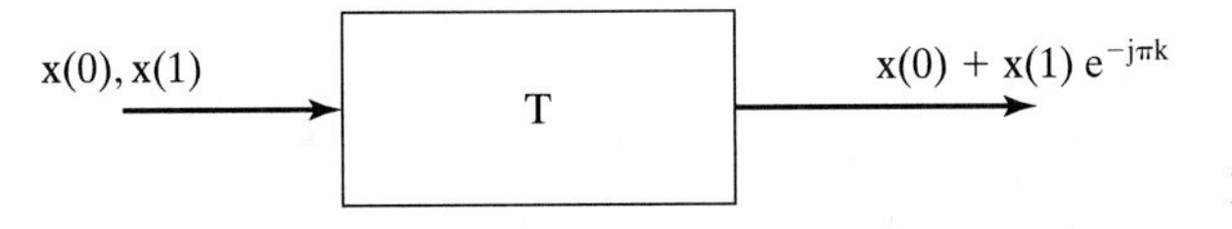

Fig. 8.39 A two-point DFT.

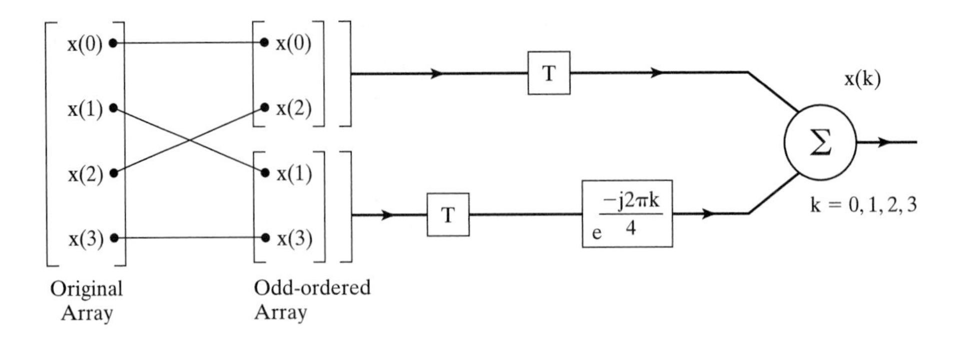

Fig. 8.40 A four-point DFT. (After Robinson and Silvia, 1978.)

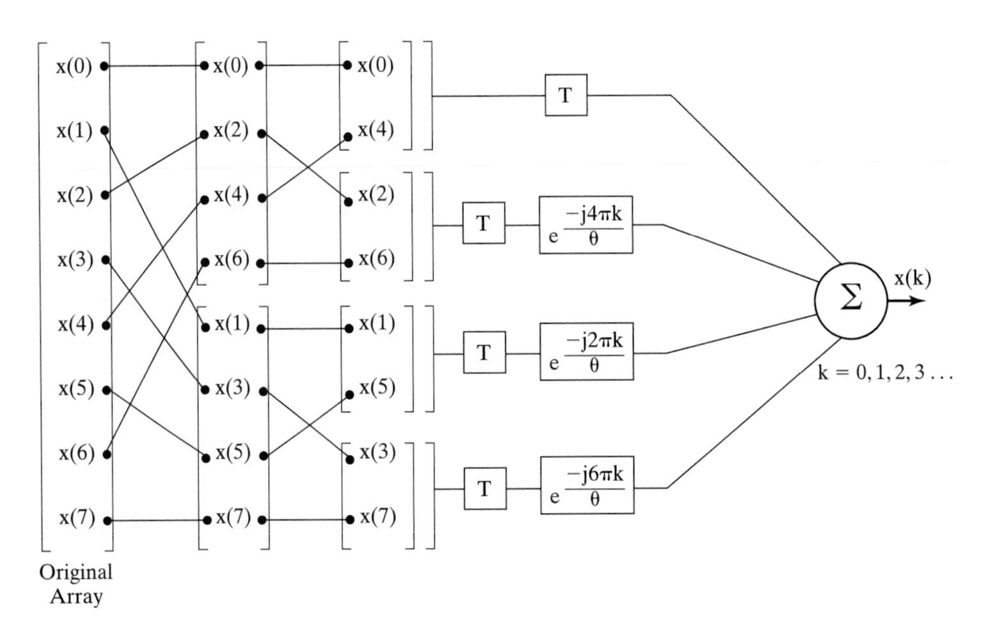

Fig. 8.41 An eight-point DFT. (After Robinson and Silvia, 1978.)

8.10 Windowing

In an application such as the monitoring of a rotating shaft, we may have a signal coming continuously from an accelerometer. In that case, we need only to study a portion of the signal. Assume that the signal has been digitized by sampling at a suitable rate, as described in Section 8.8, then the issue is how to truncate the sequence. The obvious way to do this is to multiply by a rectangular "window" function. The process is illustrated diagramatically in Fig. 8.42. But there is a problem: The DFT of x(n) as shown in Fig. 8.42b will not be the same as the DFT of w(n)x(n). Multiplication in the time domain is equivalent to convolution in the frequency domain. We have already seen in Section 8.7 that a box-car filter has a sinc function frequency response, as shown in Fig. 8.36. Hence, a pure tone input at ω_0 will have a frequency response obtained by convolving a unit sample in

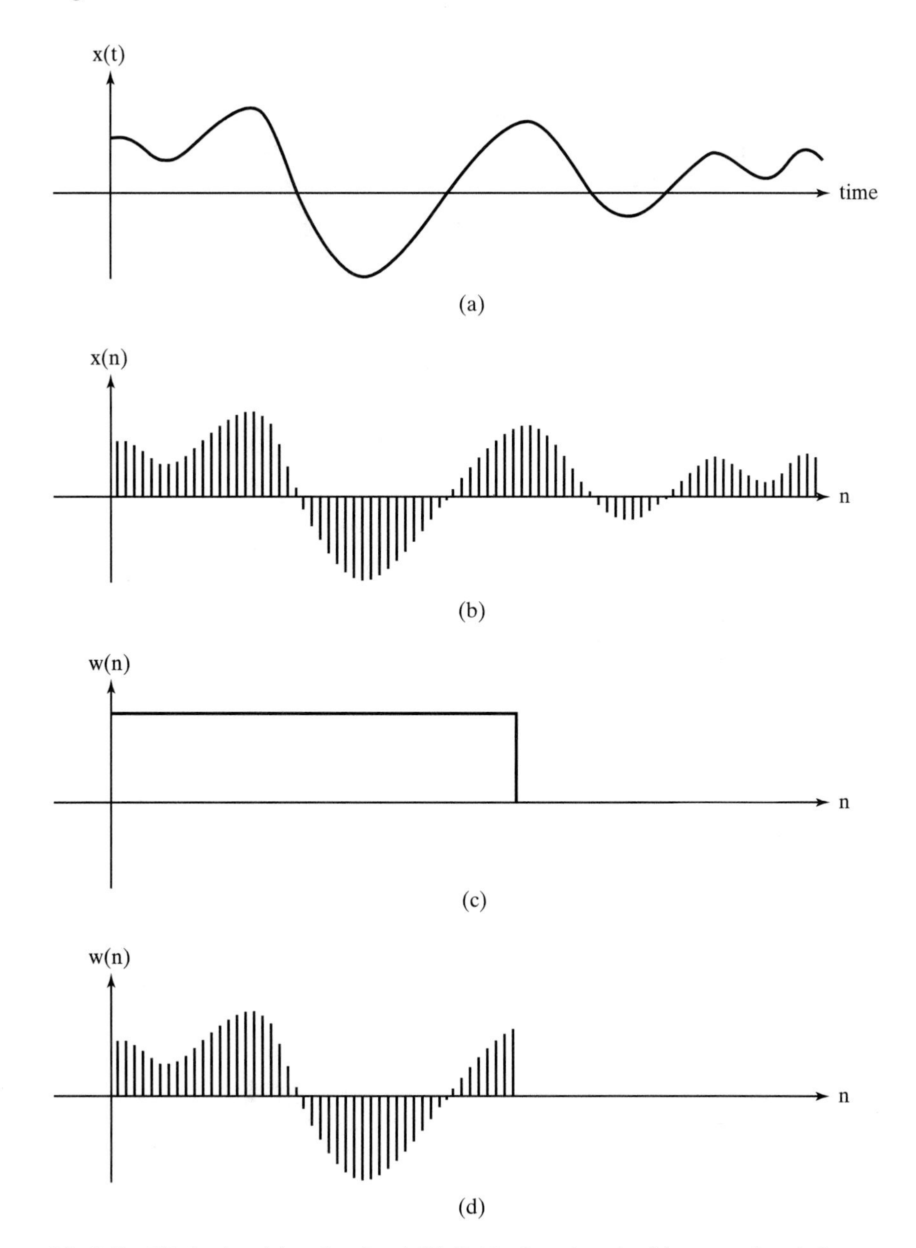

Fig. 8.42 Windowing: (a) analog signal; (b) digitized version of a; (c) rectangular window function; (d) product w(n)x(n).

the frequency domain with a sinc function, as indicated in Fig. 8.43. This response will have a maximum at ω_0, but there will also be considerable energy "leakage" to other frequencies because of the side lobes in the sinc function. The truncation operation thus results in distortion of the spectrum. To reduce leakage, it is necessary to employ a time domain truncation function with side-lobe characteristics that are smaller in magnitude

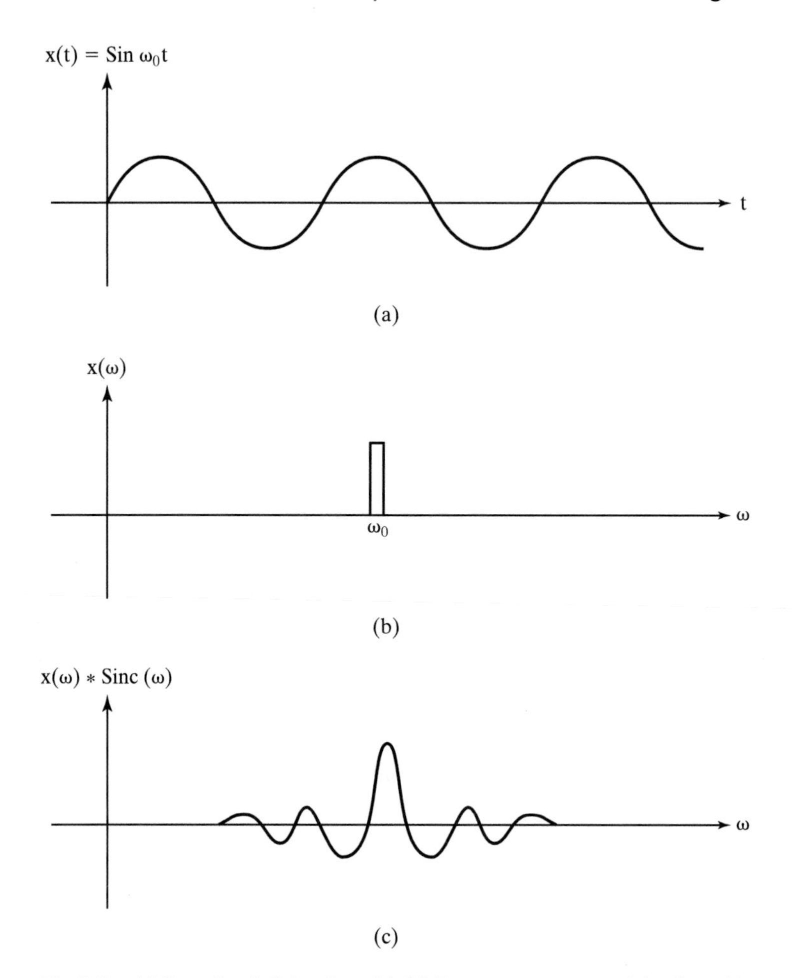

Fig. 8.43 (a) Part of an infinite sinusoid. (b) Frequency response of the sinusoid. (c) Frequency response of the same sinusoid truncated with a box-car window.

than those of the sinc (ω) function. The smaller the side lobes, the less leakage will affect the results of the DFT. Two particularly good truncation functions are the Hanning and Hamming windows given by

$$W(n) = \frac{1}{2}\left[1 - \cos\left(\frac{2\pi n}{N - 1}\right)\right] \quad \text{(Hanning)} \tag{8.10.1}$$

and

$$W(n) = 0.54 - 0.46\cos\left(\frac{2\pi n}{N - 1}\right) \quad \text{(Hamming)} \tag{8.10.2}$$

Further details on these functions can be found in Oppenheim and Schafer (1975) and in Robinson and Silvia (1978).

8.11 Correlation Functions

Correlation functions are another form of time domain operation originally developed for treating economic and other statistical data. A classical example is the correlation between rainfall and the price of grain (see Fig. 8.44). The use of correlation was introduced by Goff (1955) in acoustics for such applications as locating a noise source, measuring attenuation by a panel, and identifying a pure tone in noise.

The cross correlation of the functions x(t) and y(t) is given by

$$c_{xy}(t) = \int_{t=-\infty}^{\infty} x(t)y(t+\tau)\,dt \tag{8.11.1}$$

If x(t) and y(t) are real and have Fourier transforms $X(\omega)$ and $Y(\omega)$, we can show that the Fourier transform of $c_{xy}(t)$ is given by

$$c_{xy}(\omega) = X^*(\omega)Y(\omega) \tag{8.11.2}$$

where $X^*(\omega)$ is the conjugate of $X(\omega)$ or

$$X^*(\omega) = X(-\omega)$$

$C_{xy}(\omega)$ is called the spectral cross correlation of signals x(t) and y(t). The discrete cross correlation is defined by

$$C_{xy}(n) = \sum_{m=0}^{L-1} x(m)y(n+m) \tag{8.11.3}$$

where

$$y(n) = y(n+M)$$
$$x(n) = x(n+N)$$
$$M \leq N \text{ and } L = N \tag{8.11.4}$$

The discrete cross correlation of the discrete finite duration functions x(n) and y(n) of finite duration N and M, respectively, is given by

$$c_{xy}(n) = \sum_{m=0}^{L=N+M-2} y(m)x(n+m) \tag{8.11.5}$$

In Eqs. (8.11.1), (8.11.3), and (8.11.5), if y(t) = x(t) or y(n) = x(n), then $c_{xx}(t)$ or $c_{xx}(n)$ is called the autocorrelation or discrete autocorrelation of the function x(t) or x(n). Once the FFT of a function has been calculated, correlation functions can be readily obtained.

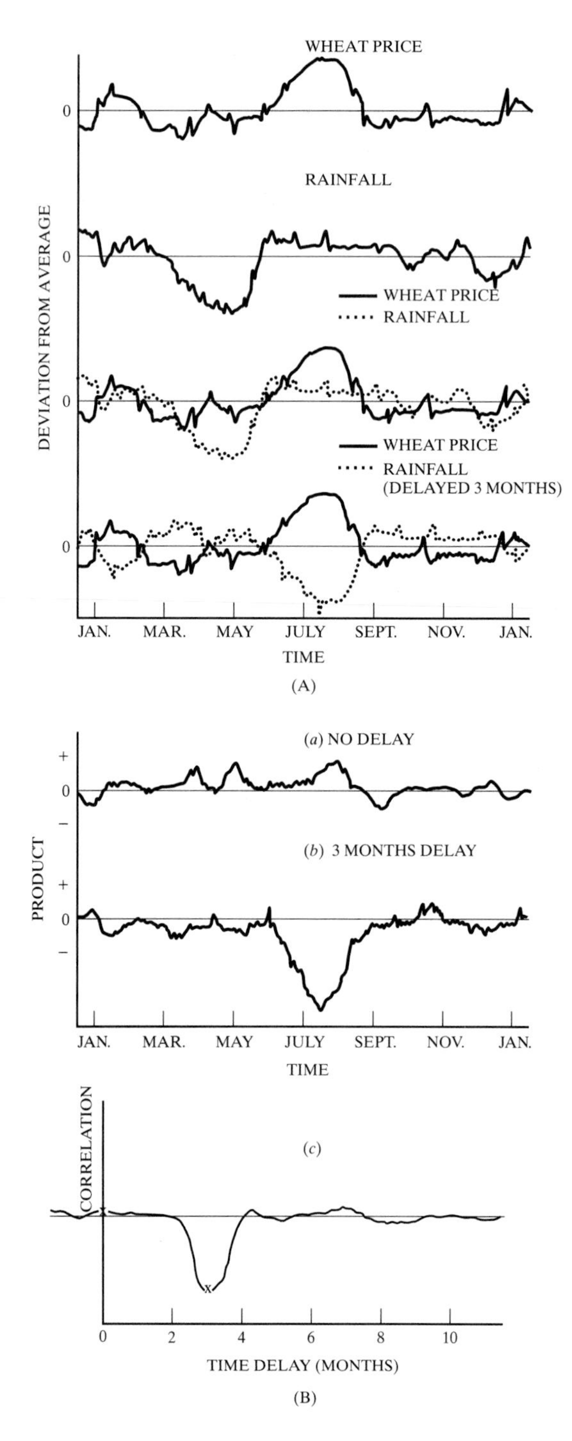

Fig. 8.44 Illustration of correlation. (A) Comparisons between hypothetical fluctuations in the price of wheat and rainfall over the area in which the wheat was grown. (B) Steps in computing the crosscorrelation function between wheat price and rainfall. Curves (*a*) and (*b*) give the product of wheat price and rainfall fluctuations with and without time delay. Curve (*c*) gives average values of curves such as (*a*) and (*b*) *versus* time delay. (After Goff, 1955.)

8.12 Homomorphic Digital Signal Processing

Homomorphic signal processing is based on theories developed by Oppenheim. The idea behind it is that the signals to be compared are subjected to a nonlinear transformation to another domain and are then added, instead of being convolved in the original domain. These additive representations are then processed by classical linear filters and returned to the original domain by the inverse of the nonlinear operation.

The complex cepstrum is an outgrowth of homomorphic systems theory. The cepstrum is defined as the Fourier transform of the logarithm of the spectral energy density. The cepstrum is sometimes loosely referred to as the spectrum of the spectrum. The cepstrum is useful in separating families of spectral peaks. The term "cepstrum" is an anagram of spectrum, and the parameter of the cepstrum is termed quefrency (an anagram of frequency). The units of quefrency are the same as those of time. A lifter is a windowing operation in the quefrency domain. A short pass filter is a low pass windowing operation in the quefrency domain. Rahmonics are integer multiples of a quefrency value corresponding to a peak amplitude in the quefrency domain.

The spectrum of a real even function is real and even. Thus the power cepstrum should be obtained from the two-sided power spectrum in order to retain sign information. Then the digital power cepstrum is obtained as

$$C_p(n) = \sum_{k=0}^{N-1} \log |X(k)|^2 \cos \frac{2\pi kn}{N} \tag{8.12.1}$$

where $\log|X(k)|$ is the log spectrum with no DC component. Note that it is advantageous to remove this DC component or mean log-spectrum value before calculating the cepstrum in order to optimize the signal-to-noise ratio in the cepstrum.

The complex cepstrum may be defined as follows:

$$C_c(n) = \sum_{k=0}^{N-1} \log|X(k)|^2 \exp \frac{2j\pi kn}{N} \tag{8.12.2}$$

where $X(k)$ is the complex cepstrum of $x(n)$:

$$X(k) = |X(k)|\, e^{j\theta(k)} \tag{8.12.3}$$

From Eq. (8.12.2), the complex logarithm of $X(k)$ is given by

$$\log X(k) = \log|X(k)| + j\theta(k) \tag{8.12.4}$$

If $x(n)$ is real, then $X(k)$ is conjugate even, from which it follows that $|X(k)|$ is even, $\log|X(k)|$ is even, and $\theta(k)$ is odd.

Convolution of two signals in the time domain corresponds in the frequency domain to a multiplication of their respective Fourier spectra. On taking the logarithm of the spectrum, the effect of an echo in a signal becomes additive, and the periodicity in the logarithmic spectrum by inverse Fourier transform becomes a series of rahmonics in the quefrency domain. Thus, the delay time of echoes is much easier to establish in the cepstrum than in the autocorrelation function. The other advantage over the autocorrelation function is that because the echoes are additive, they can be removed by subtracting delta functions from the cepstrum. In addition, if the signal containing an echo in the time domain is considered to be a periodic signal, then the shape of the autocorrelation function will be sensitive to circular shifting but the cepstrum will not be.

From this analysis it is clear that the cepstrum has advantages over the autocorrelation function in locating echoes, measuring their delay time, and measuring the reflection surface impulse response and the insensitivity of cepstrum analysis to circular shifting of the time domain signal.

REFERENCES AND FURTHER READING

ANSI S.16—1984, "American National Standard Preferred Frequencies, Frequency Levels, and Band Numbers for Acoustical Measurements."

Beckwith, T.G., N.L. Buck, and R.D. Marangoni. 1982. *Mechanical Measurements*. 3rd ed. Addison Wesley.

Cooley, J.W., and J.W. Tukey. 1965. "An Algorithm for the Machine Calculation of Complex Fourier Series." 19 *Math. Comput.* 297–301.

Cooley, J.W., P.A.W. Lewis, and P.D. Welch. 1967. "Historical Notes on the Fast Fourier Transform." Au-15 *IEEE Trans. Audio Electroacoust.* 76–79.

Goff, K.W. 1955. "The Application of Correlation Techniques to Some Acoustic Measurements." 27 *J. Acoust. Soc. Am.* 236–246.

Oppenheim, A.V., and R.W. Schafer. 1975. *Digital Signal Processing*. Prentice Hall.

Robinson, E.A., and M.T. Silvia. 1978. *Digital Signal Processing and Time Series Analysis*. Holden Day.

PROBLEMS

8.1 Explain what is meant by an electronic filter. Then design filters to pass frequencies

a. Above 60 Hz

b. Below 1000 Hz

c. Between 60 Hz and 1000 Hz

8.2 What are the advantages of processing and storing acoustic signals in digital form? Describe the methods of converting digital-to-analog signals and vice versa. How are acoustic signals stored digitally on a compact disc?

8.3 What is a convolution sum? Sketch the signal that results from convolving each pair of the following functions:

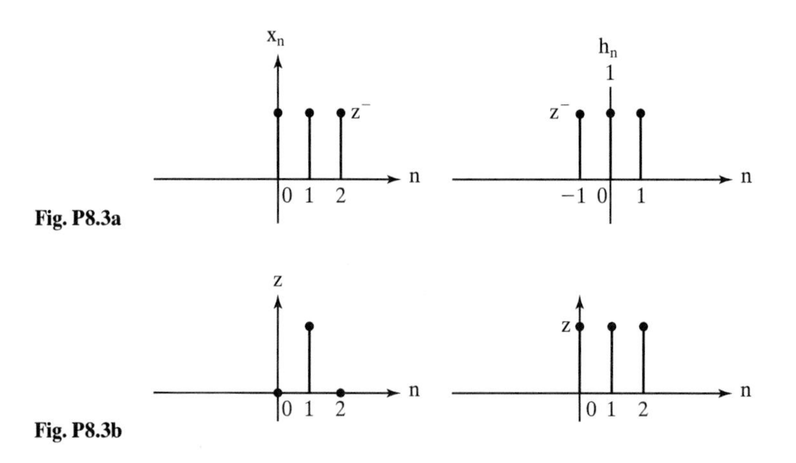

Fig. P8.3a

Fig. P8.3b

8.4 Compute by hand the FFT of the sequence

$$x_n = \cos n\pi/8, n = 0, 1, 2, 3, \ldots.$$

8.5 It is desired to make a Fourier analysis of an 80 ms duration signal. The frequency range of interest is 0−7.5 kHz. At what rate should the signal be sampled? What would be the consequence of sampling at too slow a rate?

8.6 The Fourier series for a waveform v(t) produces three spectral components with rms coefficient amplitudes of 3, 5, and 2 volts, respectively. What power would be dissipated if this signal were to be applied to a 1 ohm resistor?

8.7 Determine the Fourier series coefficients and the power density spectrum of the periodic signal shown in Fig. P8.7.

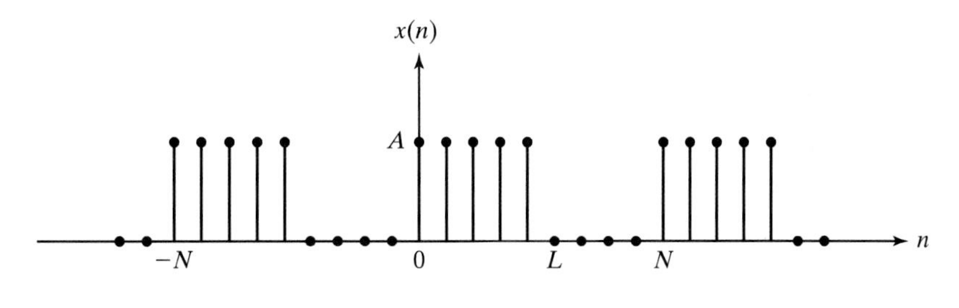

Fig. P8.7

8.8 A modified comb filter is described by the difference equation as follows:

$$y[n] = 5x[n - 2] - 5x[n - 8]$$

Show that the frequency of the filter is given by:

$$H(\omega) = 10 \sin(3\omega)e^{j\left(\frac{\pi}{2}-5\omega\right)}$$

a. Determine the impulse response of the filter.
b. Determine the step response of the filter.
c. Determine the response of the filter to the signal:

$$x_1[n] = 3u[n - 1] - 2u[n - 3]$$

8.9 Consider a rectangular pulse defined by:

$$g(t) = \begin{cases} A, & 0 \leq t \leq T \\ 0, & \text{otherwise} \end{cases}$$

It is proposed to approximate the matched filter for g(t) by an ideal low pass filter of bandwidth B; maximization of the peak pulse signal-to-pulse ratio is the primary objective. Determine the value of B for which the ideal low pass filter provides the best approximation to the matched filter.

Chapter 9

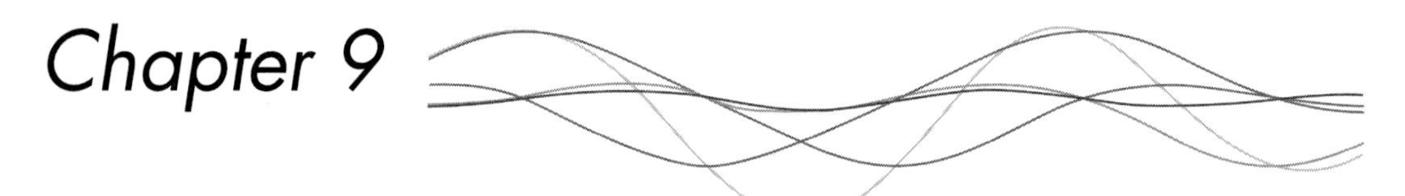

Basic Acoustic Measurements

9.1 Acoustic Parameters

We are frequently presented with the problem of measuring the characteristics of a sound or vibrational field, such as the sound pressure or particle acceleration, velocity, or displacement. In other cases, we may need to measure some associated property, such as pressure gradient, density or temperature fluctuations, or acoustic intensity. The transducer then serves as a scientific measuring instrument whose response must be an observable quantity related to the characteristics of interest in the acoustic field. The calibration of the transducer then defines the functional relationship between this response and the relevant acoustic characteristic, arrived at on an empirical basis. A similar situation arises when it is necessary to verify experimentally the response of a transducer designed theoretically. Because acoustic pressure is the characteristic most easily measured, we will first discuss the calibration of pressure-sensitive sound receivers (i.e., microphones).

Many acoustical laboratories possess a standard microphone, so the calibration of the unknown device becomes a relatively simple matter of comparing the responses of the standard and the unknown to the same sound field. This comparison circumvents the issue of the calibration of the standard. *Absolute* calibration methods do not rely on presupposed measurements.

The earliest way of obtaining any quantitative information on the strength of a sound field was to use singing flames (a singing flame is a narrow diameter, high velocity gas jet burning in air and showing a high sensitivity to sound). But the theory behind these flames was complex and hard to relate to the parameters of the sound field. Later, a way to measure the particle velocity in a sound field was devised and is known as the Rayleigh disc (Fig. 9.1). The torsion wire is twisted to produce a couple, which opposes the couple due to the sound field. The theory was

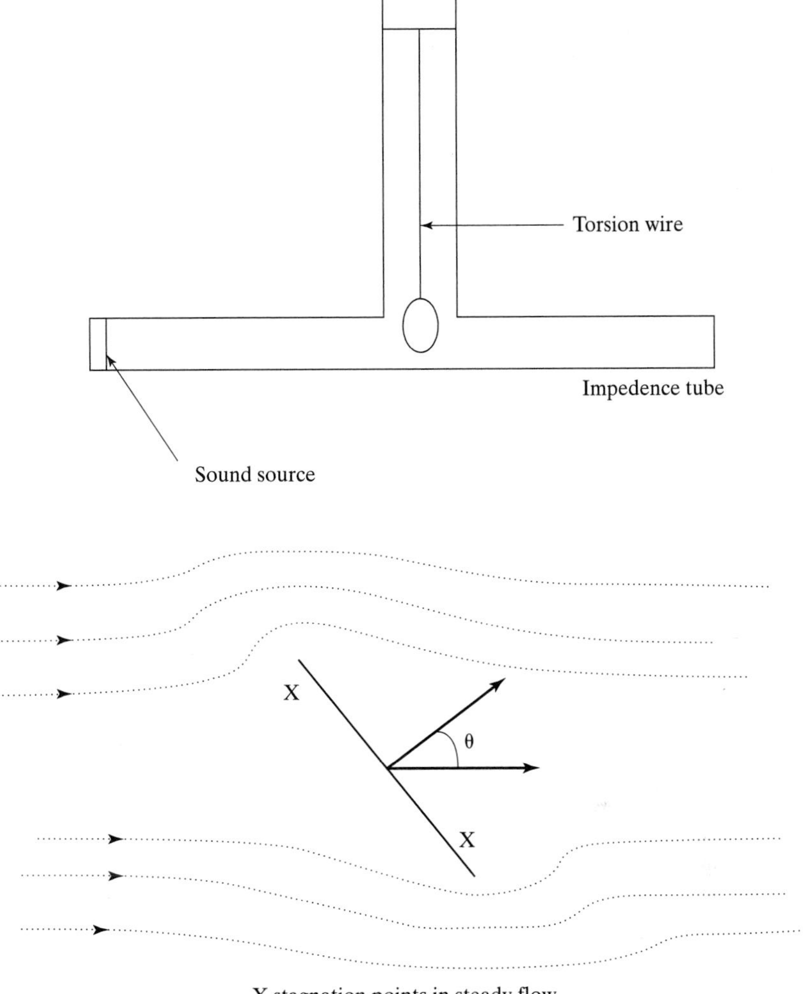

Fig. 9.1 The Rayleigh disc.

worked out by Konig for an alternating flow, and the moment of the couple was shown to be for $\theta = 45°$,

$$C = \frac{4}{3}\rho a^3 v_0^2 - \frac{4}{3}\rho a^3 v_0^3 \frac{\cos 2\omega t}{2} \tag{9.1.1}$$

The second term is usually negligible. The technique was cumbersome and limited to cases where the disc diameter was much less than a wavelength.

The Rayleigh disc was superceded as the primary standard by the development of the reciprocity technique. Another absolute calibration method consists of placing the microphone in a chamber in which a piston oscillates at a known frequency and displacement. The pressure amplitude can then be calculated. Most sound level meters are supplied with such piston phones.

9.2 Reciprocity Calibration

To use the reciprocity calibration technique as illustrated in Fig. 9.2, it is necessary to have a transmitter, T; the unknown microphone, X, which is to be calibrated; and a reversible microphone, R. A small back-enclosed loudspeaker can be used for this purpose. All the devices should be small in comparison with the sound wavelength. In the first two experiments, the same pressure p_1 is produced by T, first at R and then at X. Thus

$$p_1 = \frac{V_R}{M_R} = \frac{V_x}{M_x} \tag{9.2.1}$$

where M_R and M_x are the voltage sensitivities of R and X.

But

$$V_R = T_{em} v_1 \tag{9.2.2}$$

where T_{em} is the transduction coefficient, as discussed in Section 2.12 in Chapter 2. For example, $T_{em} = Bl$ for a loudspeaker. Now

$$v_1 = \frac{p_1 A}{Z_m} \tag{9.2.3}$$

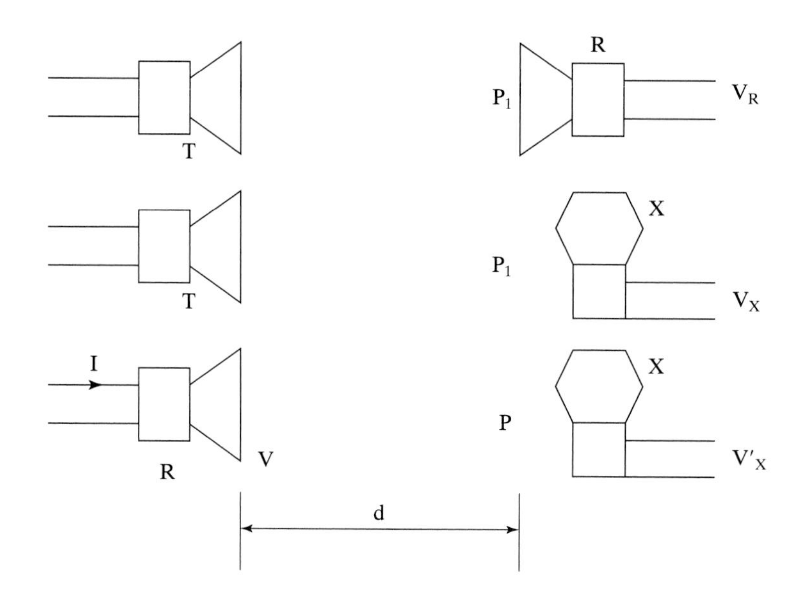

Fig. 9.2 Reciprocity calibration of a microphone of unknown sensitivity.

where A is the area of R, and Z_m is its mechanical impedance. Hence

$$\frac{V_R}{p_1} = \frac{T_{em}A}{Z_m} = M_R \tag{9.2.4}$$

In the third experiment in Fig. 9.2, the velocity produced at R for a current I is

$$v = \frac{T_{me}I}{Z_m} \tag{9.2.5}$$

Let the transfer function of the field be F; that is,

$$F = \frac{p}{v} \tag{9.2.6}$$

Usually we have a free-field condition, and the radiation is in the form of spherical waves. From Eqs. (9.2.5) and (9.2.6):

$$p = \frac{FT_{me}I}{Z_m} \tag{9.2.7}$$

Hence, from Eqs. (9.2.4) and (9.2.7)

$$p = \frac{FIM_R}{A} \tag{9.2.8}$$

Now p can also be expressed in terms of the unknown sensitivity; that is,

$$p = \frac{V'_x}{M_x} = \frac{FIM_R}{A} \tag{9.2.9}$$

Since from Eq. (9.2.1)

$$M_x = \frac{V_x}{V_R}M_R \tag{9.2.10}$$

when M_R is eliminated between Eqs. (9.2.9) and (9.2.10), we get

$$M_x = \sqrt{\frac{V_x V'_x}{IV_R}\frac{A}{F}} \tag{9.2.11}$$

For spherical waves of strength Q (=vA), the far-field pressure

$$p = \frac{\rho ckQ}{4\pi d} \quad \text{(see Eq. 3.4.15)}$$

so that

$$F = \frac{p}{v} = \frac{\rho ckA}{4\pi d} = \frac{\rho cA}{2\lambda d}$$

and hence

$$M_x = \sqrt{\frac{V_x V'_x}{IV_R} \cdot \frac{2\lambda d}{\rho c}} \tag{9.2.12}$$

It was pointed out by MacLean (1940) that if the unknown microphone and the reversible one were identical, then all three experiments in the reciprocity calibration procedure would produce the same resultant voltage; in other words,

$$V_R = V_x = V'_x$$

and in that case

$$M = \sqrt{\frac{V_R}{I} \frac{A}{F}} \tag{9.2.13}$$

In this way it is possible to calibrate two identical reversible microphones with only one electrical measurement.

Finch et al. (1964) found an interesting application of this idea, in measuring the cavitation threshold of liquid helium. They used the apparatus shown in Fig. 9.3, which does not conform to free-field conditions. In this circumstance, a standing wave is set up in the volume between the transducers. However, since the driver has a high impedance, the particle velocity v_1 will not differ appreciably from the free-field particle

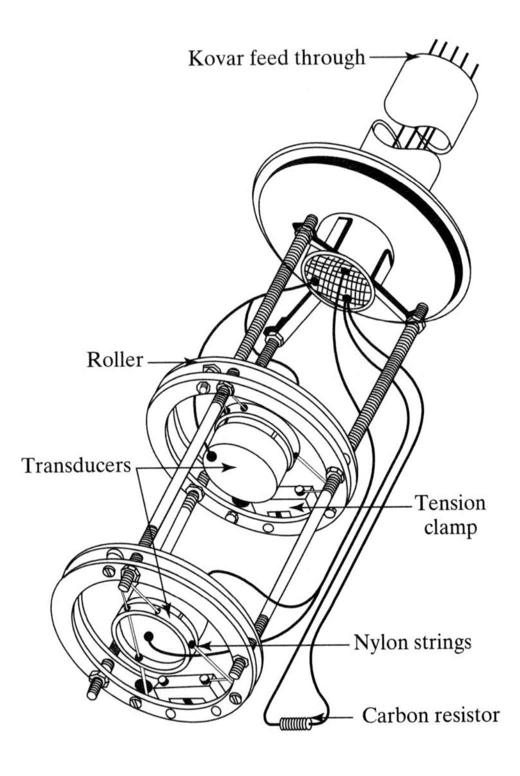

Fig. 9.3 Driver and microphone suspension. The PZT-4 transducers were 1 in. in diameter and $\frac{1}{2}$ in. in thickness. A carbon resistor, used to determine temperature stability, is seen below the lower transducer. (After Finch et al., 1964.)

velocity v_F. Suppose that due to the nondirectionality of the beam, the pressure in a progressive wave is reduced by a factor r in travelling once between the transducers. The pressure at the microphone, P, will then be given by

$$P = 2P_F r/(1 - r^2) \tag{9.2.14}$$

where P_F is the free-field pressure at the driver. Thus, in this case,

$$F = (\rho c/A)[2r/(1 - r^2)] \tag{9.2.15}$$

where ρ and c are the density and sound velocity of liquid helium, respectively. Hence, the microphone could be calibrated and the cavitation threshold determined in the same experiment.

9.3 Directivity

The response of a microphone usually depends on the direction from which the incoming sound arrives. Similarly, the radiation from a source is direction dependent. Most microphones and sources are usually increasingly directional at higher frequencies. The usual method of measuring the directivity of a microphone is to mount it on a turntable in an anechoic room and plot the response as the microphone receives sound from a fixed source. When measuring the directivity of a source, the positions of the source and the microphone are reversed. The results are usually rendered as a polar plot of sensitivity or intensity level in dB versus angle (see Fig. 9.4). The directivity factor $D(\theta)$ of a microphone is defined as the ratio of the power sensitivity $M(\theta)$ to the value M_{av} that would result from having the same sensitivity in all directions if the total power received were the same. For a symmetric receiver, it may be shown that

$$D(\theta) = \frac{M(\theta)}{M_{av}} = \frac{2M(\theta)}{\int_0^n M(\theta)\sin\theta\, d\theta} \tag{9.3.1}$$

It also follows that the total power received is $4\pi r^2 M_{av}$. When the device is used as a source, the generated intensities $I(\theta)$ and I_{av} replace $M(\theta)$ and M_{av} in Eq. (9.3.1). The directivity index is defined by

$$DI = 10 \log_{10} D(\theta) \tag{9.3.2}$$

For a piston in a baffle (see Section 3.3), it may be shown that the directivity factor is

$$DF = \frac{k^2a^2}{1 - \frac{2J_1(2ka)}{2ka}} \tag{9.3.3}$$

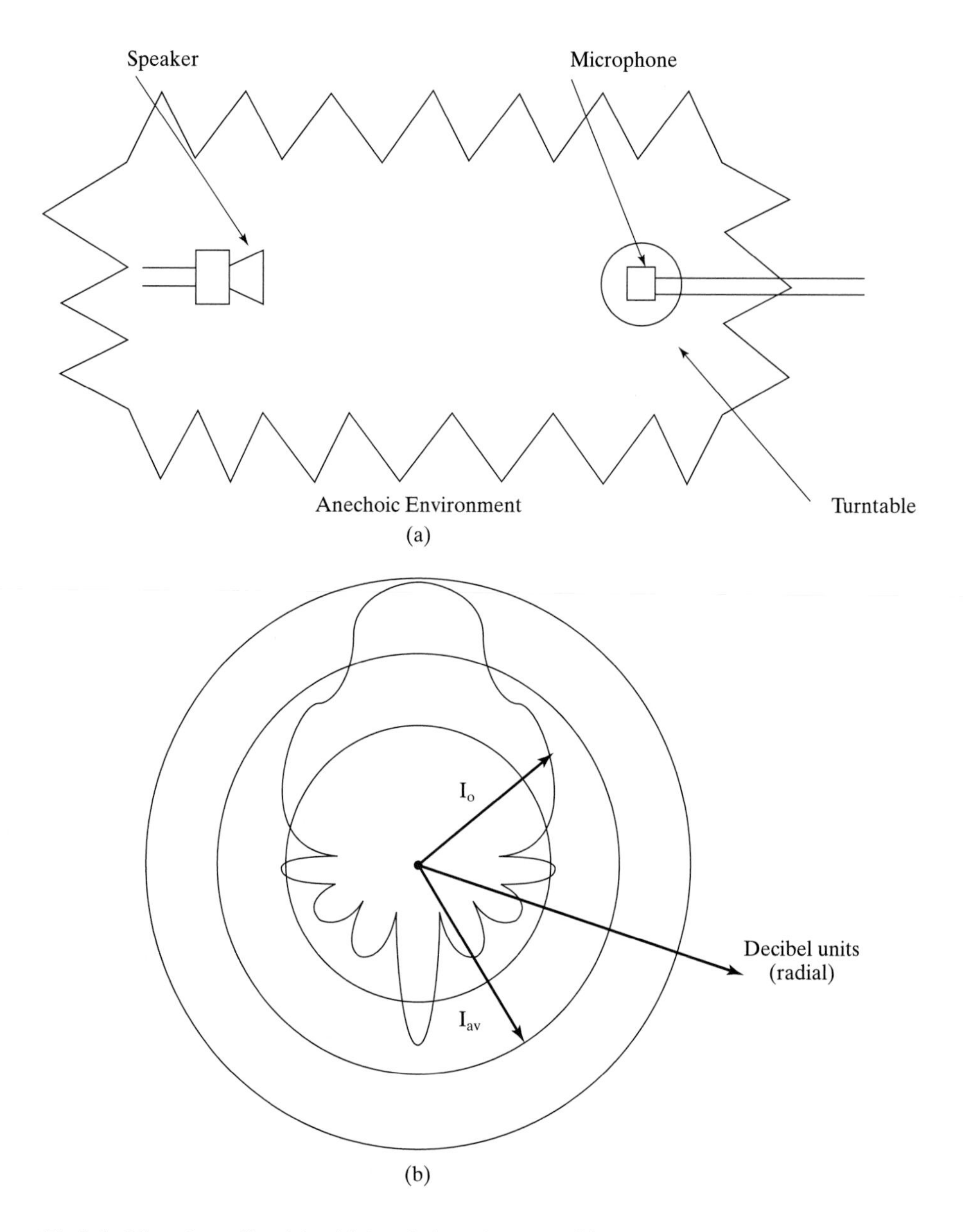

Fig. 9.4 Microphone directivity. (a) Anechoic environment. (b) Typical plot.

9.4 Particle Velocity

Particle velocity can be measured by means of a hot wire anemometer. This device is commonly used in aerodynamic measurements and operates best at low frequencies. Another method of measuring particle velocity is to use two pressure microphones at a short separation and hence determine the pressure gradient. Hence, for the component of particle velocity in the x direction, we have

$$\rho\frac{\partial v}{\partial t} = -\frac{\partial p}{\partial x} \tag{9.4.1}$$

For a harmonic wave travelling in the positive x direction

$$j\rho\omega v = +jkp$$

so that

$$v = \frac{p}{\rho c} \tag{9.4.2}$$

In general, velocity has three components and thus four microphones are required.

Another type of microphone whose output is proportional to particle velocity is the ribbon microphone, which consists of a metallic ribbon loosely suspended between the pole pieces of a magnet. It is the pressure difference between the two faces of the ribbon that causes it to move, and since the pressure difference is proportional to velocity (provided the thickness of the ribbon is small compared to a wavelength), such a device is often termed a velocity microphone. Olson (1957) provides further details on velocity microphones.

9.5 Density Fluctuation

Density fluctuations are also associated with sound fields, and density measurements are particularly important for shock wave investigations. The density and pressure fluctuations in the field are directly proportional, as given by $\delta p = c^2\delta\rho$. The most common methods for determining density fluctuations directly use optical techniques that depend on the variation of refractive index with density. The main optical methods use ultrasonic scattering (the Debye-Sears effect) and interferometry. The refractive index of a material is defined by

$$\mu = \frac{c_0}{c} \tag{9.5.1}$$

where c_0 = velocity of light in vacuo and c = velocity of light in the medium of concern. In gases, the difference between c and c_0 is small and proportional to density, to a first approximation. Hence

$$\mu = \frac{c_0}{c_0 - A\rho} = 1 + \beta\frac{\rho}{\rho_s} \tag{9.5.2}$$

where ρ_s is a reference density, usually taken at STP. For air under sodium D light (λ = 5893Å) at STP, β = .000292.

The density variation in different parts of the field causes refraction towards the region of increasing density and a relative phase shift between rays passing through different parts of the field. The refractive effect is used in scattering (Schlieren and shadowgraph methods), and the phase shift is used in interferometric methods.

An arrangement for studying ultrasonic light scattering is shown in Fig. 9.5. Light from source S (an aperture or a slit) is converted to a parallel beam by lens L_1 and passes through the test cell where an ultrasonic beam is being produced by a transducer. Some of the light is scattered by the sound field, which acts as a diffraction grating. The second lens L_2 serves to focus the light emerging from the test cell over its focal plane. There will be several diffracted orders, and the intensity of the nth order was shown by Raman and Nath (1935) to be given by the square of the nth order Bessel function:

$$I_n = J_n^2(\nu) \tag{9.5.3}$$

where

$$\nu = \text{the Raman-Nath parameter}$$
$$= 2\pi\kappa pL/\lambda_L \tag{9.5.4}$$

and

$$\kappa = \text{piezo-optic coefficient}$$
$$= \frac{\partial\mu}{\partial p}$$
$$p = \text{sound pressure amplitude}$$
$$L = \text{the width of the sound field}$$
$$\lambda_L = \text{the wavelength of the light}$$

Section 11.3 contains an exposition on Bessel functions.

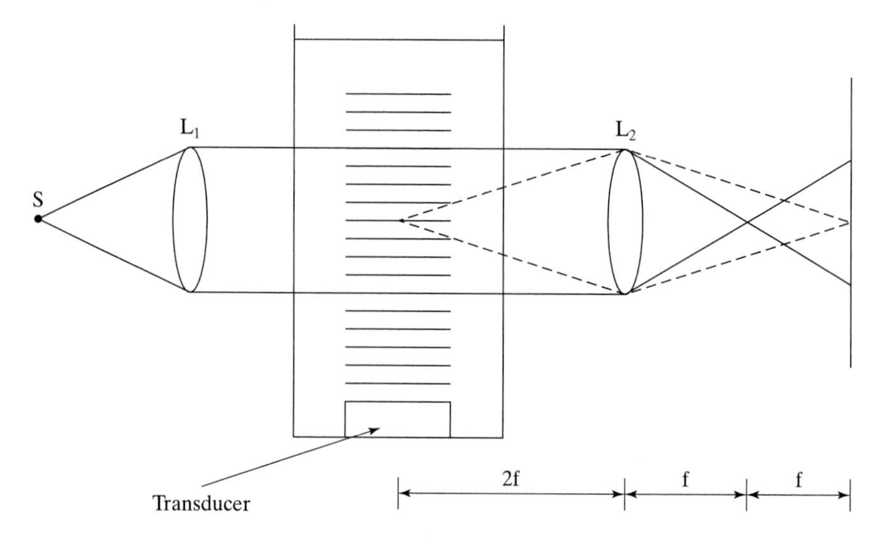

Fig. 9.5 Apparatus for studying ultrasonic light scattering (Schlieren system).

If the zeroth order diffraction pattern is now blocked off by an optical stop, only the scattered light will pass on. But this will result in the formation of an image of the sound field if L_2 and the imaging screen are set at 2f and 4f from the center of the test cell. By measuring the scattered intensity at the focal plane of L_2 with a photomultiplier, the actual density fluctuations can be deduced. At much higher frequencies, so-called Bragg diffraction is obtained. The pressure maxima in the sound field then act like atomic layers in a crystal. A unified approach to ultrasonic light diffraction, which treated Raman-Nath scattering at low frequencies and Bragg scattering at high frequencies, was first set out by Cook and Klein (1967).

9.6 Intensity

The intensity of a sound wave is the energy crossing unit area in the sound field. This energy is equivalent to the work being done by the field at that point so that the instantaneous intensity

$$I_i = R_e[p] \times R_e[v] \tag{9.6.1}$$

Consequently, if pressure and velocity are measured, the intensity can be calculated. Until recently, the normal approach was to obtain time averages of pressure and velocity (from the pressure gradient) by using two microphones. Recent developments started by Schultz, Smith, and Malme (1964), by Elko (1984), and by Tichy (1984) have pointed out that time averaging results in a loss of much valuable information. This loss may be explained by supposing that the pressure in the sound field is given by

$$p(r,t) = P(r)e^{-j\phi(r)}e^{j\omega t} \tag{9.6.2}$$

then

$$\begin{aligned} v(r,t) &= \frac{-\nabla p(r,t)}{j\omega\rho} \\ &= \frac{1}{\omega\rho}[p(r,t)\nabla\phi(r) + j\nabla P(r)\, e^{j\varphi(r)}\, e^{j\omega t}] \end{aligned} \tag{9.6.3}$$

Hence, from Eqs. (9.6.1) and (9.6.3)

$$\begin{aligned} I_i\,(r,t) = \frac{1}{2\omega\rho}[&P^2(r)\nabla\varphi(r) + P^2(r)\nabla\phi(r)\cos 2\,(\omega t - \phi) \\ &- P(r)\,\nabla\,P(r)\sin 2(\omega t - \phi)] \end{aligned} \tag{9.6.4}$$

Now it can be seen that the *average* intensity

$$\bar{I}(r) = \frac{1}{T}\int_0^T I_i(r,t)dt = \frac{P^2(r)}{2\omega\rho}\nabla\varphi(r) \tag{9.6.5}$$

Schultz, Smith, and Malme (1964) invented the concept of complex intensity in

$$I_c(r) = \bar{I}(r) + j\bar{Q}(r) = \frac{p(r)v^*(r)}{2}$$

$$= \frac{1}{2\omega\rho}[P^2(r)\,\nabla\varphi(r) - jP(r)\,\nabla P(r)] \tag{9.6.6}$$

Here the first term is the average intensity. The second term, the imaginary part of the complex intensity, is called the reactive intensity. Both $\bar{I}(r)$ and $\bar{Q}(r)$ are time independent, depending only on position. It follows that

$$\bar{Q}(r) = \frac{1}{2}I_m[p(r)\cdot v^*(r)]$$

$$= -\frac{1}{2\omega\rho}P(r)\nabla P(r)$$

$$= -\frac{1}{4\omega\rho}\nabla P^2(r) \tag{9.6.7}$$

The instantaneous intensity can now be written as follows:

$$I_i\,(r,t) = \bar{I}\,(r) + \bar{I}\,(r)\cos 2\,(\omega t - \varphi) + \bar{Q}\,(r)\sin 2\,(\omega t - \varphi) \tag{9.6.8}$$

Hence, the instantaneous power transport depends on a time independent intensity $\bar{I}(r)$ and two other vectors that oscillate with a quarter period time shift. One is collinear with I(r), and the other is in general in a different direction.

For plane progressive waves, P(r) is constant everywhere, $Q(r) = 0$, and

$$\varphi(r) = kx \tag{9.6.9}$$

$$\bar{I}(x) = \frac{P^2(x)\bar{i}}{2\rho c} \tag{9.6.10}$$

and

$$I_i(x) = \bar{I}(x)\,[1 + \cos 2\,(\omega t - kx)] \tag{9.6.11}$$

For progressive spherical waves,

$$p(r) = P(r)e^{-j\varphi(r)} = \frac{A}{r}e^{-jkr} \tag{9.6.12}$$

that is,

$$P(r) = \frac{A}{r}, \quad \varphi(r) = kr$$

so that

$$\bar{I}(r) = \frac{1}{2\omega\rho}\cdot\frac{A^2}{r^2}\cdot k\bar{e}_r = \frac{A^2}{2\rho c r^2}\bar{e}_r \qquad (9.6.13)$$

and

$$\overline{Q}(r) = -\frac{1}{4\omega\rho}\nabla\left(\frac{A^2}{r^2}\right) = \frac{A^2\bar{e}_r}{2k\rho c r^3} \qquad (9.6.14)$$

Both $\bar{I}$ and $\overline{Q}$ are directed radially. The instantaneous power flow

$$\begin{aligned} I_i(r,t) &= \bar{I} + \bar{I}\cos 2\,(\omega t - kr) + \overline{Q}\sin 2\,(\omega t - kr) \\ &= \frac{A^2}{2\rho c r^2}[1 + \cos 2\,(\omega t - kr) + \frac{1}{kr}\sin 2\,(\omega t - kr)]\,\bar{e}_r \end{aligned} \qquad (9.6.15)$$

Plane and spherical waves are special cases, and in general the flows are more complicated.

Measurements are made using two microphones at a short separation. The pressure is obtained as the average of the two signals, and the velocity from the pressure gradient. Three microphone pairs are required to obtain the full intensity vector (see Fig. 9.6). Digital processing can then be used to calculate the intensity (average and complex), the energy densities (kinetic and potential), and the specific acoustic impedance. Plotting the intensity vector can be used to locate sound sources (see Fig. 9.7). The viability of the method was improved dramatically with the advent of digital processing, although it is still limited to about 7 kHz in air due to the need to keep microphones at a separation that is small compared with a wavelength.

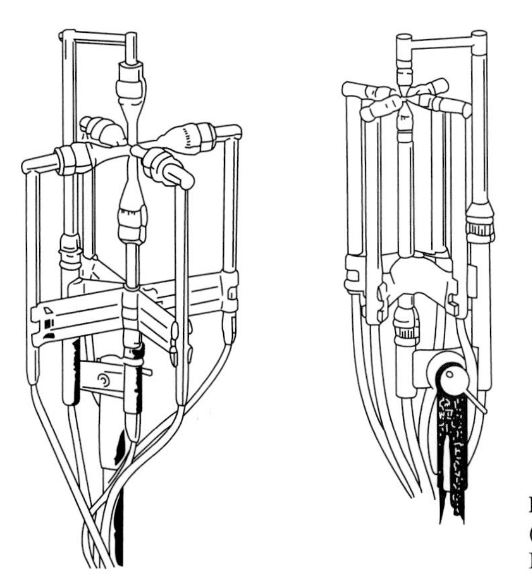

Fig. 9.6 (a) 50 mm (WA0491) intensity vector probe. (b) 12 mm (WA0447) intensity vector probe. (After Brüel & Kjaer.)

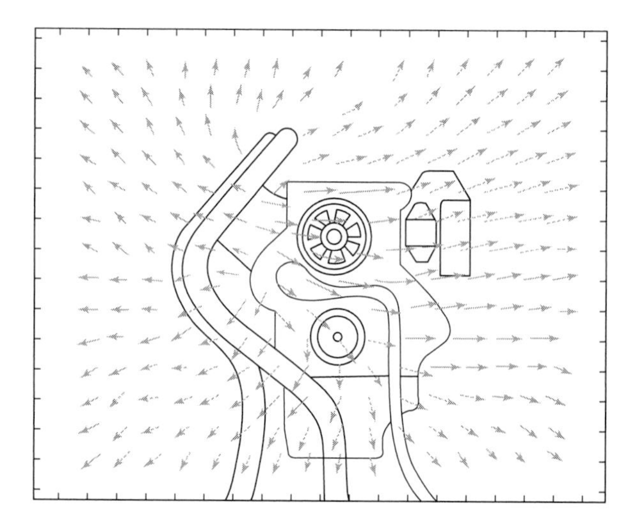

Fig. 9.7 Intensity vector plot for V8 diesel engine. (After Brüel & Kjaer.)

9.7 Speed of Sound

The earliest determinations of the speed of sound used "time of flight" measurements. These measurements were clearly of limited accuracy, a situation that was greatly improved with the invention of the standing wave tube. As we saw in Chapter 4, the pressure maxima and minima are separated by a distance of $\lambda/2$. In 1866, Kundt discovered that powder particles would collect at nodes in a standing wave tube, thus permitting a new technique to measure the velocity of sound in air or whatever gas was contained in the tube. The frequency of the sound waves was determined by exciting the sound wave with longitudinal vibrations in a rod. The rod was excited by stroking it with a resin-impregnated cloth (see Fig. 9.8). The rod is then resonant so that its length corresponds to one wavelength:

$$\lambda_R = \frac{c_R}{f} = L \tag{9.7.1}$$

Thus

$$f = \frac{c_R}{L} = \frac{1}{L}\sqrt{\frac{Y}{\rho}} \tag{9.7.2}$$

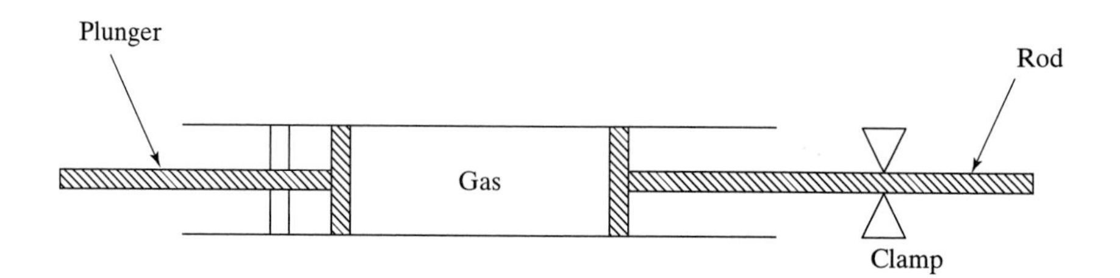

Fig. 9.8 Kundt's tube method for finding the velocity of sound in a gas.

The position of the plunger was adjusted until the gas column resonated at the same frequency as the rod.

Although an advance on the earliest techniques, this method still had limitations. The movement of particles, for example, is a complex nonlinear phenomenon. Large particles sometimes gather at velocity antinodes, and sometimes the separation of the heaps of powder is not exactly at half wavelength separations. More accurate measurements were possible by using electronically controlled piezoelectric drivers and probe microphones to locate the pressure nodes. With these refinements, the technique could be extended to the ultrasonic frequency range. It was also realized that the standing wave tube could be used to measure the impedance of a termination, as described in Section. 9.8.

A very accurate way of measuring the velocity of sound in a liquid, known as the *sing-around technique*, was developed by Greenspan and Tschiegg (1957). A pulse is sent from the driving transducer into the liquid, as shown in Fig. 9.9. The pulse repetition rate is set at slightly less than is to be expected in the circuit. The received pulse is then amplified and used to synchronize the pulse generator. Hence, the pulse repetition frequency f is given by

$$\frac{1}{f} = \tau + \frac{l}{c} \tag{9.7.3}$$

f can be measured to great accuracy, as can l; the sum of the delays in the circuit, τ, can be determined by calibration with a liquid of known velocity. Velocimeters of this type are used by the navy to measure the velocity of sound in seawater.

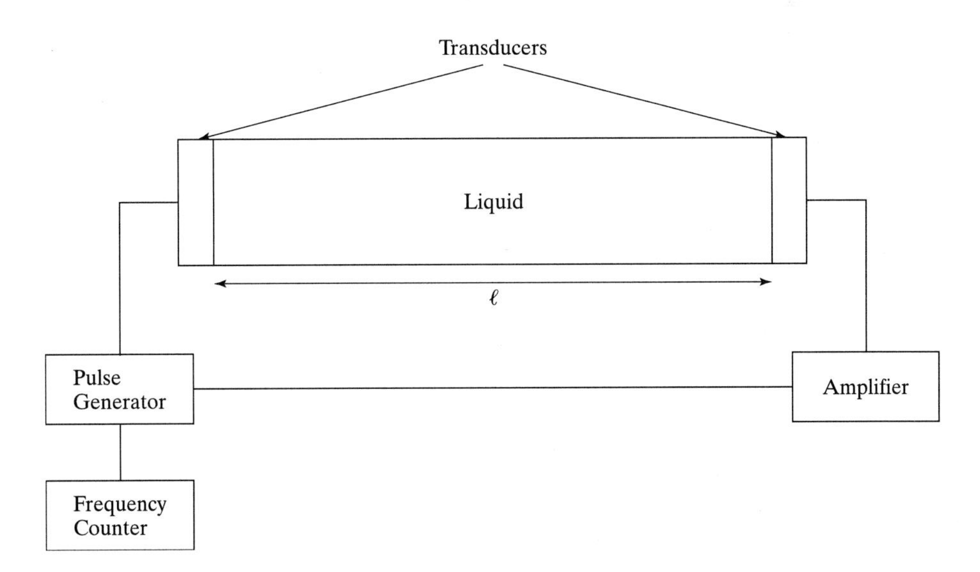

Fig. 9.9 Apparatus to measure sound velocity by the sing-around technique.

9.8 Acoustic Impedance

Reflection (or absorption coefficients) and acoustic impedance of materials can be measured using an impedance tube. A standard method for doing this is described by the American Society of Testing of Materials (1990) and the apparatus is illustrated in Fig. 9.10. It may be shown that

$$r_I = \left[\frac{\log_{10}^{-1}(A/20) - 1}{\log_{10}^{-1}(A/20) + 1}\right]^2 \tag{9.8.1}$$

and the mechanical resistance, R, and reactance, X, are given by

$$\frac{R}{\rho c} = \frac{1 - r_I}{1 + r_I + 2\sqrt{r_I}\cos\dfrac{2\pi D_1}{D_2}} \tag{9.8.2}$$

and

$$\frac{X}{\rho c} = \frac{2\sqrt{r_I}\sin\left(\dfrac{l\pi D_1}{D_2}\right)}{1 + r_I + 2\sqrt{r_I}\cos\dfrac{2\pi D_1}{D_2}} \tag{9.8.3}$$

where

$$\begin{aligned} r_I &= \text{intensity reflection coefficient} \\ &= r_p^2 \end{aligned}$$

where r_p is the pressure reflection coefficient, and where A is the difference in dB between the first maximum and first minimum of sound pressure level as measured from the test termination (see Fig. 9.11).

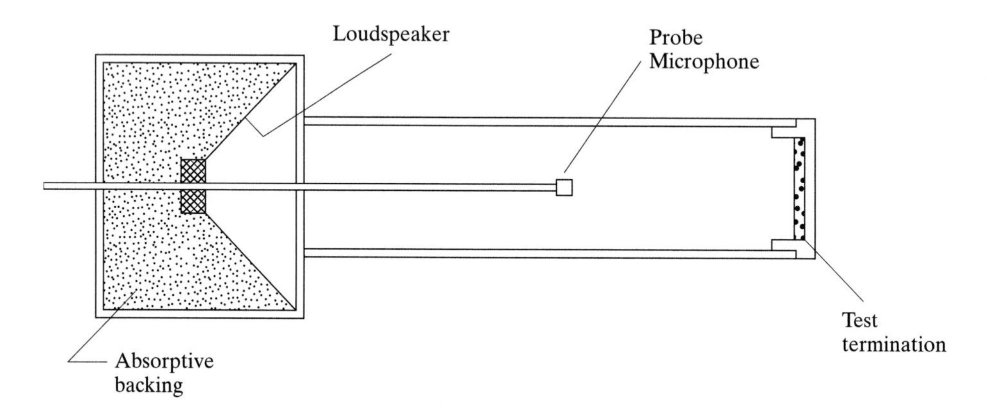

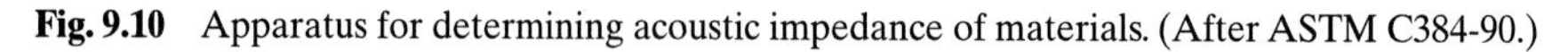
Fig. 9.10 Apparatus for determining acoustic impedance of materials. (After ASTM C384-90.)

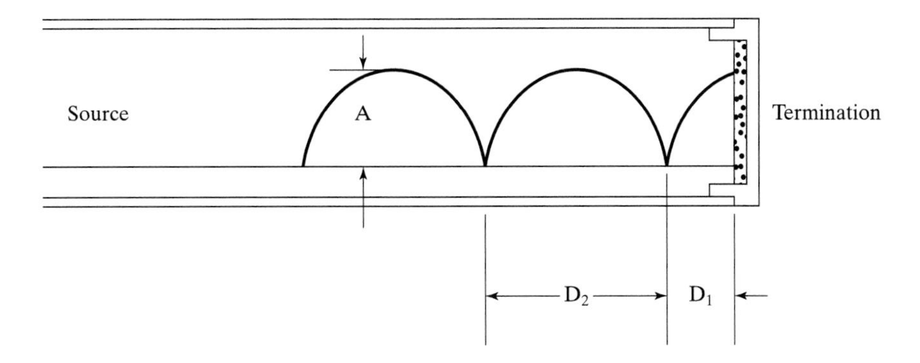

Fig. 9.11 Standing wave parameters.

9.9 Attenuation

The principal methods of measuring sound attenuation involve mechanical, optical, or electrical techniques. Some measurements have been made using thermal effects, but they are not very reliable. Different methods are often used for different frequency ranges.

1. Mechanical Methods
   Mechanical methods are the most frequently used, based on radiation pressure (see Fig. 9.12). Radiation pressure is twice the energy density of an incident sound wave. This method is for use solely on confined liquids and gases. Since

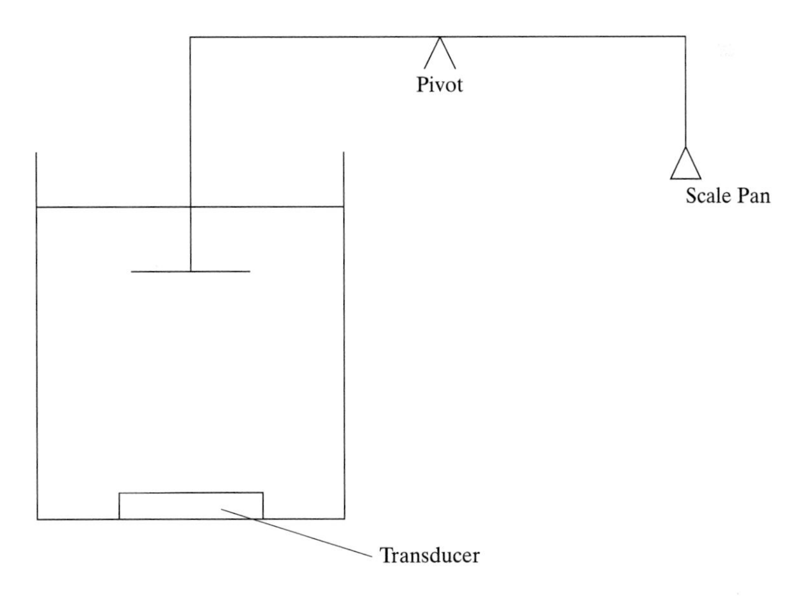

Fig. 9.12 Apparatus for measuring radiation pressure and hence attenuation.

$I = I_0 e^{-2\alpha x}$, the radiation pressure is measured at different lengths x to find α. There are a number of disadvantages to using this method:

i. Acting surface tension forces may easily be approximately 10 dynes (i.e., of the order of the forces being measured).
ii. Reflection from container walls will occur if the container is too narrow, causing standing waves to be generated and grossly incorrect results to be obtained.
iii. If a higher intensity is used, higher radiation pressure is obtained but cavitation occurs. Thus, the excess pressure must not be higher than one atmosphere (3 watts/cm^2 is the maximum intensity that should be used).
iv. At lower frequency, we may fail to intersect the entire beam so the basic relation will not apply.
v. In terms of hydrodynamic flow, the quartz wind can itself exert a pressure. This problem can be eliminated by gauze or muslin. But the radiation pressure gradient still causes flow as seen in shock waves. The shock reunites on the other side of the gauze.

For most liquids, careful measurements can be made above 10 Mz to an accuracy of 5 Mz to 10 Mz. The technique is not that accurate except for liquids of fairly high absorption.

2. Optical Methods

The Biquard method depends on the Debye-Sears effect: Successive compressions and rarefractions produce varying density, and the medium acts as a diffraction grating producing fringes on a screen. As we saw in Section 9.5, the greater the intensity the more the light is diffracted away from the main beam (i.e., the central maximum). Biquard used a photocell along the x-axis. If J_0 = intensity of light in the absence of sound and J = intensity of light in the presence of sound, then

$$1 - \frac{J}{J_0} = kI_0 e^{-2\alpha x} \tag{9.9.1}$$

Assume the output θ from the photocell is proportional to the light intensity, then

$$1 - \frac{\theta}{\theta_0} = kI_0 e^{-2\alpha x} \tag{9.9.2}$$

and

$$\log\left(\frac{\theta_0 - \theta}{\theta_0}\right) = \log(kI_0 e^{-2\alpha x}) \tag{9.9.3}$$

This is the equation of a straight line with slope $= -2\alpha$.

The optical method also has its problems. It is limited to transparent liquids, and light and sound beams have to be accurately perpendicular. Low-frequency divergence of the beam occurs. There also may be multiple reflections, as in the case of mechanical methods. High intensities produce higher order images because light is diffracted back into the main beam so that light intensity will not be proportional to sound intensity. Hydrodynamic flow also alters optical properties, which can sometimes be a problem.

The method can be enhanced by the use of photomultiplier tubes. The frequency range for optical methods is similar to that of mechanical methods, and it produces fair agreement with other methods.

3. Interferometric Method

A reflector is set parallel to a transducer face, as in Fig. 9.13. If the reflected wave is π out of phase with the incident wave, then the transducer output is reduced to zero. Because this occurs for every half wavelength movement of reflector, the effect on the driving oscillator is a good method to find sound velocity. By registering the current at two successive maxima, it is possible to find the attenuation coefficient. The interferometer is the standard instrument for gases. One difficulty with this method is that a very accurate alignment of the transducer and the reflecting surface is required. There is also a problem with the departure of the sound beam from the plane wave configuration. Lack of accurate knowledge of the reflection coefficient is also a limitation.

4. Direct Method

A microphone is placed on the axis of the beam. A knowledge of the near and far field regions, as discussed in Sections 3.12 and 3.13, is essential. The linear dimensions of the microphone should be small compared with the wavelength, or the microphone will disturb the neighboring field. Standing waves are a problem that can be overcome by using absorbing materials on the walls. The technique is more difficult for gases than for liquids. The useful frequency range in water is usually 1 mHz to 4 mHz. In gases, the frequency range is lower (20 kHz to 150 kHz). In a variation due to Knudsen and Fricke (1938), one first measures α by microphone in a chamber filled with nitrogen and then with the test gas, as a correction for wall absorption.

5. Pulse Method

The pulse method was developed to overcome standing wave problems. Sound waves introduce energy and temperature changes causing density changes and sound refraction. Pulses overcome these problems. Suitable parameters are 1000 pulses/cm with width 1 μsec at a frequency of 15 Hz and power 1/10 of

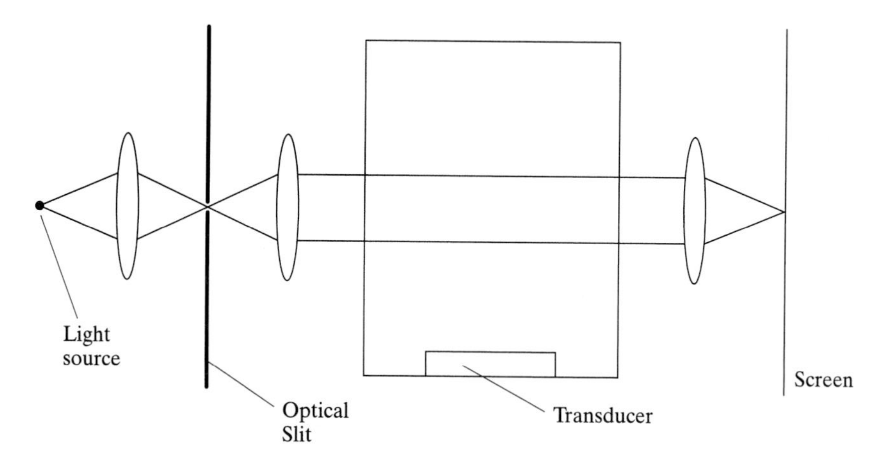

Fig. 9.13 Apparatus used by Biquard to measure absorption.

continous waves. The theory of the measurement is similar to that of the direct method. The use of narrow pulses increases the spread of the frequencies in a Fourier analysis. Simple calculations show that in pulses at 15 mHz, this error is no greater than 1 part in 150. The pulse method is the most accurate, depending on the refinement of techniques, in the range 1 Hz to 200 mHz.

6. Reverberation Methods

The general principle behind the reverberation methods is to produce a sound field in an enclosure, shut off the source, and measure the decay of intensity with time. For example, a brass sphere contained in a liquid can be excited in a radial mode by a transducer mounted on its surface (see Fig. 9.14). When the sound is shut off, the same transducer can be used to measure decay. There is absorption in both the medium and in the walls. The radial mode is used. If the radius of the liquid (R in Fig. 9.14) is an odd number of quarter wavelengths, then the energy absorbed by the walls is neglible. We use the formula

$$\alpha = \frac{1}{ct}\log\frac{I_0}{I_t} \tag{9.9.4}$$

where c = sound velocity and t = time. This method is useful in the range 24 kHz to 200 kHz.

Absorption in air can be measured from the reverberation time in a room. In the absence of furnishings, in a room with concrete walls, there is a high reflection coefficient. Under these circumstances we may use the same formula as in Eq. (9.9.4) for the spherical resonator. Some interesting phenomena involving humidity, which were discovered in this way by Knudsen, are summarized in Chapter 14. Knudsen solved the problem of the lack of dispersion over frequency among the modes of the room by using a paddle. With a large reverberation chamber, this technique can be used down to frequencies low in the audio range.

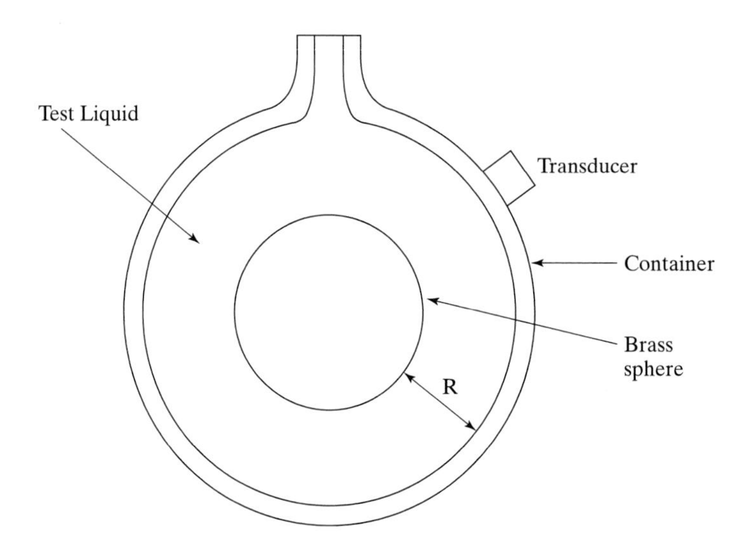

Fig. 9.14 Apparatus for measuring attenuation in a liquid.

REFERENCES AND FURTHER READING

American Society of Testing of Materials. 1990. Standard C 384–90.

Beranek, L.L. 1988. *Acoustical Measurements*. Rev. ed. Acoustical Society of America.

Brüel & Kjaer. *Measurement Microphones*. (Primer.)

Cook, B.D., and W.R. Klein. 1967. "Unified Approach to Ultrasonic Light Diffraction." SU-14 *IEEE Trans. Sonics & Ultrasonics* 123–124.

Elko, G.E. 1984. *Frequency Domain Estimation of the Complex Acoustic Intensity and Energy Density*. Ph.D. thesis, Pennsylvania State University.

Finch, R.D., R. Kagiwada, M. Barmatz, and I. Rudnick. 1964. "Cavitation in Liquid Helium." A134 *Phys. Rev.* 1425–1438.

Greenspan, M., and C.E. Tschigg. 1957. "Speed of Sound in Water by a Direct Method." 59 *J. Res. of the Nat. Bur. Stands*. 249–254.

Knudsen, V.O., and E. F. Fricke. 1938. "The Absorption of Sound in Carbon Dioxide and Other Gases." 10 *J. Acous. Soc. Am.* 89–97.

MacLean, W.R. 1940. "Absolute Measurement of Sound Without a Primary Standard." 12 *J. Acous. Soc. Am.* 140–146.

Miller, H.B., ed. 1982. "Acoustical Measurements: Methods and Instrumentation." Vol. 16 in *Benchmark Papers in Acoustics*. Hutchinson Ross Publishing Co.

Olson, H.F. 1957. *Acoustical Engineering*. Van Nostrand.

Raman, C.V., and N.S.N. Nath. 1935. "The Diffraction of Light by High Frequency Sound Waves: Part 1." 2 *Proc. Ind. Acad. Sci.* 406–412.

Schultz, T.J., P.W. Smith, Jr., and C.I. Malme. 1964. "Measurements in Nearfields and Reverberant Spaces." Bolt, Beranek, and Newman, Inc. Rep. No. 1135.

Tichy, J. 1984. "Acoustic Intensity Measurements: A Review." *Proc. AIAA/NASA 9th Aeroacoustic Conference*.

PROBLEMS

9.1 Describe how the Rayleigh disc may be used to measure particle velocity. How much further would a torsion wire have to be rotated if the velocity amplitude of a sound field were to be doubled during the use of a Rayleigh disc?

9.2 Describe how you might use two identical piezoelectric disks to measure the cavitation threshold of liquid nitrogen.

9.3 An experiment to measure the cavitation threshold of liquid nitrogen is undertaken. Given the free field pressure at the driving disk, derive an expression for the pressure at the microphone in terms of the pressure reduction factor, r, in travelling once between the transducers. Also derive an expression for the transfer function of the acoustic field in terms of the area of the disk and the specific acoustic impedance of liquid nitrogen.

9.4 A piston phone comprises a tube that may be fitted tightly over a circular microphone. A piston at the other end is driven electrodynamically with displacement amplitude η. If the cross section of the tube is A and the enclosed length is l, what pressure amplitude will be developed in air at STP? Assume the driving angular frequency is ω and that the microphone stiffness is very high.

9.5 Derive an expression for the directivity factor of a piston in a baffle.

9.6 Discuss the measurement of acoustic intensity.

9.7 An automobile manufacturer wants to identify the primary causes of noise in a test engine. Describe how this might be accomplished. Can you think of techniques other than acoustic intensity that might be employed?

9.8 Explain the operation of the sing-around technique for determining the velocity of sound in seawater.

9.9 Prove the results given in Eqs. (9.8.1), (9.8.2), and (9.8.3).

9.10 Explain the use of the formula given by Eq. (9.9.4).

9.11 How could you measure the attenuation of ultrasound in living tissue? Why would such a measurement be important?

Chapter 10

Plane Waves in Large Enclosures

10.1 Modes of a Tube with Rigid Caps

We discussed the propagation of plane waves along lines in Chapter 1 and in pipes in Chapter 4. In this chapter we will show how plane wave propagation can be used to describe the acoustic behavior of rectangular ducts and rooms. We will begin by discussing sound in a tube closed with rigid caps. This topic is simpler to understand than the case of propagation in a pipe with a general end condition, which we studied in Chapter 4. We are returning to the subject to discuss the importance of modes in analyzing the acoustics of ducts and rooms.

The general solution of the one-dimensional wave equation for velocity is

$$\begin{aligned} v &= Ae^{j(\omega t+kx)} + Be^{j(\omega t-kx)} \\ &= e^{j\omega t}[C\cos(kx) + D\sin(kx)] \end{aligned} \tag{10.1.1}$$

Now the boundary conditions for a tube with rigid caps are

$$v = 0, \text{ when } x = 0 \tag{10.1.2}$$

so that

$$C = 0 \tag{10.1.3}$$

and then

$$v = De^{j\omega t}\sin(kx) \tag{10.1.4}$$

But we also have the boundary condition

$$v = 0, \text{ when } x = \ell \quad (10.1.5)$$

so that

$$\sin(k\ell) = 0 \quad (10.1.6)$$

and

$$k\ell = n\pi, \text{ or } \ell = n\lambda/2 \quad (10.1.7)$$

In other words, the only possible solutions occur when the length of the tube is an integral number of half wavelengths. The frequencies at which these modes of oscillation occur are

$$f_n = \frac{nc}{2\ell}, \qquad n = 1, 2, 3, \ldots \quad (10.1.8)$$

The modes are illustrated in Fig. 10.1.

A similar case involves a tube with one end open and the other rigidly capped or closed. As a first approximation to this problem, which is encountered with organ pipes,

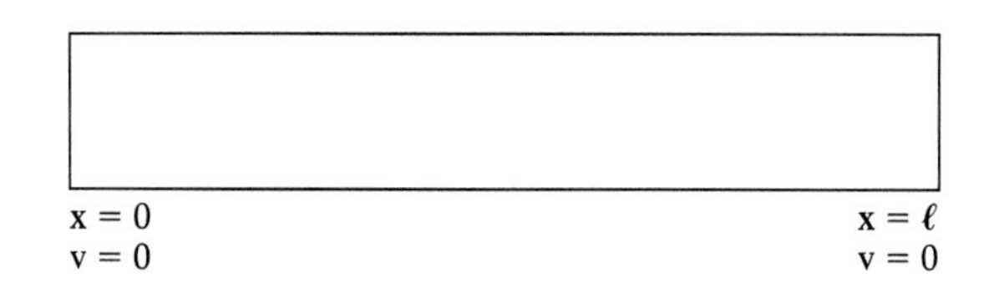

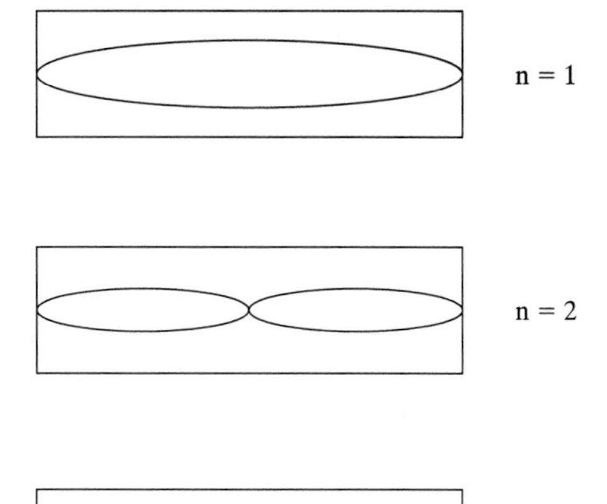

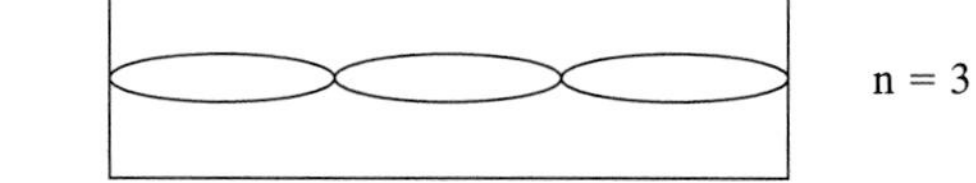

Fig. 10.1 Velocity amplitudes for the first three modes of a capped tube.

for example, the open end may be regarded as having no mass, so that the pressure will be zero at that point. Using the condition of zero velocity at $x = 0$, we have

$$v = v_o e^{j\omega t}\sin(kx) \tag{10.1.9}$$

Since

$$\frac{\partial v}{\partial t} = -\frac{1}{\rho}\frac{\partial p}{\partial x} \tag{10.1.10}$$

we obtain

$$j\omega v = -\frac{1}{\rho}\frac{\partial p}{\partial x} \tag{10.1.11}$$

and

$$\begin{aligned} p &= j\rho\omega\int v dx \\ &= j\rho\frac{\omega}{k}v_o e^{j\omega t}\cos(kx) \\ &= \rho c v_o e^{j(\omega t+\pi/2)}\cos(kx) \end{aligned} \tag{10.1.12}$$

Now, applying the boundary condition

$$p = 0, \text{ when } x = \ell$$

gives

$$k\ell = (2n + 1)(\pi/2), n = 1, 2, 3, \ldots \tag{10.1.13}$$

or

$$\ell = (2n + 1)(\lambda/4) \tag{10.1.14}$$

Thus, the natural frequencies are given by

$$f_n = \frac{(2n + 1)c}{4\ell} \tag{10.1.15}$$

The pressure antinodes occur at the same positions as the velocity nodes and vice versa, as was the case with the rigidly capped tube (see Fig. 10.2).

It was found in practice that the resonance frequencies of pipes are lower than predicted by Eq. (10.1.15). This finding can be understood by imagining the air column to have a greater effective length than that of the tube, since the surrounding air participates in the oscillation of the column to some extent. The first theoretical estimate of

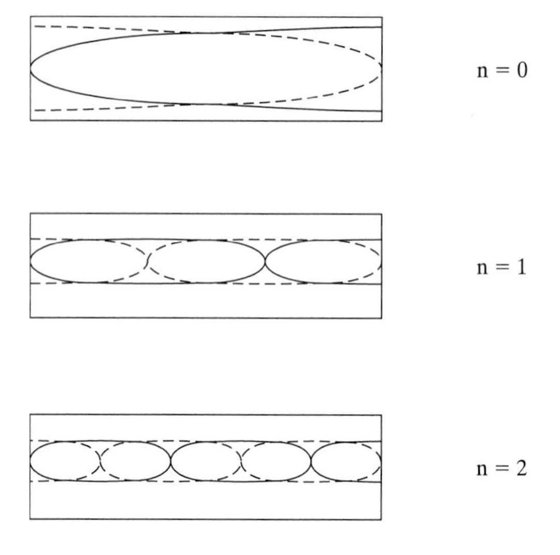

Fig. 10.2 Velocity amplitudes (——) and pressure amplitudes (– – –) for the first three modes of an open-closed pipe.

this end correction was made by Rayleigh, who considered the oscillation of the air at the mouth of the tube to be equivalent to a plane piston radiator in an infinite rigid baffle. This procedure leads to an end correction at low frequencies of 0.82R, where R is the tube radius. Levine and Schwinger (1948) showed that the end correction for a tube without the rigid baffle at low frequencies was 0.61R, which agrees with experimental findings. The theory of the piston radiator was discussed in Chapter 3.

10.2 Energy in Plane Standing Waves

We discussed energy transport in progressive waves in Chapter 1. Now we will see what happens to the energy in a standing wave. Consider an element of the medium in a plane parallel to the wavefront and having thickness dx. The kinetic energy of this volume is

$$\mathrm{KE} = \frac{1}{2}\rho v^2 S dx \tag{10.2.1}$$

where S is the area of the volume element. Now the potential energy associated with the wave is due to work done in compressing the medium; that is,

$$\delta(\mathrm{PE}) = -p\delta V \tag{10.2.2}$$

where p is the acoustic pressure. Hence

$$\mathrm{PE} = -\int p\delta V \tag{10.2.3}$$

But

$$\frac{\delta V}{V_0} = -\frac{\delta p}{\rho c^2} \tag{10.2.4}$$

and thus

$$\delta V = -\frac{V_0 \delta p}{\rho c^2} \tag{10.2.5}$$

Consequently,

$$\begin{aligned} PE &= \int \frac{pV_0}{\rho c^2} \delta p \\ &= \frac{1}{2} \frac{p^2}{\rho c^2} \end{aligned} \tag{10.2.6}$$

Thus, the energy density

$$\varepsilon = \frac{1}{2} \rho \left(v^2 + \frac{p^2}{\rho^2 c^2} \right) \tag{10.2.7}$$

For a plane wave travelling in the positive x direction,

$$p = \rho c v \tag{10.2.8}$$

Therefore

$$\varepsilon = \rho v^2 \tag{10.2.9}$$

or

$$\varepsilon = \frac{p^2}{\rho c^2} \tag{10.2.10}$$

We notice that since p and v are time dependent, the energy density must be time dependent. The time averaged energy density is thus

$$\bar{\varepsilon} = \frac{\rho v_0^2}{2} = \frac{p_0^2}{2\rho c^2} \tag{10.2.11}$$

where v_0 and p_0 are the velocity and pressure amplitudes.

For a progressive wave, the spatially averaged energy density is the same as the time averaged value. The acoustic intensity, I, of such a wave is the average flow of energy per unit cross-sectional area in the direction of propagation. Now the instantaneous flow of energy per unit cross-sectional area

$$\frac{dE}{dt} = \varepsilon c \tag{10.2.12}$$

Thus

$$I = \frac{d\overline{E}}{dt} = \overline{\varepsilon}c$$
$$= \frac{\rho c v_0^2}{2} = \frac{p_0^2}{2\rho c} \tag{10.2.13}$$

The time averaged energy density in a standing wave that is obtained with an open-ended tube is

$$\overline{\varepsilon} = \frac{\rho}{2}\left(\frac{v_0^2}{2}\sin^2 kx + \frac{v_0^2}{2}\cos^2 kx\right)$$
$$= \frac{\rho v_0^2}{4}, \text{ a constant} \tag{10.2.14}$$

10.3 Transmission Through Three Media

Suppose that waves generated in medium 1, with specific acoustic impedance $\rho_1 c_1$, impinge on a layer of medium 2, which in turn is in contact with a semi-infinite medium 3, as in Fig. 10.3. Assume that the acoustic impedances of media 2 and 3 are $\rho_2 c_2$ and $\rho_3 c_3$, respectively, and that the boundaries are located at $x = 0$ and $x = \ell$. The boundaries are neither rigid nor pressure release, as in the cases we have just considered in Section 10.1, since neither pressure nor velocity is necessarily zero at either boundary. However, it is obvious that if the displacement (and velocity and acceleration) is not the same on either side of a boundary, then the materials will either separate or interpenetrate. Neither of these possibilities is physically admissible, so we may state as a boundary condition that velocity must be continuous. The other boundary condition is established by considering a vanishingly thin layer of material located right at the boundary. The net force on this element is proportional to the difference of pressure on either side of the boundary. But the mass of the thin layer is vanishingly small; thus,

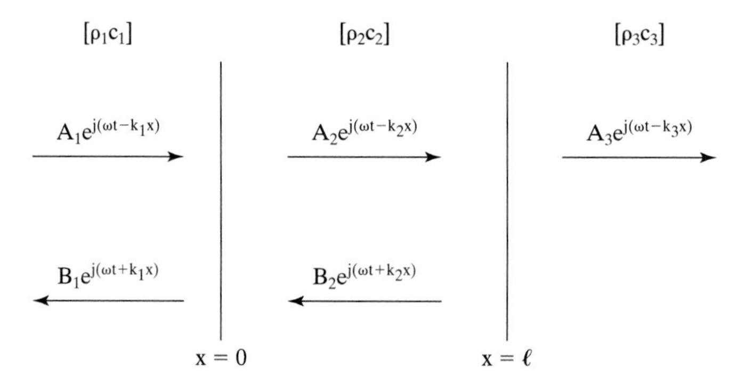

Fig. 10.3 Transmission through three media.

from Newton's second law, to avoid infinite accelerations we must stipulate continuity of pressure across the boundary.

Let the arbitrary constants in the solutions of the wave equation for pressure for the various media be as shown in Fig. 10.3. Then it follows from continuity of pressure at $x = 0$ that

$$A_1 + B_1 = A_2 + B_2 \tag{10.3.1}$$

and at $x = \ell$ that

$$A_2 e^{-jk_2\ell} + B_2 e^{jk_2\ell} = A_3 e^{-jk_2\ell} \tag{10.3.2}$$

Now, with harmonic time dependence, the relationship between pressure and velocity is

$$v = \frac{j}{\rho\omega}\frac{\partial p}{\partial x} \tag{10.3.3}$$

Thus, for forward travelling waves

$$v = \frac{p}{\rho c} \tag{10.3.4}$$

and for backward travelling waves

$$v = -\frac{p}{\rho c} \tag{10.3.5}$$

It then follows from continuity of velocity at $x = 0$ that

$$\frac{A_1}{\rho_1 c_1} - \frac{B_1}{\rho_1 c_1} = \frac{A_2}{\rho_2 c_2} - \frac{B_2}{\rho_2 c_2} \tag{10.3.6}$$

and at $x = \ell$ that

$$\frac{A_2 e^{-jk_2\ell}}{\rho_2 c_2} - \frac{B_2 e^{+jk_2\ell}}{\rho_2 c_2} = \frac{A_3 e^{-jk_3\ell}}{\rho_3 c_3} \tag{10.3.7}$$

Of particular interest is the power transmission coefficient (i.e., the ratio of the transmitted to incident intensities). This quantity is:

$$T = \frac{\dfrac{|A_3|^2}{2\rho_3 c_3}}{\dfrac{|A_1|^2}{2\rho_1 c_1}} = \frac{\rho_1 c_1}{\rho_3 c_3}\left|\frac{A_3}{A_1}\right|^2 \tag{10.3.8}$$

Thus, in order to determine T, it is necessary to solve Eqs. (10.3.1), (10.3.2), (10.3.6), and (10.3.7) for A_3 in terms of A_1. Although this is most readily done by using the determinant method, the manipulation is tedious, so we quote only the result:

$$T = \frac{4r_{13}}{(r_{13}+1)^2\left[1 - \frac{(r_{23}^2-1)(r_{12}^2-1)}{(r_{13}+1)^2}\sin^2 k_2 l\right]} \tag{10.3.9}$$

where

$$\begin{aligned} r_{13} &= \rho_3 c_3/\rho_1 c_1 \\ r_{23} &= \rho_3 c_3/\rho_2 c_2 \\ r_{12} &= \rho_2 c_2/\rho_1 c_1 \end{aligned} \tag{10.3.10}$$

We will now examine some special cases.

1. No intervening medium ($\ell = 0$)
In this case,

$$T = T_0 = \frac{4r_{13}}{(r_{13}+1)^2} \tag{10.3.11}$$

This result could be derived much more simply by having only one boundary. Notice that the transmission is zero if r_{13} is either zero or infinity; in other words, $\rho_3 c_3$ is either zero or infinity (see Fig. 10.4). These extremes correspond to a perfectly pressure relieving or a perfectly rigid termination of medium 1. Between the extremes, T has a maximum value determined by setting

$$\frac{dT_0}{dr_{13}} = \frac{4}{(r_{13}+1)^2} - \frac{8r_{13}}{(r_{13}+1)^3} = 0 \tag{10.3.12}$$

so that

$$r_{13} = 1 \tag{10.3.13}$$

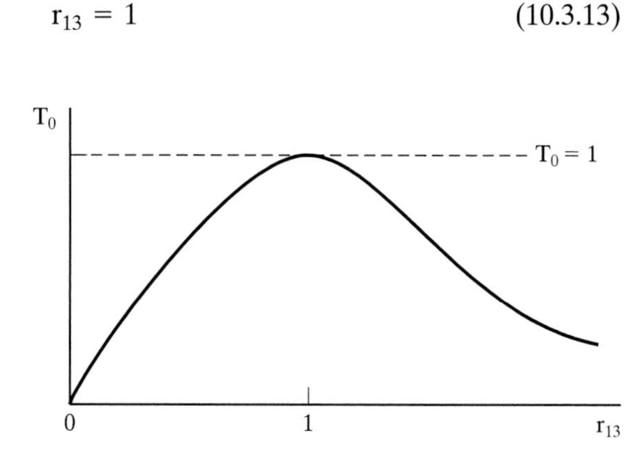

Fig. 10.4 Transmission with no intervening medium. Maximum occurs when specific acoustic impedances are matched.

T_0 is then unity. So when the specific acoustic impedances match exactly, the boundary is "invisible" to sound. Underwater transducers are frequently encapsulated with rubber to provide mechanical protection and electrical insulation. The properties of this rubber are chosen so that it has a specific acoustic impedance that is the same as that of seawater, and it is known as ρc (rho-c) rubber.

2. Thin intervening medium ($k_2\ell \ll 1$)

In this case, $\sin(k_2\ell) \to k_2\ell$. The problem is encountered with the transmission of sound through a wall, so that we will take $\rho_1 c_1 = \rho_3 c_3$. Then

$$T = \left[1 - \frac{(r_{21}^2)(r_{12}^2 - 1)}{4} k_2^2 1^2\right]^{-1} \tag{10.3.14}$$

Since we are usually interested in transmission through a solid wall in air, we may further assume that $r_{12} \gg 1$, and $r_{21} \ll 1$, then

$$T = \frac{1}{1 + \frac{r_{12}^2}{4} k_2^2 \ell^2}$$

$$= \frac{1}{1 + \left(\frac{M\omega}{2\rho_1 c_1}\right)^2} \tag{10.3.15}$$

where M = mass of wall/unit area, and ω = angular frequency of the sound. This result is usually expressed as a transmission loss in dB (TL).

$$TL = 10 \log_{10} \frac{I_{inc.}}{I_{ref.}} - 10 \log_{10} \frac{I_{trans.}}{I_{ref.}}$$

$$= 10 \log_{10} \frac{I_{inc.}}{I_{trans.}} = 10 \log_{10}\left(\frac{1}{T}\right)$$

$$= 10 \log_{10}\left[1 + \left(\frac{M\omega}{2\rho c}\right)^2\right] \tag{10.3.16}$$

This result can be used to obtain an approximate estimate of the TL for a wall. A correction has to be made for the random incidence of sound. Further, the derivation leading to Eq. (10.3.16) does not make any allowance for the stiffness of the wall and is thus known as the "limp mass" law. When this stiffness or elasticity is present, there will be certain frequencies at which so-called plate modes will be excited. In addition, there will be one frequency (called the coincidence frequency) at which the wavelength of sound in air and the wavelength of flexural waves in the panel coincide, resulting in increased transmission and a marked dip in the TL curve. A typical TL curve for a test specimen is shown in Fig. 10.5.

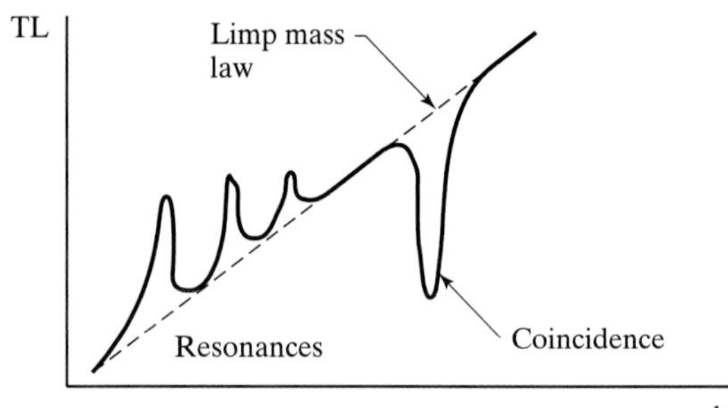

Fig. 10.5 Transmission loss through a wall. Slope is 6 dB/octave for the limp mass law.

3. Thick walls

Since from Eq. (10.3.9) we may write in general that

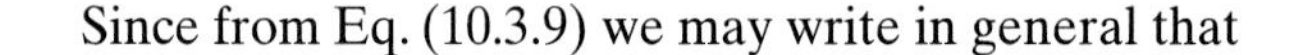

$$T = \frac{T_0}{\left[1 - \frac{(r_{23}^2 - 1)(r_{12}^2 - 1)}{(r_{13} + 1)^2} \sin^2 k_2 1\right]} \tag{10.3.17}$$

there emerges the interesting possibility that if

$$(r_{23}^2 - 1)(r_{12}^2 - 1) > 0$$

then

$$T > T_0$$

In other words, the transmission will be greater than if medium 2 were to be absent. For this to happen, $r_{23} > 1$ and $r_{12} > 1$, or $r_{23} < 1$ and $r_{12} < 1$; that is,

$$\rho_3 c_3 > \rho_2 c_2 > \rho_1 c_1$$

or

$$\rho_3 c_3 < \rho_2 c_2 < \rho_1 c_1 \tag{10.3.18}$$

If the impedance of the intervening medium lies between that of the first and third media, the transmission is increased, and if not, it is decreased.

Variation of the thickness of the intervening medium also affects the transmission through the dependence on $\sin^2(k_2\ell)$. To maximize transmission, set

$$\sin^2(k_2\ell) = 1 \tag{10.3.19}$$

that is,

$$k_2\ell = (2n + 1)(\pi/2)$$

or

$$\ell = (2n + 1)(\lambda/4) \qquad (10.3.20)$$

n is usually 0 in practice, and the thickness is one-quarter wavelength. In such a case,

$$T = \frac{4\,r_{13}}{(r_{13} + 1)^2 - (r_{23}^2 - 1)(r_{12}^2 - 1)} \qquad (10.3.21)$$

Now it can be seen that perfect transmission is possible if $T = 1$; that is,

$$4\,r_{13} = r_{13}^2 + 2\,r_{13} + 1 - r_{13}^2 + r_{12}^2 + r_{23}^2 - 1$$

so that

$$(r_{23} - r_{12})^2 = 0$$

and

$$r_{23} = r_{12}$$

or

$$\rho_1 c_1 \rho_3 c_3 = (\rho_2 c_2)^2 \qquad (10.3.22)$$

An example of such a situation is the use of a rubber sheet as a coupling between air and water. The situation is analogous to the optical coating of a lens, in which case the coating has to have a refractive index given by $(\mu_1\mu_2)^{1/2}$ where μ_1 and μ_2 are the refractive indices of air and glass, respectively.

Note that if $\rho_2 c_2$ is not intermediate between $\rho_1 c_1$ and $\rho_3 c_3$, then the quarter wave thickness will result in a minimal transmission and the half wavelength, a maximum. If $\rho_2 c_2$ is intermediate, then the half wavelength results in a minimal transmission.

10.4 Plane Waves in Three Dimensions

For the general case of propagation in three dimensions, we seek a separable solution to the wave equation including dependence on all three spatial coordinates; that is,

$$p = X(x)Y(y)Z(z)T(t) \qquad (10.4.1)$$

Substituting into the wave equation, one obtains:

$$YZTX'' + XZTY'' + XYTZ'' = \frac{XYZT''}{c^2} \qquad (10.4.2)$$

Dividing by p:

$$\frac{X''}{X} + \frac{Y''}{Y} + \frac{Z''}{Z} = \frac{1}{c^2}\frac{T''}{T} \tag{10.4.3}$$

Since the right side is a function of time only, while the left side is a function of spatial coordinates, both sides must be equal to the same constant, $-k^2$ say. If the constant is taken to be positive, we would not obtain oscillatory solutions, which is what we want. Treating the right side in the same manner as in Section 1.6.2 we obtain

$$T = Ae^{j\omega t} \tag{10.4.4}$$

We now take the left side of Eq. (10.4.3) and by rearranging obtain

$$\frac{Y''}{Y} + \frac{Z''}{Z} + k^2 = -\frac{X''}{X} \tag{10.4.5}$$

Now the right side of Eq. (10.4.5) is a function of x only, while the left side is a function of y and z, so, both sides must be equal to the same constant, ${k_x}^2$ say. Thus

$$X'' = -{k_x}^2 X \tag{10.4.6}$$

which has a solution

$$X = A_x e^{jk_x x} + B_x e^{-jk_x x} \tag{10.4.7}$$

Similarly,

$$Y = A_y e^{jk_y y} B_y e^{-jk_y y} \tag{10.4.8}$$

and

$$Z = A_z e^{jk_z z} + B_z e^{-jk_z z} \tag{10.4.9}$$

From Eq. (10.4.5) it follows that

$${k_x}^2 + {k_y}^2 + {k_z}^2 = k^2 \tag{10.4.10}$$

For a *progressive* wave in the positive x, y, and z directions as shown in Fig. 10.6,

$$A_x = A_y = A_z = 0 \tag{10.4.11}$$

and the solution is

$$\begin{aligned} p &= p_0 e^{j\omega t} e^{-j(k_x x + k_y y + k_z z)} \\ &= p_0 e^{j\omega t} e^{-jkr} \end{aligned} \tag{10.4.12}$$

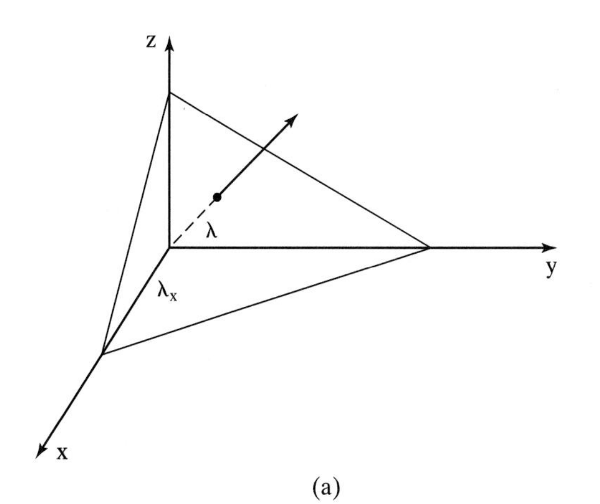

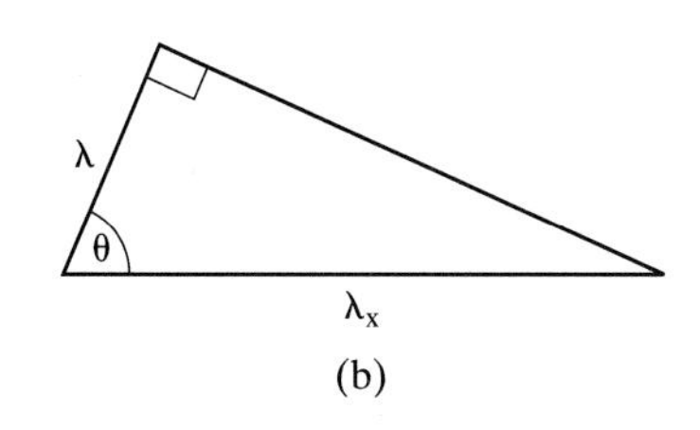

Fig. 10.6 (a) Propagation of a plane wave in an arbitrary direction. (b) Relationship between wavelength and its projection in x direction (trace wavelength).

where

$$\mathbf{k} = \mathbf{i}k_x + \mathbf{j}k_y + \mathbf{k}k_z \tag{10.4.13}$$

and

$$\mathbf{kk} = k_x^2 + k_y^2 + k_z^2 = k^2 \tag{10.4.14}$$

Now

$$k_x = 2\pi/\lambda_x = (2\pi/\lambda)\cos\theta = k\cos\theta \tag{10.4.15}$$

where λ_x is termed the trace wavelength in the x direction. Thus **k** may be regarded as a vector quantity whose components are k_x, k_y, and k_z, and the solution may be written

$$p = p_0 e^{j(\omega t - kr)} \tag{10.4.16}$$

10.5 Rectangular Waveguide

Consider a rigid walled rectangular cross-section waveguide semi-infinite in length in the z direction (Fig. 10.7). Again considering the solution of the wave equation for

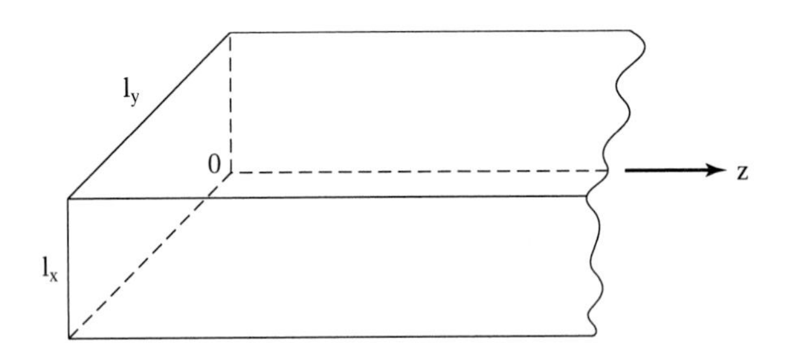

Fig. 10.7 Semi-infinite rectangular waveguide.

pressure by separation of variables, we may write

$$X = C_x \cos(k_x x) + D_x \sin(k_x x) \tag{10.5.1}$$

But

$$\frac{\partial v}{\partial t} = -\frac{1}{p} \nabla p \tag{10.5.2}$$

so

$$j\omega v_x = -\frac{1}{p}\frac{\partial p}{\partial x} \tag{10.5.3}$$

Thus, the boundary conditions for rigid walls are expressed as

$$\frac{\partial X}{\partial x} = 0 \tag{10.5.4}$$

when $x = 0$ and ℓ_x. Now

$$\frac{\partial X}{\partial x} = -C_x k_x \sin(k_x x) + k_x D_x \cos(k_x x) \tag{10.5.5}$$

and thus using the boundary condition for $x = 0$, it follows that

$$D_x = 0 \tag{10.5.6}$$

Similarly,

$$D_y = 0 \tag{10.5.7}$$

Now when $x = \ell_x$

$$\frac{\partial X}{\partial x} = 0$$

and hence

$$k_x \ell_x = m\pi$$

or

$$k_x = m\pi/\ell_x \tag{10.5.8}$$

Similarly,

$$k_y = n\pi/\ell_y \tag{10.5.9}$$

where m and n are integers, not necessarily equal. So the final solution is

$$\begin{aligned} p &= p_0\cos(k_x x)\cos(k_y y)e^{j(\omega t - k_z z)} \\ &= p_0 \cos(m\pi x/\ell_x)\cos(m\pi y/\ell_y)e^{j(\omega t - k_z z)} \end{aligned} \tag{10.5.10}$$

However

$$k_x^2 + k_y^2 + k_z^2 = k^2 = \omega^2/c^2 \tag{10.5.11}$$

By putting constraints on k_x and k_y we have put a constraint on k_z. To illustrate, let us consider some of the various modes of propagation.

a. Zero-order mode

$$m = n = 0$$

and

$$k_z = k \tag{10.5.12}$$

and there is no dependence on x and y. Plane waves can travel down the guide at any frequency.

b. 0-1 mode

$$n = 0, m = 1$$

$$p = p_0\cos(m\pi y/\ell_y)e^{j(\omega x - k_z z)} \tag{10.5.13}$$

and

$$k_z^2 = \omega^2/c^2 - (\pi/\ell_x)^2 \tag{10.5.14}$$

Now if

$$\frac{\omega^2}{c^2} < \left[\frac{\pi}{\ell_x}\right]^2 \tag{10.5.15}$$

k_z^2 is negative, k_z is imaginary, and the mode will not propagate. Motion below the cutoff frequency $\omega_c = c\pi/\ell_x$ will be evanescent.

Although there will be cutoff frequencies for other higher order modes, we will not explore them in detail here. However, some of the characteristics of these higher

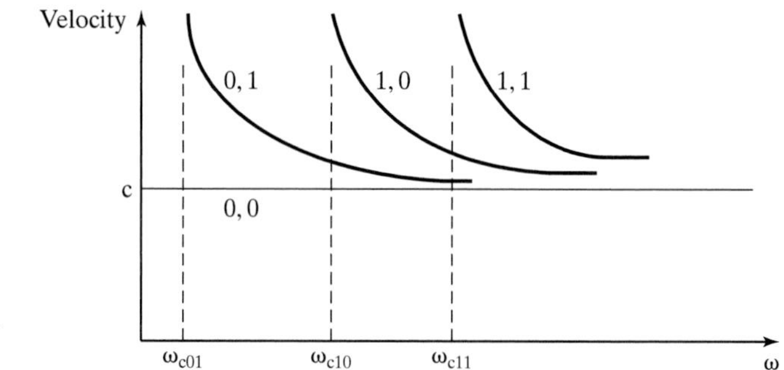

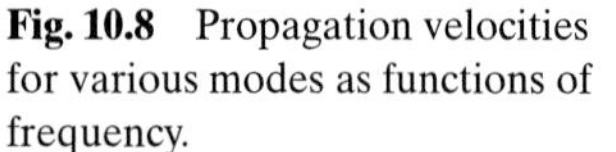

Fig. 10.8 Propagation velocities for various modes as functions of frequency.

order modes can be illustrated using the 0-1 mode as an example. First, note that the velocity of progragation of this mode is given by

$$c_{01} = \frac{\omega}{k_{z01}} = \frac{\omega}{\sqrt{\frac{\omega^2}{c^2} - \left(\frac{\pi}{\ell_x}\right)}}$$

$$= \frac{c}{\sqrt{1 - \left(\frac{c\pi}{\omega\ell_x}\right)^2}} = \frac{c}{\sqrt{1 - \left(\frac{\omega_c}{\omega}\right)^2}} \tag{10.5.16}$$

If $\omega < \omega_c$, c_{01} is imaginary, as we just discussed. If $\omega = \omega_c$, $c_{01} = \infty$. As ω is increased above ω_c, c_{01} declines from ∞ and at very high frequencies approaches c asymptotically.

The variation is shown in Fig. 10.8. Similar behavior is found for other higher order modes.

10.6 Modes of a Rectangular Enclosure

The modes of a rectangular enclosure are important in such diverse areas as room acoustics and the quantum acoustical theory of specific heats of condensed matter, as proposed by Debye. The result is obtained in a manner similar to that used for the rectangular waveguide except that an added restriction is placed on k_z. Assuming a rigid walled box so that the normal velocity is zero at $x = 0, \ell_x$; $y = 0, \ell_y$; and $z = 0, \ell_z$, we obtain the solution for the pressure as follows:

$$p = p_o \cos(k_x x) \cos(k_y y) \cos(k_z z) e^{j\omega t} \tag{10.6.1}$$

where

$$k_x = n_x\pi/\ell_x,\ k_y = n_y\pi/\ell_y, \text{ and } k_z = n_z\pi/\ell_z \tag{10.6.2}$$

The natural modes then occur at frequencies given by

$$\omega^2 = c^2\left[\left(\frac{n_x\pi}{\ell_x}\right)^2 + \left(\frac{n_y\pi}{\ell_y}\right)^2 + \left(\frac{n_z\pi}{\ell_z}\right)^2\right] \tag{10.6.3}$$

For example, the modes of an air-filled room with dimensions 10 feet × 15 feet × 30 feet between 0 Hz and 200 Hz were calculated by Bolt (1939) and are shown in Fig. 10.9. We can see that the density of the modes increases as the frequency increases. Now the equation for the natural frequencies may be rewritten

$$f = c^2\left[\left(\frac{n_x\pi}{\ell_x}\right)^2 + \left(\frac{n_y\pi}{\ell_y}\right)^2 + \left(\frac{n_z\pi}{\ell_z}\right)^2\right] \tag{10.6.4}$$

If we let

$$\frac{c}{2\ell_x} = f_x$$

$$\frac{c}{2\ell_y} = f_y$$

and

$$\frac{c}{2\ell_z} = f_z$$

then

$$f = [(n_x f_x)^2 + (n_y f_y)^2 + (n_z f_z)^2]^{\frac{1}{2}} \tag{10.6.5}$$

A lattice may be constructed in a rectangular coordinate system with lattice spacings f_x, f_y, and f_z. Then the length of a vector from the origin to any point in the lattice gives

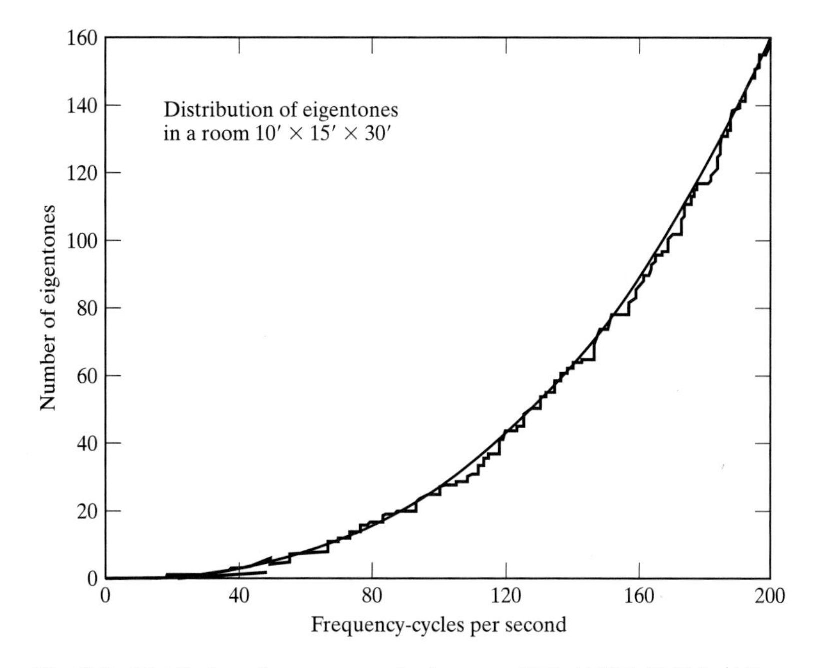

Fig. 10.9 Distribution of resonance modes in a room 10 ft. × 15 ft. × 30 ft. (After Bolt, 1939.)

one of the eigenfrequencies. Now at high frequencies, f is much greater than f_x, f_y, or f_z, so N, the number of modes occurring at frequencies less than f, will be the ratio of the octant of radius f to the volume of a unit cell, both in frequency space. This concept is illustrated in Fig. 10.10. Thus

$$N = \frac{\text{volume of octant}}{\text{volume of unit cell}} = \frac{\frac{1}{8}\frac{4}{3}\pi f^3}{f_x f_y f_z}$$

$$= \frac{\frac{\pi}{6} f^3}{\frac{1}{8}\frac{c^3}{\ell_x \ell_y \ell_z}}$$

$$= \frac{4}{3}\frac{\pi f^3}{c^3} V \tag{10.6.6}$$

where V is the volume of the box.

Now the additional number of modes introduced by an increment δf in frequency is given by

$$\delta N = \frac{4\pi f^2}{c^3} V \delta f \tag{10.6.7}$$

$$\text{and the modal density} = \frac{4\pi}{c^3} V f^2 \tag{10.6.8}$$

The actual modal density for the room used for the calculations of Fig. 10.9 is shown in Fig. 10.11, and the approximate values from Eq. (10.6.8) are shown for comparison. At

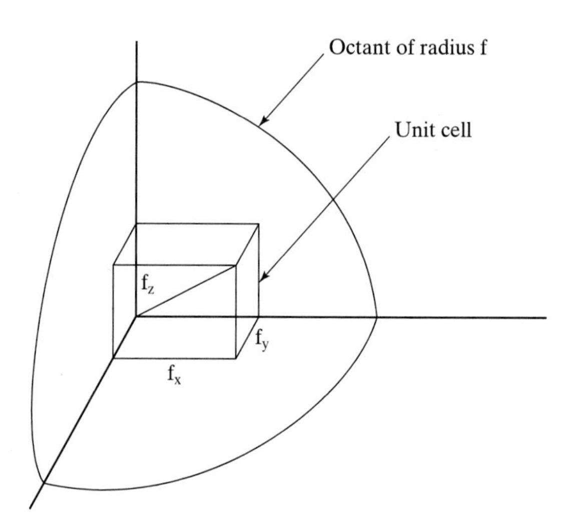

Fig. 10.10 The number of modes up to frequency f is given by the number of unit cells contained in an octant of radius f.

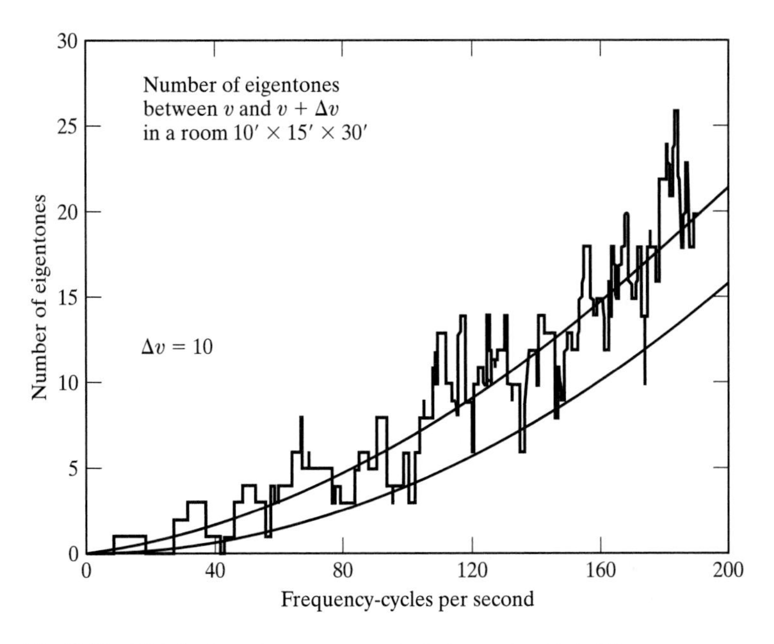

Fig. 10.11 Distribution of resonance modes between f and f + df in a room 10 ft. × 15 ft. × 30 ft. (After Bolt.)

higher frequencies the modes are so dense that sound of any frequency will be transmitted throughout the room. A similar analysis could be made, in principle, for a room of any shape, although the mathematics are too complex for any but very simple shapes. Experience tells us that regardless of shape, the room response is typically governed by a few prominent line resonances at low frequencies and a broad passband in the higher range. The low frequency resonances can produce a noticeable "booming" effect in the room response if they are irregularly spaced in frequency. Bolt (1946) made calculations for a range of room sizes and concluded that the smoothest frequency responses would be obtained from rooms with relative dimensions of about 1:1.5:2.

The sounds produced by speech and music are transient and the response of a room will then be heavily influenced by the absorption by people, furnishings, and wall and floor coverings. Sound is said to reverberate within a room, a topic we will discuss further in Chapter 16.

REFERENCES AND FURTHER READING

Bolt, R.H. 1939. "Frequency Distribution of Eigentones in a Three-Dimensional Continuum. 10 *J. Acoust. Soc. Am.* 228–234.

Bolt, R.H. 1946. "Note on Normal Frequency Statistics for Rectangular Rooms." 18 *J. Acoust. Soc. Am.* 130–133.

Kuttruff, H. 1991. *Room Acoustics*, 3rd ed. Elsevier.

Levine, H., and J. Schwinger. 1948. "On the Radiation of Sound From an Unflanged Circular Pipe." 73 *Phys. Rev.* 383–406.

PROBLEMS

10.1 A pipe with a circular internal cross section 10 cm in diameter is used for its "organ pipe" resonance by closing it off at 1 m from the mouth, which is free to radiate into the surrounding air. If the mouth of the pipe is set in a flat baffle, find the frequency at which the pipe will resonate. Estimate the frequency at which higher order modes will propagate in the pipe.

10.2 A waveguide with rigid walls is filled with air at 20°C and atmospheric pressure. If the cross section is rectangular with dimensions 6 cm × 12 cm and the guide is 2 m in length, determine the lowest six frequencies for mode propagation. Sketch the pressure and velocity profiles for each of these modes.

10.3 A circular cross-section waveguide with rigid walls is filled with air at 20°C and atmospheric pressure. If the cross section is 6 cm in diameter and the guide is 2 m in length, determine the lowest six frequencies for mode propagation. Sketch the pressure and velocity profiles for each of these modes.

10.4 Find expressions for the energy density and the intensity for plane waves in a fluid medium. A plane wave of frequency 1024 Hz in nitrogen has a displacement amplitude of 0.0024 mm. Assuming conditions of standard temperature and pressure (STP), calculate the energy density and intensity.

10.5 Plane waves propagating in air are incident normally on a water surface. Calculate the transmission coefficient for sound of frequency 1000 Hz entering the water. How would it be possible to achieve a transmission coefficient approaching unity at this frequency?

10.6 A closed tank is filled with water in its lower half. The upper half contains nitrogen gas at STP. The interface between the two halves is covered with a sheet of rho-c rubber. How thick should the sheet be in order to permit maximum acoustic transmission at 1000 Hz?

10.7 Plane waves propagating in air are incident normally on a glass window that is 3 mm in thickness. Calculate the transmission loss in dB through the window at 100 Hz, 200 Hz, 400 Hz, 800 Hz, and 1600 Hz, assuming the window acts as a limp mass. How would the results be affected if the window was treated as a plate?

10.8 Plane waves propagating in air are incident normally on a double-glazed window that consists of two glass panes each 1/8 inch in thickness. The two panes are separated by a space of 1/2 inch. Calculate the transmission loss in dB through the window at 100 Hz. What assumptions do you make in your analysis?

10.9 A rectangular room is 8 m × 10 m × 4 m in dimensions. Calculate the frequencies of the lowest six modes at atmospheric pressure and a temperature of 20°C.

10.10 A certain theory of specific heats assumes that thermal energy is stored in a solid object by sound waves that are trapped in the form of the plane wave's natural modes of the object. The energy storage depends on the number of modes in each frequency range δf. Explain why the energy storage increases as f^2. How might the theory take into account the propagation of both shear and longitudinal waves in solids?

Chapter 11

Series Solutions and Scattering

11.1 The Method of Frobenius

Suppose that a solution is required for a differential equation of the form

$$Py'' + Qy' + Ry = 0 \tag{11.1.1}$$

where P, Q, and R are polynomials in x and $y' = dy/dx$, etc. A solution may be found by assuming a series solution:

$$\begin{aligned} y &= x^t(a_0 + a_1x + a_2x^2 + \dots + a_nx^n + \dots) \\ &= \sum_{n=0}^{\infty} a_n x^{n+t} \end{aligned} \tag{11.1.2}$$

Then

$$y' = \sum_{n=0}^{\infty} a_n(n + t)x^{(n+t-1)} \tag{11.1.3}$$

and

$$y'' = \sum_{n=0}^{\infty} a_n(n + t)(n + t - 1)x^{(n+t-2)} \tag{11.1.4}$$

These values are substituted into Eq. (11.1.1), and the coefficients of each power of x are set equal to zero.

11.2 Legendre's Equation

An equation that is conveniently solved using a series solution is Legendre's equation:

$$(1 - x^2)y'' - 2xy' + l(l + 1)y = 0 \tag{11.2.1}$$

where l is generally a positive integer. Assume the series solution given in Eq. (11.1.2), hence

$$\left[\sum a_n(n + t)(n + t - 1)x^{n+t-2}\right]$$
$$- \sum\left[(n + t)(n + t - 1) + 2(n + t) - l(l + 1)\right]a_n x^{n+t} \tag{11.2.2}$$

The coefficient of x^{t-2} must be zero, and thus

$$a_0 t(t - 1) = 0 \tag{11.2.3}$$

The coefficient of x^{t-1} must also be zero, and thus

$$a_1(t + 1)t = 0 \tag{11.2.4}$$

It follows that if

$$a_0 \neq a_1 \neq 0, \text{ then } t = 0$$

Hence, from the coefficient of x^{n+t-2}:

$$a_n(n)(n - 1) - [(n - 2)(n - 3) + 2(n - 2) - l(l + 1)]a_{n-2} = 0$$

and thus

$$a_n = \frac{(n - 1)(n - 2) - l(l + 1)}{n(n - 1)} a_{n-2} \tag{11.2.5}$$

Therefore

$$a_2 = \frac{-l(l + 1)}{2.1} a_0$$

$$a_3 = \frac{[2.1 - l(l + 1)]}{3.2} a_1$$

$$a_4 = \frac{[3.2 - l(l + 1)]}{4.3} a_2$$

$$= \frac{-[3.2 - l(l + 1)]l(l + 1)}{4!} a_0$$

$$a_5 = \frac{[4.3 - l(l+1)]}{5.4} a_3$$
$$= \frac{[4.3 - l(l+1)][2.1 - l(l+1)]}{5!} a_1, \text{ etc.}$$

Hence, we can gather the terms into two series:

$$y = a_0\left\{1 - \frac{l(l+1)}{2!}x^2 - \frac{[3.2 - l(l+1)]l(l+1)}{4!}x^4 - \dots\right\}$$
$$+ a_1\left\{x + \frac{[2.1 - l(l+1)]}{3!}x^3 + \frac{[4.3 - l(l+1)][2.1 - l(l+1)]}{5!}x^5 - \dots\right\} \quad (11.2.6)$$

If $l = 0$,

$$y = a_0 + a_1\left[x + \frac{2.1x^3}{3!} + \dots\right]$$

If $l = 1$,

$$2.1 - l(l+1) = 0$$

and

$$y = a_1 x + a_0(\dots)$$

If $l = 2$,

$$3.2 - l(l+1) = 0$$

and

$$y = a_0(1 - 3x^2) + a_1(\dots)$$

If $l = 3$,

$$4.3 - l(l+1) = 0$$

and

$$y = a_1\left(x - \frac{5}{3}x^3\right) + a_0(\dots)$$

In general, if $l = n - 2$, it follows from Eq. (11.2.5) that one of the two series reduces to a polynomial. Hence, we obtain bounded solutions when a_0 has a value and a_1 is zero, or vice versa. When these polynomials are multiplied by a factor, they are termed

Legendre polynomials, or zonal harmonics $P_l(x)$. We will use these functions in the theory of scattering from a sphere in Section 11.5.

For values of l greater than unity, the multiplying factor is

$$(-1)^{l/2}\frac{1.3.5\ldots(l-1)}{2.4.6\ldots l},\ l \text{ even}$$

and

$$(-1)^{(l-1)/2}\frac{1.3.5.\ldots(l)}{2.4.6\ldots(l-1)},\ l \text{ odd}$$

which yields the following values for the Legendre polynomials:

$$P_0(x) = 1 \tag{11.2.7}$$

$$P_1(x) = x \tag{11.2.8}$$

$$P_2(x) = \frac{1}{2}(3x^2 - 1) \tag{11.2.9}$$

$$P_3(x) = \frac{1}{2}(5x^3 - 3x) \tag{11.2.10}$$

$$P_4(x) = \frac{1}{8}(35x^4 - 30x^2 + 3) \tag{11.2.11}$$

$$P_5(x) = \frac{1}{8}(63x^5 - 70x^3 + 15x) \tag{11.2.12}$$

11.3 Bessel's Equation

Another important equation with a convenient series solution is Bessel's equation of order l:

$$x^2y'' + xy' + (x^2 - l^2)y = 0 \tag{11.3.1}$$

Any solution of this equation is called a Bessel function. We have encountered Bessel functions previously in discussing the piston radiator and the measurement of density fluctuation (see Sections 3.13 and 9.5). We are now in a position to appreciate the mathematics in more detail.

Again assume a series solution as in Eq. (11.1.2). On substituting into Bessel's equation, we obtain

$$\sum a_n[(n + t)^2 - l^2]x^{n+t} + \sum a_n x^{n+t+2} = 0 \tag{11.3.2}$$

The coefficient of x^t must be zero, and thus

$$(t^2 - l^2)a_0 = 0 \tag{11.3.3}$$

The coefficient of x^{t+1} must also be zero, and thus

$$a_1[(t + 1)^2 - l^2] = 0 \tag{11.3.4}$$

if $a_0 \neq 0$, $t = \pm l$, and $a_1 = 0$. The coefficient of x^{n+t} in Eq. (11.3.2) must be zero, so that

$$[(n + t)^2 - l^2]a_n + a_{n-2} = 0$$

or

$$a_n = \frac{-a_{n-2}}{(n + t)^2 - l^2} = \frac{-a_{n-2}}{(n + t - l)(n + t + l)} \tag{11.3.5}$$

Considering various values of n then yields

$$a_2 = \frac{-a_0}{(2 + t + l)(2 + t + l)}$$

$$a_4 = \frac{-a_2}{(4 + t + l)(4 + t - l)}$$

$$= \frac{a_0}{(4 + t + l)(4 + t - l)(2 + t + l)(2 + t - l)}$$

and since a_1 is zero, $a_3 = a_5 = \ldots = 0$.
But

$$y = x^t(a_0 + a_2x^2 + a_4x^4 + \ldots)$$

Letting $t = l$,

$$y = a_0x^l\left\{1 - \frac{x^2}{2(2l + 2)} + \frac{x^4}{2.4(2l + 2)(2l + 4)} - \ldots\right\} \tag{11.3.6}$$

This series is called a Bessel function of the first kind of order l, $J_l(x)$, if

$$a_0 = \frac{1}{2^l\Gamma(l + 1)} \tag{11.3.7}$$

where the gamma function

$$\Gamma(l + 1) = \int_0^\infty x^l e^{-x} dx$$

which is l if l is an integer but is called $l!$ even if l is not an integer. The reason for this choice of a_0 is that it permits a simple form for the general term a_n.
Letting $t = -l$,

$$y = b_0 x^{-l}\left\{1 - \frac{x^2}{2(2-2l)} + \frac{x^4}{2.4(2-2l)(4-2l)} - \dots\right\} \tag{11.3.8}$$

This series is called a Bessel function of the first kind of order $-l$, $J_{-l}(x)$, if

$$b_0 = \frac{2^l}{\Gamma(1-l)} \tag{11.3.9}$$

If l is *not* an integer, both series are convergent, different, and valid for all values of x, and a general solution of Bessel's equation is

$$y = AJ_l(x) + BJ_{-l}(x) \tag{11.3.10}$$

If l is an integer, only one of the series is convergent. For example, suppose $l = +2$, then

$$2l + 4 = 0$$

and $J_{-2}(x)$ has infinite coefficients, so we would have to set B in Eq. (11.3.10) to zero. Another complete solution of Bessel's equation can be obtained by taking the combination

$$y = AJ_l(x) + BN_l(x) \tag{11.3.11}$$

where $N_l(x)$ is another independent solution of Bessel's equation, sometimes called a Bessel function of the second kind, or a Neumann function. It is defined by

$$N_l(x) = \frac{J_l(x)\cos(l\pi) - J_{-l}(x)}{\sin(l\pi)} \tag{11.3.12}$$

Bessel and Neumann functions are shown in Fig. 11.1 a and b. Another family of functions, Bessel functions of the third kind, or Hankel functions, are defined by

$$\begin{aligned} H_1^{(1)}(x) &= J_l(x) + iN_l(x) \\ H_1^{(2)}(x) &= J_l(x) - iN_l(x) \end{aligned} \tag{11.3.13}$$

These functions are related to J_l and N_l in the same way that e^{ix} is related to sin(x) and cos(x).

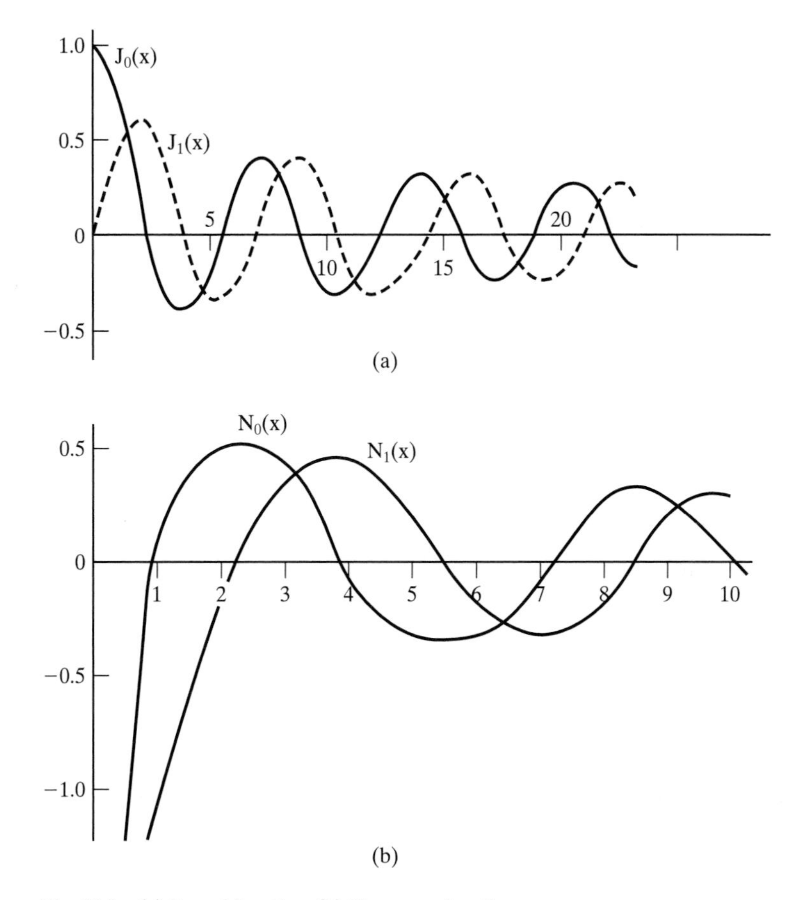

Fig. 11.1 (a) Bessel function. (b) Neumann function.

11.4 Series Solutions of the Wave Equation

With these mathematical preliminaries behind us we can now see their application in acoustics. In spherical polar coordinates, the wave equation is

$$\frac{1}{r^2}\left(r^2\frac{\partial p}{\partial r}\right) + \frac{1}{r^2 \sin\theta}\frac{\partial}{\partial\theta}\left(\sin\frac{\partial p}{\partial\theta}\right) + \frac{1}{r^2\sin^2\theta}\frac{\partial^2 p}{\partial\varphi^2} = \frac{1}{c^2}\frac{\partial^2 p}{\partial t^2} \tag{11.4.1}$$

Assume a separable harmonic solution:

$$P = R(r)\Theta(\theta)\phi(\varphi)e^{j\omega t} \tag{11.4.2}$$

then

$$\frac{\Theta\phi}{r^2}\left(r^2\frac{dR}{dr}\right) + \frac{R\phi}{r^2\sin\theta}\frac{d}{d\theta}\left(\sin\theta\frac{d\Theta}{d\theta}\right) + \frac{R\Theta}{r^2\sin^2\theta}\frac{d^2\phi}{d\varphi^2} = -\frac{\omega^2}{c^2}R\Theta\phi \tag{11.4.3}$$

Dividing by $R\Theta\phi$ and rearranging:

$$\frac{1}{R}\frac{d}{dr}\left(r^2\frac{dR}{dr}\right) + k^2r^2 = -\frac{1}{\Theta \sin\theta}\frac{d}{d\theta}\left(\sin\theta\frac{d\Theta}{d\theta}\right) - \frac{1}{\phi\sin^2\theta}\frac{d^2\phi}{d\varphi^2} \tag{11.4.4}$$

Now the left side is a function of r only while the right side is a function of θ and φ, so both sides must be equal to the same constant, say A. Hence

$$\frac{1}{r^2}\frac{d}{dr}\left(r^2\frac{dR}{dr}\right) + \left(k^2 - \frac{A}{r^2}\right)R = 0 \tag{11.4.5}$$

and

$$-\frac{\sin\theta}{\Theta}\frac{d}{d\theta}\left(\sin\Theta\frac{d\Theta}{d\theta}\right) - A\sin^2\theta = \frac{1}{\phi}\frac{d^2\phi}{d\varphi^2} \tag{11.4.6}$$

By a similar argument, both sides of Eq. (11.4.6) must be equal to a constant, say $-n^2$. Then

$$\frac{d^2\phi}{d\varphi^2} = -n^2\phi \tag{11.4.7}$$

and

$$\frac{1}{\sin\theta}\frac{d}{d\theta}\left(\sin\theta\frac{d\Theta}{d\theta}\right) + \left(A - \frac{n^2}{\sin^2\theta}\right)\Theta = 0 \tag{11.4.8}$$

The solution of Eq. (11.4.7) is

$$\phi = \cos(n\,\varphi) \text{ or } \sin(n\,\varphi) \tag{11.4.9}$$

But φ must be single valued and thus n must be an integer. Frequently there is axial symmetry and then $n = 0$. In these cases, Eq. (11.4.8) becomes

$$\frac{1}{\sin\theta}\frac{d}{d\theta}\left(\sin\theta\frac{d\Theta}{d\theta}\right) + A\Theta = 0 \tag{11.4.10}$$

which can be reduced to Legendre's equation by substituting

$$\eta = \cos\theta \text{ and } \frac{d\eta}{d\theta} = -\sin\theta$$

since Eq. (11.4.10) may be rewritten

$$\frac{\cos\theta}{\sin\theta}\frac{d\Theta}{d\theta} + \frac{d^2\Theta}{d\theta^2} + A\Theta = 0 \tag{11.4.11}$$

and then

$$\frac{d\Theta}{d\theta} = \frac{d\Theta}{d\eta}\frac{d\eta}{d\theta} = -\sin\theta\frac{d\Theta}{d\eta}$$

and

$$\frac{d^2\Theta}{d\theta^2} = \sin\theta\cos\theta\frac{d\theta}{d\eta}\frac{d\Theta}{d\eta} + \sin^2\theta\frac{d^2\Theta}{d\eta^2}$$

Thus, substituting into Eq. (11.4.11)

$$(1-\eta^2)\frac{d^2\Theta}{d\eta^2} - 2\eta\frac{d\Theta}{d\eta} + A\Theta = 0 \qquad (11.4.12)$$

This is Legendre's equation if the constant $A = l(l+1)$. Consequently, the solutions are

$$\begin{aligned}
l = 0, A = 0, \quad & P_0(\eta) = 1 \\
& P_0(\cos\theta) = 1 \\
l = 1, A = 2, \quad & P_1(\eta) = \eta \\
& P_1(\cos\theta) = \cos\theta \\
l = 2, A = 6, \quad & P_2(\eta) = \frac{1}{2}(3\eta^2 - 1) \\
& P_2(\cos\theta) = \frac{1}{4}(3\cos 2\theta + 1) \\
l = 3, A = 12 \quad & P_3(\eta) = \frac{1}{2}(5\eta^3 - 3\eta) \\
& P_3(\cos\theta) = \frac{1}{8}(5\cos 3\theta + \cos\theta)
\end{aligned}$$

We will now return to the solution of Eq. (11.4.5), which becomes

$$\frac{1}{r^2}\frac{d}{dr}\left(r^2\frac{dR}{dr}\right) + \left[k^2 - \frac{l(l+1)}{r^2}\right]R = 0 \qquad (11.4.13)$$

Let $\zeta = kr$, and therefore

$$\frac{d^2R}{d\zeta^2} + \frac{2}{\zeta}\frac{dR}{d\zeta} + [\zeta^2 - l(l+1)]\frac{R}{\zeta^2} = 0 \qquad (11.4.14)$$

This equation is very similar to Bessel's equation and is one of a family of equations whose solutions can be found in terms of Bessel's functions. The solutions to Eq. (11.4.14) are called spherical Hankel functions and thus

$$R = Ah_l^{(1)}(\zeta) + Bh_l^{(2)}(\zeta) \tag{11.4.15}$$

where

$$h_l^{(1)} = j_l(\zeta) + in_l(\zeta)$$

and

$$h_l^{(2)} = j_l(\zeta) - in_l(\zeta)$$

and

$$j_l(\zeta) = \sqrt{\frac{\pi}{2\zeta}} J_{l+1/2}(\zeta) \tag{11.4.16}$$

and

$$n_l(\zeta) = \sqrt{\frac{\pi}{2\zeta}} N_{l+1/2}(\zeta)$$

These functions are used for such problems as radiation from a general spherical source, from a point on a sphere, from a zone on a sphere, or for propagation in a conical horn. A good example of the use of series solutions is the problem of scattering from a sphere.

11.5 Scattering from a Sphere

The problem of scattering from a rigid sphere was first solved by Rayleigh. If the sphere is located at the origin of coordinates, as shown in Fig. 11.2, the scattered wave must be a solution of the wave equation in polar coordinates. It will have no dependence on φ if the incident wave is directed along the z-axis, and hence the velocity potential of the scattered wave can be expressed as a sum of the solutions of Eq. (11.4.10) or

$$\psi_s = \sum_{l=0}^{\infty} a_l P_l(\eta) h_l^{(1)}(\zeta) e^{j\omega t} \tag{11.5.1}$$

Now the incident wave is

$$\psi_i = Ae^{j(\omega t - kz)} \tag{11.5.2}$$

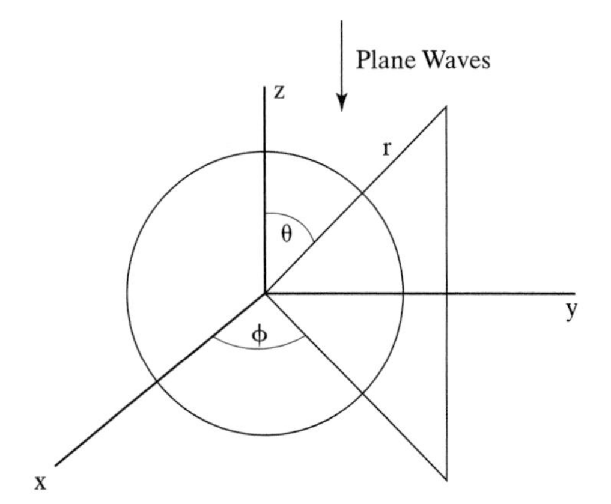

Fig. 11.2 The geometry for scattering of plane waves by a sphere.

We would be able to describe the scattered wave in terms of the incident wave if we could express the coefficients a_l in terms of the incident amplitude A. To do this we must use the boundary condition that the velocity is zero over the surface of the sphere. We therefore expand the incident wave in terms of Legendre polynomials and Bessel's functions.

Since the incident waves are both "incoming" and "outgoing," the appropriate solution is

$$R = h_l^{(1)}(\zeta) + h_l^{(2)}(\zeta) = j_l(\zeta) \tag{11.5.3}$$

in other words,

$$\begin{aligned}\psi_i &= A\sum_{l=0}^{\infty} b_l P_l(\eta) j_l^{(1)}(\zeta) e^{j\omega t} \\ &= Ae^{j(\omega t - kz)} \\ &= Ae^{j\omega t}e^{-jkr\cos\theta} \\ &= Ae^{j\omega t}e^{-j\zeta\eta}\end{aligned} \tag{11.5.4}$$

Hence

$$e^{-j\zeta\eta} = \sum_{l=0}^{\infty} b_l P_l(\eta) j_l(\zeta) \tag{11.5.5}$$

Multiply both sides of Eq. (11.5.5) by $P_l(\eta)$ and integrate with respect to η from -1 to $+1$:

$$\int_{-1}^{+1} e^{j\zeta\eta} P_l(\eta) d\eta = b_l j_l(\zeta) \frac{2}{2l + 1} \tag{11.5.6}$$

Since

$$\left[\int_{-1}^{+1} P_n(\eta) P_m(\eta) d\eta = \begin{cases} 0, & n \neq m \\ \dfrac{2}{2m + 1}, & n = m \end{cases}\right]$$

Rayleigh proved that

$$\begin{aligned}\int_{-1}^{+1} e^{j\zeta\eta} P_l(\eta) d\eta &= 2i^l \sqrt{\frac{\pi}{2\zeta}} J_{+1/2}(\zeta) \\ &= 2i^l j_l(\zeta)\end{aligned} \tag{11.5.7}$$

and hence

$$b_l = i^l(2l + 1)$$

and

$$\psi_i = A\sum_{l=0}^{\infty} i^l(2l + 1)P_l(\eta)j_l(\zeta)e^{j\omega t} \tag{11.5.8}$$

To satisfy the boundary condition, at $r = a$,

$$\frac{\partial\psi_s}{\partial r} + \frac{\partial\psi_i}{\partial r} = 0 \tag{11.5.9}$$

or

$$\sum_{l=0}^{\infty} a_l P_l(\eta)\frac{dh^{(1)}}{d\zeta}(\zeta) + A\sum_{l=0}^{\infty} i^l(2l + 1)P_l(\eta)\frac{dj_l(\zeta)}{d\zeta} = 0 \tag{11.5.10}$$

But for j_m, n_m, or h_m,

$$\frac{d}{d\zeta}[j_m(\zeta)] = \frac{1}{2m + 1}[mj_{m-1}(\zeta) - (m + 1)j_{m+1}(\zeta)]$$

and thus

$$\frac{a_l}{2l + 1}[lh^{(1)}{}_{l-1}(\zeta) - (l + 1)h^{(1)}_{l+1}(\zeta)] + Ai^l[lj_{l-1}(\zeta) - (l + 1)j_{l+1}(\zeta)] = 0 \tag{11.5.11}$$

Hence, the coefficients in the scattered radiation are given by

$$a_l = \frac{-Ai^l(2l + 1)[lj_{l-1}(\zeta) - (l + 1)j_{l+1}(\zeta)]}{lh^{(1)}{}_{l-1}(\zeta) - (l + 1)h^{(1)}_{l+1}(\zeta)} \tag{11.5.12}$$

It can be shown that

$$mn_{m-1}(\zeta) - (m + 1)n_{m+1}(\zeta) = (2m + 1)B_m \cos\delta_m$$

and

$$(m + 1)j_{m+1}(\zeta) - mj_{m-1}(\zeta) = (2m + 1)B_m \sin\delta_m \tag{11.5.13}$$

where B_m and δ_m are functions of ζ. Thus

$$a_l = Ai^{l+1}(2l + 1)e^{-i\delta_l} \sin\delta_l \tag{11.5.14}$$

and hence, from Eq. (11.5.1)

$$\psi_s = A\sum_{i=0}^{\infty} i^{l+1}(2l + 1)e^{-i\delta_l} \sin\delta_l P_l(\eta)h_l{}^{(1)}(\zeta)e^{j\omega t} \tag{11.5.15}$$

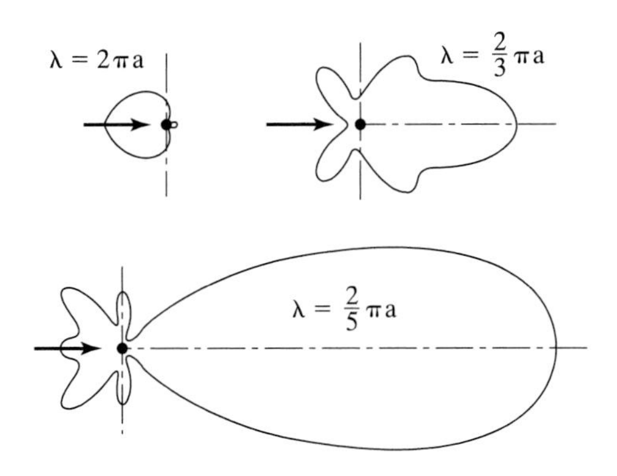

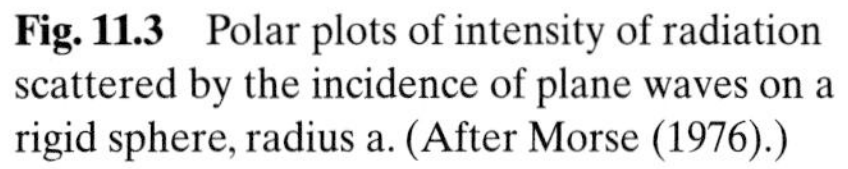
Fig. 11.3 Polar plots of intensity of radiation scattered by the incidence of plane waves on a rigid sphere, radius a. (After Morse (1976).)

Now the pressure and intensity can be calculated. Polar plots of intensity given by Morse are shown in Fig. 11.3.

For small ka,

$$\frac{I_s}{I_i} = \frac{16\pi^4 a^6}{9\lambda^4 r^2}(1 - 3\cos\theta)^2 \tag{11.5.16}$$

This is the celebrated case of Rayleigh scattering (1871), in which the scattered intensity depends inversely on the fourth power of wavelength.

11.6 Scattering by Soft Particles

In general, the scattering particles cannot be assumed to be rigid, and to solve this problem, a phase shift analysis—as first applied in quantum mechanics—is employed. This method may be illustrated by a simplified example in which only the zero order ($l = 0$) harmonic element in Eq. (11.5.8) is considered. Then since

$$\sqrt{\frac{\pi}{2kr}}J_{1/2}(kr) = \frac{\sin kr}{kr} \text{ at large r, and } P_0(\cos\theta) = 1$$

$$\psi_i = \frac{\sin kr}{kr}e^{i\omega t} \tag{11.6.1}$$

or

$$\psi_i = \frac{e^{j\omega t}}{2jkr}(e^{jkr} - e^{-jkr}) \tag{11.6.2}$$

that is, the incident plane wave is replaced by two spherical waves, one converging on the particle and the other diverging from it. The presence of the particle will produce a phase shift of, say 2δ, in the diverging component of the departing wave, whose velocity potential will be

$$\frac{e^{j\omega t}}{2jkr}(e^{jkr} - e^{-j(kr - 2\delta)})$$

Thus, the scattered wave, being the difference of the departing and incident waves, will have velocity potential

$$\psi_s = e^{j\omega t}(2jkr)^{-1}(e^{-jkr} - e^{-j(kr-2\delta)})$$

So the intensity of the scattered wave will be

$$\psi^2 = \frac{|e^{2i\delta} - 1|^2}{|2ikr|^2} \tag{11.6.3}$$

The scattering cross section is defined as the total scattered power per unit incident intensity, and the absorption cross section, as the total power absorbed per unit incident intensity. From Eq. (11.6.3), it may be shown that the scattering cross section is

$$\sigma_s = \frac{4\pi}{k^2}\sin^2\delta \tag{11.6.4}$$

and the absorption cross section is

$$\sigma_s = \frac{\pi}{k^2}(1 - |e^{2i\delta}|^2) \tag{11.6.5}$$

We can see immediately from Eq. (11.6.5) that there will be no absorption unless δ is imaginary. If it is not imaginary, 2δ simply corresponds to the phase shift suffered by the spherical waves in passing through the particle and is therefore given by $2(k - k_1)\,a$, where k_1 is the propagation constant in the material of the particle. The phase shift in the case of Rayleigh scattering, where it is assumed that the velocity of sound in the particle is close to that in the surrounding medium, is thus small, being the difference of two small quantities. Even if the particle diameter is smaller than the wavelength in the surrounding medium, but is comparable with the wavelength in the particle, there will be a large phase shift. In other words, if the sound velocity in the particle is smaller than the sound velocity in the surrounding medium, there will be a large phase change. Furthermore, if the density of the particle is less than that of the surrounding medium, due to the consequent acoustic mismatch, there is a possibility of the formation of standing waves within the particle. These will result in an increase of phase shift and scattering cross section and the particle will be caused to pulsate. Such a situation is known as resonance. A particle causing a small phase change (e.g., solid particle in liquid) is said to be hard; a particle causing a large phase shift is said to be soft (e.g., gas bubble in liquid).

The basis of the analogy involving acoustical, optical, and quantum mechanical scattering is the fact that in the theory of each is a variable that obeys the wave equation. In acoustics, it is the velocity potential. In optics, both the electric and magnetic vectors of a photon satisfy the equation. In quantum mechanics, the variable is called the wave function, but its physical significance lies in the fact that its squared modulus is the probability of finding a particle at any particular point. In optics, the variable is a vector, and the waves are called vector waves, whereas in acoustics and quantum mechanics the variable is a scalar and the waves are called scalar waves. Thus, in optics, polarization can arise due to scattering, a phenomenon without analogy in the scattering of scalar waves.

In optics and quantum mechanics, the terms "hardness" and "softness" of particles have analogous interpretations. The parameter determining hardness in optics is the refractive index, and in quantum mechanics, it is the potential field of the scattering center. These parameters determine the magnitudes of the phase shifts of the scattered waves. Rayleigh, who first solved both the acoustic and optical cases, assumed these phase shifts were small. In quantum mechanics, the equivalent of Rayleigh's assumption is called the Born approximation.

11.7 The Behavior of Rigid Particles Under Radiation Pressure

The effect of back scattering of plane progressive waves by spheres of radius that are small compared with the sound is that a force, known as radiation pressure, acts on the sphere in the direction of the incident radiation.

As in Rayleigh's analysis, King (1934) expanded the velocity potentials of the incident and scattered waves in terms of Legendre polynomials and Bessel functions, and then took as boundary conditions: first, that the sphere was capable of vibration, and second, that the fluid remained in contact with the sphere.

The coefficients in the expansions for velocity potentials of incident and scattered waves having been determined by the boundary conditions, and with appropriate assumptions regarding the nature of the wave (i.e., progressive or stationary plane waves), King found the time averaged radiation pressure on a small ($ka \ll 1$) rigid sphere in a plane progressive wave to be

$$\bar{p} = 2\pi\rho_0|A|^2k^6a^6\frac{\left\{1 + \frac{2}{9}\left(1 - {}^{\rho_0}/_{\rho_1}\right)\right\}^2}{\left(2 + {}^{\rho_0}/_{\rho_1}\right)} \tag{11.7.1}$$

and in a plane stationary wave to be

$$\bar{p} = \pi\rho_0|A|^2\sin(2kh)k^3a^3\frac{\left\{1 + \frac{2}{3}\left(1 - {}^{\rho_0}/_{\rho_1}\right)\right\}}{\left(2 + {}^{\rho_0}/_{\rho_1}\right)} \tag{11.7.2}$$

where A is the complex amplitude of the incident radiation, and h is the distance of the sphere from a fixed plane of reference.

In comparing Eqs. (11.7.1) and (11.7.2), we can see that the force on a sphere in a progressive wave depends on $(ka)^6$ whereas in a standing wave it depends on $(ka)^3$. As (ka) has been assumed to be small, the magnitude of the force in a standing wave will be much greater than in a progressive wave.

We can also see that in a standing wave there is a sinusoidal variation in the force. A good quantitative agreement with these results was found by Rudnick (1951) who suspended a 1 mm radius cork sphere (by a vertical thread) in a standing wave field whose axis of symmetry was horizontal. The force on the sphere was determined by the deflection from the vertical position.

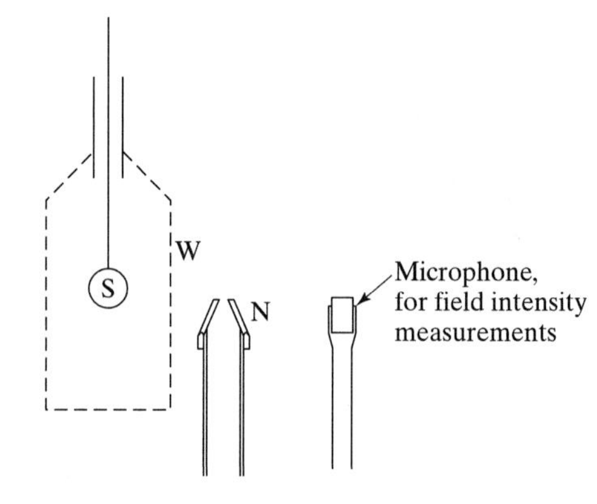

Fig. 11.4 Embleton's apparatus for measuring radiation scattering of a spherical wave.

King thought that radiation pressure played an important part in the formation of Kundt's tube figures, as mentioned in Section 9.7 in Chapter 9. From the foregoing theory, and taking into account the effects of viscosity and the inertia of the particles, King established an equation of motion and was able to evaluate the period of oscillation of particles across the nodes or antinodes. For cork spherules in air, he predicted that if the particles are initially uniformly distributed, they would be swept towards the antinodes in about 10 seconds.

King's theory has been extended by Embleton (1954) for spherical waves and tested experimentally with the apparatus illustrated in Fig. 11.4, consisting of a hollow glass sphere, of size ranging from 0.2 cm to 3 cm in radius. W was a windshield. The point source of sound was the brass nozzle N, which gave a horizontally acting radiation pressure, causing displacement that was measured with a telescope. Embleton reported no serious discrepancy between theory and experiment.

11.8 Scattering by Bubbles

Gas bubbles play an important role in underwater sound propagation because they cause much greater attenuation than do small solid particles. This phenomenon may be explained in terms of the acoustic mismatch between the gas and the liquid as compared with the closer match between solid and liquid. A sound field incident on a bubble causes it to pulsate, so that the system of scattered waves is more intense than in the case of a solid particle. It is customary, as was done for rigid particles, to consider first an ideal case in which there is no degradation of acoustic energy. Such an ideal bubble is considered to have a radius that is small compared with the sound wavelength. This limitation is imposed so that the pressure in and on the bubble can be taken to be uniform, and spherical symmetry can be assumed for the system. Only sinusoidal pulsations of small amplitude compared with the bubble radius are to be considered. Modifications have to be made to account for the degradation of acoustic energy. However, nearly all the absorption due to soft particles arises from volume pulsation, whereas with rigid particles the energy degradation occurs mainly in the surrounding medium.

The problem of the vibration of a single gas bubble was first investigated by Minnaert (1933), who wanted to understand the origin of sound from a babbling brook. We will rederive Minnaert's result by regarding the bubble (radius R) as a pneumatic spring on which the liquid bounces. From Eq. (1.2.23), we have for the stiffness, assuming adiabatic conditions,

$$S = \frac{\gamma P A^2}{v} = 4\pi\,\gamma\,P\,3R \tag{11.8.1}$$

where P is the equilibrium gas pressure inside the bubble. The effective mass of the pulsating liquid is given by Eq. (3.2.25), assuming the radius is small compared with the wavelength of the scattered sound:

$$M = 4\pi R^3 \rho \tag{11.8.2}$$

Here ρ is the density of the liquid. Hence, the resonant frequency is

$$f_0 = \frac{1}{2\pi}\sqrt{\frac{S}{M}} = \frac{1}{2\pi R}\sqrt{\frac{3\gamma P}{\rho}} \tag{11.8.3}$$

Minnaert verified this prediction of a resonant bubble frequency by a series of experiments in which the notes emitted by bubbles, issued from a jet into water, were compared with notes from a set of standard tuning forks.

Smith (1935) included the effect of surface tension in the bubble and approached the problem from the viewpoint of a modified expression for the resonant frequency:

$$f_0 = \frac{1}{2\pi R}\sqrt{\frac{3\gamma(P_0 + 2T/R)}{\rho}} \tag{11.8.4}$$

This expression reduces to that obtained by Minnaert when T, the surface tension, is neglected. Usually the surface tension is much smaller than the hydrostatic pressure, so Minnaert's expression is sufficiently accurate. Smith showed that the bubble oscillation would suffer damping due to radiation of acoustic energy into the surrounding liquid. Using standard expressions for the radiation resistance of a pulsating sphere and for the effective mass of the coupled liquid, Smith found the logarithmic decrement of the bubble pulsation to be

$$\Delta = \pi^2 f_r d/c \tag{11.8.5}$$

where c = the sound velocity in the liquid, and where d = 2R, the bubble diameter. Now it can be seen from Eq. (11.8.3) that

$$(f_0 d) = \frac{1}{\pi}\sqrt{\frac{3P\gamma}{\rho}}$$

and is thus a constant for bubbles of a given gas in a given liquid at a given pressure (e.g., for air bubbles in water, $(f_0 d) = 657$ cm. sec.$^{-1}$). Thus Δ, being proportional to

(f_0d), is also constant under these stated conditions, and for air bubbles in water,

$$\Delta = \pi^2 \frac{(f_0d)}{c} \approx \frac{10 \times 657}{1500 \times 10^2} \approx 0 \cdot 043$$

Smith pointed out that this decrement arising from radiation was much smaller than the observed values, which may vary between 0.045 and 0.10, as found by Meyer and Skudrzyk (1953). Smith therefore suggested another possible source of damping: heat conduction, which has since proved to be of importance. Meyer and Skudrzyk realized that viscous losses for a spherical resonator would also be important. Pfriem (1940) first investigated theoretically the energy loss due to thermal conduction. The process can be easily visualized: As the bubble is compressed, the temperature rises and heat starts to flow out of the bubble, which tends to cool the bubble before the expansion begins and thus the maximum temperature will be reached before the maximum pressure. This phase shift is maintained throughout the cycle, thus introducing a loss of energy that appears as a net flow of heat into the liquid. In purely adiabatic conditions or purely isothermal conditions, there is no energy lost, as both are reversible thermodynamic processes. By solving the heat conduction equation for the gas space, Pfriem was able to show that, under certain circumstances, the loss of energy per cycle could considerably exceed the radiation loss.

The nature of the viscous loss can be explained in the following manner. If a small element of the liquid outside the bubble is considered, as shown in Fig. 11.5, then when this element is displaced from r_1 to r_2, its shape will change, but to a first approximation its volume will remain constant. It can in fact be demonstrated that it is being extended in one direction and compressed in the other, which is equivalent to shearing and gives rise to an energy loss.

MacPherson (1957) calculated the magnitude of the damping due to the more important processes, and these are shown in Fig. 11.6. Values predicted for the total damping decrement, including radiation conduction and viscous losses, were verified experimentally over a wide range of frequencies and for several different gases by a succession of German and U.S. workers during and after World War II.

11.9 Scattering in Multiple Bubble Systems

The first reference to the acoustic behavior of large quantities of bubbles was by Mallock (1910), who considered the loss of musical tone of a glass tumbler when it contained froth or aerated water instead of bubble-free water. Mallock did not consider

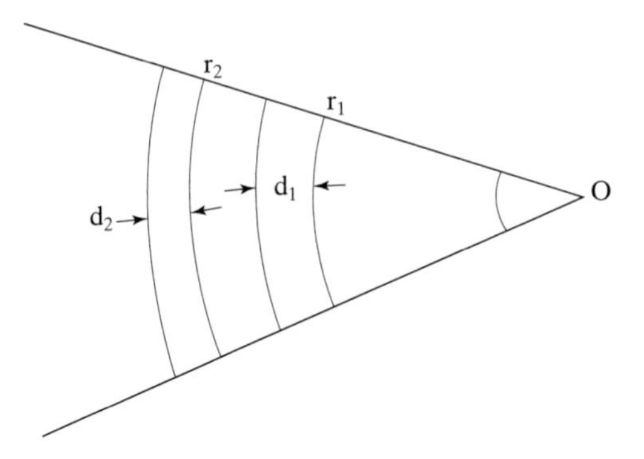

Fig. 11.5 Construction to illustrate viscous damping of a bubble.

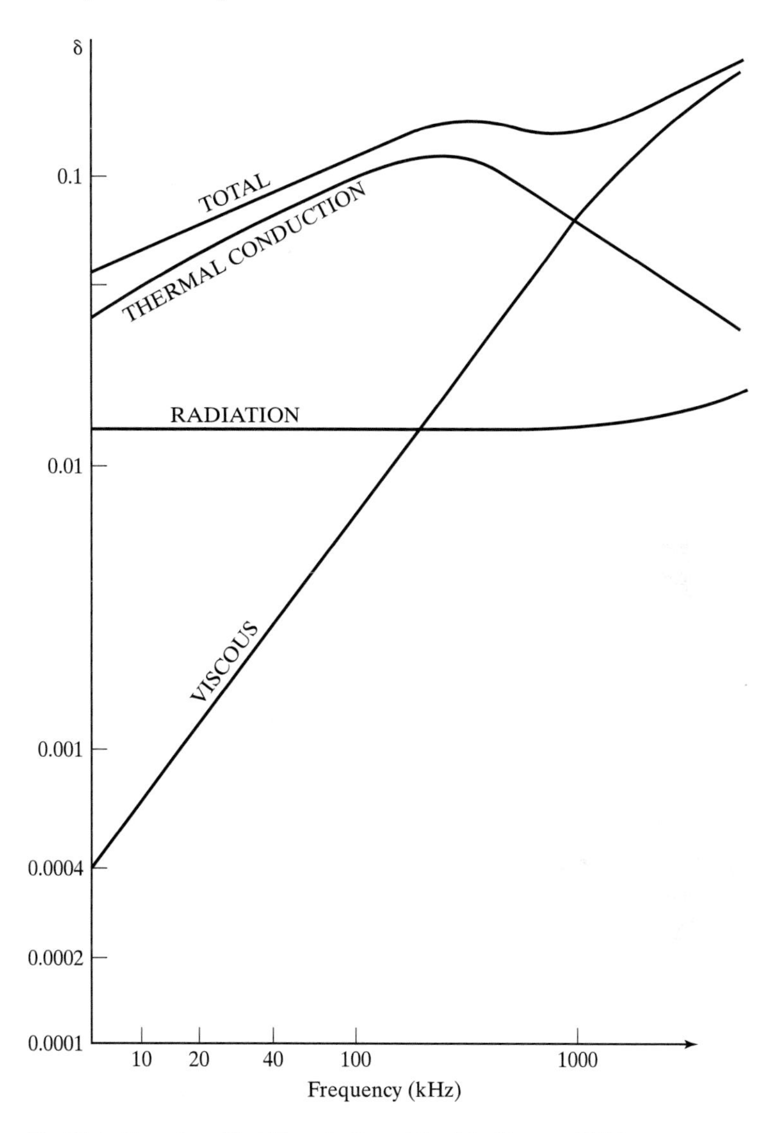

Fig. 11.6 Damping of bubble pulsations. After MacPherson (1957).

the possibility of bubble resonances or damping due to the presence of the bubbles, but considered the bubbles (density ρ_1 and elasticity E_1) and the liquid (density ρ_0 and elasticity E_0) to form a homogeneous fluid of density ρ and elasticity E. Then if x is the proportion of bubbles by volume:

$$\rho = x\rho_1 + (1 - x)\rho_0$$

and

$$\frac{1}{E} = \frac{x}{E_1} + \frac{(1 - x)}{E_0} \tag{11.9.1}$$

so that

$$E = \frac{E_0E_1}{xE_0 + (1 - x)E_1} \tag{11.9.2}$$

Thus, from Eqs. (11.9.1) and (11.9.2), the mean velocity

$$c = \sqrt{\frac{E}{\rho}} = \sqrt{\frac{E_0E_1}{\{xE_0 + (1 - x)E_1\}\{x\rho_1 + (1 - x)\rho_0\}}} \tag{11.9.3}$$

Fig. 11.7 shows c as a function of x. Wood (1941) used Eq. (11.9.3) to evaluate the specific acoustic impedance of the mixture:

$$\rho_0 = \sqrt{E\rho} = \sqrt{\frac{E_0E_1\{x\rho_1 + (1 - x)\rho_0\}}{\{xE_0 + (1 - x)E_1\}}} \tag{11.9.4}$$

Using this result, the reflection coefficient of a relatively thin bubble layer may be easily calculated, and this was done by Wood and is shown in Fig. 11.8. Wood was the first to point out, on the basis of his theory, the large reduction of intensity of the noise of a ship's propeller by the bubbly water in its wake.

The first experimental work on "clouds" of bubbles was carried out during World War II by Meyer and Skudrzyk (1953), who derived theoretically a complex compressibility for a liquid-gas mixture. In making this derivation, they took into account resonance and the various losses mentioned in Section 11.8. From their calculation of the complex compressibility, they deduced expressions for the phase velocity and attenuation in the medium.

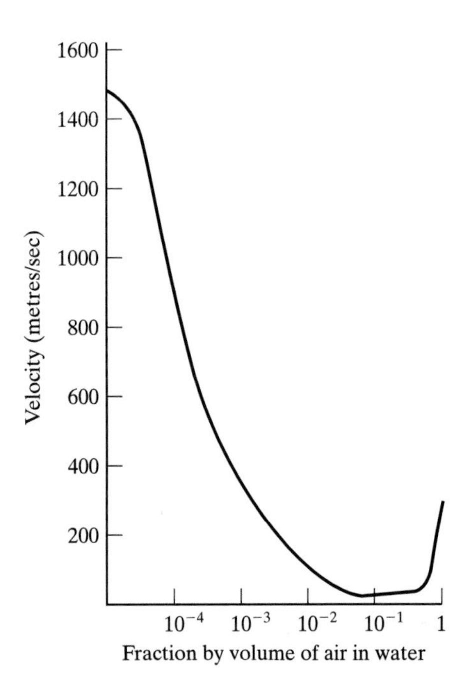

Fig. 11.7 Wood's theory for sound velocity of bubbly water.

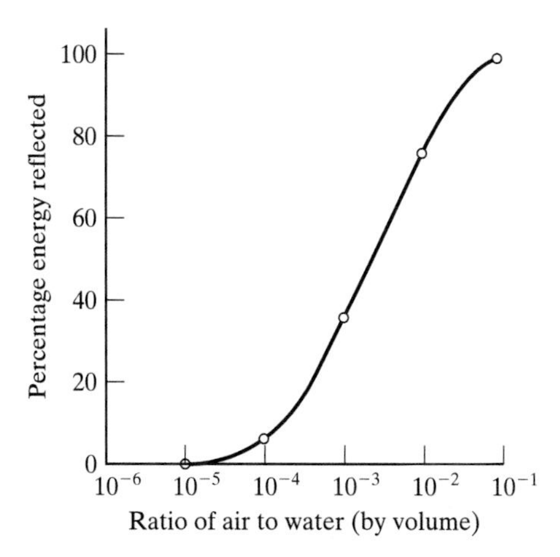

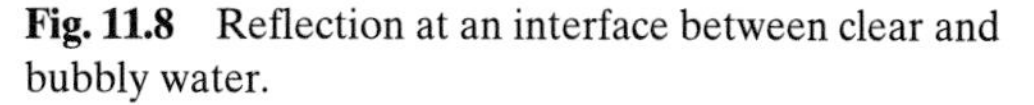
Fig. 11.8 Reflection at an interface between clear and bubbly water.

Meyer and Skudrzyk used three different methods to produce bubbles experimentally: (1) electrolysis using plates or grids as electrodes; (2) blowing air through porous plates; and (3) a pumping system in which air was taken into a stream of water at a low pressure point and then forced into the bulk of the liquid. The measurements were made in a large water tank (20 meters × 20 meters × 4 meters) and receiving and transmitting transducers were placed in the bubble field. The wavelength and attenuation were measured by varying the position of one transducer with respect to the other. The results show only an order of magnitude agreement with the theoretical predictions, one of the major difficulties being the lack of control over the number and size of bubbles present at any particular time.

At about the same time as Meyer and Skudrzyk were carrying out their investigations in Germany, Spitzer (1943) in the United States produced a theory of scattering by bubbles that predicted a behavior similar to Meyer and Skudrzyk's results. Silberman (1957) tested Spitzer's theory experimentally, using vertical pipes containing bubbly water. A beryllium copper diaphragm at the base of the tube generated waves that were reflected from the water surface at the top of the tube, thus producing standing waves. The phase velocity was obtained by locating pressure nodes and antinodes, and the attenuation constant, by measuring the relative amplitudes of the maxima and minima of the standing wave system. When the data had been interpreted and various corrections (for the effects of the pipe wall and variations in concentration and bubble size) had been applied, results of the type shown in Figs. 11.9 and 11.10 were obtained, showing a good agreement between theory and experiment.

Foldy (1945) published a general mathematical treatment for the multiple scattering of waves by randomly distributed scatterers, which was applied to the problem of the behavior of bubbles of various sizes in liquids. Foldy also showed that in general the part of the liquid containing the bubbles can be treated as a homogeneous medium with a complex propagation constant, but in certain cases it is also necessary to consider the effect of the presence of a scattered or incoherent wave system.

Detection of distant submerged vessels at more than about 1 km in seawater is possible only by using sound. The invention of submarines capable of staying submerged for long periods therefore prompted a search for suitable sound-absorbing coverings as protection against detection. One possibility that was considered consisted of

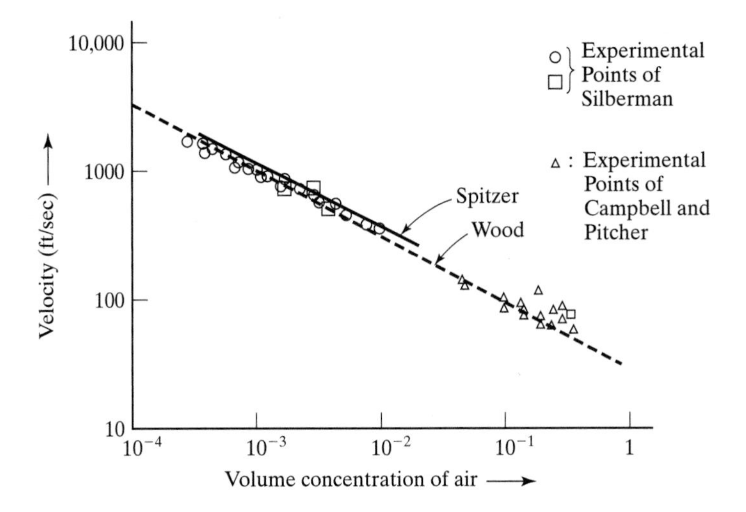

Fig. 11.9 Velocity of sound in water with air bubbles versus concentration.

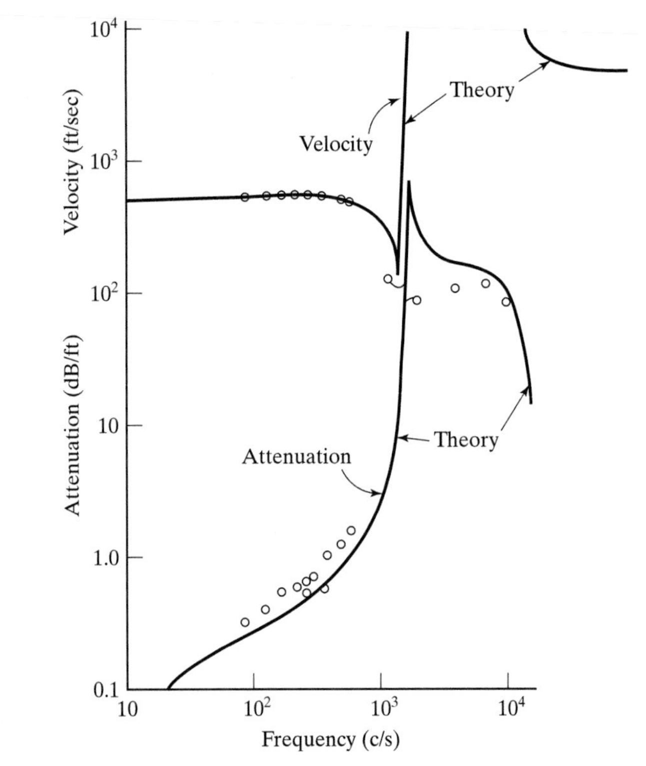

Fig. 11.10 Velocity of sound and attenuation of water with air bubbles versus frequency.

enclosing the submarine in a swarm of resonant bubbles, which would be difficult to attach to the submarine. In addition, the swarm of bubbles could not be too thick, or it would act as a sound reflector itself; nor could it be too thin, or the submarine would still reflect sound. Investigators in Germany tried several different methods to overcome these difficulties and eventually found the most efficient "carrier" of the bubbles was artificial rubber, which is similar to water in its acoustical properties. This approach led to the development of the resonant absorber.

11.10 Radiation Pressure on Bubbles

King (1934), at the time he published his theory of radiation pressure on incompressible spheres, believed it to be applicable to bubbles and cited evidence that small cavitation bubbles ($\rho_0/\rho_1 > 2 \cdot 5$) went to the nodes of a standing wave system. However, Sollner and Bondy (1936), working with emulsions, empirically arrived at the conclusion that particles for which $\rho_0/\rho_1 > 1$ (i.e., particles less dense than the medium) went to the nodes, and particles of which $\rho_0/\rho_1 < 1$ (i.e., particles more dense than the medium) went to the antinodes. Yosiaka and Kawasima (1955) observed that the radiation pressure on certain bubbles was 10^5 to 10^8 times as great as that given by King's theory. This serious discrepancy prompted them to work out the theory of radiation pressure on compressible spheres.

For small spheres (i.e., $(k_0a)^2$, $(k_1a)^2 \ll 1$) and for ρ_0/ρ_1 approximately unity (i.e., an emulsion), they showed that the radiation force in a plane progressive wave was always in the direction of propagation. They found that, in a standing wave for particles of a given size and at a given frequency, the radiation pressure was proportional to the function

$$\frac{\rho_0/\rho_1 + \left[2(\rho_1/\rho_0 - 1)/3\right]}{1 + 2\rho_1/\rho_0} - \frac{1}{3(\rho_1/\rho_0)(c_1^2/c_0^2)} \tag{11.10.1}$$

Thus, for any value of (c_1/c_0) there is a critical value of ρ_1/ρ_0, which will make the expression Eq. (11.10.1) zero (in other words, for these values the radiation pressure will be zero). For a given value of c_1/c_0, Yosiaka and Kawasima also showed that below the critical value of ρ_1/ρ_0, particles travel to the nodes, and above the critical value, particles travel to the antinodes.

For bubbles with $\rho_1/\rho_0 \ll 1$, Yosiaka and Kawasima predicted that in a plane progressive wave, the radiation pressure would show a sharp resonance at the resonant frequency of the bubble pulsation. They found that in plane stationary waves, the radiation pressure was dependent on a function ρ_1/ρ_0, c_1/c_0, k_1, and a. It follows that bubbles smaller than the resonant size travel to the nodes, and that bubbles larger than resonant size travel to the antinodes, while the radiation pressure on a resonant bubble is zero.

REFERENCES AND FURTHER READING

Embleton, T.F.W. 1954. "Mean Force on a Sphere in a Spherical Sound Field." 26 *J. Acoust. Soc. Am.* 40–50.
Foldy, L.L. 1945. "The Multiple Scattering of Waves." 67 *Phys. Rev.* 107–119.
King, L.V. 1934. "Forces on Spheres in Sound Fields. 147 *Proc. Roy. Soc. A.* 212.

MacPherson, J.D. 1957. "The Effect of Gas Bubbles on Sound Propagation in Water." 70 *Proc. Phys. Soc. B.* 85–92.

Mallock, A. 1910. 84 *Proc. Roy. Soc. A.* 391.

McLachlan, N.W. 1961. *Bessel Functions for Engineers*. Oxford Univ. Press.

Meyer, E., and E. Skudrzyk. 1953. 3 *Akust. Beih.* 434.

Minnaert, M. 1933. "On Musical Air Bubbles and the Sounds of Running Water." 16 *Phil. Mag.* 235.

Morse, P.M. [1936] 1976. Vibration and Sound, 2nd ed. American Institute of Physics.

Pfriem, H. 1940. 5 *Akust. Zeits.* 202.

Rayleigh, Lord. 1871. "On the Light from the Sky, Its Polarization and Colour." 4 *Phil. Mag.* 41, 107–120.

Rayleigh, Lord. 1872. "Investigation of the Disturbance Produced by a Spherical Obstacle on the Waves of Sound." 4 *Proc. Lond. Math. Soc.* 253–283.

Rudnick, I. 1951. "Measurement of the Acoustic Radiation Pressure on a Sphere in a Standing Wave Field." 23 *J. Acoust. Soc. Am.* 633–634.

Silberman, E. 1957. "Sound Velocity and Attenuation in Bubbly Mixtures Measured in Standing Wave Tubes." 29 *J. Acoust. Soc. Am.* 925–933.

Smith, F.D. 1935. "On the Destructive Mechanical Effects of Gas Bubbles Liberated by the Passage of Intense Sound Through a Liquid." 19 *Phil. Mag.* 1147.

Sollner, K., and C. Bondy. 1936. "The Mechanism of Coagulation by Ultrasonic Waves." 32 *Trans. Faraday Soc.* 616–624.

Spitzer, L., Jr. 1943. NDRC Report No. 6, 1-sr 20–918.

Strasberg, M. 1954. "Damping of Pulsating Air Bubbles in Water. 4 *Acustica* 518.

Wood, A.B. [1930] 1941. *A Textbook of Sound*. MacMillan.

Wylie, C.R., Jr. 1960. *Advanced Engineering Mathematics*. 2nd ed. McGraw Hill.

Yosiaka, K., Y. Kawasima and H. Hirano. 1955. "Acoustic Radiation Pressure on Bubbles and Their Logarithmic Decrement." 5 *Acustica* 173–178.

PROBLEMS

11.1 Find solutions in series for the following equations and examine their convergence.

a. $y'' = xy$

b. $4x^2y'' + (4x + 1)y = 0$

11.2 Show that one solution of the Hermite equation of order p

$$y'' - xy' + py = 0$$

is a polynomial when p is a positive integer.

11.3 Express the wave equation (a) in cylindrical coordinates, and (b) in spherical polar coordinates.

11.4 Show that

$$P_5(x) = \frac{1}{8}(63x^5 - 70x^3 + 15x)$$

11.5 Show that

$$J_n(x) = \sum_{p=0}^{\infty} \frac{(-1)^p}{p!\Gamma(p + n + 1)}\left(\frac{x}{2}\right)^{n+2p}$$

11.6 From the result of Problem 11.5, find the ratio of a_n/a_0 in the expansion of $J_n(x)$.

11.7 Show that

$$J_{n+1}(x) = \frac{2n}{x} J_n(x) - J_{n-1}(x)$$

11.8 Show that

$$J'_n(x) = \frac{1}{2}[J_{n-1}(x) - J_{n+1}(x)]$$

11.9 Prove the result given in Eq. (11.5.7).

11.10 Explain how a series expansion in terms of Legendre polynomials might be used to calculate the acoustic intensity of sound radiated from a zonal radiator; in other words, a sphere over whose surface the velocity distribution is given by

$$u(\theta) = \begin{cases} u_0 & (0 \le \theta < \theta_0) \\ 0 & (\theta_0 < \theta < \pi) \end{cases}$$

Chapter 12

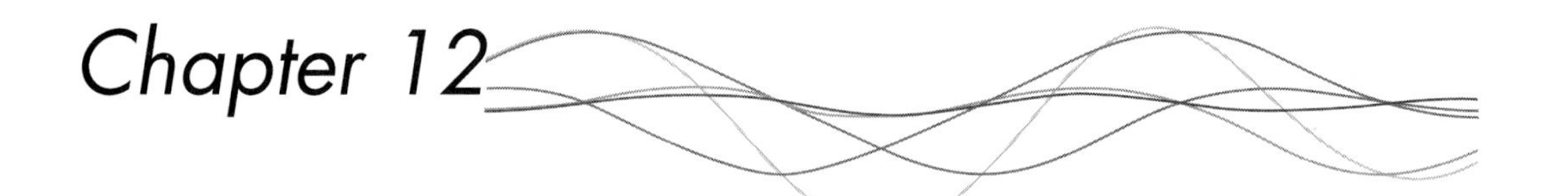

Vibration of Structural Elements

12.1 Historical Notes

The study of the vibrations of strings is perhaps the oldest topic in the history of acoustics. Pythagoras (c. 570–497 B.C.) is credited with beginning such investigations. Unfortunately, it is not easy to learn exactly what Pythagoras did because he and the members of his school believed in keeping their discoveries and deliberations secret. We rely on much later writers, including Boethius (480–524 A.D.), for accounts of the early work. It seems that Pythagoras was the first to show that the pitch of the note emitted by a vibrating string was inversely related to its length. According to Boethius, he arrived at this conclusion by using a monochord (i.e., a string stretched over knife edges which could be excited into vibration by plucking), thus began the preoccupation of the ancient Greeks with harmony and concordance. Some of their efforts produced useful outcomes, as in the improvement of musical instruments and the definition of a musical scale, but there was also a certain amount of speculation, crossing into the realm of the mystical.

The Arab Safi al-Din made the next significant contribution in the area with his proposal in about 1252 for a melodic doctrine based on the division of the octave into sixteen intervals. Galileo carried out a number of experiments on vibrating strings, which he described in his book *Discourses on Two New Sciences*, published in 1638, although it is believed that he performed his experiments earlier in his life. He discovered the relationship of pitch to frequency; explained consonance and dissonance; found the frequency ratios corresponding to musical intervals; discussed resonance and sympathetic vibrations; and determined the quantitative dependence of the frequency of vibration of a string on its length, diameter, density, and tension.

Mersenne's studies on the same subject were published in 1636, although it is believed that they were performed after Galileo's. In order to determine the number of vibrations made by a string in unit time, Mersenne used very long spinet and lute strings, of 100 to 120 feet in length. These made a return in about one second, while a string half the length made two vibrations in a second. Each time the length was halved, the number of vibrations was doubled, although as he noted, once the number of vibrations exceeded about ten per second, it became impossible to resolve them. He also did experiments with a hemp rope 90 feet long and a brass "string" 138 feet long. The theoretical explanation of these discoveries is based on an understanding of the nature of elasticity, so it is appropriate to review the early experiments in this area. For more details on the history of acoustics, a good account of this topic can be found in Hunt (1978).

12.2 Measurement of Elasticity

The experimental study of the strength of materials and elasticity dates back to the time of Leonardo da Vinci. The earliest scientific investigations on the subject were performed by Sir Robert Hooke, who measured the extensions of metal wires under loading. The typical behavior of such a wire is shown in Fig. 12.1. In the linear region, the applied stress (force F divided by area A) was found to be proportional to the strain in the wire (elongation $\delta\ell$ divided by original length ℓ), a result known as Hooke's law:

$$\frac{F}{A} = Y\frac{\delta\ell}{\ell} \tag{12.2.1}$$

where Y is a constant, called Young's modulus.

A rod subjected to load in this way usually undergoes a lateral contraction, and the ratio of lateral to longitudinal strains is found to be constant for a given material in the linear region. We define

$$-\frac{\text{lateral strain}}{\text{longitudinal strain}} = \nu \tag{12.2.2}$$

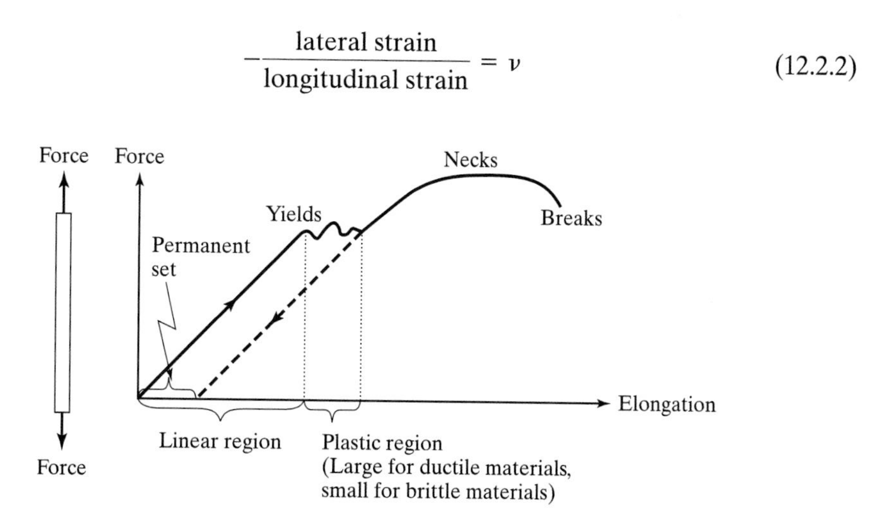

Fig. 12.1 The elongation of a wire under load.

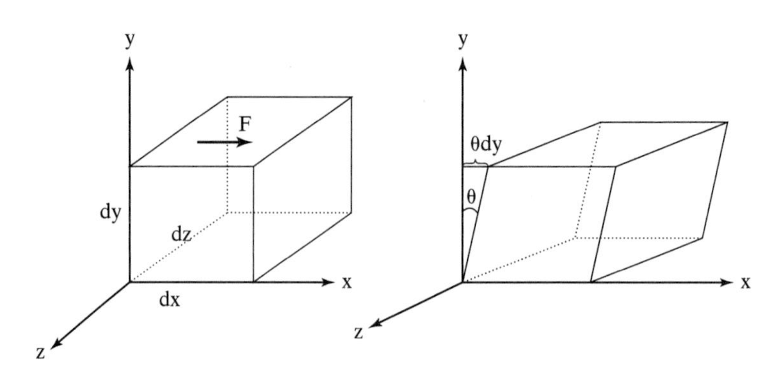

Fig. 12.2 Shear of an element.

where ν is known as Poisson's ratio, which has a value of 0.29 for steel and is normally in the range 0.2 to 0.5 for other solids. Solids can also be sheared (see Fig. 12.2), and in this case the appropriate elastic modulus—called the shear modulus—is given by

$$\mu = \frac{\text{shear stress}}{\text{shear strain}} = \frac{F/A}{\theta^{\delta y}/_{\delta y}} = \frac{F/A}{\theta} \tag{12.2.3}$$

We may also define a "bulk" modulus, B, which relates the volume strain in an element to an applied hydrostatic stress, thus

$$B = \frac{\text{hydrostatic pressure}}{-\text{volume strain}} = \frac{p}{^{-\delta V}/_{V}} \tag{12.2.4}$$

The negative sign is used because a compressive stress results in a volume decrement.

These four elastic moduli (Y, ν, μ, and B) provide engineers with a practical theory of the elasticity of most common materials. The displacements of most structures under load can be calculated with Eqs. (12.2.1) to (12.2.4). So can the modes of vibration of simple structural elements, as we shall see in subsequent sections. In one sense, a string (or wire, cable, or rope) may be regarded as the simplest of the structural elements. An ideal string can be used to transmit a force only if it is in tension and it has no ability to withstand compression.

12.3 The Wave Equation for a String

The theory of the vibration of a string was first worked out by d'Alembert (1747). The vibrations of a string are of course transverse, and the string will not vibrate unless it is under tension. We can assume that the string must have been stretched significantly to be under tension, so that an additional small displacement will have only a negligible effect on the tension. We neglect gravity and assume that the string is uniform. Fig. 12.3 shows such a string displaced from its equilibrium position, and Fig. 12.4 shows a magnified view of an element of the string. The displacement is exaggerated in the figure and in fact may be assumed to be very small. Thus, the components of force acting on the string in the x direction

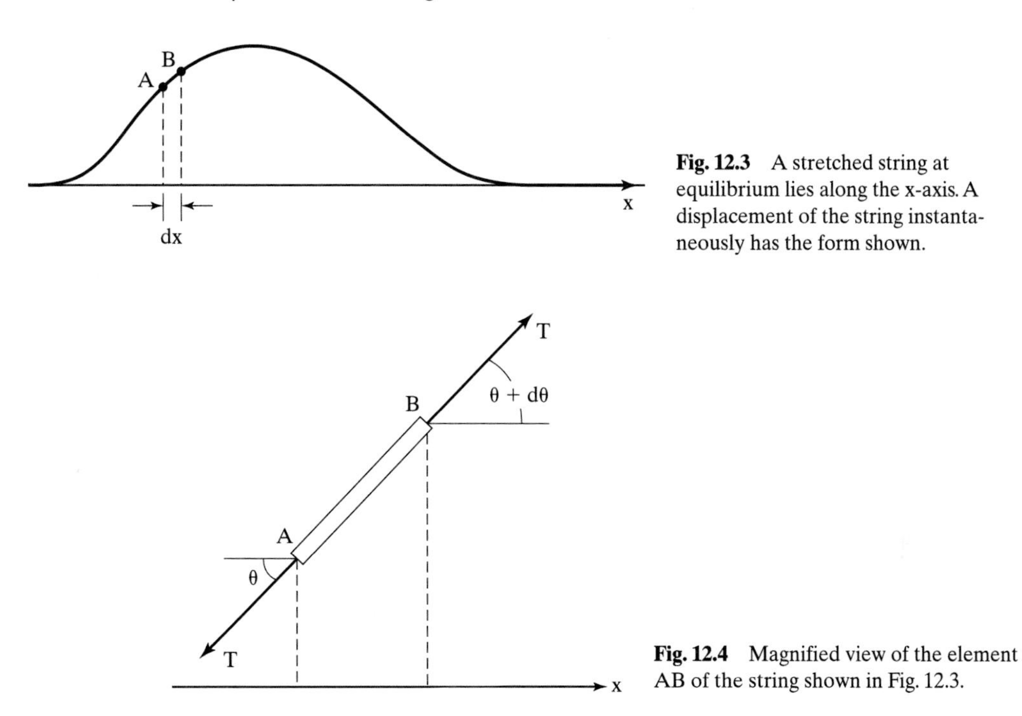

Fig. 12.3 A stretched string at equilibrium lies along the x-axis. A displacement of the string instantaneously has the form shown.

Fig. 12.4 Magnified view of the element AB of the string shown in Fig. 12.3.

add to zero, to a first approximation. In the y direction, the components at A are equal to $T \sin \theta = T \tan \theta = Tdy/dx$. So, the difference between the y components at A and B is proportional to the rate of change of dy/dx, and the net force on the element is

$$T\frac{\partial^2 y}{\partial x^2}\delta x = m\delta x\frac{\partial^2 y}{\partial t^2} \tag{12.3.1}$$

where m is the mass of unit length of the string. The length of the element is approximately equal to δx, since the displacements are small. Hence, we obtain the equation governing the motion of the string, which is a wave equation:

$$T\frac{\partial^2 y}{\partial x^2} = m\frac{\partial^2 y}{\partial t^2} \tag{12.3.2}$$

We see that the propagation velocity is given by

$$c = \sqrt{\frac{T}{m}} \tag{12.3.3}$$

The solutions of this equation are similar to those we have studied in earlier chapters. The string is a good way to demonstrate the properties of one-dimensional waves because the transverse nature of the displacement is readily visualized. As we have already seen, the general solution of the equation is

$$y = f(x - ct) + g(x + ct) \tag{12.3.4}$$

which represents a superposition of waves travelling in the positive and negative x directions, respectively. The exact forms of the functions will be determined by the initial conditions on the string. An interesting example is a string initially plucked in the triangular shape as shown in Fig. 12.5a by being clamped at points P, Q, and R. The string is assumed to extend a great distance to both the right and the left in the figure. What would we expect to happen when the constraints at these three points are removed? There is nothing in the initial conditions to suggest that waves should move preferentially in either the positive or negative directions, so a solution of the equation will result if we superimpose two "partial" waves each with amplitude equal to one half of the initial displacement, as shown in Fig. 12.5b. Now when the constraints are released, the two partial waves will move off in opposite directions, giving rise to displacements as shown in Figs. 12.5c and 12.5d. But what will happen when a progressive wave reaches a rigid termination, T, on the string, as shown in Fig. 12.6? In this case, since the displacement has to add to zero at T at all times, the solution can be constructed by superimposing a backward travelling wave with the opposite amplitude, as shown in the figure.

The forced transverse vibration of an infinite string can be treated in a manner similar to the case of longitudinal oscillations in a line, as given in Section 1.6.2 in Chapter 1. If the driving force is harmonic and the string is driven at one end, then we may presume that progressive waves will run away from this end in the direction of the other end, which is assumed to be a great distance away. Thus, the solution will be of the form:

$$y = Ae^{j(\omega t - kx)} \tag{12.3.5}$$

At any point along the string, an applied force must be just able to overcome the tension so that, by inspecting Fig. 12.4, we see that a vertical applied force must be $-T \sin \theta = -T dy/dx = f$. Hence

$$f = jkTAe^{j(\omega t - kx)} \tag{12.3.6}$$

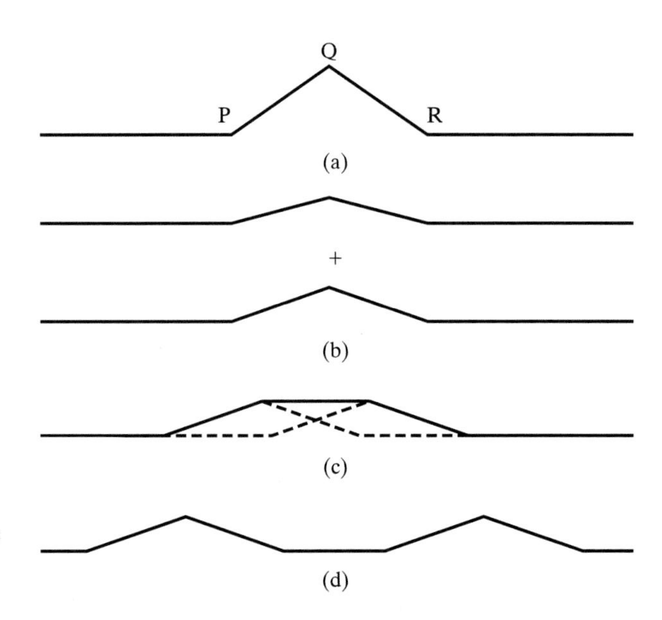

Fig. 12.5 The initial plucking of a string results in two partial waves that travel in opposite directions.

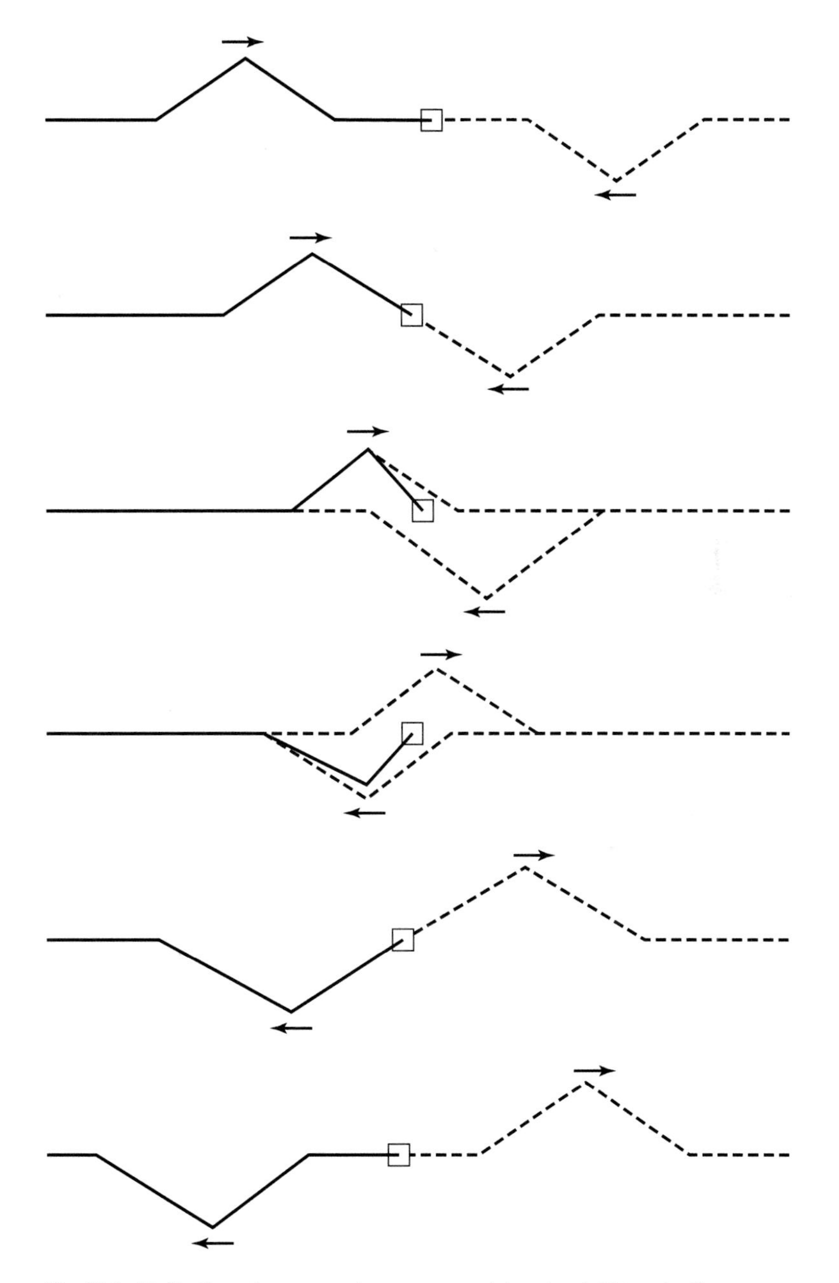

Fig. 12.6 Reflection of a progressive wave on a string at a rigid termination.

and the impedance is obtained by dividing by the particle velocity dy/dt,

$$\frac{f}{v} = T\frac{k}{\omega} = \frac{T}{c} = mc \tag{12.3.7}$$

If the force is applied at the end of the string $x = 0$, then

$$f = Fe^{j\omega t} = jkTAe^{j\omega t} \tag{12.3.8}$$

and we determine the arbitrary constant $A = F/jkT$.

12.4 A String of Finite Length

A string of infinite length is of academic interest only; it is more useful to study a string of finite length. A situation that is frequently encountered is a string fixed at its two ends and given some initial displacement, in which case it will vibrate in a characteristic manner. Consider a string of fixed length, say ℓ, subject to a harmonic excitation at angular frequency ω. In this case, the general solution of Eq. (12.3.2) can be expressed in terms of forward and backward travelling harmonic waves, as explained in Section 1.6.2. Thus

$$y = Ae^{j(\omega t - kx)} + Be^{j(\omega t + kx)} \tag{12.4.1}$$

Assuming that the ends of the string are rigidly clamped, then the boundary conditions may be expressed by

$$y = 0, \qquad x = 0 \text{ and } \ell \tag{12.4.2}$$

From these conditions, we obtain

$$A + B = 0 \qquad \text{and} \qquad Ae^{-jk\ell} + Be^{jk\ell} = 0 \tag{12.4.3}$$

Hence

$$\sin k\ell = 0 \tag{12.4.4}$$

or

$$k\ell = n\pi \qquad \text{or} \qquad f_n = \frac{\omega_n}{2\pi} = \frac{nc}{2\ell} \tag{12.4.5}$$

The conclusion is that there are a series of frequencies f_n at which the string may vibrate, which are integral multiples of the lowest or fundamental value:

$$f_1 = \frac{c}{2\ell} = \frac{l}{2\ell}\sqrt{\frac{T}{m}} \tag{12.4.6}$$

In this result, we see the explanation for the findings of Galileo and Mersenne that the fundamental resonance frequency is inversely proportional to the length, proportional to the square root of the tension, and inversely proportional to the square root of the mass of unit length. Since the mass per unit length is in turn proportional to the cross-sectional area, it follows that the fundamental is inversely proportional to the diameter of the string.

In general, any initial displacement of the string can be decomposed into a combination of natural modes corresponding to the allowable resonance frequencies given by Eq. (12.4.5). The relative weights of the various modes can be determined by a Fourier analysis. Any mode whose frequency is related to the fundamental by an integral multiple is said to be a harmonic. The fundamental is sometimes called the first harmonic. The higher resonance frequencies are also sometimes described as overtones. Thus, the first overtone is the second harmonic.

12.5 The Wave Equation for a Membrane

The theory of the vibration of a membrane is very similar to that of a string. Again, the membrane will vibrate only if it is under tension, which implies that it must be stretched. We may imagine that the stretched membrane consists of a large number of strings parallel to, say, the x-axis. But normally the membrane will also be stretched in a perpendicular direction, so we must also postulate that the membrane consists of an assembly of strings stretched in the y direction. This situation is as shown in Fig. 12.7, assuming that the displacement is in the z direction. Using reasoning similar to that for the string, we see that the net restoring force in the z direction will be

$$T\left(\frac{\partial^2 z}{\partial x^2} + \frac{\partial^2 z}{\partial y^2}\right)\delta x \delta y = m\delta x \delta y \frac{\partial^2 z}{\partial t^2} \tag{12.5.1}$$

where T is now interpreted as the tension per unit length in the membrane and m is the mass of unit area. Thus, we obtain for the membrane a two-dimensional wave equation:

$$\left(\frac{\partial^2 z}{\partial x^2} + \frac{\partial^2 z}{\partial y^2}\right) = \frac{m}{T}\frac{\partial^2 z}{\partial t^2} \tag{12.5.2}$$

The propagation velocity is given by

$$c = \sqrt{\frac{T}{m}} \tag{12.5.3}$$

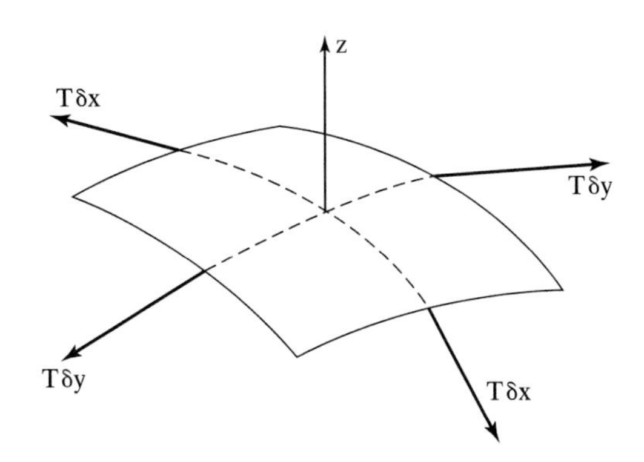

Fig. 12.7 An element of a membrane under tension T per unit length.

There are a number of practical instances involving circular membranes in the capacitor microphone, for example, and in drums. The tympanum that separates the outer and middle ear may also be treated as a membrane. Of greatest interest in such a case are the natural vibration frequencies, which can be obtained by solving Eq. (12.5.3) with the appropriate boundary conditions. It is most convenient to use plane polar coordinates:

$$\frac{\partial^2 y}{\partial r^2} + \frac{1}{r}\frac{\partial y}{\partial r} + \frac{1}{r^2}\frac{\partial^2 y}{\partial \theta^2} = \frac{1}{c^2}\frac{\partial^2 y}{\partial t^2} \tag{12.5.4}$$

Suppose that the motion is harmonic at angular frequency ω, then assuming a separable solution:

$$y(r,\theta,t) = R(r)\Theta(\theta)e^{j\omega t} \tag{12.5.5}$$

Substituting into Eq. (12.5.4) yields

$$\Theta\frac{d^2R}{dr^2} + \frac{\Theta}{r}\frac{dR}{dr} + \frac{R}{r^2}\frac{d^2\Theta}{d\theta^2} + k^2R\Theta = 0 \tag{12.5.6}$$

Multiplying by $r^2/(\Theta R)$ and separating terms in r and θ to different sides then yields

$$\frac{r^2}{R}\frac{d^2R}{dr^2} + \frac{r}{R}\frac{dR}{dr} + k^2r^2 = -\frac{1}{\Theta}\frac{d^2\Theta}{d\theta^2} \tag{12.5.7}$$

Since the right side of Eq. (12.5.7) is a function of r only and the left side is a function of θ only, it follows that both sides must be equal to a constant, m^2 say. Then

$$\frac{d^2\Theta}{d\theta^2} = -m^2\Theta \tag{12.5.8}$$

and the harmonic solution is

$$\Theta = \cos(m\theta + \alpha) \tag{12.5.9}$$

where α is an arbitrary phase angle. Since the displacement must be single valued with increase of 2π in θ, it follows that m must be an integer. Hence, Eq. (12.5.7) becomes

$$r^2\frac{d^2R}{dr^2} + r\frac{dR}{dr} + (k^2r^2 - m^2)R = 0 \tag{12.5.10}$$

This is Bessel's equation, which we encountered in Chapter 11 as Eq. (11.3.1), if we substitute $x = kr$ and $m = l$. Consequently, following Eq. (11.3.11), a general solution can be written:

$$y = AJ_m(kr) + BN_m(kr) \tag{12.5.11}$$

But, as we can see in Fig. 11.1a, the Neumann function becomes infinite at $kr = 0$, a physically inadmissable situation, so that $B = 0$ and

$$y = AJ_m(kr) \tag{12.5.12}$$

Suppose that the radius of the membrane is a, then the boundary condition is

$$J_m(ka) = 0 \tag{12.5.13}$$

If the values of the argument of $J_m(\zeta)$ that cause the function to go to zero are designated as ζ_{mn}, then

$$J_m(\zeta_{mn}) = 0 \tag{12.5.14}$$

and there are only certain allowed values of k, namely

$$k_{mn} = \frac{\zeta_{mn}}{a} \tag{12.5.15}$$

The natural frequencies are thus

$$f_{mn} = \frac{ck_{mn}}{2\pi} = \frac{c\zeta_{mn}}{2\pi a} \tag{12.5.16}$$

Hence, the normal modes of the membrane are given by

$$y_{mn}(r,\theta,t) = A_{mn}J_m(k_{mn}r)\cos(m\theta + \gamma_{mn})e^{j\omega_{mn}t} \tag{12.5.17}$$

The displacement is given by the real part of this, namely

$$y_{mn}(r,\theta,t) = A_{mn}J_m(k_{mn}r)\cos(m\theta + \gamma_{mn})\cos(\omega_{mn}t + \phi_{mn}) \tag{12.5.18}$$

The azimuthal phase angle γ_{mn} is one of the arbitrary constants of the solution. It determines the location of the nodal lines and depends on the way in which the membrane is excited. Some modes are shown in Fig. 12.8.

12.6 Forced Vibrations of a Membrane

Consider a circular membrane driven by a uniformly distributed pressure on one side:

$$p = p_0e^{j\omega t} \tag{12.6.1}$$

Now we must modify the equation of motion, Eq. (12.5.1), which will become

$$T\left(\frac{\partial^2 z}{\partial x^2} + \frac{\partial^2 z}{\partial y^2}\right)\delta x\delta y + p_0e^{j\omega t}\delta x\delta y = m\delta x\delta y\frac{\partial^2 z}{\partial t^2} \tag{12.6.2}$$

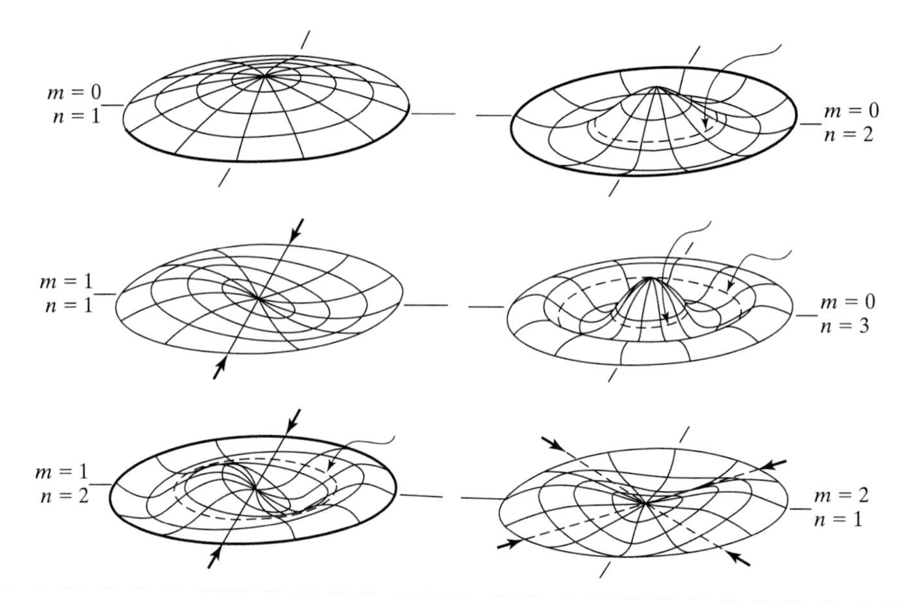

Fig. 12.8 Shapes of some of the normal modes of vibration of the circular membrane. Arrows point to the nodal lines. (After Morse (1948).)

and consequently the wave equation becomes

$$c^2\nabla^2 z = \frac{\partial^2 z}{\partial t^2} - \frac{p_0}{m}e^{j\omega t} \tag{12.6.3}$$

Assume a steady-state solution of the form

$$z = z_0 e^{j\omega t} \tag{12.6.4}$$

then

$$\nabla^2 z_0 + k^2 z_0 = -\frac{p}{mc^2} = -\frac{p}{T} \tag{12.6.5}$$

where $k = \omega/c$. The complete solution of Eq. (12.6.5) consists of a complementary function and the particular solution

$$z_{0P} = -\frac{p}{k^2T} \tag{12.6.6}$$

Thus, the complete solution is

$$z_0 = AJ_0(kr) - \frac{p}{k^2T} \tag{12.6.7}$$

Applying the boundary condition $z_0 = 0$ at $r = a$ gives

$$A = \frac{p_0}{k^2 T} \frac{1}{J_0(ka)} \tag{12.6.8}$$

Hence, from Eqs. (12.6.4), (12.6.7), and (12.6.8)

$$z = \frac{p_0}{k^2 T}\left[\frac{J_0(kr)}{J_0(ka)} - 1\right]e^{j\omega t} \tag{12.6.9}$$

and the amplitude at any point on the membrane is

$$z_0 = \frac{p_0}{k^2 T}\left[\frac{J_0(kr) - J_0(ka)}{J_0(ka)}\right] \tag{12.6.10}$$

Note that z_0 is proportional to p_0 and inversely proportional to T. Whenever the driving frequency corresponds to a free oscillation frequency, $J_0(ka)$, an infinite amplitude is theoreticallly obtained. In practice, the amplitude will be limited by damping.

The most important application of the theory of a membrane is to the diaphragm of a capacitor microphone, as discussed in Section 6.4 in Chapter 6. The capacitance between this diaphragm and the backing plate generates a voltage that depends on the average membrane displacement. This average displacement is

$$\bar{y} = \frac{e^{j\omega t}}{\pi a^2}\int \frac{p_0}{k^2 T}\left[\frac{J_0(kr) - J_0(ka)}{J_0(ka)}\right]2\pi r dr = \frac{p_0}{k^2 T}\frac{J_2(ka)}{J_0(ka)}e^{j\omega t} \tag{12.6.11}$$

At low frequencies, ka is much less than 1, and

$$J_0 \rightarrow 1 - \frac{k^2 a^2}{4} \tag{12.6.12}$$

while

$$J_2 \rightarrow \frac{k^2 a^2}{8}\left[1 - \frac{k^2 a^2}{12}\right] \tag{12.6.13}$$

so that

$$\bar{y} = \frac{p_0 a^2}{8T}\left[1 + \frac{k^2 a^2}{6}\right]e^{j\omega t} \tag{12.6.14}$$

Thus, if ka is small, the output does not depend on frequency. The limiting frequency is given by $ka \leq 1$, or

$$f \leq \frac{c}{2\pi a} = \frac{1}{2\pi a}\sqrt{\frac{T}{m}} \tag{12.6.15}$$

Hence, if we define the stiffness of the diaphragm as the ratio of the net force to the average displacement, then

$$S = \frac{p_0 \pi a^2}{\bar{y}} = 8\pi T \tag{12.6.16}$$

Eq. (12.6.16) then proves the result assumed previously in Section 6.4.

12.7 Longitudinal Waves in a Rod

The problem of longitudinal waves in a rod was first investigated experimentally by Chladni in 1787 and the theory was supplied by Navier in 1824. It is similar to the case of propagation in a fluid-filled pipe.

Consider a long thin rod in which the stresses are not in equilibrium, as shown in Fig. 12.9a, where the free-body diagram for an element of length δx is shown. Here, σ_{xx} is the component of stress in the x direction, acting on a face perpendicular to the x direction. The stress balance equation for the element shown in Fig. 12.9a is

$$A \frac{\partial \sigma_{xx}}{\partial x} \delta x = A \rho \delta x \frac{\partial^2 \xi}{\partial t^2} \tag{12.7.1}$$

where A = cross-sectional area, ρ = density, and ξ = displacement in the x direction. Now in this situation, from Eq. (12.2.1)

$$\sigma_{xx} = Y \frac{\partial \xi}{\partial x} \tag{12.7.2}$$

and thus, from Eq. (12.7.1)

$$Y \frac{\partial^2 \xi}{\partial x^2} = \rho \frac{\partial^2 \xi}{\partial t^2} \tag{12.7.3}$$

So longitudinal waves propagate along the rod with what is known as the bar velocity:

$$c_b = \sqrt{\frac{Y}{\rho}} \tag{12.7.4}$$

a well-known result in elementary acoustics. As in fluid-filled pipes and strings of finite length, there will be a series of resonance frequencies determined by the end conditions,

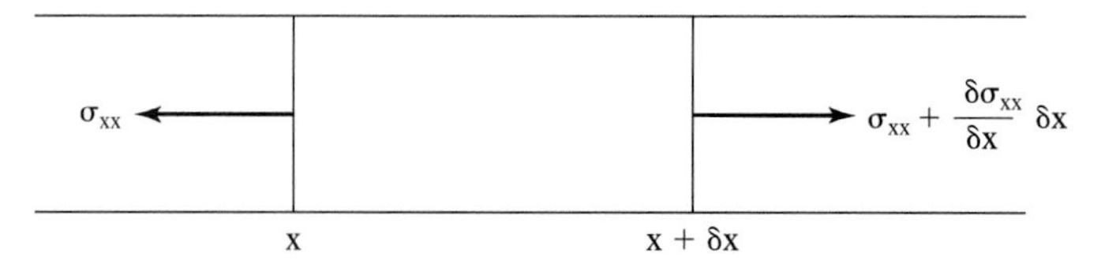

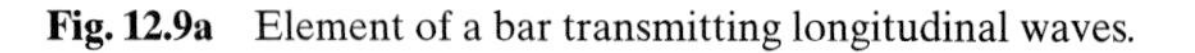
Fig. 12.9a Element of a bar transmitting longitudinal waves.

whether they are clamped or free. Determining the values of these frequencies is a worthwhile student exercise.

12.8 Shear and Torsional Waves in Rods

The theory of shear waves in a bar, first studied by Bernoulli in 1735, is very similar to that of longitudinal waves. Again we consider the balance of forces on an element of the bar (Fig. 12.9b), where σ_{xy} is the component in the y direction of stress acting on a face perpendicular to the x direction. The net force in this case is

$$A\frac{\partial\sigma_{xy}}{\partial x}\delta x = A\rho\delta x\frac{\partial^2\eta}{\partial t^2} \tag{12.8.1}$$

where η is now the displacement in the y direction. But

$$\sigma_{xy} = \mu\frac{\partial\eta}{\partial x} \tag{12.8.2}$$

hence

$$\mu\frac{\partial^2\eta}{\partial x^2} = \rho\frac{\partial^2\eta}{\partial t^2} \tag{12.8.3}$$

so that the transverse displacement propagates as a wave in the x direction with velocity

$$c_T = \sqrt{\frac{\mu}{\rho}} \tag{12.8.4}$$

The credit for the derivation of the complete torsional wave equation is usually given to Saint-Venant for his 1849 paper. Torsional vibrations of a rod occur when sections rotate about the rod's axis while remaining in the same plane (see Fig. 12.10). If the couple at x is T, and at x + dx is $T + \frac{\partial T}{\partial x}\delta x$, then the net couple acting on the section is

$$\frac{\partial T}{\partial x}\delta x = I\frac{\partial^2\theta}{\partial t^2} \tag{12.8.5}$$

where θ is the mean angle of rotation and I is the moment of inertia of the section. Now if two opposing couples, each of magnitude T, act at opposite ends of a cylinder of

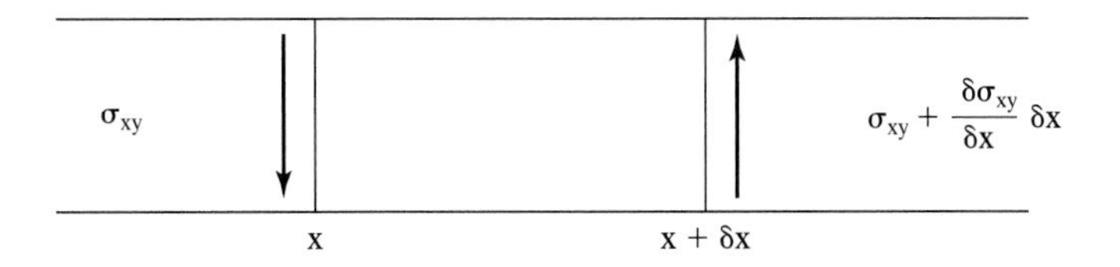

Fig. 12.9b Element of a bar transmitting shear waves.

length ℓ radius r, they produce a relative rotation between the end faces, θ, where

$$T = \frac{\pi}{2}\mu r^4 \frac{\theta}{\ell} \tag{12.8.6}$$

Now since

$$\mu = \frac{\text{stress}}{\text{strain}} = \frac{(\delta F/\rho)\delta\alpha\delta\rho}{\rho\theta/\ell} \tag{12.8.7}$$

it follows that

$$\delta F = \mu\rho^2\delta\alpha\delta\rho\ \theta/\ell \tag{12.8.8}$$

is the force on the small element shown in Fig. 12.11. So the couple on a ring or a cylindrical shell at radius ρ is obtained by integrating around α:

$$\delta T = 2\pi\mu\rho^3\delta\rho\ \theta/\ell \tag{12.8.9}$$

Thus, if the relative rotation between x and x + dx is $d\theta$, we obtain the total couple acting on a cross-sectional face by integrating ρ from zero to r:

$$T = \frac{\pi}{2}\mu r^4 \frac{d\theta}{dx} \tag{12.8.10}$$

Hence, since the net torque on the element is $(\partial T/\partial x)\ \delta x$,

$$\frac{\pi}{2}\mu r^4 \frac{\partial^2\theta}{\partial x^2}\delta x = I\frac{\partial^2\theta}{\partial t^2} \tag{12.8.11}$$

But for a cylinder of length δx

$$I = \rho\frac{\pi}{2}\mu r^4\delta x \tag{12.8.12}$$

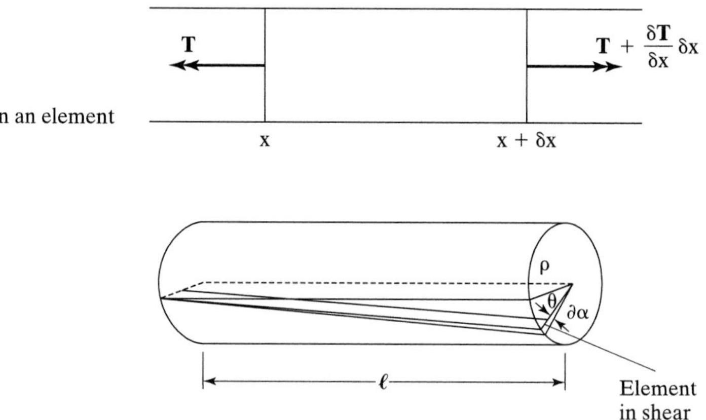

Fig. 12.10 Torsional couple on an element of a rod.

Fig. 12.11 Twisting of a rod.

and therefore

$$\frac{\partial^2 \theta}{\partial x^2} = \frac{\rho}{\mu} \frac{\partial^2 \theta}{\partial t^2} \tag{12.8.13}$$

so that the twisting propagates as a wave with the same velocity as shear waves.

12.9 The Statics of Bending of Beams

Most structures are composed of beams, which are intended to carry lateral, as opposed to axial, loads. A major consideration in the design or selection of beams is the amount of bending caused by a given load. (The bending of beams was first treated by Euler and Bernoulli, using a straight beam with a uniform cross section and a longitudinal plane of symmetry.) The beam is further assumed to be made of homogeneous isotropic material. Suppose a load with magnitude q(x) is distributed as in Fig. 12.12a. This load will tend to bow the beam downwards, and it is said to be in a state of positive bending (so that it will assume a shape so as to collect water). Notice that the load is assigned a positive sign when directed upwards, although actual loads are usually downwards and thus negative. Now consider the elementary portion of the beam shown in Fig. 12.12b. The bending moment, magnitude M(x), acting on the left side of the element must be in the negative z direction (i.e., clockwise using the right-hand rule) and the shear force V(x) must be in the positive y direction. On the right side of the element, the reactions of the remainder of the beam will be a shear force and moment increased slightly in magnitude but reversed in direction, as compared to the left side. Applying the static equilibrium conditions leads to

$$\frac{dV}{dx} = q \tag{12.9.1}$$

and

$$\frac{dM}{dx} = V \tag{12.9.2}$$

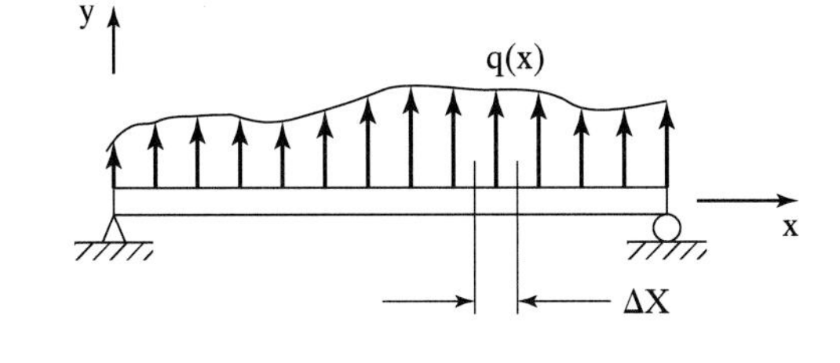

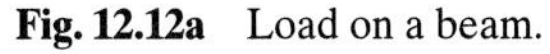

Fig. 12.12a Load on a beam.

Fig. 12.12b Free-body diagram of a beam element.

These equations can produce what are called shear and bending moment diagrams, used by engineers as the first step in analyzing the effects of static loads on a beam. The next step is to estimate the stresses caused as a result of a particular load distribution. Suppose that there is a constant bending moment along some region of the beam, which might result from having

i. equal opposite couples applied at the ends of the beam, or
ii. in other ways, at least over a portion of the beam, as shown in Fig. 12.13.

Such a situation is called pure bending, and we assume that the beam has a positive radius of curvature R. Under these conditions, planes perpendicular to the axis remain perpendicular to the axis and rotate relative to one another. Deformation is related to strain by considering a small element and thinking of the beam as comprised of an assembly of longitudinal fibers.

In positive bending, the fibers in the upper portion of the beam are shortened, while those in the lower portion are elongated (see Fig. 12.14). In between, there is a surface where the fibers do not change in length, which is called the neutral axis. Its intersection with the beam cross section is called the bending axis (see Fig. 12.15). Beam cross sections rotate about the bending axis. Choose the x-axis to coincide with the neutral axis. Consider the element AB (Fig. 12.15) whose new length, A′B′, will be (R-y).

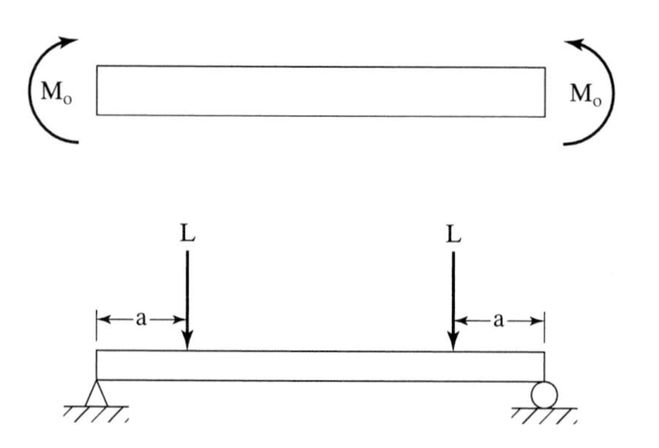

Fig. 12.13 Ways of achieving pure bending in a beam.

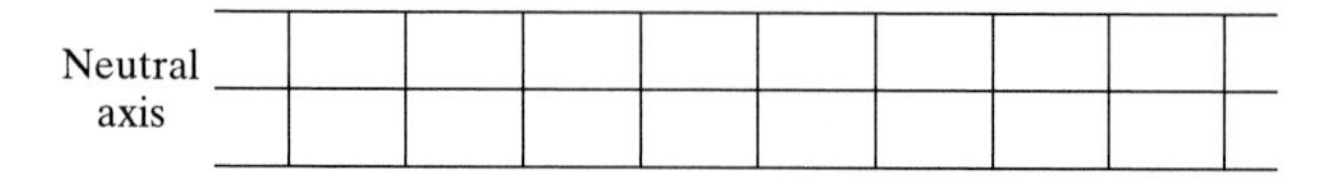

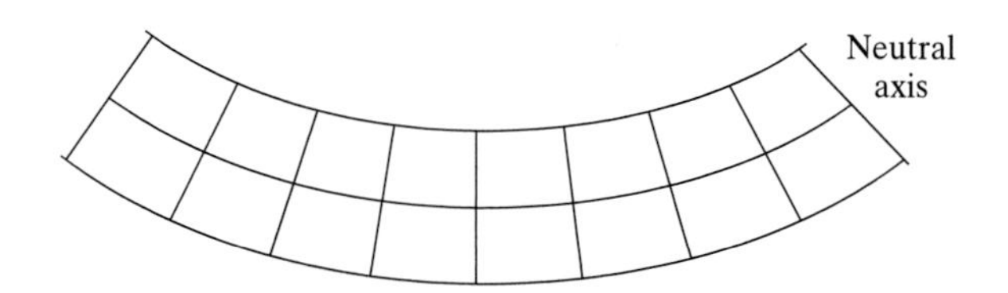

Fig. 12.14 Deformation due to bending.

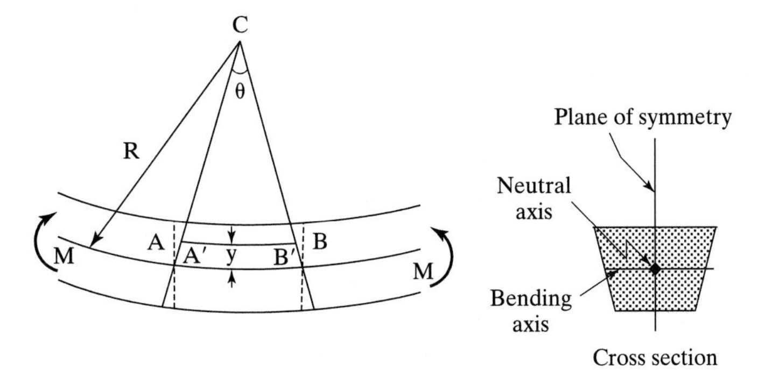

Fig. 12.15 Strains in bending.

Hence, the strain on AB will be

$$\epsilon_{AB} = \frac{A'B' - AB}{AB} = -\frac{y}{R} \tag{12.9.3}$$

Assume we are in the linearly elastic region. The stress distribution must be proportional to the strain distribution through Hooke's law as shown in Fig. 12.16. Thus

$$\sigma_{AB} = Y\epsilon_{AB} = -\frac{Y_y}{R} \tag{12.9.4}$$

System 1 in Fig. 12.17 must be equipollent with System 2, and thus

$$\int \sigma dA = 0 \tag{12.9.5}$$

and

$$\int y\sigma dA = M \tag{12.9.6}$$

Assuming R does not vary throughout the beam, from Eq. (12.9.5)

$$-\frac{Y}{R}\int ydA = 0 \tag{12.9.7}$$

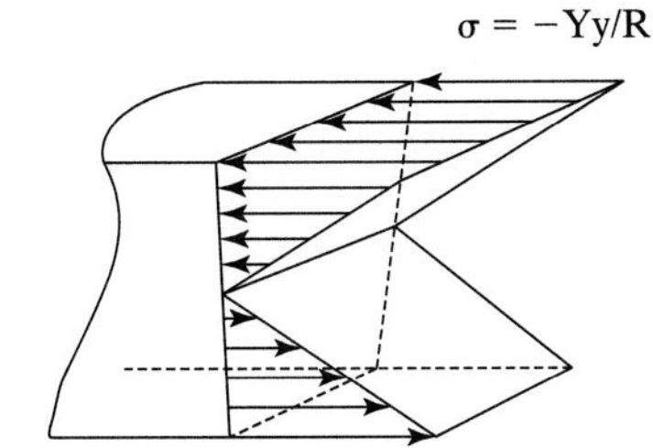

Fig. 12.16 Distribution of stress in bending.

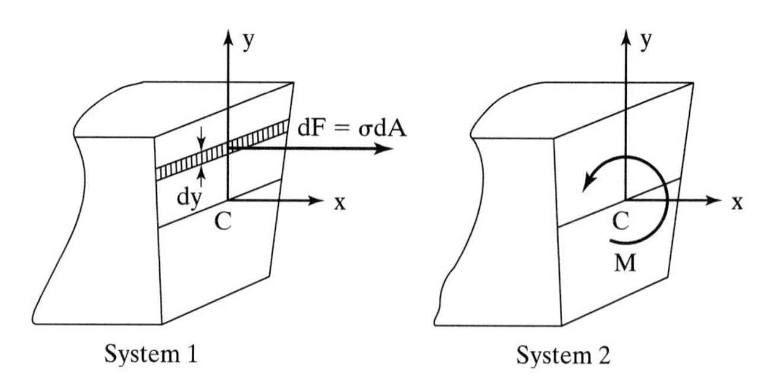

Fig. 12.17 Force distribution in bending.

But

$$-\frac{Y}{R}\int y dA = 0 \tag{12.9.8}$$

where $\bar{y}$ is the distance from the centroid. Thus, $\bar{y}$ is zero, and we can conclude that the neutral plane must pass through the centroid. From Eq. (12.9.6)

$$\sigma_{AB} = -\frac{Y}{R}\int y^2 dA \tag{12.9.9}$$

But the integral in Eq. (12.9.9) is the second moment of area about the bending axis, I, and hence

$$M = -\frac{YI}{R} \tag{12.9.10}$$

Eq. (12.9.10) is called the moment-curvature relation; the moment-stress relation is

$$\sigma = \frac{My}{I} \tag{12.9.11}$$

Eq. (12.9.11) shows that, for a given value of y, the smaller the stress, the greater is I. This is the reason for using I beams, which have a maximum moment of area for a minimum amount of material. Eqs. (12.9.1), (12.9.2), (12.9.10), and (12.9.11) enable a designer to find the stresses in a beam and to calculate its deflection for a given load distribution and end conditions.

12.10 The Dynamics of Bending: Flexural Vibrations

Consider a beam as shown in Fig. 12.18. In this nonequilibrium or dynamic situation, there will be a slight variation in bending moment along the section. If the bending is slight, we may regard the section as experiencing an unbalanced force in the y direction, then

$$\rho A\frac{\partial^2\eta}{\partial t^2} = -\frac{\partial V}{\partial x} \tag{12.10.1}$$

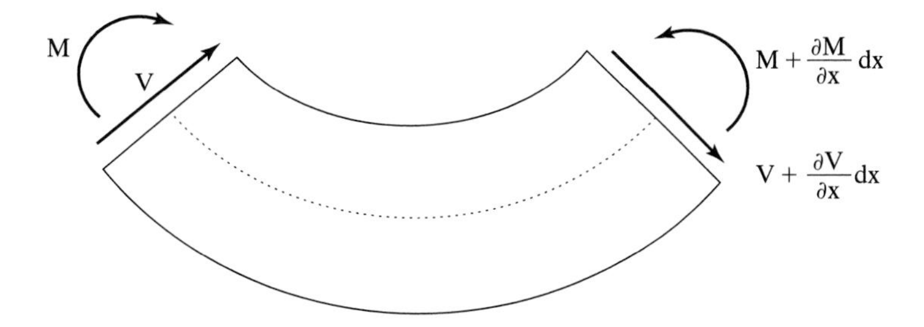

Fig. 12.18 Beam in bending: dynamic case.

where η is the displacement of the axis of the beam. Using Eqs. (12.9.1) and (12.9.10) gives

$$V = \frac{\partial}{\partial x}\left(\frac{YI}{R}\right) \tag{12.10.2}$$

Now, in general, the radius of curvature, R, of a curve $\eta = f(x)$ is given by

$$\frac{1}{R} = \frac{d^2\eta/dx^2}{\left\{1 + (d\eta/dx)^2\right\}^{3/2}} \tag{12.10.3}$$

In our case, $d\eta/dx$ is small and thus

$$\frac{1}{R} = \frac{d^2\eta}{dx^2} \tag{12.10.4}$$

Thus, substituting into Eq. (12.9.10), we obtain

$$M = -YI\frac{\partial^2\eta}{\partial x^2} \tag{12.10.5}$$

and substituting into Eq. (12.10.2),

$$V = YI\frac{d^3\eta}{dx^3} \tag{12.10.6}$$

Similarly, from Eqs. (12.10.1), (12.10.2), and (12.10.4)

$$\rho A\frac{\partial^2\eta}{\partial t^2} = -\frac{\partial^2}{\partial x^2}\left[YI\frac{\partial^2\eta}{\partial x^2}\right] \tag{12.10.7}$$

For a beam of constant cross section, the inertia is constant and may be represented by K^2A, where K is called the radius of gyration:

$$\frac{\partial^2\eta}{\partial t^2} = -\frac{YK^2}{\rho}\frac{\partial^4\eta}{\partial x^4} = -c_b^2K^2\frac{\partial^4\eta}{\partial x^4} \tag{12.10.8}$$

where c_b is the bar wave velocity from Eq. (12.7.4). Eq. (12.10.8) is a fourth-order equation and not a wave equation. However, this is not to say that wavelike solutions cannot be obtained. Assume that it is possible for a sinusoidal wave to propagate along the beam:

$$\eta = \eta_0 e^{j(\omega t - kx)} \tag{12.10.9}$$

Then

$$\frac{\omega^2}{k^2} = c^2 = c_b^2 K^2 k^2 \tag{12.10.10}$$

Thus, the phase velocity

$$c = c_b k = \frac{2\pi c_b K}{\lambda} \tag{12.10.11}$$

In other words, the phase velocity depends on the wavelength (or frequency).

This result is unsatisfactory in that at short wavelength (or high frequency) the velocity approaches infinity. The reason for this inadequacy is that the derivation assumed purely translational motion in the y direction, ignoring rotational effects. It was also assumed that rectangular sections remained rectangular, which is not true in the event the length of the beam is comparable with the thickness. These two defects were corrected by Rayleigh (1894) and Timoshenko (1921), respectively. However, for many purposes, the Euler-Bernoulli theory is accurate enough, and for simplicity we will continue to use it in our discussions.

12.11 Natural Flexural Modes of a Beam

To find the natural flexural modes of a beam, we have to solve Eq. (12.10.7), which can be done by assuming a separable solution:

$$\eta = X(x)T(t) \tag{12.11.1}$$

When Eq. (12.11.1) is substituted into Eq. (12.10.7), the result is

$$\frac{\partial^4 X/\partial x^4}{X} = -\frac{\partial^2 T/\partial t^2}{c_b^2 K^2 T} \tag{12.11.2}$$

Now one side of Eq. (12.11.2) is a function of x while the other side is a function of t. Since x and t are independent variables, it follows that both sides must equal a constant, β^4 say. Hence

$$\frac{\partial^4 X}{\partial x^4} = \beta^4 X \tag{12.11.3}$$

and

$$\frac{\partial^2 T}{\partial t^2} = -\beta^2 c_b^2 K^2 T \tag{12.11.4}$$

The solution of Eq. (12.11.4) is

$$T = Ae^{j\omega t} \tag{12.11.5}$$

where

$$\omega^2 = \beta^4 c_b^2 K^2 \tag{12.11.6}$$

or

$$\beta^4 = \frac{\omega^2}{c_b^2 K^2} = \frac{\omega^2 \rho}{YK^2} = \frac{\omega^2 \rho A}{YI} = \frac{\omega^2 m}{YI} \tag{12.11.7}$$

where m is the mass per unit length of the beam. We obtain the solution of Eq. (12.11.3) by expressing it in the differential notation, thus

$$(D^4 - \beta^4)X = 0 \tag{12.11.8}$$

or

$$(D^2 - \beta^2)(D^2 + \beta^2)X = 0 \tag{12.11.9}$$

Either

$$(D^2 - \beta^2)X = 0 \quad \text{or} \quad (D^2 + \beta^2)X = 0 \tag{12.11.10}$$

when either

$$X = Ae^{\beta x} + Be^{-\beta x} \quad \text{or} \quad X = Ae^{j\beta x} + Be^{-j\beta x} \tag{12.11.11}$$

We combine these to obtain the general solution

$$X = C_1 \cos \beta x + C_2 \sin \beta x + C_3 \cosh \beta x + C_4 \sinh \beta x \tag{12.11.12}$$

To proceed further we need to stipulate boundary conditions. If the end of a beam is pinned, then its displacement at that point will be zero, and the bending moment will also be zero, since a pin cannot supply a rotational constraint. From Eq. (12.10.5), it follows that the condition of zero moment is equivalent to the second derivative being zero. For a clamped or fixed-end condition, the displacement and its slope will be zero. Finally, for a free-end condition, the bending moment and shear force will have to be zero. From Eq. (12.10.6), it follows that the condition of zero shear is equivalent to the third derivative being zero. We can now proceed to find the natural modes with some common types of supports.

Consider a *simply supported beam*, for which the boundary conditions at $\pi = 0, \ell$ can be stated as

$$X = 0 \text{ and } \frac{d^2X}{dx^2} = 0 \tag{12.11.13}$$

Note that these boundary conditions, and thus the modes of vibration, will be the same for a pinned-pinned beam. From Eq. (12.11.12), using the conditions at $x = 0$,

$$C_1 + C_3 = 0 \tag{12.11.14}$$

and

$$-\beta^2 C_1 + \beta^2 C_3 = 0 \tag{12.11.15}$$

It follows that

$$C_1 = C_3 = 0 \tag{12.11.16}$$

Using the conditions for $x = \ell$,

$$C_2 \sin \beta\ell + \beta^2 C_4 \sinh \beta\ell = 0 \tag{12.11.17}$$

and

$$-\beta^2 C_2 \sin \beta\ell + \beta^2 C_4 \sinh \beta\ell = 0 \tag{12.11.18}$$

The only way we can obtain a nontrivial solution to Eqs. (12.11.17) and (12.11.18) is to set

$$C_4 = \sin \beta\ell = 0 \tag{12.11.19}$$

Hence

$$\beta\ell = n\pi \tag{12.11.20}$$

and from Eq. (12.11.7),

$$\omega_n^2 = \left(\frac{n\pi}{\ell}\right)^4 \frac{YI}{\rho A} \tag{12.11.21}$$

and the mode shapes are given by

$$\eta = A_n \sin \frac{n\pi x}{\ell} e^{j\omega_n t} \tag{12.11.22}$$

As a second example we will consider a *clamped-clamped beam* (or *fixed-fixed beam*), for which the boundary conditions at $x = 0, \ell$ can be stated as

$$X = 0, \frac{dX}{dx} = 0 \tag{12.11.23}$$

Using the conditions at $x = 0$

$$C_1 + C_3 = 0 \tag{12.11.24}$$

and

$$\beta(C_2 + C_4) = 0 \tag{12.11.25}$$

It follows that

$$C_1 = -C_3 \tag{12.11.26}$$

and

$$C_2 = -C_4 \tag{12.11.27}$$

and, using the end conditions at $x = \ell$,

$$\frac{\cosh \beta\ell - \cosh \beta\ell}{\sinh \beta\ell + \sin \beta\ell} = \frac{\sin \beta\ell - \sinh \beta\ell}{\cos \beta\ell - \cosh \beta\ell} \tag{12.11.28}$$

which leads to the frequency equation

$$\cosh \beta\ell \cos \beta\ell = 1 \tag{12.11.29}$$

so that $\beta_1\ell = 4.73$, $\beta_2\ell = 7.853$, etc. The natural mode shapes for the clamped-clamped beam are shown in Fig. 12.19. The natural modes for other supporting conditions, including cantilevers, have been worked out, and there are tabulations available (see

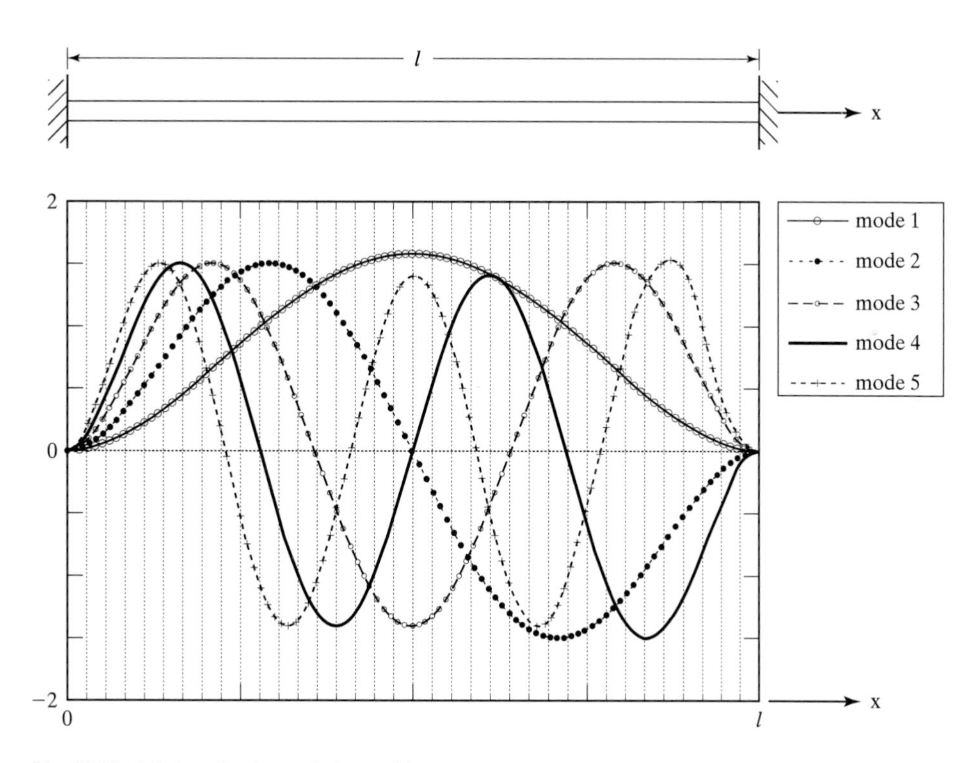

Fig. 12.19 Modes of a clamped-clamped beam.

Gorman (1975), Warburton (1976), and Weaver et al. (1990)). It is also possible to solve for the natural modes of beams where the end conditions do not conform to the cases we just discussed. Warburton (1976) shows how to treat general end conditions where a lumped mass is present and where linear and rotational motions are constrained by springs rather than rigid fixtures.

12.12 Response of a Beam to an Applied Force

The vibration of a structure is usually the result of a forced oscillation. If an applied force, which is a function of time f(t) and which acts in the y direction, is distributed along the length of the beam, the equation of motion of an element will include an additional term p(x)f(t), equal to the applied force per unit length, added to Eq. (12.10.7); in other words,

$$\rho A\frac{\partial^2 v}{\partial t^2} + \frac{\partial^2}{\partial x^2}\left(YI\frac{\partial^2 v}{\partial x^2}\right) = p(x)f(t) \qquad (12.12.1)$$

This result is valid for uniform and nonuniform beams. In Eq. (12.12.1), it is assumed that the applied force varies with time in the same way for all the points on the beam. Before solving Eq. (12.12.1), it will be modified to include damping.

For the elastic beam, the stress σ_x and the strain ϵ_x in the longitudinal direction at a point distant y from the center line are related by Hooke's law; that is,

$$\sigma_x = Y\epsilon_x \qquad (12.12.2)$$

In order to include damping, we consider a viscoelastic material with a stress-strain relation given by

$$\sigma_x = Y\left(\epsilon_x + \alpha\frac{\partial \epsilon_x}{\partial t}\right) \qquad (12.12.3)$$

The strain in a fiber at a distance y from the neutral axis is

$$\epsilon_x = -\frac{y}{R} = -y\frac{\partial^2 \eta}{\partial x^2} \qquad (12.12.4)$$

where R is the radius of curvature. The relation between the bending moment M and stress σ_x is

$$M = \int \sigma_x b(y)y\,dy \qquad (12.12.5)$$

where b(y) is the breadth of the beam at y and the integration is performed over the depth. Thus, the effect of including damping is to modify the bending moment-curvature relation from Eq. (12.10.5) to

$$M = -YI\left(\frac{\partial^2 \eta}{\partial x^2} + \alpha\frac{\partial^3 \eta}{\partial x^2 \partial t}\right) \qquad (12.12.6)$$

Using Eq. (12.12.6), the equation of motion is

$$\rho A \frac{\partial^2 \eta}{\partial t^2} + \frac{\partial^2}{\partial x^2}\left[YI\left(\frac{\partial^2 \eta}{\partial x^2} + \alpha \frac{\partial^3 \eta}{\partial x^2 \partial t}\right)\right] = p(x)f(t) \tag{12.12.7}$$

A solution of Eq. (12.12.7) will be sought in terms of the normal modes, $\Phi_r(x)$; each mode shape is associated with a frequency ω_r, and

$$\eta = \sum_r A_r \Phi_r(x) \sin(\omega_r t + \beta_r) \tag{12.12.8}$$

is the general solution for free vibrations of the beam (i.e., the solution of Eq. (12.10.7) or of Eq. (12.12.8) with the right side equal to zero and $\alpha = 0$). The general solution of Eq. (12.12.7) will be of the form

$$\eta = \sum_r \Phi_r(x) q_r(t) \tag{12.12.9}$$

where $q_r(t)$ is a function of time.

Substituting for η from Eq. (12.12.9) in Eq. (12.12.7), multiplying by $\Phi_s(x)$, and integrating with respect to x over the length of the beam gives

$$\int_0^\ell \rho A \Phi_S \Sigma_r(\Phi_r \ddot{q}_r)dx + \int_0^\ell \Phi_s \frac{d^2}{dx^2}\left[YI\Sigma_r\left(\frac{d^2\Phi_r}{dx^2} q_r + \alpha \frac{d^2\Phi_r}{dx^2}\dot{q}_r\right)\right]dx \tag{12.12.10}$$

$$= \int_0^\ell p(x)\Phi_s \cdot f(t)dx$$

Now it can be shown that if $r \neq s$, then

$$\int_0^\ell \Phi_r(x)\Phi_s(x)dx = 0 \tag{12.12.11}$$

$$\int_0^\ell YI \frac{d^2\Phi_r}{dx^2}(x)\frac{d^2\Phi_s}{dx^2}(x) = 0 \tag{12.12.12}$$

and

$$\int_0^\ell \Phi_s(x)\frac{d^2}{dx^2}\left[YI\frac{d^2\Phi_r}{dx^2}(x)\right]dx = 0 \tag{12.12.13}$$

If $r = s$, then

$$\omega_r^2 \int_0^\ell \rho A \Phi_r^2(x)dx = \int_r^\ell \Phi_r(x)\frac{d^2}{dx^2}\left[YI\frac{d^2\Phi_r}{dx^2}(x)\right]dx \tag{12.12.14}$$

These results are known as orthogonality conditions. The proof of the orthogonality conditions is a worthwhile student exercise. The result is true for uniform and nonuniform beams with simple end conditions.

Using the orthogonality relation for the beam in Eq. (12.12.10), we have

$$\ddot{q}_s + \alpha\omega_s^2\dot{q}_s + \omega_s{}^s q_s = \frac{f(t)\displaystyle\int_0^{\ell} p(x)\Phi_s(x)dx}{\displaystyle\int_0^{\ell} \rho A\left[\Phi_s(x)\right]^2 dx} \tag{12.12.15}$$

If the applied force is located at one point along the beam, as shown in Fig. 12.20, where a force pf(t) acts at x = a, with the distribution of p(x) as shown (i.e., a delta function), $p(a)\Delta x \rightarrow P$ as $\Delta x \rightarrow 0$. Thus, the upper integral in Eq. (12.12.15) becomes $P\Phi_s$ (a) and

$$\ddot{q}_s + \alpha\omega_s^2\dot{q}_s + \omega_s{}^s q_s = \frac{P\Phi_s(a)f(t)}{\displaystyle\int_0^{\ell} \rho A[\Phi_s(x)]^2 dx} \tag{12.12.16}$$

If the mode shapes are normalized so that

$$\int_0^{\ell} \rho A[\Phi_s(x)]\, dx = m, s = 1, 2, 3\ldots \tag{12.12.17}$$

where m is the mass of the beam, then for an applied force pf(t) at x = a, we get

$$\ddot{q}_s + \alpha\omega_s{}^2\dot{q}_s + \omega_s{}^2 q_s = \frac{P\Phi_s(a)f(t)}{m} \tag{12.12.18}$$

Note that for the special case of a uniform beam, the normalizing condition of Eq. (12.12.17) reduces to

$$\int_0^{\ell} [\Phi(x)]^2 dx = \ell \tag{12.12.19}$$

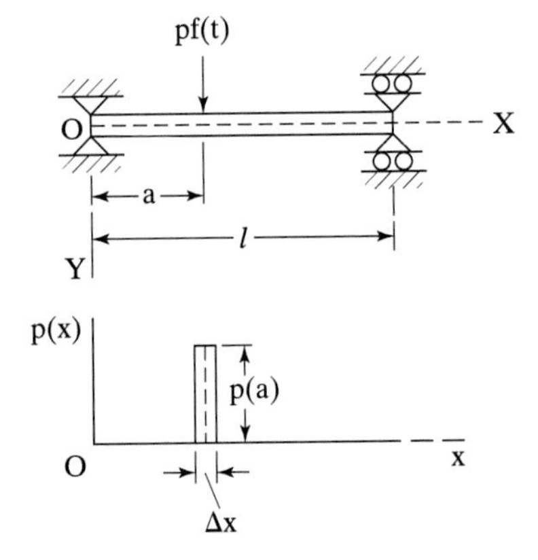

Fig. 12.20 Beam with an applied force at x = a. (After Warburton (1976)).

where ℓ is the length of the beam, as in this case $m = \rho A\ell$. Eq. (12.12.18) is of the same form as Eq. (12.9.1), and we can proceed to a solution in the same way as in that case by taking the Laplace transform of both sides. Assuming zero initial values, we have from Eq. (12.9.5).

$$L[q_s] = \frac{(p/m)\Phi_s(a)L[f(t)]}{(s^2 + x\omega_s^2 s + \omega_s^2)} \tag{12.12.20}$$

Now let

$$\alpha\omega_s^2 = 2a \text{ and } \omega_s^2 = a^2 + b^2 \tag{12.12.21}$$

that is,

$$L[q_s] = \frac{p/m\,\Phi_s(a)}{(s + a)^2 + b^2} L[f(t)] \tag{12.12.22}$$

But

$$\frac{L}{(s + a)^2 + b^2} = L\left[\frac{e^{-at}\sin bt}{b}\right] \tag{12.12.23}$$

Hence

$$L[q_s] = (p/m)\Phi_s(a)L\left[\frac{e^{-at}\sin bt}{b}\right]L[f(t)] \tag{12.12.24}$$

Thus, by the convolution theorem, Eq. (2.8.24):

$$q_s = \frac{p\Phi_s(a)}{m} f(t) * \frac{e^{-at}\sin bt}{b} = \frac{p\Phi_s(a)}{mb} \cdot \int_0^t f(\tau)e^{-a(\tau-\tau)}\sin b(t - \tau)d\tau \tag{12.12.25}$$

and finally

$$q_s = \frac{p\Phi_s(a)}{m\omega_s'} \int_0^t f(\tau)\exp\left[-\frac{1}{2}\alpha\omega_s^2(t - \tau)\right]\sin\omega_s'(t - \tau)d\tau \tag{12.12.26}$$

where

$$\omega_s' = \omega_s(1 - \alpha^2\omega_s^2/4)^{1/2}] \tag{12.12.27}$$

Therefore

$$\eta(x,t) = \sum_s \left[\frac{p\Phi_s(a)\Phi_s(x)}{m\omega_s'} \int_0^\ell f(\tau)\exp\left[-\frac{1}{2}\alpha\omega_s^2(t - \tau)\right]\sin\omega_s'(t - \tau)d\tau\right] \tag{12.12.28}$$

In other words, in response to a forcing function, all the natural modes of the beam are excited with amplitudes proportional to that of each mode at the excitation point. One interesting consequence is that if the beam is excited at x = a and if this is the nodal point of a particular mode, we then have

$$\Phi_s(a) = 0 \tag{12.12.29}$$

and consequently from Eq. (12.12.28)

$$q_s(t) = 0 \tag{12.12.30}$$

in other words, the response is zero for all values of t. Furthermore, we see from Eq. (12.12.9) that for this mode

$$\eta_s(x, t) = 0 \tag{12.12.31}$$

for all values of x and t.

12.13 Structural Acoustics

The subject of the statics and dynamics of beams, started early in the eighteenth century by Euler and Bernoulli, has grown steadily in importance over the years. This was partly due to the innovation of steelmaking and to the use of steel beams in bridges and buildings. More recently, there has been increased use of reinforced concrete, and the subject of the dynamic response of structures made of this material becomes of paramount importance in earthquake engineering. The extraction of oil from below the seabed has necessitated the construction of enormous trusses to sustain the drilling equipment and to deliver the oil to the surface. These offshore platforms are generally made of tubular members, which also bend in a manner described by the Euler-Bernoulli beam theory. Wave action on offshore platforms can give rise to vibrational responses, which can be disastrous unless they are taken into account.

The developments in civil engineering were paralleled in transportation by the advent of the railroads. Rail is itself a beam and here again Euler and Bernoulli have served us well, since their theory can be readily adapted to encompass a beam on an elastic foundation. We can then investigate the problem of a moving load on such a beam. (Warburton (1976) provides further details on this topic.) There is a critical velocity at which the mass may travel, which preferentially generates flexural vibrations: a situation the railroad engineer will want to avoid. We already discussed the railroad wheel in Chapter 1 in connection with finite element analysis. But there is another approach to the vibrations of the wheel, first discussed by Stappenbeck (1954), who applied the theory of a vibrating ring to the problem (see Love (1944)). Once again, the ring is treated as a beam, although now in a circular form. The two ends of the beam are joined, and the "boundary" conditions become a requirement that the displacement must repeat itself with every complete circuit of the wheel. It is the thick rim of the wheel that approximates the ring. The plate of the wheel serves as a stiffener. Stappenbeck found that the series of resonant frequencies observed in the radiated sound spectrum of the wheel do in fact closely follow the ratios predicted by the ring theory. Haran (1983) expanded on the idea by showing that the wheel could be regarded as a series of concentric rings, which, when connected together suitably, quite accurately predict the observed resonances.

The same flexural vibrations that cause failure when they are of high amplitude can be used for nondestructive testing and monitoring at low amplitude. High-speed rotating equipment, as evidenced in turbines and other applications, can be subject to flexural vibrations in the presence of unbalanced loads. The vibrations thus generated can be detected using accelerometers mounted on the bearings of the machines, thus serving as an early warning of incipient problems. A full review of work in this area has been given by Dimarogonas and Paipetis (1983). Research work on the use of structural beam vibrations to detect cracks has been reviewed by Man, et al. (1994).

Flexural vibrations have also been put to service in the flextensional transducer. Imagine a device similar to the tonpilz transducer, with head and tail masses that have been connected by beams: that is a picture of the flextensional transducer. It is resonant at relatively low frequencies and can produce high power outputs. These two characteristics make it suitable for use with the deep-sea channels in underwater communication. A review of the electroacoustics of these devices has been presented by Woollett (1989).

The basic concept of the Euler-Bernoulli theory is that the bending of the beam produces a stretching of fibers, which can then be related to the assumptions of elasticity using Hooke's law. A similar presupposition can be used to derive a theory for plates and shells. The elucidation of the problem has an interesting history. The first, and incorrect, attack on it was undertaken by James Bernoulli, a nephew of the famous Daniel, by trying to represent the plate as a superposition of crossed beams. The modes of vibration of plates are well visualized by the patterns taken up by sand or other particulate scattered on the plate. The resulting patterns are often called Chladni figures for the scientist who first observed them and popularized the subject with demonstrations at the courts of Europe. This topic caught the attention of Napoleon, who offered a prize for the first correct theory of plate vibrations. This prize was eventually won by Sophie Germaine in 1815, although some points were not finally resolved until 1850 by Kirchoff. The treatment uses concepts developed in the general theory of elasticity derived in Chapter 13 and is quite elaborate. Consequently, we will simply state that disturbances in the plate obey a fourth-order differential equation, and solutions can be found analytically for the simple cases of rectangular and circular plates that agree well with experimental findings. An extensive literature has been accumulated on vibrating plates, and Leissa (1993) has put together a compendium of this knowledge.

Equations similar to those that govern the behavior of plates are found for a vibrating shell. This subject has been of interest historically in connection with the vibration of bells. Contributions to the theory were made by Sophie Germaine, Aron, and Rayleigh, although Love (1888) is given credit for the first general solution extending the assumptions made for beams and plates. Love's theory is couched in general curvilinear coordinates from which solutions for beams, arches, rings, and cylinders emerge as special cases. A modern presentation using such an approach is given by Soedel (1981). A direct treatment of the vibration of plates and cylindrical shells may be found in Warburton (1976) and a further compendium on vibration of shells has been presented by

Leissa (1993). During the Cold War era, the United States and the Soviet Union put great efforts into the development of passive sonar (i.e., systems that listen for and recognize enemy submarines). To a first approximation, the hull of a submarine can be treated as a cylindrical shell, which, through its vibrations, radiates sound into the water. An important aspect of that problem involves the radiation loading of the surrounding medium on the shell, for which Junger and Feit (1993) give a good introduction.

REFERENCES AND FURTHER READING

d'Alembert, J.L. 1747. "Investigation of the Curve Formed by a Vibrating String." 3 *Hist. Acad. Sci.* 214–219. (See translation by R.B. Lindsay in *Acoustics: Historical and Philosophical Development*. 1973. Dowden, Hutchinson and Ross.)

Dimarogonas, A.D., and S.A. Paipetis. 1983. *Analytical Methods in Rotor Dynamics*. Applied Science Publishers.

Gorman, D.J. 1975. *Free Vibration Analysis of Beams and Shafts*. Wiley-Interscience.

Haran, S., and R.D. Finch. 1983. "Ring Model of Railroad Wheel Vibrations. 74 *J. Acoust. Soc. Am.* 1433–1440.

Harari, A. 1995. *Structural Acoustics*. Technical Digest. Naval Undersea Warfare Center Division, Newport, R.I.

Hunt, F.V. 1978. *Origins in Acoustics*. Yale Univ. Press. Republished: Acoustical Society of America (1992).

Junger, M.D., and D. Feit. 1933. *Sound, Structures, and their Interaction*, 2nd ed. Acoustical Society of America.

Leissa, A. 1993. *Vibration of Plates*. Acoustical Society of America.

Leissa, A. 1993. *Vibration of Shells*. Acoustical Society of America.

Love, A.E.H. 1888. "On the Small Free Vibrations and Deformations of Thin Elastic Shells." 179A *Phil. Trans. Roy. Soc.* 491–546.

Love, A.E.H. 1944. *A Treatise on the Mathematical Theory of Elasticity*, 4th ed. Dover.

Man, X.C., et al. 1994. "Slot Depth Resolutions in Vibration Monitoring of Beams Using Frequency Shift." 95 *J. Acoust. Soc. Am.* 2029–2037.

Morse, P.M. 1948. *Vibration and Sound*, 2nd ed. McGraw Hill.

Soedel, W. 1981. *Vibrations of Shells and Plates*. Marcel Dekker.

Stappenbech, H. von. 1954. "Das Kurvengerausch Der Strassenbahn." 96 *Z. Ver. Dtsch. Ing.* 171–175.

Strutt, J.W. (Baron Rayleigh) [1894] 1945. *The Theory of Sound*. Vol. 1. 2nd ed. Reprinted: Dover Publications.

Timoshenko, S. 1921. "On the Correction for Shear of the Differential Equation for Transverse Vibrations of Prismatic Bars." 41 *Philos. Mag.* 744–746.

Warburton, G.B. 1976. *The Dynamical Behavior of Structures*, 2nd ed. Pergamon.

Weaver, W.J., S. Timoshenko, and D.H. Young. 1990. *Vibration Problems in Engineering*. Wiley.

Woollett, R.S. 1989. *The Flexural Bar Transducer*. Naval Underwater Systems Center, Newport, R.I.

PROBLEMS

12.1 $y = 0.04 \cos(3t - 2x)$ is the equation of a wave travelling on a string of linear density 0.2 kg/m, where x and y are in meters and t is in seconds.

a. What are the wave's amplitude, phase speed, frequency, wavelength, and wave number?
b. What is the particle velocity of an element at $x = 0$ and at $t = 0$?

12.2 A semi-infinite string ($x < 0$) with linear density ρ and under tension T is joined at $x = 0$ to a second semi-infinite string ($x > 0$) under the same tension but with linear density 3ρ. If a wave of angular frequency ω and amplitude a is travelling in the positive x direction on the first string, find the amplitude of the associated wave travelling on the second string.

12.3 A flexible cable is fixed at its upper end and hangs in such a way that it can oscillate under gravity. If the lateral displacement is y at distance x from the lower end, show that the equation of motion is given by

$$\frac{\partial^2 y}{\partial t^2} = g\left[x\frac{\partial^2 y}{\partial x^2} + \frac{\partial y}{\partial x}\right]$$

12.4 A bar of length L undergoes longitudinal vibrations. The bar is rigidly fixed at $x = 0$ and free to move at $x = L$.

a. Show that only odd harmonic overtones will occur.
b. Determine the fundamental frequency of the bar if it is composed of steel and has a length of 0.7 m.

12.5 A rectangular membrane has width w and length 2w. Find the ratios of the first four overtone frequencies relative to the fundamental.

12.6 A beam is 1 m in length and has modulus of elasticity $Y = 2 \times 10^{11}$ Pa, shear modulus $G = 1 \times 10^{11}$ Pa, density $r = 7800$ kg/m^3, rectangular cross section with height $h = 0.1$ m and width $w = .05$ m, and simply supported ends. Estimate the lowest natural frequency first using the Euler-Bernoulli theory and then using the Rayleigh-Timoshenko theory.

12.7 An aluminum beam of 1 m length with circular cross section of 2 cm diameter is simply supported at both ends. For transverse vibrations,

a. Show that the normal modes are the same as for the fixed-fixed string.
b. Find the frequencies of the normal modes.
c. Are the overtones harmonics as they are for the fixed-fixed string?

12.8 An aluminum cantilever beam of 1 m length with square cross section of 1 cm $\times$ 1 cm dimensions is clamped at one end and free at the other.

a. Find the frequency of the lowest mode of transverse vibration.
b. If the free end has a displacement amplitude of 3 cm in this mode, determine all the constants in the equation for transverse displacement.
c. Plot the displacement amplitude of the beam at this frequency.

12.9 A uniform beam of length l and mass M_b is loaded with a mass M_o at one end and clamped at the other. What are the boundary conditions? Find the frequency equation.

12.10 Estimate the frequencies for natural vibrations of a ring of radius R and of uniform cross section A, density ρ, and Young's modulus Y by treating it as a free-free beam whose ends are joined. What will be the difference in the frequencies of the in-plane versus the out-of-plane vibrations.

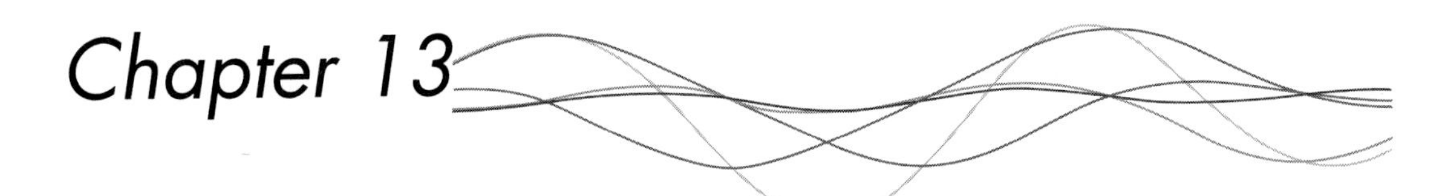

Chapter 13 Propagation in Solids

13.1 Elasticity of Extended Media

If we want to consider wave propagation in extended elastic media, such as the Earth for example, it is clear that the approaches we have taken in Chapters 11 and 12 cannot be applied. For longitudinal waves, we studied the motion of an element whose lateral dimension was the same as the bar itself, based on the assumption that the motion of all parts of the cross section are the same. Obviously, this assumption cannot hold true if the thickness of the bar becomes comparable to a wavelength. Again, the assumptions of the Euler-Bernoulli theory of flexural vibration (namely, that fibers along the beam are either stretched or compressed) do not make sense when applied to infinite media. For an extended medium, we will consider the motion of an element whose dimensions are all small. The procedure is similar to the one we followed in Chapter 3, where we treated radiation in three dimensions in a fluid medium. But now we will modify that approach for a general elastic medium, meaning one that regains its original form after the removal of applied stresses. Although the main reason for doing this is to be able to describe what happens in solids, we will see that the technique is also useful for fluids when we discuss the effect of viscosity in Chapter 14. The main difference between fluids and solids is that solids can sustain shearing deformation, whereas fluids cannot, and in order to take this into account we need a more sophisticated description of deformation than we have used so far. We need to establish this new description and state Hooke's law in a generalized form before getting to the heart of the matter: the dynamics of a small element.

Consider any point P in an elastic medium, whose position vector, referred to as a Cartesian coordinate system, is

$$\mathbf{r} = \mathbf{i}x + \mathbf{j}y + \mathbf{k}z \tag{13.1.1}$$

Suppose now that the medium moves so that the particle that was at P suffers a displacement Δ, and finishes at P′ (see Fig. 13.1). Such a displacement might be the result

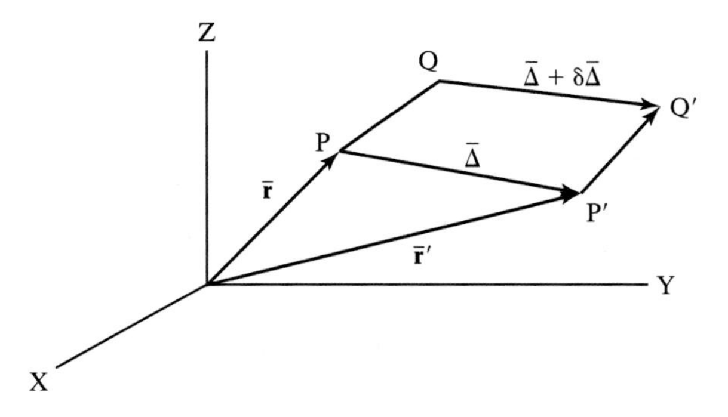

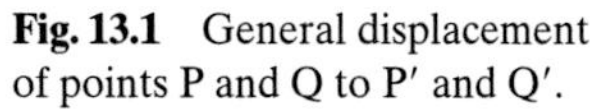
Fig. 13.1 General displacement of points P and Q to P′ and Q′.

of a simple translation or rotation of the medium, without deformation. In order to tell if the medium has been deformed, we have to consider the displacement of a second particle, which moves from Q to Q′ say, with displacement $\Delta + \delta\Delta$. If Q is close to P, then PQ may be written $\delta\mathbf{r}$, and

$$\delta\Delta = \delta(\delta\mathbf{r}) \tag{13.1.2}$$

Let

$$\Delta = \mathbf{i}\xi + \mathbf{j}\eta + \mathbf{k}\zeta \tag{13.1.3}$$

then

$$\delta\Delta = \mathbf{i}(\xi_x\delta x + \xi_y\delta y + \xi_z\delta z) + \mathbf{j}(\eta_x\delta x + \eta_y\delta y + \eta_z\delta z) + \mathbf{k}(\zeta_x\delta x + \zeta_y\delta y + \zeta_z\delta z) \tag{13.1.4}$$

where

$$\xi_x = \frac{\partial\xi}{\partial x}, \quad \xi_y = \frac{\partial\xi}{\partial y}, \text{ etc.} \tag{13.1.5}$$

Rearranging Eq. (13.1.4), we obtain

$$\begin{aligned}\delta\Delta = \mathbf{i}&\left[\xi_x\delta x + \frac{1}{2}(\eta_x + \xi_y)\delta y + \frac{1}{2}(\xi_z + \zeta_x)\delta z\right]\\ &+ \mathbf{j}\left[\frac{1}{2}(\xi_y + \eta_x)\delta x + \eta_y\delta y + \frac{1}{2}(\zeta_y + \eta_z)\delta z\right]\\ &+ \mathbf{k}\left[\frac{1}{2}(\xi_z + \zeta_x)\delta x + \frac{1}{2}(\eta_z + \zeta_y)\delta y + \xi_z\delta z\right]\\ &+ \mathbf{i}\left[\frac{\mathbf{1}}{\mathbf{2}}(\xi_y - \eta_x)\delta y + \frac{1}{2}(\xi_z - \zeta_x)\delta z\right]\end{aligned}$$

$$+ \mathbf{j}\left[\frac{1}{2}(\eta_x - \xi_y)\delta x + \frac{1}{2}(\eta_z - \xi_y)\delta z\right]$$

$$+ \mathbf{k}\left[\frac{1}{2}(\xi_x - \zeta_z)\delta x + \frac{1}{2}(\xi_y - \eta_z)\delta y\right] \quad (13.1.6)$$

Now the second of the vectors in Eq. (13.1.6) represents a purely rotational movement, as we will see. Consequently, the first part of the expression for $\delta\Delta$ must be due to deformation only, since a purely translational motion would not produce a difference in displacement of the various particles in the medium.

To prove the point about the second part of the right side of Eq. (13.1.6), consider the rotation of a perfectly rigid body through a small angle α, as in Fig. 13.2. The displacement of a point P would then be

$$\Delta_r = \alpha \times r \quad (13.1.7)$$

But notice an interesting property:

$$\nabla \times \Delta_r = \nabla \times \alpha \times r$$

$$= \nabla \times [\mathbf{i}(\alpha_y z - \alpha_z y) + \mathbf{j}(\alpha_z x - \alpha_x z) + \mathbf{k}(\alpha_x y - \alpha_y x)]$$

$$= \begin{bmatrix} \mathbf{i} & \mathbf{j} & \mathbf{k} \\ \dfrac{\partial}{\partial x} & \dfrac{\partial}{\partial y} & \dfrac{\partial}{\partial z} \\ (\alpha_y z - \alpha_z y) & (\alpha_z x - \alpha_x z) & (\alpha_x y - \alpha_y x) \end{bmatrix}$$

$$= \mathbf{i}\left[\frac{\partial}{\partial y}(\alpha_x y - \alpha_y x) - \frac{\partial}{\partial z}(\alpha_z x - \alpha_x z)\right] + \mathbf{j}[\;] + \mathbf{k}[\;]$$

$$= \mathbf{i}[\alpha_x + \alpha_x] + \mathbf{j}2\alpha_y + \mathbf{k}2\alpha_z \quad (13.1.8)$$

since α is independent of x, y, and z and since x, y, and z are independent variables. Thus

$$\nabla \times \Delta_{\mathbf{r}} = 2\alpha \quad (13.1.9)$$

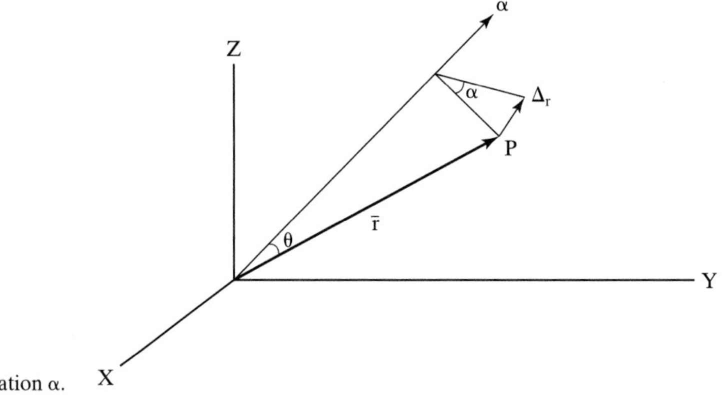

Fig. 13.2 A rigid-body rotation α.

or

$$\alpha = \frac{1}{2}\nabla \mathrm{x}\, \Delta_{\mathbf{r}} = \frac{1}{2}\nabla \mathrm{x}(\alpha \mathrm{ x } \mathrm{r}) \tag{13.1.10}$$

But

$$\nabla \mathrm{x}\, \Delta = \mathbf{i}(\xi_y - \eta_z) + \mathbf{j}(\xi_z - \zeta_x) + \mathbf{k}(\eta_x - \xi_y) \tag{13.1.11}$$

Thus

$$\frac{1}{2}\Delta \mathrm{ x } \nabla \mathrm{ x } \mathrm{d}\, \mathrm{r} = \frac{1}{2}\begin{bmatrix} \mathbf{i} & \mathbf{j} & \mathbf{k} \\ (\zeta_y - \eta_z) & (\xi_z - \zeta_x) & (\eta_x - \xi_y) \\ \mathrm{dx} & \mathrm{dy} & \mathrm{dz} \end{bmatrix}$$

$$= \mathbf{i}\left[\frac{1}{2}(\xi_z - \zeta_x)\,\mathrm{dz} + \frac{1}{2}(\xi_y - \eta_x)\,\mathrm{dy}\right] + \mathbf{j}[\;] + \mathbf{k}[\;] \tag{13.1.12}$$

This is the second vector in Eq. (13.1.6). Thus, the deformation is represented by the first vector of Eq. (13.1.6), whose coefficients may be written in a square array matrix:

$$D = \begin{bmatrix} \xi_x & \frac{1}{2}(\eta_x + \xi_y) & \frac{1}{2}(\zeta_x + \xi_z) \\ \frac{1}{2}(\xi_y + \eta_x) & \eta_y & \frac{1}{2}(\xi_y + \eta_z) \\ \frac{1}{2}(\zeta_x + \xi_z) & \frac{1}{2}(\eta_x + \xi_y) & \xi_z \end{bmatrix} \tag{13.1.13}$$

which is known as the strain matrix. Hence, $\xi_x = \partial\xi/\partial x$ = increase in length per unit length in the x direction. η_y and ξ_z may be understood in a similar manner. Now consider the other terms. By definition, the shear modulus = shearing stress/strain = $F/A/\theta$, where the shear strain is the shear angle. Consider a section of a rectangular block in the x-z plane, as in Fig. 13.3. Suppose the shears are caused by couples with moments parallel to the y-axis. For small angles, the total shear strain will be $\alpha_1 + \alpha_2 = AA'/OA + CC'/OC$. But

$$AA' = \frac{\partial\zeta}{\partial x}\delta x$$

and

$$CC' = \frac{\partial\xi}{\partial z}\delta z \tag{13.1.14}$$

Thus, the total shear strain about the y-axis = $\xi_z + \zeta_x$. Similarly, the shear about the z-axis is $\xi_y + \eta_z$. Notice that these strains are given by the cyclic rotations $x \rightarrow y \rightarrow z \rightarrow x$ and $\xi \rightarrow \eta \rightarrow \zeta \rightarrow \xi$.

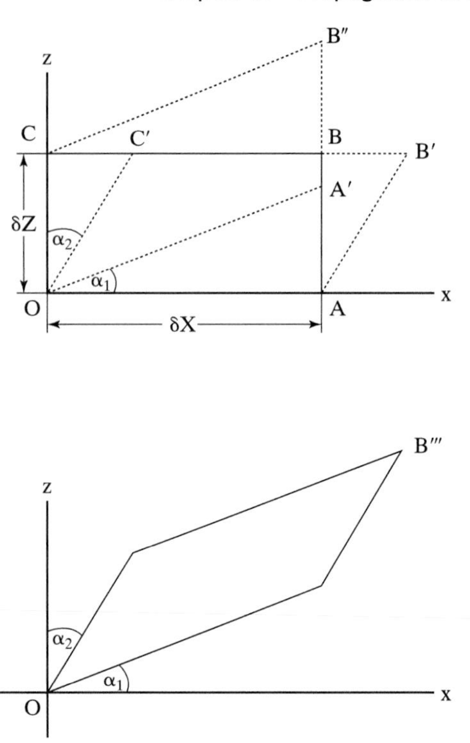

Fig. 13.3 Shears caused by couples with moments parallel to the y-axis.

Now for pure shear there will be no rotational displacement, and hence from Eq. (13.1.12)

$$\xi_z = \zeta_x, \xi_y = \eta_x, \text{etc.} \tag{13.1.15}$$

Thus, the terms on the right side of the diagonal of the strain matrix, Eq. (13.1.13), are one half of the total shear strain. Note that each of these terms is mirrored by an equal term across the diagonal; for example,

$$1/2(\eta_x + \xi_y) = 1/2(\xi_y + \eta_x) \tag{13.1.16}$$

We may rewrite the strain matrix, D, thus

$$D = \begin{bmatrix} \epsilon_{xx} & \epsilon_{xy} & \epsilon_{xz} \\ \epsilon_{yx} & \epsilon_{yy} & \epsilon_{yz} \\ \epsilon_{zx} & \epsilon_{zy} & \epsilon_{zz} \end{bmatrix} \tag{13.1.17}$$

where $\epsilon_{xx} = \xi_x$, etc.; $\epsilon_{xy} = 1/2(\eta_x + \xi_y)$, etc.; and $\epsilon_{xy} = \epsilon_{yx}, \epsilon_{yz} = \epsilon_{zy}, \epsilon_{zx} = \epsilon_{xz}$. The conclusion is that a continuous medium can be deformed only with three independent linear strains and three independent shear strains.

13.2 Strain Tensor Ellipsoid

It is now convenient to introduce the concept of a tensor ellipsoid. Consider a radius vector

$$\mathbf{r} = \mathbf{i}x + \mathbf{j}y + \mathbf{k}z \tag{13.2.1}$$

and a second vector

$$\mathbf{w} = \mathbf{i}w_x + \mathbf{j}w_y + \mathbf{k}w_z \tag{13.2.2}$$

where

$$\begin{aligned} w_x &= a_{11}x + a_{12}y + a_{12}z \\ w_y &= a_{21}x + a_{22}y + a_{23}z \\ w_z &= a_{31}x + a_{32}y + a_{33}z \end{aligned} \tag{13.2.3}$$

These coefficients may be written as a matrix

$$\begin{bmatrix} a_{11} & a_{12} & a_{13} \\ a_{21} & a_{22} & a_{23} \\ a_{31} & a_{32} & a_{33} \end{bmatrix} \tag{13.2.4}$$

Now if the matrix is symmetric

$$a_{21} = a_{12}, a_{31} = a_{13}, a_{32} = a_{23} \tag{13.2.5}$$

then

$$\mathbf{r} \cdot \mathbf{w} = a_{11}x^2 + a_{22}y^2 + a_{33}z^2 + 2a_{12}xy + 2a_{23}yz + 2a_{13}zx = p \tag{13.2.6}$$

Thus

$$\frac{\partial p}{\partial x} = 2a_{11}x + 2a_{12}y + 2a_{13}z = 2w_x \tag{13.2.7}$$

Similarly,

$$\frac{\partial p}{\partial y} = 2w_y \text{ and } \frac{\partial p}{\partial z} = 2w_z \tag{13.2.8}$$

Hence

$$w = \frac{1}{2}\text{grad } p = \frac{1}{2}\nabla p \tag{13.2.9}$$

Given the coefficients a_{11}, a_{12}, etc., then each value of p substituted into Eq. (13.2.6) defines a quadratic surface or ellipsoid, as illustrated in Fig. 13.4a. Each point (x, y, z) in space

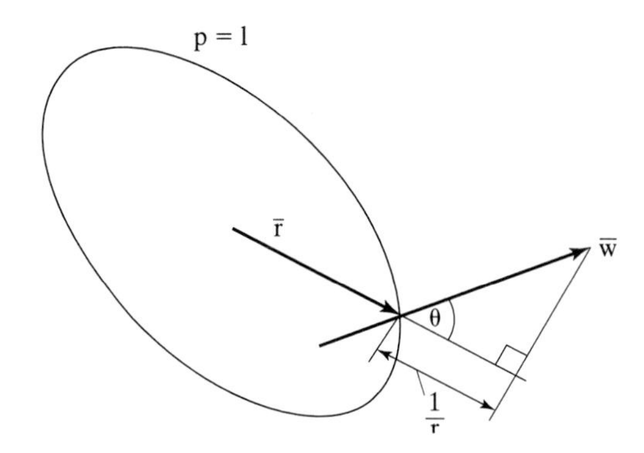

Fig. 13.4a Ellipsoid for tensor $\bar{w}$.

will lie on one of these ellipsoidal surfaces, and the magnitude and direction of **w** at any point (x, y, z) will be given by 1/2 grad p. For example, if $p = 1$, since $p = \mathbf{r} \cdot \mathbf{w} = rw \cos\theta$,

$$w \cos\theta = \frac{1}{r} \tag{13.2.10}$$

so the component of w in the direction of the radius vector is equal in magnitude to the reciprocal of the magnitude of the radius vector.

Let us now consider the application of this theorem to the case of strains by referring again to Fig. 13.1. Transfer the origin of coordinates to P and use r, x, y, and z for δr, δx, δy, and δz to determine the position of point Q, remembering that x, y, and z will then have to be small. Now the difference in displacement of P and Q is the displacement of Q relative to P. That part of this relative displacement due to strain, **d** say, is related to x, y, and z through the strain matrix. Since the strain matrix is a 3×3 symmetric matrix, we may substitute **d** for **w** in our general theorem. The strains ϵ_{xx}, ϵ_{xy}, etc. now correspond to the coefficients a_{11}, a_{12}, etc. Thus, we may define a strain tensor ellipsoid whose equation will be

$$\epsilon_{xx}x^2 + \epsilon_{yy}y^2 + \epsilon_{zz}z^2 + 2\epsilon_{xy}xy + 2\epsilon_{yz}yz + 2\epsilon_{zx}zx = \text{constant} \tag{13.2.11}$$

Consider one such surface, passing through Q as in Fig. 13.4b. Now the displacement of Q, **d** will be a normal to the surface, so there will be three directions along the principal axes in which the displacement lies in the direction of **r**. Consequently, in these directions

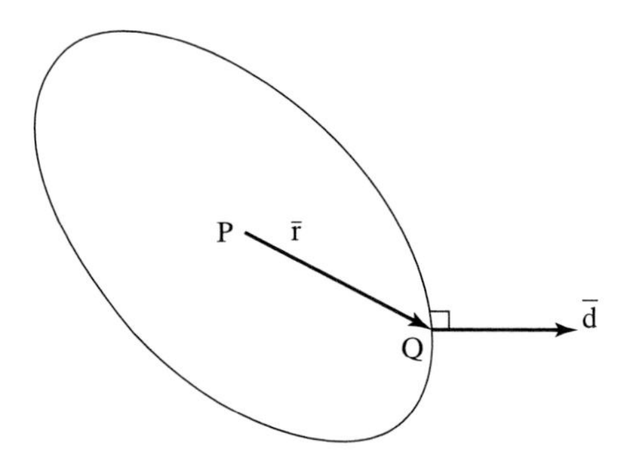

Fig. 13.4b Strain tensor ellipsoid. Displacement is normal to surface.

the strain is linear (i.e., there is no shear along the principal axes). Thus, the state of strain in the medium around P may be specified by the three linear strains in the directions of the principal axes. For example, if the principal axes are x′, y′, and z′, the equation of the ellipsoid may be rewritten

$$\epsilon_{x'x'}x'^2 + \epsilon_{y'y'}y'^2 + \epsilon_{z'z'}z'^2 = \text{constant} \tag{13.2.12}$$

since

$$\epsilon_{x'y'} = \epsilon_{y'z'} = \epsilon_{z'x'} = 0 \tag{13.2.13}$$

To interpret this result physically, consider a small spherical surface in the unstrained medium. Application of linear strains in the xyz directions deforms the sphere to an ellipsoid. Subsequent applications of shear strains ϵ_{xy}, ϵ_{yz}, and ϵ_{zx} merely rotate this ellipsoid. The same final deformation could have been achieved by the application of linear strains only in the direction of the principal axes.

13.3 Transformation of Coordinates

The results of the previous section are in the frame of the principal axes and are important to subsequent development of our argument, so that we need to know how to carry out transformations between these and other coordinate axes. Let α_i, β_i, and γ_i be direction cosines between the ith principal axis and the x-, y-, and z-axes. Now the position vector **r** is the same in both coordinate systems (see Fig. 13.5); that is, $\mathbf{r} = \mathbf{r}'$. Expressing both position vectors in their respective systems:

$$\mathbf{r} = \mathbf{i}x + \mathbf{j}y + \mathbf{k}z \tag{13.3.1}$$

and

$$\mathbf{r}' = \mathbf{i}'x' + \mathbf{j}'y' + \mathbf{k}'z' \tag{13.3.2}$$

Now

$$\mathbf{i}' \cdot \mathbf{r}' = \mathbf{i}' \cdot \mathbf{r} \tag{13.3.3}$$

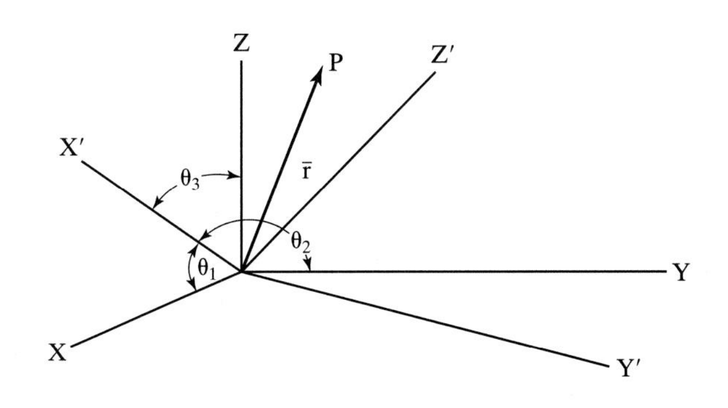

Fig. 13.5 Transformation of coordinate axes.

hence

$$x' = \mathbf{i}' \cdot \mathbf{i}x + \mathbf{i}' \cdot \mathbf{j}y + \mathbf{i}' \cdot \mathbf{k}z \tag{13.3.4}$$

But

$$\begin{aligned} \mathbf{i}' \cdot \mathbf{i} &= \cos\theta_1 = \alpha_1 \\ \mathbf{i}' \cdot \mathbf{j} &= \cos\theta_2 = \beta_1 \\ \mathbf{i}' \cdot \mathbf{k} &= \cos\theta_3 = \gamma_1 \end{aligned} \tag{13.3.5}$$

Thus

$$x' = \alpha_1 x + \beta_1 y + \gamma_1 z \tag{13.3.6}$$

and similarly,

$$y' = \alpha_2 x + \beta_2 y + \gamma_2 z \tag{13.3.7}$$

and

$$z' = \alpha_3 x + \beta_3 y + \gamma_3 z \tag{13.3.8}$$

Now the principal axes quadratic form is

$$\sum_i \epsilon_i \mathbf{x}'^2_i = \text{const} = p \tag{13.3.9}$$

But

$$\begin{aligned} \sum_i \epsilon_i \mathbf{x}'^2_i &= \sum_i \epsilon_i(\alpha_i x + \beta_i y + \gamma_i z)^2 \\ &= \sum_i \epsilon_i \alpha_i^2 x^2 + \sum_i \epsilon_i \beta_i^2 y^2 + \sum_i \epsilon_i \gamma_i^2 z^2 \\ &\quad + 2\sum_i \epsilon_i \alpha_i \beta_i xy + 2\sum_i \epsilon_i \beta_i \gamma_i yz \\ &\quad + 2\sum_i \epsilon_i \gamma_i \alpha_i xz \end{aligned} \tag{13.3.10}$$

Comparing with Eq. (13.2.11),

$$\begin{aligned} \epsilon_{xx} &= \sum_i \epsilon_i \alpha_i^2, & \epsilon_{xy} &= \sum_i \epsilon_i \alpha_i \beta_i \\ \epsilon_{yy} &= \sum_i \epsilon_j \beta_i^2, & \epsilon_{yz} &= \sum_i \epsilon_i \beta_i \gamma_i \\ \epsilon_{zz} &= \sum_i \epsilon_i \gamma_i^2, & \epsilon_{zx} &= \sum_i \epsilon_i \gamma_i \alpha_i \end{aligned} \tag{13.3.11}$$

So all strains in the xyz coordinate system can be expressed in terms of ϵ_1, ϵ_2, and ϵ_3, the linear strains of the principal axes coordinate system.

But we know that these systems are both Cartesian, and if the origin of both frames is the same, then

$$x^2 + y^2 + z^2 = x'^2 + y'^2 + z'^2 \tag{13.3.12}$$

But from Eq. (13.3.6)

$$\begin{aligned} x'^2 &= \alpha_1^2 x^2 + \beta_1^2 y^2 + \gamma_1^2 z^2 \\ &+ 2\alpha_1\beta_1 xy + 2\beta_1\gamma_1 yz + 2\gamma_1\alpha_1 zx \end{aligned} \tag{13.3.13}$$

and hence

$$\begin{aligned} x'^2 + y'^2 + z'^2 &= x^2\sum_i \alpha_i^2 + y^2\sum_i \beta_i^2 \\ &+ z^2\sum_i \gamma_i^2 + 2xy\sum_i \alpha_i\beta_i + 2yz\sum_i \beta_i\gamma_i \\ &+ 2zx\sum_i \gamma_i\alpha_i \\ &= x^2 + y^2 + z^2 \end{aligned} \tag{13.3.14}$$

Thus, by comparing coefficients

$$\begin{aligned} \sum \alpha_i^2 &= 1, \quad \sum \alpha_i\beta_i = 0 \\ \sum \beta_i^2 &= 1, \quad \sum \beta_i\gamma_i = 0 \\ \sum \gamma_i^2 &= 1, \quad \sum \gamma_i\alpha_i = 0 \end{aligned} \tag{13.3.15}$$

13.4 Generalized Form of Hooke's Law

Strains are caused by stresses, stress being defined as the applied force divided by the area of application. Various stresses may be applied to an infinitesimal element, as shown in Fig. 13.6, and we need a convention to name them. We will use σ as a general symbol for stress. The stresses on OGFE in Fig. 13.6a are σ_{xx}, σ_{xy}, and σ_{xz}, where the first subscript denotes the axis perpendicular to the plane of application, and the second subscript indicates the direction of the applied force. Now there must be oppositely directed stresses on ABCD if there is to be translational equilibrium. Note that the directions of the stresses are chosen to be positive so that the linear stresses (σ_{xx}, σ_{yy}, σ_{zz}) cause positive dilatations, and the shear stresses cause positive shear strains. Suppose that the stresses on the front faces differ from the rear face stresses by small amounts. In that case, the forces on the front faces will be as shown in Fig. 13.6b.

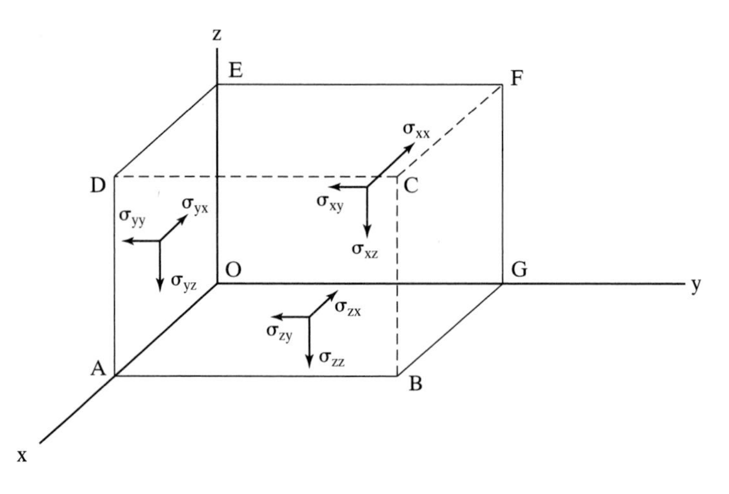

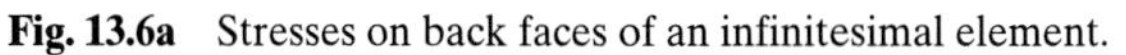
Fig. 13.6a Stresses on back faces of an infinitesimal element.

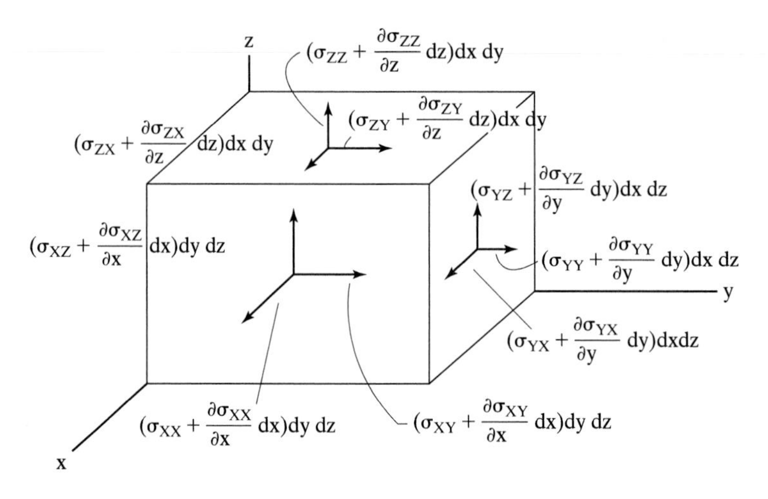

Fig. 13.6b Forces on front faces of an infinitesimal element.

Hence, taking moments about a z-axis through the center of the element:

$$\sigma_{xy}\, dy\, dz \frac{dx}{2} - \left(\sigma_{yx} + \frac{\partial \sigma_{yx}}{\sigma_y} dy \right) dx\, dz \frac{dy}{2}$$
$$+ \left(\sigma_{xy} + \frac{\partial \sigma_{xy}}{\partial x} dx \right) dy\, dz \frac{dx}{2} - \sigma_{yx}\, dx\, dz \frac{dy}{2} = 0 \qquad (13.4.1)$$

Dividing by the volume element, and letting the remaining terms in dx and dy go to zero, yields

$$\sigma_{xy} = \sigma_{yx} \qquad (13.4.2)$$

By similar arguments, $\sigma_{yz} = \sigma_{zy}$ and $\sigma_{zx} = \sigma_{xz}$.

The condition of translational equilibrium leads to the realization that there are only nine independent strains, and the condition of rotational equilibrium further reduces this to six. The stresses may also be written in tensor form:

$$X = \begin{bmatrix} \sigma_{xx} & \sigma_{xy} & \sigma_{xz} \\ \sigma_{yz} & \sigma_{yy} & \sigma_{yz} \\ \sigma_{zx} & \sigma_{zy} & \sigma_{zz} \end{bmatrix} \tag{13.4.3}$$

The most general form of Hooke's law states that stress is a linear function of the strains

$$\begin{aligned}
\sigma_{xx} &= c_{11}\epsilon_{xx} + c_{12}\epsilon_{yy} + c_{13}\epsilon_{zz} + c_{14}\epsilon_{yz} + c_{15}\epsilon_{zx} + c_{16}\epsilon_{xy} \\
\sigma_{yy} &= c_{21}\epsilon_{xx} + c_{22}\epsilon_{yy} + \\
\sigma_{zz} &= c_{31}\epsilon_{xx} + c_{32}\epsilon_{yy} + c_{33}\epsilon_{zz} + \cdots \\
\sigma_{yz} &= c_{41}\epsilon_{xx} + \cdots + \cdots \quad + c_{44}\epsilon_{yz} + \cdots \\
\sigma_{zx} &= c_{51}\epsilon_{xx} + \\
\sigma_{xy} &= c_{61}\epsilon_{xx} + \cdots \qquad + c_{16}\epsilon_{xy}
\end{aligned} \tag{13.4.4}$$

where c_{11}, etc. are constants. In matrix notation, we have

$$\begin{bmatrix} \sigma_{xx} \\ \sigma_{yy} \\ \sigma_{zz} \\ \sigma_{yz} \\ \sigma_{zx} \\ \sigma_{xy} \end{bmatrix} = \begin{bmatrix} c_{11} & c_{12} & c_{13} & c_{14} & c_{15} & c_{16} \\ c_{21} & c_{22} & & & & \\ c_{31} & & c_{33} & & & \\ c_{41} & & & c_{44} & & \\ c_{51} & & & & c_{55} & \\ c_{61} & & & & & c_{66} \end{bmatrix} \begin{bmatrix} \epsilon_{xx} \\ \epsilon_{yy} \\ \epsilon_{zz} \\ \epsilon_{yz} \\ \epsilon_{zx} \\ \epsilon_{xy} \end{bmatrix} \tag{13.4.5}$$

The constants c_{11}, etc. are said to be elastic moduli. There is no reason why strains should not be written in terms of stresses. In this case, we might use constants p_{ik} instead of c_{ik}. The coefficients p_{ik} would then be termed inverse moduli. In summary, to state Hooke's law in the most general form we need thirty-six independent constants. But fortunately things are not usually that complex, as we will see in subsequent sections from our discussion of energy and symmetry.

13.5 Energy of Elastic Deformation

Consider a unit cube of material. Suppose the strains ϵ_{ik} on the material are increased by small amounts to $\epsilon_{ik} + \Delta\epsilon_{ik}$. Now, for normal stresses, the work done in causing these changes = stress × area × extension. But for a unit cube, the area is unity and the extension is equal to the change of strain (see Fig. 13.7). So for example,

$$\Delta U = \sigma_{xx}\,\Delta\epsilon_{xx} \tag{13.5.1}$$

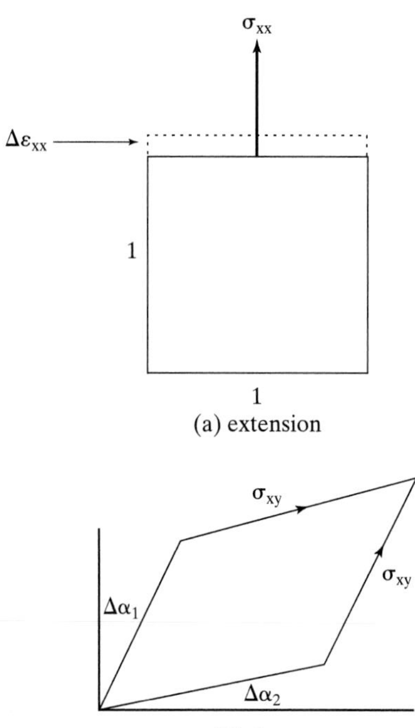

Fig. 13.7 Doing work on a unit cube.

Similarly, for shearing,

$$\Delta U = \sigma_{xy}(\Delta\alpha_1 + \Delta\alpha_2) = \sigma_{xy}\,\Delta(\alpha_1 + \alpha_2) = 2\sigma_{xy}\,\Delta\epsilon_{xy} \tag{13.5.2}$$

Hence, for all possible deformations,

$$\Delta U = \sigma_{xx}\,\Delta\epsilon_{xx} + \sigma_{yy}\,\Delta\epsilon_{yy} + \sigma_{zz}\,\Delta\epsilon_{zz} + 2\sigma_{yz}\,\Delta\epsilon_{yz} + 2\sigma_{zx}\,\Delta\epsilon_{zx} + 2\sigma_{xy}\,\Delta\epsilon_{xy} \tag{13.5.3}$$

But

$$U = \text{total energy} = U\,(\epsilon_{xx}, \epsilon_{xy}, \epsilon_{yx}, \dots) \tag{13.5.4}$$

Thus

$$\begin{aligned}\Delta U &= \frac{\partial U}{\partial \epsilon_{xx}}\Delta\epsilon_{xx} + \cdots + \frac{\partial U}{\partial \epsilon_{xy}}\Delta\epsilon_{xy} + \frac{\partial U}{\partial \epsilon_{yx}}\Delta\epsilon_{yx} \\ &= \frac{\partial U}{\partial \epsilon_{xx}}\Delta\epsilon_{xx} + \cdots + 2\frac{\partial U}{\partial \epsilon_{xy}}\Delta\epsilon_{xy} + \cdots \end{aligned} \tag{13.5.5}$$

Hence, from Eqs. (13.5.3) and (13.5.5)

$$\sigma_{xx} = \frac{\partial U}{\partial \epsilon_{xx}} \text{ etc. and } \sigma_{zx} = \frac{\partial U}{\partial \epsilon_{zx}} \tag{13.5.6}$$

Then from Eqs. (13.4.4) and (13.5.6),

$$c_{11} = \frac{\partial \sigma_{xx}}{\partial \epsilon_{xx}} = \frac{\partial^2 U}{\partial \epsilon_{xx}^2} \tag{13.5.7}$$

and

$$c_{15} = \frac{\partial \sigma_{xx}}{\partial \epsilon_{zx}} = \frac{\partial}{\partial \epsilon_{zx}}\left(\frac{\partial U}{\partial \epsilon_{xx}}\right) \tag{13.5.8}$$

But

$$c_{51} = \frac{\partial \sigma_{zx}}{\partial \epsilon_{xx}} = \frac{\partial}{\partial \epsilon_{xx}}\left(\frac{\partial U}{\partial \epsilon_{zx}}\right) \tag{13.5.9}$$

and hence $c_{15} = c_{51}$ and, in general, $c_{ij} = c_{ji}$. So the strain matrix is symmetric about the diagonal:

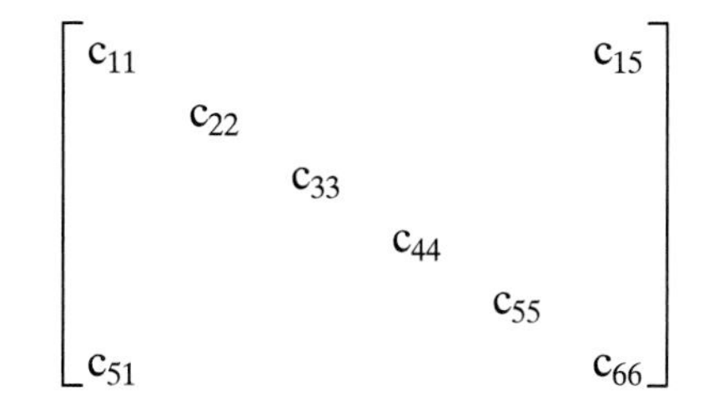

Thus, we have reduced the thirty-six constants to six on the diagonal and thirty off the diagonal, of which only half (fifteen) are independent, so there are $6 + 15 = 21$ independent constants in the most general case.

It also follows from Eq. (13.5.3) and Eqs. (13.5.7) to (13.5.9) that

$$\begin{aligned}\Delta U = {} & (c_{11}\epsilon_{xx} + c_{12}\epsilon_{yy} + c_{13}\epsilon_{zz} + c_{14}\epsilon_{yz} + c_{15}\epsilon_{zx} + c_{16}\epsilon_{xy})\Delta\epsilon_{xx} \\ & + \cdots + 2(c_{41}\epsilon_{zx} + \cdots + c_{44}\epsilon_{yz} + \cdots)\,\Delta\epsilon_{yz} + \cdots \end{aligned} \tag{13.5.10}$$

Note that Eq. (13.5.10) contains thirty-six terms.

By integrating, we obtain an expression for the energy:

$$\begin{aligned}U = {} & \frac{c_{11}\epsilon_{xx}^2}{2} + c_{12}\epsilon_{yy}\epsilon_{xx} + c_{13}\epsilon_{zz}\epsilon_{xx} + c_{14}\epsilon_{yz}\epsilon_{xx} \\ & + c_{15}\epsilon_{zx}\epsilon_{xx} + c_{16}\epsilon_{xy}\epsilon_{xx} \\ & + \cdots + 2\left(c_{41}\epsilon_{xx}\epsilon_{yz} + \cdots + \frac{c_{44}}{2}\epsilon_{yz}^2 + \cdots\right)\end{aligned} \tag{13.5.11}$$

However, since $c_{14} = c_{41}$, etc., the thirty-six terms in this expression can be reduced to twenty-one, thus

$$
\begin{aligned}
U = {} & \frac{c_{11}\epsilon_{xx}^2}{2} + 2c_{12}\epsilon_{yy}\epsilon_{xx} + 2c_{13}\epsilon_{zz}\epsilon_{xx} + 3c_{14}\epsilon_{yz}\epsilon_{xx} \\
& + 3c_{15}\epsilon_{zx}\epsilon_{xx} + 3c_{16}\epsilon_{xy}\epsilon_{xx} + \frac{c_{22}}{2}\epsilon_{yy}^2 \\
& + 2c_{23}\epsilon_{zz}\epsilon_{yy} + 3c_{24}\epsilon_{yz}\epsilon_{yy} + 3c_{25}\epsilon_{zx}\epsilon_{yy} \\
& + 3c_{26}\epsilon_{xy}\epsilon_{yy} + \frac{c_{33}}{2}\epsilon_{zz}^2 + 3c_{34}\epsilon_{yz}\epsilon_{zz} + \cdots \\
& + 3c_{35}\epsilon_{zx}\epsilon_{zz} + 3c_{36}\epsilon_{xy}\epsilon_{zz} + c_{44}\epsilon_{yz}^2 \\
& + 4c_{45}\epsilon_{zx}\epsilon_{yz} + 4c_{46}\epsilon_{xy}\epsilon_{yz} + c_{55}\epsilon_{zx}^2 \\
& + 4c_{56}\epsilon_{xy}\epsilon_{zx} + c_{66}\epsilon_{xy}^2
\end{aligned}
\tag{13.5.12}
$$

13.6 Isotropic Media

Depending on the physical nature of the medium concerned, the twenty-one independent constants may be reduced further in number. Consider the very important case of an isotropic medium. Any state of strain may be represented by three orthogonal linear strains acting along the principal axes (i.e., $\epsilon_{x'x'}$, $\epsilon\epsilon_{y'y'}$, and $\epsilon_{z'z'}$, which we may call ϵ_1, ϵ_2, and ϵ_3 for simplification). Now in the frame of the principal axes, the stresses that caused the strains can only be functions of ϵ_1, ϵ_2, and ϵ_3; so, for example

$$\sigma_{x'x'} = a\epsilon_1 + b\epsilon_2 + c\epsilon_3 \equiv \sigma_1 \tag{13.6.1}$$

But for an isotropic medium, stress in the x direction must cause the same strain in the y or z directions, hence $b = c$. Thus

$$\sigma_1 = a\epsilon_1 + b(\epsilon_2 + \epsilon_3) \tag{13.6.2}$$

Again

$$\sigma_2 = a'\epsilon_1 + b'\epsilon_2 + c'\epsilon_3 \tag{13.6.3}$$

Repeating the above argument, $a' = c'$. And if σ_1 and σ_2 are numerically equal, for an isotropic medium the strains ϵ_1 and ϵ_2 would have to be numerically equal, so $b' = a$ and similarly $a' = c' = b$; thus

$$\sigma_2 = b\epsilon_1 + a\epsilon_2 + b\epsilon_3 \tag{13.6.4}$$

and finally

$$\sigma_3 = b\epsilon_1 + b\epsilon_2 + a\epsilon_3 \tag{13.6.5}$$

Since there is no shear strain in the frame of the principal axes, it may be assumed that there is no shear stress in that frame. Thus, to describe the relation between strain and stress for an isotropic medium requires only two independent constants, at least if the frame of reference is the principal axes of the strain tensor ellipsoid. Now we shall show that the result is true for any set of axes. But first a small rearrangement:

$$\sigma_1 = (a - b)\epsilon_1 + b(\epsilon_1 + \epsilon_2 + \epsilon_3)$$
$$\sigma_2 = (a - b)\epsilon_2 + b(\epsilon_1 + \epsilon_2 + \epsilon_3)$$

and

$$\sigma_3 = (a - b)\epsilon_3 + b(\epsilon_1 + \epsilon_2 + \epsilon_3) \tag{13.6.6}$$

Let $b = \lambda$ and $(a - b) = 2\mu$, where λ and μ are the so-called Lame constants. Thus

$$\sigma_1 = 2\mu\epsilon_1 + \lambda(\epsilon_1 + \epsilon_2 + \epsilon_3)$$
$$\sigma_2 = 2\mu\epsilon_2 + \lambda(\epsilon_1 + \epsilon_2 + \epsilon_3)$$

and

$$\sigma_3 = 2\mu\epsilon_3 + \lambda(\epsilon_1 + \epsilon_2 + \epsilon_3) \tag{13.6.7}$$

which is the conventional way of stating Hooke's law for an isotropic medium using the principal axes as a reference frame. Now we will discuss the statement of the law in any reference frame.

The stress tensor, being a 3×3 symmetric matrix, has a stress tensor ellipsoid, whose equation in the frame of the principal axes must be

$$\sum_i \sigma_i x'^2_i = \text{constant} \tag{13.6.8}$$

The same tranformations for change of coordinates must apply for stresses as for strains; that is

$$\sigma_{xx} = \sum_i \sigma_i \alpha_i^2, \qquad \sigma_{xy} = \sum_i \sigma_i \alpha_i \beta_i$$
$$\sigma_{yy} = \sum_i \sigma_i \beta_i^2, \qquad \sigma_{yz} = \sum_i \sigma_i \beta_i \gamma_i,$$
$$\sigma_{zz} = \sum_i \sigma_i \gamma_i^2, \qquad \sigma_{zx} = \sum_i \sigma_i \gamma_i \alpha_i. \tag{13.6.9}$$

Thus

$$\sigma_{xx} = \alpha_1^2[2\mu\epsilon_1 + \lambda(\epsilon_1 + \epsilon_2 + \epsilon_3)] + \alpha_2^2[2\mu\epsilon_2 + \lambda(\epsilon_1 + \epsilon_2 + \epsilon_3)]$$
$$+ \alpha_3^2[2\mu\epsilon_3 + \lambda(\epsilon_1 + \epsilon_2 + \epsilon_3)] \tag{13.6.10}$$

But from Eq. (13.3.14), $\sum_i \alpha_i^2 = 1$, and hence

$$\sigma_{xx} = \lambda(\epsilon_1 + \epsilon_2 + \epsilon_3) + 2\mu[\epsilon_1\alpha_1^2 + \epsilon_2\alpha_2^2 + \epsilon_3\alpha_3^2] \tag{13.6.11}$$

From Eq. (13.3.11) it is easy to see that

$$\epsilon_{xx} + \epsilon_{yy} + \epsilon_{zz} = \sum \epsilon_i(\alpha_i^2 + \beta_i^2 + \gamma_i^2) \tag{13.6.12}$$

But α_1, β_i, and γ_i are the direction cosines between the ith principal axis and the x-, y-, and z-axes; thus

$$\alpha_i^2 + \beta_i^2 + \gamma_i^2 = 1 \tag{13.6.13}$$

Hence, from Eqs. (13.6.12) and (13.6.13)

$$\epsilon_{xx} + \epsilon_{yy} + \epsilon_{zz} = \epsilon_1 + \epsilon_2 + \epsilon_3 \tag{13.6.14}$$

and from Eqs. (13.6.11), (13.6.14), and (13.3.11)

$$\sigma_{xx} = 2\mu\epsilon_{xx} + \lambda(\epsilon_{xx} + \epsilon_{yy} + \epsilon_{zz}) \tag{13.6.15}$$

Similarly,

$$\sigma_{yy} = 2\mu\epsilon_{yy} + \lambda(\epsilon_{xx} + \epsilon_{yy} + \epsilon_{zz})$$

and

$$\sigma_{zz} = 2\mu\epsilon_{zz} + \lambda(\epsilon_{xx} + \epsilon_{yy} + \epsilon_{zz}) \tag{13.6.16}$$

Now

$$\begin{aligned}
\sigma_{xy} &= \sum \sigma_i\alpha_i\beta_i \\
&= \alpha_1\beta_1\sigma_1 + \alpha_2\beta_2\sigma_2 + \alpha_3\beta_3\sigma_3 \\
&= \alpha_1\beta_1[2\mu\epsilon_1 + \lambda(\epsilon_1 + \epsilon_2 + \epsilon_3)] + \alpha_2\beta_2[2\mu\epsilon_3 + \lambda(\epsilon_1 + \epsilon_2 + \epsilon_3)] \\
&+ \alpha_3\beta_3[2\mu\epsilon_3 + \lambda(\epsilon_1 + \epsilon_2 + \epsilon_3)] \\
&= 2\mu\left[\epsilon_1\alpha_1\beta_1 + \epsilon_2\alpha_2\beta_2 + \epsilon_3\alpha_3\beta_3\right] \\
&+ \lambda(\epsilon_1 + \epsilon_2 + \epsilon_3)(\alpha_1\beta_1 + \alpha_2\beta_2 + \alpha_3\beta_3)
\end{aligned} \tag{13.6.17}$$

But from Eq. (13.3.14)

$$\alpha_1\beta_1 + \alpha_2\beta_2 + \alpha_3\beta_\times = 0 \tag{13.6.18}$$

and from Eq. (13.3.11)

$$\epsilon_{xy} = \sum_i \epsilon_i \alpha_i \beta_i \tag{13.6.19}$$

Thus

$$\sigma_{xy} = 2\mu\epsilon_{xy}$$
$$\sigma_{xz} = 2\mu\epsilon_{xy}$$

and

$$\sigma_{yz} = 2\mu\epsilon_{yz} \tag{13.6.20}$$

So, we have finally proved that only two independent constants are required to state Hooke's law for an isotropic medium.

13.7 Young's Modulus, Poisson's Ratio, and Bulk Modulus

It is now convenient to derive the elastic moduli that are used in engineering practice—and which we have used in earlier chapters—in terms of the Lame constants. Young's modulus is the ratio of stress and strain along the length of a bar, as shown in Fig. 13.8; that is

$$Y = \frac{\sigma_{xx}}{\epsilon_{xx}} = \frac{\lambda(\epsilon_{xx} + \epsilon_{yy} + \epsilon_{zz}) + 2\mu\epsilon_{xx}}{\epsilon_{xx}}$$
$$= \lambda\left(1 + \frac{\epsilon_{yy}}{\epsilon_{xx}} + \frac{\epsilon_{zz}}{\epsilon_{xx}}\right) + 2\mu \tag{13.7.1}$$

Now for a homogeneous isotropic medium under stress, σ_{xx}, $\epsilon_{yy} = \epsilon_{zz}$. But Poisson's ratio, υ, is the negative ratio of lateral to longitudinal strain:

$$\upsilon = -\frac{\epsilon_{yy}}{\epsilon_{xx}} = -\frac{\epsilon_{zz}}{\epsilon_{xx}} \tag{13.7.2}$$

Thus, from Eqs. (13.7.1) and (13.7.2)

$$Y = \lambda(1 - 2\upsilon) + 2\mu = \lambda + 2\mu - 2\upsilon\lambda \tag{13.7.3}$$

But in the case under consideration

$$\sigma_{yy} = \sigma_{zz} = 0 \tag{13.7.4}$$

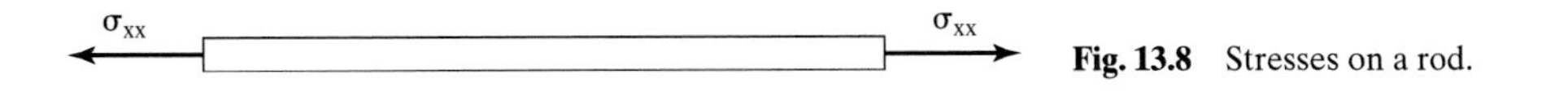

Fig. 13.8 Stresses on a rod.

Thus

$$2\mu\frac{\epsilon_{yy}}{\epsilon_{xx}} + \lambda\left(1 + \frac{\epsilon_{yy}}{\epsilon_{xx}} + \frac{\epsilon_{zz}}{\epsilon_{xx}}\right) = 0 \tag{13.7.5}$$

so that

$$-2\mu\upsilon + \lambda(1 - 2\upsilon) = 0 \tag{13.7.6}$$

and

$$\upsilon = \frac{\lambda}{2(\mu + \lambda)} \tag{13.7.7}$$

Substituting for υ into Eq. (13.7.3),

$$Y = \lambda + 2\mu - \frac{2\lambda^2}{2(\mu + \lambda)} = \frac{\mu(3\lambda + 2\mu)}{\mu + \lambda} \tag{13.7.8}$$

The bulk modulus

$$B = -\frac{\text{Pressure}}{dV/V} \tag{13.7.9}$$

where dV/V is the volume strain. Consider a small cube, with sides dz, dy, and dx. Now a change in length in the x direction produces ϵ_{xx} dxdydx as a change in volume; that is

$$dV_x = (\epsilon_{xx}\, dx)\, dy\, dz \tag{13.7.10}$$

or

$$(dV_x/V) = \epsilon_{xx} \tag{13.7.11}$$

and thus

$$\begin{aligned}\left(\frac{dV}{V}\right)_{\text{total}} &= \epsilon_{xx} + \epsilon_{yy} + \epsilon_{zz}\\ &= \frac{\partial\xi}{\partial x} + \frac{\partial\eta}{\partial y} + \frac{\partial\zeta}{\partial z}\\ &= \nabla\cdot\Delta\end{aligned} \tag{13.7.12}$$

Now substituting into Eq. (13.7.9) and recognizing that pressure is a negative stress,

$$B = \frac{\sigma_{xx}}{\epsilon_{xx} + \epsilon_{yy} + \epsilon_{zz}} \tag{13.7.13}$$

and for hydrostatic pressure

$$\sigma_{xx} = \sigma_{yy} = \sigma_{zz} \tag{13.7.14}$$

Thus, from Eqs. (13.6.15) and (13.7.14)

$$\begin{aligned}\sigma_{xx} &= 2\mu\epsilon_{xx} + \lambda(\epsilon_{xx} + \epsilon_{yy} + \epsilon_{zz}) \\ &= \sigma_{yy} = 2\mu\epsilon_{yy} + \lambda(\epsilon_{xx} + \epsilon_{yy} + \epsilon_{zz})\end{aligned} \tag{13.7.15}$$

Hence, $\epsilon_{xx} = \epsilon_{yy}$ and similarly $\epsilon_{yy} = \epsilon_{zz}$. Consequently, in this case

$$\sigma_{xx} = 2\mu\epsilon_{xx} + 3\lambda\epsilon_{xx} \tag{13.7.16}$$

and finally, from Eq. (13.7.13)

$$B = \frac{2\mu\epsilon_{xx} + 3\lambda\epsilon_{xx}}{3\epsilon_{xx}} = \lambda + \frac{2}{3}\mu \tag{13.7.17}$$

We must emphasize that Eq. (13.7.17) applies only for hydrostatic pressure.

In summary, Young's modulus, Poisson's ratio, and the bulk modulus may be expressed in terms of the Lame constants:

$$Y = \frac{\mu(3\lambda + 2\mu)}{\mu + \lambda} \tag{13.7.18}$$

and

$$\upsilon = \frac{\lambda}{2(\mu + \lambda)} \tag{13.7.19}$$

and for hydrostatic loads,

$$B = \lambda + \frac{2}{3}\mu \tag{13.7.20}$$

which implies that any one of these "engineering" moduli may be expressed in terms of the other two. Thus, multiplying Eq. (13.7.20) by υ, Eq. (13.7.19) by 1/3, and subtracting yields

$$\lambda = \frac{3\upsilon B}{1 + \upsilon} \tag{13.7.21}$$

while from Eqs. (13.7.20) and (13.7.21),

$$\mu = \frac{3}{2}\,\frac{B(1 - 2\upsilon)}{1 + \upsilon} \tag{13.7.22}$$

Now substitute for μ and λ into Eq. (13.7.18) to obtain

$$Y = 3B(1 - \upsilon) \tag{13.7.23}$$

or eliminate B from Eqs. (13.7.22) and (13.7.23)

$$Y = 2\mu(1 + \upsilon) \tag{13.7.24}$$

or finally, eliminate υ from Eqs. (13.7.23) and (13.7.24)

$$Y = 9B\mu/(\mu + 3B) \tag{13.7.25}$$

13.8 Dynamics of an Elastic Medium: The Navier Equation

In the preceding sections we discussed the statics of an element in an extended medium, but in general the state of stress will be different at different points in the medium. Consequently, there will be unbalanced forces and motion within the medium. It is these motions that we now want to consider. Suppose a small cube, as in Fig. 13.9, has the net forces acting in the x direction due to variations in stress across it. For example, due to variation in σ_{xx}, we have a net force

$$F_{xx} = \frac{\partial \sigma_{xx}}{\partial x} dxdydz \tag{13.8.1}$$

Thus, the force in the x direction accounting for all stress variations in the x direction is

$$F_x = \left(\frac{\partial \sigma_{xx}}{\partial x} + \frac{\partial \sigma_{yx}}{\partial y} + \frac{\partial \sigma_{zx}}{\partial z} \right) dxdydz \tag{13.8.2}$$

Let $\theta = \epsilon_{xx} + \epsilon_{yy} + \epsilon_{zz} =$ the volume strain, then, from Eq. (13.6.15)

$$\frac{\partial \sigma_{xx}}{\partial x} = \frac{\partial}{\partial x}(\lambda\theta + 2\mu\epsilon_{xx}) + \lambda\frac{\partial \theta}{\partial x} + 2\mu\frac{\partial^2 \xi}{\partial x^2} \tag{13.8.3}$$

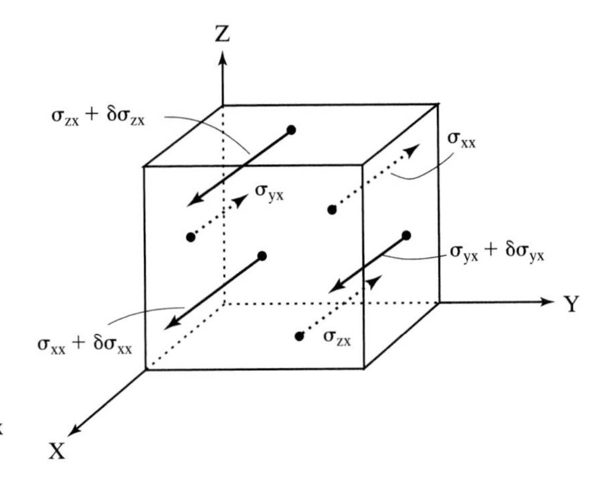

Fig. 13.9 Components of stress in the x direction acting on an element.

From Eq. (13.6.20)

$$\frac{\partial \sigma_{yx}}{\partial y} = 2\mu \frac{\partial}{\partial y}(\epsilon_{yx}) = \mu(\xi_{yy} + \eta_{xy}) \tag{13.8.4}$$

and similarly

$$\frac{\partial \sigma_{yx}}{\partial y} = 2\mu \frac{\partial}{\partial y}(\epsilon_{yx}) = \mu(\xi_{yy} + \eta_{xy}) \tag{13.8.5}$$

Thus, substituting into Eq. (13.8.2)

$$F_x = \left\{\lambda \frac{\partial \theta}{\partial x} + 2\mu \frac{\partial^2 \xi}{\partial x^2} + \mu \frac{\partial^2 \xi}{\partial y^2} + \mu \frac{\partial^2 \eta}{\partial x \partial y} + \mu \frac{\partial^2 \xi}{\partial z^2} + \mu \frac{\partial^2 \zeta}{\partial x \partial z}\right\} dxdydz \tag{13.8.6}$$

But

$$\frac{\partial \theta}{\partial x} = \frac{\partial^2 \xi}{\partial x^2} + \frac{\partial^2 \eta}{\partial x \partial y} + \frac{\partial^2 \zeta}{\partial x \partial z} \tag{13.8.7}$$

thus

$$F_x = \left\{(\lambda + \mu)\frac{\partial \theta}{\partial x} + \mu \nabla^2 \xi\right\} dxdydz \tag{13.8.8}$$

Similarly,

$$F_y = \left\{(\lambda + \mu)\frac{\partial \theta}{\partial y} + \mu \nabla^2 \eta\right\} dxdydz \tag{13.8.9}$$

and

$$F_z = \left\{(\lambda + \mu)\frac{\partial \theta}{\partial x} + \mu \nabla^2 \zeta\right\} dxdydz \tag{13.8.10}$$

Thus, the total force acting on the cube:

$$F = \{(\lambda + \mu)\,\nabla\theta + \mu\nabla \cdot \nabla\Delta\}\, dxdydz \tag{13.8.11}$$

Now the motion of the cube will be governed by Newton's second law: F = mass $\times$ acceleration. If the velocity in the medium is $\mathbf{v}(x, y, z, t)$, then the rate of change of velocity at any point (x, y, z) is $\partial\mathbf{v}/\partial t$. But the cube may move in a time interval δt from (x, y, z) to $(x + \delta x, y + \delta y, z + \delta z)$, where the velocity is different, say $\mathbf{v}'$, so to be precise, its acceleration will not be just $\partial\mathbf{v}/\partial t$. We see that

$$\mathbf{v}' - \mathbf{v} = \frac{\partial \mathbf{v}}{\partial t}\delta t + \frac{\partial \mathbf{v}}{\partial x}\delta x + \frac{\partial \mathbf{v}}{\partial y}\delta y + \frac{\delta \mathbf{v}}{\partial z}\delta z \tag{13.8.12}$$

If the limit as δt approaches zero,

$$\delta x = v_x \delta t; \ \delta_y = v_y \delta t; \ \delta z = v_z \delta t \tag{13.8.13}$$

Thus

$$\underset{\delta t \to 0}{\text{Lim}} \left(\frac{\mathbf{v}' - \mathbf{v}}{\delta t} \right) = \frac{\partial \mathbf{v}}{\partial t} + v_x \frac{\partial \mathbf{v}}{\partial x} + v_y \frac{\partial \mathbf{v}}{\partial y} + v_z \frac{\partial \mathbf{v}}{\partial z} = \frac{D\mathbf{v}}{Dt} \tag{13.8.14}$$

by definition, where $D\mathbf{v}/Dt$ is called the hydrodynamic derivative, or in vector form

$$\frac{D\mathbf{v}}{Dt} = \frac{\partial \mathbf{v}}{\partial r} + \mathbf{v} \cdot \text{grad } \mathbf{v} \tag{13.8.15}$$

Thus, applying Newton's law,

$$\mathbf{F} = \rho dxdydz \frac{D\mathbf{v}}{Dt} \tag{13.8.16}$$

assuming ρ does not change greatly with position, and finally

$$\rho \frac{D\mathbf{v}}{Dt} = (\lambda + \mu) \nabla\theta + \mu \nabla \cdot \nabla \Delta \tag{13.8.17}$$

Actually, it turns out for sound waves that $\partial \mathbf{v}/\partial t >> \mathbf{v}$. grad $\mathbf{v}$, and we may safely say that

$$\frac{D\mathbf{v}}{Dt} \cong \frac{\partial \mathbf{v}}{\partial t} = \frac{\partial^2 \Delta}{\partial t^2} = \ddot{\Delta} \tag{13.8.18}$$

Thus

$$\rho \ddot{\Delta} = (\lambda + \mu) \nabla\theta + \mu \nabla \cdot \nabla \Delta \tag{13.8.19}$$

This result is called the Navier-Stokes equation. Now, there is a vector identity that states

$$\nabla x \ \nabla x \ A = \nabla(\nabla \cdot A) - (\nabla \cdot \nabla A) \tag{13.8.20}$$

But

$$\nabla \cdot \Delta = \xi_x + \eta_y + \xi_z = \theta \tag{13.8.21}$$

and thus

$$\nabla(\nabla \cdot \Delta) = \nabla\theta \tag{13.8.22}$$

thus from Eq. (13.8.19)

$$\rho \ddot{\Delta} = (\lambda + 2\mu) \nabla\theta - \mu \nabla x \nabla x \Delta \tag{13.8.23}$$

We immediately recognize two cases of special interest:

1. If Δ is irrotational; that is

$$\nabla x \Delta = 0 \tag{13.8.24}$$

then from Eqs. (13.8.22), (13.8.23), and (13.8.20)

$$\rho\ddot{\Delta} = (\lambda + 2\mu)\nabla\nabla\cdot\Delta = 0 \tag{13.8.25}$$

But in this special case

$$\nabla\nabla\cdot\Delta = \nabla\cdot\nabla\Delta \tag{13.8.26}$$

and thus

$$\rho\ddot{\Delta} = (\lambda + 2\mu)[\mathbf{i}\nabla^2\xi + \mathbf{j}\Delta^2\eta + \mathbf{k}\nabla^2\xi] \tag{13.8.27}$$

which decomposes into three scalar wave equations with propagation velocity

$$c_i = \sqrt{\frac{\lambda + 2\mu}{\rho}} = \sqrt{\frac{B + \frac{4}{3}\mu}{\rho}} \tag{13.8.28}$$

where we have used Eq. (13.7.17).

2. If Δ is solenoidal; that is

$$\nabla\cdot\Delta = 0 \tag{13.8.29}$$

then from Eq. (13.8.21)

$$\Delta\theta = 0 \tag{13.8.30}$$

Hence, from Eq. (13.8.19)

$$\rho\ddot{\Delta} = \mu\nabla\cdot\nabla\Delta \tag{13.8.31}$$

which decomposes into three wave equations with velocity of propagation:

$$c_s = \sqrt{\frac{\mu}{\rho}} \tag{13.8.32}$$

Now there is a theorem in vector analysis which states that any vector may be expressed as the sum of the gradient of a scalar and the curl of a vector:

$$\Delta = \mathrm{grad}\Phi + \mathrm{curl}\mathbf{A} \tag{13.8.33}$$

Now if Δ is known, Φ and the three components of **A** are unknown, so that Eq. (13.8.33) really represents three equations with four unknowns. We may find Φ easily using the vector identity

$$\nabla \cdot \nabla \mathrm{x} \mathbf{A} = 0 \tag{13.8.34}$$

so that

$$\nabla \cdot \Delta = \nabla^2 \Phi \tag{13.8.35}$$

Similarly

$$\nabla \mathrm{xgrad} \Phi = 0 \tag{13.8.36}$$

thus

$$\nabla \mathrm{x}\, \Delta = \nabla \mathrm{x}\, \nabla \mathrm{x} \mathrm{A} = \nabla(\nabla \cdot \mathrm{A}) - \nabla^2 \mathrm{A} \tag{13.8.37}$$

So, rewriting Eq. (13.8.32) in general as

$$\Delta = \Delta_s + \Delta_i \tag{13.8.38}$$

where

$$\Delta_i = \mathrm{grad}\Phi \tag{13.8.39}$$

and since Δ_i is irrotational

$$\nabla \mathrm{x}\, \Delta_i = 0 \tag{13.8.40}$$

Similarly

$$\Delta_s = \mathrm{curl}\mathbf{A} \tag{13.8.41}$$

and since Δ_s is soleniodal

$$\nabla \cdot \Delta_s = 0 \tag{13.8.42}$$

This possible division into irrotational and solenoidal components may be understood physically as follows. An irrotational displacement, since $\nabla \mathrm{x}\, \Delta = 2\alpha$ (see Section 13.1), is such that there is no solid body rotation, so both shear and dilatation may be involved with such a wave motion. On the other hand, a solenoidal displacement, since

$$\nabla \cdot \Delta = \frac{dV}{V} \tag{13.8.43}$$

is such that there is no dilatation involved in the motion, so there may be shear and rotation. To find any further information, the various possibilities (e.g., plane, spherical, or cylindrical waves) must be examined in detail. Note also that a fluid cannot support a shear stress (i.e., $\mu = 0$). Consequently, it is impossible to propagate solenoidal waves in a fluid. Moreover, the velocity of irrotational waves in a fluid reduces simply to

$$c_i = \sqrt{\frac{B}{\rho}} \tag{13.8.44}$$

which agrees with the result for the sound velocity in a fluid derived in Chapter 3.

13.9 Plane Waves in an Infinite Solid Isotropic Medium

1. Irrotational waves obey the equation

$$\rho\ddot{\Delta} - (\lambda + 2\mu)\nabla^2\Delta = 0 \tag{13.9.1}$$

or

$$\rho\left[\mathbf{i}\ddot{\xi} + \mathbf{j}\ddot{\eta} + \mathbf{k}\ddot{\xi}\right] = (\lambda + 2\mu)\left[\mathbf{i}\nabla^2\xi + \mathbf{j}\nabla^2\eta + \mathbf{k}\nabla^2\zeta\right] \tag{13.9.2}$$

Now consider a wave travelling in the x direction. Since wavefronts are surfaces over which the displacement is the same, it follows that ξ, η, and ζ are functions of x and t only. There can be no acceleration in the y and z direction, hence Eq. (13.9.2) reduces to

$$\rho\ddot{\xi} = (\lambda + 2\mu)\frac{\partial^2\xi}{\partial x^2} \tag{13.9.3}$$

We can conclude that only ξ propagates as a wave (i.e., a plane irrotational wave is longitudinal). One of the most important properties of an irrotational displacement is that it may be represented as the gradient of a scalar function, which is known as a displacement potential. It is customary, however, to work with the velocity potential. From the condition of irrotationality, as in Eq. (13.8.23), it follows that

$$\nabla \mathrm{x} \mathbf{v} = 0 \tag{13.9.4}$$

Define

$$-\mathrm{grad}\Phi \equiv \mathbf{v} \tag{13.9.5}$$

where Φ is the velocity potential. Now

$$\ddot{\Delta} = c_i^2\nabla^2\Delta \tag{13.9.6}$$

so

$$\ddot{\mathbf{v}} = c_i^2\nabla^2\mathbf{v} \tag{13.9.7}$$

Hence from Eqs. (13.9.5) and (13.9.7)

$$-\nabla\ddot{\Phi} = c_i^2\nabla^2(-\nabla\Phi) = c_i^2\nabla\nabla^2\Phi \tag{13.9.8}$$

and consequently

$$\ddot{\Phi} = c_s^2 \nabla^2 \Phi \tag{13.9.9}$$

2. Solenoidal waves obey the equation

$$\rho \ddot{\Delta} = \mu \nabla^2 \Delta \tag{13.9.10}$$

Now

$$\nabla \cdot \Delta_s = \frac{\partial \xi}{\partial x} + \frac{\partial \eta}{\partial y} + \frac{\partial \xi}{\partial z} = 0 \tag{13.9.11}$$

But for a plane wave, ξ, η, and ζ are functions of x and t only. Thus

$$\frac{\partial \xi}{\partial x} = 0 \tag{13.9.12}$$

and

$$\frac{\partial^2 \xi}{\partial x^2} = 0 \tag{13.9.13}$$

and hence the Navier equation for plane solenoidal waves reduces to two scalar equations:

$$\rho \ddot{\eta} = c_i^2 \frac{\partial^2 \eta}{\partial x^2} \tag{13.9.14}$$

and

$$\rho \ddot{\xi} = c_i^2 \frac{\partial^2 \xi}{\partial x^2} \tag{13.9.15}$$

In other words, the two transverse components of displacement progagate as waves.

13.10 Propagation in Bounded Solids

In Section 13.8 we saw that disturbances in an isotropic medium obey the Navier equation. In an unbounded medium, there are a variety of possible solutions (irrotational or solenoidal, plane, spherical, or cylindrical waves). The actual wave generated in a certain physical situation will depend on the source of the disturbance (e.g., plane piston, plane shearer, point source, and line source); in other words, it will depend on the conditions at some boundary in the isotropic medium.

Suppose that we add a second boundary to such a problem. The resulting solution is not then so obvious. For instance, suppose a plane piston radiates into an infinite medium producing dilatational waves. Then if the infinite medium is restricted by an infinite plane boundary, the solution will be different and we may exploit physical intuition to guess that the differences will center on what happens at the boundary. In reality, there are no such things as infinite plane pistons (to produce purely dilatational waves) nor infinite free boundaries, and even if there were we could not have both at once, unless they were parallel. The best we may hope for in what we are about to do is that it will resemble reality under some circumstances.

Assume it is possible to have a purely dilatational (or irrotational) wave incident on an infinite free surface, say the x-y plane. As shown in Fig. 13.10, the propagation direction is normal to the z-axis. The question is: What happens? A first guess might be that it will be reflected as a dilatational wave. We should then be able to satisfy the boundary conditions by choosing the amplitude and direction of the reflected wave.

Let the incident wave have amplitude A_1, angular frequency ω, and angle of incidence α_1, as in Fig. 13.10, where the solid diamond is a symbol for a dilatational wave. Then the displacement normal to the wavefront,

$$\varphi_1 = A_1 \sin(\omega t + f_1 x + g_1 y) \tag{13.10.1}$$

where

$$f_1 = \frac{2\pi}{\lambda}\cos\alpha_1 \equiv k\cos\alpha_1 \equiv \frac{\omega}{c_1}\cos\alpha_1 \tag{13.10.2}$$

where c_1 = velocity of dilatational waves and where

$$g_1 = k\sin\alpha_1 = (\omega/c_1)\sin\alpha_1 \tag{13.10.3}$$

This is illustrated in Fig. 13.11a. Then, as shown in Fig. 13.11b, the displacement in the x direction

$$\xi_1 = -\varphi_1 \cos\alpha_1 \tag{13.10.4}$$

and the displacement in the y direction

$$\eta_1 = -\varphi_1 \sin\alpha_1 \tag{13.10.5}$$

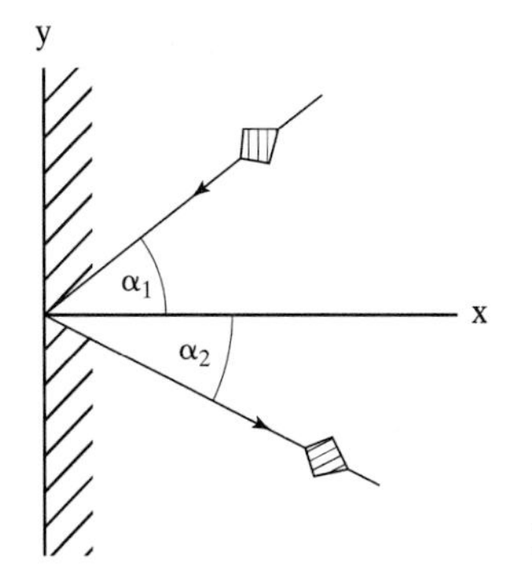

Fig. 13.10 A dilatational wave is incident on a plane boundary.

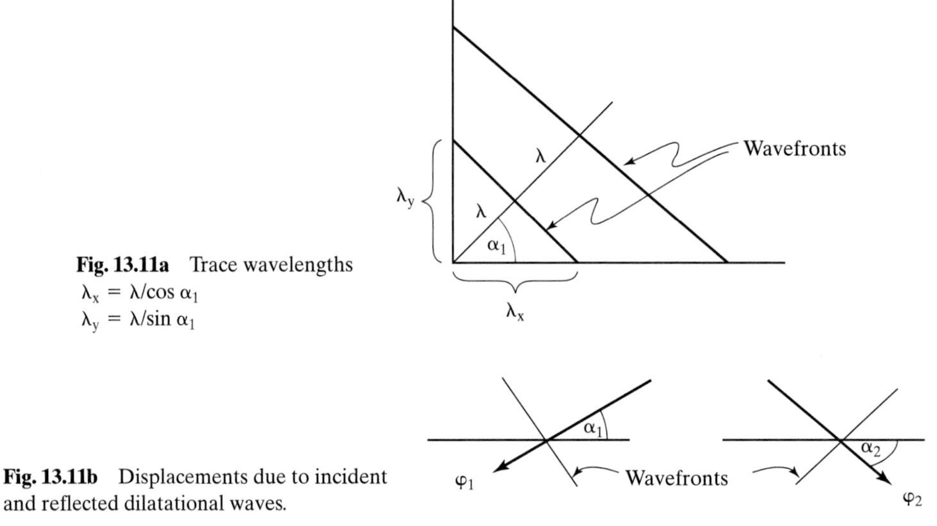

Fig. 13.11a Trace wavelengths
$\lambda_x = \lambda/\cos\alpha_1$
$\lambda_y = \lambda/\sin\alpha_1$

Fig. 13.11b Displacements due to incident and reflected dilatational waves.

Similarly, the postulated reflected wave has displacement

$$\varphi_2 = A_2 \sin(\omega t - f_2 x + g_2 y + \delta) \tag{13.10.6}$$

where δ is an arbitrary phase change. We assume the frequency, and consequently the propagation constant, does not change on reflection. Similarly

$$\xi_2 = \varphi_2 \cos\alpha_2 \text{ and } \eta_2 = -\varphi_2 \sin\alpha_2 \tag{13.10.7}$$

Now at the boundary the stresses must be zero (i.e., $\sigma_{xx} = \sigma_{xz} = 0$). Thus

i.
$$\sigma_{xz} = 2\mu\epsilon_{xz} = \mu\left(\frac{\partial\xi}{\partial z} + \frac{\partial\zeta}{\partial x}\right) = 0 \tag{13.10.8}$$

But $\zeta = 0$, hence ξ must be independent of z, which is a trivial result.

ii.
$$\sigma_{xx} = \lambda\left(\frac{\partial\xi}{\partial x} + \frac{\partial\eta}{\partial y} + \frac{\partial\zeta}{\partial z}\right) + 2\mu\frac{\partial\xi}{\partial x} \tag{13.10.9}$$

Letting

$$\xi = \xi_1 + \xi_2,\ \eta = \eta_1 + \eta_2, \text{ and } \zeta = 0 \tag{13.10.10}$$

we have

$$\begin{aligned}\sigma_{xx} &= \lambda\left(\frac{\partial\xi_1}{\partial x} + \frac{\partial\eta_1}{\partial y}\right) + 2\mu\frac{\partial\xi_1}{\partial x} + \lambda\left(\frac{\partial\xi_2}{\partial x} + \frac{\partial\eta_2}{\partial y}\right) + 2\mu\frac{\partial\xi_2}{\partial x}\\ &= -[\lambda(f_1\cos\alpha_1 + g_1\sin\alpha_1) + 2\mu f_1\cos\alpha_1]\varphi_1'\\ &\quad - [\lambda(f_2\cos\alpha_2 + g_2\sin\alpha_2) + 2\mu f_2\cos\alpha_2]\varphi_2'\end{aligned} \tag{13.10.11}$$

where

$$\varphi_1' = A_1 \cos(\omega t + f_1 x + g_1 y) \tag{13.10.12}$$

and

$$\varphi_2' = A_2 \cos(\omega t - f_2 x + g_2 y + \delta) \tag{13.10.13}$$

Substituting for f_1, f_2, g_1, and g_2, and letting $x = 0$

$$\sigma_{xx} = -k[(\lambda + 2\mu \cos^2 \alpha_1)\varphi_1' + (\lambda + 2\mu \cos^2 \alpha_2)\varphi_2'] = 0 \tag{13.10.14}$$

Hence

$$A_1(\lambda + 2\mu \cos^2 \alpha_1) \cos(\omega t + g_1 y) + A_2(\lambda + 2\mu \cos^2 \alpha_2) \cos(\omega t + g_2 y + \delta) = 0 \tag{13.10.15}$$

In other words, at any value of y, we have two simple harmonic motions of the same period, whose sum is zero. Thus, we may say that either

a. they have the same phase, but equal and opposite amplitude, or
b. they have the same amplitude, but are π out of phase.

These two cases are in reality indistinguishable. In the first case

$$g_1 y = g_2 y + \delta \tag{13.10.16}$$

and

$$A_1(\lambda + 2\mu \cos^2 \alpha_1) = -A_2(\lambda + 2\mu \cos^2 \alpha_2) \tag{13.10.17}$$

But δ cannot be a function of y (i.e., the phase change on reflection cannot depend on the position of incidence). Thus, δ is zero and from Eq. (13.10.16)

$$g_1 = g_2 \tag{13.10.18}$$

and consequently, from Eq. (13.10.3)

$$\alpha_1 = \alpha_2 \text{ and thus } A_1 = -A_2 \tag{13.10.19}$$

In the second case

$$g_1 y + \pi = g_2 y + \delta \tag{13.10.20}$$

and

$$A_1(\lambda + 2\mu \cos^2 \alpha_1) = +A_2(\lambda + 2\mu \cos^2 \alpha_2) \tag{13.10.21}$$

Thus,

$$\delta = \pi \text{ and } g_1 = g_2, \alpha_1 = \alpha_2, \text{ and } A_1 = A_2. \tag{13.10.22}$$

Now a and b are entirely equivalent statements. We may conclude from studying the boundary condition $\sigma_{xx} = 0$ that an incident dilatational wave would reflect as a dilatational wave with the angle of reflection equal to the angle of incidence, and with the same displacement amplitude, but with a phase change of π. But there is also the boundary condition $\sigma_{xy} = 0$, and we must check that this gives a consistent result. Thus

$$\begin{aligned}
\sigma_{xy} &= \mu\left(\frac{\partial \eta}{\partial x} + \frac{\partial \xi}{\partial y}\right) \\
&= -\mu\left[\frac{\partial}{\partial x}(\varphi_1 \sin \alpha_1 + \varphi_2 \sin \alpha_2) + \frac{\partial}{\partial y}(\varphi_1 \cos \alpha_1 - \varphi_2 \cos \alpha_2)\right] \\
&= -k\mu[(\sin 2\alpha_1)\varphi_1' - (\sin 2\alpha_2)\varphi_2'] = 0, \text{ when } x = 0
\end{aligned} \tag{13.10.23}$$

Consequently

$$A_1 \cos(\omega t + g_1 y) \sin 2\alpha_1 = A_2 \cos(\omega t + g_2 y + \delta) \sin 2\alpha_2 \tag{13.10.24}$$

Thus, we have two equal simple harmonic motions, and

$$g_1 = g_2, \alpha_1 = \alpha_2, \delta = 0, \text{ and } A_1 = A_2 \tag{13.10.25}$$

This implies that the wave reflects with angle of incidence equal to angle of reflection, the same displacement amplitude, and *no* phase change. This result is inconsistent with the result from Eq. (13.10.22), and we are forced to conclude that the boundary conditions cannot be met with a reflected dilatational wave alone. The solution to the conundrum is to add a shear wave, as shown in Fig. 13.12:

$$\varphi_3 = A_3 \sin(\omega t - f_3 x + g_3 y + \delta_T) \tag{13.10.26}$$

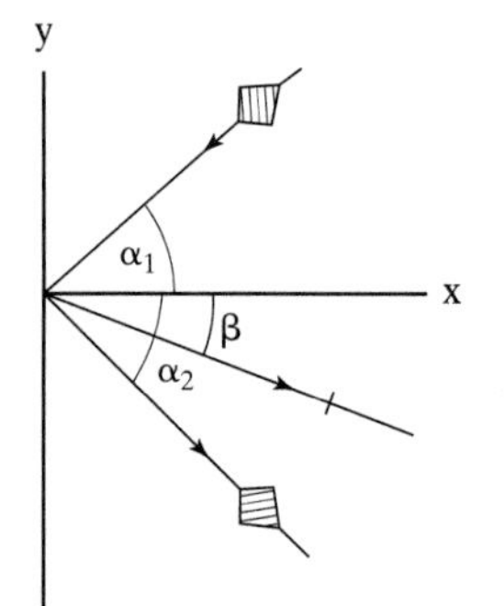

Fig. 13.12a Generation of shear wave on reflection of a dilatational wave.

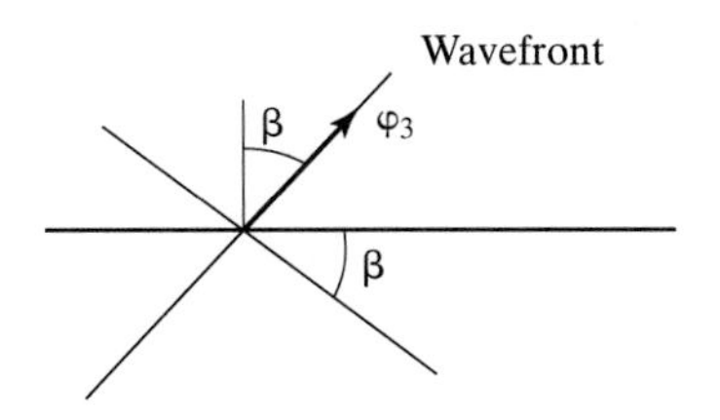

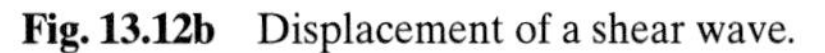
Fig. 13.12b Displacement of a shear wave.

where φ_3 lies in the x-y plane and

$$f_3 = \frac{\omega}{c_T}\cos\beta;\ g_3 = \frac{\omega}{c_T}\sin\beta \tag{13.10.27}$$

and

$$\xi_3 = \varphi_3 \sin\beta;\ \eta_3 = \varphi_3 \cos\beta \tag{13.10.28}$$

Now repeating the procedure used before, from the condition for vanishing shear stress at the boundary:

$$[f_1 \sin\alpha_1 + g_1 \cos\alpha_1]\varphi_1' - [f_2 \sin\alpha_2 + g_2 \cos\alpha_2]\varphi_2' + [g_3 \sin\beta - f_3 \cos\beta]\varphi_3' = 0 \tag{13.10.29}$$

We now have three simple harmonic motions whose sum is zero. Thus

$$\delta = \delta_T = 0 \tag{13.10.30}$$

and

$$g_1 = g_2 = g_3 \tag{13.10.31}$$

Hence,

$$\alpha_1 = \alpha_2 \tag{13.10.32}$$

and

$$\frac{\sin\alpha_1}{c_\ell} = \frac{\sin\alpha_2}{c_\ell} = \frac{\sin\beta}{c_T} \tag{13.10.33}$$

so that

$$\frac{\sin\alpha_1}{\sin\beta} = \frac{c_\ell}{c_T} \tag{13.10.34}$$

and

$$\frac{\omega}{c_\ell}[A_1 - A_2]\sin 2\alpha_1 = \frac{\omega}{c_T}A_3 \cos 2\beta \tag{13.10.35}$$

We must again check that $[\sigma_{xx}]_{x=0}$ is zero. Following the earlier analysis, we now have in place of Eq. (13.10.15):

$$A_1\frac{\omega}{c_\ell}(\lambda + 2\mu \cos^2 \alpha_1) \cos(\omega t + g_1 y) + A_2\frac{\omega}{c_\ell}(\lambda + 2\mu \cos^2 \alpha_2) \cos(\omega t + g_2 y + \delta)$$

$$- A_3\frac{\omega}{c_T}\mu \sin 2\beta \cos(\omega t + g_3 y + \delta_T) = 0 \tag{13.10.36}$$

Hence,

$$\delta = \delta_T = 0 \quad \text{and} \quad g_1 = g_2 = g_3 \tag{13.10.37}$$

implying that

$$\alpha_1 = \alpha_2 (=\alpha) \quad \text{and} \quad \frac{c_\ell}{c_T} = \frac{\sin \alpha}{\sin \beta} \tag{13.10.38}$$

Further

$$\frac{(A_1 + A_2)}{c_\ell}(\lambda + 2\mu \cos^2 \alpha_1) = \frac{A_3 \mu}{c_T} \sin 2\beta \tag{13.10.39}$$

But

$$\left(\frac{c_\ell}{c_T}\right)^2 = \frac{\lambda + 2\mu}{\mu} \tag{13.10.40}$$

so that with some manipulation

$$(A_1 + A_2)\frac{c_\ell}{c_T}(1 - 2 \sin^2 \beta) + A_3 \sin 2\beta = 0 \tag{13.10.41}$$

Using Eq. (13.10.38), we finally obtain

$$(A_1 + A_2) \sin \alpha_1 \cos 2\beta + A_3 \sin \beta \sin 2\beta = 0 \tag{13.10.42}$$

It is possible to solve Eqs. (13.10.35) and (13.10.42) to find A_2/A_1 and A_3/A_1; that is, it is possible to meet the boundary conditions with the addition of a reflected shear wave. For example

$$\frac{A_2}{A_1} = \frac{\begin{vmatrix} \cos 2\beta & -2 \sin \beta \cos \alpha_1 \\ \sin \beta \sin 2\beta & \sin \alpha_1 \cos 2\beta \end{vmatrix}}{\begin{vmatrix} 2 \sin \beta \cos \alpha_1 & \cos 2\beta \\ \sin \alpha_1 \cos 2\beta & \sin \beta \sin 2\beta \end{vmatrix}}$$

$$= \frac{2 \sin^2 \beta \sin 2\beta \cos \alpha_1 + \sin \alpha_1 \cos^2 2\beta}{2 \sin^2 \beta \sin 2\beta \cos \alpha_1 - \sin \alpha_1 \cos^2 2\beta} \tag{13.10.43}$$

A similar expression is found for A_3/A_1. For the case

$$\sigma = \frac{1}{3}, \quad \frac{c_\ell}{c_T} = 2$$

we can obtain the results shown in Fig. 13.13. Note in particular that for normal and grazing incidence, no shear wave is reflected. The shear wave, in this case, has a maximum amplitude at 48 degrees for incidence of the dilatational wave, and the reflected dilatational wave has its minimum amplitude at 65 degrees.

In the case of reflection of shear waves, there are two cases of interest:

i. when the direction of displacement is parallel to the z-axis, and
ii. when the direction of displacement is perpendicular to the z-axis.

Again, it is necessary to satisfy the boundary condition of vanishing stress at the boundary. Using similar procedures to those already employed, the first case is satisfied by having a reflected shear wave only, with the angle of reflection equal to the angle of incidence. In the second case, however, a dilatational wave is also generated, except for the case of normal incidence, and again Snell's law (see Section 3.7.2) is obeyed, as given by Eq. (13.10.38).

Some of the earliest corroboration of the continuum mechanical theory of sound came in the field of seismology. It was observed that earthquakes consisted usually of two distinct tremors, known as the primary (P) and secondary (S) waves. The primary wave has been shown to be essentially longitudinal, and the secondary wave, essentially transverse. It was often observed that a third tremor followed shortly after the secondary wave, and this was a puzzle until Rayleigh advanced the theory of surface waves on solids in 1887, which we will discuss in Section 13.11.

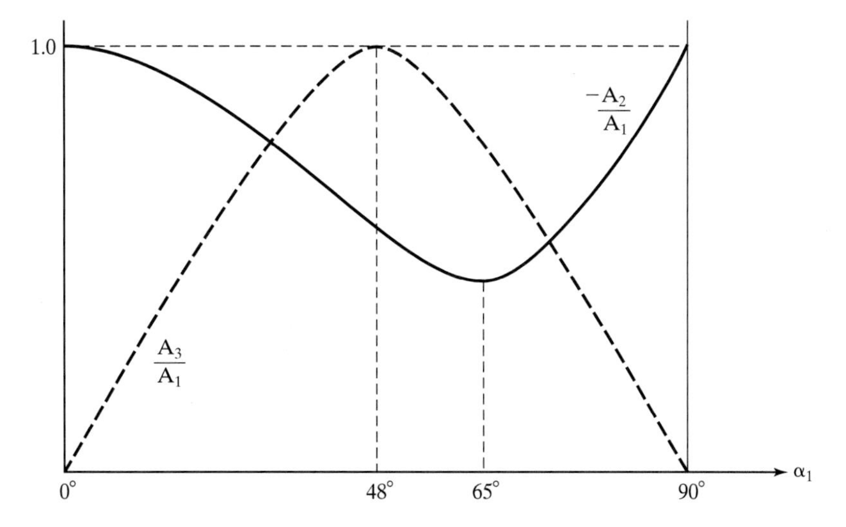

Fig. 13.13 Fractional amplitudes of reflected dilatational (A_2/A_1) and generated shear waves (A_3/A_1) at various angles of incidence of a dilatational wave.

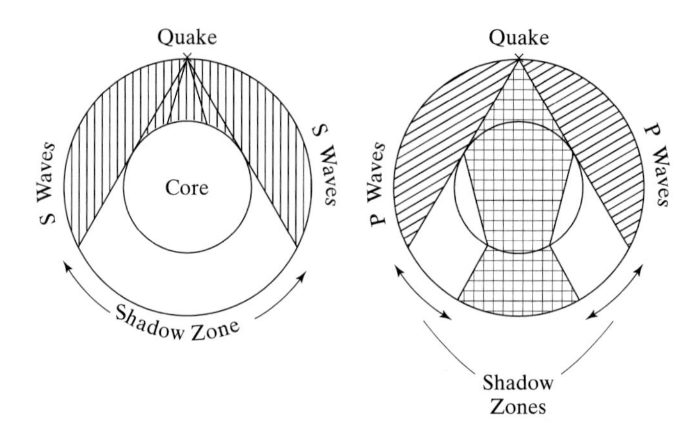

Fig. 13.14 The distribution of shear (S) waves and compressional (P) waves over the Earth's surface from an earthquake near the surface. Shear waves cannot penetrate the molten core.

Accurate quantitative data on earthquakes has been collected since Milne invented the seismograph in 1897. An outcome of this systematic data collection was the realization that there were shadow zones in the distribution of the P and S waves over the Earth's surface (see Fig. 13.14). Thus, S waves from a quake at the North Pole would be completely cut off below a certain latitude in the Southern Hemisphere. To explain this, it was postulated that the Earth possesses a liquid core. The shadow zones in the case of the P waves were explained if this core had a high density, refracting the incident waves. From the arrival times of the waves at various positions on the surface it is possible to deduce the position of the source. Seventy percent of earthquakes originate within thirty-seven miles of the Earth's surface. Such studies have also led to the conclusion that the Earth has a crust, some twenty to thirty miles thick, in which the velocity of P waves is some 3 km/s, and a lower solid layer, the mantle, in which the P waves travel at about 8 km/s. The velocities of S waves are about 2 km/s and 6 km/s, respectively, in these two layers.

13.11 Surface Waves

To explain the occurrence of tertiary tremors in earthquakes, Rayleigh postulated the existence of surface waves on solids by analogy with surface waves on water, which travel more slowly than sound waves. We know that, in general, a displacement may be expressed as the sum of solenoidal and irrotational parts (see Section 13.8); in other words,

$$\Delta = \Delta_i + \Delta_s \tag{13.11.1}$$

where the irrotational part of the displacement is

$$\Delta_i = \nabla\varphi \tag{13.11.2}$$

and the solenoidal part of the displacement is

$$\Delta_s = \nabla \times \mathbf{A} \tag{13.11.3}$$

Since both Δ_i and Δ_s propagate as waves independently (see Section 13.8), it follows that

$$\ddot{\Delta}_i - c_\ell^2 \nabla^2 \Delta_i = 0 \tag{13.11.4}$$

and

$$\ddot{\Delta}_s - c_T^2 \nabla^2 \Delta_s = 0 \tag{13.11.5}$$

It also follows that if Δ_i and Δ_s propagate as waves, φ and $\mathbf{A}$ should propagate as waves and thus

$$\ddot{\varphi} - c_\ell^2 \nabla^2 \varphi = 0 \tag{13.11.6}$$

and

$$\ddot{\mathbf{A}} - c_T^2 \nabla^2 \mathbf{A} = 0 \tag{13.11.7}$$

We will now consider a half space where the boundary lies in the x-y plane and the z-axis points into the continuum (see Fig. 13.15). Suppose a wave travels in the x direction, and allow for a possible variation in amplitude and phase with depth by specifying that Δ is a function of x and z only (i.e., Δ is not a function of y). Similarly, $\Delta_i = \Delta_i(x,z)$ and $\Delta_s = \Delta_s(x,z)$, so $\nabla\varphi$ is a function of x and z only; in other words

$$\Delta_i = \mathbf{i}\frac{\partial \varphi}{\partial x} + \mathbf{k}\frac{\partial \varphi}{\partial z} \tag{13.11.8}$$

and $\nabla \times \mathbf{A}$ is a function of x and z only; that is

$$\Delta_s = \mathbf{i}\left(\frac{\partial A_z}{\partial y} - \frac{\partial A_y}{\partial z}\right) + \mathbf{k}\left(\frac{\partial A_y}{\partial x} - \frac{\partial A_x}{\partial y}\right) \tag{13.11.9}$$

But $\mathbf{A}$ also propagates as a wave in the x direction, so $\mathbf{A}$ and its components are not functions of y; then

$$\Delta_s = -\mathbf{i}\frac{\partial A_y}{\partial z} + \mathbf{k}\frac{\partial A_y}{\partial x} \tag{13.11.10}$$

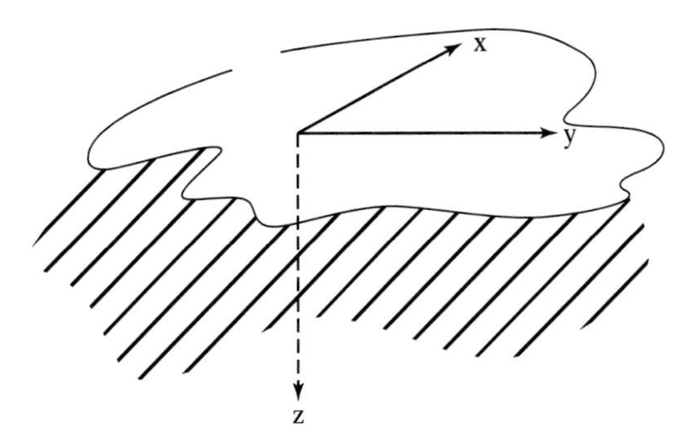

Fig. 13.15 Geometry of Rayleigh wave discussion.

The conclusion is that we may specify Δ_s in terms of a single component of A; namely, A_y or ψ, let us say. Hence

$$\xi = \frac{\partial \varphi}{\partial x} - \frac{\partial \psi}{\partial z} \qquad (13.11.11)$$

and

$$\zeta = \frac{\partial \varphi}{\partial z} + \frac{\partial \psi}{\partial x} \qquad (13.11.12)$$

Also from Eqs. (13.11.6) and (13.11.7), we obtain

$$\frac{\ddot{\varphi}}{c_\ell^2} = \frac{\partial^2 \varphi}{\partial x^2} + \frac{\partial^2 \varphi}{\partial z^2} \qquad (13.11.13)$$

and

$$\frac{\ddot{\psi}}{c_T^2} = \frac{\partial^2 \psi}{\partial x^2} + \frac{\partial^2 \psi}{\partial z^2} \qquad (13.11.14)$$

Now we are looking for solutions of the form

$$\varphi = C_1(z) e^{j(\omega t - k_1 x)} \qquad (13.11.15)$$

and

$$\psi = C_2(z) e^{j(\omega t - k_2 x)} \qquad (13.11.16)$$

Substituting into Eq. (13.11.13) yields

$$-\frac{\omega^2}{c_\ell^2}\varphi = -k_1^2 \varphi + e^{j(\omega t - k_1 x)} \frac{\partial^2 C_1}{\partial z^2} \qquad (13.11.17)$$

so that

$$\frac{\partial^2 C_1}{\partial z^2} = \left[k_1^2 - \frac{\omega^2}{c_\ell^2}\right] C_1 \qquad (13.11.18)$$

and similarly

$$\frac{\partial^2 C_2}{\partial z^2} = \left[k_2^2 - \frac{\omega^2}{c_T^2}\right] C_2 \qquad (13.11.19)$$

Now we already know about the solutions

$$k_1^2 = \frac{\omega^2}{c_\ell^2} \quad \text{and} \quad k_2^2 = \frac{\omega^2}{c_T^2} \qquad (13.11.20)$$

which correspond to longitudinal and shear waves propagating in the x direction. What Rayleigh realized was that there was another solution in which, say

$$k_1 = k_2 = k \tag{13.11.21}$$

Then

$$\frac{\partial^2 C_1}{\partial z^2} = \alpha_1^2 C_1 \tag{13.11.22}$$

and

$$\frac{\partial^2 C_2}{\partial z^2} = \alpha_2^2 C_2 \tag{13.11.23}$$

where

$$\alpha_1 = \sqrt{k^2 - \frac{\omega^2}{c_\ell^2}} \quad \text{and} \quad \alpha_2 = \sqrt{k^2 - \frac{\omega^2}{c_T^2}} \tag{13.11.24}$$

General solutions of Eqs. (13.11.22) and (13.11.23) are

$$C_1 = \beta_1 e^{-\alpha_1 z} + \gamma_1 e^{\alpha_1 z} \tag{13.11.25}$$

and

$$C_2 = \beta_2 e^{-\alpha_2 z} + \gamma_2 e^{\alpha_2 z} \tag{13.11.26}$$

Clearly, $\gamma_1 = \gamma_2 = 0$ or the waves would have infinite displacement at infinite depth and

$$C_1 = \beta_1 e^{-\alpha_1 z} \tag{13.11.27}$$

and

$$C_2 = \beta_2 e^{-\alpha_2 z} \tag{13.11.28}$$

To determine the remaining arbitrary constants β_1 and β_2, we must specify the boundary conditions (i.e., the stress must vanish at the boundary):

$$\sigma_{zz} = \sigma_{zy} = \sigma_{zx} = 0, \quad \text{when} \quad z = 0 \tag{13.11.29}$$

Hence

$$\sigma_{xx} = \lambda\theta + 2\mu\epsilon_{zz} = \lambda\nabla\cdot\Delta + 2\mu\frac{\partial\zeta}{\partial z} = \lambda\nabla^2\varphi + 2\mu\frac{\partial}{\partial z}(\Delta_{iz} + \Delta_{sz})$$

$$=\lambda\nabla^2\varphi + 2\mu\frac{\partial^2\varphi}{\partial z^2} + 2\mu\frac{\partial^2\psi}{\partial z\partial x} = \lambda\frac{\partial^2\varphi}{\partial x^2} + (\lambda + 2\mu)\frac{\partial^2\varphi}{\partial z^2} + 2\mu\frac{\partial^2\psi}{\partial z\partial x} = 0 \tag{13.11.30}$$

Since

$$\psi = \beta_1 e^{-\alpha_1 z} e^{j(\omega t - kx)} \tag{13.11.31}$$

and

$$\varphi = \beta_2 e^{-\alpha_2 z} e^{j(\omega t - kx)} \tag{13.11.32}$$

on substituting into Eq. (13.11.30) we obtain

$$\sigma_{zz} = -\lambda k^2 \varphi + (\lambda + 2\mu)\alpha_1^2 \varphi + jk\alpha_2 2\mu\varphi = 0 \tag{13.11.33}$$

Now when $z = 0$, $\sigma_{zz} = 0$, and thus

$$-\lambda k^2 \beta_1 + (\lambda + 2\mu)\alpha_1^2 \beta_1 + jk\alpha_2 2\mu\beta_2 = 0 \tag{13.11.34}$$

or

$$\beta_1[\lambda({\alpha_1}^2 - k^2) + 2\mu{\alpha_1}^2] + j2\mu k\alpha_2\beta_2 = 0 \tag{13.11.35}$$

Similarly

$$\sigma_{xz} = \mu\left(\frac{\partial \xi}{\partial z} + \frac{\partial \zeta}{\partial x}\right) = \mu\left(2\frac{\partial^2 \varphi}{\partial z \partial x} + \frac{\partial^2 \psi}{\partial x^2} - \frac{\partial^2 \psi}{\partial z^2}\right) \tag{13.11.36}$$

Substituting for φ and ψ gives

$$\sigma_{xz} = \mu(2jk\alpha_1\varphi - k^2\psi - \alpha_2^2\psi) \tag{13.11.37}$$

Now

$$\sigma_{xz} = 0 \text{ when } z = 0$$

and thus

$$j2k\alpha_1\beta_1 - k^2\beta_2 - \alpha_2^2\beta_2 = 0 \tag{13.11.38}$$

Eliminating β_1 and β_2 from Eqs. (13.11.35) and (13.11.38) finally results in a relation between ω and k, thus

$$4\mu k^2\alpha_1\alpha_2 = ({\alpha_2}^2 + k^2)[\lambda({\alpha_1}^2 - k^2) + 2\mu{\alpha_1}^2] \tag{13.11.39}$$

Now the velocity of propagation of the surface disturbance is, say, $(\omega/k) = c_R$, and thus from Eq. (13.11.24)

$$\alpha_1 = k\sqrt{1 - \frac{c_R^2}{c_\ell^2}} \quad \text{and} \quad \alpha_2 = k\sqrt{1 - \frac{c_R^2}{c_T^2}} \tag{13.11.40}$$

Substitution of Eq. (13.11.40) into Eq. (13.11.39) and squaring both sides yields

$$16\mu^2\left(1-\frac{c_R^2}{c_\ell^2}\right)\left(1-\frac{c_R^2}{c_T^2}\right)=\left(2-\frac{c_R^2}{c_T^2}\right)^2\left[2\mu\left(1-\frac{c_R^2}{c_\ell^2}\right)-\lambda\frac{c_R^2}{c_\ell^2}\right]^2 \quad (13.11.41)$$

Dividing by μ^2 and using Eq. (13.7.7) (which we repeat):

$$\upsilon=\frac{\lambda}{2(\mu+\lambda)} \quad (13.11.42)$$

yields

$$16\left(1-\frac{c_R^2}{c_\ell^2}\right)\left(1-\frac{c_R^2}{c_T^2}\right)=\left(2-\frac{c_R^2}{c_T^2}\right)^2\left[2-\frac{c_R^2}{c_\ell^2}\left(\frac{2-2\upsilon}{1-2\upsilon}\right)\right]^2 \quad (13.11.43)$$

If λ and μ are known, Eq. (13.11.43) can be solved. For example, if $\sigma = 1/4$ (for steel $\sigma = 0.29$), then from Eq. (13.11.42)

$$2\mu+2\lambda=4\lambda \text{ and } \mu=\lambda \quad (13.11.44)$$

In addition, we have

$$c_\ell=\sqrt{\frac{3\lambda}{\rho}},\ c_T=\sqrt{\frac{\lambda}{\rho}}\left(\text{i.e., } c_\ell=\sqrt{3}c_T\right) \quad (13.11.45)$$

and in Eq. (13.11.43) the term

$$\left(\frac{2-2\upsilon}{1-2\upsilon}\right)=\frac{2-\frac{1}{2}}{1-\frac{1}{2}}=3 \quad (13.11.46)$$

Thus, substituting into Eq. (13.11.43) and letting $(c_R/c_T)^2 = x$

$$16\left(1-x/3\right)(1-x)=(2-x)^2(2-x)^2 \quad (13.11.47)$$

leading to

$$3x^3-24x^2+56x-32=0 \quad (13.11.48)$$

and

$$x=4 \text{ or } \left(2+2/\sqrt{3}\right) \text{ or } \left(2-2/\sqrt{3}\right) \quad (13.11.49)$$

When $x = 4$,

$$\alpha_1=k\sqrt{1-\frac{c_R^2}{3c_T^2}}=k\sqrt{1-\frac{4}{3}} \quad (13.11.50)$$

which is imaginary and $\alpha_2 = k\sqrt{1 - 4}$, which is also imaginary. The waves would not then be confined to the surface. When $x = 2 + 2/\sqrt{3}$,

$$\alpha_1 = k\sqrt{1 - \frac{1}{3}\left(2 + 2/\sqrt{3}\right)} = k\sqrt{0.33 - 0.6} \tag{13.11.51}$$

which is imaginary. Further

$$\alpha_2 = k\sqrt{1 - (2 + 2/\sqrt{3})} \tag{13.11.52}$$

which is also imaginary, and the waves do not correspond to the type postulated. Finally, when $x = 2 - 2/\sqrt{3}$,

$$\alpha_1 = k\sqrt{1 - \frac{2}{3} + \frac{2\sqrt{3}}{9}} = k\sqrt{0.33 + 0.37} \tag{13.11.53}$$

This is real and

$$\alpha_2 = k\sqrt{1 - 2 + \frac{2\sqrt{3}}{3}} = k\sqrt{\frac{3.4}{3} - 1} = k\sqrt{0.1} \tag{13.11.54}$$

which is also real. Thus, there is a solution of the Navier equation which predicts that plane waves restricted to the surface can propagate. These are called Rayleigh waves and their velocity of propagation

$$c_R = \sqrt{2 - \frac{2}{\sqrt{3}}}c_T = 0.92c_T \tag{13.11.55}$$

We have finally verified on theoretical grounds that the Rayleigh waves will travel more slowly than shear waves.

13.12 Propagation in a Plate

The problem of propagation in a plate was first tackled by Lamb in 1917. If an infinite plate is parallel to the x-y plane, the thickness is in the z direction, and waves are propagated in the x direction, then by the same arguments used in the discussion of surface waves, the components of displacement will be

$$\xi = \frac{\partial \varphi}{\partial x} - \frac{\partial \psi}{\partial z}, \eta = 0 \tag{13.12.1}$$

and

$$\zeta = \frac{\partial \varphi}{\partial z} + \frac{\partial \psi}{\partial x} \tag{13.12.2}$$

Similarly, the potentials φ and ψ obey wave equations

$$\ddot{\varphi} = c_\ell^2 \nabla^2 \varphi \tag{13.12.3}$$

and

$$\ddot{\psi} = c_T^2 \nabla^2 \psi \tag{13.12.4}$$

Assume solutions of the form

$$\varphi = f_1(z) e^{j(\omega t - kx)} \tag{13.12.5}$$

and

$$\psi = f_2(z) e^{j(\omega t - kx)} \tag{13.12.6}$$

Assuming plane waves can propagate along the plate, on substituting Eq. (13.12.5) into Eq. (13.12.3)

$$-\omega^2 \varphi = c_\ell^2 \left(-k^2 \varphi + \frac{\varphi}{f_1(z)} \frac{\partial^2 f_1}{\partial z^2} \right) \tag{13.12.7}$$

and

$$\frac{\partial^2 f_1}{\partial z^2} = \left[k^2 - \frac{\omega^2}{c_\ell^2} \right] f_1 \tag{13.12.8}$$

Consider the case for which

$$\left[k^2 - \frac{\omega^2}{c_\ell^2} \right] > 0 \tag{13.12.9}$$

in other words,

$$\frac{\partial^2 f_1}{\partial z^2} = \alpha_1^2 f_1 \tag{13.12.10}$$

Solutions for Eq. (13.12.10) are

$$f_1 = A_1 \sinh \alpha_1 z + B_1 \cosh \alpha_1 z \tag{13.12.11}$$

and

$$f_2 = A_2 \sinh \alpha_2 z + B_2 \cosh \alpha_2 z \tag{13.12.12}$$

We can recognize two subcases: one for symmetrical displacement and one for asymmetrical displacement, as illustrated in Fig. 13.16. The first is the case of dilatational Lamb waves; the second, the case of flexural or shear Lamb waves. In the first case, $\Delta(z) = -\Delta(-z)$; in

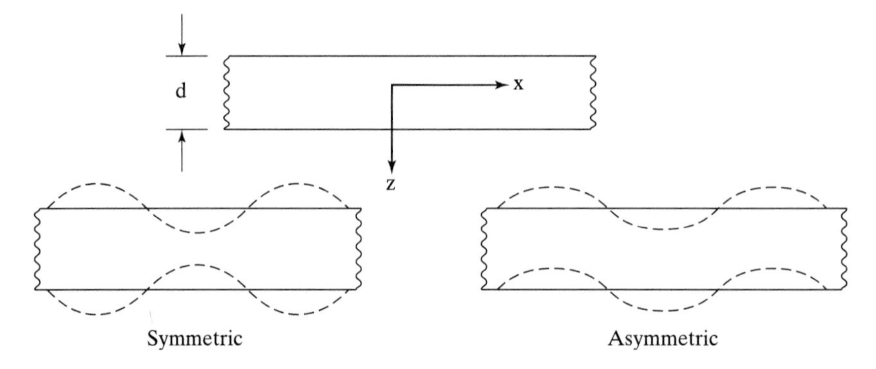

Fig. 13.16 Two types of Lamb wave motion.

the second, $\Delta(z) = +\Delta(-z)$, so for symmetry $\xi(z) = -\xi(-z)$. But

$$\xi(z) = \frac{\partial f_1}{\partial z} e^{j(\omega t + kx)} + jkf_2 e^{j(\omega t + kx)} \tag{13.12.13}$$

Thus

$$\frac{\partial f_1}{\partial z} = A_1\alpha_1 \cosh \alpha_1 z + B_1\alpha_1 \sinh \alpha_1 z \tag{13.12.14}$$

which must be symmetric; in other words, A_1 must be zero and f_2 must be symmetric, that is B_2 must be zero. Thus, the solutions must be of the form

$$\varphi = B_1 \cosh(\alpha_1 z) e^{j(\omega t - kx)} \tag{13.12.15}$$

and

$$\psi = A_2 \sinh(\alpha_2 z) e^{j(\omega t - kx)} \tag{13.12.16}$$

To delineate the remaining arbitrary constants of the solution (A, B, and k), we must invoke the boundary conditions of vanishing stress on the faces of the plate. Suppose these faces lie at $z = \pm d/2$. The stresses in the z direction are

$$\sigma_{xz} = 2\mu\epsilon_{xz} = \mu\left(\frac{\partial \zeta}{\partial x} + \frac{\partial \xi}{\partial z}\right) \tag{13.12.17}$$

$$\sigma_{yz} = 2\mu\epsilon_{yz} = \mu\left(\frac{\partial \eta}{\partial z} + \frac{\partial \zeta}{\partial y}\right) = 0 \tag{13.12.18}$$

and

$$\sigma_{zz} = \lambda(\epsilon_{xx} + \epsilon_{yy} + \epsilon_{zz}) + 2\mu\epsilon_{zz} \tag{13.12.19}$$

From Eqs. (13.12.17), (13.12.2), (13.12.15), and (13.12.16)

$$\sigma_{xz} = \mu\left[2\frac{\partial^2\varphi}{\partial x\partial z} + \frac{\partial^2\psi}{\partial x^2} - \frac{\partial^2\psi}{\partial z^2}\right]$$
$$= \mu[2B_1(-jk)\alpha_1 \sinh(\alpha_1 z)e^{j(\omega t-kx)} - k^2\psi - \alpha_2^2\psi] \quad (13.12.20)$$

When $z = \pm d/2$, $\sigma_{xz} = 0$ and thus

$$\left\{A_2(k^2 + \alpha_2^2)\sinh\left(\frac{\alpha_2 d}{2}\right) + j2B_1k\alpha_1\sinh\left(\frac{\alpha_1 d}{2}\right)\right\} = 0 \quad (13.12.21)$$

From Eqs. (13.12.19), (13.12.2), (13.12.15), and (13.12.16)

$$\sigma_{zz} = \lambda\left(\frac{\partial\xi}{\partial x} + \frac{\partial\zeta}{\partial z}\right) + 2\mu\frac{\partial\zeta}{\partial z}$$
$$= \lambda\left(\frac{\partial^2\varphi}{\partial x^2} - \frac{\partial^2\psi}{\partial x\partial z}\right) + (2\mu + \lambda)\left(\frac{\partial^2\varphi}{\partial z^2} - \frac{\partial^2\psi}{\partial z\partial x}\right)$$
$$= \lambda(-k^2\varphi + jk\alpha_2 A\cosh(\alpha_2 z)e^{j(\omega t-kx)})$$
$$+ (2\mu + \lambda)(\alpha_1^2\varphi - jk\alpha_2 A\cosh(\alpha_2 z)e^{j(\omega t-kx)}) \quad (13.12.22)$$

When $z = \pm d/2$, $\sigma_{zz} = 0$, and thus

$$[\alpha_1^2(2\mu + \lambda) - k^2\lambda]B_1\cosh\left(\alpha_1\frac{d}{2}\right) - 2\mu jk\alpha_2 A_2\cosh\left(\alpha_2\frac{d}{2}\right) = 0 \quad (13.12.23)$$

But from Eq. (13.11.24)

$$\alpha_1^2 = k^2 - \frac{\omega^2\rho}{2\mu + \lambda} \quad (13.12.24)$$

and thus

$$\alpha_1^2(2\mu + \lambda) - k^2\lambda = 2\mu k^2 - \omega^2\rho \quad (13.12.25)$$

or from Eq. (13.11.24)

$$\alpha_1^2(2\mu + \lambda) - k^2\lambda = \mu k^2 + \mu\left(k^2 - \frac{\omega^2}{c_T^2}\right) = \mu(k^2 + \alpha_2^2) \quad (13.12.26)$$

Substituting from Eq. (13.12.26) into Eq. (13.12.23)

$$(k^2 + \alpha_2^2)B_1\cosh\left(\alpha_1\frac{d}{2}\right) - 2jkA_2\alpha_2\cosh\left(\alpha_2\frac{d}{2}\right) = 0 \quad (13.12.27)$$

From Eqs. (13.12.21) and (13.12.27), we obtain

$$\frac{\tanh\left(\frac{\alpha_2 d}{2}\right)}{\tanh\left(\frac{\alpha_1 d}{2}\right)} = \frac{4k^2\alpha_1\alpha_2}{(k^2 + \alpha_2^2)^2} \tag{13.12.28}$$

Solutions are obtained most readily for the extreme cases of very thin and very thick plates. For very thin plates,

$$\tanh\left(\frac{\alpha_2 d}{2}\right) \to \frac{\alpha_2 d}{2}, \text{ etc.} \tag{13.12.29}$$

and Eq. (13.12.28) yields

$$(k^2 + \alpha_2^2)^2 - 4k^2\alpha_1^2 = 0 \tag{13.12.30}$$

or using Eq. (13.12.24)

$$k^4 + 2k^4\left(1 - \frac{c_L^2}{c_T^2}\right) + k^4\left(1 - \frac{c_L^2}{c_T^2}\right)^2 - 4k^4\left(1 - \frac{c_L^2}{c_\ell^2}\right) = 0 \tag{13.12.31}$$

where $c_L^2 = \dfrac{\omega^2}{k^2}$ then

$$-4\frac{c_L^2}{c_T^2} + \frac{c_L^2}{c_T^4} + 4\frac{c_L^2}{c_\ell^2} = 0 \tag{13.12.32}$$

Hence

$$\begin{aligned} c_L^2 &= c_T^4\left[\frac{4}{c_T^2} - \frac{4}{c_\ell^2}\right] = 4\frac{\mu^2}{\rho^2}\left[\frac{\rho}{\mu} - \frac{\rho}{2\mu + \lambda}\right] \\ &= 4\frac{\mu}{\rho}\frac{(2\mu + \lambda - \mu)}{(2\mu + \lambda)} = \frac{4\mu(\mu + \lambda)}{\rho(2\mu + \lambda)} \end{aligned} \tag{13.12.33}$$

and

$$c_L = 2c_T\sqrt{\frac{\mu + \lambda}{2\mu + \lambda}} \tag{13.12.34}$$

Similarly, for thick plates

$$\tanh\left(\frac{\alpha_2 d}{2}\right) \to 1, \text{ etc.} \tag{13.12.35}$$

Thus, from Eq. (13.12.28)

$$(k^2 + \alpha_2^2)^2 - 4k^2\alpha_1\alpha_2 = 0 \tag{13.12.36}$$

which may be rewritten using Eq. (13.11.24) as

$$\begin{aligned}4\mu k^2\alpha_1\alpha_2 &= (\alpha_2^2 + k^2)\left[\mu k^2 + \mu k^2\left(1 - \frac{\omega^2}{k^2c_T^2}\right)\right] \\ &= (\alpha_2^2 + k^2)[2\mu k^2 - \omega^2\rho] \\ &= (\alpha_2^2 + k^2)\left[2\mu k^2 - (2\mu + \lambda)\frac{\omega^2}{c_\ell^2}\right] = (\alpha_2^2 + k^2)[2\mu\alpha_1^2 + \lambda(\alpha_1^2 - k^2)]\end{aligned} \tag{13.12.37}$$

in other words

$$4\mu k^2\alpha_1\alpha_2 = (\alpha_2^2 + k^2)[\lambda(\alpha_1^2 - k^2) + 2\mu\alpha_1^2] \tag{13.12.38}$$

which was precisely the expression obtained for Rayleigh waves, Eq. (13.11.39), as is to be expected. Waves of wavelength that are short in comparison with plate thickness propagate with the velocity of Rayleigh waves. In between the extremes, the velocity varies. The solution for the asymmetrical case follows a similar procedure.

REFERENCES AND FURTHER READING

Achenbach, J.D. 1973. *Wave Propagation in Elastic Solids*. North Holland Publishing.
Kolsky, H. 1963. *Stress Waves in Solids*. Dover.

PROBLEMS

13.1 For steel, Young's modulus is 19.5×10^{10} Pa and the value of Poisson's ratio is 0.29. The density is 7700 Kg/m^3. Given that for an isotropic medium Hooke's law states

$$\sigma_{xx} = \lambda\theta + 2\mu\epsilon_{xx}, \text{ etc.}$$

where the symbols have their usual meanings, derive expressions for Young's modulus (Y), Poisson's ratio (ν), and the bulk modulus (B), in terms of the Lame constants λ and μ. Determine the value of the shear modulus of steel.

13.2 An elastic rod of Young's modulus Y and density ρ has length l and is clamped at end $x = l$. The rod is initially at rest. At time $t = 0$, the end $x = 0$ is subject to a force F(t). Assuming that a linear one-dimensional stress wave propagates, determine the reflection at end $x = l$.

13.3 A very long unstressed rod is travelling in the positive x direction with velocity V. A distant section of the rod is then brought to rest in a short, but not zero, time and then clamped. Sketch the particle velocity wave $f(t - x/c)$ that runs along the rod at time t, before reflection. Note that $c = (Y/\rho)^{1/2}$.

13.4 Starting with the assumption of Hooke's law for an isotropic medium in terms of the Lame constants, show that there are only two possible velocities of sound propagation if the medium is unbounded. Determine the velocity of irrotational wave motion in steel.

13.5 A dilatational wave is incident on a free flat surface. Show that the boundary conditions cannot be satisfied by assuming that only a dilatational wave is reflected. State briefly how the boundary conditions may be satisfied. Describe briefly how plane shear waves are reflected at flat surfaces. What is meant by "critical angle" in this connection?

13.6 A thin layer of a limp substance is stuck on the surface of an elastic half space. The mass density of the layer is ρ_s. A plane harmonic wave is incident on the surface at angle θ_0. Determine the influence of the covering on the amplitudes of the reflected waves.

13.7 Discuss the propagation of surface waves at an interface of a solid and a fluid.

13.8 Show that when a wave travels in the x direction along a free surface defined by the x−y plane, the z direction being directed into a homogeneous, isotropic material, that its displacement is given by

$$\Delta = i\xi + k\zeta$$

where

$$\xi = \frac{\partial\varphi}{\partial x} - \frac{\partial\psi}{\partial z}$$

and

$$\zeta = \frac{\partial\psi}{\partial z} + \frac{\partial\varphi}{\partial x}$$

Explain briefly the difference between Rayleigh and Lamb waves.

13.9 Show that the velocity of propagation of dilatational Lamb waves, c_L, is given by

$$c_L^2 = \frac{4\mu(\mu + \lambda)}{\rho(2\mu + \lambda)}$$

in the case of propagation in a thin plate of material with Lame constants μ and λ and density ρ.

13.10 The matrix of elastic constants for a cubic crystal is

$$\begin{vmatrix} c_{11} & c_{12} & c_{12} & & & \\ c_{12} & c_{11} & c_{12} & & 0 & \\ c_{12} & c_{12} & c_{11} & & & \\ & & & c_{44} & & \\ & 0 & & & c_{44} & \\ & & & & & c_{44} \end{vmatrix}$$

Show that plane waves may be propagated in the direction of a crystal axis with velocity

$$\sqrt{\frac{c_{11}}{\rho}} \quad or \quad \sqrt{\frac{c_{44}}{\rho}}$$

where ρ is the density. Show that the waves are longitudinal and transverse, respectively.

Chapter 14

Damping, Attenuation, and Absorption

14.1 The Navier-Stokes Equation

The discussion of the continuum mechanical theory of acoustic propagation in the previous chapters ignored the occurrence of energy loss. When such energy loss occurs in a vibrating system, the phenomenon is usually referred to as damping. The gradual decay of the amplitude of a progressive wave is called attenuation, while the energy loss on reflection from a wall is called absorption. These three terms are used rather loosely and are all manifestations of the same underlying phenomenon, namely, the irreversible conversion of acoustic energy, usually into heat. We will start by introducing damping for the case of a fluid, since the situation is simplified by the absence of a shear modulus. We will first modify the generalized theory of elasticity by introducing the concept of viscosity. It has been known for a long time that in order to maintain a flow at constant velocity, it is necessary to apply a shearing stress. To a good first approximation, the required shearing stress was found to be proportional to the velocity gradient; that is, for flow in the x direction, with a velocity gradient in the y direction, the shear stress is

$$\sigma_{yx} = M\frac{\partial \dot{\xi}}{\partial y} = 2M\dot{\epsilon}_{yx} \tag{14.1.1}$$

where M is known as the coefficient of shear viscosity. Stokes (1845) was the first to suggest that a complete set of resistive stresses in a fluid could be related to the rates of change of strain. In general, the rates of change of strain will be

$$\dot{\epsilon}_{xx} = \frac{\partial \dot{\xi}}{\partial x} \varepsilon_{xy} = \frac{1}{2}\left(\frac{\partial \dot{\eta}}{\partial x} + \frac{\partial \dot{\xi}}{\partial y}\right)$$

$$\dot{\epsilon}_{yy} = \frac{\partial \dot{\eta}}{\partial y} \varepsilon_{yz} = \frac{1}{2}\left(\frac{\partial \dot{\xi}}{\partial y} + \frac{\partial \dot{\eta}}{\partial z}\right)$$

and

$$\dot{\epsilon}_{zz} = \frac{\partial \dot{\xi}}{\partial z} \varepsilon_{zx} = \frac{1}{2}\left(\frac{\partial \dot{\xi}}{\partial x} + \frac{\partial \dot{\xi}}{\partial z}\right) \tag{14.1.2}$$

Assuming in general a linear relationship, we may write for the resistive stresses

$$\begin{bmatrix} \sigma'_{xx} \\ \sigma'_{yy} \\ \sigma'_{zz} \\ \sigma'_{yz} \\ \sigma'_{zx} \\ \sigma'_{yx} \end{bmatrix} = \begin{bmatrix} c'_{11} & c'_{12} & c'_{13} & c'_{14} & c'_{15} & c'_{16} \\ c'_{21} & c'_{22} & & & & \\ c'_{31} & & c'_{33} & & & \\ c'_{41} & & & c'_{44} & & \\ c'_{51} & & & & c'_{55} & \\ c'_{61} & & & & & c'_{66} \end{bmatrix} = \begin{bmatrix} \dot{\epsilon}_{xx} \\ \dot{\epsilon}_{yy} \\ \dot{\epsilon}_{zz} \\ \dot{\epsilon}_{yz} \\ \dot{\epsilon}_{zx} \\ \dot{\epsilon}_{yx} \end{bmatrix} \tag{14.1.3}$$

Now for small changes in the strains, the work done against all the possible resistive stresses will be

$$\begin{aligned} \Delta U &= \sigma'_{xx}\,\Delta\dot{\epsilon}_{xx} + \sigma'_{yy}\,\Delta\dot{\epsilon}_{yy} + \sigma'_{zz}\,\Delta\dot{\epsilon}_{zz} \\ &\quad + 2\sigma'_{xy}\,\Delta\dot{\epsilon}_{xy} + 2\sigma'_{yz}\,\Delta\dot{\epsilon}_{yz} + 2\sigma'_{zx}\,\Delta\dot{\epsilon}_{zx} \end{aligned} \tag{14.1.4}$$

Thus

$$\begin{aligned} \frac{\Delta U}{\Delta t} &= \sigma'_{xx}\,\dot{\epsilon}_{xx} + \sigma'_{yy}\,\dot{\epsilon}_{yy} + \sigma'_{zz}\,\dot{\epsilon}_{zz} \\ &\quad + 2\sigma'_{xy}\,\dot{\epsilon}_{xy} + 2\sigma'_{yz}\,\dot{\epsilon}_{yz} + 2\sigma'_{zx}\,\dot{\epsilon}_{zx} \end{aligned} \tag{14.1.5}$$

Now

$$c'_{11} = \frac{\partial \sigma'_{xx}}{\partial \dot{\epsilon}_{xx}} = \frac{\partial^3 U}{\partial t\, \partial^2 \dot{\epsilon}_{xx}} \tag{14.1.6}$$

and

$$c'_{15} = \frac{\partial \sigma'_{xx}}{\partial \dot{\epsilon}_{xx}} = \frac{\partial^3 U}{2\partial t \partial \dot{\epsilon}_{zx} \partial \dot{\epsilon}_{xx}} = \frac{\partial \sigma'_{zx}}{\partial \dot{\epsilon}_{xx}} = c'_{51} \tag{14.1.7}$$

Thus, we can conclude that the matrix is symmetric with only twenty-one constants.

Obviously, a fluid is isotropic and homogeneous, and since there will be a set of coordinates in which the state of flow may be expressed in terms of three mutually perpendicular velocity components, parallel to the coordinate axes, we may reduce these twenty-one constants to two, thus

$$
\begin{aligned}
\sigma'_{x'x'} &= A\dot{\epsilon}_{x'x'} + B\dot{\epsilon}_{y'y'} + B\dot{\epsilon}_{z'z'} \\
\sigma'_{y'y'} &= B\dot{\epsilon}_{x'x'} + A\dot{\epsilon}_{y'y'} + B\dot{\epsilon}_{z'z'} \\
\sigma'_{z'z'} &= B\dot{\epsilon}_{x'x'} + B\dot{\epsilon}_{x'x'} + A\dot{\epsilon}_{z'z'}
\end{aligned}
\tag{14.1.8}
$$

and

$$\sigma'_{y'z'} = \sigma'_{z'x'} = \sigma'_{x'y'} = 0 \tag{14.1.9}$$

where x′y′z′ is the principal coordinate system. Now if we let $A - B = 2M$ and $B = A$ in the same way as in Section 13.6, then

$$
\begin{aligned}
\sigma'_{x'x'} &= \Lambda\dot{\theta}' + 2M\dot{\epsilon}_{x'x'} \\
\sigma'_{y'y'} &= \Lambda\dot{\theta}' + 2M\dot{\epsilon}_{y'y'}
\end{aligned}
$$

and

$$\sigma'_{z'z'} = \Lambda\dot{\theta}' + 2M\dot{\epsilon}_{z'z'} \tag{14.1.10}$$

Transforming to a general set of coordinates

$$
\begin{aligned}
\sigma'_{xx} &= \Lambda\dot{\theta} + 2M\dot{\epsilon}_{xx} \\
\sigma'_{y'y'} &= \Lambda\dot{\theta} + 2M\dot{\epsilon}_{yy} \\
\sigma'_{z'z'} &= \Lambda\dot{\theta} + 2M\dot{\epsilon}_{zz}
\end{aligned}
\tag{14.1.11}
$$

and

$$\sigma'_{yz} = 2M\dot{\epsilon}_{yz};\ \sigma'_{zx} = 2M\dot{\epsilon}_{zx};\ \sigma'_{xy} = 2M\dot{\epsilon}_{xy} \tag{14.1.12}$$

If the comparison is now made with the conventional definition of viscosity, we see that M is the coefficient of shear viscosity. But what is Λ? To answer this question, first note that there must be a coefficient of bulk viscosity β, analogous to the bulk modulus of elasticity, which by comparison with Eq. (13.7.20) in Chapter 13 must be given by

$$\beta = \Lambda + \frac{2}{3}M \tag{14.1.13}$$

Now β is the stress per rate of change of volume strain. It was shown by Maxwell that for a perfect gas $\beta = 0$, and so for a first approximation Stokes set

$$\Lambda = -\frac{2}{3}M \tag{14.1.14}$$

implying that only one independent constant should be sufficient to describe resistive stresses.

We may also carry over a large part of the former analysis in dealing with the dynamics of a viscous fluid. Thus, as we saw in Eq. (13.8.6), the net force on an element in a dynamic situation was given by

$$F = [(\lambda + \mu) \nabla\theta + \mu\nabla \cdot \nabla\Delta] \, dxdydz \tag{14.1.15}$$

Thus, the resistive forces must be

$$F = [(\Lambda + M) \nabla\dot{\theta} + M\nabla \cdot \nabla\mathbf{v}] \, dxdydz \tag{14.1.16}$$

Now for a fluid, the stress is due to hydrostatic pressure, and thus

$$\sigma_{xx} = \lambda\theta = \sigma_{yy} = \sigma_{zz} = -p \tag{14.1.17}$$

Consequently, from Eq. (14.1.15), remembering that the shear modulus vanishes, and from Eq. (14.1.17)

$$\frac{F}{dV} = \lambda\nabla\theta = -\nabla p \tag{14.1.18}$$

Thus, the total force on an element of a viscous fluid

$$\mathbf{F} + \mathbf{F}' = \left[-\nabla p + \frac{M}{3}\nabla\nabla\cdot\mathbf{v} + M\nabla\cdot\nabla\mathbf{v}\right] dxdydz \tag{14.1.19}$$

Now using Newton's second law

$$\frac{\partial\mathbf{v}}{\partial t} + \mathbf{v}\cdot\nabla\mathbf{v} = -\frac{1}{\rho}\nabla p + \frac{\upsilon}{3}\nabla\nabla\cdot\mathbf{v} + \upsilon\nabla\cdot\nabla\mathbf{v} \tag{14.1.20}$$

where υ, the kinematic viscosity, is defined as M/ρ. Eq. (14.1.20) is known as the Navier-Stokes equation. An alternative form results from using the identity

$$\nabla\text{x}\,\nabla\text{x}\mathbf{v} = \nabla\nabla\cdot\mathbf{v} - \nabla\cdot\nabla\mathbf{v} \tag{14.1.21}$$

yielding

$$\frac{\partial\mathbf{v}}{\partial t} + \mathbf{v}\cdot\nabla\mathbf{v} = -\frac{1}{\rho}\nabla p + \frac{4\upsilon}{3}\nabla\nabla\cdot\mathbf{v} - \upsilon\nabla\text{x}\,\nabla\text{x}\mathbf{v} \tag{14.1.22}$$

14.2 Viscous Attenuation of Sound

We now want to describe the effect of viscosity on wave propagation, the theory of which was first derived by Stokes.

For irrotational flow

$$\nabla x \mathbf{v} = \mathbf{i}\left(\frac{\partial\dot{\zeta}}{\partial \mathbf{y}} - \frac{\partial\dot{\eta}}{\partial z}\right) + \mathbf{j}\left(\frac{\partial\dot{\xi}}{\partial z} - \frac{\partial\dot{\zeta}}{\partial x}\right) + \mathbf{k}\left(\frac{\partial\dot{\eta}}{\partial \mathbf{x}} - \frac{\partial\dot{\xi}}{\partial y}\right) = 0 \qquad (14.2.1)$$

Consider a plane wave travelling in the x direction, then

$$\dot{\eta} = \dot{\zeta} = 0 \qquad (14.2.2)$$

and if the wave is irrotational, it follows from Eq. (14.2.1) that

$$\frac{\partial\dot{\xi}}{\partial z} = \frac{\partial\dot{\xi}}{\partial y} = 0 \qquad (14.2.3)$$

Substituting Eqs. (14.2.1) , (14.2.2), and (14.2.3) into the Navier-Stokes equation (14.1.22), and making the usual approximation that the convective part of the acceleration is negligible, we obtain

$$\frac{\partial\dot{\xi}}{\partial t} = -\frac{1}{\rho}\frac{\partial p}{\partial x} + \frac{4}{3}\upsilon\frac{\partial^2\dot{\xi}}{\partial x^2} \qquad (14.2.4)$$

The equation of continuity of mass gives

$$\frac{\partial\rho}{\partial t} + \rho_o \nabla\cdot\mathbf{v} = 0 \qquad (14.2.5)$$

since ρ varies very little. We may write a thermodynamic equation of state as

$$p_e = A\rho_e \qquad (14.2.6)$$

since the excess pressure is proportional to the excess density. Therefore, from Eqs. (14.2.5) and (14.2.6)

$$\frac{\partial\rho_e}{\partial t} = \frac{1}{A}\frac{\partial p_e}{\partial t} = -\rho_o \nabla\cdot\mathbf{v} \qquad (14.2.7)$$

Thus

$$p_e = -A\rho_o \int \nabla\cdot\mathbf{v}dt = -A\rho_o \int \frac{\partial\dot{\xi}}{\partial x}dt \qquad (14.2.8)$$

and consequently

$$\frac{1}{\rho_o}\frac{\partial p_e}{\partial x} = -A\frac{\partial^2\xi}{\partial x^2} \qquad (14.2.9)$$

Substituting this result in the one-dimensional Navier-Stokes equation (14.2.4):

$$\frac{\partial^2\xi}{\partial t^2} = A\frac{\partial^2\xi}{\partial x^2} + \frac{4}{3}\upsilon\frac{\partial^2\dot{\xi}}{\partial x^2} \qquad (14.2.10)$$

Now if $\mathbf{v}$ is zero, we obtain the one-dimensional wave equation for an inviscid fluid, with $c^2 = A$. Thus

$$\frac{\partial^2 \xi}{\partial t^2} = c^2 \frac{\partial^2 \xi}{\partial x^2} + \frac{4}{3} \upsilon \frac{\partial^2 \dot{\xi}}{\partial x^2} \tag{14.2.11}$$

If we now assume sinusoidal time dependence; that is

$$\xi = \xi(x) e^{j\omega t} \tag{14.2.12}$$

thus

$$\frac{\partial^2 \xi}{\partial t^2} = \left[c^2 + j\omega \frac{4}{3} \upsilon \right] \frac{\partial^2 \xi}{\partial x^2} \tag{14.2.13}$$

which is a wave equation with complex propagation velocity, say c', then

$$c'^2 = c^2 + j\omega \frac{4}{3} \upsilon = c^2 \left[1 + j\omega \frac{4\upsilon}{3c^2} \right] \tag{14.2.14}$$

and

$$c' = c \left[1 + j\omega \frac{4\upsilon}{3c^2} \right]^{\frac{1}{2}} \tag{14.2.15}$$

If the damping is small

$$c' = c \left[1 + j \frac{2\omega\upsilon}{3c^2} \right] \tag{14.2.16}$$

Now a solution of the equation

$$\frac{\partial^2 \xi}{\partial t^2} = c'^2 \frac{\partial^2 \xi}{\partial x^2} \tag{14.2.17}$$

is

$$\xi = \xi_o e^{j(\omega t - k'x)} \tag{14.2.18}$$

where

$$k' = \frac{\omega}{c'} = \frac{\omega}{c\left[1 + j \frac{2\omega\upsilon}{3c^2} \right]} = \frac{\omega}{c} \left[1 - j \frac{2\omega\upsilon}{3c^2} \right]$$

$$= k - j \frac{2\omega^2 \upsilon}{3c^3} = k - j\alpha \tag{14.2.19}$$

Thus

$$\xi = \xi_o e^{-\alpha x} e^{j(\omega t - kx)} \tag{14.2.20}$$

where α is called the coefficient of attenuation due to viscosity. Hence

$$\alpha = \frac{2\omega^2 \upsilon}{3c^3} = \frac{8\pi^2 f^2 M}{3\rho c^3} \tag{14.2.21}$$

Note the dependence of α on f^2. Note also that this value will be correct only if $2\omega\upsilon/3c^2 \ll 1$, (i.e., at low frequencies).

14.3 Heat Conduction as a Source of Attenuation in Fluids

Compression of a fluid raises its temperature and not all of this heat is returned as acoustic energy, some of it being lost through thermal conduction. It is this source of attenuation that we now want to consider. The theory we are about to discuss derives from the work of Kirchoff (1868), and it involves a reconsideration of the basic theory of wave propagation in a fluid, as was given in Chapter 3, to allow for heat conduction. The equations of motion and continuity are unaffected, but the equation of state has to be modified. To see how this is done, we must take an excursion into thermodynamics. From the first law

$$dQ = dU + pdV \tag{14.3.1}$$

Thus, the specific heat at constant volume

$$C_v = \left(\frac{\partial Q}{\partial T}\right)_v = \left(\frac{\partial U}{\partial T}\right)_v = \left(\frac{\partial U}{\partial p}\right)_v \Big/ \left(\frac{\partial U}{\partial P}\right)_v \tag{14.3.2}$$

But

$$dp = \left(\frac{\partial p}{\partial T}\right)_v dT + \left(\frac{\partial p}{\partial V}\right)_T dV \tag{14.3.3}$$

Thus, at constant pressure

$$\left(\frac{\partial p}{\partial T}\right)_v + \left(\frac{\partial p}{\partial V}\right)_T \left(\frac{\partial V}{\partial T}\right)_p = 0 \tag{14.3.4}$$

and therefore

$$\left(\frac{\partial p}{\partial T}\right)_v = -V\left(\frac{\partial p}{\partial V}\right)_T \frac{1}{V}\left(\frac{\partial V}{\partial T}\right)_p = B_T \beta \tag{14.3.5}$$

where B_T = the isothermal bulk modulus, and β = the coefficient of thermal expansion.

Hence, from Eqs. (14.3.2) and (14.3.5)

$$\left(\frac{\partial U}{\partial p}\right)_p = C_v\left(\frac{\partial T}{\partial p}\right)_v = \frac{C_v}{B_T\beta} \tag{14.3.6}$$

Now it follows from Eq. (14.3.1) that

$$\left(\frac{\partial U}{\partial V}\right)_p = \left(\frac{\partial U}{\partial T}\right)_p\left(\frac{\partial T}{\partial V}\right)_p = \left[C_p - p\left(\frac{\partial V}{\partial T}\right)_p\right]\left(\frac{\partial T}{\partial V}\right)_p$$

$$= \frac{C_p}{\left(\frac{\partial V}{\partial T}\right)_p} - p = \frac{C_p}{V\beta} - p \tag{14.3.7}$$

Hence, from Eqs. (14.3.1), (14.3.6), and (14.3.7)

$$dQ = \left(\frac{\partial U}{\partial p}\right)_v dp + \left(\frac{\partial U}{\partial V}\right)_p dV + pdV$$

$$= \frac{C_v}{B_T\beta}dp + \frac{C_p}{V\beta}dV \tag{14.3.8}$$

Consequently, for time variation we have

$$\frac{dQ}{dt} = \frac{1}{\beta}\left(\frac{C_v}{B_T}\frac{dp}{dt} + \frac{C_p}{V}\frac{dV}{dt}\right) \tag{14.3.9}$$

or

$$\dot{Q} = \frac{1}{\beta}\left(\frac{C_v}{B_T}\dot{p} + \frac{C_p}{V}\dot{V}\right) \tag{14.3.10}$$

If $\dot{Q} = 0$ (i.e., if the process is adiabatic)

$$\dot{p} = B_T\frac{C_p}{C_v}\frac{\dot{V}}{V} \tag{14.3.11}$$

or

$$dp = -B_T\frac{C_p}{C_v}\cdot\frac{dV}{V} \tag{14.3.12}$$

or

$$B_s = \gamma B_T \tag{14.3.13}$$

where γ is the ratio of specific heats, and B_s is the adiabatic bulk modulus. Now, in general,

$$\rho_e = \rho - \rho_o = \frac{M}{V} = \frac{M}{V_o} \cong -M\frac{dV}{V_o^2} = -\rho_o\frac{dV}{V} \tag{14.3.14}$$

where M is the fluid mass and thus

$$dp = p_e = -B\frac{dV}{V} = \frac{B}{\rho_o}\rho_e \tag{14.3.15}$$

in other words, the excess pressure is proportional to the excess density, which was our assumption in Section 14.2 and in Chapter 3.

We can now see that if the process is neither adiabatic nor isothermal, this simple assumption of proportionality is wrong, since from Eqs. (14.3.8) and (14.3.14)

$$\begin{aligned} p_e = dp &= \frac{B_T\beta}{C_v}dQ + \frac{C_pB_T}{C_v\rho_0}\rho_e \\ &= \frac{B_s}{\rho_o}\rho_e + \frac{B_T\beta}{C_v}dQ \end{aligned} \tag{14.3.16}$$

Now for a plane wave propagating in the x direction, the temperature gradients will be in the x direction. If the heat flows are Q and Q + dQ across the z-y planes at x and x + dx, in time interval dt, then the net heat flow out of the enclosed slab

$$dQ = \kappa A\frac{\partial^2 T}{\partial x^2}\cdot dxdt \tag{14.3.17}$$

where

κ = the thermal conductivity
A = the slab area
T = the temperature

But Adx = V = the volume of material in the slab. If we consider one mole of material (M), then its volume will be

$$V = \frac{M}{\rho_o} \tag{14.3.18}$$

Thus

$$\frac{dQ}{dt} = \kappa\frac{M}{\rho_o}\frac{\partial^2 T}{\partial x^2} \tag{14.3.19}$$

Hence, from Eqs. (14.3.16) and (14.3.19)

$$\frac{dp}{dt} = \frac{B_s}{\rho_o}\frac{d\rho_e}{dt} + \frac{B_T\beta}{C_v}\cdot\frac{\kappa M}{\rho_o}\frac{\partial^2 T}{\partial x^2} \tag{14.3.20}$$

It is now advisable to eliminate T, so as to obtain an equation involving only p and ρ. Since

$$dT = \left(\frac{\partial T}{\partial p}\right)_V dp + \left(\frac{\partial T}{\partial V}\right)_p dV \tag{14.3.21}$$

it follows from Eq. (14.3.5) that

$$dT = \frac{dp}{B_T\beta} + \frac{dV}{V\beta} = \frac{dp}{B_T\beta} - \frac{d\rho}{\rho_0\beta} \tag{14.3.22}$$

Hence

$$\frac{\partial T}{\partial x} = \frac{1}{B_T\beta}\frac{\partial p}{\partial x} - \frac{1}{\rho_0\beta}\frac{\partial \rho}{\partial x} \tag{14.3.23}$$

and

$$\frac{\partial^2 T}{\partial x^2} = \frac{1}{B_T\beta}\frac{\partial^2 p}{\partial x^2} - \frac{1}{\rho_0\beta}\frac{\partial^2 \rho}{\partial x^2} \tag{14.3.24}$$

Thus, rewriting Eq. (14.3.20) using Eq. (14.3.24), we have

$$\dot{p}_e - \frac{B_s}{\rho_o}\dot{\rho}_e = \frac{M\kappa}{\rho_o C_v}\left(\frac{\partial^2 p_e}{\partial x^2} - \frac{B_T}{\rho_o}\frac{\partial^2 \rho_e}{\partial x^2}\right) \tag{14.3.25}$$

We will now make the assumption that κ is small, so that $p_e = c^2\rho_e$, then

$$\dot{p}_e \cong \frac{B_s}{\rho_o}\dot{\rho}_e = c^2\dot{\rho}_e \tag{14.3.26}$$

and using Eqs. (14.3.13) and (14.3.26), Eq. (14.3.25) becomes

$$\begin{aligned}\dot{p}_e - c^2\dot{\rho}_e &\cong \frac{M\kappa}{\rho_o C_v}\left(c^2 - \frac{c^2}{\gamma}\right)\frac{\partial^2 \rho_e}{\partial x^2} \\ &\cong \frac{M\kappa}{\rho_o C_p}(\gamma - 1)\frac{\partial^2 \rho_e}{\partial t^2}\end{aligned} \tag{14.3.27}$$

But the equations of motion and continuity will be

$$\rho_o\frac{\partial^2 \xi}{\partial t^2} = -\frac{\partial p_e}{\partial x} \tag{14.3.28}$$

and

$$\rho_o\frac{\partial \dot{\xi}}{\partial x} = -\frac{\partial \rho_e}{\partial t} \tag{14.3.29}$$

Hence, Eqs. (14.3.27) and (14.3.29) yield

$$\dot{p}_e = -c^2\rho_o\frac{\partial\xi}{\partial x} - \frac{M\kappa}{\rho_o C_p}(\gamma - 1)\rho_o\frac{\partial\ddot{\xi}}{\partial x} \tag{14.3.30}$$

and

$$-\frac{1}{\rho_o}\frac{\partial\dot{p}}{\partial x} = c^2\frac{\partial^2\dot{\xi}}{\partial x^2} + \frac{M\kappa(\gamma - 1)}{\rho_o C_p}\frac{\partial^2\ddot{\xi}}{\partial x^2} \tag{14.3.31}$$

or using Eq. (14.3.28)

$$\frac{\partial^2\dot{\xi}}{\partial t^2} = c^2\frac{\partial^2\dot{\xi}}{\partial x^2} + \frac{M\kappa(\gamma - 1)}{\rho_o C_p}\frac{\partial^2\ddot{\xi}}{\partial x^2} \tag{14.3.32}$$

which may be compared with the wave equation involving viscosity Eq. (14.2.11). Consequently, we may use the analysis of Section 11.2 to see immediately that the attenuation coefficient will be

$$\alpha_{th} = \frac{M\kappa(\gamma - 1)\omega^2}{2c^3\rho_o C_p} \tag{14.3.33}$$

Note that here C_p is the specific heat/mole.

If we cannot make the assumption of small thermal conductivity, the analysis is considerably more complicated and will not be discussed in detail here. The resulting attenuation coefficient is found to be

$$\alpha = \frac{\rho_o\omega^3\tau(\gamma - 1)}{2k\kappa_s[1 + \omega^2\tau^2]} = \frac{\rho_o V_p(\gamma - 1)\omega^2\tau}{2B_s(1 + \omega^2\tau^2)} \tag{14.3.34}$$

where

$$\frac{M\kappa k^2}{C_p\rho_o\omega^2} = \tau \tag{14.3.35}$$

V_p = the phase velocity = ω/k, and k = the propagation constant. Thus, using Eq. (14.3.26)

$$\alpha = \frac{V_p}{2c^2}(\gamma - 1)\frac{\omega^2\tau}{1 + \omega^2\tau^2} \tag{14.3.36}$$

At low frequencies $\omega\tau \ll 1$ and $V_p \approx c$. Thus

$$\alpha = \frac{(\gamma - 1)}{2c}\omega^2\frac{M\kappa}{C_p\rho_o c^2} = (\gamma - 1)\frac{M\kappa}{2\rho_o C_p c^3}\omega^2 \tag{14.3.37}$$

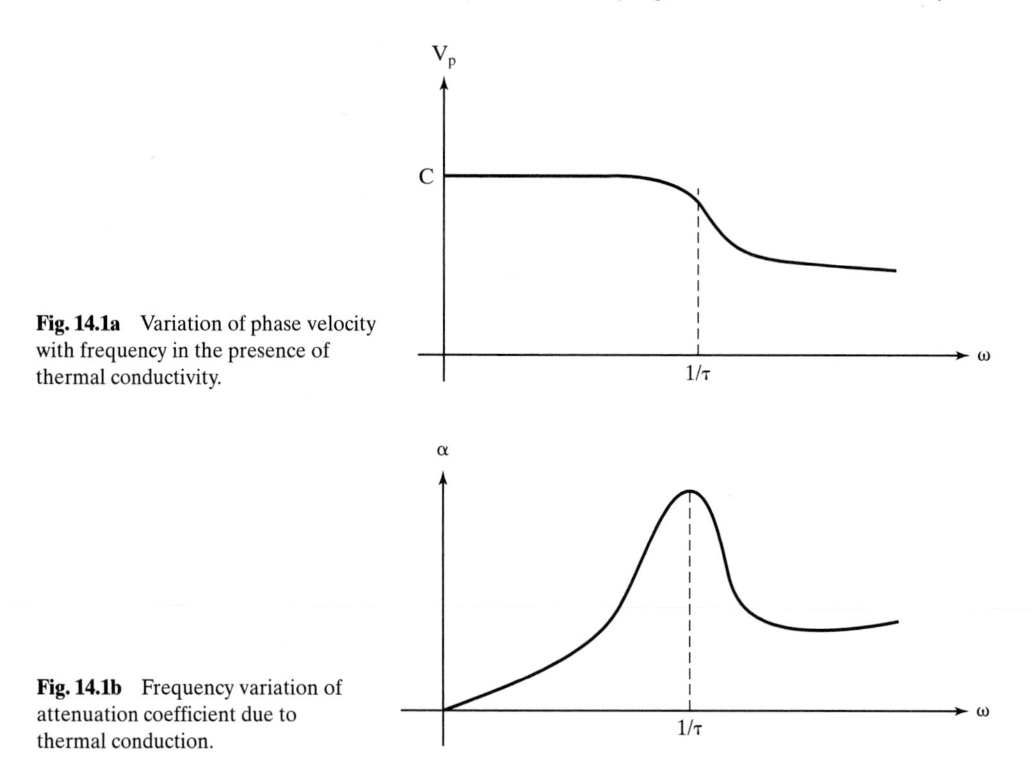

Fig. 14.1a Variation of phase velocity with frequency in the presence of thermal conductivity.

Fig. 14.1b Frequency variation of attenuation coefficient due to thermal conduction.

which agrees with the result for low conductivity in Eq. (14.3.33). At high frequencies, it is possible to show that the phase velocity varies with frequency according to

$$V_p = c\sqrt{\frac{1 + \omega^2\tau^2}{1 + \gamma\omega^2\tau^2}} \tag{14.3.38}$$

Such a variation is shown in Fig. 14.1a. It is also possible to show that at high frequencies

$$\alpha = \text{constant} \times v_p^5 \frac{\omega^2\tau^2}{1 + \omega^2\tau^2} \tag{14.3.39}$$

as shown in Fig. 14.1b. These variations are typical of a relaxation phenomenon. At low frequencies, there is sufficient time for heat conduction to occur, and the propagation will be isothermal. At very high frequencies, the time available in a cycle for the thermal diffusion from the high to the low temperature zones will be very short, and the propagation is adiabatic. In an intermediate frequency range, the temperature gradient will increase due to the smaller wavelength, and there will be a maximum energy loss.

14.4 Sound Absorption by the Atmosphere

To recapitulate: When an acoustic wave propagates in a medium, there is a dissipation of acoustic energy. This loss of energy from a sound beam is called attenuation, which is divided into two parts: absorption (or conversion of acoustic into thermal energy) and deflection (or scattering of acoustic energy out of the beam). Losses in the medium are divided into three basic types: viscous losses, heat conduction losses, and (as we shall see) losses due to molecular exchange of energy. These mechanisms cause the acoustic energy to degrade slowly into heat. Sound waves propagating within a homogeneous medium are attenuated only by absorption. The presence of any discontinuity or inhomogenity in the medium (e.g., a suspended particle, thermal micro cells of different temperature, or a region of different classic properties) will scatter acoustic energy out of the beam. We alluded to scattering in Chapter 11. Acoustic energy is also lost from a sound beam in an inhomogeneous medium because at the boundaries of suspended particles there are additional viscous and heat conduction losses.

Based on the theory discussed in Sections 11.2 and 11.3, we would expect the attenuation coefficient in a fluid to be given by

$$\alpha = \frac{2M\omega^2}{3\rho_o c^3} + \frac{(\gamma - 1)\kappa\omega^2}{2\rho_o C_p c^3} \tag{14.4.1}$$

The first experimental investigation of atmospheric absorption was made in Hyde Park in London by Tyndall (1874) using guns and organ pipes. This work was followed by Duff (1898) on the St. John River in New Brunswick, Canada. Duff used whistles for sources, which were sounded on one side of the river with observers on the other side. It was found that the classical absorption mechanisms given by Eq. (14.4.1) were not sufficient to explain the value actually measured. Duff surmised that there must be a third factor in the attenuation. This third factor was postulated by Herzfeld and Rice (1928) to be the slow rate of exchange of energy between the translational movement and the internal degrees of freedom of the air molecules. Knudsen (1931) investigated the effect of humidity and temperature on the absorption of air by the reverberation method, which used two model chambers of different volumes to eliminate the effect of surface absorption. Kneser (1933) believed that Knudsen's investigation could not be explained on the basis of the classical theory of Stokes, Rayleigh, and Kirchhoff. He proposed that the mechanism of atmospheric absorption was vibrational relaxation of oxygen. This mechanism was supposedly dependent on the rate of adjustment of the internal equilibrium between vibrating and nonvibrating oxygen within the molecules. The kinetic energy of the molecules is converted into other forms of energy, such as potential energy, internal rotational and vibrational energy, and energy of association and dissociation of ionized solutions. Kneser accounted for the phenomenon of the effect of humidity by assuming the relaxation time to be a function of the water content. Sivian (1947) carried out his own measurements of the absorption of ultrasonic waves in gases inside cylindrical tubes. The average values of atmospheric conditions were a pressure of 76 cm of mercury, temperature of 26.5°C, and relative humidity of 37%. His results were of the order of 50% greater than the values given by the classical theory. Other researchers measured the intensity attenuation coefficient of air of various humidities in reverberation chambers.

Knowledge of the absorption mechanism of sound in the air has improved considerably since the late 1960s as a result of a number of investigations. Bass et al. (1995) summarize this work. Modern values of the absorption coefficient are shown in Fig. 14.2.

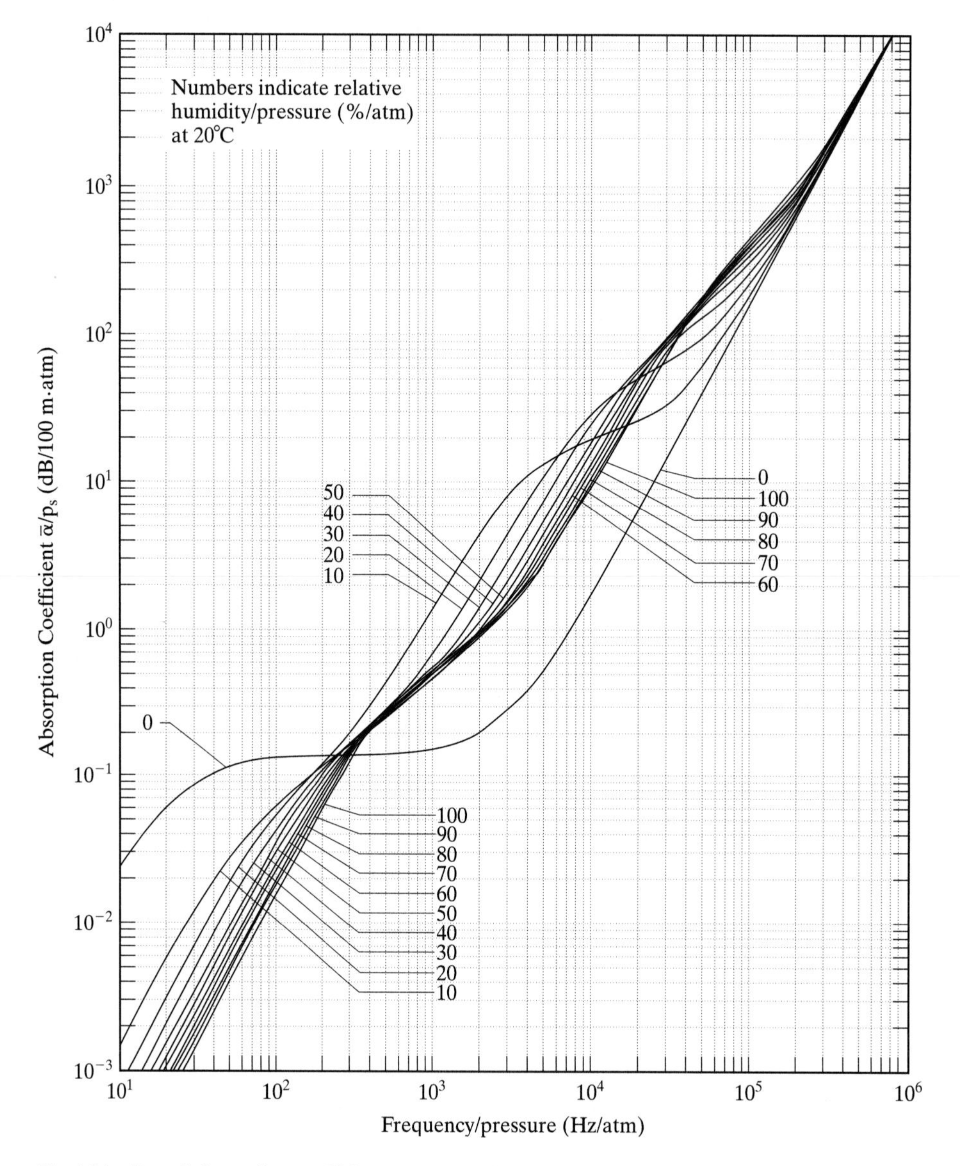

Fig. 14.2 Sound absorption coefficient per atmosphere, SI units, for air at 20°C. The abscissa is frequency/pressure, and the parameter is relative humidity/pressure in the range 0 to 100%/atm. (From Bass et al (1995).)

14.5 Attenuation in Water

The theory of attenuation in water is the same as in the case of air for the viscous and thermal conduction mechanisms. For fresh water, the theoretical attenuation coefficient is $8.1 \times 10^{-15} f^2$ due to viscosity and $3 \times 10^{-18} f^2$ due to thermal conduction. The first measurements of attenuation in seawater were made in the early 1930s by transmitting pulses from a surface ship to a submarine. These were followed by measurements in various parts of the ocean, and during World War II, extensive additional data were collected. The measured value in fresh water was 24×10^{-15} f^2, about three times the expected value. Liebermann's (1949) explanation for this "excess" absorption is the effect of volume viscosity. When a correction is made to the theory to take into account volume viscosity, the absorption coefficient becomes

$$\alpha = \frac{16\pi^2}{3\rho c^3}\left(M + \frac{3\beta}{4}\right) f^2 \tag{14.5.1}$$

The ratio M/β is 2.81 for water. Laboratory studies were made by Wilson and Leonard (1954) using the resonator technique described in Section 9.9. They showed that the excess absorption below 100 kHz is due to ionic relaxation of magnesium sulfate, even though this salt comprises only 4.7% by weight of the total dissolved salts in seawater. At frequencies below 5 kHz, still more excess absorption was found, which was shown to be due to a chemical relaxation of boric acid. An equation for the absorption coefficient in seawater was first given by Fisher and Simmons (1977), and a comprehensive review of the subject was given by Francois and Garrison (1982). The equation for the absorption coefficient at frequency f in seawater is usually written as follows:

$$\alpha = \frac{A_1P_1f_1f^2}{f_1^2 + f^2} + \frac{A_2P_2f_2f^2}{f_2^2 + f^2} + A_3P_3f^2 \tag{14.5.2}$$

The first two terms are due to the relaxation effects of boric acid and magnesium sulfate where the relaxation frequencies are f_1 and f_2. The third term represents the absorption of pure water. A_1, A_2, and A_3 were originally intended as constants, but it now seems that some slight variations occur. P_1, P_2, and P_3 are included to allow for dependence on pressure. The variations of α with frequency are shown in Fig. 14.3. Near the surface of the ocean, there is additional attenuation due to scattering by bubbles and marine life. Clay and Medwin (1977) provide a more detailed account of this topic.

14.6 Absorption of Sound in Fluid-Filled Pipes

When an acoustic wave travels in a pipe, an exchange of energy occurs between the medium and the wall of the pipe. This exchange is caused by the viscous resistance and the heat conduction between the medium and the wall of the pipe. Kirchhoff (1868) presented the following formulae for the absorption coefficient and phase speed of the acoustic wave propagating in a circular pipe:

$$\alpha_w = \frac{1}{ac_m}\sqrt{\frac{\eta_e\omega}{2}} \tag{14.6.1}$$

$$c = c_m\left[1 - \frac{1}{a}\sqrt{\frac{\eta_e}{2\omega}}\right] \tag{14.6.2}$$

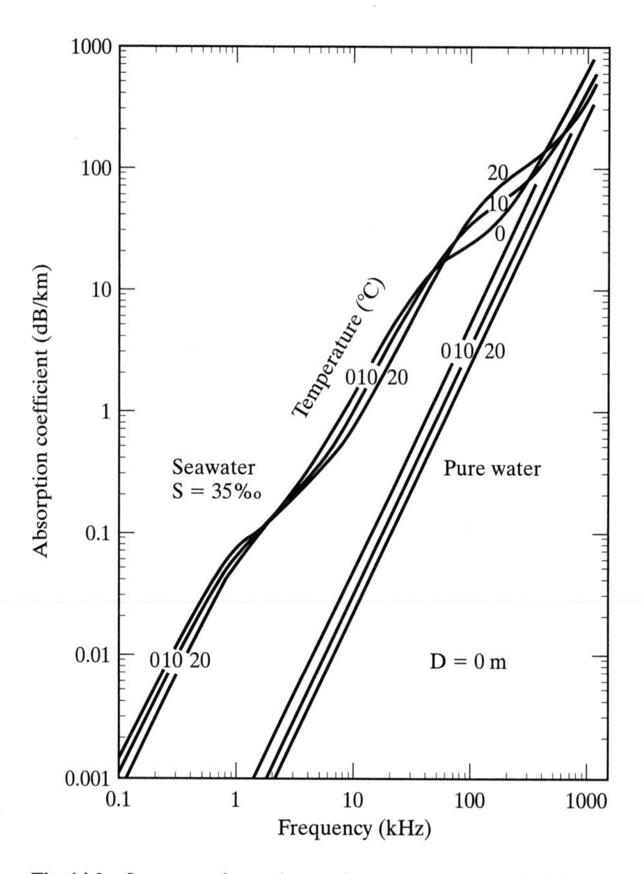

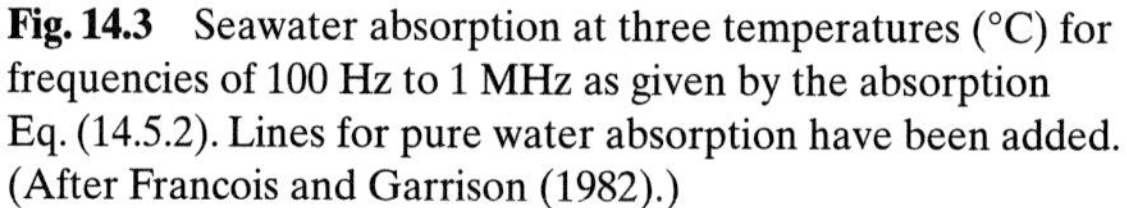

Fig. 14.3 Seawater absorption at three temperatures (°C) for frequencies of 100 Hz to 1 MHz as given by the absorption Eq. (14.5.2). Lines for pure water absorption have been added. (After Francois and Garrison (1982).)

where

$$\eta_e = \frac{M}{\rho}\left[1 + (\gamma - 1)\sqrt{\frac{\kappa}{C_p M}}\right]^2$$

α_w = absorption coefficient between the medium and the wall of the pipe
c = phase velocity of sound
a = radius of the pipe
c_m = sound velocity in the medium
ρ = density of the medium
ω = angular frequency
M = coefficient of dynamic shear viscosity
γ = ratio of specific heats
κ = thermal conductivity of the medium
C_p = specific heat at constant pressure

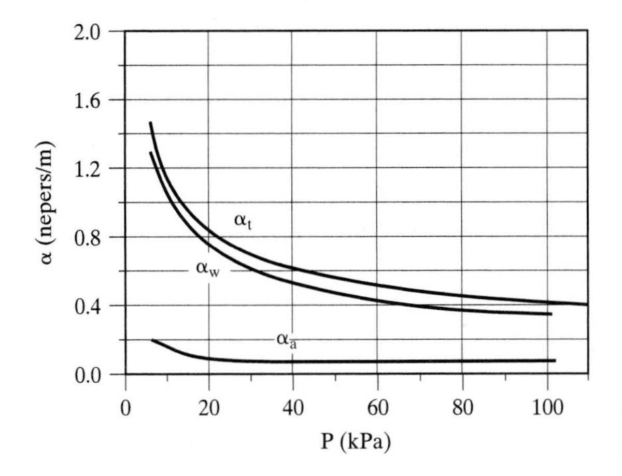

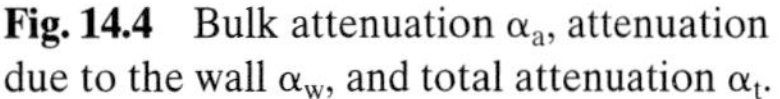
Fig. 14.4 Bulk attenuation α_a, attenuation due to the wall α_w, and total attenuation α_t.

These formulae can be applied to a homogeneous medium in a circular tube with a smooth surface. The temperature of the medium at the wall of the tube is assumed to be constant, and there is little variation of temperature gradient at the wall. For example, a tube of radius 1.27×10^{-2} m filled with air at atmospheric pressure and 20°C can be analyzed as follows: We have $M = 1.875 \times 10^{-5}$ Pa · s, $\rho = 1.21$ kg/m^3, $\gamma = 1.403$, and $\kappa/(C_P\mu) = 1.242$, and the effective coefficient of shear viscosity η_e is 3.2551×10^{-5} m^2/sec. The attenuation coefficient is 0.3364 nepers/m. If the pressure of the air is reduced to 6.9 kPa but the temperature is kept constant, the density of the air changes to 0.0823 kg/m^3. The dynamic viscosity and the thermal conductivity of the air, which are functions of temperature, remain the same. The attenuation coefficient will then be 1.2897 nepers/m. The calculations of the attenuation coefficient due to the pipe wall, using Eq. (14.6.1) and the numerical values just given, were made in the range of air pressure of 6.9 kPa and 101.4 kPa. The attenuation coefficient in the fluid α_w is plotted against air pressure in Fig. 14.4. The attenuation coefficient in the fluid body α_a in the same range of air pressure is thus much smaller than the attenuation coefficient between the medium and the wall α_w. The variations of α_a in the range of 10% to 100% humidity are as small as in open air in the same range of air pressure. Therefore, the total attenuation coefficient α_t in the closed tube is the sum of α_w and the bulk attenuation α_a, as shown in Fig. 14.4.

14.7 Sound Absorption in Rooms

Sound absorption in a room is a crucial topic in architectural acoustics. The finding that absorption in a tube is much greater than in unrestricted air leads to the realization that the cause of absorption in a room is the impinging of sound on carpeting, furnishings, and finishes containing trapped air. As a practical matter, it is the movement of the air in the interstices of these materials that determines the absorption in the room. This phenomenon was first realized by Wallace Clement Sabine (1898), and the following is a derivation of Sabine's famous reverberation formula.

Suppose there is a sound source with a constant power output in the room. When the source is first activated, the energy in the room will accumulate, rapidly at first and then more slowly, finally reaching a steady-state condition. How long does it take to reach this condition? What is the intensity in the room when it does? How long would it take for the sound to completely die away if the source were to be stopped?

Assume that in the steady state, the sound energy density is uniform throughout the room and the energy travels equally in all directions (disregard standing wave effects). As we just saw, the dissipation is assumed to be confined to the bounding surfaces.

Fig. 14.5 shows an element of area ds on a wall of the room. Let the energy density be ε per unit volume. Hence, the energy in volume dV is ε dV, and the portion of it that will reach dS will be the fraction travelling in the solid angle $d\omega = dS \cos\theta / r^2$. Thus, the quantity of energy that will reach dS from dV is

$$\varepsilon\, dV \frac{d\omega}{4\pi} = \varepsilon\, dV\, ds \frac{\cos\theta}{4\pi r^2} \tag{14.7.1}$$

All the volume elements in the torus of cross section hatched in Fig. 14.5 will contribute similarly to the energy arriving at ds. The volume of this torus is $2\pi r^2 \sin\theta\, dr\, d\theta$; the total energy received in time dt comes from a hemisphere of radius c and is therefore given by

$$\begin{aligned} &\int\int \varepsilon 2\pi r^2 \sin\theta\, dr d\theta ds \frac{\cos\theta}{4\pi r^2} \\ &= \frac{\varepsilon\, ds}{2} \int_0^{cdt} dr \int_0^{\pi/2} \sin\theta \cos\theta\, d\theta \\ &= \frac{\varepsilon c}{4} ds\, dt \end{aligned} \tag{14.7.2}$$

Thus, the total energy falling on unit area in unit time is $\varepsilon c/4$.

We can now discuss the decay of sound. Sabine's approach to this phenomenon was to suppose the walls to be perfectly reflecting except for an open window of area A.

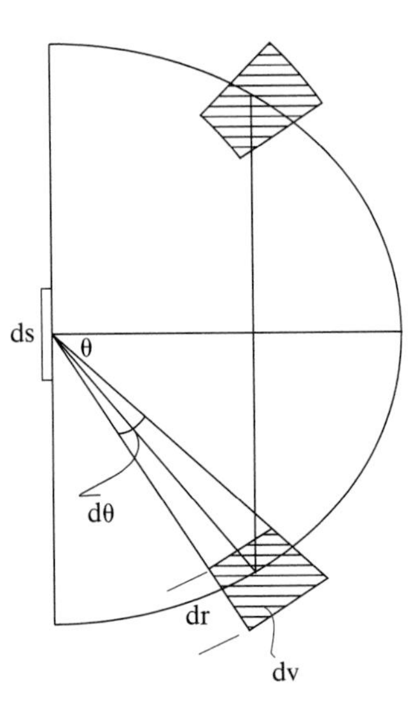

Fig. 14.5 Energy incident on a wall element.

The rate of energy loss from the room will be $\varepsilon cA/4$, assuming all the energy falling on the window passes through it. Thus, if the source is cut off at $t = 0$, then after time t the energy remaining will be $V\varepsilon$, where V is the room volume and ε is presumed to be a function of time. Hence

$$-\frac{d(V\varepsilon)}{dt} = \frac{\varepsilon cA}{4} \tag{14.7.3}$$

and consequently

$$\varepsilon(t) = \varepsilon_o e^{-\frac{cA}{4V}t} \tag{14.7.4}$$

The growth of sound in a room containing the source will be given by

$$W - \frac{\varepsilon cA}{4} = \frac{d(V\varepsilon)}{dt} \tag{14.7.5}$$

Let

$$\frac{cA}{4V} = h \tag{14.7.6}$$

then

$$hdt = \frac{hVd\varepsilon}{W - h\varepsilon V} \tag{14.7.7}$$

Integrating Eq. (14.7.7), with the initial condition $\varepsilon(0) = 0$, gives

$$e^{-ht} = \frac{W - h\varepsilon V}{W} \tag{14.7.8}$$

Rearranging gives

$$\varepsilon = \frac{W}{hV}\left(1 - e^{-ht}\right) = \frac{4W}{cA}\left(1 - e^{-\frac{cA}{4V}t}\right) \tag{14.7.9}$$

The reverberation time T is defined as the time for the sound energy density to decay by 60 dB, or to 10^{-6} of its original value. It follows from Eq. (14.7.4) that

$$T = \frac{13.8 \times 4V}{cA} = \frac{0.161V}{A} \tag{14.7.10}$$

where V is in m^3 and A is in m^2. The absorption in a room is rarely all concentrated in a single open window, but is usually spread over several surfaces whose areas $S_1, S_2, \ldots, S_n$ have absorption coefficients $\alpha_1, \alpha_2, \ldots, \alpha_n$ defined as the ratio of energy absorbed by the surface to the energy incident on it. In that case, we set

$$A = S\alpha = S_1\alpha_1 + S_2\alpha_2 + \cdots + S_n\alpha_n \tag{14.7.11}$$

where now S is that total surface area, α is the average absorption coefficient, and A is termed the absorption of the room in sabins.

In 1898, Sabine first obtained the result in Eq. (14.7.10) experimentally using organ pipes and a stopwatch. He plotted T versus A, and is reported to have shouted from his study: "Mother, it's a hyperbola!" This famous moment in the history of architectural acoustics marked the beginning of the scientific design of spaces for acoustics. The absorption coefficients of building materials and furnishings have been measured and tabulated (see Appendix I), thus permitting the reverberation time to be estimated at various frequencies. The reverberation time is the major factor in determining the acoustics of a room, and reverberation times suitable for various activities are also shown in Appendix I. Kuttruff (1979) provides further details on the science of room acoustics, and advice on design and construction is given in Knudsen and Harris (1950).

14.8 Damping in Solids

Damping, or "internal friction," is important in determining the vibrational response of structural members and for what it can reveal about internal structural conditions during vibration monitoring. The macroscopic behavior of gases can be predicted from the simple assumptions of molecular behavior made in the kinetic theory. But the behavior of solids is far more complex, involving yield, flow, and failure. Unlike the case for a gas, the past history of a solid influences its present state, as in slip, aging, work hardening, plastic deformations, etc. Because of this, the most successful approach to the study of the deformation and flow of solid matter (rheology) has been the phenomenological stance of solid mechanics. Here, we will first measure the properties of a unit volume of the material and next formulate rheological, or "constitutive," equations that can be used to predict the response of members (i.e., items made entirely of the given material, excluding interfaces, joints, or connectors such as struts, springs, torsion bars, shafts, beams, and turbine blades). The theories of elasticity and viscoelasticity are examples of this phenomenological approach. Lazan (1968) provides further details on this topic.

Suppose a force F is applied to unit volume of a material, thus producing a displacement x. For a linear material, the behavior could in general be described by the equation

$$\left[a_o + a_1\frac{\partial}{\partial t} + a_2\frac{\partial^2}{\partial t^2} + \cdots\right]F = \left[b_o + b_1\frac{\partial}{\partial t} + b_2\frac{\partial^2}{\partial t^2} + \cdots\right]x \qquad (14.8.1)$$

The various simplifications of this equation can be represented by different spring and dashpot combinations, as shown in Fig. 14.6. More complex combinations have been postulated to model creep and other phenomena. For linear elasticity, modeled by a spring, only a_o and b_o are nonzero in Eq. (14.8.1) and

$$k = \frac{F}{\eta} = \frac{b_o}{a_o} \qquad (14.8.2)$$

Similarly, for viscoelasticity

$$M = \frac{F}{\dot{\eta}} = \frac{b_1}{a_o} \qquad (14.8.3)$$

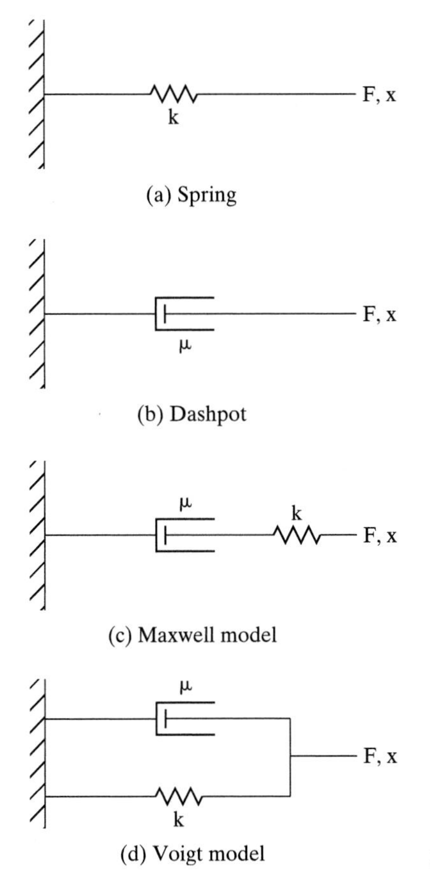

Fig. 14.6 Simple models of unit volume of a material.

In the so-called Maxwell model, the constitutive equation is

$$a_o F + a_1 \frac{dF}{dt} = b_o x + b_1 \frac{dx}{dt} \tag{14.8.4}$$

or

$$\frac{dF}{dt} + \frac{k}{M} F = k \frac{dx}{dt} \tag{14.8.5}$$

where $a_o = 1$, $a_1 = M/k$, $b_o = 0$, and $b_1 = M$. But the Maxwell model does not simulate real materials realistically. The Voigt model overcomes some of these limitations with the constitutive equation

$$\frac{dx}{dt} + \frac{k}{M} \eta = \frac{F}{M} \tag{14.8.6}$$

14.9 Damping Nomenclature for Solids

The load applied to the unit volume or the test member is usually cyclic and may be considered sinusoidal. The work done on the damping force per cycle is

$$W_d = \int F_d \, dx \tag{14.9.1}$$

For a member modeled as a spring with viscous damping, or a Voigt model unit volume, the damping force is $M\dot{x}$, and in the steady state the displacement and velocity are

$$x = x_o \cos(\omega t - \alpha)$$

and

$$\dot{x} = -x_o\omega \sin(\omega t - \alpha) \tag{14.9.2}$$

Hence, the energy dissipated per cycle becomes

$$\begin{aligned} W_d &= M\dot{x}\, dx = M\dot{x}^2\, dt \\ &= M\omega^2 x_o^2 \int_0^{2\pi/\omega} \sin^2(\omega t - \alpha)\, dt = \pi M \omega x_o^2 \end{aligned} \tag{14.9.3}$$

The velocity may be written in the form

$$\begin{aligned} \dot{\eta} &= -\omega x_o \sin(\omega t - \alpha) = \pm\omega x_o \sqrt{1 - \cos^2(\omega \tau - d)} \\ &= \pm\omega\sqrt{x_o^2 - x^2} \end{aligned} \tag{14.9.4}$$

The damping force thus can be written as

$$F_d = M\dot{x} = \pm M\omega\sqrt{x_o^2 - x^2} \tag{14.9.5}$$

We may rearrange Eq. (14.9.5) as follows:

$$\left(\frac{F_d}{\mu\omega\eta_o}\right)^2 + \left(\frac{x}{x_o}\right)^2 = 1 \tag{14.9.6}$$

which is an ellipse, as shown in Fig. 14.7a. The energy dissipated per cycle is the area enclosed by the ellipse. If we add the spring force $k\eta$ to F_d, then the ellipse is rotated to the position given in Fig. 14.7b, which is the hysteresis loop of the linear Voigt model. At high amplitudes in metals, the loop departs from the elliptical shape. The loss coefficient is defined as the ratio of damping energy loss per radian to the peak potential energy U:

$$\eta = W_d/2\pi U \tag{14.9.7}$$

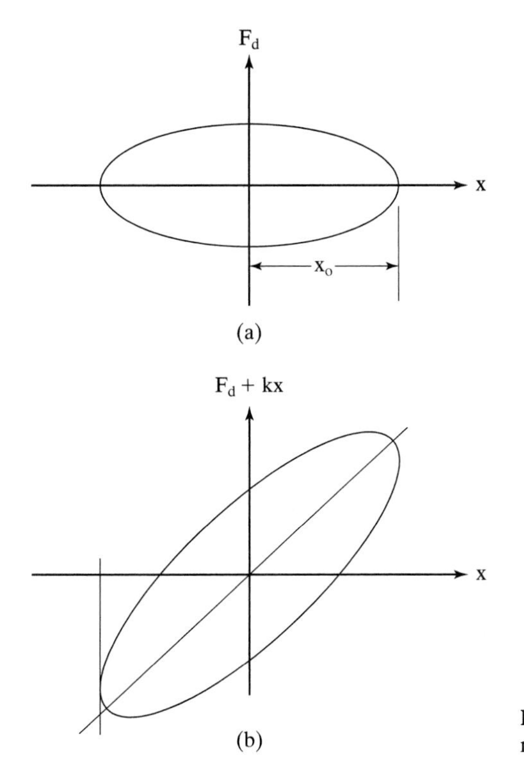

Fig. 14.7 The hysteresis loop of the linear Voigt model.

The quality factor is

$$Q = \frac{1}{\eta} \tag{14.9.8}$$

Experiments have shown that for most structural metals, such as steel or aluminum, the energy dissipated per cycle is independent of frequency over a wide range, and proportional to the square of the vibration amplitude. The loss coefficient is then a constant, independent of the strain rate.

The damping is often specified in terms of a complex elastic modulus, which is related to the properties of a member giving rise to a complex spring constant, k*, thus

$$k^* = k + jk' \tag{14.9.9}$$

where k is the elastic spring constant and k′ is the loss constant. For a bar under axial loading, the stress is

$$\sigma = \frac{F}{A} = E\varepsilon = \frac{Ee}{\ell} = \frac{ek}{A} \tag{14.9.10}$$

Hence

$$E = (\ell/A)\kappa \text{ and } E' = (\ell/A)k' \tag{14.9.11}$$

E is Young's modulus and E′ is the loss modulus.

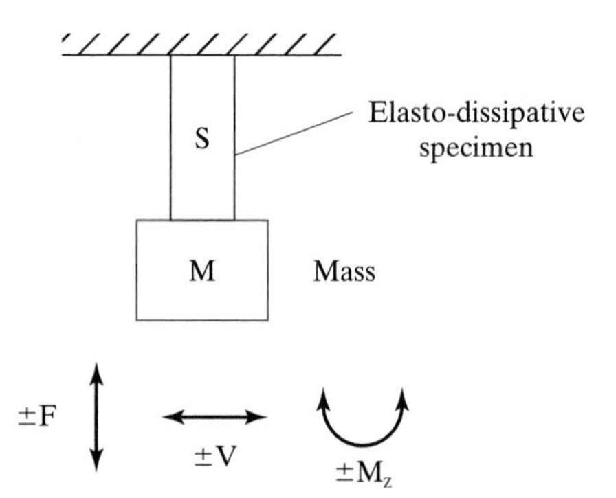

Fig. 14.8 Test setup for measuring damping in axial load, shear, or torsion.

Material properties are usually measured in some macroscopic test setup, as indicated in Fig. 14.8. From the measured macroscopic values of spring constant and damping, the unit elasticity and loss constant can be deduced using the method just illustrated. Such methods are employed both by structural engineers and physicists. Engineers use the measurements to make predictions of structural response to vibration; physicists to elucidate damping micromechanisms. Lazan (1968) and Mason (1958) provide further details on this topic. Similar phenomenology has been found in both metals and polymers. In certain frequency ranges, sharp peaks are found in the values of the loss coefficient. These peaks are due to various relaxation mechanisms, including grain boundary shear, micro- and macrothermoelasticity, eddy current effects, and the movement of dislocations.

REFERENCES AND FURTHER READING

Bass, H.E., L.C. Sutherland, A.J. Zuckerwar, D.T. Blackstock, and D.M. Hester. 1995. "Atmospheric Absorption of Sound: Further Developments." 97 *J. Acoust. Soc. Am.* 680–683.

Clay, C. S., and H. Medwin. 1977. *Acoustical Oceanography: Principles and Applications*. Wiley.

Duff, A.W. 1898. "The Attenuation of Sound and the Constant of Radiation of Air." 6 *Phys. Rev.* 129–139.

Fisher, F.H., and V. P. Simmons. 1977. "Sound Absorption in Seawater." 62 *J. Acoust. Soc. Am.* 558–564.

Francois, R.E., and G.R. Garrison. 1982. "Sound Absorption Based on Ocean Measurements." 72 *J. Acoust. Soc. Am.* 896–907, 1879–1890.

Herzfeld, K.F., and T.A. Litovitz. 1959. *Absorption and Dispersion of Ultrasonic Waves*. Academic Press.

Herzfeld, K.F., and F.O. Rice. 1928. "Dispersion and Absorption of High Frequency Sound Waves." 31 *Phys. Rev.* 691–695.

Kirchoff, G. 1868. 134 *Ann. Physik.* 177–193.

Kneser, H.O. 1933. 16 *Ann. Physik.* 337.

Knudsen, V.O. 1931. "The Effect of Humidity Upon the Absorption of Sound in a Room, and a Determination of the Coefficients of Absorption of Sound in Air." 3 *J. Acoust. Soc. Am.* 126–138.

Knudsen, V.O., and C.M. Harris. 1950. *Acoustical Designing in Architecture*. Wiley.

Kuttruff, H. 1979. *Room Acoustics*, 2nd ed. Applied Science Publishers.

Lazan, B.J. 1968. *Damping of Materials in Structural Mechanics*. Pergamon.

Liebermann, L.N. 1949. "The Second Viscosity of Liquids." 75 *Phys. Rev.* 1415–1422.

Mason, W.P. 1958. *Physical Acoustics and the Properties of Solids*. Van Nostrand.
Sabine, Wallace Clement, 1898. "Reverberation," The American Architect and The Engineering Record.
Sivian, L.J. 1947. "High Frequency Absorption in Air and Other Gases." 19 *J. Acoust. Soc. Am.* 914–916.
Stokes, G.G. 1845. "On the theories of the Internal Friction of Fluids in Motion etc." 8 *Trans. Cambridge Phil. Soc.* 287.
Stokes, G.G. 1880–1905. *Mathematical and Physical Papers*. 5 vols. Cambridge Univ. Press.
Tyndall, J. 1874. 164 *J. Phil. Trans. Roy. Soc.* 183.
Wilson, O.B., and R.W. Leonard. 1954. "Measurements of Sound Absorption in Aqueous Salt Solutions by a Resonator Method." 26 *J. Acoust. Soc. Am.* 223–226.

PROBLEMS

14.1 Show that the coefficient of attenuation (α) of sound in a liquid due to shear viscosity (η) is given at low frequency (f) by

$$\alpha = \frac{8\pi^2 f^2 \eta}{3\rho c^3}$$

where ρ is the density and c is the sound velocity.

14.2 A damper is attached to an instrument pointer that moves along a line. The damper consists of a metal square 3 cm on the side that moves between two parallel plates, leaving a 0.1 mm gap on either side. Find the damping constant if the gap is filled with oil of viscosity 15 mPa · s.

14.3 Compare the wave equations for propagation with viscosity, Eq. (14.2.11), and for propagation with heat conduction, Eq. (14.3.32). Show that the attenuation coefficient due to thermal conduction will be

$$\alpha = \frac{M\kappa(\gamma - 1)\omega^2}{2\rho c^3 C_p}$$

Identify the various symbols.

14.4 Find values for attenuation by the atmosphere due to combined viscosity, thermal conduction, and molecular rotational relaxation at standard temperature and pressure at 20 kHz and 200 kHz. Assume the relative humidity is 40%.

14.5 Estimate the reduction in sound pressure level in dB in seawater at 10 Hz and 1 kHz over distances of 1 km and 1000 km. Assume the temperature is 10°C and the salinity is 35%.

14.6 Estimate the reduction in sound pressure level in dB in the atmosphere at 1 Hz and 1 kHz over distances of 1 km and 20 km. Assume the temperature is 20°C and the relative humidity is 40%.

14.7 A circular air-filled pipe is used as a resonator in a fog signal. Assuming that the pipe is 10 cm inside diameter, calculate the attenuation coefficient due to viscous drag at the wall at atmospheric pressure and 20°C at 100 Hz, 400 Hz, and 1 kHz.

14.8 Estimate the reverberation time in a rectangular room 20 m × 10 m × 5 m in dimensions at 30 Hz, 100 Hz, 300 Hz, and 1000 Hz. Assume the floor is covered in carpet and that the walls and ceiling are covered in plaster. Will the room be suitable for playing chamber music? If not, suggest how it may be remedied.

14.9 A spherical resonator is 25 cm in diameter. If it is filled with water at room temperature, what will be its lowest natural frequency? Assume the walls are of negligible thickness and that the neck is small. Estimate the Q value of this lowest mode.

14.10 Describe how the internal damping in a solid specimen might be measured. What factors would you expect to influence the result?

Nonlinear Acoustics

15.1 Nonlinear Oscillators

All the phenomena we have discussed so far were assumed to be linear; that is, the displacements of particles, whether in lumped constant or distributed systems, have been presumed to be in the linear, elastic region. Although this is often a good assumption, leading to insightful and useful results, there are many situations in which displacements go beyond the elastic limit, resulting in nonlinear behavior that may demonstrate effects that are different from any we have discussed thus far. Some of these phenomena—such as shock waves, subharmonics, solitons, combination tones, cavitation, and chaos—are of great practical importance. In this chapter, we will discuss nonlinearity in three principal examples: the simple oscillator, the propagation of a pressure disturbance along a tube, and cavitation.

We begin with a simple point mass oscillator without damping but having a nonlinear spring. We know that the kinetic energy of such a system depends on the velocity of the mass and its potential energy depends on the displacement. (See Section 1.1.4 in Chapter 1). It follows that the state of the system can be represented by a point on a phase plane for which the abscissa is the particle's position and the ordinate is its velocity. This geometrical approach to the problem of nonlinear oscillation was pioneered by Poincaré and Liapunov. For a linear system the equation of motion is given by

$$M\ddot{x} + Sx = 0 \tag{15.1.1}$$

A solution of Eq. (15.1.1) may be written as

$$x = x_0 \cos(\omega_0 t + \alpha) \tag{15.1.2}$$

where

$$\omega_0^2 = S/M \tag{15.1.3}$$

so that

$$v = \dot{x} = -\omega_0 x_0 \sin(\omega t + \alpha) \qquad (15.1.4)$$

Eq. (15.1.2) gives the abscissa of the phase diagram and Eq. (15.1.4) gives its ordinate. Eliminating t between these two equations then leads to the equation of the path

$$\frac{x^2}{x_0^2} + \frac{v^2}{\omega_0^2 x_0^2} = 1 \qquad (15.1.5)$$

Thus, the paths are seen to be a family of ellipses, as shown in Fig. 15.1. The point representing the state will always move around the ellipse in a clockwise manner, since a positive velocity is in the positive x direction.

Now consider the case where the spring is nonlinear:

$$M\ddot{x} + f(x) = M\ddot{x} + Sx + bx^3 \qquad (15.1.6)$$

The nonlinearity is introduced through the constant b, which, when assumed to be greater than zero, results in the restoring force being greater than in the linear case. The spring is then said to be "hardening." If b is negative, on the other hand, the spring is said to be "softening." Since

$$\frac{d^2x}{dt^2} = \frac{dv}{dt} = \frac{dv}{dx} \cdot \frac{dx}{dt} = \frac{vdv}{dx} \qquad (15.1.7)$$

it follows that

$$Mv\frac{dv}{dx} = -(Sx + bx^3) \qquad (15.1.8)$$

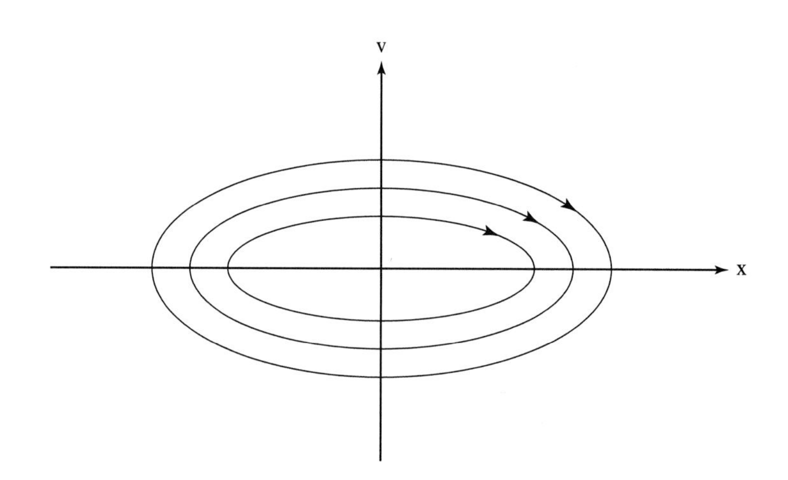

Fig. 15.1 Velocity-displacement plane for a linear oscillator.

Integration yields

$$M\frac{v^2}{2} + \frac{Sx^2}{2} + \frac{bx^4}{4} = E \tag{15.1.9}$$

where E is the total energy of the system. When x and v are both small, the nonlinear term vanishes and the equations are again ellipses. The maximum displacement, a, occurs when v is zero and can thus be shown from Eq. (15.1.9) to satisfy the expression

$$a^2 = \frac{-S + \sqrt{S^2 + 4bE}}{b} \tag{15.1.10}$$

Only the positive sign of the radical gives a valid solution when $b > 0$, since E and a^2 are both necessarily positive and b is small. When $b < 0$, however, both positive and negative values of the radical are valid. The phase curves are symmetric, and thus the period of the motion can be obtained from an integral:

$$T = \oint \frac{dx}{v} = 4\int_0^b \frac{dx}{\sqrt{2\frac{E}{M} - \left(\omega_0^2 + \frac{bx^2}{2M}\right)x^2}} \tag{15.1.11}$$

In the linear case when b is zero, it can be shown that Eq. (15.1.11) yields $T = 2\pi/\omega_0$, as expected. When b is positive (hardening spring), the result of the integral, namely T, is smaller, since ω_0 is effectively replaced by a larger value that depends on the amplitude a. For a softening spring, the period is greater and the resonance frequency is less. This result is represented in Fig. 15.2.

It is also interesting to discuss the phase curves given by Eq. (15.1.9). For the hardening spring, the curves are closed, as shown in Fig. 15.3. For the softening spring, however, the behavior is more interesting, as seen in Fig. 15.4. If we set

$$b = -\rho^2 \tag{15.1.12}$$

then Eq. (15.1.9) yields

$$v^2 = \frac{2E}{M} - \omega_0^2 x^2 + \frac{\rho^2 x^4}{2M} \tag{15.1.13}$$

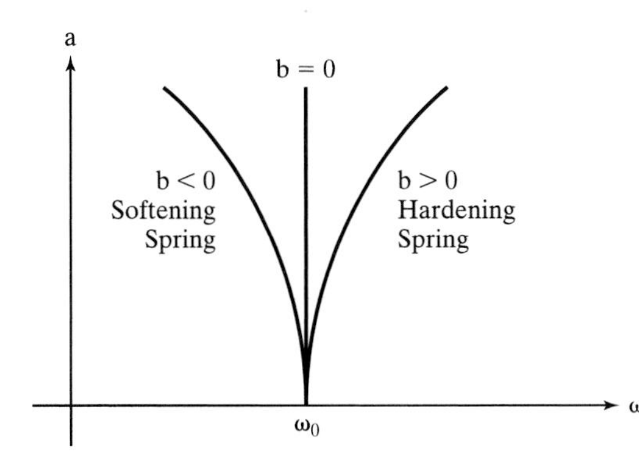

Fig. 15.2 Dependence of angular resonance frequency on amplitude a for linear and nonlinear oscillators.

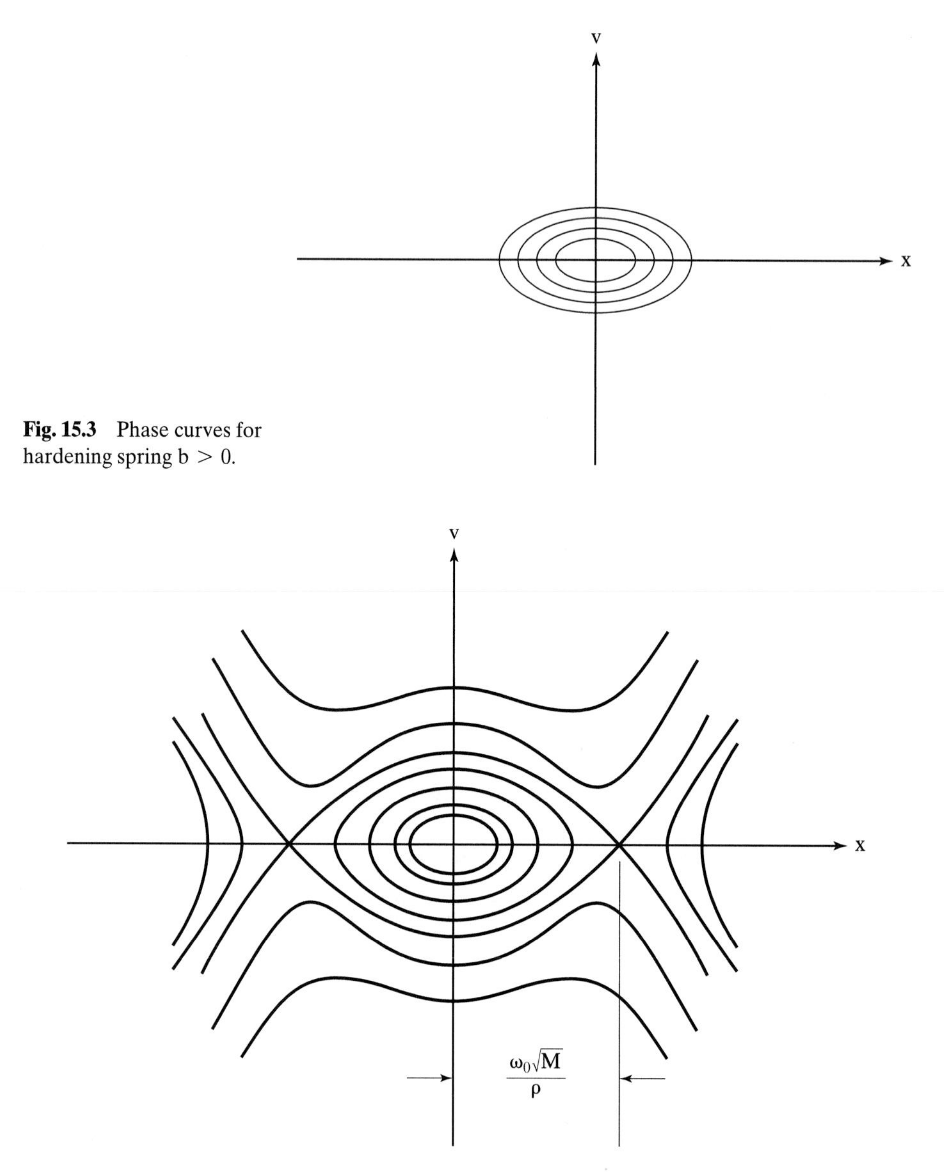

Fig. 15.3 Phase curves for hardening spring $b > 0$.

Fig. 15.4 Phase curves for softening spring $b < 0$.

Near the origin, the curves again approximate ellipses. By considering the minimum of Eq. (15.1.13), it can be shown that its right side is always positive if

$$\omega_0^4 - 2E\rho^2/M^2 < 0 \tag{15.1.14}$$

Thus, the transition from curves that cross the x-axis to curves that do not occurs for

$$E = \frac{\omega_0^2 M^2}{2\rho^2} \tag{15.1.15}$$

For this curve, the velocity at zero displacement, v_0, must be given by

$$v_0^2 = \frac{\omega_0^2 M}{\rho^2} \tag{15.1.16}$$

The transition phase curve crosses the x-axis at

$$x = \pm\omega_0\sqrt{M}/\rho = \pm(S/b)^{1/2} \tag{15.1.17}$$

which is the equilibrium position for which, from Eq. (15.1.6),

$$f(x) = Sx - bx^3 = 0 \tag{15.1.18}$$

The origin is a point of stable equilibrium, while the crossing points are unstable positions. If damping is included in the problem, the equation becomes

$$M\ddot{x} + \varphi(\dot{x}) + f(x) = 0 \tag{15.1.19}$$

The damping term can assume the following forms:

a. Coulomb damping

$$\varphi(\dot{x}) = \begin{cases} F, \dot{x} > 0 \\ -F, \dot{x} < 0 \end{cases} \tag{15.1.20}$$

b. Viscous damping

$$\varphi(\dot{x}) = R\dot{x} \tag{15.1.21}$$

c. Self-sustained oscillations

$$\varphi(\dot{x}) = -Av + Bv^3/3 \tag{15.1.22}$$

Stoker (1950) discusses these rather specialized topics in greater detail. In the first two cases, the damping causes energy loss and the phase curve will spiral inward, as shown in Fig. 15.5. In the third case, the negative damping term in Eq. (15.1.22) implies that energy can be delivered to the system, thus permitting the buildup of an oscillation. However, the term in v^3 limits the buildup to a finite value and the phase curve is then said to be a limit cycle, as shown in Fig. 15.6. Systems of this sort were first studied by van der Pol (1927). Such self-excited oscillations can be seen with a mass held by a spring on a belt with friction, on a spring-suspended aerofoil, and in certain electronic amplifying circuits.

15.2 Forced Nonlinear Oscillations

Again consider a mass M with nonlinear spring and damping but now under a forcing function:

$$M\ddot{x} + \varphi(\dot{x}) + f(x) = F(t) \tag{15.2.1}$$

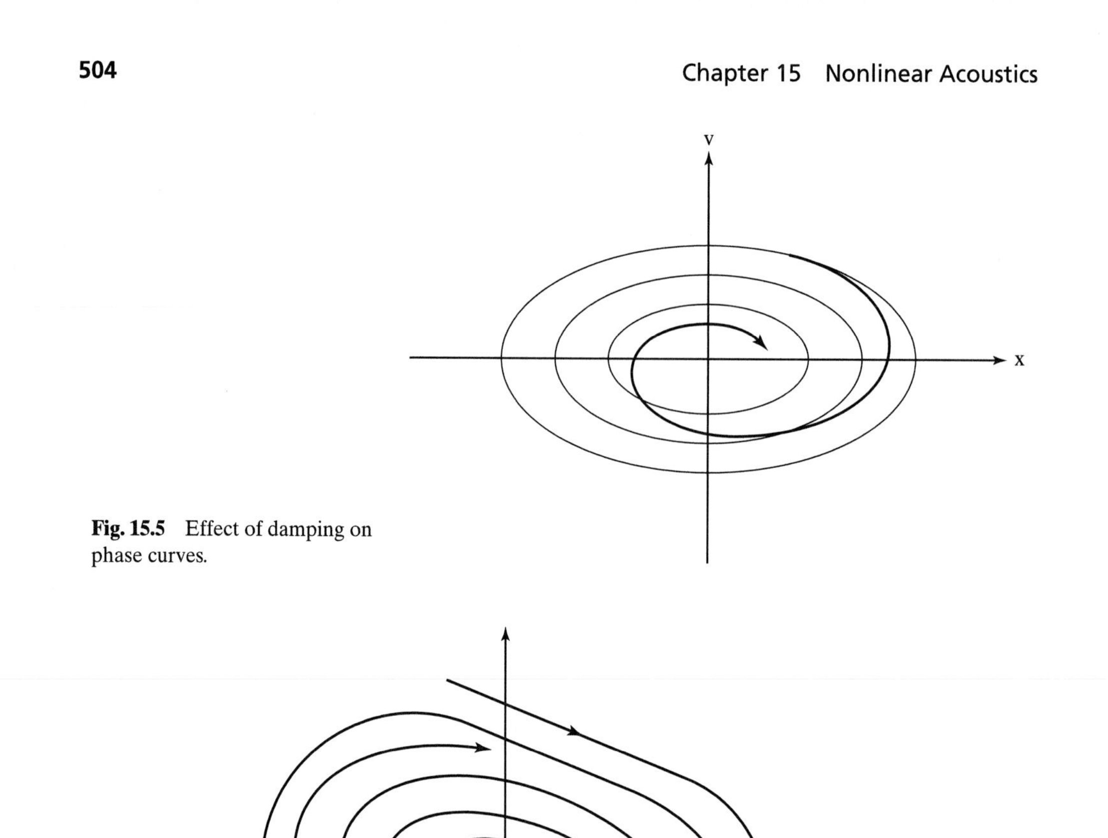

Fig. 15.5 Effect of damping on phase curves.

Fig. 15.6 Self-sustained oscillation showing limit cycle. (After van der Pol, 1927.)

This is known as the Duffing equation. With viscous damping and nonlinearity in the spring as studied previously:

$$M\ddot{x} + R\dot{x} + (Sx + bx^3) = F(t) \tag{15.2.2}$$

Suppose the damping is zero and we have a harmonic driving force, then

$$M\ddot{x} = -Sx - bx^3 + F_0 \cos \omega t \tag{15.2.3}$$

There are both periodic and nonperiodic solutions of the Duffing equation. The periodic ones were the first to receive attention, since approximation methods can easily be used to obtain qualitative information on their character. Either perturbation or iterative schemes can be used, and we may either work with the differential equation directly, or, since the solutions are assumed periodic, we may use a Fourier expansion. Such an expansion will contain a harmonic oscillation, at the same frequency as the driving force and subharmonics with periods equal to multiples of the period of the driving force. Duffing used an iteration method.

A first-order solution is obtained thus

$$x_1 = A \cos \omega t \tag{15.2.4}$$

A may be regarded as the value of the first Fourier coefficient of the oscillation. If this value is substituted on the right side of Eq. (15.2.3), we obtain

$$M\ddot{x}_2 = -SA \cos \omega t + F_0 \cos \omega t - bA^3 \cos^3 \omega t \tag{15.2.5}$$

Hence, using the trigonometric expansion of $\cos^3 \omega t$ and dividing by M, we obtain a second-order approximation:

$$\ddot{x}_2 = \left(-\frac{3bA^3}{4M} - \frac{SA}{M} + \frac{F_0}{M} \right) \cos \omega t - \frac{bA^3}{4M} \cos 3\omega t \tag{15.2.6}$$

Integrating Eq. (15.2.6) twice with respect to time yields

$$x_2 = \frac{1}{\omega^2} \left(\frac{3bA^3}{4M} + \frac{SA}{M} - \frac{F_0}{M} \right) \cos \omega t + \frac{bA^3}{36M\omega^2} \cos 3\omega t \tag{15.2.7}$$

Let

$$x_2 = A_1 \cos \omega t + \frac{bA^3}{36M\omega^2} \cos 3\omega t \tag{15.2.8}$$

Duffing then argued that since b must be small, comparing Eqs. (15.2.4) and (15.2.8), it follows that A_1 must be approximately equal to A. Thus

$$\omega^2 = \frac{S}{M} + \frac{3}{4} \frac{bA^2}{M} - \frac{F_0}{MA} \tag{15.2.9}$$

To obtain the response curves, we sketch in an ω^2, A plane, the curve:

$$\omega^2 = \frac{S}{M} + \frac{3}{4} \frac{bA^2}{M} \tag{15.2.10}$$

and the set of curves $\omega^2 = -\frac{F_0}{MA}$, for various values of F_0. Such curves are shown in Fig. 15.7a for a hardening spring (i.e., $b > 0$). When the abscissas of these curves are added to obtain the right side of Eq. (15.2.9), the curves are as given in Fig. 15.7b. It is common practice to plot the modulus of the amplitude as shown in Fig. 15.8a. Fig. 15.8b shows the corresponding response for a softening spring.

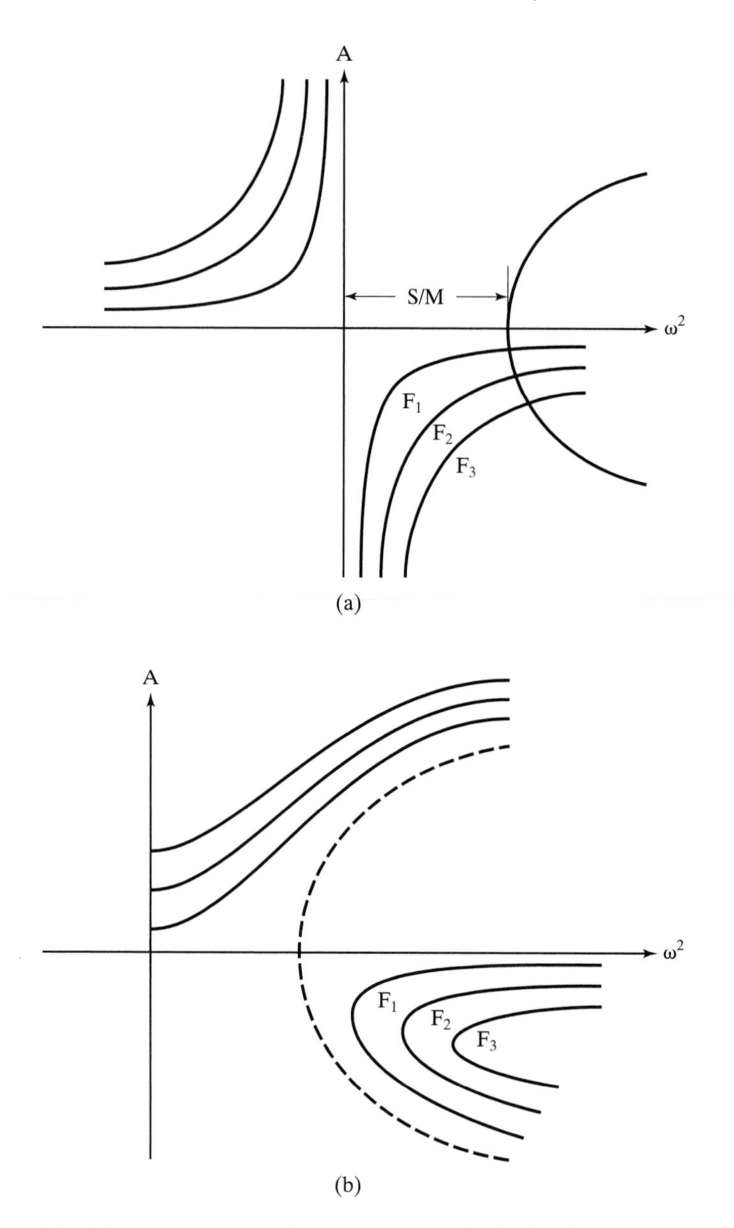

Fig. 15.7 Steps in constructing response curves for the Duffing oscillator.

The Duffing equation can also be solved with a viscous damping term present, although we will not discuss the details here. Suffice it to say that the effect of the damping is to cause a limitation of the response curves at the resonance, so that these curves then resemble those shown in Figs. 15.9a and b. The curves may be thought of as similar to those for an undampened oscillator, only bent to the right for

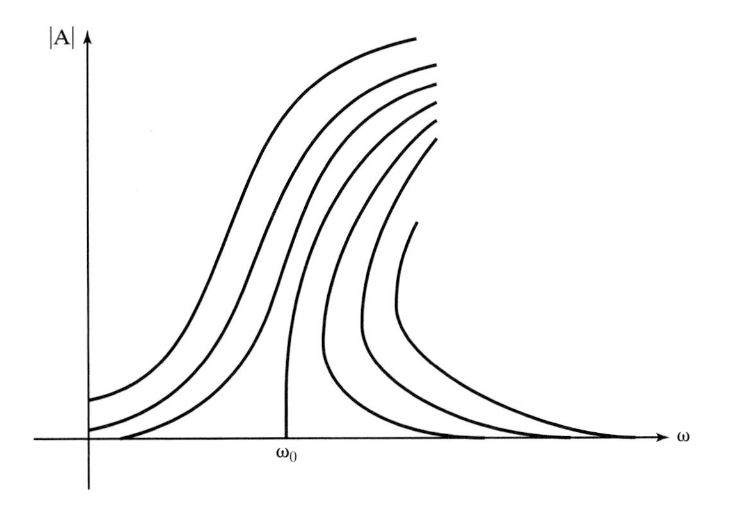

Fig. 15.8a Response curve for a hardening spring under forced oscillation.

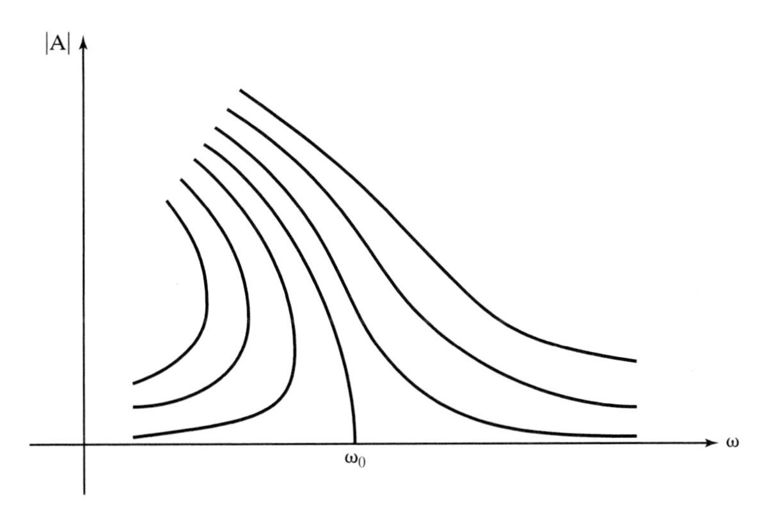

Fig. 15.8b Response curve for a softening spring under forced oscillation.

the hardening spring and to the left in the softening case. For a given force amplitude, we see that at some frequencies the solutions can take three possible amplitude values. This curious circumstance gives rise to what is termed the "jump phenomenon," as illustrated in Fig. 15.10. Consider a system with a hardening spring being driven at a high frequency at point 1. As the driving frequency is lowered with a constant force amplitude, eventually point 2 is reached, at which the response curve turns back on itself. Then, if the frequency is lowered further, there has to be a jump to point 3 on the other branch of the curve. Such jumps have been observed for both hardening and softening springs.

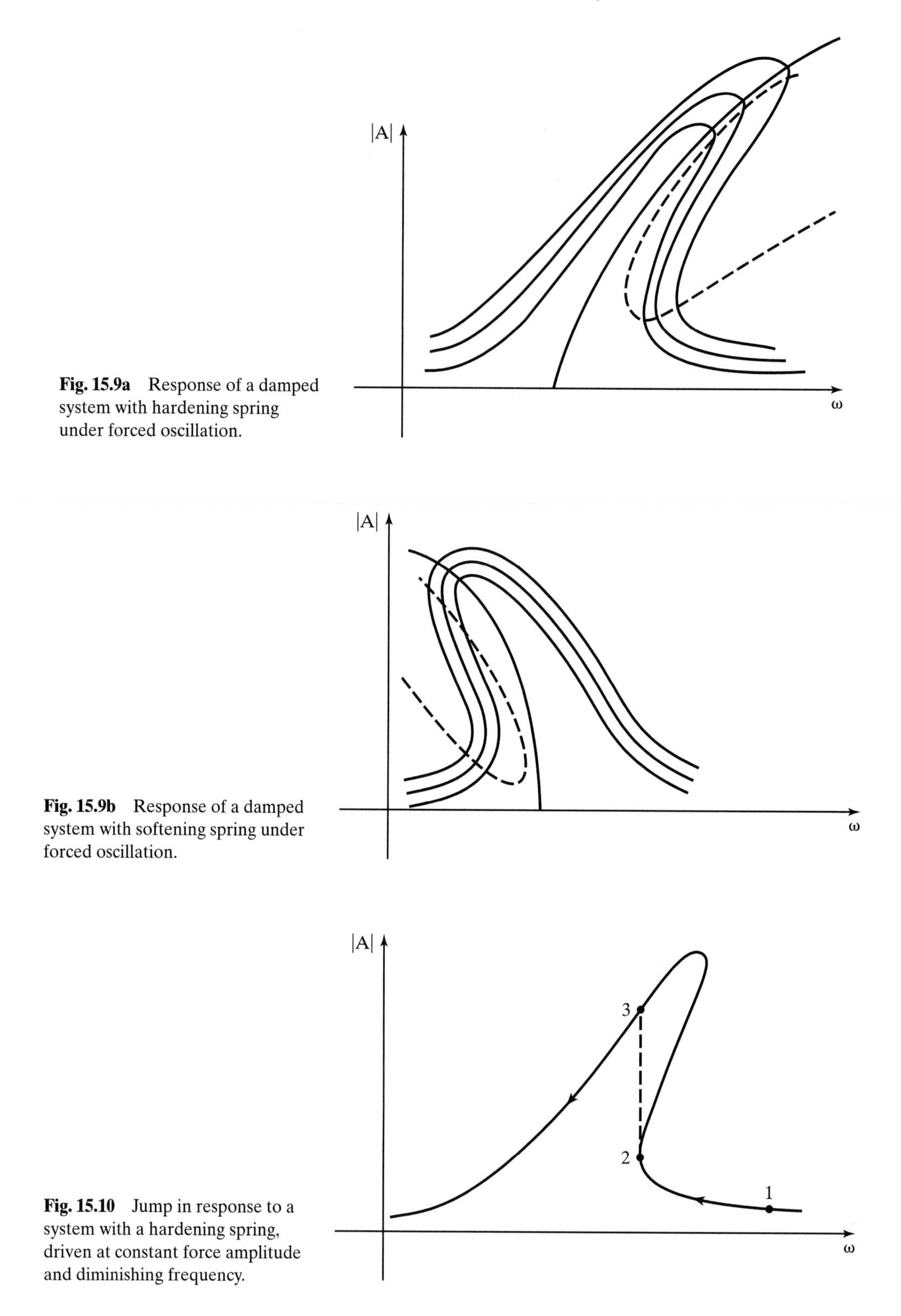

Fig. 15.9a Response of a damped system with hardening spring under forced oscillation.

Fig. 15.9b Response of a damped system with softening spring under forced oscillation.

Fig. 15.10 Jump in response to a system with a hardening spring, driven at constant force amplitude and diminishing frequency.

15.3 Nonlinear Effects in Wave Motion

We will turn now to nonlinear effects in wave motion. In the linear acoustic theory presented in Chapters 1, 3, and 4, it was assumed that the amplitudes of disturbances from equilibrium were small. However, there are phenomena that have been known since ancient times for which such assumptions are anything but true. The high-speed impact of two bodies, the movement of the tip of a whip, or an explosion are examples of such phenomena and they may be associated with sudden, local pressure changes comparable to or greater than an atmosphere in magnitude. These disturbances can then travel through the surrounding air as what are known as shock waves. Since in linear acoustics we assume that all disturbances are small, it is surprising that such large amplitude effects still propagate as waves. We might ask how this might be so, and at what velocity the waves will travel. The first theoretical treatment of the subject was provided by Earnshaw (1860), who considered a one-dimensional arrangement that we would now call a shock tube. He proved that the local velocity of propagation was higher in the compressive part of a disturbance so that the density gradient should be expected to steepen with time and eventually produce a discontinuous pressure front.

As we saw in Sections 1.6.1, 3.1, and 4.1, small adiabatic disturbances propagate through a fluid at speed c where

$$c^2 = \left(\frac{\partial p}{\partial \rho}\right)_s \tag{15.3.1}$$

For a perfect gas under adiabatic conditions

$$PV^\gamma = \text{constant or } P = \text{constant } \rho^\gamma \tag{15.3.2}$$

and

$$c^2 = \frac{\gamma P}{\rho} = \gamma \Re T \tag{15.3.3}$$

For a flowing fluid, we introduce the concept of the Mach number

$$M = v/c \tag{15.3.4}$$

which is a most important parameter in gas dynamics. The principles governing the production of waves are the same as in the linear case. First, we have continuity of mass. The continuity equation will be

$$\frac{\partial}{\partial t}(\rho A) + \frac{\partial}{\partial x}(\rho v A) = 0 \tag{15.3.5}$$

For steady flow, $d(\rho v A)$ is zero, and thus

$$\frac{d\rho}{\rho} + \frac{du}{u} + \frac{dA}{A} = 0 \tag{15.3.6}$$

If the flow is incompressible, $d\rho$ will be zero, and

$$\frac{dv}{v} = \frac{-dA}{A} \tag{15.3.7}$$

For compressible flow, we use Euler's formula

$$vdv = \frac{-dP}{\rho} = \frac{-dP}{d\rho}\frac{d\rho}{\rho} = -c^2\frac{d\rho}{\rho} \tag{15.3.8}$$

and hence

$$\frac{d\rho}{\rho} = -M^2\frac{dv}{v} \tag{15.3.9}$$

Eqs. (15.3.6) and (15.3.9) finally yield

$$\frac{dv}{v} = \frac{-dA/A}{1 - M^2} \tag{15.3.10}$$

The following special cases are of interest:

1. Low velocity, $M \rightarrow 0$: Then a decrease in area gives a proportionate increase in velocity.
2. $0 < M < 1$: A decrease in area produces an increase in velocity, even greater than for incompressible flow, since the denominator of Eq. (15.3.10) is less than unity.
3. $M > 1$ (i.e., at supersonic speeds): Now an increase in area yields an increase in velocity.
4. $M = 1$: Then $\frac{dv}{v}$ can only be finite if dA/A is zero. For a tube in which v increases continuously from zero, passing through $M = 1$, the tube must converge in the subsonic region, diverge in the supersonic, and have a throat where $M = 1$.

For a constant cross-sectional area, continuity of mass reduces to

$$\frac{\partial\rho}{\partial t} + \rho\frac{\partial v}{\partial x} + v\frac{\partial\rho}{\partial x} = 0 \tag{15.3.11}$$

If we consider two constant cross sections, designated by subscripts 1 and 2, which are on either side of a nonequilibrium region, (see Fig. 15.11), then for steady flow, we have from Eq. (15.3.11)

$$\rho_1 v_1 = \rho_2 v_2 \tag{15.3.12}$$

since ρv is then constant.

Second, conservation of energy must apply. From the first law of thermodynamics for unit mass of fluid, we have

$$du = dq - pdV \tag{15.3.13}$$

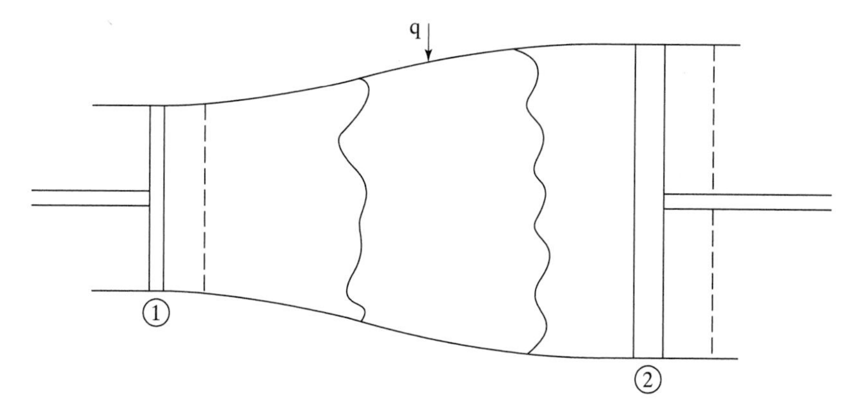

Fig. 15.11 Energy balance on a nonequilibrium region of a flow.

where du is the change in internal energy, dq is the heat supplied, and pdV is the work done on the fluid whose volume is V. Hence, the specific heat at constant volume

$$C_v = \left(\frac{\partial q}{\partial T}\right)_v = \left(\frac{\partial u}{\partial T}\right)_v \tag{15.3.14}$$

and the specific heat at constant pressure

$$C_p = \left(\frac{\partial q}{\partial T}\right)_p = \left(\frac{\partial u}{\partial T}\right)_p + p\left(\frac{\partial V}{\partial T}\right)_p \tag{15.3.15}$$

The enthalpy, h, is defined by

$$h = u + pV \tag{15.3.16}$$

Hence

$$C_p = \left(\frac{\partial h}{\partial T}\right)_p \tag{15.3.17}$$

If C_p and C_v are constants, then

$$h = C_pT + \text{const}$$

and

$$u = C_vT + \text{const} \tag{15.3.18}$$

In general, as seen from Fig. 15.11,

$$q = h_2 - h_1 + \frac{v_2^2}{2} - \frac{v_1^2}{2} \tag{15.3.19}$$

For an adiabatic process, $q = 0$, from Eqs. (15.3.18) and (15.3.19)

$$C_pT_2 + \frac{v_2^2}{2} = C_pT_1 + \frac{v_1^2}{2} \tag{15.3.20}$$

We obtain some useful results from Eq. (15.3.20). For a perfect gas, we have

$$C_pT + \frac{v^2}{2} = C_pT_0 \tag{15.3.21}$$

when T_0 is a reservoir temperature. From Eq. (15.3.3), we have

$$c^2 = \frac{\gamma P}{\rho} = \gamma \Re T = \gamma(C_p - C_v)T = (\gamma - 1)C_pT \tag{15.3.22}$$

Thus, from Eqs. (15.3.21) and (15.3.22)

$$\frac{c^2}{\gamma - 1} + \frac{v^2}{2} = \frac{c_0^2}{\gamma - 1} \tag{15.3.23}$$

Hence, multiplying by $(\gamma - 1)/c^2$

$$\frac{c_0^2}{c^2} = \frac{T_0}{T} = 1 + \frac{(\gamma - 1)}{2}M^2 \tag{15.3.24}$$

Then, using the adiabatic relations, we obtain

$$\frac{P_0}{P} = \left(1 + \frac{\gamma - 1}{2}M^2\right)^{\gamma/(\gamma-1)} \tag{15.3.25}$$

and

$$\frac{\rho_0}{\rho} = \left(1 + \frac{\gamma - 1}{2}M^2\right)^{1/(\gamma-1)} \tag{15.3.26}$$

Instead of the reservoir, we may use any other point in the flow to evaluate the constant in the energy equation. A useful point is a throat where $M = 1$. The flow variables are then said to be sonic and designated by an asterisk, and $v^* = c^*$. The energy equation then gives

$$\frac{v^2}{2} + \frac{c^2}{\gamma - 1} = \frac{v^{*2}}{2} + \frac{c^{*2}}{\gamma - 1} = \frac{(\gamma + 1)c^{*2}}{2(\gamma - 1)} \tag{15.3.27}$$

Hence, the relation between the speed of sound in a throat and a reservoir is

$$\frac{c^{*2}}{c_0^2} = \frac{2}{1 + \gamma} = \frac{T^*}{T_0} \tag{15.3.28}$$

Thus, for a given fluid the sonic and reservoir temperatures are in a fixed ratio. For air

$$\frac{T^*}{T_0} = 0.833 \quad \text{and} \quad \frac{c^*}{c_0} = 0.913 \tag{15.3.29}$$

It is not necessary for a throat to actually exist for the sonic values to be used as a reference. Let

$$M^* = u/c^* \tag{15.3.30}$$

although this is not in accord with our convention. We would expect M^* to be unity, but it is a convenient symbol to use.

The relation between M^* and M can be obtained by dividing the energy equation Eq. (15.3.27) by v^2, yielding

$$M^{*2} = \frac{(\gamma + 1)M^2}{2 + (\gamma - 1)M^2} = \frac{\gamma + 1}{\dfrac{2}{M^2} + \gamma - 1} \tag{15.3.31}$$

and

$$M^2 = \frac{2}{\dfrac{(\gamma + 1)}{M^{*2}} - (\gamma - 1)} \tag{15.3.32}$$

Third, we have conservation of momentum, as illustrated in Fig. 15.12. From Newton's second law, the applied force equals the rate of change of momentum, and hence

$$\rho_2 v_2^2 A_2 - \rho_1 v_1^2 A_1 = p_1 A_1 - p_2 A_2 \tag{15.3.33}$$

For constant area

$$p_2 + \rho_2 v_2^2 = p_1 + \rho_1 v_1^2 \tag{15.3.34}$$

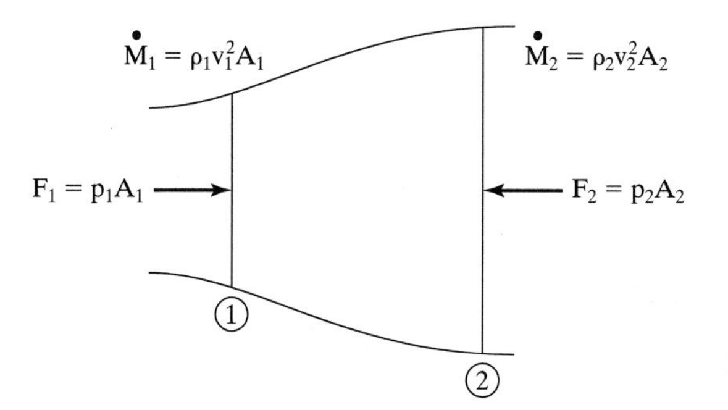

Fig. 15.12 Momentum balance on a non-equilibrium region in a shock tube.

15.4 Shock Waves

We will now restate the governing equations for adiabatic flow through a constant area pipe with a nonequilibrium region, as illustrated in Fig. 15.13.

Continuity of mass

$$\rho_1 v_1 = \rho_2 v_2 \tag{15.4.1}$$

Conservation of energy

$$C_p T_1 + \frac{v_1^2}{2} = C_p T_2 + \frac{v_2^2}{2} \tag{15.4.2}$$

Conservation of momentum

$$p_1 + \rho_1 v_1^2 = p_2 + \rho_2 v_2^2 \tag{15.4.3}$$

Perfect gas law

$$p = \rho \Re T \tag{15.4.4}$$

Dividing Eq. (15.4.3) by Eq. (15.4.1) and using Eq. (15.3.3), we have

$$v_1 - v_2 = \frac{p_2}{\rho_2 v_2} - \frac{p_1}{\rho_1 v_1} = \frac{c_2^2}{\gamma v_2} - \frac{c_1^2}{\gamma v_1} \tag{15.4.5}$$

Now, using Eqs. (15.3.20), (15.4.4), (15.4.5), and (15.3.3), we obtain

$$\frac{v_1^2}{2} + \frac{c_1^2}{\gamma - 1} = \frac{v_2^2}{2} + \frac{c_2^2}{\gamma - 1} \tag{15.4.6}$$

For a flow speed equal to the velocity of sound such as occurs at a throat when $v = c^*$, Eq. (15.4.6) yields

$$\frac{v_1^2}{2} + \frac{c_1^2}{\gamma - 1} = \frac{v_2^2}{2} + \frac{c_2^2}{\gamma - 1} = \frac{c^{*2}(\gamma + 1)}{2(\gamma - 1)} \tag{15.4.7}$$

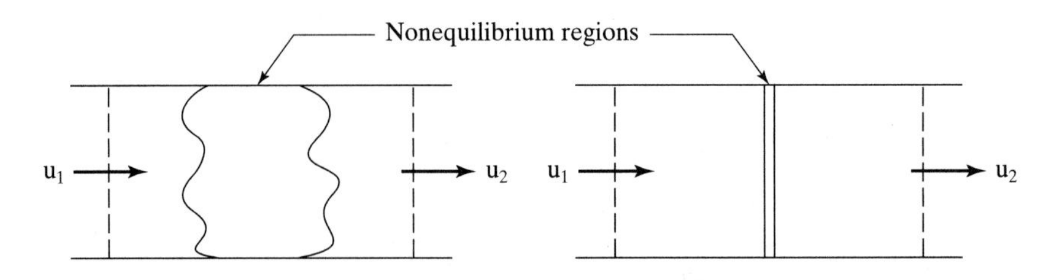

Fig. 15.13 Flow in a tube of constant area.

Combining Eqs. (15.4.5) and (15.4.7) after some manipulation yields

$$v_1 v_2 = c^{*2} \tag{15.4.8}$$

which is known as the Prandtl relation. Alternatively, it may be written

$$M_2^* = 1/M_1^* \tag{15.4.9}$$

where M_2^* is given by Eq. (15.3.30).

We can obtain the relationship between the Mach numbers by combining Eqs. (15.3.31) and (15.4.7), thus

$$M_2^2 = \frac{1 + \dfrac{\gamma - 1}{2} M_1^2}{\gamma M_1^2 - \dfrac{\gamma - 1}{2}} \tag{15.4.10}$$

We also have for the ratio of the velocities

$$\frac{v_1}{v_2} = \frac{v_1^2}{v_1 v_2} = \frac{v_1^2}{c^{*2}} = M_1^{*2} \tag{15.4.11}$$

which in turn leads to other useful results. For instance, in combination with the continuity equation Eq. (15.3.25), we obtain

$$\frac{\rho_2}{\rho_1} = \frac{v_1}{v_2} = \frac{(\gamma + 1) M_1^2}{(\gamma - 1) M_1^2 + 2} \tag{15.4.12}$$

where we also used Eq. (15.3.31). To obtain the pressure ratio, we use the continuity equation together with the momentum equation Eq. (15.4.3), thus

$$p_2 - p_1 = \rho_1 v_1^2 - \rho_2 v_2^2 = \rho_1 v_1 (v_1 - v_2) \tag{15.4.13}$$

Hence

$$\frac{p_2 - p_1}{p_1} = \frac{\rho_1 v_1^2}{p_1}\left(1 - \frac{v_2}{v_1}\right) \tag{15.4.14}$$

From Eqs. (15.4.12) and (15.3.3), we obtain the pressure jump

$$\frac{p_2 - p_1}{p_1} = \frac{\Delta p}{p_1} = \frac{2\gamma}{\gamma + 1}(M_1^2 - 1) \tag{15.4.15}$$

Finally, the pressure ratio is

$$\frac{p_2}{p_1} = 1 + \frac{2\gamma}{\gamma + 1}(M_1^2 - 1) \tag{15.4.16}$$

The temperature ratio is obtained from $T_2/T_1 = (p_2/p_1)(\rho_1/\rho_2)$, so that

$$\frac{T_2}{T_1} = 1 + \frac{2(\gamma - 1)}{(\gamma - 1)^2}\frac{\gamma M_1^2 + 1}{M_1^2}(M_1^2 - 1) = \frac{c_2^2}{c_1^2} \tag{15.4.17}$$

Eqs. (15.4.12), (15.4.16), and (15.4.17) give the ratios of density, pressure, and temperature in terms of the Mach number. However, it is frequently more convenient to express all these parameters in terms of the pressure ratio. It follows from Eq. (15.4.16) that

$$M_1^2 = \frac{(p_2/p_1 - 1)(\gamma + 1) + 2\gamma}{2\gamma} \tag{15.4.18}$$

Hence, from Eq. (15.4.12)

$$\frac{v_1}{v_2} = \frac{(\gamma + 1)M_1^2}{(\gamma - 1)M_1^2 + 2} = \frac{1 + \left(\dfrac{\gamma + 1}{\gamma - 1}\right)\dfrac{p_2}{p_1}}{\dfrac{\gamma + 1}{\gamma - 1} + \dfrac{p_2}{p_1}} = \frac{\rho_2}{\rho_1} = \frac{p_2}{p_1}\frac{T_1}{T_2} \tag{15.4.19}$$

Eq. (15.4.19) is known as the Rankine-Hugoniot relations.

Thus far in our analysis we have imagined that the fluid flows through a stationary pressure discontinuity, or shock front, as shown in Fig. 15.14a. The physics would be unchanged if the flow direction were reversed, as in Fig. 15.14b. Now suppose we impose a velocity v_1 on the fluid towards the right. The fluid ahead of the shock would then be brought to rest and the shock could be thought of as moving with velocity v_1 through the fluid. The fluid behind the shock now has velocity $v_1 - v_2$ to the right (as seen in Fig. 15.14c. This would be the situation if the fluid were being driven by a piston with velocity $v_p = v_1 - v_2$, and the situation is as shown in Fig. 15.14d. Assuming the piston can be started impulsively at $x = 0$, $t = 0$, the shock and piston paths are as shown in Fig. 15.14e. The density, pressure, and temperature ratios are not affected by the transformation and are given by the Rankine-Hugoniot relations. Using the pressure ratio as the basic parameter, from Eq. (15.4.18)

$$v_5 = M_1 c_1 = c_1\left[\left(\frac{p_2}{p_1} - 1\right)\frac{\gamma + 1}{2\gamma} + 1\right]^{1/2} = c_1\left[\frac{p_2}{p_1}\left(\frac{\gamma + 1}{2\gamma}\right) + \frac{\gamma - 1}{2\gamma}\right]^{1/2} \tag{15.4.20}$$

The fluid velocity in front of the piston is

$$v_p = v_1 - v_2 = v_s(1 - v_2/v_1) \tag{15.4.21}$$

From Eq. (15.4.20) and the Rankine-Hugoniot relation from Eq. (15.4.19),

$$v_p = \frac{c_1}{\gamma}\left(\frac{p_2}{p_1} - 1\right)\left\{\frac{\dfrac{2\gamma}{\gamma + 1}}{\dfrac{p_2}{p_1} + \dfrac{\gamma - 1}{\gamma + 1}}\right\}^{1/2} \tag{15.4.22}$$

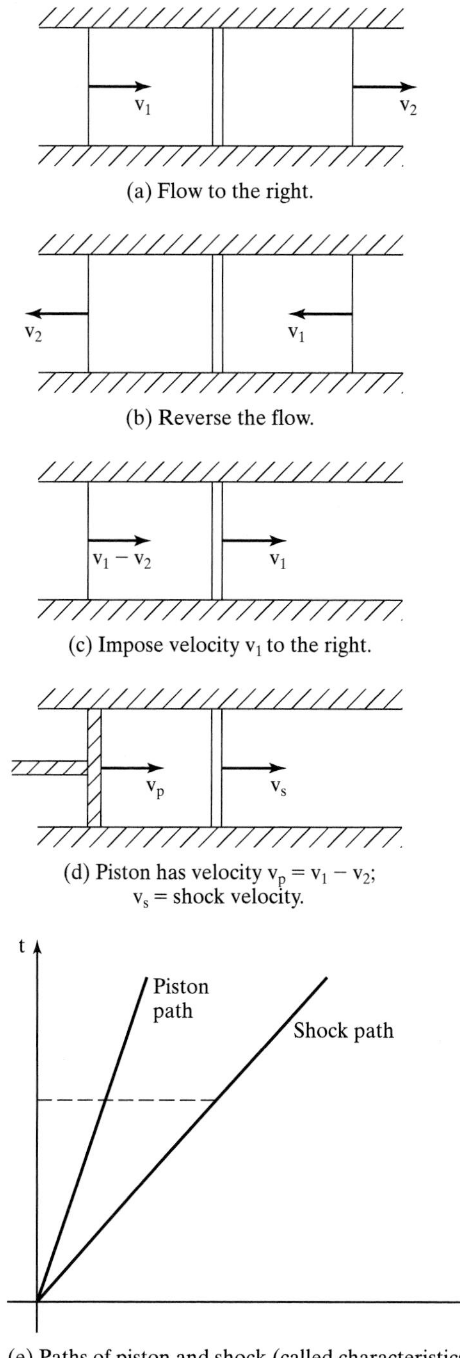

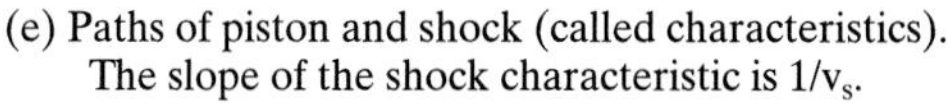
(e) Paths of piston and shock (called characteristics). The slope of the shock characteristic is $1/v_s$.

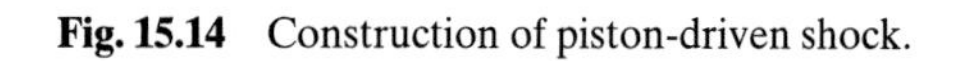
Fig. 15.14 Construction of piston-driven shock.

For a weak shock

$$\frac{\Delta p}{p_1} = \frac{p_2 - p_1}{p_1} << 1 \tag{15.4.23}$$

The other parameters then also have small disturbances, and thus we have from Eqs. (15.3.2) and (15.4.22)

$$\frac{\Delta \rho}{\rho} = \frac{\rho_2}{\rho_1} - 1 = \frac{1}{\gamma}\frac{\Delta p}{p_1} \approx \frac{v_p}{c_1} \tag{15.4.24}$$

and using a first-order approximation in Eq. (15.4.21)

$$v_s \approx c_1\left(1 + \frac{\gamma + 1}{4\gamma} \cdot \frac{\Delta p}{p_1}\right) \tag{15.4.25}$$

Thus, we see that with a weak shock the shock velocity only barely exceeds the sound velocity in the quiescent fluid. The situation will not differ from the linear acoustic case. For a strong shock, however, p_2/p_1 is large and we then have from the Rankine-Hugoniot relations

$$\frac{\rho_2}{\rho_1} \to \frac{\gamma + 1}{\gamma - 1} \tag{15.4.26}$$

$$\frac{T_2}{T_1} \to \frac{p_2}{p_1}\left(\frac{\gamma - 1}{\gamma + 1}\right) \tag{15.4.27}$$

$$v_s \to c_1\left(\frac{\gamma + 1}{2\gamma}\frac{p_2}{p_1}\right)^{1/2} \tag{15.4.28}$$

and

$$v_p \to c_1\left(\frac{2}{\gamma(\gamma + 1)}\frac{p_2}{p_1}\right)^{1/2} \tag{15.4.29}$$

In this case, we can conclude that a piston driving into a stationary gas raises its pressure, density, and temperature and generates a shock that will propagate into the quiescent gas at a speed exceeding the sound velocity in the quiescent state.

Most laboratory work is carried out using a shock tube. High and low pressure regions in a tube are separated by a diaphragm that is punctured, which results in an essentially instantaneous creation of a pressure front in the gas. The theory behind this situation is not handled as easily as in the case of the piston-driven shock because it cannot be derived from a stationary case. We have to use the nonlinear-nonstationary-governing equations.

1. Continuity equation, from Eq. (15.3.11)

$$\frac{\partial \rho}{\partial t} + \rho\frac{\partial v}{\partial x} + v\frac{\partial \rho}{\partial x} = 0 \tag{15.4.30}$$

2. Momentum equation

$$-\frac{\partial p}{\partial x} = \rho\frac{\partial v}{\partial t} + \rho v\frac{\partial v}{\partial x} \tag{15.4.31}$$

Note that no damping terms are included and that isentropic conditions exist, since no external heat is added. Eq. (15.4.31) is known as Euler's formula.

3. Equation of state from Eq. (15.3.3)

$$\frac{\partial p}{\partial x} = c^2\frac{\partial \rho}{\partial x} \tag{15.4.32}$$

The continuity and momentum equations can now be rewritten as follows:

$$\rho_1\frac{\partial s}{\partial t} + \rho_1\left(\frac{\partial v}{\partial x} + s\frac{\partial v}{\partial x}\right) + \rho_1 v\frac{\partial s}{\partial_x} = 0 \tag{15.4.33}$$

and

$$\frac{\partial v}{\partial t} + v\frac{\partial v}{\partial x} + \frac{c^2}{1 + s}\frac{\partial s}{\partial t} = 0 \tag{15.4.34}$$

where s is the condensation and is given by

$$s = \frac{\rho - \rho_1}{\rho_1} \tag{15.4.35}$$

The principal difficulty in solving these equations is due to the presence of nonlinear terms, such as $v(\partial v/\partial x)$ and $s(\partial v/\partial x)$, in which the independent variables appear as coefficients of their derivatives. The equations can be linearized by assuming small disturbances, which leads to the usual acoustic equations

$$\frac{\partial s}{\partial t} + \frac{\partial v}{\partial x} = 0 \tag{15.4.36}$$

and

$$\frac{\partial v}{\partial t} + c^2\frac{\partial s}{\partial x} = 0 \tag{15.4.37}$$

These equations yield the following wave equations:

$$\frac{\partial^2 v}{\partial t^2} = c^2\frac{\partial^2 v}{\partial x^2} \tag{15.4.38}$$

and

$$\frac{\partial^2 s}{\partial t^2} = c^2\frac{\partial^2 s}{\partial x^2} \tag{15.4.39}$$

Consider the shock tube in this linearized approximation. A diaphragm at $x = 0$ divides the tube into two chambers in which the pressures are different. When the diaphragm is broken, a wave motion is set up in the tube, a situation illustrated in Fig. 15.15. A compression wave is created in the expansion chamber, and an expansion wave results in the compression chamber.

A solution of Eqs. (15.4.38) and (15.4.39) is

$$s(x, t) = F(x + ct) + G(x - ct)$$

and

$$v(x, t) = cF(x + ct) + cG(x - ct) \tag{15.4.40}$$

as illustrated in Fig. 15.15.

If the acoustic assumptions cannot be made, the nonlinear equations have to be solved. A physical approach might be as follows: From the viewpoint of an observer moving with the local particle velocity, the acoustic theory applies locally. Let there be a wavelet superimposed on a finite amplitude wave at $x = x_n$. Relative to an observer moving with the local fluid velocity v_n, the wavelet propagation velocity is

$$(c_n') = \left(\frac{dp}{d\rho}\right)_n^{1/2} \tag{15.4.41}$$

Hence

$$c_n = c_n' + v_n \tag{15.4.42}$$

but

$$(c_n')^2 = \frac{\gamma P}{\rho} \tag{15.4.43}$$

and

$$PV^\gamma = P_1 V_1^\gamma \tag{15.4.44}$$

Thus

$$(c_n')^2 = c_1^2\left(\frac{\rho}{\rho_1}\right)^{\gamma-1} \tag{15.4.45}$$

Hence, for an acoustic wave it may be shown that

$$v = \frac{\Delta p}{\rho c} = \frac{c^2 \Delta \rho}{\rho c} = c\left(\frac{\Delta \rho}{\rho}\right) \tag{15.4.46}$$

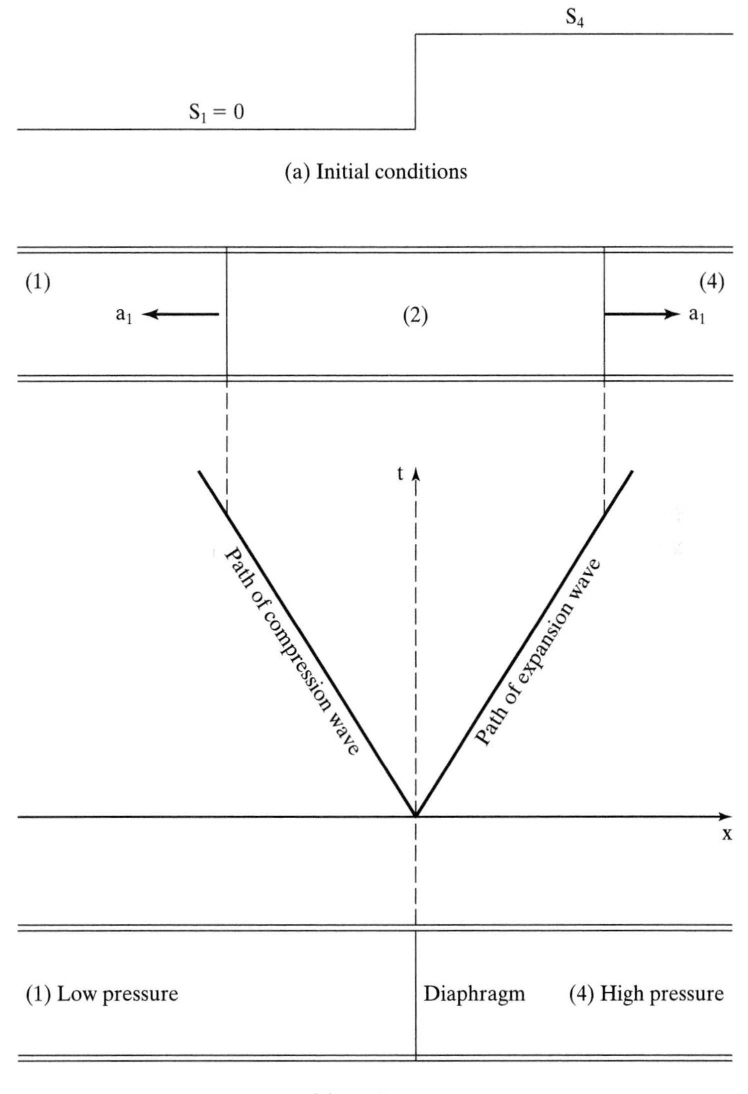

(b) x-t diagram

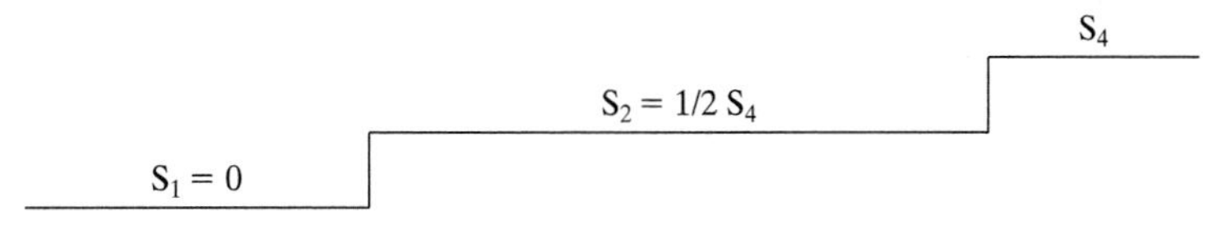

(c) Conditions at time t

Fig. 15.15 Acoustic model of shock tube.

Thus, locally the particle velocity is related to density by

$$dv = \pm c_n'\left(\frac{d\rho}{\rho}\right) \tag{15.4.47}$$

$$= \pm c_1\left(\frac{\rho}{\rho_1}\right)^{\frac{\gamma-1}{2}}\frac{d\rho}{\rho} \tag{15.4.48}$$

Integrating

$$\begin{aligned} v_u &= \pm\int_{\rho_1}^{\rho} c_n^1\frac{d\rho}{\rho} = \pm\int_{\rho_1}^{\rho} c_1\left(\frac{\rho}{\rho_1}\right)^{\frac{\gamma-1}{2}}\frac{d\rho}{\rho} \\ &= \pm\frac{c_1}{(\rho_1)^{\frac{\gamma-1}{2}}}\left[\frac{\rho^{\frac{\gamma-1}{2}}}{(\gamma-1)/2}\right]_{\rho_1}^{\rho} \\ &= \pm\frac{2[c_n' - c_1]}{(\gamma-1)} \end{aligned} \tag{15.4.49}$$

From Eqs. (15.4.42) and (15.4.49), we obtain

$$c_n = c_1 \pm \left(\frac{\gamma+1}{2}\right)v_n \tag{15.4.50}$$

From Eqs. (15.4.42), (15.4.45), and (15.4.49)

$$c_n = c_1\left[1 \pm \frac{\gamma+1}{\gamma-1}\left\{\left(\frac{\rho}{\rho_1}\right)^{\frac{\gamma-1}{2}} - 1\right\}\right] \tag{15.4.51}$$

These two results show that the wave propagation speed depends on the local particle velocity or the local density. Where $\rho > \rho_1$, in regions of condensation the wave speed is higher than c_1, and in regions of rarefaction it is lower. Thus, the wave distorts as it propagates, as shown in Fig. 15.16. In high condensation regions, the characteristics are more inclined, since their slopes are inversely proportional to velocity. However, the slopes do not really cross, as at t_3, since this would result in the wave having three velocities at one point. As the temperature and velocity gradients steepen, viscosity and thermal conduction effects take effect and limit the further distortion of the wave. When all these effects are taken into account, we are able to predict the behavior in the shock tube, as illustrated in Fig. 15.17. The basic parameter of the shock tube is the diaphragm pressure ratio p_4/p_1. When the diaphragm bursts, a shock wave propagates into the expansion chamber with speed c_3, and a rarefaction or an expansion wave propagates into the compression chamber with speed c_4. Derivation of the velocities and shock strength is left as a student exercise.

The subject of shock waves is of great practical importance in applications from aeronautical engineering to the flow of gases in pipelines. Liepmann and Roshko (1957) and Pain and Rogers (1962) provide further information on this topic.

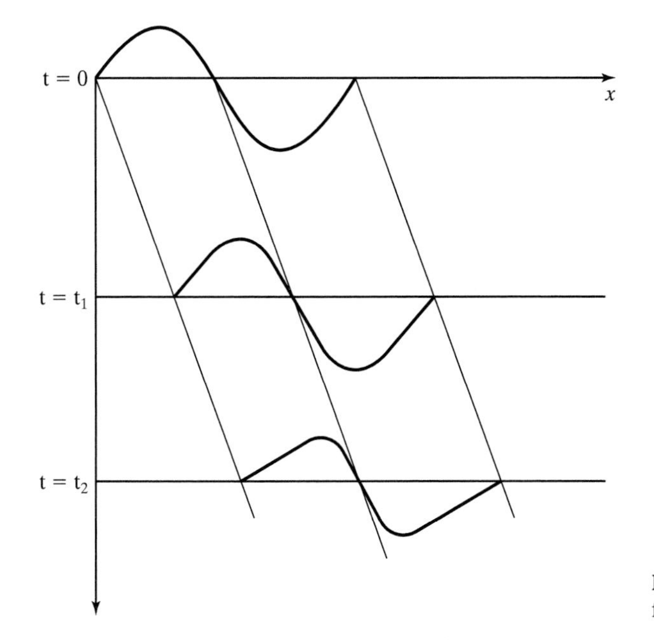

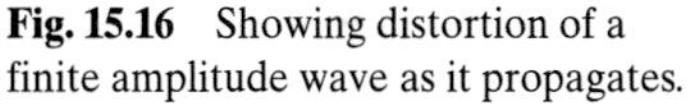
Fig. 15.16 Showing distortion of a finite amplitude wave as it propagates.

15.5 Ultrasonic Cavitation

Finch and Neppiras (1973) assumed the dynamics of a vapor bubble in a liquid to be governed by the laws of conservation of mass, momentum, and energy. The liquid was assumed to be inviscid and incompressible and the vapor to obey the ideal gas law. Heat conduction was assumed to follow Fourier's law and the temperature and pressure of the vapor to be related by the Clausius-Claperon equation. The bubble radius was assumed to be small compared with the sound wavelength and the thermal diffusion length in the vapor. The variables in the process are given in Table 15.1.

The following set of equations was found to govern the motion

1. Equation of motion

$$R\ddot{R} + \frac{2}{3}\dot{R}^2 = (1/\rho_L)(P_v - 2\sigma/R - P_s + P_0 e^{j\omega t}) \tag{15.5.1}$$

2. Energy equation

$$\begin{aligned} 3\gamma P_v\dot{R} + R\dot{P}_v = {} & 3(\gamma - 1)\xi k_L(T_L)_R \\ & - [L - C_v T - (P_v 4\pi R^3)/(3M)] \\ & x\dot{M}_v/(4\pi R^2) \end{aligned} \tag{15.5.2}$$

3. Equation of state

$$P_v \frac{4}{3}\pi R^3 = M_v \Re T_v \tag{15.5.3}$$

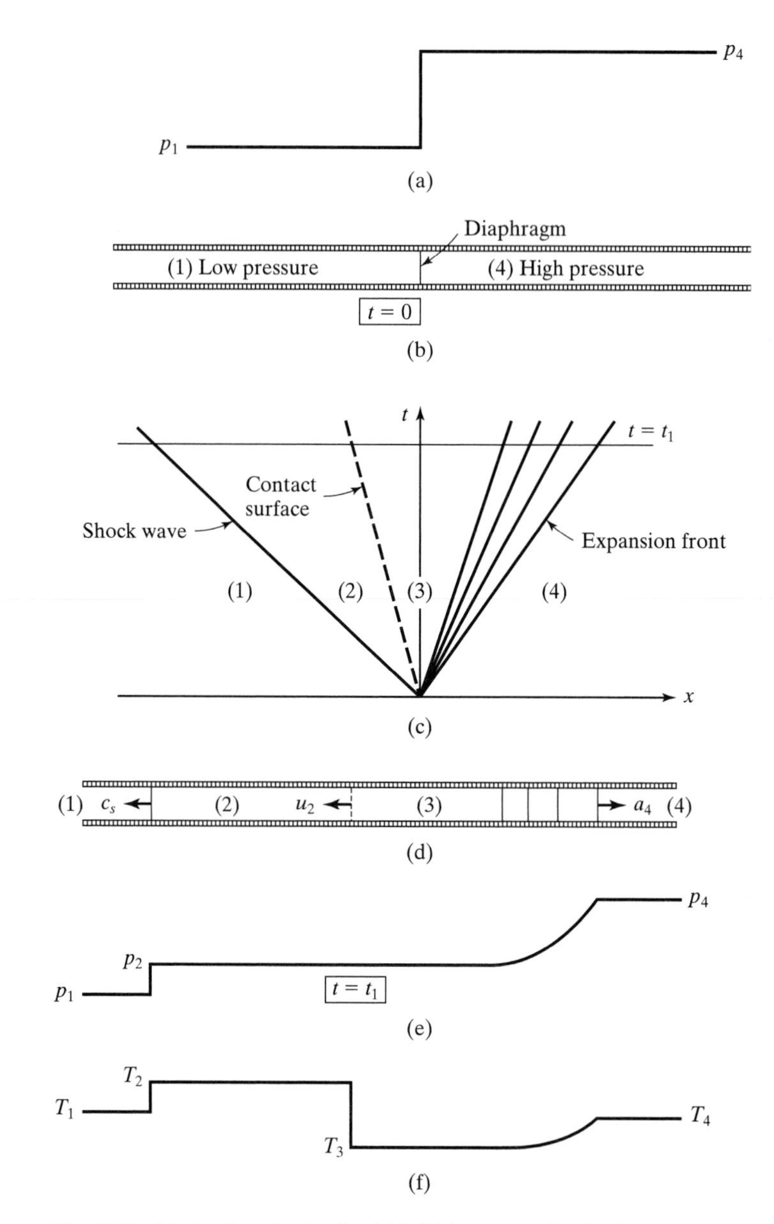

Fig. 15.17 Motion in a shock tube. (a) Initial pressure distribution. (b) Initial configuration. (c) Shock and expansion wave characteristics. (d) Configuration at t_1. (e) Pressure profile at t_1. (f) Temperature profile at t. (After Liepmann and Roshko, 1957.)

Table 15.1 Symbols for Cavitation Equations

ω_f = frequency of driving force
ω_n = natural frequency of bubble
ω_r = real roots of the characteristic equation
P_a = amplitude of driving force
P_v = vapor pressure
P_s = static pressure of the liquid
σ = surface tension
ρ_L = density of the liquid
$k = C_p/C_v$
C_p = specific heat at constant pressure
C_v = specific heat at constant volume
R = radius of bubble
s = Laplace variable
Subscripts
o = initial state at time zero
v = vapor
s = surface
max = related to the maximum natural frequency
c = related to the critical radius
T = temperature
M_v = mass of vapor inside bubble
α_L = thermal conductivity of liquid

4. Heat conduction equation

$$\frac{DT_L}{Dt} = \alpha_L \nabla^2 T_L \tag{15.5.4}$$

5. Temperature-vapor-pressure relation

$$T_v - T_{L\infty} = \Phi(P_v - P_{v\infty}) \tag{15.5.5}$$

The unknown quantities in these five equations are R, P_v, T_v, T_L, and M_v.

Some understanding of the phenomena involved can be obtained from a general linearization, using the method of the Taylor series expansion. The equations are rearranged so that the highest derivative is a function of the other variables and then all variables are assumed to take the form

$$x(t) = x_o + x(t) \tag{15.5.6}$$

that is, the sum of an equilibrium value and a time-dependent term. Thus, if the highest derivative is z(X, Y, W, ...), then assume a Taylor series expansion and linearizing

$$z(t) = \left(\frac{\partial Z}{\partial x}\right)_o x(t) + \left(\frac{\partial Z}{\partial y}\right)_o y(t) + \left(\frac{\partial Z}{\partial w}\right) w(t) + \cdots \tag{15.5.7}$$

We can illustrate the technique by rearranging the equation of motion

$$\ddot{R} = \frac{1}{R\rho_L}\left[P_v - \frac{2\sigma}{R} - P_s + p_A(t) - \frac{3}{2}\dot{R}^2\rho_L\right] \tag{15.5.8}$$

where $p_a(t)$ is an applied pressure, which is not necessarily sinusoidal. Applying the linearization scheme now yields

$$\ddot{r}(t) = (1/R_o\rho_L)(p_v(t) + p_a(t)) + (1/R_o^2\rho_L)(P_{vo} - 4\sigma/R_o - P_{so})r(t) \tag{15.5.9}$$

where it is assumed that the bubble wall is initially at rest. Similarly, we can linearize the other governing equations. Consider the simple case of a bubble with an insoluble gas content in a nonconducting liquid. The entire right side of the energy equation, Eq. (15.5.2), then reduces to zero and on linearizing yields

$$\dot{r}(t) = \frac{-R_o}{3\gamma P_{vo}}\dot{p}_v(t) \tag{15.5.10}$$

The equation of state yields

$$p_v(t) = \frac{3M_v\Re}{4\pi R_o^3}\theta(t) - \frac{9M_v\Re T_{vo}}{4\pi R_o^4}r(t) \tag{15.5.11}$$

where the temperature variation is given by

$$T(t) = T_o + \theta(t) \tag{15.5.12}$$

With our simplifying assumptions, the heat conduction equation and the temperature-vapor-pressure relation are not relevant. The relation between the equilibrium parameters is

$$P_{vo} = 2\sigma/R_o + P_{so} \tag{15.5.13}$$

We now have to solve the system of three equations—Eqs. (15.5.5), (15.5.9), and (15.5.11). This can be done by transforming to the Laplace domain. The transformed equations are

1. For motion

$$\left[s^2 + \frac{1}{R_o^2\rho_L}\left(P_{vo} - \frac{4\sigma}{R_o} - P_{so}\right)\right]r(s) = \frac{1}{R_o\rho_L}[P_v(s) + p_A(s)] \tag{15.5.14}$$

2. For energy

$$sr(s) = -\left(\frac{R_o}{3\gamma P_{vo}}\right)sp_v(s) \tag{15.5.15}$$

3. For the equation of state

$$P_v(s) = \frac{3M_v\Re}{4\pi R_o^3}\theta(s) - \frac{9M_v\Re T_{vo}r(s)}{4\pi R_o^4} \tag{15.5.16}$$

For a given driving pressure $p_A(s)$, after some simplification using Eq. (15.5.13), these three simultaneous equations can be solved to find the radial, vapor pressure, or temperature responses, $r(s)$, $P_v(s)$, or $\theta(s)$, respectively. The radial response, for example, is given by

$$r(s) = \frac{\frac{1}{R_o\rho_L}\cdot p_A(s)}{(s+\alpha)(s-\alpha)} \tag{15.5.17}$$

where

$$\alpha = \sqrt{\frac{2\sigma/R_o - 3\gamma P_o}{R_o^2\rho_L}} \tag{15.5.18}$$

Now if the bubble is driven by an applied sinusoidal sound field; that is

$$p_A(t) = p_o \sin \omega t \tag{15.5.19}$$

then

$$p_n(s) = \frac{p_o\omega}{s^2+\omega^2} \tag{15.5.20}$$

Hence

$$r(s) = \frac{\frac{p_o\omega}{R_o\rho_L}}{(s+j\omega)(s-j\omega)(s+\alpha)(s-\alpha)} \tag{15.5.21}$$

The solution in the time domain is obtained by taking the inverse Laplace transform of Eq. (15.5.21) to yield

$$r(t) = c_1 \sin(\omega t) + c_2 e^{\alpha t} + c_3 e^{-\alpha t} \tag{15.5.22}$$

where

$$c_1 = \frac{-p_o/R_o\rho_L}{\alpha^2+\omega^2} \tag{15.5.23}$$

$$c_2 = \frac{+p_o\omega/R_o\rho_L}{2\alpha(\alpha^2+\omega^2)} \tag{15.5.24}$$

and

$$c_3 = \frac{-p_o\omega/R_o\rho_L}{2\alpha(\alpha^2 + \omega^2)} \tag{15.5.25}$$

There are three regimes of bubble size in which the nature of the response is qualitatively different. In the regime where $R_o < \frac{2\sigma}{3\gamma P_{vo}}$ from Eq. (15.5.18), α is real, and due to the positive sign of c_2, the bubble will grow according to

$$r(t) = -\frac{p_o/R_o\rho_L}{\alpha^2 + \omega^2}\sin \omega t + \frac{p_o\omega/R_o\rho_L}{\alpha(\alpha^2 + \omega^2)}\sinh \alpha t \tag{15.5.26}$$

When

$$R_o > \frac{2\sigma}{3\gamma P_{vo}}$$

α becomes imaginary, thus

$$\alpha = j\sqrt{\frac{3\gamma P_{vo} - 2\sigma/R_o}{R_o\rho_L}} = j\omega_o \tag{15.5.27}$$

where ω_o is the bubble's natural resonance frequency. Then

$$r(t) = -\frac{(P_o/R_o\rho_L)}{\omega^2 - \omega_o^2}\sin \omega t + \frac{p_o\omega/R_o\rho_L}{\omega_o(\omega^2 - \omega_o^2)}\sin \omega_o t \tag{15.5.28}$$

This is an oscillatory solution with two harmonic components. If the driving frequency is less than ω_0, the component at the driving frequency predominates. If the driving frequency is greater than ω_0, then the component at ω_0 predominates.

In the case $R_o = R_c$, α is zero and the solution becomes

$$r(t) = -\frac{(p_o/R_o\rho_L)}{\omega^2 - \omega_o}\sin \omega t + \frac{p_o t}{R_o\rho_L\omega} \tag{15.5.29}$$

in other words, the motion is a sinusoid superimposed on a ramp function. The amplitude of the sinusoid is infinite, but of course this model neglects damping.

From Eq. (15.5.17)

$$r(s) = \frac{p_A(s)/R_o\rho_L}{(s + \alpha)(s - \alpha)}$$

$$= w(s)p_A(s) \tag{15.5.30}$$

We may then think of the bubble as a system with input $p_A(s)$ and output r(s), and transfer function

$$w(s) = \frac{1/R_o\rho_L}{(s+\alpha)(s-\alpha)} \tag{15.5.31}$$

It is possible to rewrite this equation as

$$w(s) = \frac{4\pi R_o^2}{4\pi R_o^3\rho_L s^2 + 4\pi R_o\left(3\gamma P_{vo} - \dfrac{2\sigma}{R_o}\right)} \tag{15.5.32}$$

We may say that the bubble has an equivalent mass

$$M = 4\pi R_o^3\rho_L \tag{15.5.33}$$

that is, an amount equal to three times the mass of liquid displaced. The stiffness

$$S = 4\pi R_o\left(3\gamma P_{ro} - \frac{2\sigma}{R_0}\right) \tag{15.5.34}$$

When R_o is less than R_c, the stiffness becomes negative and there is a positive instead of a negative feedback, resulting in collapse.

Some qualitative results are evident. The first point to be considered is the bubble's natural frequency as a function of the radius. This function is illustrated in Fig. 15.18,

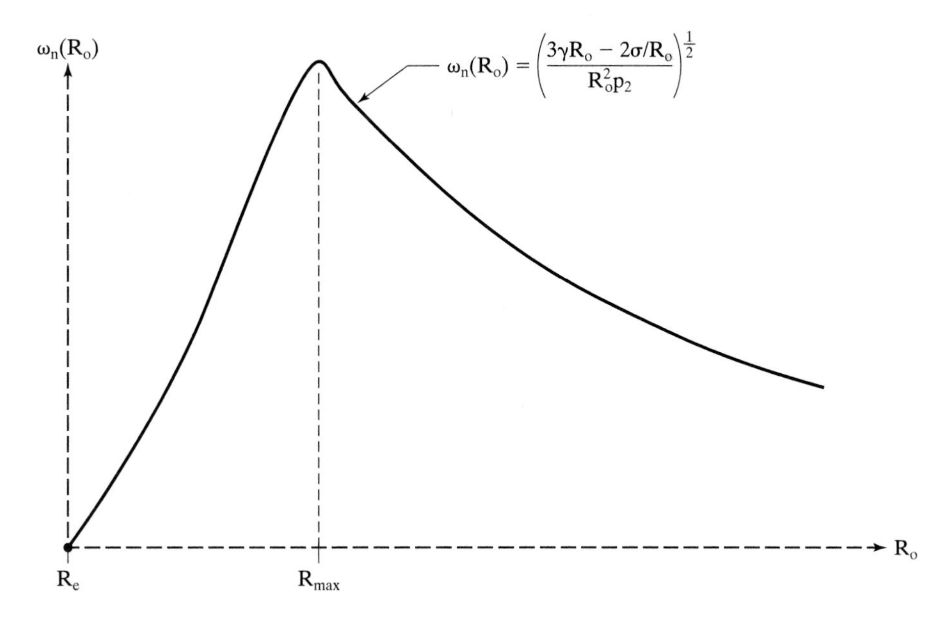

Fig. 15.18 Bubbles natural frequency as a function of R_o.

which shows that as the initial radius R_o is increased from the critical radius there is a maximum resonant frequency. The radius at this frequency is R_{max}. As the radius increases beyond R_{max}, the natural frequency decreases nonlinearly and tends towards zero as the radius approaches infinity. For radii in the range shown, the transient response is given by Eq. (15.5.28).

These solutions for the radial response can be analyzed in several ways. First, assume that a bubble pre-exists at some constant radius such that its natural frequency is large. Then, if the frequency of the forcing function is increased from zero, it is found that the amplitude of the transient response remains extremely small until the input frequency is on the order of the natural frequency. As the two frequencies draw closer together, the amplitude function displays an enormous growth. A graphical representation of this result is shown in Fig. 15.19.

Another way to analyze the solution is to fix the driving frequency at some large value and observe the coefficients as a function of the equilibrium bubble radius R_o. For values of R_o close to the critical radius (i.e., for small natural freguencies), the amplitude response will be extremely small. As R_o is increased, a vertical asymptote is found when the natural frequency reaches the input frequency; and as R_o is increased towards the maximum frequency, the amplitude response returns to small values.

In attempting to qualitatively relate the solutions of an ideal bubble to a real vapor bubble, consider the assumptions that were made. One: It was assumed the bubbles were filled with an ideal gas. The effect of replacing the gas with a condensable vapor would result in reducing the vapor pressure during a compression process. Two: It was assumed the bubble was an isolated system. Allowing heat transfer to occur at the bubble's surface will result in a reduction of internal pressure during a compression process; it will also help to reduce pressure drops during expansion processes.

Cavitation is a subject of considerable practical importance. It first became of engineering significance when high-speed propellers were introduced for marine propulsion. It was found that these propellers were subject to rapid erosion, which was proved to be due to the formation and subsequent collapse of cavitation bubbles as the propellors rotated. Rayleigh (1917) investigated the subject by positing the existence of a spherical void in the liquid. He showed that such a cavity would implode with accelerating inward velocity, thereby generating high pressures in the liquid. For representative starting conditions, he showed that such pressures might reach thousands of atmospheres. It was to these pressures that Rayleigh ascribed the erosional effects of cavitation. More recent work suggests that a bubble collapsing in the vicinity of a solid surface produces a miniature high-speed jet, and it is believed that this may be the source of damage.

A high amplitude acoustic field can also generate cavitation. Water is believed to contain a profusion of microscopic bubbles stabilized against collapse under surface tension with "skins" of surfactant molecules. If the sound pressure amplitude exceeds some threshold, then these microbubbles can act as nuclei for the occurrence of macroscopic cavitation events. In a sound field, a bubble may oscillate several times with increasing amplitude before finally undergoing a Rayleigh collapse. A numerical integration of equations similar to Eqs. (15.5.1) to (15.5.5) was obtained by Noltingk and Neppiras (1950) showing the occurrence of a collapse, as illustrated in Fig. 15.20.

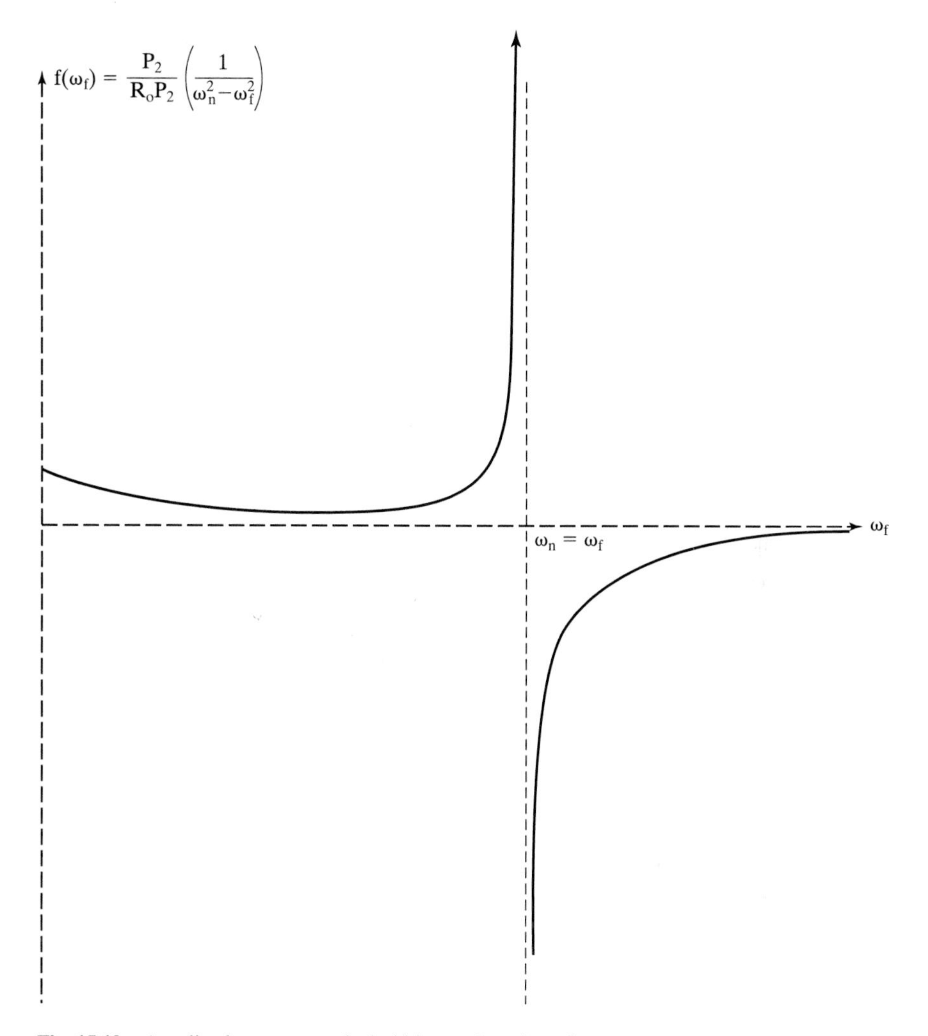

Fig. 15.19 Amplitude response of a bubble as a function of ω_f.

Over a cycle there may be a net diffusion of heat and gas into a cavitation bubble, so that bubbles can grow to macroscopic size from microscopic nuclei. The occurrence of clouds of cavitation bubbles in front of sonar transducers effectively limited their power output. The gas and vapor content of an imploding Rayleigh cavity also reaches very great pressures and temperatures, which can result in their producing an incandescence known as sonoluminescence. Many chemical reactions are spurred by the same agency. Cavitation is put to use in the process of ultrasonic cleaning. Cavitation may be the mechanism that makes ultrasonic dental cleaners and phase-emulsifiers effective. There are a number of industrial processes that also depend on ultrasonic cavitation for their operation. Frederick (1965) and Kuttruff (1991) provide further details on this topic.

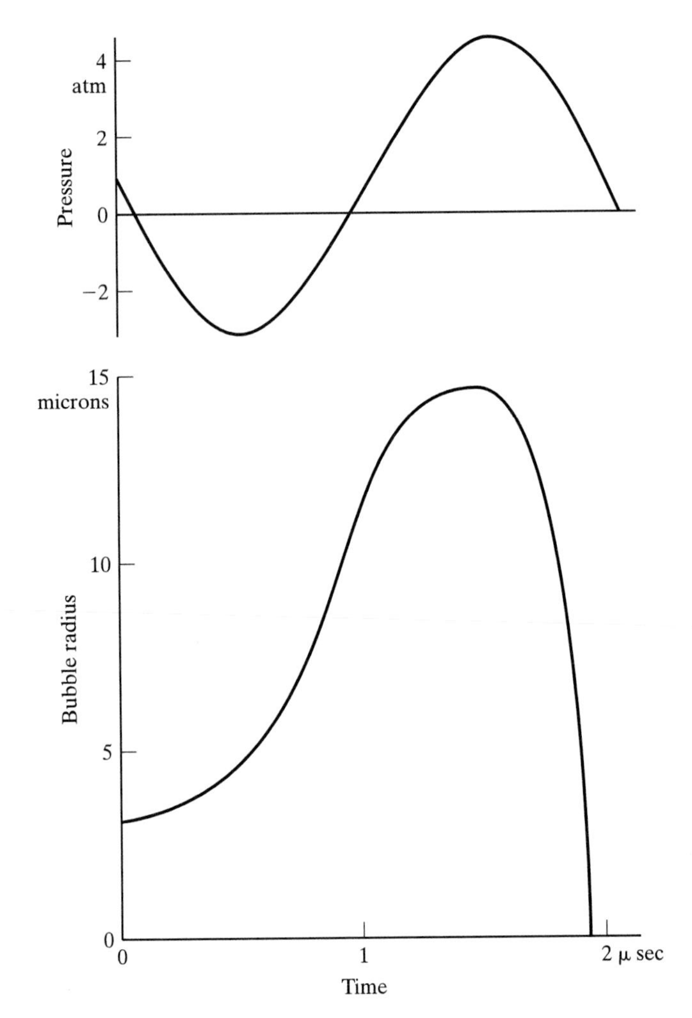

Fig. 15.20 Calculated variation of radius with time for cavitating bubble. (After B.E. Noltingk and E.A. Neppiras, 1950.)

15.6 Conclusion

A short review in an introductory text such as this cannot provide details of all the important and interesting nonlinear phenomena in the realm of acoustics. Thus, we have not discussed flow-induced sound and vibration, the source of whistles and jet noise. (Blake (1986) provides a discussion of this topic.) Nor have we discussed thermally maintained oscillations, currently being investigated for use in new forms of heat engines and refrigerators.

The advent of the digital computer has also acted as a stimulus to research in the solution of nonlinear equations, and over the past twenty years or so a vast amount of new information has been accumulated in this area, which has become known as "chaos." The term is chosen because of the unpredictable and irregular motion involved. Details of this work are covered in technical reviews by Moon (1987) and by Lauterborn and Parlitz (1988), with an excellent layman's account by Gleick (1987). For the most part, the engineer tries to avoid nonlinear vibrations, and there are no practical applications of nonlinear vibration science, although that situation may change. But it is clear that some of the findings in the area are of profound philosophical interest, and for that reason the following summary is given.

The behavior of the nonlinear system is generally described in state space, which is a generalization of the phase space description we have already encountered. The system will follow some trajectory in state space, which over a number of cycles will occupy some volume. For a dissipative system, this volume will shrink as time goes on to converge on a point, or in a closed orbit called a limit cycle (which will be traversed repeatedly), or on a toroidal surface on which the trajectory is wound. A trajectory on such a torus is a called a quasiperiodic motion. These final configurations in state space are known as "attractors." Chaotic motion follows a new type of attractor, called a strange attractor, which can exist only in a state space of at least three dimensions. Strange attractors exhibit self-similarity in that features of their structures repeat again and again on ever smaller scales. One way of visualizing the behavior of a nonlinear system is by mapping the positions of the intersections of successive orbits of trajectories on a plane in state space, which is known as a Poincare section.

A dynamic system may possess several different attractors, which may be reached starting from different points in state space. The region of state space containing points from which trajectories lead to a certain attractor are said to constitute the basin of attraction for that particular attractor. The basins of attraction can be remarkably convoluted so that neighboring points in state space can belong to quite different attractors. Very small differences in initial conditions can thus lead to very large differences in end conditions. The uncertainty in the end conditions increases as time goes on, and some researchers have related this process to the concept of entropy in information theory. The rapidity of divergence of neighboring trajectories in state space is measured by the so-called Lyapunov exponent.

If some parameter of a system is varied, there may be a change in the type of attractor operative, and such a switch is called a bifurcation point. There are three basic types of bifurcation points: the Hopf, the saddle point, and the period doubling (or pitchfork). The pitchfork bifurcation operates only on periodic orbits. A limit cycle of a given period changes to one of exactly double that period. In other words, a half frequency subharmonic appears. When one such period doubling has occurred, with further increases in the control parameter there will be more pitchfork bifurcations to four times the original period, and then eight, and so on. This is called a cascade and results eventually in periods of infinite extent (i.e., in motions that are not periodic). The system is said to have followed a period doubling route to chaos.

REFERENCES AND FURTHER READING

Beyer, R.T. 1974. *Nonlinear Acoustics*. U.S. Department of the Navy.

Blackstock, D.T. 1972. Nonlinear Acoustics (Theoretical), in *American Institutes of Physics Handbook*, 3rd ed. McGraw Hill.

Blake, W.K. 1986. *Mechanics of Flow-Induced Sound and Vibration*. Academic Press, Inc.

Duffing, G. 1918. *Erwugene Schwingungen bei Veranderlicher Eigenfrequenz*. F. Vieweg. U. Sohn.

Earnshaw, S. 1860. 150 *Proc. Roy. Soc.* 133.

Finch, R.D., and E.A. Neppiras. 1973. "Vapor Bubble Dynamics." 53 *J. Acoust. Soc. Am.* 1402–1410.

Frederick, J.R. 1965. *Ultrasonic Engineering*. Wiley.

Gleick, J. 1987. *Chaos*. Viking.
Kuttruff, H. 1991. *Ultrasonics Fundamentals and Applications*. Elsevier.
Lauterborn, W., and U. Parlitz. 1988. "Methods of Chaos Physics and Their Application to Acoustics." 84 *J. Acoust. Soc. Am.* 1975–1993.
Liepmann, H.W., and A. Roshko. 1957. *Elements of Gas Dynamics*. Wiley.
Moon, F.C. 1987. *Chaotic Vibrations*. Wiley.
Noltingk, B.E., and E.A. Neppiras. 1950. 63B *Proc. Phys. Soc.* 674.
Pain, H.J., and E.W.E. Rogers. 1962. "Shock Waves of Gases." 25 *Rep. Prog. Phys.* 287–337.
Pippard, A.B. 1985. *Response and Stability*. Cambridge Univ. Press.
Rayleigh, Lord. 1917. "On the Pressure Developed in a Liquid During the Collapse of a Spherical Cavity." 34 *Philo. Mag.* 94–98.
Stoker, T.J. 1950. *Nonlinear Vibrations in Mechanical and Electrical Systems*. Wiley-Interscience.
Van der Pol, B. 1927. "Forced oscillations in a system with nonlinear resistance." *Philo. Mag.*

PROBLEMS

15.1 A pendulum consists of a massless rod of length l and a bob of concentrated mass m. Show that the angular frequency ω is given approximately by

$$\omega^2 = \frac{g}{l}\left(1 - \frac{\theta_0^2}{8}\right)$$

where θ_0 is the initial angular displacement and g is the acceleration due to gravity.

15.2 Show that Eq. (15.1.11) yields $T = 2\pi/\omega_0$ when $b = 0$.

15.3 A block of mass M rests on a horizontally moving belt with coefficient of friction μ. The block is attached to a horizontal spring of stiffness S, as shown in Fig. P15.3 Describe what happens using x as a measure of the block's displacement.

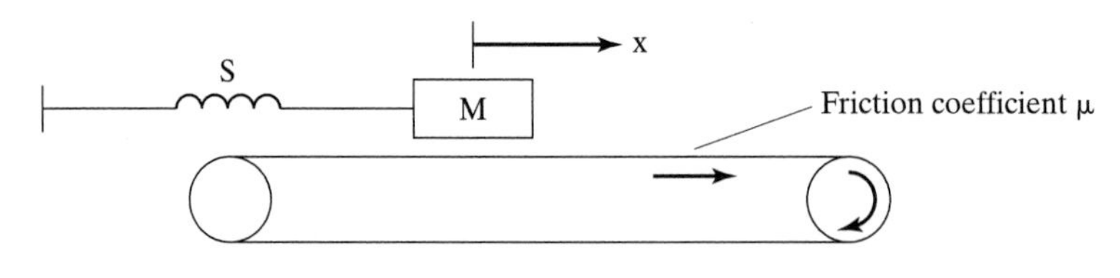

Fig. P15.3

15.4 Explain what is meant by the "jump phenomenon" in the forced oscillation of a mass on a nonlinear spring.

15.5 Derive the results given in Eqs. (15.3.31) and (15.3.32).

15.6 Give a theoretical explanation for the occurrence of subharmonics during ultrasonic cavitation.

15.7 A spherical cavity in water, with radius R_0, contains gas at STP. The cavity is expanded rapidly to a radius R_e and then collapses rapidly to radius R_c, where R_c is much less than

R_0. Assuming that no gas diffuses in or out of the cavity and that condensation and evaporation are negligible, estimate the final temperature and pressure in the cavity.

15.8 **a.** Suggest ways in which the expansion and collapse of the cavity in Problem 15.7 might be brought about.

b. Suggest some realistic values for R_0, R_e, and R_c to be obtained using the methods you might use in part a. What processes will actually determine R_e and R_c in your opinion?

15.9 It is known that collapsing cavities can produce light. Discuss the mechanisms that might be involved in such a process.

15.10 It has been hypothesized that collapsing cavities might produce thermonuclear reactions. Discuss the likelihood of such an effect occurring.

Chapter 16

Noise Control

16.1 The Auditory System

Acoustical engineers need to have a knowledge of hearing and speech for many reasons, and we have already alluded to some of them. In our earlier discussion of the production of sound and its measurement, it was understood that the instrumentation had to serve human audition and speech; in this chapter, we will discuss problems of noise control. But for certain applications, the concern is more intimately connected with the hearing and speech production process. Engineers may be responsible for the design of instrumentation used to measure speech and hearing. They may also be responsible for designing hearing aids and prostheses for all or parts of the ear or vocal tract. It is therefore crucial that engineers be acquainted with the anatomy and functioning of the relevant organs. We will begin by discussing the hearing mechanism.

The ear is an electroacoustic marvel. In a volume of approximately one cubic inch, it contains a sound system that includes an impedance matcher, a broad range frequency analyzer, and a multichannel transducer to convert mechanical energy to electrical energy—in other words, an internal two-way communication system as well as the means to maintain a delicate hydraulic balance. The human ear possesses a remarkable dynamic range in both frequency and amplitude response. It responds to oscillations over the frequency range from 20 Hz to 20 kHz, a factor of 1,000 in frequency (compared with the eye, which covers only a factor of about two). The ear can detect a displacement amplitude of less than one tenth of the diameter of a hydrogen atom, and it can still function in the presence of the noise generated during the takeoff of an Apollo rocket, an energy ratio of about 10^{12}.

The human ear consists of three parts: the outer, middle, and inner ears, as shown in Fig. 16.1. Airborne sound is received by the ear through air conduction and transmitted through the outer ear to the tympanic membrane (eardrum or tympanum) at the entrance to the middle ear. The tympanum vibrates, in turn setting the ossicles of the middle ear into vibration. There is a sequence of three ossicles in the middle ear: the malleus, the incus, and the stapes. The acoustic energy flows along this chain into the cochlea, which is a hollow spiral of about two and a half turns, filled with fluid into which the hair cells project. The energy moves through the fluid to the hair cells, where it is converted into electrochemical signals that are carried to the brain via the auditory nerve. The high sensitivity of the ear is achieved by a high-energy conversion efficiency at each stage of the transmission path.

Egyptian records of 1500 B.C. refer to the ear as an organ of hearing and respiration. The early Greeks, notably Empedocles (504–443 B.C.), conceived of sounds as vibratory movements, propagated through the air, and they were aware that hearing is a result of the passage of these vibrations into the ear. Beyond this basic fact, Empedocles and his followers had little understanding of the hearing process, and their knowledge of the ear's anatomy was limited. They knew about the drum membrane and the tympanic cavity behind it, but the other deep-lying structures escaped their notice. Galen (130–200 A.D.), took a step forward by recognizing the importance of nerve excitation in the sensory process in general, and he knew about the auditory nerve, having seen its bundle of fibers passing out of the internal auditory meatus to the brain. No further progress was made until the sixteenth century, when anatomists brought to light most of the rest of the previously unknown parts of the ear. By the middle 1500s, the essential features of the conduction apparatus were well recognized. The two larger ossicles were discovered first by Berengario da Capri in 1514. Then in 1546 Ingrassia discovered the third ossicle, the stapes, and the two windows of the cochlea. In 1561 Fallopius distinguished the two

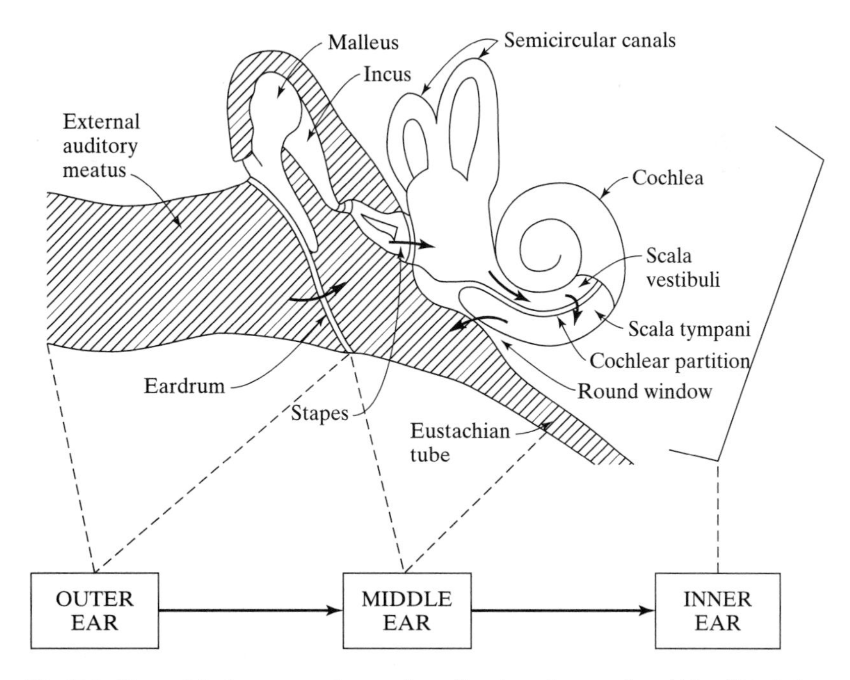

Fig. 16.1 Parts of the human ear. Arrows show direction of energy flow. (After Helmholtz, 1862.)

principal divisions of the inner ear, and gave them their present names of cochlea and labyrinth. In 1564 Eustachius discovered the tube now known by his name, connecting the tympanic cavity with the pharynx.

A well-documented science of hearing has existed only since 1862, when Helmholtz advanced a theory that continues to be influential today. He recognized some of the problems regarding the transfer of energy from air to the cochlear fluids. He attributed this energy transmission to ossicular lever action and the tympanum to stapes areal ratio. The first prosthesis for the middle ear was described by Toynbee in 1865. It consisted of a rubber disk about three quarters of an inch in diameter with an attached silver wire. Since that time, there has been much progress in understanding the mechanism of hearing, in making experimental measurements (especially with the aid of electronic instrumentation), and in treating hearing disorders. Schubert (1978) can be consulted for a more detailed history of hearing research.

We understand biological systems to the extent that we can duplicate their observed behavior with postulated artificial devices. Chapter 2 introduced the concept of system diagrams in which each component of a more complex system is represented as a block, each with its own transfer function. Our current understanding of the function performed by each of the three parts of the ear may be presented in the form of the block diagram shown in Fig. 16.2. The first block represents the outer ear. The function of the pinna is, to some extent, to focus sound at the entrance to the ear canal. The canal serves to provide the delicate equipment of the middle and inner ear with some protection. There is a great difference in impedance of air and the fluid of the inner ear. The major role of the middle ear is thus impedance matching. Without the middle ear, transmission of acoustic energy would be very poor. Within the inner ear, the motion of

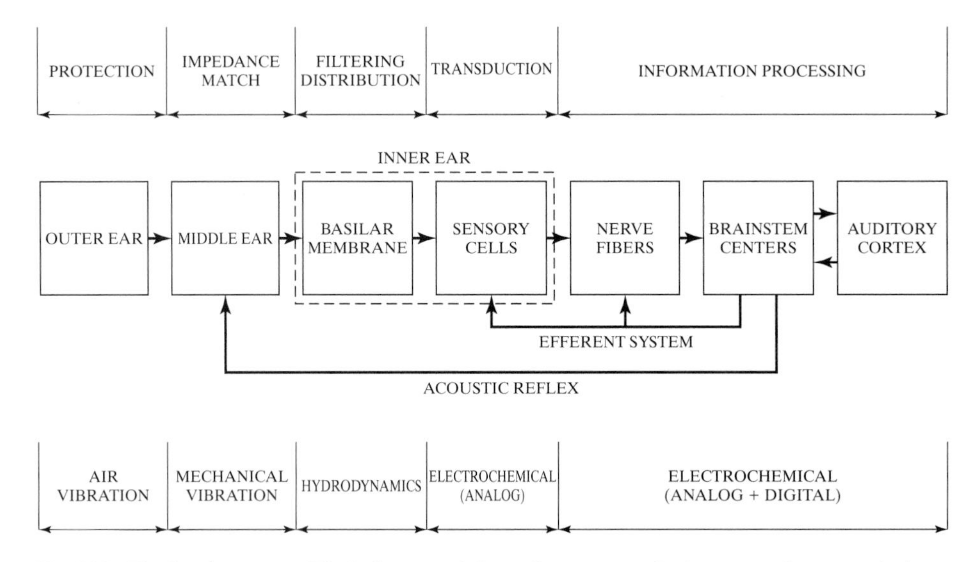

Fig. 16.2 The hearing system. Block diagram of the auditory system. In the center, the anatomical subdivisions are shown, with the arrows indicating the direction of information flow. At the top, the function of the various segments is indicated, while at the bottom the dominant mode of operation is described. (After Dallos, 1973.)

the fluid is transmitted through the basilar membrane (which acts as a frequency filter) to the hair cells (which produce the signal sent to the brain). The movement of these cells is the final mechanical link in the auditory chain. The hair cells serve as transducers. The auditory nerve passes its signals through various brainstem centers to the auditory cortex. These centers also serve to control the outgoing (efferent) nerves, which control the blood supply in the ear. They also control the acoustic reflex, a feedback mechanism that protects the inner ear by tightening the stapedial muscle to reduce the transfer function of the middle ear when the sound becomes excessively loud. The last three blocks in Fig. 16.2 may thus all be regarded as having information-processing functions. Fig. 16.2 may also be regarded either as a very abstract representation of the ear or as a very abstract representation of an artificial system. Later in this chapter, we will discuss modeling of the ear, which in effect is the process of converting the abstract functional representation of an artificial system into physically realizable devices.

Balsalla (1988) has pointed out that a strong analogy can be drawn between biological evolution and the history of technology. The acoustical engineer charged with developing hearing aids or prosthetic devices can thus expect to find ideas from a comparative study of the hearing mechanism in various species and its phylogeny (evolutionary history). The earliest vertebrates were aquatic and their tympanums were exposed directly to water. The typical amphibian tympanum is planar and consists mainly of a set of radial fibers that converge on a thick fibrous disk in the center. Evolved from the amphibians, the reptiles have a tympanum that is flat, except for a small lateral protrusion at the center, and has both radial and nonradial fibers. The avian tympanum evolved from the reptilian; it is conical and points outwards, and it also has both radial and nonradial fibers. Mammals evolved from the reptiles independently from the birds. The mammalian tympanum is conical and points inwards, and the radial and nonradial fibers are separated into two distinct layers.

Studies have been conducted on many animals to determine their thresholds of hearing, and a considerable variation of ability has been found. Some are capable of perceiving very high frequencies of up to 90 kHz (e.g., bats), while others have exceptionally good low-frequency hearing at less than 50 Hz (e.g. chinchillas and kangaroo rats). The differences in the audiograms of different species depend on the length of the meatus, the size of the tympanum, and the ossicles. Generally, there is a correlation between size and the frequency of maximum hearing accuity, with larger animals having better hearing at lower frequencies. Elephants, for example, hear well into the infrasonic range. Bekesy (1960) has argued that while a small ear may be well suited for good high-frequency reception, good low-frequency reception may be possible only with a larger ear.

The outer ear consists of a visible portion called the pinna, or auricle, and a portion inside the head, the ear canal or external auditory meatus (a meatus being a passage in the body). The pinna is a cartilaginous structure with many convoluted folds leading to the opening of the external auditory meatus. In humans, it is without useful musculature, so it remains relatively immobile with reference to the head. It serves as a horn that receives acoustic energy and directs it to the auditory canal, as well as serves as possible aid to the localization of sounds in space. In certain animals, the pinna can be turned towards the sound source, and in such instances it undoubtedly improves sensitivity by reflecting or scattering sound toward the meatal opening. The meatus begins near the center of the pinna and is an irregularly shaped canal, whose length in humans is about 30 mm. Its longitudinal axis is nearly perpendicular to the side of the head, and in cross section the height of the meatus is slightly greater than its width. Its mean diameter is about 7 mm. The outer third of the meatus has a cartilaginous wall while the rest is bony. The inner wall of the canal is lined with skin, bearing numerous hairs and wax-producing glands, which serve a protective function and keep the intrusion of foreign matter to a minimum. The meatus has a natural resonant frequency of about 3000 Hz.

The middle ear is an air-filled volume situated in a cavity that in higher mammals is surrounded by the temporal bone of the skull, but in most laboratory animals is surrounded by a thin bony compartment, the auditory bulla, attached to the skull. The parts of a normal human middle ear are shown in Fig. 16.3. The middle ear cavity extends from the tympanum on its lateral border to the bony cochlear wall on its medial extreme. The middle ear communicates with the inner ear through two openings in this bony wall—the oval window and the round window—and it also communicates with the nasopharynx via the Eustachian tube. This tube, which is closed most of the time, opens during swallowing to permit the air pressure in the middle ear to stay at the atmospheric value. The tympanum is about 69 mm^2 in area and has the shape of a flat cone, whose altitude is about 2 mm with the apex pointed inwards and its oval base held obliquely within the meatus by a bony ring, the annulus. The tympanum itself is made up of three layers. The outermost layer is continuous with the lining of the meatus, while the innermost layer is continuous with the lining of the middle ear. The structure that gives it its characteristic cone shape and its structural stability is the central fibrous layer. It is this layer that is composed of two groups of fibers, one radially oriented and the other concentrically arranged.

The tympanum is connected over a length of a few millimeters to the manubrium, a long process of the malleus, the first of the ossicles. The malleus, in turn, is connected to another ossicle, the incus. The nature of the incudomalleal joint is a matter of some contention, as we will soon discuss. The long process of the incus is connected to the third

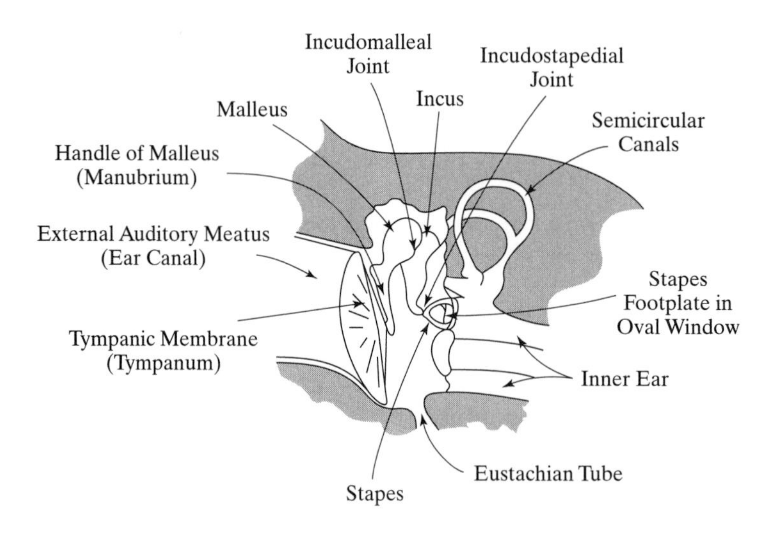

Fig. 16.3 The human middle ear.

ossicle, the stapes, through the incudostapedial joint. This joint shows greatest flexibility for movements perpendicular to the piston-like motion of the stapes. The footplate of the stapes rests in the oval window of the inner ear. The tympanum is under slight inward tension.

Some earlier researchers, including Bekesy, reported that the incudomalleal joint is rigid. That this appears to be incorrect has now been demonstrated by many researchers. Marquet (1981) in particular went to great lengths to show that a normal, healthy incudomalleal joint (see Fig. 16.4) is not rigid, but flexible. He pointed out that the incudomalleal joint has a ball-and-socket type of configuration. The malleus provides the ball element, while the incus provides the socket element. His extensive study used material from more than 3,000 temporal bone studies, as well as information acquired during ear surgeries and the results of audiological tests. He also used data derived from the use of cinematography, photography, holography, stroboscopy, and television recordings for comparisons with live and dead material. He believed that one of the main causes of error in evaluating the action of the incudomalleal joint can be attributed to the fact that most human studies of the joint are done on dead material, with the effects of rigor mortis causing the flexibility to change.

Ballenger (1977) describes how the ossicles are suspended in the middle ear cavity partly by the rigid connection of the manubrium to the tympanum and partly by several ligaments and two muscles. Three ligaments attach to the malleus (the ligamentum mallei superius, ligamentum mallei laterale, and ligamentum mallei anterius). Two ligaments attach to the incus (ligamentum incudis superius and ligamentum incudis posterius). The stapes footplate is attached to the margin of the oval window by an annular ligament (ligamentum annulare stapedis). The middle ear has two muscles: the tensor tympani and the stapedius. The tensor tympani is attached near the neck of the malleus; the stapedius is attached near the neck of the stapes. When the tensor tympani contracts, it draws the

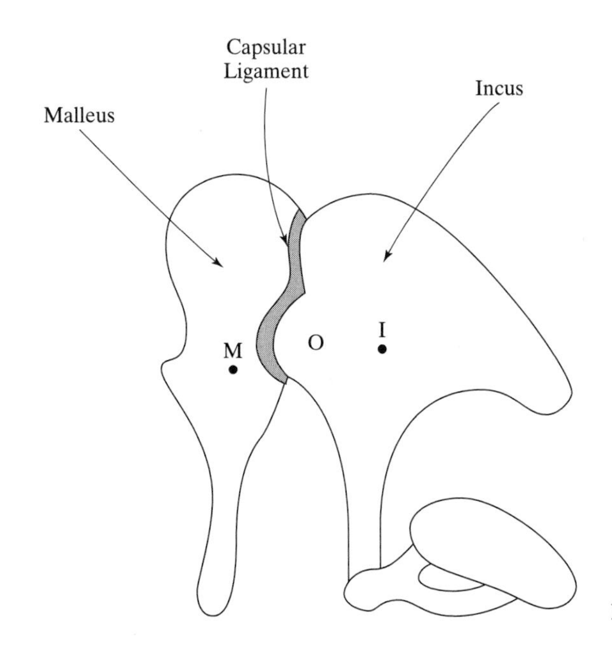

Fig. 16.4 The incudomalleal joint.

handle of the malleus inwards. This action places the tympanum under tension. Contraction of the stapedius draws the head of the stapes downwards and out of the oval window. The acoustic middle ear reflex is activated when high amplitude sounds are introduced to the middle ear. It is completely involuntary. This reflex is bilateral: When one ear is stimulated, the muscles of both ears are engaged. According to Moller (1961), contraction of the stapedial muscle lessens the mobility of the stapes. Contraction of this muscle causes the stapes to move and reduces sound transmission to the inner ear. Different views exist on which of these muscles are involved in the acoustic middle ear reflex. Bekesy (1960) claimed that both the stapedius and tensor tympani muscles are involved in the reflex. But later researchers have stated that in humans, only the stapedius is involved in this reflex.

In a review on the middle ear muscles, Borg and Counter (1989) expressed the view that the primary function of the acoustic stapedius reflex (ASR) is to protect the inner ear from the damaging effects of excessive displacement of the stapes caused by sound intensity levels that are 80 dB to 90 dB above the threshold of hearing. Not only does the ASR protect against the effects of loud external sounds, it is also active during speech, presumably to offer protection against the high sound levels speech produces close to the head. The ASR reduces transmission to the middle ear by 20 dB or more. The ASR takes between 100 ms and 200 ms to become effective, which is sufficient to protect against slowly rising sounds, such as thunder, but inadequate to muffle fast rising sounds, such as a gun shot.

The vestibule or entrance chamber of the inner ear connects with the middle ear through two membranes: the oval window and the round window. The stapes rests on the oval window. The cochlea is shaped like a snail shell enclosing a conduit of decreasing diameter, coiled about two and a half times. Uncoiled, the cochlea would be about 3.5 cm long. The conduit is divided along its length by the cochlear partition, creating upper and lower passages: the scala vestibuli and the scala tympani, respectively. The upper and lower chambers are filled with a liquid called the endolymph, which is sodium rich. The two scala are connected through an opening called the helicotrema at the apex of the coil. The cochlear partition also contains a passage, called the scala media (or sometimes the cochlear duct). The fluid in the scala media is also called endolymph, and it is potassium rich. The detail of this structure is shown in Fig. 16.5. The three fluid-filled tubes are separated from one another by two membranes: the upper being known as Reisner's membrane and the lower as the basilar membrane. It was discovered in the 1950s that there are various electrical potentials associated with the cochlea. The perilymph in the scalae is at a +3 mv DC potential relative to the bone, but it is now customary to take the level within the scala vestibuli as the reference. The potential within the scala media is a high positive value of 80 mv to 100 mv, while inside the organ of Corti the value is −40 mv. These DC voltages are present at all times, while what is known as the cochlear microphonic is an AC voltage produced in the cochlea, first observed in the 1930s. Dallos (1973) provides more details on these potentials.

Bekesy provided the first explanation for the action of the inner ear in 1948. According to his theory, any movement of the oval window produces a displacement of the endolymph. The cochlear partition is consequently deflected. Bekesy showed that each region of the cochlea will be most responsive to a particular frequency range (some of his results are shown in Fig. 16.6a). In other words, the cochlea acts as a frequency filter. Note that the higher frequencies have their maximum sensitivity closer

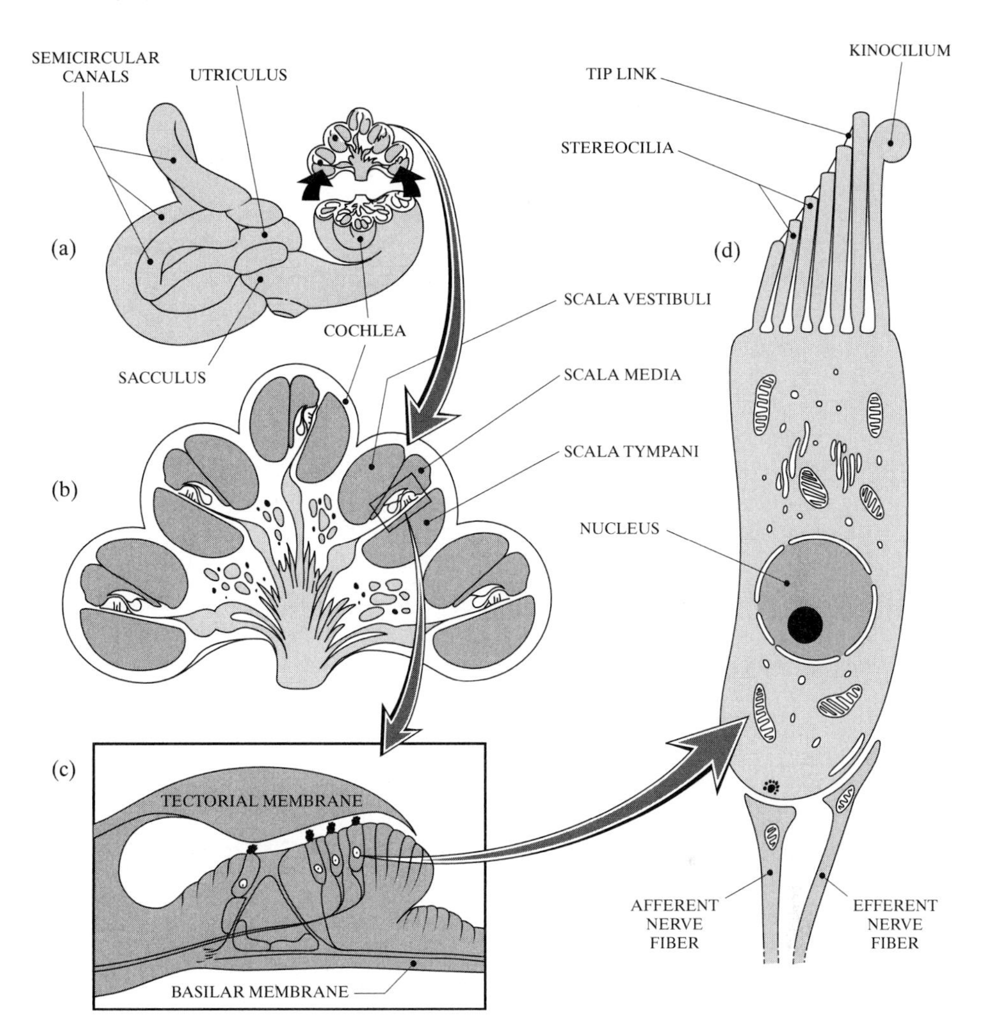

Fig. 16.5 The internal ear (a) includes the five acceleration-sensitive organs of the vestibular labyrinth and the cochlea, the detector of airborne sound. The sacculus and utriculus include, respectively, about 15,000 and 30,000 hair cells in planar sheets. The vertically oriented sacculus best detects up-and-down accelerations, while the utriculus is most sensitive to accelerations in the horizontal plane. Each of the three semicircular canals, which measure angular accelerations, consists of a fluid-filled tube interrupted by a gelatinous diaphragm into which insert some 7,000 mechanically sensitive hair bundles. The spiraling cochlea (b) comprises three fluid-filled tubes separated from one another by a pair of elastic, helical partitions. Upon the thicker of these partitions, the basilar membrane (c), sits the organ of Corti, which includes 16,000 hair cells disposed in four rows. Each hair cell (d) is anatomically and functionally divisible into two regions. The hair bundle at the cell's top is the detector for mechanical inputs. The bundle contains numerous actin-stiffened stereocilia and may include a single true cilium at its tall edge. The cell's basolateral membrane surface is specialized for electrical amplification of the receptor potential and for synaptic transmission of information to afferent nerve fibers. Efferent nerve fibers, whose activity regulates hair-cell sensitivity, also terminate on the basolateral membrane. (From *Physics Today*, 1994.)

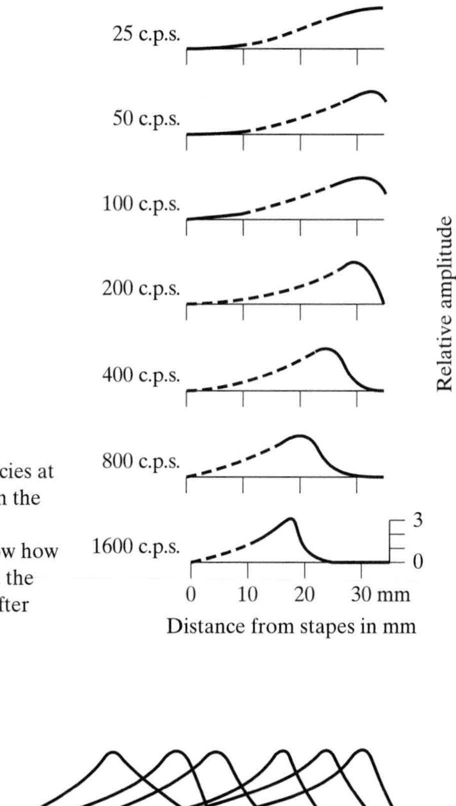

Fig. 16.6a The stapes was driven with different frequencies at a constant amplitude. After an opening had been made in the cochlea, the amplitude of the vibration of the cochlear partition was measured at various places. The figures show how the maximum amplitude of vibration is displaced toward the stapes as the frequency of the vibrations is increased. (After Bekesy (1949).)

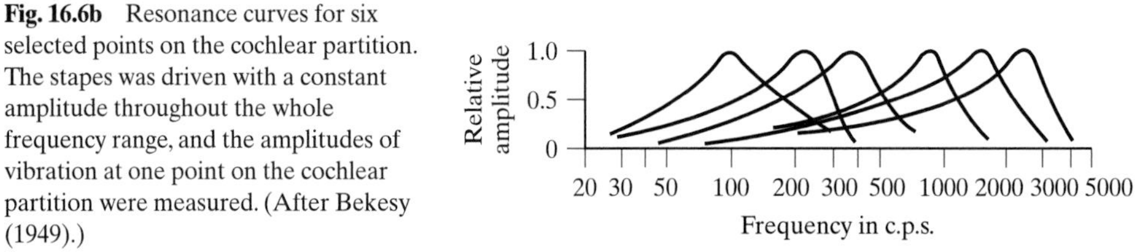

Fig. 16.6b Resonance curves for six selected points on the cochlear partition. The stapes was driven with a constant amplitude throughout the whole frequency range, and the amplitudes of vibration at one point on the cochlear partition were measured. (After Bekesy (1949).)

to the stapes. Another way of illustrating the phenomenon is to show the resonance curves at various points on the membrane as functions of frequency, as in Fig. 16.6b. These results provide a plausible explanation for a phenomenon known as masking. A pure tone of frequency f_1 and of a certain amplitude will produce a certain response all along the membrane, including the point for the maximum response at frequency f_2. A pure tone at f_2 will therefore have to exceed the response due to f_1 before it is heard. The f_2 tone is said to have been masked by the f_1 tone.

Situated on the basilar membrane is a structure known as the organ of Corti, as shown in Fig. 16.5c. It is here the hair cells that perform the transduction are located. There are three rows of outer hair cells and one row of inner hair cells for a total of 16,000. In recent years, there has been rapid progress in understanding the details of the transduction process. The sensitive element of hair cells are the so-called stereocilia, which are cylindrical in shape, having a diameter of 5 microns and a length of 35 microns. Each stereocilium consists of a protein called actin, which is very stiff. Because only a few actin filaments extend down into the body of the cell, the stereocilia are effectively hinged at their base. The stereocilia are grouped together in bundles. Hudspeth proposed

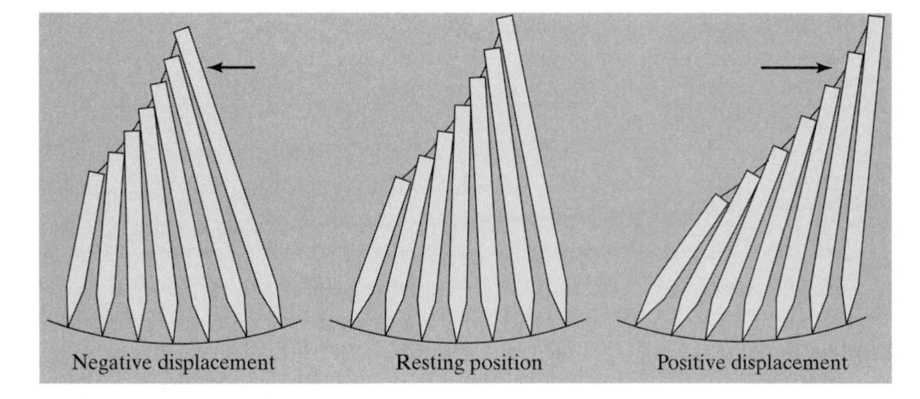

Fig. 16.7 Motion of a hair bundle when stimulated by a mechanical force indicates that the individual stereocilia pivot about their basal insertions while remaining straight. Although a hair bundle can survive displacements as great as those shown, the loudest tolerable sound would move a bundle by only one stereociliary diameter. The gating-spring model supposes that mechanoelectrical transduction results from the longitudinal shear produced between adjacent stereocilia in a deflected hair bundle. During positive stimulation the tip link that connects two contiguous stereocilia is stretched and promotes opening of ion channels by pulling on a molecular gate. Negative stimuli relax the tip link and permit the closure of some of the channels that are open at rest. (From *Physics Today*, 1994.)

that the transduction occurs when the whole bundle is displaced backwards or forwards, as shown in Fig. 16.7 (see Hudspeth and Markin (1994)). The hair cell has an electrical potential because of differences in ionic concentrations in the fluids on either side of its surface membrane. Corey and Hudspeth (1983) proposed that the mechanism for mechanoelectrical transduction is the molecular gating of certain ion channels. It is supposed that the molecules in question are members of the myosin family, one member of which permits the contraction of muscles. There are also some nonlinear effects that may be accounted for by the mechanical gating theory.

There are phenomena involved with the mechanics of the cochlea that are still unclear. One is that the motion of the basilar membrane is about 100 times greater than would be expected on the basis of most physical models, which suggests that there is also some amplifying mechanism in the cochlea. Another such phenomenon is otoacoustic emission; that is, the production of sound by the ear when it is stimulated with an acoustic click. A third evidence for an active process in the cochlea is the occurrence of spontaneous otoacoustic emissions; in other words, the continuous emission of sound at one or more frequencies, which is found with most human ears.

16.2 Measurements on the Ear

Several types of measurements are made on the ear. Audiometric measurements provide data on the response of the entire hearing system. They are subjective in the sense that the listener is asked to make a comparison judgment between sounds. Impedance measurements provide data on the impedance of the ear from the measurement point (usually the tympanum) onwards. Transfer function measurements are made across a part of the system, usually the middle ear.

Audiometric measurements are of great importance to the designer of telephone or sound reproduction equipment or of architectural spaces because these measurements serve to specify the frequency responses and sound levels to be accomplished. The audiogram of an individual can be used by an audiologist or a medical specialist to diagnose hearing deficiencies. An audiological measurement frequently made is the determination of the threshold of hearing as a function of frequency. An average of results from a number of subjects is shown in Fig. 16.8 and is labeled the minimum audible field (MAF). The audiogram typically shows a derioration in the threshold with age, an effect known as presbycusis. This occurs even in the absence of exposure to loud noise. Exposure to noise will accelerate the loss of hearing, as will certain diseases. Noise exposure usually first results in a temporary threshold shift, with a subsequent return to normal hearing, but continued exposure will result in a permanent hearing loss. We hear the sound of our own voice to a considerable degree through bone conduction, which is why we sound different to ourselves from a recorded sound source played back through the air. Audiograms can also be obtained using bone conduction, and they are important diagnostic aids.

Another related measurement is the determination of increments of loudness above the threshold. The loudness of a sound is a psychoacoustic quantity, available

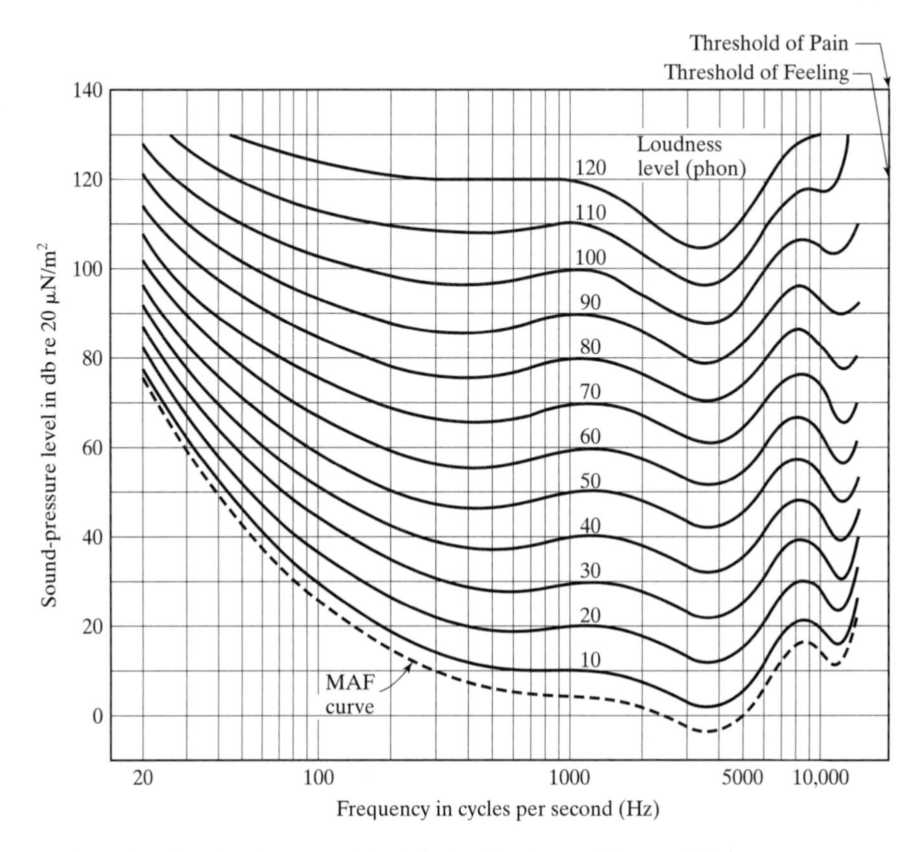

Fig. 16.8 Loudness level contours (phon). (After Fletcher and Munson, 1933.)

only within the subject's brain. The measurement is carried out by determining the sound pressure levels of pure tones of various frequencies that sound to the subject equal in loudness to a 1000 Hz reference tone. The averaged results of such determinations on a number of subjects were first given by Fletcher and Munson (1933) as the equal loudness contours given in Fig. 16.8. These contours might be thought of as the inverse of the transfer function of the entire hearing system. The units of these contours are phons, where the loudness level of a sound in phons is defined as the sound pressure level of the 1000 Hz reference tone. Another way to specify loudness is in sones, where one sone is defined as the loudness of a 1000 Hz tone having a sound pressure level of 40 dB. A sound of loudness 2 sones is twice as loud as the 1 sone sound and for the average listener will correspond to a 10 dB increment in sound pressure level.

For impedance measurements, an acoustic bridge, as shown schematically in Fig. 16.9, is used. It consists of a symmetrical electroacoustic transducer E mounted between two tubes, A and B, of identical length and internal diameter. The transducer is bridged by a third smaller tube that connects tube A to tube B. The bridging tube is Y-shaped, and its third branch leads to the examiner's ear. Tube A is connected to the unknown impedance, the tympanum, and tube B is terminated by a variable matching impedance. The transducer is fed from a generator. Vibration of the transducer diaphragm produces a wave in tube A that is propagated towards the end of the tube. There it is partially reflected and propagated in the opposite direction. After several reflections back and forth, a standing wave pattern is built up, which depends on the unknown impedance at the end of the tube. A similar process takes place in tube B and, when the matching impedance is made equal to the unknown impedance, the wave pattern becomes identical but with opposite phase to that in tube A. Thus, the variable impedance is adjusted until the sound generated becomes inaudible to the examiner. This technique was pioneered by Zwislocki (1962) to obtain the impedance of the tympanum, as shown in Fig. 16.10. It is now used as a standard diagnostic test. Van Camp et al. (1986) provide a review of this subject.

In an important measurement of the cochlear impedance, Zwislocki (1965), showed that it is essentially reactive, but not equal to that of an unrestricted water load. The actual acoustic function of the middle ear is to transform the pressure in the meatus at the tympanum to a stimulus at the oval window of the cochlea. The transfer function of the middle ear has been measured on human cadavers by several researchers, including Bekesy (1960). But Zwislocki and Feldman (1963), using impedance measurements, showed that highly significant changes in the mechanical properties of the middle ear occur after death. The only direct observations of the displacement of the

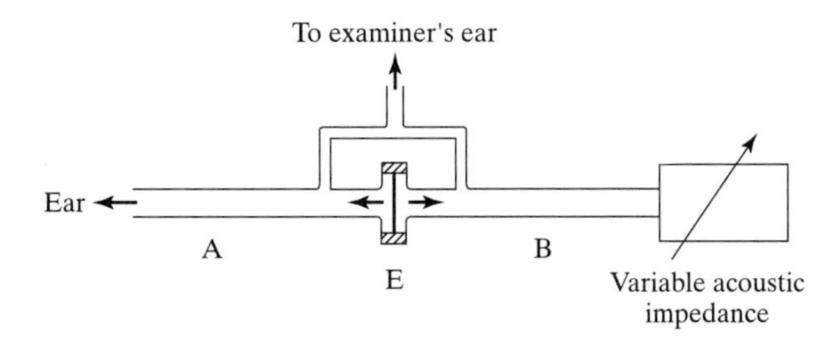

Fig. 16.9 Schematic diagram of the acoustic bridge used to measure the impedance of the tympanum.

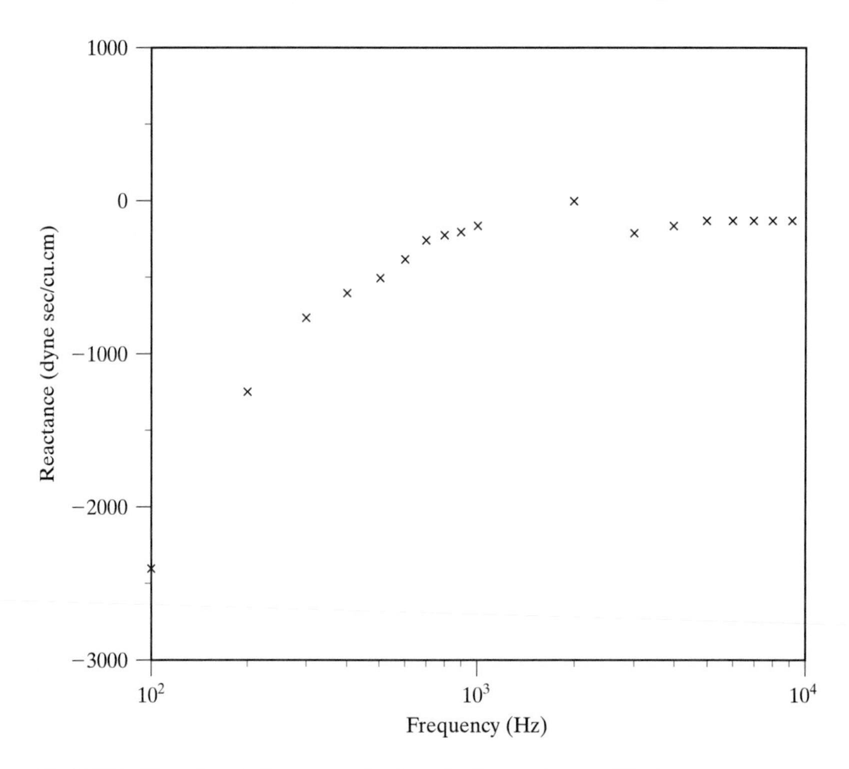

Fig. 16.10 Experimentally measured values for the reactance of the tympanum. (After Zwislocki (1962).)

stapes footplate in live animals were made by Guinan and Peake (1967) using stroboscopic illumination in cats. They made measurements on twenty-five animals and found the averaged response. Their measurements of phase angle are of considerable importance in evaluating different models of the middle ear. The most important point is that the phase difference between the pressure at the tympanum and the stapes exceeded 180 degrees at high frequencies.

16.3 Hearing Aids and Prostheses for the Ear

Certain conditions cause deterioration in the sensitivity of the hair cells. Under such conditions, the listener may benefit from amplification of the sound arriving at the tympanum. Systems to carry out this objective have been available for many years. What is required is a microphone, an amplifier, and a small loudspeaker. The sound from the speaker is delivered to the ear by a pipe moulded to the entry of the meatus. The first hearing aids used transducers developed for telephony, but they were bulky packages. Miniaturization employing transistors on integrated circuits has led to the marketing of products that can fit behind the auricle or be conveniently mounted on eyeglass frames. One problem with hearing aids is that they do not preserve the phase information in signals delivered to the two ears in the normal way. Consequently, they amplify both signal and noise so that the listener may still have difficulties interpreting speech.

The design of prostheses for the ear may be considered a problem in engineering acoustics. Such devices are necessary because certain diseases of the middle ear are treated by replacement of one or more of the three ossicles (malleus, incus, and stapes) or the tympanum. Traumatic injuries to the ossicular chain can accompany certain skull fractures, which can dislocate the ossicles. When the ossicular chain is interrupted and the tympanum is left intact, a 60 dB hearing loss can be experienced. Wolferman (1970) states that with advanced middle ear disease, it is not unusual to find the head of the malleus and the stapes footplate as the only remaining bones in the middle ear. Because of its location, the head of the malleus is often spared when middle ear infection has destroyed the rest of the ossicular chain. Infection can cause deterioration of the bones and joints of the middle ear. Congenital defects can leave the middle ear damaged or even nonexistent. It has become a widespread practice to replace one or more of the ossicles by a prosthesis, which in its simplest form is a rod, as described by Sheehy (1978). Middle ear prosthetic devices are generally grouped into two categories. One is the partial ossicular replacement prosthesis. The PORP replaces one or two of the middle ear bones. Replacing the stapes alone with a rod-like PORP has been carried out successfully for more than thirty years, as described by Robinson (1979), for example. The other category is the total ossicular replacement prosthesis. The TORP replaces all three bones in the ossicular chain. Most TORPs are mechanically equivalent to a simple rod and do not allow for the flexibility of the incudomalleal joint. Fig. 16.11 shows a TORP in place.

Most TORPs do not permanently restore standard hearing. Lesinki (1987) reports that bone transplants suffer from the disadvantages of loss of the middle ear amplification, instability of reconstruction resulting in prosthesis migration, and excessive pressure on the oval window, causing dizziness. Synthetic polymeric and metallic prostheses have generally met with long-term failure (two to three years after implantation) with consequent loss of hearing because of instability, extrusion through the tympanic membrane, or biodegradability. Lesinki reported that short-term hearing restoration was reported for TORP and PORP synthetic ossicles, but only 15% to 20% of the earlier implants were still satisfactory five years postoperatively. He also claimed that patients with the rod-type prostheses still had the equivalent of a 20 dB to 25 dB hearing loss. Although these values can be described as within the range of normal hearing, and represent a considerable improvement in the hearing of the subjects, it is clear that there remains room for further improvement.

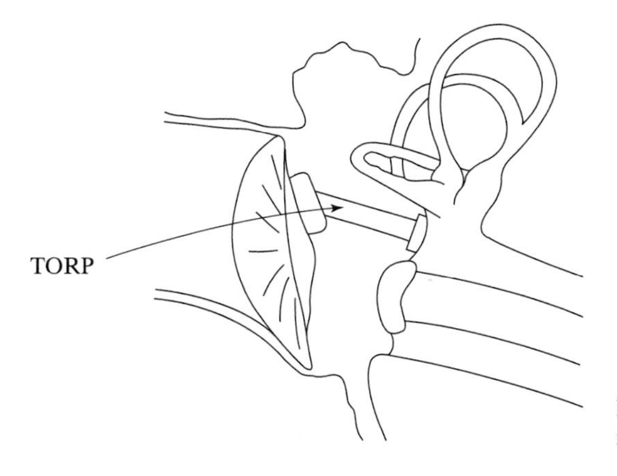

Fig. 16.11 A total ossicular replacement prosthesis.

16.4 The Voice

The other major part of the human acoustic system is the voice. Fig. 16.12a shows a cross section of the vocal tract, and Fig. 16.12b is a cross section of the larynx seen from above. Air is forced upwards, from the lung through the vocal folds. (The word "cord" has been dropped in favor of the more descriptive "fold.") Sound is produced when the two folds part, producing a gap called the glottis. Observations and measurements of the vibrating folds have been made by optical and ultrasonic techniques, as described by Kaneko et al. (1991). When a pure tone is being voiced, vibration of the folds is observed to take place at the same frequency as the acoustical output. The same conclusion is drawn from modeling studies by Kakita (1988). The mechanism of the sound generation is now generally believed to be associated with a fluttering motion of the fleshy surfaces of the folds, as illustrated in Fig. 16.13. This in turn produces pressure fluctuations in the air flow, but the flutter also causes a flow separation as the air is expelled, resulting in a jet formation. This sound source would give rise to a buzzing sound without the modifying influence of the rest of the vocal tract. The buccal and nasal cavities act as Helmholtz resonators, thus giving the sound its characteristic spectral content. The relative intensity and frequency values of these "formant" resonances can be changed by moving the tongue and changing the volume and speed of the airflow. One approach to speech processing described by Oppenheim (1978) attempts to map the acoustic signals in terms of the parameters of the vocal tract. "Helium speech,"

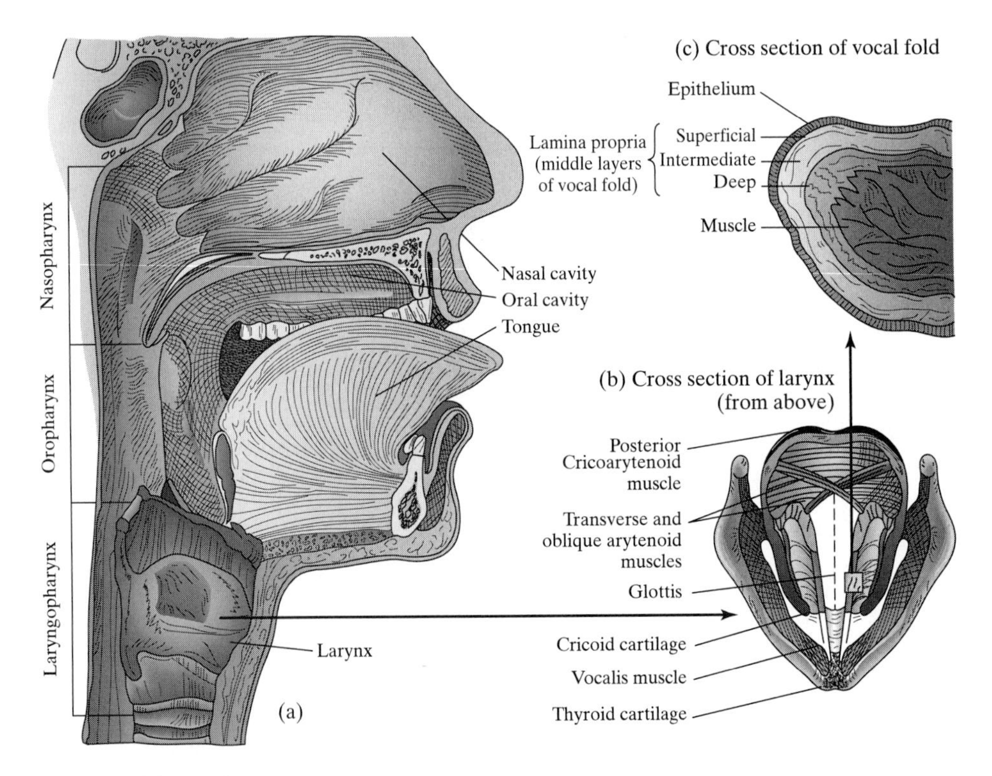

Fig. 16.12 The vocal tract.

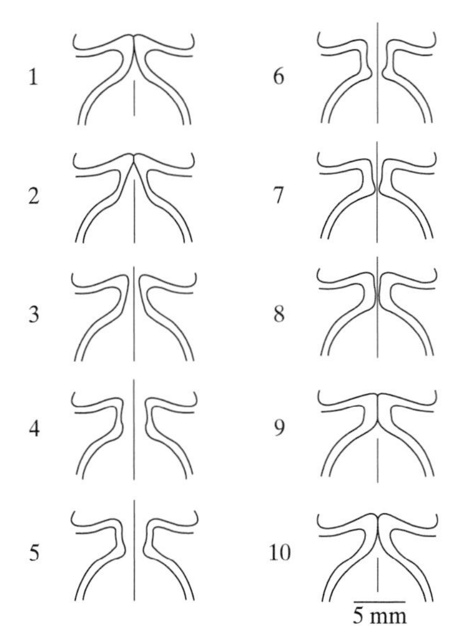

Fig. 16.13 Motion of the vocal folds, during speech (frontal view) [after Hirano (1977)].

produced by inhaling helium, makes a normal speaker sound like Donald Duck (i.e., the register is shifted to higher frequencies). It is believed that this happens because the cavity resonance frequencies are proportional to $\gamma P/\rho$; that is, they will depend on the density of the gas, since the pressure will still be essentially atmospheric.

Kirchner (1988) has presented an interesting survey on the evolution of the larynx. The simplest larynx found in the animal kingdom, in the lungfish, is a slit in the ventral wall of the pharynx with a sphincter muscle. A cricoid ring first appears in the amphibians. Airway protection during swallowing is most efficient when the glottis is high in the pharynx, which is the case with most mammals, including the great apes. It is only in humans that the larynx is as low as it is, with the consequent risk of choking. Kirchner concluded that "the evolution of the human larynx was directed as much toward communication as toward the more vital protection of this airway." Another fascinating observation is that there is a difference in the height of the larynx between *Homo sapiens* and the Neanderthal man. In the latter, the larynx was much higher in the pharynx, approaching that seen in the gorilla or some apes. Could it be that the height of the larynx prevented the Neanderthal man from speaking, thus putting him at a crucial evolutionary disadvantage from the Cro-Magnons?

16.5 Effects of Noise on Humans

Humans are affected by sound in a number of ways. First, we use sound as a means of communication; consequently it is important to avoid the masking of speech or music by noise. Masking can cause us to fail to understand a spoken message or to miss a warning signal. Such interference can annoy the listener, inducing a psychological state of stress with associated physiological conditions, such as increased pulse rate and elevated blood pressure. Sound can also interfere with sleep, without the subject being aware of it. If sound is sufficiently intense, it can cause a temporary hearing loss, which can become permanent with long exposure. Unwanted sound is called noise. The acoustical engineer is often called on to control noise, either in a preexisting situation or preferably in the design stage of a new project.

In dealing with a noise control problem, it is helpful to remember the paradigm represented in Fig. 16.14. Together, the sound source, the transmission path, and the receiver constitute a system. The first step is to identify the components of this system for the problem of concern. The source is usually a vibrating object, or it might be aerodynamic (e.g., jets or engine exhausts), or it may be a man-made transducer (e.g., a loudspeaker). The transmission path is frequently through the air, but it may involve a solid-borne link (e.g., the mounting of a machine to a floor or the passage of sound through a wall). In addition, there may be multiple sound sources involved, as well as multiple transmission paths. Once the system configuration has been established, the next step is generally to make some measurements in order to specify the magnitude of the problem. The instrument most commonly used for this purpose is the sound level meter, which we discuss in Section 16.6. Beyond making the measurements, we have to decide if the noise is acceptable, for which we need the help of appropriate rating criteria (which we will discuss shortly). The last step involves the specification of the treatment, which, in general, consists of the following measures: 1) reducing the noise at the source (usually the best remedy from the acoustical point of view); 2) blocking the transmission path or absorbing the sound along its path; or 3) providing the receiver with hearing protection (usually considered the least desirable remedy). In some cases, economic or other considerations may require a combination of these measures.

Our discussion in this chapter assumes a basic understanding of the wave nature of sound and its propagation (described in Chapter 1), of Fourier analysis (summarized in Sections 2.1 to 2.5 of Chapter 2), of spherical waves (described in Chapter 3), as well as Section 6.4 in Chapter 6 on capacitor microphones, and basic measurements in Chapter 9. Some topics in this chapter have been treated earlier as extensions of other subjects, but because they have been developed primarily in the context of noise control, they fall naturally into this chapter as well. For example, propagation of sound outdoors is in one sense an extension of the discussion in Section 14.4 on attenuation in the atmosphere. Similarly, the use of a wall or a barrier to block sound is an obvious noise control remedy, and we will discover that the theory derives from the study of diffraction, which we covered in Chapter 3. The discussion of noise in enclosures harks back to several earlier chapters, including Section 14.7 in Chapter 14 on sound absorption in rooms and to Chapter 12 on flexural vibration. Finally, the design of mufflers rests both on the theory of pipes and horns (Chapter 4) and on attenuation in tubes (Section 14.6).

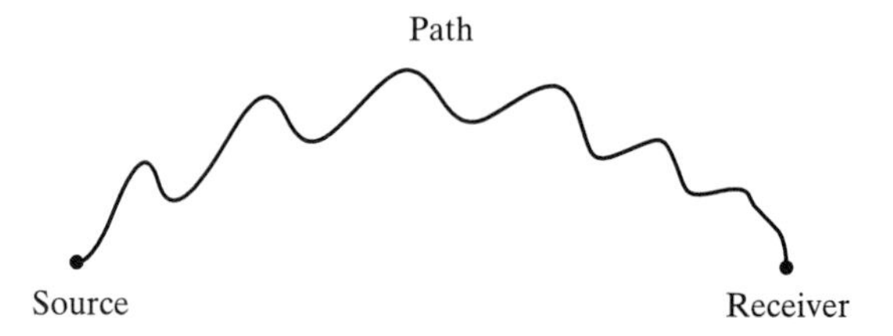

Fig. 16.14 A paradigm for noise control: Sound starts at some source and follows a transmission path to the receiver. We may treat the problem at the source, interrupt the transmission path, or protect the receiver.

16.6 Noise Measurement and Criteria

The device most frequently used for noise control measurements is the sound level meter, which generally consists of a microphone, of a capacitance-type with a membrane in the highest quality instruments, or an electret or piezoelectric transducer where the requirements are less demanding. The frequency response of the output from the microphone can be shaped by one of a number of filters, which are designed to reproduce the response of the human ear at various intensity levels. The A-weighting corresponds to the ear's response at the lowest intensity levels, and the C-weighting to that at the highest, the B-weighting being intermediate. The signal is then rectified and fed to a meter. In older instruments, this meter was a simple needle and scale, but today the readout is usually displayed as a numeric. The response can be set to fast or slow to capture transient versus steady sounds. Fig. 16.15 is a schematic of a sound level meter, and Fig. 16.16 shows the international standards for the various weighting networks. The agreed-upon classification for sound level meters is as follows:

Type 0: Laboratory reference standard (for calibrating other instruments)
Type 1: Precision sound level meter (for laboratory use)
Type 2: General-purpose sound level meter for general field use and recording
Type 3: Survey sound level meter (for preliminary readings)

The sound level meter is usually designed to be held in the hand, but where the user would interfere with the measurements being taken, it is often possible to detach the microphone, set it at the desired measurement point, and connect it by cable to the case of the instrument. When making measurements outdoors, care must be taken to prevent the wind from generating noise as it blows over the microphone. This can be done by enclosing the microphone with a windshield made of silk or foam rubber. The response of a microphone may depend on the direction of incidence of the sound, so that where this might present a problem, the manufacturer's specification should be consulted. Most microphones are insensitive to temperature over a wide range, but humidity can present a problem. If moisture has condensed in a microphone, care should be taken to dry it out by warming it under a lamp. Most sound level meters can be calibrated using a piston-phone, and it is good practice to carry out a calibration before making measurements.

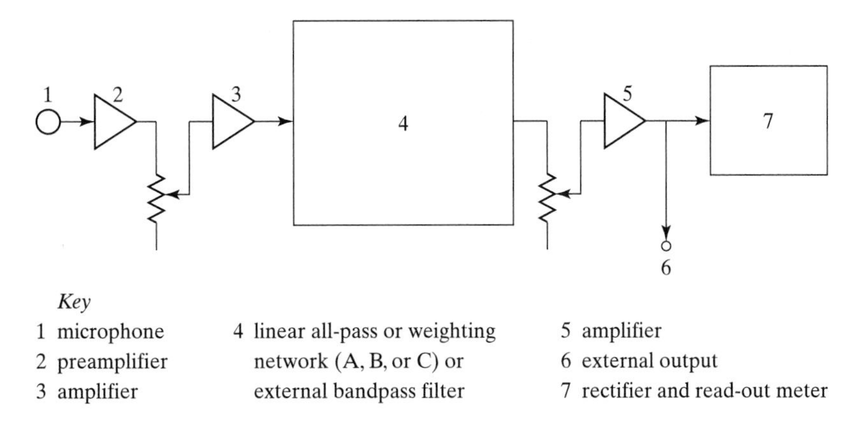

Fig. 16.15 A schematic representation of the elements of a sound level meter.

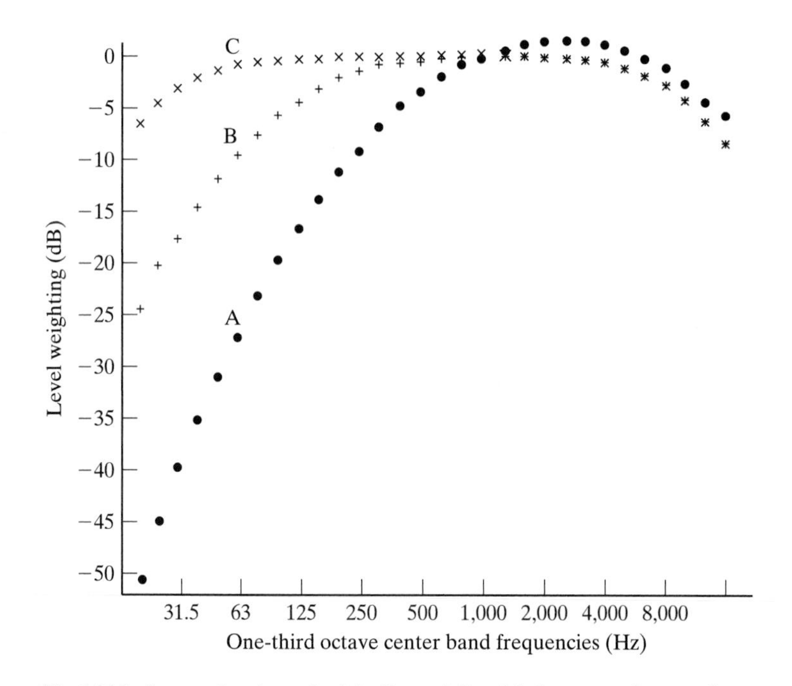

Fig. 16.16 International standard A-, B-, and C-weighting curves for sound level meters.

Sometimes other instruments are useful in noise studies. Analysis of the sound into third or tenth octave bands or narrow band analysis can sometimes yield valuable information on the source of sound. (See Chapter 8 for further details on this topic.) It is also often useful to make tape recordings of noise for subsequent analysis. A vibration meter is an accelerometer calibrated to permit direct readout of displacement, velocity, or acceleration.

Noise is evaluated for several practical concerns. If the levels are high, our problem is to ensure that the risk of damage to hearing is minimized. At lower sound levels, the problems are more in the psychoacoustic domain. We may be required to ensure that speech communication is possible or that working efficiency is not impaired. We may want to determine if a particular space will be suitable for a specific type of use (e.g., general office work, speech, music). Alternatively, it may be necessary to make sure that noise levels experienced by a community are acceptable. Over the years, specialists in various areas—audiologists, psychologists, sociologists—have established criteria to assess the quality of the noise environment or to determine what difficulties will be an issue.

Hearing damage results from both the intensity and the duration of the noise exposure. There are two main theories of hearing damage. The *equal energy theory* holds that risk of hearing damage is determined by the total amount of noise energy to which the ear is exposed each day (eight hours). If the noise energy exposure during an eight-hour period is the maximum safe level, then for a half-day exposure (four hours), to experience the same total energy the noise level can be 3 dB greater, regardless of how the exposure is distributed throughout the eight-hour period. This theory seems to be validated by data on workers exposed to similar noise levels on a daily basis for years at a time. Based on such evidence, many government agencies have established criteria for the noise dosage to which a worker may be subjected without unacceptable risk to hearing. Table 16.1 lists the maximum time exposures at various A-weighted sound levels presently permitted in the United States under the Occupational Safety and Health Act (OSHA). Further details on noise exposure levels can be found in the *Federal Register.*

The values given in Table 16.1 refer to continuous noise, but sounds of short duration can also result in permanent hearing loss if they are of sufficient amplitude. The paradigm that seems best to account for interrupted noise is the *equal temporary effect theory*. The idea behind this theory is that the risk of hearing damage is related to the temporary threshold shift (TTS) resulting from noise exposure in young adults. Studies indicate that intermittent noise is less harmful than steady noise, with a doubling of exposure being associated with a 5 dB decrease in intensity exposure. Among intermittent sounds, we distinguish between impact noise and impulse noise. Impact noise results when one body collides with another, such as when an object is dropped on a floor. The sound may show extensive ringing before it dies away. Impulse noise is a rapid-rise time-pressure pulse, such as results from a rifle shot. The military has investigated this topic extensively, and a damage risk criterion has been proposed by Coles et al. (1968). Fig. 16.17 shows schematic representations of the two types of intermittent sounds, and Fig. 16.18 shows recommended upper bounds for both.

Table 16.1 OSHA Permissible Noise Exposures

Duration per Day in Hours	Sound Level dBA, Slow Response
8	90
6	92
4	95
3	97
2	100
1-1/2	102
1	105
1/2	110
1/4 or less	115

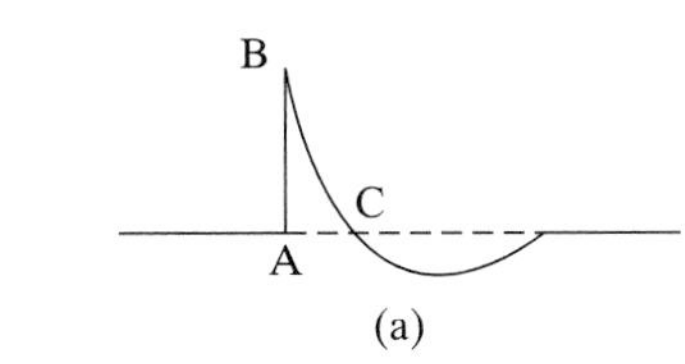

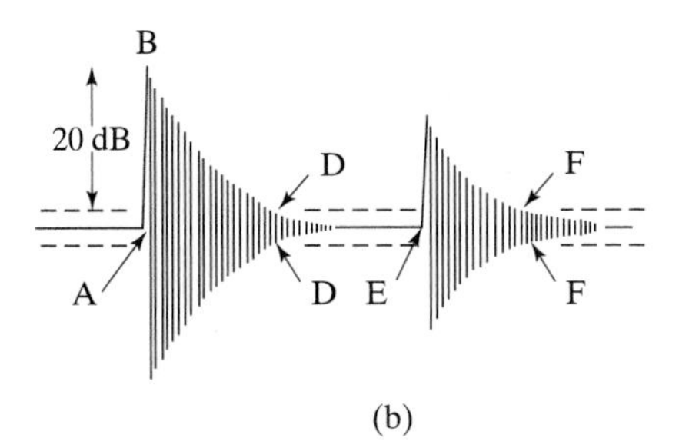

Fig. 16.17 Idealized oscilloscopic waveforms of impulse noises. Peak level: pressure difference *AB*. Rise time: time difference *AB*. (a) *A* duration: time difference *AC*. (b) *B* duration: time difference *AD* (+*EF* when a reflection is present). (After Coles et al., 1968.)

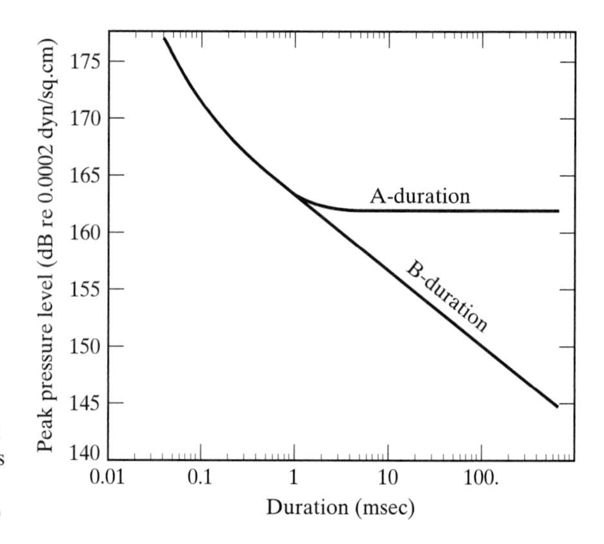

Fig. 16.18 Peak pressure level and duration limits for impulses having near-instantaneous rise times that will not produce an excessive risk of hearing loss. (After Coles et al., 1968.)

Interference with speech starts at levels lower than the criteria for hearing damage. Speech frequencies range from 200 Hz to 6000 Hz, although those frequencies most necessary for communication lie in the 500 Hz to 4000 Hz range. The peak energy output of a male voice is about 700 Hz, while that of the female is 350 Hz. The intelligibility of speech will depend on the background noise. Speakers tend to raise their voices in the presence of noise. These two effects of noise on speech are quantified in Fig. 16.19. Here, the speech interference level is computed by taking the arithmetical average of the sound pressure levels in the octave bands at 500 Hz, 1000 Hz, and 2000 Hz. An alternative, although slightly less accurate, method of specifying the background

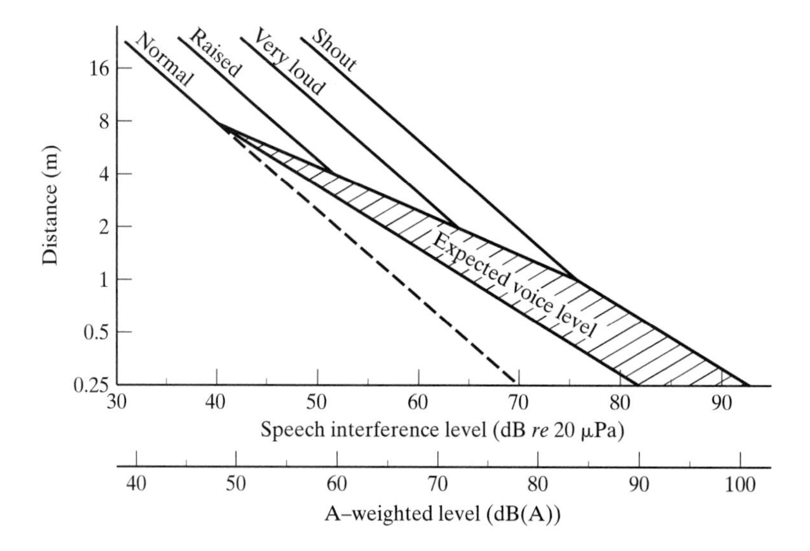

Fig. 16.19 Rating noise with respect to speech interference. (After American Standard ANSI S3.14–1977.)

noise is to use the A-weighted sound level. The voice level required for adequate communication at various distances can then be read off the nomograph. At high sound levels, the voice level will be increased. The expected level based on statistical data for 95% sentence intelligibility is shown as the cross-hatched region in Fig. 16.19. In general, where voice communication is necessary in an industrial environment, the noise level should not exceed 70 dBA. Pure tones can mask sound very well, and should therefore be avoided where speech is crucial. On the other hand, if masking is required, a pure tone at about the frequency of maximum hearing acuity (about 400 Hz) together with a number of harmonics would be useful.

The primary acoustic factors that influence the annoyance caused by noise are the sound level, the frequency, and the duration of the noise. Secondary acoustic factors are the complexity of the spectrum, fluctuations in sound level and frequency, the rise time of the noise, and the localization of the sound. Several nonacoustic factors also play a role, including the listener's physiology, adaptation, and past experience; how the listener's activity affects the noise; the predictability of when the noise will occur; the necessity of the noise; as well as individual differences and personality. Generally, the higher the sound level, the more likely a subject is to find it annoying. Frequencies in the mid range, where hearing is most acute, are found to be the most annoying. The longer a noise continues, the more annoying it will be. If the pure tones in the frequency spectrum do not combine harmoniously, they will be annoying. For example, a bell is designed to ring in a harmonious way, but a length of pipe when struck usually sounds dissonant. It has been found that sound levels that fluctuate (as with traffic) are more annoying than the equivalent continuous sound levels. Frequency fluctuations are an additional source of annoyance. Sounds with rapid rise times (explosions, sonic booms, impact) are more annoying than continuous sounds with the same energy content. Sounds whose source is difficult to localize are generally more annoying than those of known position. Molino [in Harris (1979)] provides further information on the annoyance of noise.

One of the first schemes devised to assess ambient noise was the noise criterion (NC) rating. It was originally proposed as a way to categorize the maximum acceptable air conditioning sound in different spaces, but it is used in a wider variety of applications, reflecting concerns about both speech interference and annoyance due to noise. Although the definitions of the ratings are quite similar to those of A-scale values, they have to be obtained graphically based on octave band analyzer measurements. The benefit of this procedure is that the frequency distribution of the noise is available to the engineer who may have to design for its correction. The given set of analyzer measurements should all lie just below the contour whose number specifies the NC rating, as shown in Fig. 16.20. More information on the history, particulars, and variations of the NC rating can be found in Beranek (1971) and in Harris (1979). Some ambient levels recommended by Bies and Hansen (1988) in terms of A-scale levels are given in Appendix K.

Several rating schemes have been devised to categorize the noise environment experienced by a community. Some are used for special situations, such as traffic or aircraft noise. One rating scheme that has general applicability is the equivalent noise level, usually abbreviated L_{eq}. This rating and its derivatives have been used by several government agencies to assess community noise. To arrive at the L_{eq} value, the sound

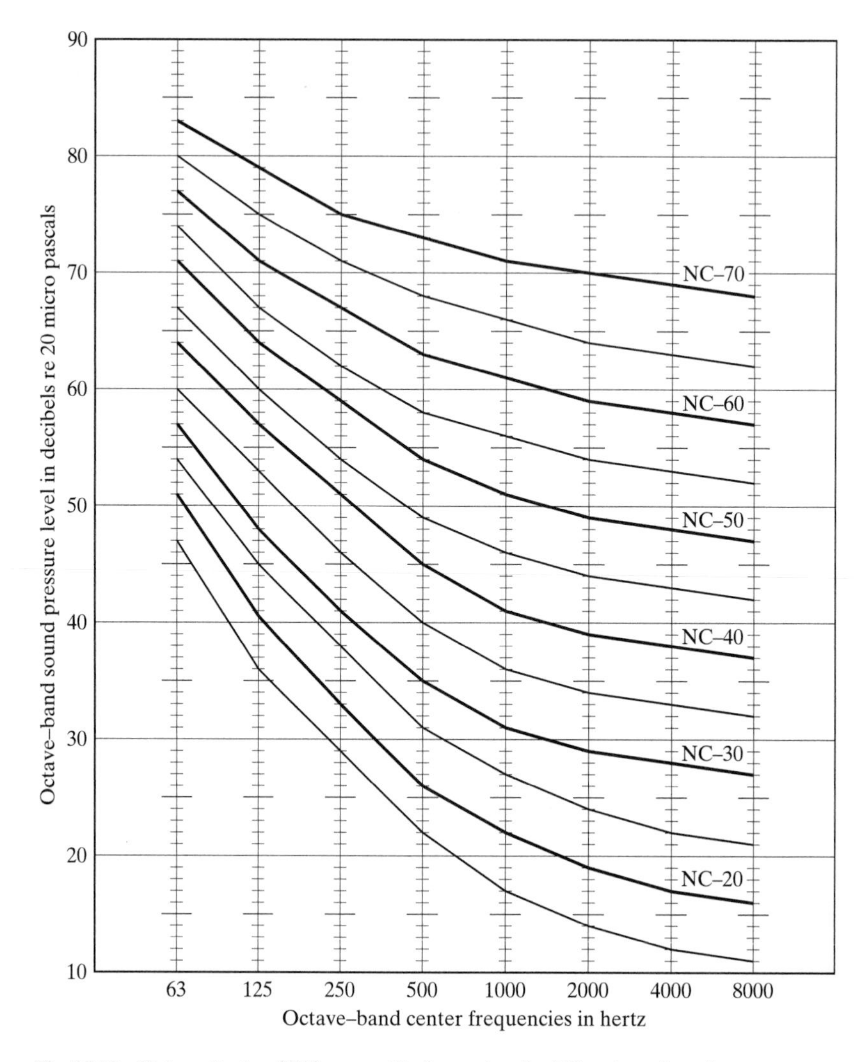

Fig. 16.20 Noise criterion (NC) curves. To determine the NC rating of a noise spectrum, first plot the octave-band sound pressure levels of the spectrum on the above chart. The NC rating is then determined by the *lowest* NC curve which just touches the plot of the octave-band spectrum.

level is sampled at regular intervals and an average value is calculated. Sound level meters with incorporated computer chips to do the computations automatically (called community noise analyzers) are now available. To allow for the added effect of interference with sleep, the "day-night level," or L_{dn}, was introduced. In calculating this number, 10 dB is added to the samples of the sound level measured between 10 p.m. and 7 a.m. Additional corrections can be made for sounds with rapid rise times. The L_{dn} value has been found to correlate well with community response, as shown in Fig. 16.21.

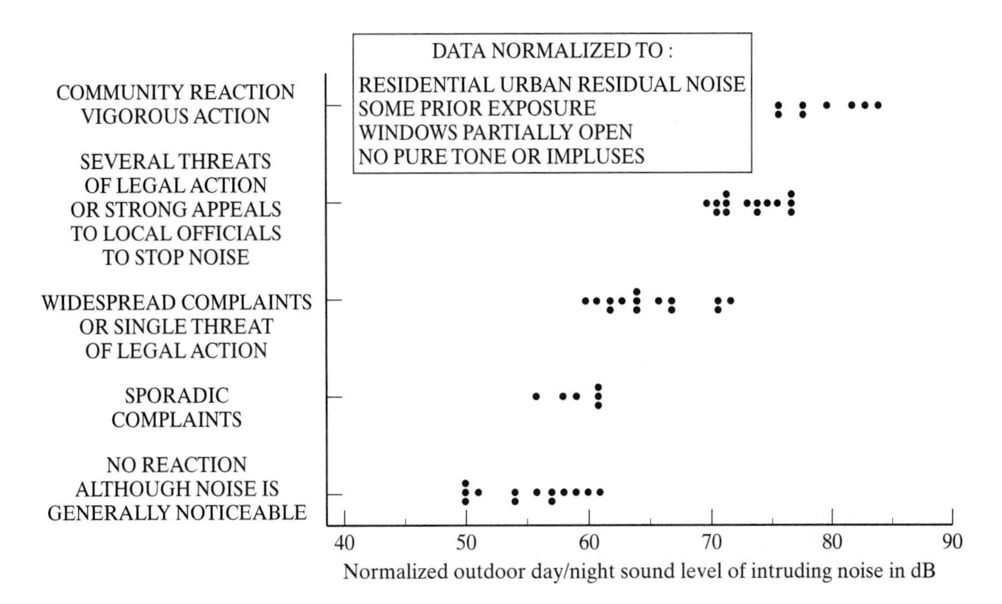

Fig. 16.21 Correlation between NL_{dn} and community reaction from 55 case histories. (EPA, 1974.)

16.7 Outdoor Sound

When it comes to understanding how sound behaves outdoors, our starting point is the basic theory covered in Chapters 1 and 3, as well as our discussion of directivity in Chapter 9 (in Section 9.3). The theory of radiation and scattering was covered at a more advanced level in Chapter 11, but for most noise control work the discussion there is not required reading. There are, however, certain peculiarities of sound propagation in the open that need to be covered here.

Suppose we want to estimate the sound pressure level at some distance r from a noise source. If we know the directivity factor of the source, then it follows from Eq. (9.3.1) that the total radiated power is

$$W = 4\pi r^2 I(\theta)/D \qquad (16.7.1)$$

where $I(\theta)$ is the intensity (the average energy crossing unit area in unit time) at an angle θ and D is the directivity factor. For a nondirective point source, the directivity factor is unity, and the intensity is proportional to the square of the sound pressure and inversely proportional to the square of the distance from the source. Thus, it follows that the sound pressure level in the vicinity of a point source falls 6 dB with doubling of distance, or 3 dB with doubling of distance from a line source. By taking logarithms of both sides of Eq. (16.7.1), we obtain

$$10 \log_{10} W - 10 \log_{10} W_{ref} = 10 \log_{10} 4\pi + 20 \log_{10} r + 10 \log_{10} I(\theta) - 10 \log_{10} I_{ref} - 10 \log_{10} D \qquad (16.7.2)$$

Hence

$$L_P = L_W - 20 \log_{10} r - 10 \log_{10} 4\pi + DI \qquad (16.7.3)$$

where L_P is the intensity level at the observation point P, L_W is the source level, and DI is the directivity index given by Eq. (9.3.2). In practice, this equation needs a number of refinements to allow for attenuation. The corrections may be made by subtracting a term AE, which is defined as the excess attenuation in dB. Then

$$AE = AA + AG + AM + AB + AF \qquad (16.7.4)$$

where AA is the air absorption, AG is the ground absorption, AM is due to meteorological effects, AB is due to barriers, and AF is due to shrubs and forests. The subject of absorption by the atmosphere, AA, was covered in Section 14.4, and attenuation coefficients in dB per foot are given in Fig. 14.2 for losses due to viscosity, thermal conduction, and molecular relaxation. These values are quite small compared with typical attenuation due to turbulence, which is a major contributor to meteorological attenuation, AM. The role of turbulence as a major sound scatterer and attenuator was first suspected by Tyndall (1874) in his research on fog signals, and the subject has received much attention since then (Tatarskii (1971), and Brown and Hall (1978) provide good reviews of this topic). Other important meteorological effects are due to wind and temperature gradients, which cause refraction effects as illustrated in Fig. 16.22. During the day, the ground gets hot and a temperature "lapse" occurs. At night, the situation is reversed and a temperature inversion occurs. Before Tyndall's time, it was frequently asserted that fog attenuated sound, but he was able to show that in fact sound would often carry better under foggy conditions than in conditions of bright sunlight, when turbulence causes heavy scattering. The excess attenuation due to most fogs, rain, or falling snow is less than 0.5 dB per Km and it may thus be neglected. Attenuation due

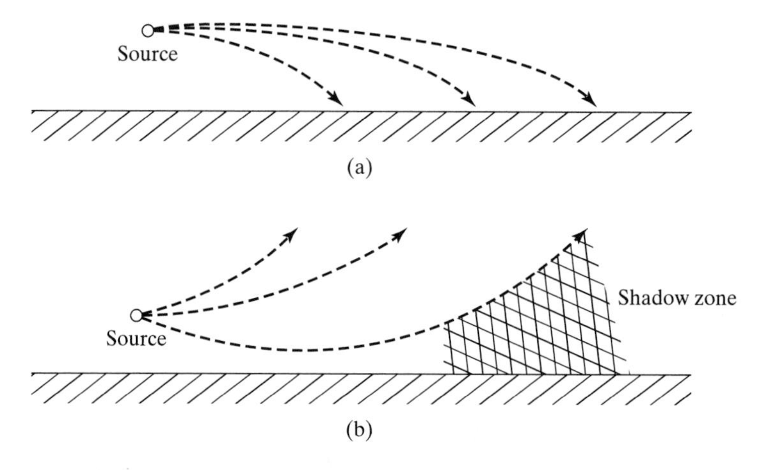

Fig. 16.22 Refraction due to wind and temperature gradients. (a) Propagation downwind or under temperature inversion. (b) Propagation upwind or under temperature lapse conditions.

to shrubs and trees, AF, is essentially zero at low frequencies. Although such foliage may block the sight of a noise source, which may have some psychological value, it has no direct effect on sound transmission except at very high frequencies. Trees may, however, have an appreciable ground effect, as we will soon see. (Attenuation by barriers is covered in Section 16.8 so only the ground effect remains to be discussed now.)

The ground can serve as both a reflector and an absorber of sound. The problem is usually posed in terms of the situation illustrated in Fig. 16.23. The theory was adapted from the electromagnetic case by Lawhead and Rudnick (1951), who showed that the reflection coefficient of the waves is given by

$$R = \frac{Z_2 \sin\varphi - Z_1\left[1 - \left({}^{k_1}\!/_{k_2}\right)^2 \cos^2\varphi\right]^{1/2}}{Z_2 \sin\varphi + Z_1\left[1 - \left({}^{k_1}\!/_{k_2}\right)^2 \cos^2\varphi\right]^{1/2}} \tag{16.7.5}$$

where Z_1 and Z_2 are the specific acoustic impedances of the atmosphere and the ground, respectively, and k_1 and k_2 are the corresponding propagation constants. The ratio k_1/k_2 will usually be small so that the reflection coefficient will approximate the plane wave reflection coefficient and will depend on the ratio Z_1/Z_2; thus

$$R = \frac{\sin\varphi - Z_1/Z_2}{\sin\varphi + Z_1/Z_2} \tag{16.7.6}$$

For a rigid ground, Z_2 will be infinite and R will be unity (i.e., all the energy will be reflected). This situation is approximated for hard surfaces such as concrete or asphalt.

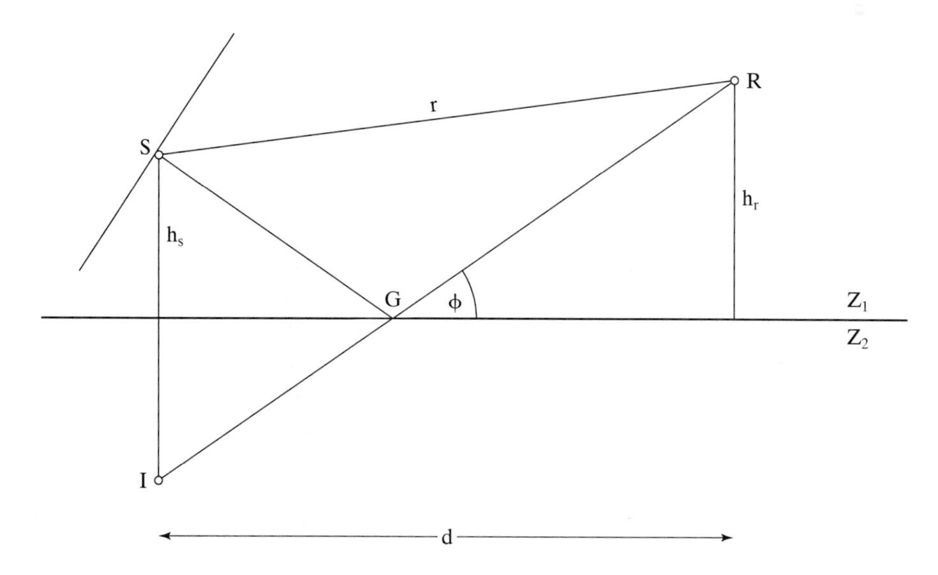

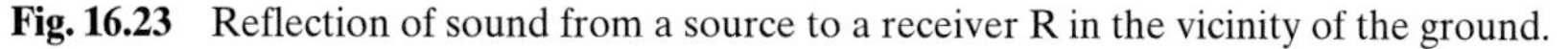

Fig. 16.23 Reflection of sound from a source to a receiver R in the vicinity of the ground.

The opposite extreme of a very soft surface is exemplified by a covering of snow or pine needles, while there is a range of impedance values in between represented by grass, roadside dirt, small stones, etc. The impedance of such grounds has a high resistive component because of the air trapped in interstices. Outdoor measurements have been made by Embleton, Piercy, and Olson (1976), who were able to show the effects of interference between the direct waves and the ground waves over hard surfaces. There is also some evidence that there is a surface wave that propagates in the air over a compliant ground.

16.8 Treatment at Source

We are now ready to discuss some of the possible treatments for noise problems. It is generally considered to be the best case if the noise can be eliminated or at least abated at the source. For our purposes, we may classify sources into two types: vibration of objects or solid surfaces, and fluid flows. Vibrating objects include machinery, structures, loudspeakers, bells, and some musical instruments; fluid flows include jets, engine exhausts, voices, whistles, sirens, and the remaining musical instruments. In many of these cases, there is little that may be done to alleviate the output at source. The problem then becomes one of interrupting the transmission path to the listeners who wish or need to have protection. There are some situations, however, where it is possible to intercede at the source, preferably at the design stage.

The control of vibration is frequently feasible for household and industrial machinery. (We have already discussed the damping of a one degree of freedom system in Chapter 1, and in Chapter 14 we covered the subject of internal damping in solids.) It is possible to increase the damping of a solid object by coating it with a viscoelastic material. The damping in this case can be determined by measuring the decay rate at the various resonance frequencies of a suspended panel. The frequency response in each vibration mode is the same as that for a one degree of freedom system, so that the decay rate is directly proportional to the damping coefficient. Various rubbers have high damping, and it has been found to be more effective if the damping layer is constrained between two plates. Ungar (1971) provides a detailed review of the damping of panels.

To control aerodynamically generated sound at the source, we must, if possible, reduce the flow speed, since the sound is actually produced by turbulent eddies in the flow and the degree of turbulence depends on the flow speed. It has been found that the sound intensity due to a jet stream is proportional to the eighth power of the velocity. If it is not possible to reduce the flow speed in a particular application, then the next resort is usually the use of a muffler. There are two main types of mufflers: reactive and dissipative. The simplest type of reactive muffler is illustrated in Fig. 16.24, which shows an expansion chamber in an exhaust line. In this case, there will be reflections of waves travelling down the line at each junction, resulting in the buildup of a standing wave in the expansion chamber. The theory is similar to the case of transmission through three media given in Section 10.3. An exercise for the student is to show that the pure tone insertion loss due to the expansion chamber is given by

$$L_{TL} = 10 \log \left[1 + \frac{1}{4}\left(m - \frac{1}{m} \right)^2 \sin^2 kl \right] \text{dB} \qquad (16.8.1)$$

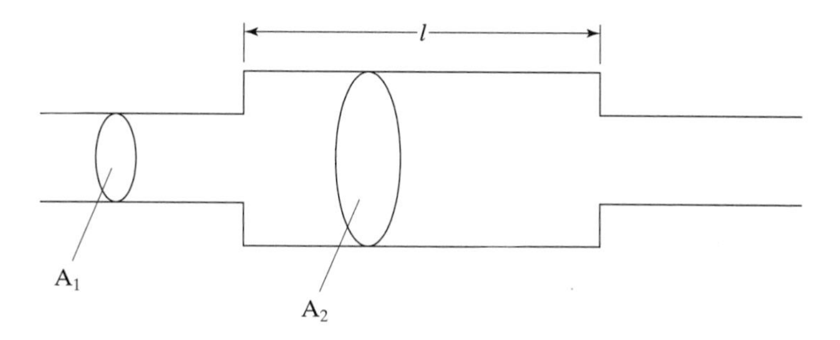

Fig. 16.24 An expansion-chamber-type reactive muffler. m = A_2/A_1.

where m is the ratio of the cross-sectional areas of the chamber and the line; k is the propagation constant; and l is the length of the chamber. Some values of transmission loss calculated using this result are shown in Fig. 16.25. Reactive mufflers employing two or more chambers may be used. When the length of the expansion chamber is much less than a wavelength, the enclosed volume acts like a Helmholtz resonator or a

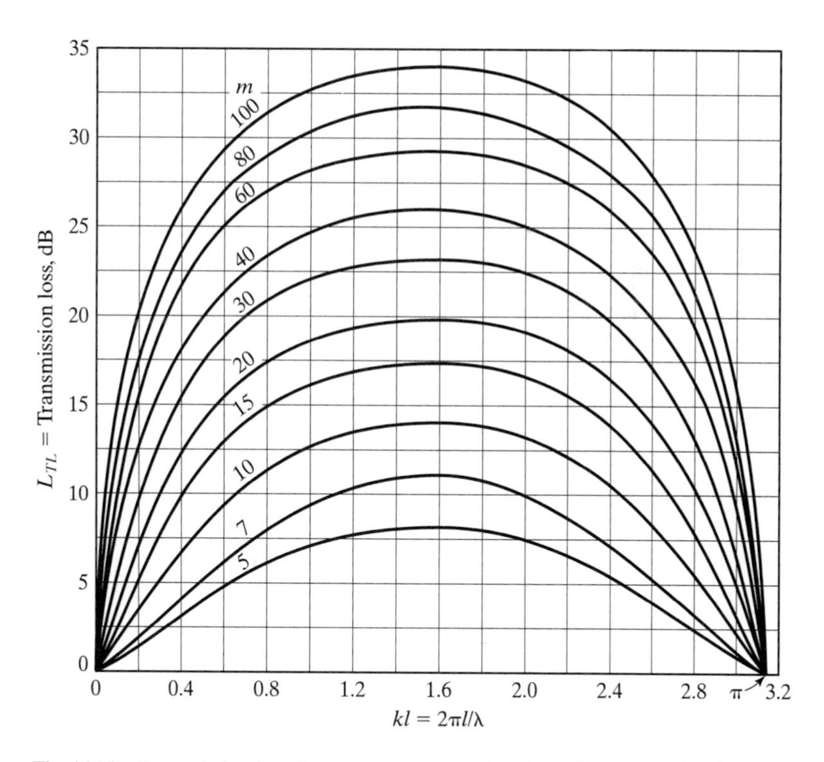

Fig. 16.25 Transmission loss L_{TL} of an expansion chamber of length l and A_2/A_1 = m (see Fig. 16.24). The cross section of the muffler need not be round, but its greatest transverse dimension should be less than 0.8 λ (approximately) for the graph to be valid. For values of kl between π and 2π, subtract π and use the scale given along the abscissa. Similarly, for values between 2π and 3π, subtract 2π; etc. Note that when $kl = \pi$, then $l = \lambda/2$; when $kl = 2\pi$, then $l = \lambda$; etc. (After Embleton, 1971.)

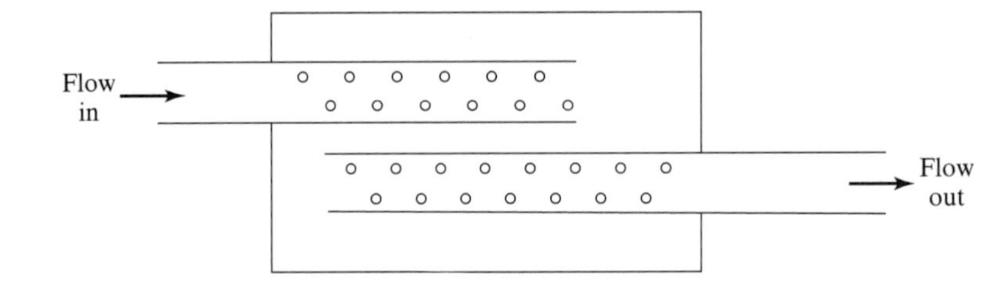

Fig. 16.26 Schematic of a tailpipe.

pneumatic spring. Such a volume resonator will have a maximum transmission loss at its resonance frequency. Mufflers of this type are used in pulsation damping of lines fed by compressors and should be tuned to the compressor operating frequency. The volume resonator is usually configured as a side branch to the main line. Another type of reactive muffler, used as an automobile tailpipe, is illustrated in Fig. 16.26. Each of the holes in the wall of the line shown in this sketch acts as a resonator. It is also possible to design a muffler by lining the walls of a pipe or a duct with a sound-absorbing material. Sabine found experimentally that, provided the dimensions of the duct are less than about one tenth of a wavelength, the attenuation A (dB/ft) of a lined duct is given by

$$A = 12.6\alpha^{1.4}\frac{P}{S} \tag{16.8.2}$$

where α is the sound absorption coefficient for the duct lining, P (inch) is the perimeter of the duct which is lined, and S (inch^2) is the open cross section of duct. The attenuation can be increased by dividing the duct with internal metal walls and lining these also. Alternatively, sound-absorbing lining can be added to any of the reactive mufflers described in this section. Embleton (1971) provides further information on this topic.

16.9 Treatment of the Transmission Path

Assuming that whatever could be done at source has been done, we come to the possibility of treating the transmission path. The first topic here will be vibration isolation of any reciprocating or rotating machinery, which we already discussed in Section 1.3.5 in terms of the transmissibility of a single degree of freedom system. Close to the resonance frequency it helps to increase the damping in the system, but at frequencies greater than $\sqrt{2}\omega_0$, the addition of damping increases the transmissibility. Metal springs of various configurations are often used, as are pads of rubber, cork, or felt. Muster and Plunkett (1971) provide more information on this topic.

Another way to reduce noise is through the use of a barrier. To be effective, the wall must be solid and massive, in which case sound can reach a receiver on the far side only by diffraction. In recent years, walls of this sort have been used increasingly to shield residential neighborhoods from highway noise. The situation in this case is illustrated in Fig. 16.27. The theory used is carried over from optics and is given in terms of the Fresnel number:

$$N = \pm\frac{2}{\lambda}(A + B - d) \tag{16.9.1}$$

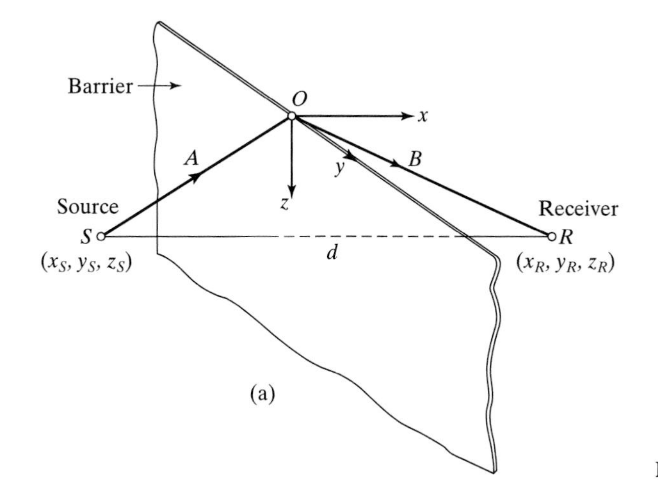

Fig. 16.27 Diffraction over a barrier.

where λ is the sound wavelength, d is the shortest direct distance between the source and the receiver, and (A + B) is the shortest distance between them measured over the wall. The positive sign is taken when the receiver is in the shadow zone, and the negative sign when the receiver is in the "bright" zone. The excess attenuation is then as given in Fig. 16.28. As a practical matter, there is a limit of 20 dB to 25 dB in the attenuation that can be achieved through the use of a barrier.

Another noise control solution is to enclose the offending sound entirely. The level inside the enclosure will then build up in the manner described in Section 14.7 and, as shown in Eq. (14.7.9), will depend on the total absorption in the room. A tabulation of absorption coefficients for various materials is given in Appendix J. How much sound escapes to the outside then depends on the transmission through the walls of the enclosure. The transmission loss through a panel was discussed in Section 10.3, where the transmission loss was shown to be (see Eq. (10.3.16))

$$\mathrm{TL} = 10 \log_{10}\left[1 + \left(\frac{M\omega}{2\rho c}\right)^2\right] \tag{16.9.2}$$

This result, as previously explained, depends on the assumption that the panel behaves as a limp mass. This assumption is clearly of limited validity, since the elasticity of the panel will result in its having various flexural modes of vibration that will influence the transmission loss, as will the effect of coincidence, producing a total picture of the sort shown in Fig. 10.5. The character of the sound transmission by a panel is often presented as a single number rating known as the sound transmission class (STC). The STC is determined by taking the transmission loss values at sixteen standard third octave band center frequencies and comparing them with the contours shown in Fig. 16.29. The given TL curve should not fall below the matching contour by more than an average of 2 dB, and no individual value should be more than 8 dB below the contour. Windows and doors are often the weakest links in the transmission from an enclosure, so they are usually given such STC ratings. Any gaps in the walls of an enclosure can result in the sound being able to flank otherwise

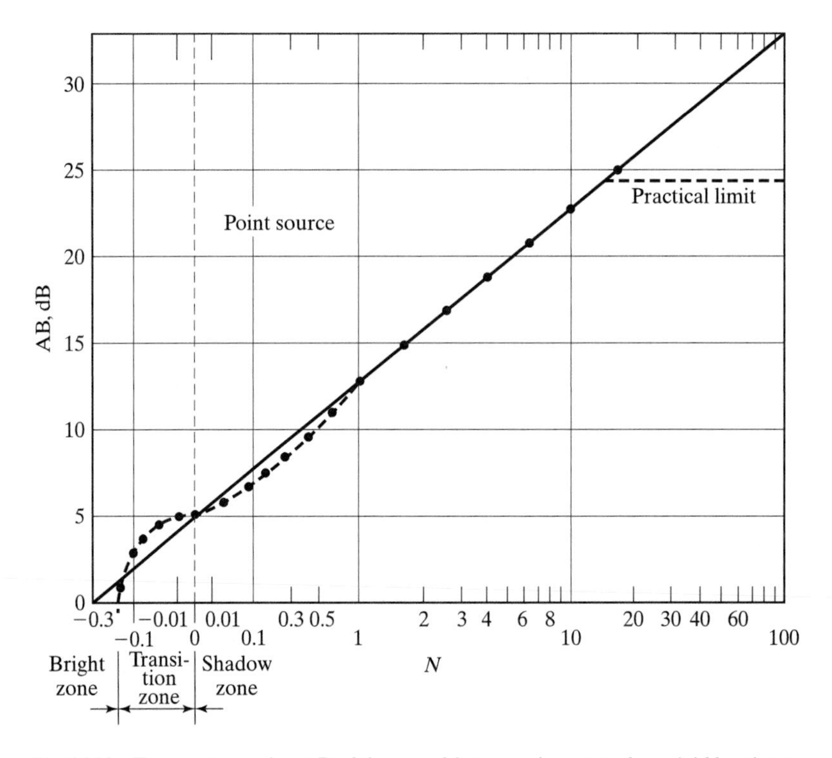

Fig. 16.28 Excess attenuation AB of the sound from a point source by a rigid barrier as a function of Fresnel number N (see Eq. 16.9.1). A negative N refers to the case where the receiver is able to see the source. (After Beranek, 1971.)

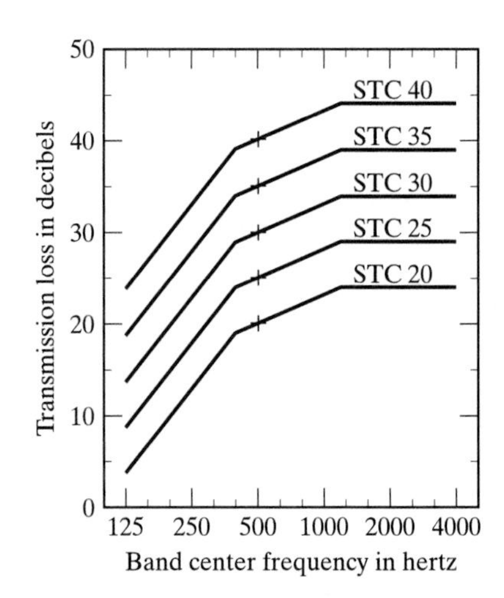

Fig. 16.29 Sound transmission class (STC) contours used for rating airborne sound insulation. Contour labels correspond to the transmission loss at 500 Hz, and are in increments of 1 dB. (After Northwood, Warnock, and Quirt in Harris (1979).)

excellent insulation. Particular care must be taken to ensure that windows and doors are sealed, that pipes passing between rooms are caulked, and that the spaces above rooms with suspended ceilings are insulated from one another. An extensive treatment of the topic of sound insulation is given by Northwood, Warnock, and Quirt (1979).

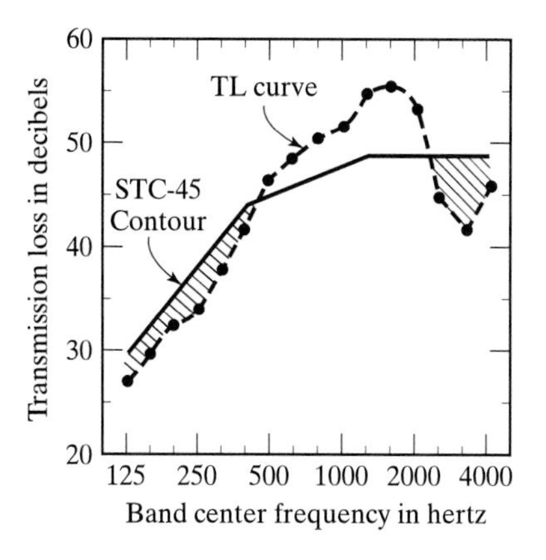

Fig. 16.30 Derivation of the sound transmission class (STC) of a partition from transmission loss (TL) data. The shaded portion lying between the TL curve and the matched STC contour must not exceed an average (over all bands) of 2 dB, and no individual TL should be more than 8 dB below the contour. (After Northwood, Warnock, and Quirt in Harris 1979.)

16.10 Hearing Protection

When engineering measures to control noise at its source or along its transmission path are not feasible, remedies have to be sought at the listener's end of the system. There are two possible approaches here. The first is the so-called administrative solution in which a worker's total exposure to noise is limited by prescribing the hours he or she may work in different environments at different sound levels. The plant in which the listener works will first be surveyed to ascertain the sound levels in different places and a work schedule will be planned so that the total exposure does not exceed the limits given in Table 16.1. To monitor the worker's actual exposure, instruments known as noise dosimeters can be carried on the person to measure the fraction of the total daily permissable noise exposure the worker has enountered. Such a hearing conservation program should include the worker being given regular audiograms to ascertain that there is no deterioration in hearing.

The second approach to hearing protection is to physically intervene with the sound reaching the ear. A simple way to do this is to insert plugs in the ear canal. It is important that the plug fit closely in the ear canal, as even a small gap can almost completely vitiate its effectiveness. If fitted properly, plugs can usually provide about a 10 dB reduction in the level at the tympanum at low frequencies and up to 40 dB at high frequencies. Ear muffs generally have a liquid-filled seal that is held against the head by tension in the headband. Again, it is important that this seal be completely effective. Ear muffs are generally not as effective as plugs at low frequencies but are slightly better at the high end of the range. A combination of plugs and muffs improves the effectiveness of either one alone. A more recent approach to the problem uses active noise control. The head is fitted with an ear muff that contains a microphone to sense the noise and a small loudspeaker to produce the exact pressure variation of the noise but out of phase with it, resulting in cancellation of the noise.

REFERENCES AND FURTHER READING

Ballenger, J.J. 1977. *Diseases of the Nose, Throat, and Ear*. Lea & Febiger.

Balsalla, G. 1988. *The Evolution of Technology*. Cambridge Univ. Press.

Bekesy, G.V. 1948. "On the Elasticity of the Cochlear Partition." 20 *J. Acoust. Soc. Am.* 227–241.

Bekesy, G.V. 1949. "The Vibration of the Cochlear Partition. 21 *J. Acoust. Soc. Am.* 233–245.

Bekesy, G.V. 1960. *Experiments in Hearing*. McGraw Hill.

Beranek, L.L. ed. 1971. *Noise and Vibration Control*. McGraw Hill.

Bies, D.A., and C.H. Hansen. 1988. *Engineering Noise Control*. Unwin Hyman.

Borg, E., and S.A. Counter. 1989. "The Middle-Ear Muscles." 261 *Sci. Am.* 74–80.

Brink, W.R. 1992. *Modeling of the Middle Ear for Prosthetic Devices*. M.S. thesis, University of Houston.

Brown, E.H., and F.F. Hall, Jr. 1978. "Advances in Atmospheric Acoustics." 16 *Rev. Geophys. Space Phys.* 47–110.

Coles, R.R.A., G.R. Garinther, D.C. Hodge, and C.G. Rice. 1968. "Hazardous Exposure to Impulse Noise." 43 *J. Acoust. Soc. Am.* 336–343.

Corey, D.P., and A.J. Hudspeth. 1983. 3 *J. Neurosci.* 962.

Dallos, P. 1973. *The Auditory Periphery*. Academic Press.

Embleton, T.F.W. 1971. "Mufflers," in Beranek, *Noise and Vibration Control*. McGraw Hill.

———. 1996. "Tutorial on Sound Propagation Outdoors." 100 *J. Acoust. Soc. Am.* 31–48.

Embleton, T.F.W., J.E. Piercy, and N. Olson. 1976. "Outdoor Propagation Over Ground of Finite Impedance." 59 *J. Acoust. Soc. Am.* 267–277.

English, G.M. 1976. *Otolaryngology, A Textbook*. Harper & Row.

Environmental Protection Agency. 1974. *Information on Levels of Environmental Noise Requisite to Protect Public Health and Welfare With an Adequate Margin of Safety*. EPA Doc. 550/9-74-004.

Fay, R.R. 1988. *Hearing in Vertebrates: A Psychophysics Databook*. Hill-Fay Assocs. Winnetka, Ill.

Fletcher, H., and W.A. Munson. 1933. "Loudness, Its Definition, Measurement, and Calculation." 5 *J. Acoust. Soc. Am.* 82–108.

Guinan, J.J., Jr., and W.T. Peake. 1967. "Middle Ear Characteristics of Anesthetized Cats." 41 *J. Acoust. Soc. Am.* 1237–1261.

Harris, C.M. 1979. *Handbook of Noise Control*. McGraw Hill.

Heffner, H., and B. Masterson. 1980. "Hearing in Glires: Domestic Rabbit, Cotton Rat, Feral House Mouse, and Kangaroo Rat." 68 *J. Acoust. Soc. Am.* 1584–1599.

Helmholtz, H. [1862] 1954. *On the Sensations of Tone*. Dover.

Hirano, M. 1977. "Structure and Vibratory Behavior of the Vocal Folds." In M. Sawashima and F.S. Cooper, eds., *Dynamic Aspects of Speech Production*. Univ. of Tokyo Press.

Hudspeth, A.J., and V.S. Markin. Feb. 1994. "Mechanoelectrical Transduction by Hair Cells." *Phys. Today*.

Kakita, Y. 1988. "Simultaneous Observation of the Vibratory Pattern, Sound Pressure, and Airflow Signals Using a Physical Model of the Vocal Folds." In *Vocal Physiology: Voice Production, Mechanisms, and Functions*, edited by O. Fujimura. Raven Press.

Kaneko, T., et al. 1991. "Ultrasound Laryngography: Multiple Simultaneous Recording of Vocal Fold Vibration." In *Vocal Fold Physiology*, edited by J. Gauffin and B. Hammarberg. Singular Publishing Group.

Karjalainen, S., R. Harma, and J. Karja. 1983. "Results of Stapes Operations With Preservation of the Stapedius Muscle Tendon." 96 *Acta Otol.* 113–117.

Kirchner, J.A. 1988. "Functional Evolution of the Human Larynx: Variations Among the Vertebrates." In *Vocal Physiology: Voice Production, Mechanisms, and Functions*, edited by O. Fujimura. Raven Press.

Kobrack, H. 1959. *The Middle Ear*. Univ. of Chicago Press.

Kringlebotn, M., and T. Gundersen. 1985. "Frequency Characteristics of the Middle Ear." 77 *J. Acoust. Soc. Am.* 159–164.

Kryter, K.D. 1970. *The Effects of Noise on Man*. Academic Press.

Lawhead, R.B., and I. Rudnick. 1951. "Acoustic Wave Propagation Along a Constant Normal Impedance Boundary." 23 *J. Acoust. Soc. Am.* 546–549.

Lenkauskas, E. 1986. U.S. Patent No. 4,624,672.

Lesinki, S.G. 1987. U.S. Patent No. 4, 655,776.

Marquet, J. 1981. "The Incudo-Malleal Joint." 95 *J. Laryn. & Otol.* 543–565.

Moller, A.R. 1961. "Network Model of the Middle Ear." 33 *J. Acoust. Soc. Am.* 168–176.

Moller, A.R. 1963. "Transfer Function of the Middle Ear." 35 *J. Acoust. Soc. Am.* 1526–1534.

Moller, A.R. 1983. *Auditory Physiology*. Academic Press.

Molino, J.R. 1979. "Annoyance and Noise" in Harris (1979), 16-1–16-9.

Muster, D., and R. Plunkett. 1971. "Isolation of Vibrations," in Beranek.

Nassef, A. 1987. *Models for Prostheses of the Middle Ear*. Ph.D. dissertation, University of Houston.

Northwood, T.D., A.C.C. Warnock, and J.D. Quirt, "Airborne Sound Insulation," in Harris (1979), 22-1–22-21.

Oppenheim, A.V. 1978. *Applications of Digital Signal Processing*. Prentice Hall.

Popper, A.N., and R.R. Fay. 1980. *Comparative Studies of Hearing in Vertebrates*. Springer Verlag.

Robinson, M. 1979. "The Robinson Stapes Prosthesis: A 15-Year Study." 87 *Otol. Head Neck Surg.* 60–65.

Sataloff, R.T. 1992. "The Human Voice." 267 *Sci. Am.* 108–115.

Schubert, E.D. 1978. "History of Research on Hearing." In *Hearing*, Vol. 4 of *Handbook of Perception*, edited by E.C. Carterette and M.P. Friedman. Academic Press.

Schuknecht, H.F. 1974. *Pathology of the Ear*. Harvard Univ. Press.

Shaw, E.A.G., and M.R. Stinson. 1983. "The Human External and Middle Ear: Models and Concepts," in *Mechanics of Hearing*, edited by E. de Boer and M.A. Viergever. Delft Univ. Press.

Shea, J.J., Jr. 1976. U.S. Patent No. 3, 931,648.

Sheehy, J.L. 1978. "TORPS and PORPS in Tympanoplasty. 3 *Clin. Otol.* 451.

Tatarskii, V.I. 1971. *The Effects of the Turbulent Atmosphere on Wave Propagation*. U.S. Department of Commerce, National Oceanic and Atmospheric Administration.

Tonndorf, J., and S.M. Khanna. 1966. "Some Properties of Sound Transmission in the Middle and Outer Ears of Cats." 41 *J. Acoust. Soc. Am.* 513–521.

Tyndall, J. 1874. "Sound propagation through fogs," 164 *Phil Trans. Roy. Soc.* 183

Ungar, E.E. 1971. "Damping of Panels," in Beranek.

Van Camp, K.J. et al. 1986. *Principles of Tympanometry*. ASHA Monographs No. 24. American Speech-Language-Hearing Association.

Wente, E.G., and A.L. Thuras. "Moving-coil Telephone Receivers and Microphones. 3 *J. Acoust. Soc. Am.* 44–55.

Wever, E.G., and M. Lawrence. 1954. *Physiological Acoustics*. Princeton Univ. Press.

Wolferman, A. 1970. *Reconstructive Surgery of the Middle Ear*. Grune & Stratton.

Zwislocki, J. 1957. "Some Impedance Measurements on Normal and Pathological Ears." 29 J. Acoust. Soc. Am. 1312–1317.

Zwislocki, J. 1962. "Analysis of the Middle-Ear Function. Part 1: Input Impedance." 34 *J. Acoust. Soc. Am.* 1514–1523.

Zwislocki, J. 1965. "Analysis of Some Auditory Characteristics." In Vol. 3 of *Handbook of Mathematical Psychology*, edited by R. Luce, R. Bush, and E. Galanter. Wiley.

Zwislocki. J.J., and A.S. Feldman. 1963. "Post-mortem Acoustic Impedance of Human Ears." 35 *J. Acoust. Soc. Am.* 104–107.

PROBLEMS

16.1 Two identical speakers driven by a pure tone signal through the same amplifier channel each produce 75 dB at a microphone location equidistant from each speaker. What would the sound level be if both speakers operated together? What would be the effect of introducing a 60° phase shift in one channel?

16.2 The pinna of the human ear is about 70 mm in length, while that of the mouse is about 7 mm. Assuming the two ears are otherwise scaled in the same proportions, what might this suggest about the frequency range of hearing in the mouse?

16.3 Why was the quantity L_{eq} devised to describe the level of a time varying noise? Calculate the L_{eq} of a noise that is 85 dB(A) for 15 minutes; 72 dB(A) for 150 minutes; 90 dB(A) for 120 minutes; 100 dB(A) for 5 minutes; and 80 dB(A) for 4 hours.

16.4 A pure tone of 300 Hz varied over 8 hours with an rms pressure given as a function of time by $p = (t^2 + 6t + 3) \times 10^{-2}$ Pa where t is in hours. Calculate the A-weighted L_{eq} value.

16.5 A warning hooter is sounded n times per hour, producing a constant sound pressure level of H dBA for a duration measured to be T seconds. If the background sound level is negligible, find an expression for the one-hour equivalent sound level. Hence, derive an expression for the day-night level.

16.6 Calculate the daily noise dose for a worker who is employed for 8 hours per day and spends 30% of his time in an environment of 86 dB(A); 25% at 88 dB(A); 25% at 90 dB(A); and 20% at 92 dB(A). How much should his workday be shortened to meet OSHA regulations, assuming the percentage of his time in each environment remains the same?

16.7 A steel panel is 4 m square and 1.5 mm thick. Calculate its fundamental resonance frequency and its critical frequencies. Estimate the transmission loss as a function of frequency. Assume the panel loss factor is 0.001.

16.8 A large meeting room in a conference center is 5 m high, 18 m long, and 10 m wide. The floor is covered with carpet on foam rubber, the ceiling is an acoustic tile of mineral fiber, and 70% of the floor space is occupied with upholstered chairs, but there are no people. The front wall and the front 10 m of the side walls are painted cement. The back wall and the back 8 m of the side wall are covered with one-inch polyurethane foam. For frequencies of 500 Hz, 1000 Hz, and 2000 Hz, find the average absorption coefficient for the room.

16.9 The noise level at the ear of a machine operator must be reduced below 90 dBA to meet the requirements of the health and safety executive. An investigation of the machine reveals that a flat panel 0.09 m^2 vibrates in resonance at 500 Hz. The measured average displacement is 0.1 mm and the whole surface of the panel vibrates in phase. To establish whether treatment of the panel is worthwhile, estimate the dBA contribution (at the operator's ear) due to the vibration. Base your estimate on radiation from a sphere of equivalent surface area. Check whether it is reasonable to make the equivalent sphere assumption. Take into account the efficiency of radiation for the equivalent sphere by considering the real and imaginary parts of the radiation impedance. The A-weighting correction at 500 Hz is −3.5 dB.

16.10 A piece of equipment used for constructing the new mechanical engineering building produces 95 dB-A sound pressure level at the front door of the administration building 100 feet away. In response to complaints by faculty and students, the contractor moves it 200 feet away. What will be the resulting sound pressure level?

16.11 Consider a roof-mounted compressor, with a sound power level of 120 dB-A at 120 Hz. A resident 800 m away, and 10 m lower in elevation, is complaining about the noise from this unit. A 5 m tall noise barrier is proposed, as depicted in Fig. P16.11.

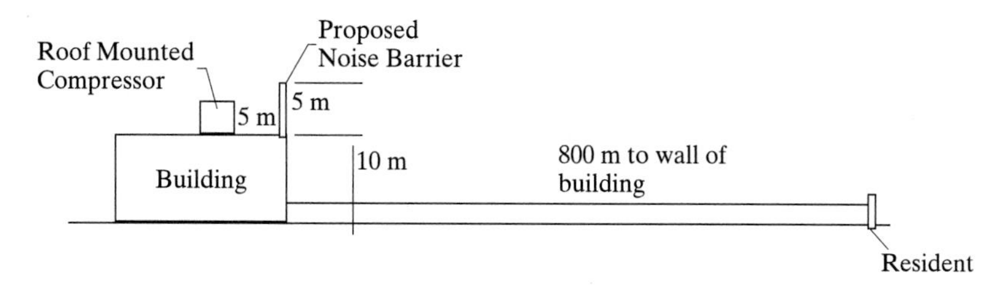

Fig. P16.11

a. What is the sound power level at the residence, without the barrier in place?

b. What is the level at the residence, with the barrier in place?

Chapter 17

Acoustic Systems

17.1 Systems Theory

The word "system" has been used since the time of the ancient Greeks. It is used frequently in scientific and engineering work, sometimes to denote a piece of hardware and sometimes to refer to a method of performing a task. The word is commonly used in other senses as well. In this chapter, we will discuss the concept underlying the use of the word. We will also formulate a taxonomy of systems (i.e., a classification) by finding the common properties and by examining how the various classes of the taxonomy differ from one another. The intention of this exercise is both to improve our understanding of the world and to provide guidance for engineering practice.

We will begin our discussion with the underlying concept—partly to provide a working definition, but also to develop a deeper understanding. Philosophers have discussed the acquisition of human knowledge for centuries, a subject known as epistemology. Our investigation is a part of epistemology. The wide scope of systems theory is illustrated in Fig. 17.1, which is inspired by the three worlds cosmology of Pierce and Popper. The domains of natural, human, and formal systems are equivalent to these philosophers' physical, mental, and man-made worlds. Systems theory might also be regarded as an extension of Dewey's critique of technology as summarized by Hickman (1990). Dewey, a member of the American school of pragmatic philosophy, maintained that human artifacts or instruments are constructions that we use to serve our purposes. In his epistemological view, our entire knowledge consists of such instrumentalities, thus Hickman claims that technology is central to Dewey's view. The same argument might be applied to Popper's philosophy.

Many definitions of the word "system" have been given, reflecting both analytical and holistic approaches, as well as descriptive and normative approaches. Bertalanffy (1968) defined a system as "a set of elements standing in interrelation among themselves and with the environment." Since this definition puts some emphasis on the components of

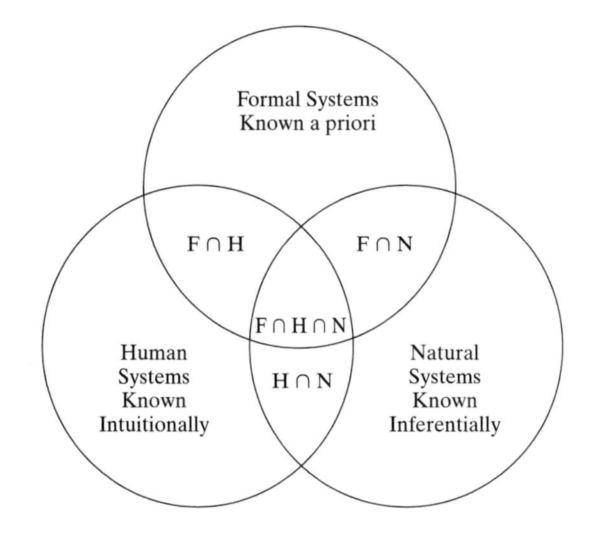

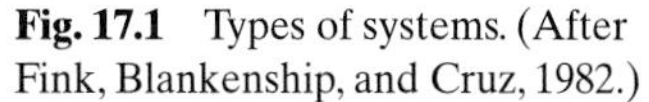
Fig. 17.1 Types of systems. (After Fink, Blankenship, and Cruz, 1982.)

a system, it might be said to be analytic in nature. A holistic definition, given by Rapoport (1986), is that a system is that which remains recognizably itself. Most definitions imply that there is some invariance which underlies a system. Very often it is possible to define this invariance precisely and to state it in mathematical terms. A spring balance, for example, can be specified in terms of two variables: the unloaded length of the spring, and its stiffness. An organism, such as a human being, is specified in terms of the chemical constitution of its genome. The city of Rome can be defined in terms of its geographical location, infrastructure, and the culture of its inhabitants. The system may be subject to change, but we can usually continue to recognize it intuitively and to pinpoint the underlying invariance.

We sometimes distinguish between the mental processes of analysis and synthesis. In analysis, we examine an object's constituent parts. For example, when we describe a method we usually break it down into different steps. In synthesis, we seek to grasp or recognize a whole. Both of these mental processes are essential to cognition. The holistic and analytic approaches to knowledge are encountered in everyday life. An example of the holistic approach is the recognition of a face; an example of the analytic approach is the sorting of objects into categories. These two approaches to knowledge are found in many academic disciplines. Deductive mathematics was originally concerned with intuitively recognized configurations until Descartes introduced analytical geometry. In biology, followers of the holistic approach are taxonomists who study organisms, species, genera, families, etc. The analytic approach, on the other hand, is conducive to physiology and genetics. In psychology, there are the gestaltists versus the behaviorists. In engineering, there are engineering scientists as opposed to designers. Another dichotomy exists in the descriptive versus normative approaches to knowledge. In the descriptive approach, we are concerned with understanding "how" something works; in the normative approach, we ask "what for?" Engineers design systems to serve goals, as well as to analyze systems descriptively.

A system may be defined in terms of a statement of a boundary or a demarcation (e.g., the so-called free-body diagram used in mechanics or the designation of a thermodynamic system as opposed to a control volume encountered in thermodynamics).

In some simple systems, the invariance is stated as the constant of proportionality between cause (input) and effect (output), which is sometimes known as a transfer function. In more complex systems, there may be multiple inputs and outputs to consider. In nonlinear systems, the invariance cannot be specified in terms of a simple constant of proportionality, and we will need to state the differential equation governing the behavior. In the case of a computer, the invariance may be embedded in the form of a program with a lengthy sequence of instructions. The genome of an organism may be thought of as a program for the manufacture of proteins. In each case, we see that there is a "schema" that remains invariant as long as the system preserves its identity. In more complex systems, the schema requires more information for its specification. We can still recognize the continuity of a system with a slowly changing, or even an adaptive, schema. The evolution of schema may also be regarded as a domain of system science.

Natural systems can be classified to show a progression from the simple to the more complex, and it is believed that this is a result of schematic evolution. The basic constituents of nature are the elementary particles, which can be classified and explained using the so-called standard model. Combinations of elementary particles give rise to the atoms of the chemical elements and these in turn to molecules. The special structure of the carbon atom enables it to form the organic molecules that are essential to life. Astronomers study the evolution of the largest scale systems: galaxies, stars, and planets. There are many other systems discovered in the evolution of planetary bodies that are the concern of geologists. Quite often the intriguing scientific issues center around explaining how such systems perpetuate themselves. Another great scientific challenge is to explain how life started on Earth.

Once started, life forms also show a progression from the simple to the complex. Viruses are among the simplest life forms, to the extent that they are indeed living, which has been questioned. They show behavior resembling that of robots (i.e., they respond only to certain limited stimuli). Unicellular forms, such as bacteria, are already systems displaying a great complexity in their organization. There exist a vast number of plant and animal species whose classification shows a tree-like arrangement, explained by Darwin's theory of natural selection and its genetic corollary: namely, that the occurrence of numerous mutations in the genome allowed an organism to adapt to changes in the environment. Those that adapted were the fittest to survive.

Human beings have an additional means of adaption to the environment: a brain of sufficient complexity as to enable us to create artificial systems. The emergence of tool-making and language is speculated to be the crucial evolutionary advantage of Homo Sapiens over their rivals. Human artifacts also show a ramifying classification reminiscent of the tree of life. Languages, for example, can be grouped into families and traced back to common ancestors. Perhaps, at one time, there was just one proto-language. Most manufactured objects show evolutionary development. Mathematics may be regarded as a human artifact, a precise form of language. Science may be regarded as a collection of schemata or theories by which we model the world.

Human beings are social animals who have also created much greater and richer social systems than other animals: We have political, legal, economic, business, religious, ethical, familial, and personal systems giving structure to our lives. The branches of knowledge are reflected in the organization of the universities and the professions, a phenomenon sometimes called the socialization of knowledge. The design of a complex system, such as a battleship or a space station, may itself be accomplished by an organization of specialists. Systems theory has also been applied to the question of how best to organize social systems.

The issues here are of great concern to both analysts and synthesists. How can we explain the remarkable ability of the human brain to create new systems? If we could explain this amazing phenomenon, perhaps we could improve the artificial intelligence of machines. Another pressing question: Is it possible to construct a general theory of systems, in some way analogous to physics, but going beyond the traditional limits of physics to the discussion of systems as yet unrealized?

Bertalanffy, who was among the first to propose the development of a general theory of systems, was most interested in the organization of biological systems. He developed his ideas during the 1930s and 1940s. Norbert Wiener, a celebrated mathematician, was thinking along parallel lines. In the late 1940s, he published his ideas and invented the word "cybernetics" to designate the field of systems science. There is now an accepted field of systems engineering, whose domains are illustrated in Fig. 17.2.

An interesting aspect of Fig. 17.2, from a review of *Systems Engineering* by Fink, Blankenship, and Cruz (1982), is that it omits some important areas of systems engineering, including the analysis of mechanical structures, thermodynamics, and computer science. Thermodynamics arose historically from the emergence of the steam engine. The practical problems of designing new systems often provide the impetus for the growth of new branches of science. The term "entropy" was invented and the second law of thermodynamics was formulated to state that in a closed system, entropy would increase. This increasing of entropy contrasts with the behavior of biological systems, which have shown evolutionary trends to increasing complexity. Systems scientists now believe the difference is due to the fact that biological systems are not closed; in the thermodynamic sense, they are open. It appears that there is, in general, a close correlation between goal-directedness in a system and its being open. As we have already

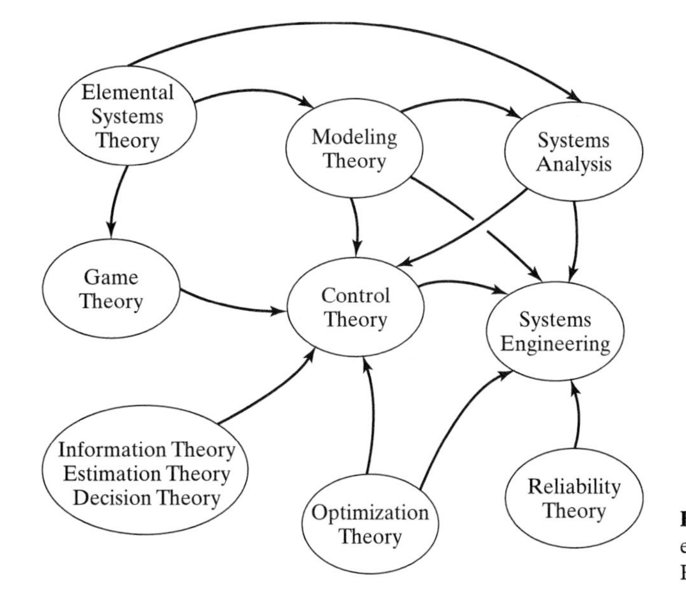

Fig. 17.2 Domains of systems engineering. (After Fink, Blankenship, and Cruz, 1982.)

argued, computer science is an important area (which in fact appears to be subsuming many other engineering fields), and together with cognitive science and artificial intelligence, seems poised to make important contributions to our understanding of the brain and an explanation of consciousness.

Much of this book has been concerned with the analysis of simple and complex acoustic systems, starting with a discussion of those having one degree of freedom (such as a mass on a spring) and followed by a discussion of two degree and multidegree of freedom systems. Since these were all supposed to be discrete element systems, we extended the discussion to distributed systems. The distribution of material could be along a line (as with strings, rods, or fluid-filled pipes), or over a surface (as with a membrane, plate, or shell), or throughout a three-dimensional space (e.g., a parallelopiped, a sphere, or an essentially limitless space, as in the atmosphere, the ocean, or the Earth). We have also discussed the electrical analogs of these mechanical systems and simple combinations of electrical and mechanical components, as in loudspeakers or other transducers. Another aspect of linear system theory that we investigated is the response of a system to excitation. Frequently, the response is stable and our concern is then with the transmission efficiency of the system or noise control. However, there are systems that have an unstable response and we need to identify them (transient cavitation bubbles, for instance). Sometimes our concern is to control the oscillations of a dynamic system. We also encounter mechanical systems with nonlinear or even chaotic responses. Some of the phenomena we have discussed can be found in nature, but for the most part the systems we discussed have been artificial.

We have also encountered some systems in connection with the study of noise control, where we pointed out the usefulness of the sound-path-receiver paradigm as an approach to noise control. If we include the element of signal processing for the benefit of a human observer and allow for the possible completion of a feedback loop to the source, then we have a more general picture of an acoustic system, as seen in Fig. 17.3. Measurement instruments, such as sound level meters and noise dosimeters, are examples of this type of system. Communication systems, including the human speech/hearing

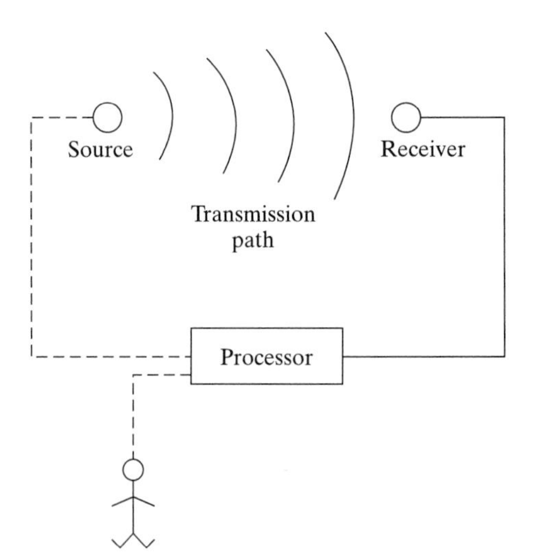

Fig. 17.3 Schematic of an *acoustic system*.

system, and sound systems for residences, offices, theaters, etc. are very similar. The major design consideration in these is to secure a good signal-to-noise ratio.

We will now discuss more complex artificial systems, in which engineers have used an understanding of acoustics for increasingly sophisticated purposes. We will examine ranging, imaging, monitoring, recognition, speech processing, and music systems in the hope that the taxonomy of acoustic systems presented here will illustrate the relevance of systems theory to engineering in general. We will also discuss the evolution of acoustic systems and their related social systems.

17.2 Ranging and Detection Systems

The main purpose of a ranging system is to determine the distance to an object of interest, as illustrated in Fig. 17.4a. This type of system was originally proposed by Langevin during World War I. For underwater sound, the object of interest might be fish or a submarine. In medical ultrasonics, the concern is to determine the location of interfaces among the organs of the body. In nondestructive testing, we want to locate flaws or gauge thicknesses. In seismic prospecting, the concern is to locate oil-bearing strata. One way to analyze the problem is to think of the system as comprised of three components: a source, a transmission path, and a target, as in Fig. 17.4b. We might then proceed by finding the transfer function of each component. An alternative approach, which we will use, follows the energy flow through the system. The directionality of the

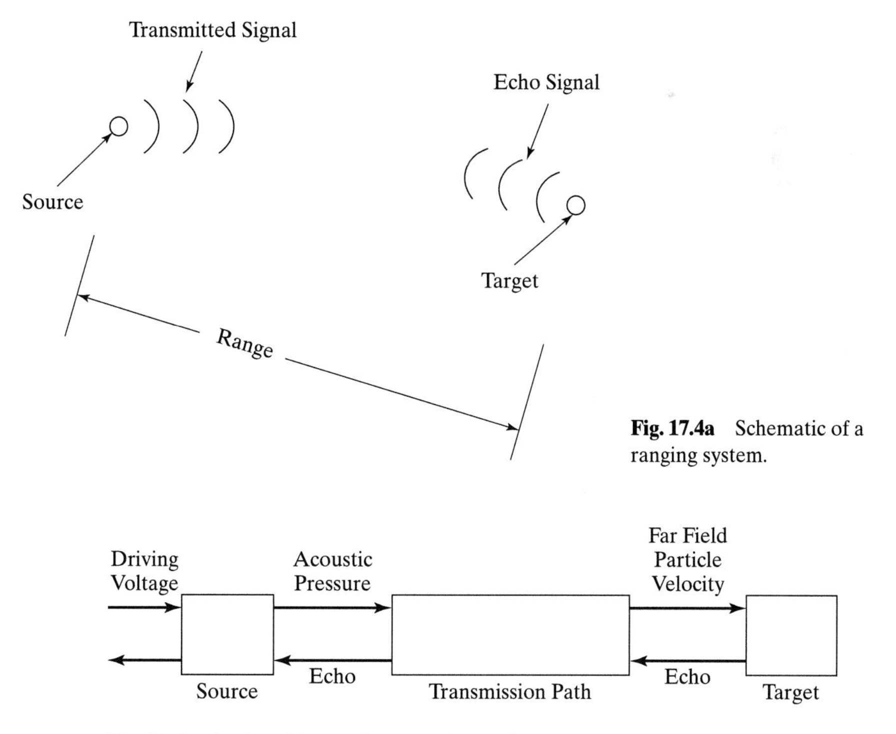

Fig. 17.4a Schematic of a ranging system.

Fig. 17.4b System diagram for a ranging system.

source is defined by the directivity factor (DF):

$$DF = I_o/I_{av} \tag{17.2.1}$$

where I_{av} is the average value of the intensity and I_o is its value in the forward direction. Thus, if W is the total power radiated by the source, then at a distance r, the average intensity will be

$$I_{av} = W/4\pi r^2 \tag{17.2.2}$$

If the intensity at an angle θ to the axis is I_θ, then we can show that

$$DF = \frac{2I_o}{\int_0^\pi I_\theta \sin\theta \, d\theta} \tag{17.2.3}$$

From Eqs. (17.2.1) and (17.2.2), we have

$$I_o = \frac{W}{4\pi r^2} \times DF \tag{17.2.4}$$

In the case of a sectoral radiator (i.e., one that radiates a uniform intensity through a solid angle Ω, and nothing in any other direction),

$$DF = \frac{4\pi}{\Omega} \quad \text{and} \quad I_o = \frac{W}{\Omega r^2} \tag{17.2.5}$$

In general, however, we must use Eq. (17.2.4). The intensity at the source is defined as the intensity at unit distance (1 meter) and the source level is given by

$$SL = 10 \log_{10}\frac{(\text{intensity at source})}{(\text{reference intensity})} \tag{17.2.6}$$

$$= 10 \log_{10}\frac{I_1}{I_{ref}} = 10 \log_{10}\left[\frac{W}{4\pi} \times \frac{DF}{I_{ref}}\right] \tag{17.2.7}$$

It follows that the intensity at the target will be given by

$$I_T = I_1\frac{W_T}{W_1 R^2} \tag{17.2.8}$$

where R is the range, and allowing for attenuation

$$I_T = \frac{I_1}{R^2}e^{-\alpha R} \tag{17.2.9}$$

Now the quantity

$$TL = -10 \log_{10}(e^{-\alpha R}/R^2) \tag{17.2.10}$$

is called the transmission loss. Some of the energy will be absorbed by the target and some will be scattered. Let the fraction scattered in the backward direction be f_s, so that if the scattered intensity is measured at 1 meter,

$$f_s = \frac{I_s}{I_T} \tag{17.2.11}$$

and

$$I_s = f_s I_T = f_s \frac{I_1}{R^2} e^{-\alpha R} \tag{17.2.12}$$

Taking logarithms

$$\begin{aligned} 10 \log_{10} \frac{I_s}{I_{ref}} &= 10 \log_{10} \frac{I_1}{I_{ref}} + 10 \log_{10} f_s + 10 \log_{10} \frac{e^{-\alpha R}}{R^2} \\ &= SL + TS - TL \end{aligned} \tag{17.2.13}$$

where TS = target strength = $10 \log_{10} f_s$.

Finally, the scattered wave has to travel back to the transmitter, and the intensity of the echo then received will be

$$I_E = \frac{I_s}{R^2} e^{-\alpha R} \tag{17.2.14}$$

Thus, the echo level is obtained from Eqs. (17.2.10), (17.2.13), and (17.2.14), yielding

$$EL = 10 \log_{10} \frac{I_E}{I_{ref}} = SL + TS - 2TL \tag{17.2.15}$$

This signal is then "transduced" by the transmitter, now acting as a microphone, to yield a voltage that must exceed the noise due to both the acoustic environment and the intrinsic noise in the microphone. Equation (17.2.15) is called the sonar equation.

17.3 Imaging Systems

Imaging systems are intended to produce an optical display that can be interpreted by a human observer using the well-developed visual signal processing capabilities of the brain. The optical image may be a shadowgraph or a Schlieren field (such as we discussed in Chapter 9), a drawing, a photograph, a trace on a cathode ray screen, a two-dimensional section called a tomogram, or a three-dimensional representation, such as a hologram. The power of acoustic systems in these applications lies in the fact that sound can propagate through media that are highly attenuative to electromagnetic waves in general and optical waves in particular. Provided acoustic radiation can be generated in the medium, made to interact with the region of interest, and then sensed in some accessible field, an optical image can be produced. It is this realization that allows us to "see" with sound and has led to the advent of movies of babies in utero, as well as pictures of defects deep inside metal castings, of sunken ships in the ocean, and of the interiors of the Earth, of other planets, and even the Sun.

One form of imaging system may be regarded as an extension of the pulse echo ranging system described in Section 17.2. Suppose that a pulse is radiated in a certain direction and an echo results from some interface. In the simplest systems, the echo is received by the transmitting transducer and the voltage on the transmitter is displayed on an oscillograph screen. This is sometimes called an A-scan and will have the form shown in Fig. 17.5. The ellapsed time t will be proportional to the distance to the reflecting interface. Depending on the nature of the medium, there may be more than one echo discernable in the trace. This type of display was the earliest used in ultrasonics for both medical and nondestructive testing applications. By rotating the transmitter or by rotating the emitted beam, it is possible to obtain echoes from different directions. The amplitude of the reflected signal can then be displayed as the brightness of the oscilloscope spot and the trace is made at the same angle to a reference direction as that of the beam so as to obtain a scan or sweep of the field. The beam sweeping can be accomplished electronically using an array of transducers, as described in Chapter 3. In this way, we can build up an image of the reflecting surface, known as a B-scan, as shown in Fig. 17.6a. The resolution is a critical issue in systems of this type. In order to visualize internal structure, the pulse width has to be minimized. Since the pulse width is determined by the sound wavelength in the medium, it is necessary to use as high a frequency as possible, the limitation being the increasingly high attenuation as frequency increases. Further improvements in clarity have been achieved by using data processing to extract phase information and thus refocus the image. Fig. 17.7 shows images obtained from such a system.

Another approach to imaging was initiated by Pohlman (1937). Pohlman described a cell that consisted of a water sample contained between a glass wall and a thin membrane, transparent to ultrasound. Metal particles were suspended in the cell. When the cell was irradiated with ultrasound from the membrane side, these particles would align perpendicular to the sound in the manner of Rayleigh discs. Thus, the insonified region of the cell could be rendered visible by scattering light through the glass wall. The Pohlman cell has served as the progenitor of a number of systems using the scattering of sound from particles for imaging or other purposes. Doppler shifts can be used to determine the velocity of blood flow, which can then be displayed in color to enhance images of the heart, for example. Cespedes (1993) and his advisor Ophir used ultrasonic scattering from particles in tissue to measure strain in a technique known as "elastography."

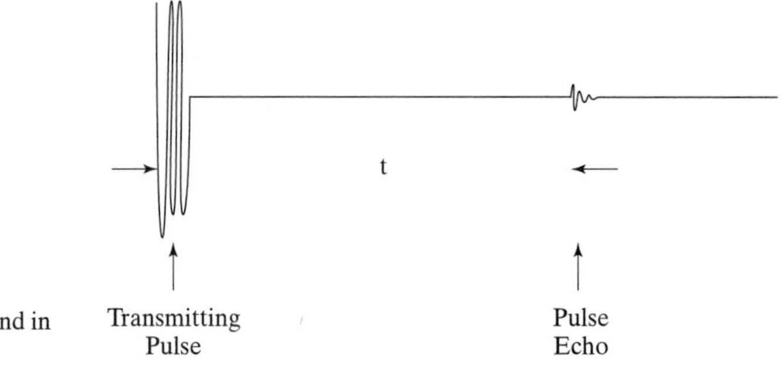

Fig. 17.5 An A-scan display found in the pulse echo technique.

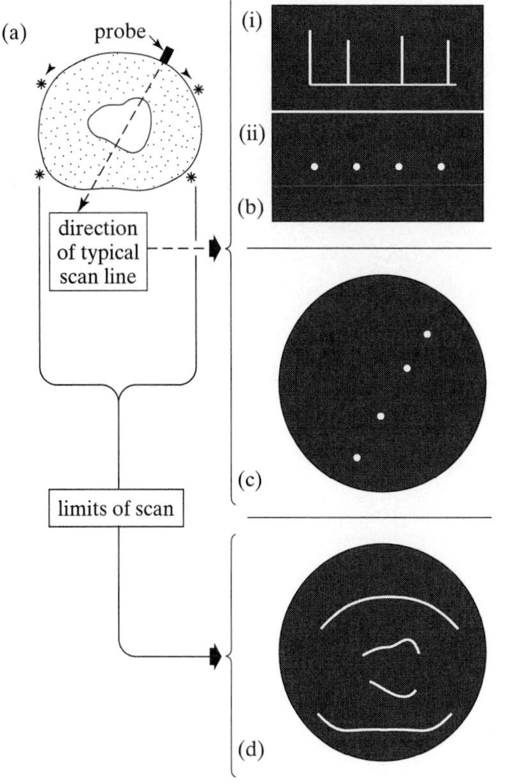

Fig. 17.6 A-scope and B-scope methods for displaying ultrasonic pulse-echo information. (a) Schematic representation of a section through a patient; (b) (i) A-scope presentation of a typical scan line; (ii) B-scope presentation of the same scan line; (c) B-scan as in (b)(ii), but with the direction of the timebase linked to the direction of the ultrasonic beam; (d) compound B-scan, integrated from many individual scans, each one similar to (c). (After Wells, 1977.)

Imaging can also be effected through the use of lenses. We are most familiar with this technique in optics, but as we saw in Section 3.7 the same laws must govern ray propagation in acoustics. The use of lenses in acoustics has received the most serious attention in regard to the ultrasonic microscope, first proposed by Sokolov (1935). In principle, the device can be understood as a lens used as a magnifier, as in Fig. 17.8a. In the optical realization, the image is either viewed by eye or captured on a photographic emulsion. In the ultrasonic version, some other means has to be found to convert the acoustic field variations into a visible image. Cook and Werchan (1971) used heat-sensitive liquid crystals, while Quate and co-workers used a thin deposit of piezoelectric material (see Lemons and Quate (1979)). One of Quate's designs for an ultrasonic microscope is shown in Fig. 17.8b. Sound is generated on the piezoelectric film in response to the input signal. The use of a thin film transducer implies that the sound is generated at a very high frequency (0.2 GHz to 2 GHz), thus permitting a correspondingly high resolution. The sound propagates as a plane beam through the sapphire (low attenuation) support. The curved surface on the left acts as a lens and serves to focus sound energy on the object. The second lens produces the magnified image on the second piezoelectric film from which a signal is taken to a CRO tube display. The object is scanned and the CRO display is kept in synchronization with the scan to build up a two-dimensional image. Such microscopes have been used to view biological material (e.g., the interiors of cells) and have the advantage that they produce images with better

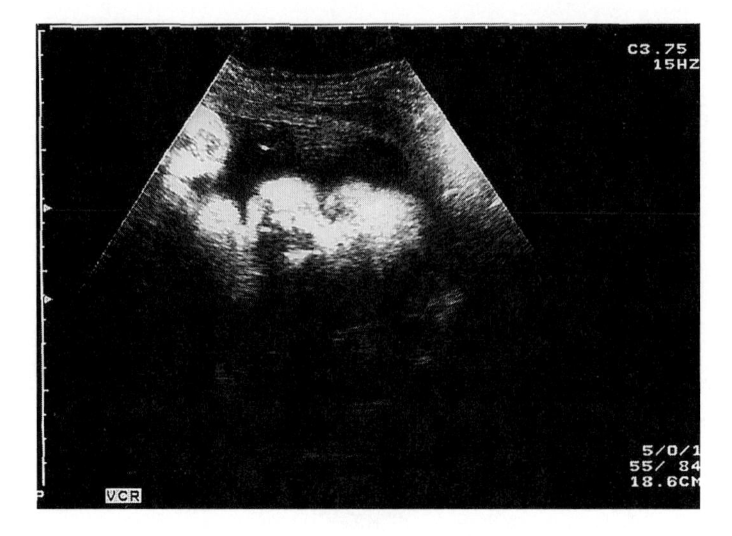

Fig. 17.7 Pulse echo sonogram of a baby in uterus.

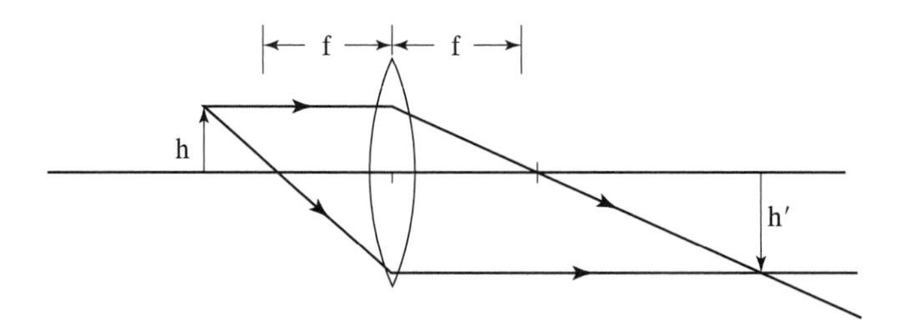

Fig. 17.8a Formation of magnified image by a lens. Magnification = h'/h.

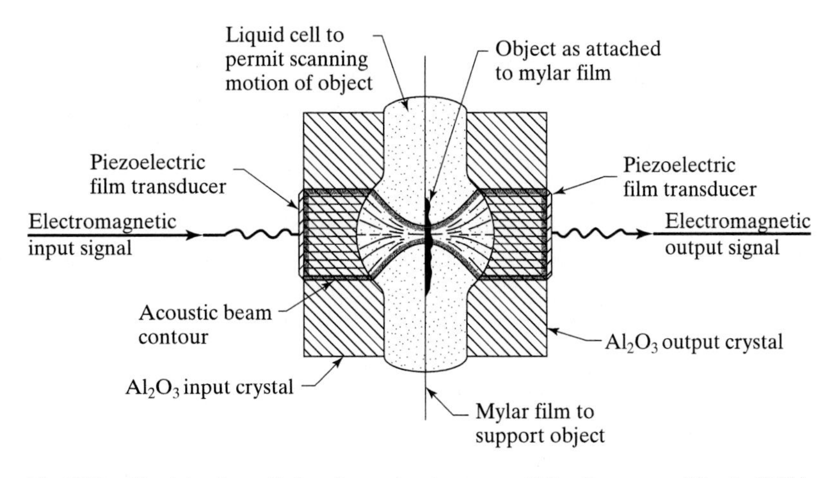

Fig. 17.8b Quate's schematic for ultrasonic microscope. (After Lemons and Quate, 1979.)

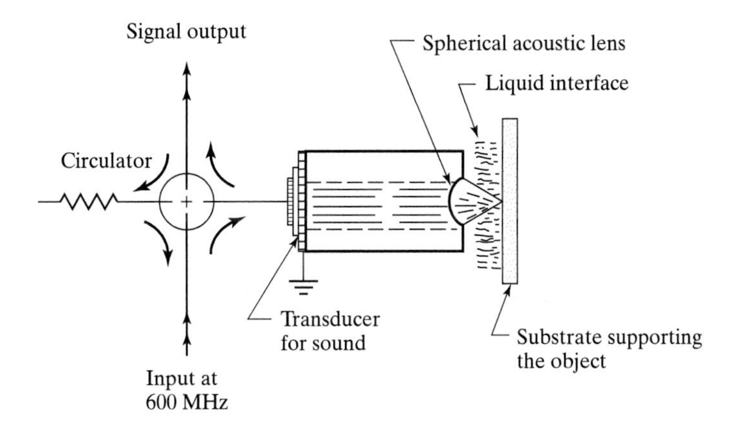

Fig. 17.8c Quate's diagrammatic scheme for the reflection mode. The substrate is scanned mechanically to obtain the image. (After Lemons and Quate, 1979.)

contrast than optical microscopes, as illustrated in Fig. 17.9. They have also been used in nondestructive evaluation of composites, claddings, coatings, and printed circuit boards. Another design by Quate uses the same lens for both collimating and receiving, as shown in Fig. 17.8c. A circulator is used to keep the input and output signals separated, since they are of much different strengths.

Another form of imaging is holography. The optical hologram, first demonstrated by Gabor (1948), became commonplace with the development of the laser. This development in turn precipitated work on acoustic holography, although it appears that an arrangement anticipating many of the features of the most successful acoustic holography systems was actually constructed for flaw detection by Sokolov in 1929. We will begin by reviewing the general theory of the hologram, which is actually an interference pattern produced when a monchromatic (or in the acoustical case, monotonic) beam scattered from some object meets a coherent reference beam. Such a situation is shown in Fig. 17.10. The hologram is used when a beam with exactly the same properties as the reference beam is shone through it. The desired image is then formed by focusing the reconstructed scattered wave. The image will be three dimensional if the original object is three dimensional.

To understand what happens, we will consider a simple case first. Suppose that the monotonic object beam of Fig. 17.10a has been scattered from a point at a great distance so that it is essentially a plane wave. The reference beam is also a plane wave of the same frequency. For optical holography, we would have to be very careful that the object and reference beams are "coherent"; in other words, that they maintain the same constant phase difference. This is usually done by deriving the two beams from the same laser source by using a beam splitter. Assuming the coherence condition is met, the two beams will produce an interference pattern that will be a series of straight fringes, which may be recorded in the optical case using a photographic plate. The hologram is played back by shining a laser beam of the same wavelength through the photographed fringes. But this series of straight fringes constitutes a diffraction grating and thus the transmission consists of a zero-order wave, two first-order waves, and perhaps some very weak higher order waves (see Section 3.8). We will prove shortly that the first-order wave travels in the same direction as the original object wave and is in fact a reconstruction of it.

For a more complicated object, the situation may be represented as shown in Fig. 17.11. Here O is the object wave and R is the reference wave. The hologram is

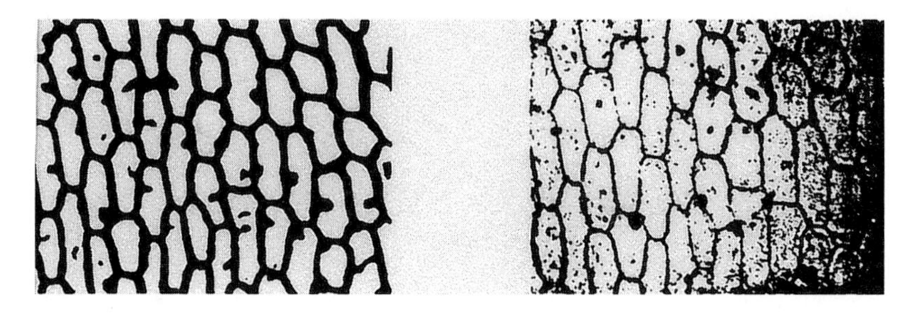

Fig. 17.9 Comparison of the acoustic and optical images of onion cells. (After Lemons and Quate, 1979.)

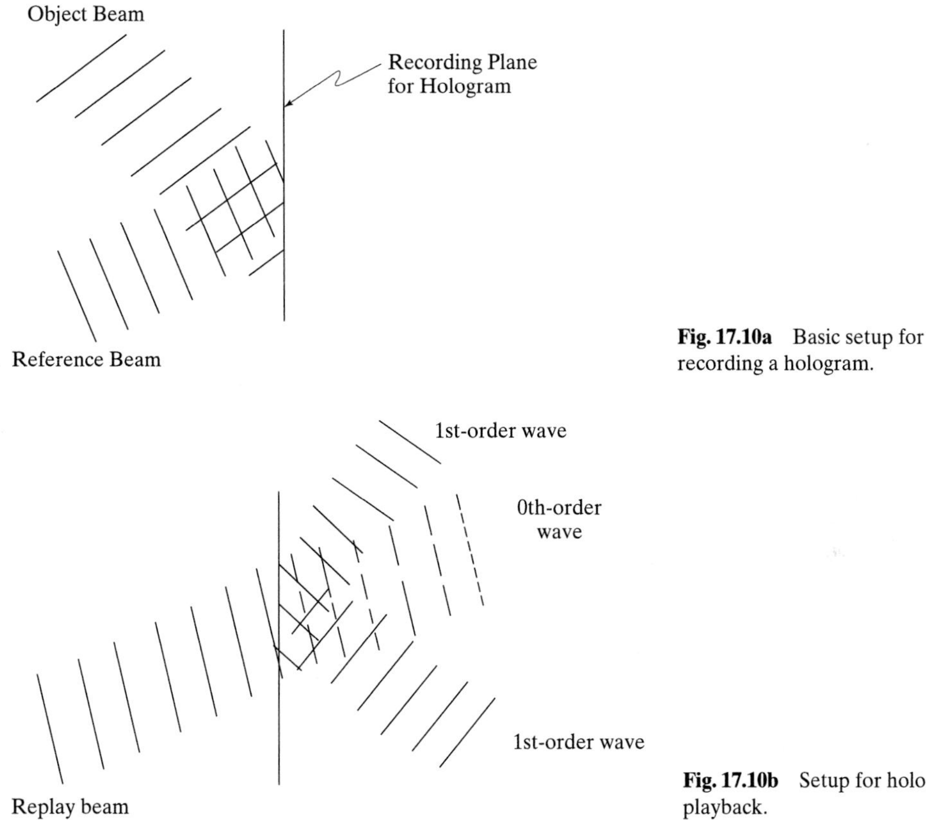

Fig. 17.10a Basic setup for recording a hologram.

Fig. 17.10b Setup for hologram playback.

recorded on the plane H. In the optical case, the darkening of the photographic plate records the amplitude of the object wave while the fringes are determined by its phase. The total field at H is O + R. The photographic plate responds to the intensity and assuming that on playback the process is linear, the transmittance will be given by

$$t(x) = A|O + R|^2 = A(|O|^2 + |R|^2 + OR^* + O^*R) \tag{17.3.1}$$

where * denotes a complex conjugate. When the hologram is replayed using reference wave R, the transmitted field will be

$$\varphi(x) = R(x)t(x) = A(R|O|^2 + R|R|^2 + |R|^2O + R^2O^*) \tag{17.3.2}$$

If $|R|^2$ is a constant, then the third term is proportional to the original object wave. If the reference wave is plane and we think of the object wave as being composed of a collection of plane waves, each one of which interferes with R to form a grating, then the replayed hologram will produce a reconstructed object wave. The other terms in Eq. (17.3.2) represent the zero-order wave and the conjugate first-order wave, all of which are filtered out.

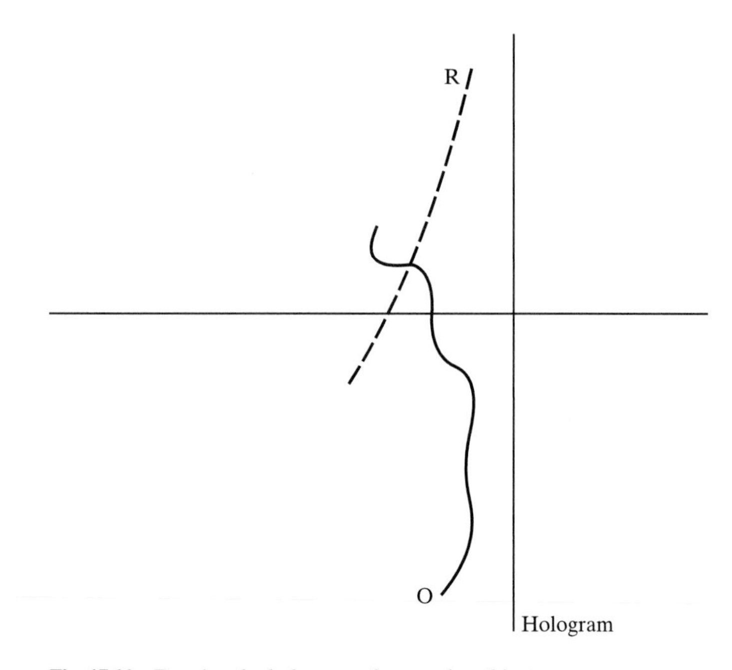

Fig. 17.11 Forming the hologram of a complex object. O = object wave; R = reference wave.

In acoustic holography, the problem is finding a way to record the hologram in place of photography. One way, illustrated in Fig. 17.12a, is by scanning the holographic plane with a microphone. The object modulated signal is multiplied with the signal originally sent to the object, a process that duplicates the effect of interference. An optical hologram can be produced by using the multiplied signals to modulate a light source, which is then scanned over a photographic plate in synchronization with the moving microphone. This hologram can then be replayed using a laser to produce an optical image. However, the best-quality images have been produced using liquid surface holography, as illustrated in Fig. 17.12b. The object is submerged in a tank of water and the hologram is a ripple pattern produced by interference of the object and reference beams on the surface. The ripple pattern can be illuminated for image reconstruction, as shown in Fig. 17.12b. Both scanned and liquid surface holography operate at frequencies in the range 0.1 MHz to 1 MHz.

Our next example of imaging comes from the field of geophysics. Where an oil well already exists, it is possible to determine the velocity of sound in the surrounding rock by sending an ultrasonic pulse for a short distance down the well. However, when the objective is seismic prospecting from the surface, other approaches have to be taken. The attenuation below ground is very severe indeed, and the distances that have to be covered are very great so that use of ultrasonic frequencies is out of the question. It follows that the signals we need to employ are of very high amplitude and very low frequency, such as those produced by explosives (onshore) and air guns (offshore). The upper layer of the ground has the highest attenuation, so it is common to place the source below this

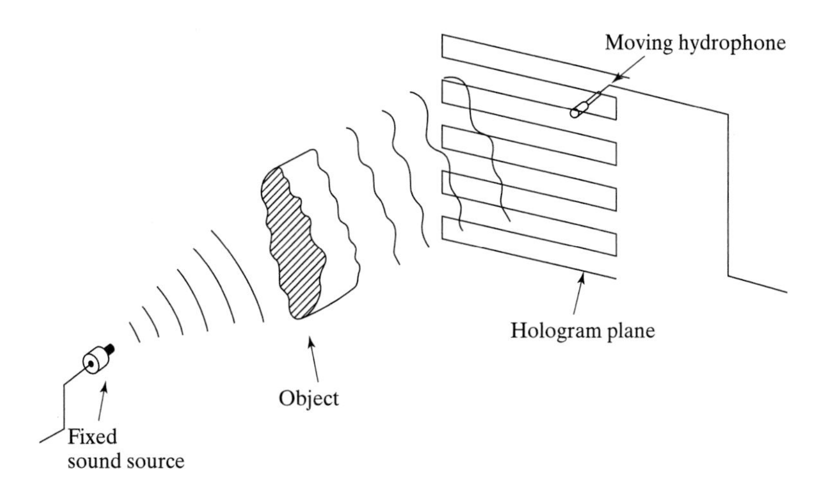

Fig. 17.12a Scanned holography.

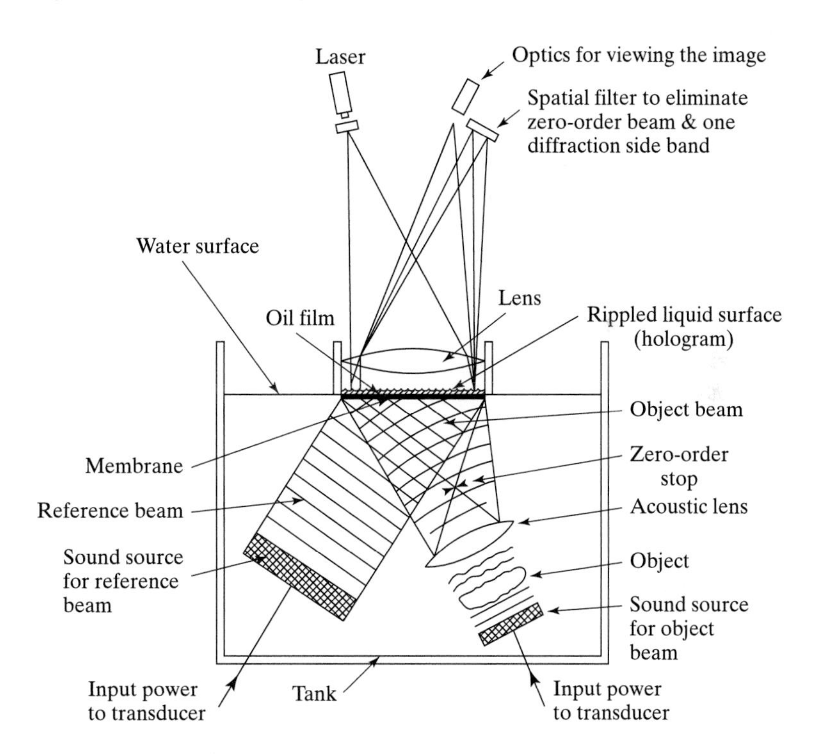

Fig. 17.12b Liquid-surface holography. (After Wade, 1976.)

layer. The source sends dilatational waves into the Earth, and echoes from subsurface strata are received by seismometers at intervals along the ground. Sets of seismometers like this are often referred to as arrays, but they should not necessarily be thought of as operating in the same manner as the arrays we discussed in connection with ultrasonic applications. The procedure is known as the amplitude-versus-offset (AVO) method. It is usual to "stack" the data by superimposing amplitude traces of signals from transducers at successive offset positions, as shown in Fig. 17.13. A horizontal interface between strata will thus give rise to a line across the traces, as shown in the figure. There are a number of data processing approaches that have been developed to improve the situation. It has become standard practice to blacken the area below the excursions to one side of the trace. Unfortunately, placing the source below the topmost layer usually gives rise to multiple echoes, or "ghosts." Eventually, it was realized that the data received at any given seismometer are a convolution of the source signal and the response of the ground and that the ghosts could be removed by deconvolving the data with what is known as a Wiener filter. The arrangements of lines in the stack can be used to generate hypotheses on the nature of the strata below the surface (e.g., Are they flat or tilted? Are there synclines or anticlines present?). The data can then be "migrated" (i.e., converted into what is in effect an image of the subsurface strata). Fig. 17.14 shows an example of such an image obtained from a salt dome below the Gulf of Mexico. The degree of interpretation involved in seismic processing begins to approach that of a recognition problem, as described in Section 17.4.

17.4 Recognition Systems

We have seen that a system may be defined as a constancy amid change. But how is such a constancy recognized? We are familiar with the intuitive abilities of humans to recognize faces or handwriting or sentences in the English language. To write a computer program to carry out one of these tasks is still challenging at the present time. In each of these cases, we presume that some underlying constancy determines the signals we receive. Thus, the structure of a face is defined by the genome of the individual person. Handwriting is determined by muscular movements, which are in turn determined by neuronal impulses, and these processes are in turn determined by both learned and genetic factors. The English language can be defined by sets of rules. However, the expression of the underlying constancies is mitigated by uncertainties in each case, and the processes producing the final results are thus said to be statistical or stochastic. Recognition systems attempt to classify received signals according to the systemic constancies that generated them.

Acoustic monitoring systems are examples of the general class of recognition systems. By "acoustic monitoring," we mean the detection of sound or vibration signals with a view to gaining some knowledge about the object from which the signals emanated. Many structure and pieces of equipment—such as railroad wheels and rail, rotating shafts, turbine blades, piping, and all manner of trusses and frames used in bridges and buildings—can develop dangerous defects. We would like to assess the safety of these structures and equipment without having a human inspector conduct an inch-by-inch time-consuming examination. Acoustic monitoring aims to do just this by sensing the vibrations of the structure and detecting any changes that occur. Fig. 17.15 shows a schematic of a monitoring system. Suppose a test object has some condition of interest, such as a crack. When the test object is excited by an input

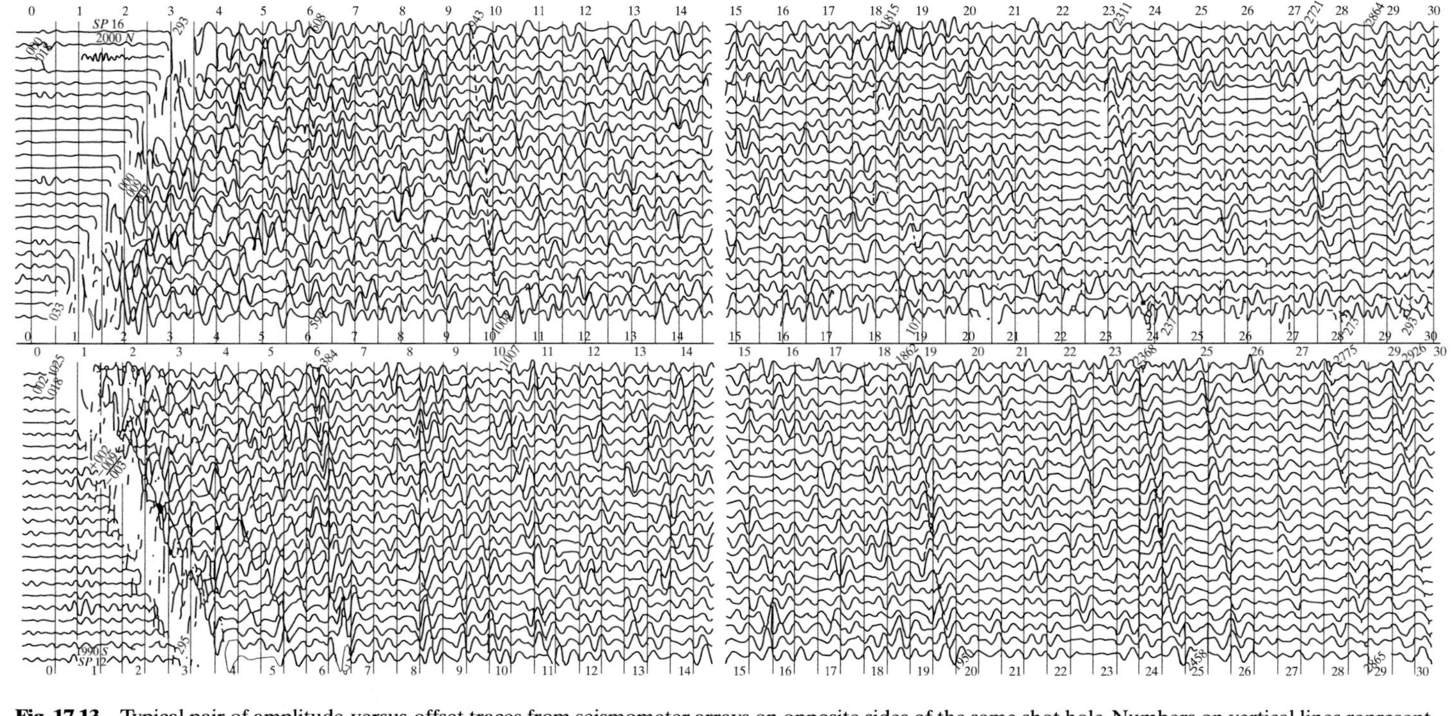

Fig. 17.13 Typical pair of amplitude-versus-offset traces from seismometer arrays on opposite sides of the same shot hole. Numbers on vertical lines represent 1 min after explosion in tenths of a second. (National Geophysical Co., published in *Geophysics*, 1947.)

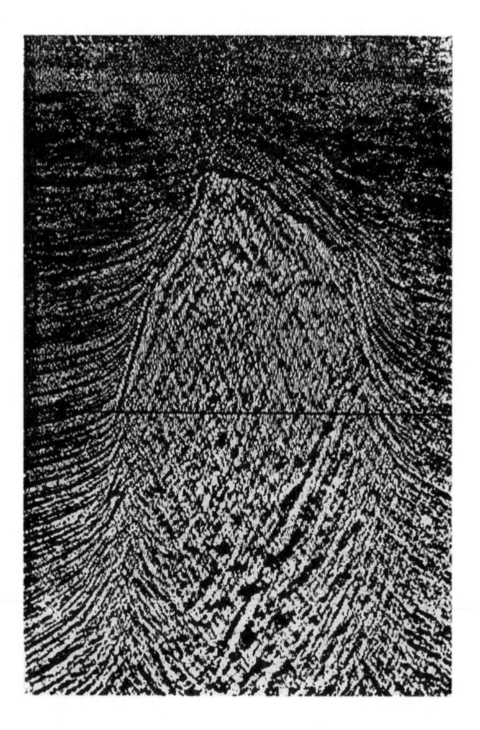

Fig. 17.14 Seismic image of a salt dome below the Gulf of Mexico.

force, its response can be detected by a transducer to yield a signal. If there is a defect in the object, then this signal will be different, all other conditions remaining the same. The system is therefore completed by some sort of device for recognition of the difference in the received signal. The limitation of such a system lies in its uncertainties: variability in the exciting force, geometrical and material variations, and finally noise at the reception end. These various uncertainties may produce differences in the received signal as great as those due to the condition of interest, thus masking its presence.

The use of acoustic monitoring probably dates back to antiquity, since tapping on an object can generate sounds whose differences can be heard when an object is defective. The input force can be made more reproducible if a coin or a hammer is used to do the tapping. The use of sound and the human ear to do the signal transmission and recognition relies on a rather sophisticated natural system whose artificial duplication has only recently been contemplated. Thus, the first fully automated acoustic monitoring system was a simpler version than the tapping test in which a rotating shaft was monitored for unbalance by taking a vibration signal from its bearing. The recognition of the out-of-balance condition depends on the signal amplitude exceeding a predetermined level. In more advanced systems, a frequency analysis of the signal is made. This

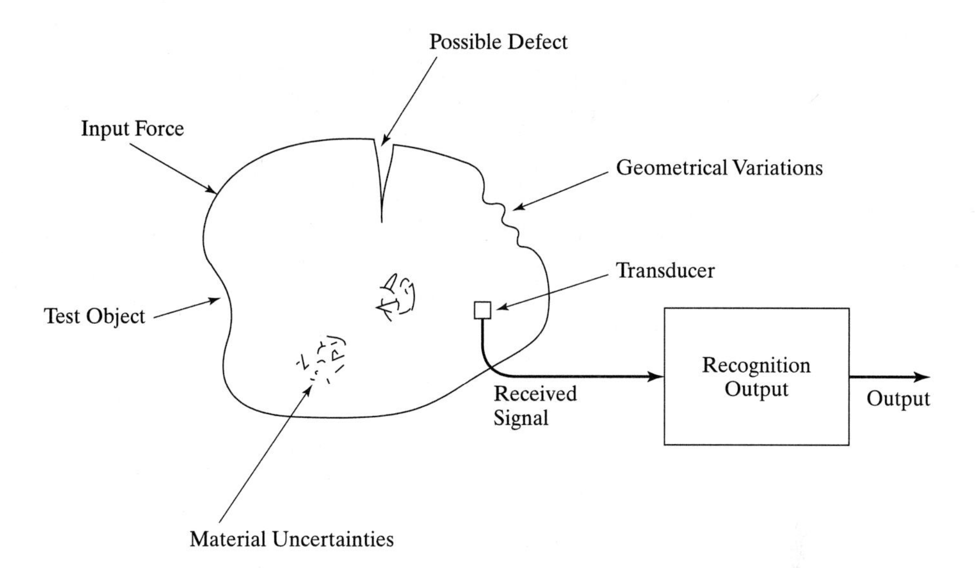

Fig. 17.15 Schematic of a monitoring system.

author was involved in research on such a system intended for monitoring railroad wheels, and a brief description of this research will illustrate some general design considerations. (See Nagy, Dousis, and Finch (1978).)

The main failure categories that need to be found in railroad wheels are cracked plates, rims, and flanges, as illustrated in Fig. 17.16. The elements of an acoustic monitoring system consist of an automatic hammer, a microphone, the means for frequency analysis, and some decision-making capability. Tapping the wheels with a hammer has

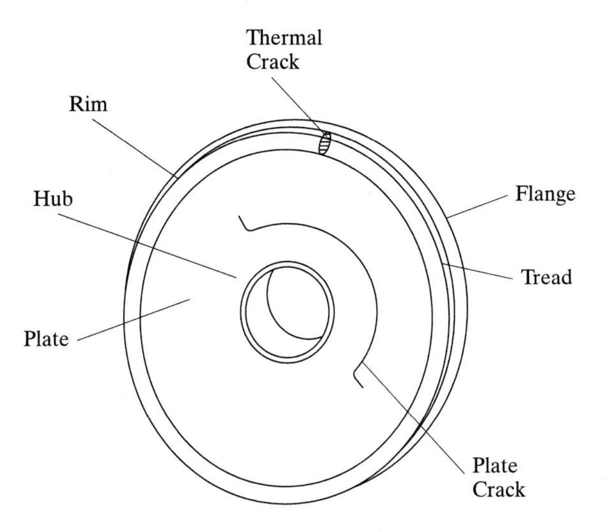

Fig. 17.16 Typical defects on a wheel. (After Nagy et al., 1978.)

been a long-standing practice in the railroad industry, so the task might be regarded as the automation of this art. In the first embodiments of the concept, the frequency analysis was performed on analog signals and the outputs of the various frequency channels were digitized and fed to an early-generation minicomputer. In later versions, the signals were digitized directly from the microphone, and a fast Fourier transform was made as the initial step in data processing by a microcomputer. Fig. 17.17 shows the narrowband spectrum of impact noise from a good wheel struck with a bar. Note that the spectrum is rich in resonances, like a bell. The acoustic signal is dependent on the wheel type and size (there are about forty wheel variations in use in the United States). In addition, the signal depends on the wear on the wheel, the load it is carrying, and its surface condition (some wheels are covered in a thick layer of grease). This posed the question: Which "template" should be used to compare with the sound from a given wheel?

It was decided that the best comparison would be with the sound of the other member of the wheelset. The wheels are always mounted and changed in pairs, so that they are nominally of the same size and type, and by the same token should have similar histories of usage and loading. Thus a significant difference in the "acoustic signatures" of the two members of a wheelset should indicate the presence of a defect or an unusual condition in one of the wheels. Fig. 17.18 shows a comparison of the spectra from two good wheels mounted on the same axle. The signal amplitudes differ somewhat, but the values of the resonance frequencies are very closely reproduced. Fig. 17.19 shows the comparison of two wheels of the same nominal size but from different manufacturers. There are some pronounced differences in the resonance frequencies. Even more differences in the resonance frequencies are found in Fig. 17.20, which contrasts the spectra of two wheels of the same manufacturer, mounted on the same axle, but one being defective. Finally, Fig. 17.21 shows the comparison of the spectra of two good wheels, with one having a heavy layer of grease. The most interesting aspects of the spectrum of the greasy wheel are increased damping at higher frequencies and the complete absence of sound above about 4 kHz in this sample. It was also clear from experimental findings that cracked wheels are more highly damped than their unbroken counterparts.

Based on such findings, a system was designed and operated for several years during the 1980s to gather data from trains in revenue service (See Haran, Jansen, and

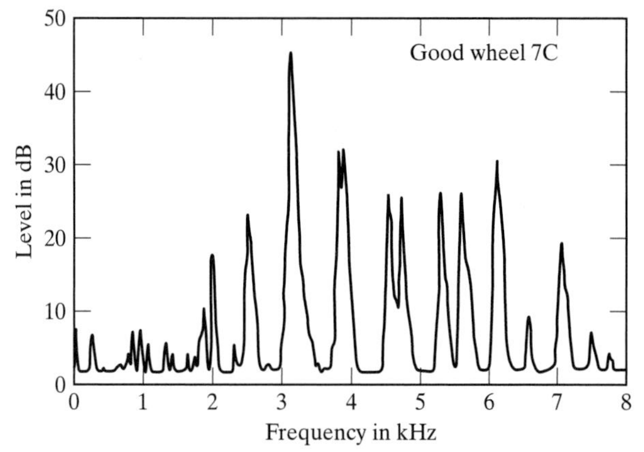

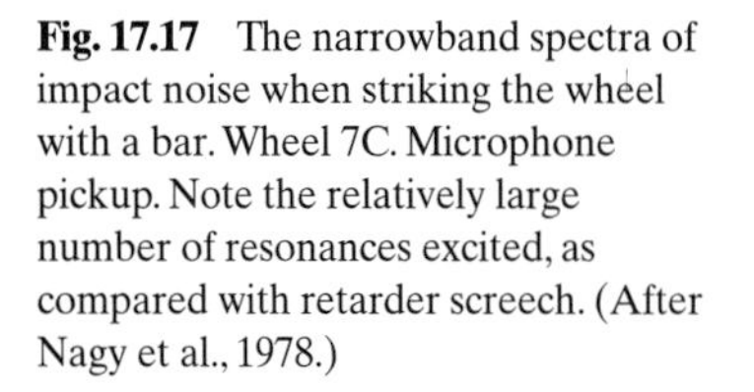
Fig. 17.17 The narrowband spectra of impact noise when striking the wheel with a bar. Wheel 7C. Microphone pickup. Note the relatively large number of resonances excited, as compared with retarder screech. (After Nagy et al., 1978.)

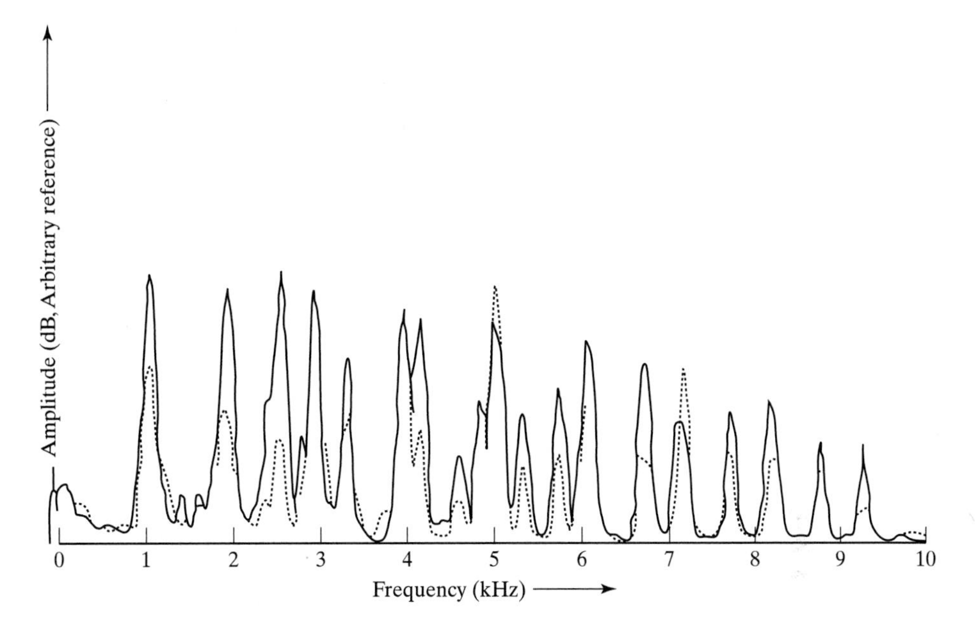

Fig. 17.18 Fixed band spectra of impact on two good wheels on either end of the same axle. (After Nagy et al., 1978.)

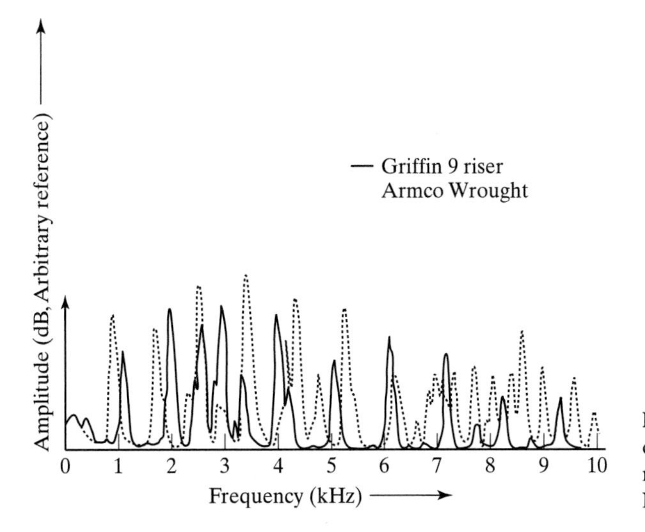

Fig. 17.19 Fixed band spectra of impact on two good 33 in. wheels, on a Griffin 9 riser, one an Armco Wrought. (After Nagy et al., 1978.)

Finch (1989).) The system configuration is shown in Fig. 17.22. Ideally, the railroad companies would like to inspect each wheel on a moving train as it arrives at a classification yard in order to know if any given wheel is defective. There are a number of interesting challenges in the design of such a system, including the hammer, which has to be able to impact each wheel with the same force in the same locality regardless of train speed, and the microphone, which has to be able to operate in all weathers.

For our discussion, the most interesting issues concern the software required to report a "good/defective" decision in near real time. There are two features of the sound

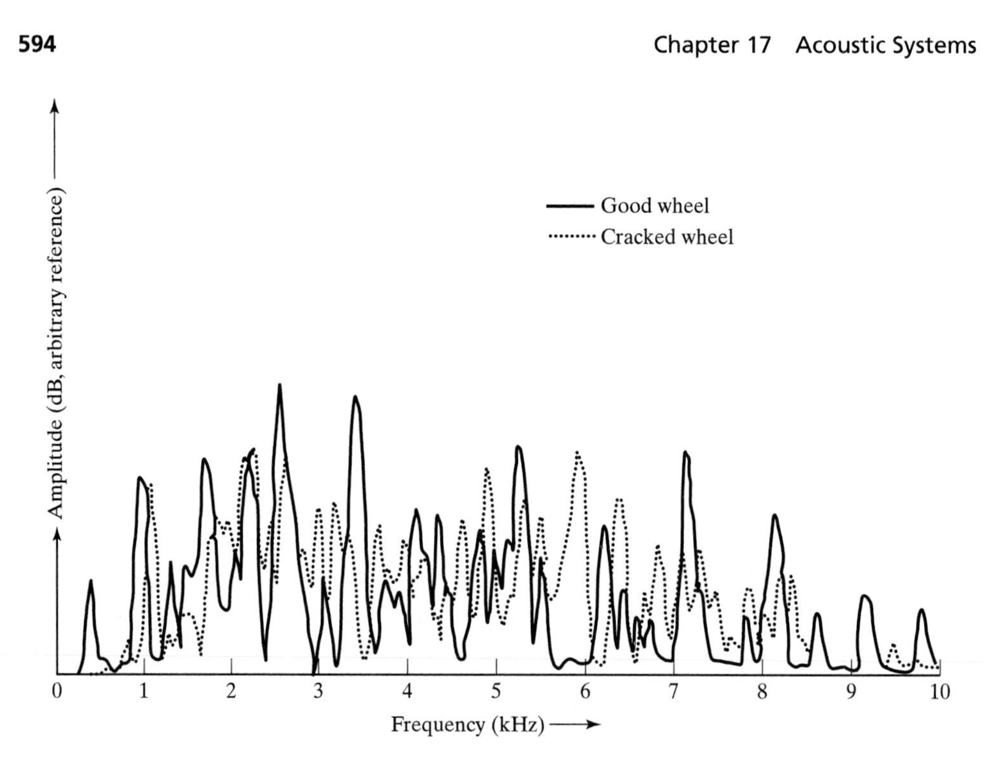

Fig. 17.20 Fixed band spectra of impact on two wheels, one with a large thermal crack. (After Nagy et al., 1978.)

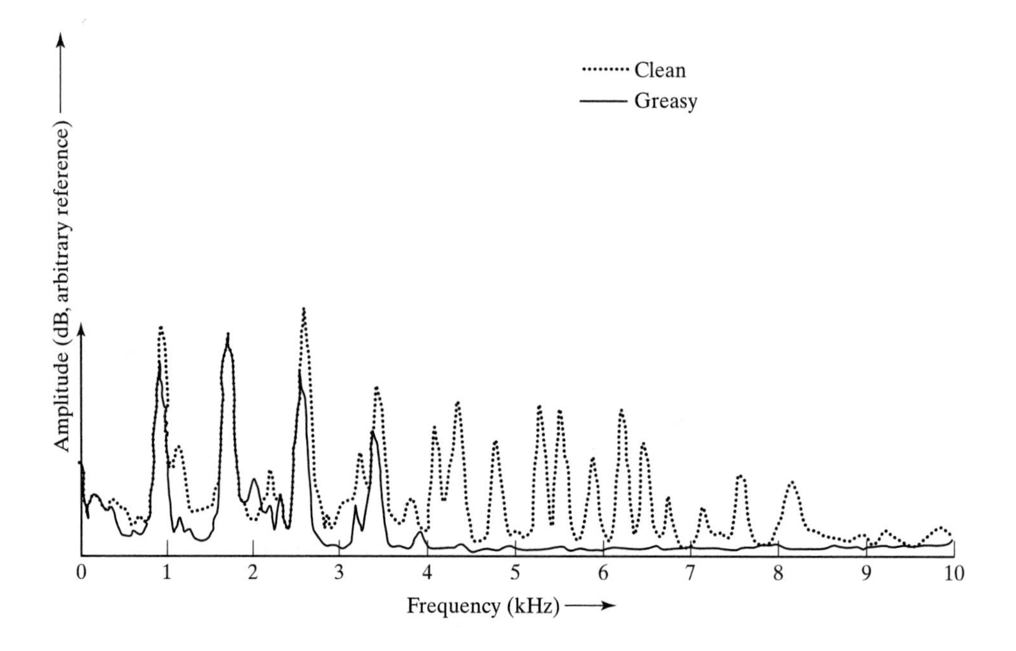

Fig. 17.21 Fixed band spectra of impact on two 33″ wheels, one with a thick layer of grease. (After Nagy et al., 1978.)

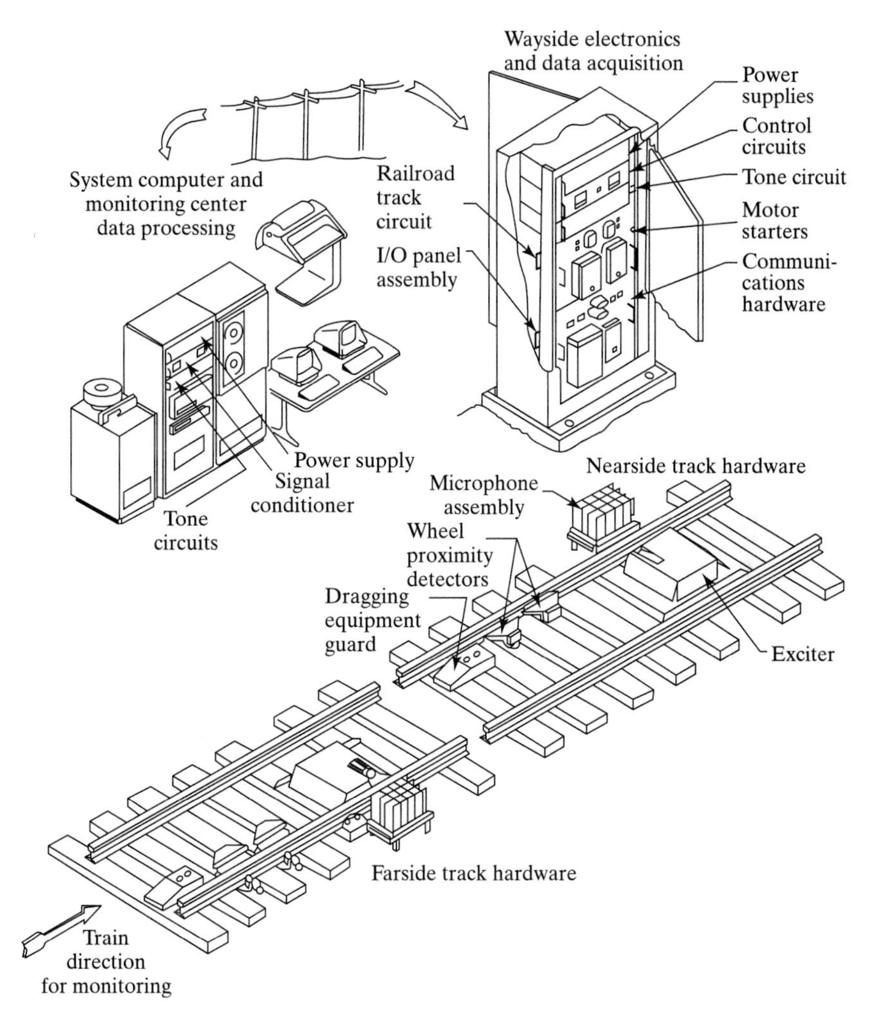

Fig. 17.22 Acoustic signature inspection—wheel-monitoring system. (After Nagy et al., 1978.)

spectra of a wheelset that betray the presence of a crack in a wheel: the noncoincidence of the resonance frequencies and a difference in the damping rates. The first is determined by the mass and stiffness matrices, and the second by the energy absorption. These two characteristics can be quantized by counting the number of resonances occuring at the same frequency (see Fig. 17.23) within a certain bandwidth, and by calculating the difference in energy between the two spectra. These two numbers can be plotted for each wheelset on a two-dimensional "scattergram" as shown in Fig. 17.24. For a wheelset with no defect, we would expect the number of common resonances to be relatively high and the damping difference to be relatively low. The situation should be reversed for a wheelset with a defect. Fig. 17.25 shows the results obtained using a test with some defective wheels mounted on it. It is possible to draw a discriminant line across the plane to separate the test population into "good" and "bad" wheelsets.

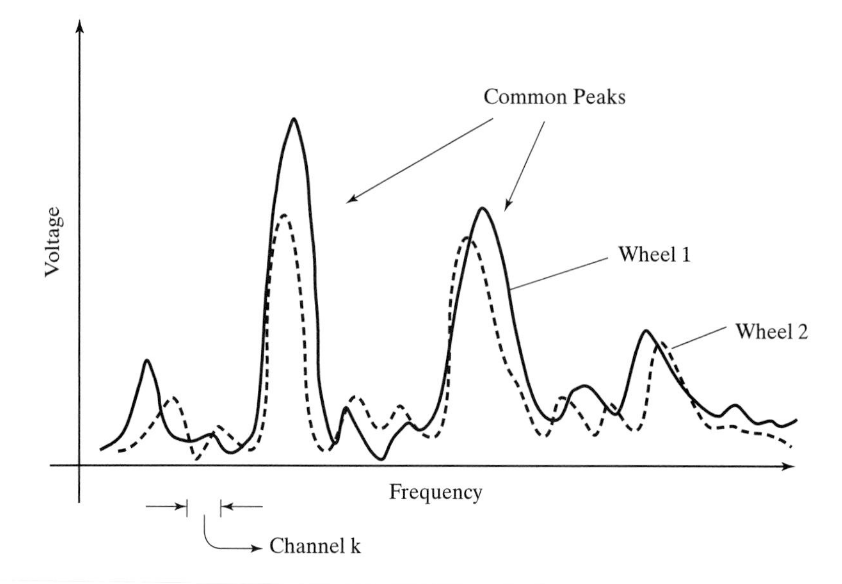

Fig. 17.23 Fourier transform of signals from a pair of wheels (wheels 1 and 2 on the same axle), showing common peaks.

The perpendicular distance of a point representing a given wheelset to the discriminating line could be used as a single number measure of the similarity of the sounds of the two wheelsets. We might also think of this process as the projection of the points onto a line perpendicular to the discriminant. (If we had some other independent feature of the difference in the sounds between the wheelsets, then it might be used as a third dimension of the pattern space, and if there were additional features, then they would make the space multidimensional.) Suppose we reduce the data to a single number representation for each wheelset using the procedure just described. Then it is easy to show that this number, DI, will be given by

$$\mathrm{DI} = c_1\mathrm{DA} - c_2\mathrm{NC} \tag{17.4.1}$$

where DA and NC are the energy difference measure and the common resonance count as we just described. DI is now a single number representation of the difference in the sounds of the two wheels. If DI is a relatively high number, then there will be a reason to believe that one of the wheels may be defective. But there is a difficulty in knowing where to set the discriminant value between the two populations. If the discriminant is set too low, then we may detect a larger number of the defective wheels but at the expense of declaring a large number of false alarms. On the other hand, if the discriminant is set too high, then we may fail to detect a large number of defective wheels. Finding the optimum setting for the discriminant is a problem in decision theory. It is necessary to determine the statistical distribution of the DI values for the good wheel population, as well as the range of values for defective wheels. But defective wheels are quite rare and thus their DI values have to be obtained in the laboratory. A comparison of the distribution of good wheel DI values and some values from defective

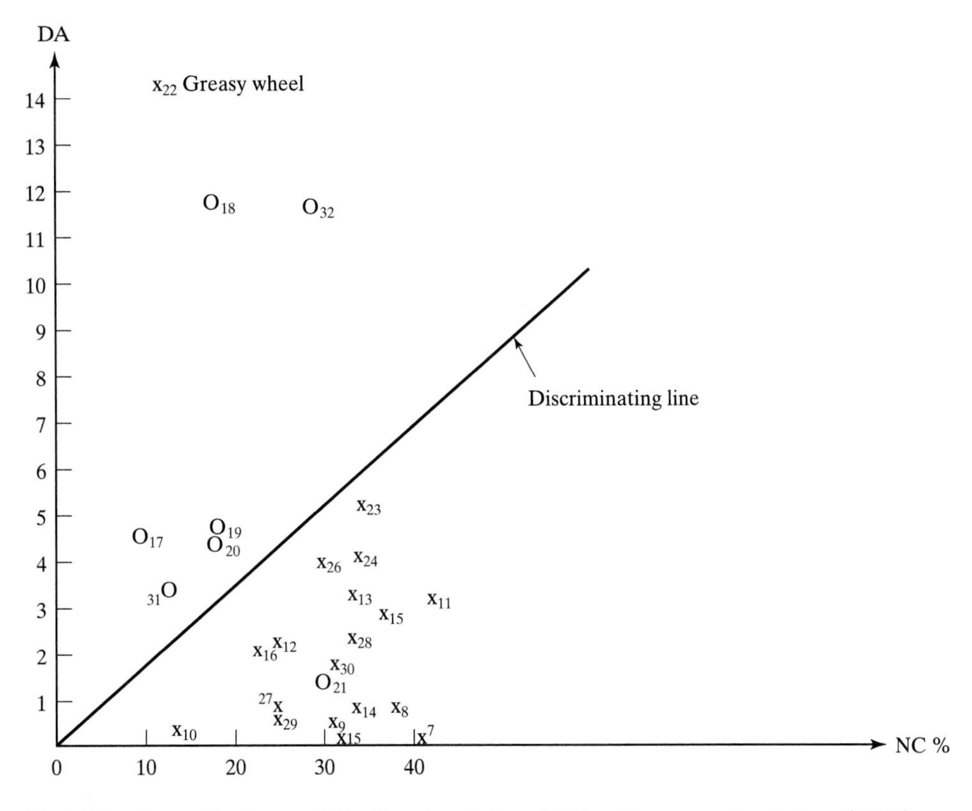

Fig. 17.24 Recognition by combining time domain term (DA) and frequency domain term (NC%). "O" indicates wheels with crack or flange cut. Numbers refer to axles. (After Nagy et al., 1978.)

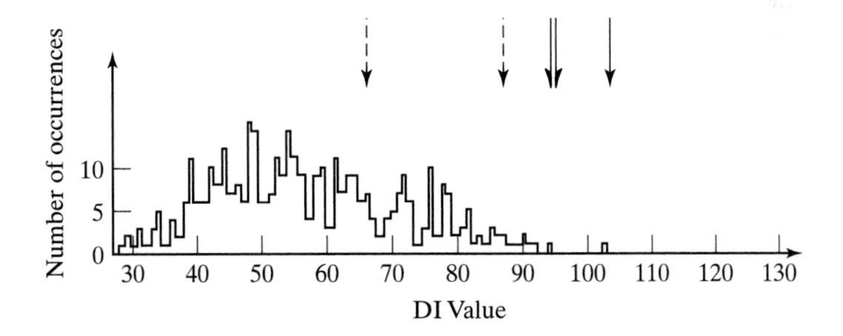

Fig. 17.25 Histogram of DI values after correction for system errors. DI = C_1SD. ← Indicates DI value for wheelsets with a defective member, measured in laboratory. ←–– Indicates DI value for good wheelsets, measured in laboratory. Sample size: 369. Frequency range: 0–7200 Hz, with correction for timing error and missing timing pulse. (After Nagy et al., 1978.)

wheels is shown in Fig. 17.25. An algorithm to find the discriminant that maximizes the separation distances is known as the Fisher linear discriminant. Statistical procedures have also evolved to divide data into a given number of subclasses or to recognize the existence of subclasses. These procedures are known as cluster analysis.

17.5 Recognition Using Artificial Neural Nets

Artificial neural nets (ANNs) were first conceived as a way to model biological nervous systems by McCulloch and Pitts (1943). Other early researchers included Hebb (1949), Rosenblatt (1959), and Widrow and Hoff (1960). It was the work of Hopfield (1982), Rummelhart and McClelland (1990), and Grossberg (1986) during the 1980s that developed into a burgeoning new field, sometimes called connectionist, or parallel distributed, processing. ANNs appear to have great potential for recognition tasks. Instead of performing sequential instructions as in the von Neumann computer architecture, ANNs employ massively parallel nets with many computational elements connected by links of various weights. They tend to be more robust than von Neumann machines because damage to a few nodes has less influence.

Fig. 17.26 shows a node in a neural network. The inputs are $x_0, x_1, \ldots, x_{n-1}$ and the output y is a function of the variable α where

$$\alpha = \sum_{i=0}^{n-1} w_i x_1 - \vartheta \tag{17.5.1}$$

and where w_i is the weightings and ϑ is an offset value. Common selections for $f(\alpha)$ are given in Figs. 17.26 b, c, and d, all of which are nonlinear. By adjusting the weights, it is possible to "train" the net. It is usual for the weighting of the inputs to be continuously changed with time to permit improvement in performance, which permits a faster adaptation than with the statistical techniques mentioned in Section 17.4 where all the training data have to be processed simultaneously before a change can be made. It is important to realize that a neural net may also be regarded as a content-addressable or associative memory.

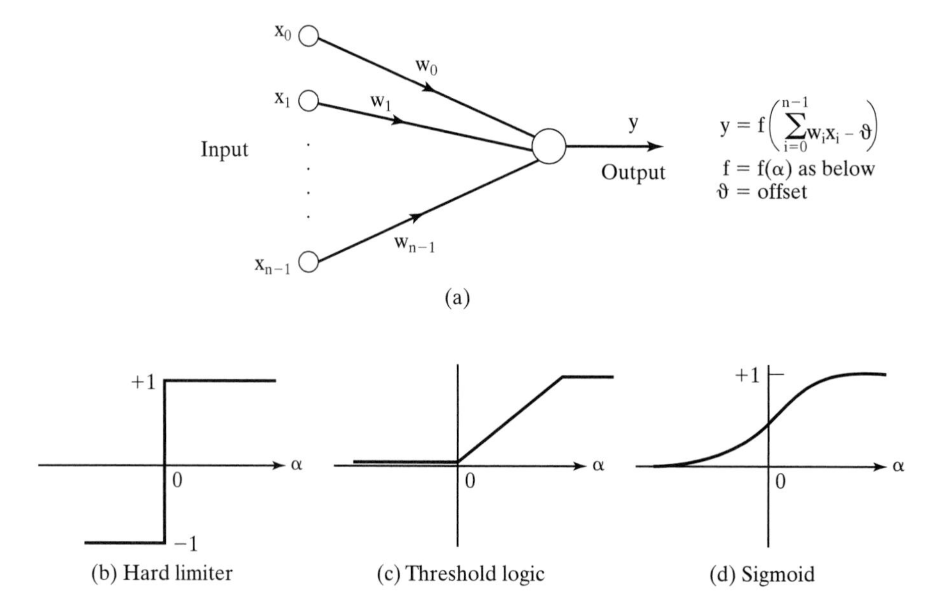

Fig. 17.26 A node in a neural network.

The neural net that excited great interest initially was Rosenblatt's "perceptron," a single-layered device of the form illustrated in Fig. 17.25 using a hard limiter at the node. As a simple example, suppose we want to decide whether a pattern belongs to either class A or class B based on just two inputs x_0 and x_1. In this case

$$y = f(\alpha) = f(w_0x_0 + w_1x_1 - \vartheta) \tag{17.5.2}$$

The node computes the weighted sum of the two inputs, subtracts a threshold, and passes the result to a hard limiter to give an output that is either +1 or −1. Fig. 17.27, which may be regarded as a generalized version of Fig. 17.24, is a scatter diagram of the input occurrences. Suppose we use a decision boundary that is a straight line with equation

$$x_0 = mx_1 + c \tag{17.5.3}$$

The perpendicular distance d of a given point (x_1, x_0) from the line is then

$$d = \frac{x_0 - mx_1 - c}{\sqrt{1 + m^2}} \tag{17.5.4}$$

and if we let

$$m = -\frac{w_1}{w_0} \quad \text{and} \quad c = \frac{\vartheta}{w_0} \tag{17.5.5}$$

then

$$d = \frac{x_0 + \dfrac{w_1}{w_0}x_1 - \dfrac{\vartheta}{w_0}}{\sqrt{1 + m^2}} = \frac{\alpha/w_0}{\sqrt{1 + m^2}} \tag{17.5.6}$$

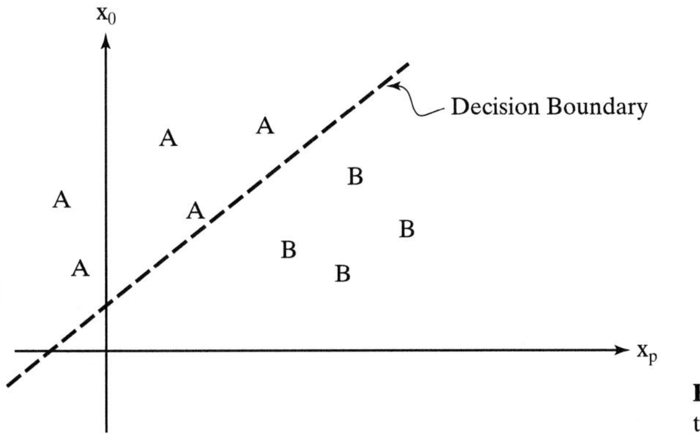

Fig. 17.27 Deciding if an input belongs to class A or class B using two inputs.

Thus, we see that α is proportional to the distance from the discriminating line. We obtain the equation of the discriminating line itself by setting α equal to zero, and then

$$x_0 = -\frac{w_1}{w_0}x_1 + \frac{\vartheta}{w_0} \tag{17.5.7}$$

The equation depends on the connection weights and the threshold. This simple ANN can therefore perform the same discriminating task as the statistical method we discussed in Section 17.4. If there are more than two inputs, we will have a multidimensional space divided by a hyperplane.

The original perceptron training algorithm used by Rosenblatt (1959) is as follows:

1. Initialize weights and threshold to small random values.
2. Present the new input and the desired output $(x_0, x_1, \ldots, x_{N-1}, d(t))$. For example, in the two input case just mentioned, $d(t)$ is $+1$ if the input is from class A and -1 if from class B.
3. Calculate the actual output

$$y(t) = f\left(\sum_{i=0}^{N-1} w_i(t)x_i(t) - \vartheta\right) \tag{17.5.8}$$

4. Adapt the weights

$$w_i(t+1) = w_i(t) + \eta[d(t) - y(t)]x_i(t) \tag{17.5.9}$$

   Here η, called the gain, lies between 0 and 1 and is chosen to give a compromise between rapid adaptation and stable weight estimates.
5. Repeat by going back to Step 2.

A problem was discovered with the perceptron training procedure in that decision boundaries may oscillate when the distributions of the input classes overlap. Such problems can occur in speech recognition using formant frequencies. This difficulty was used by Minsky and Papert (1969) to illustrate the weakness of the perceptron. Later, it became clear that these problems can be overcome by using multilayered perceptrons, as illustrated in Fig. 17.28. The additional layers are often referred to as "hidden" layers.

It is also possible to recognize acoustic signatures for monitoring purposes using a neural network. In some experiments performed at the University of Houston (see Robin et al. (1994)), vibration signals were obtained from clamped-clamped steel beams of rectangular cross section. Some of the beams had saw cuts to simulate damage. The digitized data were used to train the ANN to predict future values of the signal. The prediction errors obtained by presenting the trained ANN with test data from other beams were compared with those obtained from the same beam. This approach resulted in a measure that was useful to estimate the slot depth, and was sensitive enough to detect slots as shallow as 0.1 inch. The ANNs were trained by adjusting the connection weights until a desired output was generated for a given input pattern. An

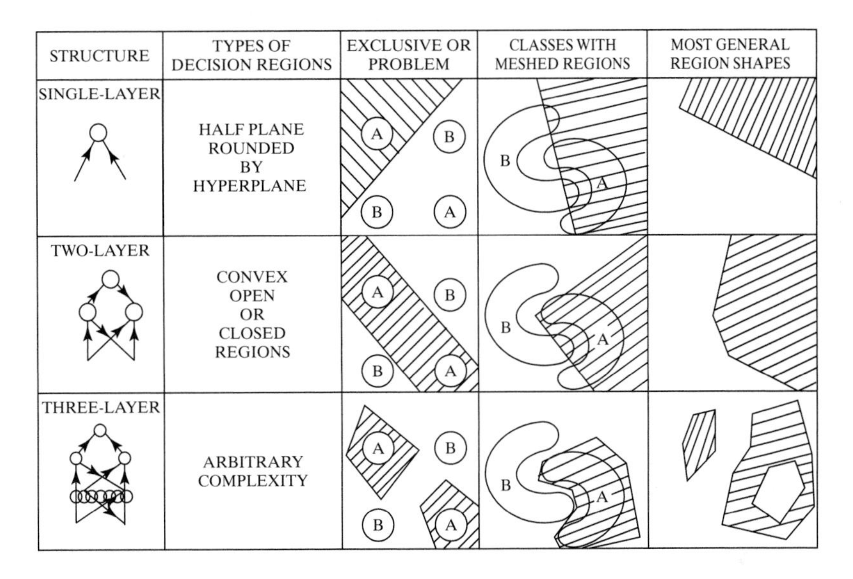

Fig. 17.28 Types of decision regions that can be formed by single- and multilayer perceptions with one and two layers of hidden units and two inputs. Shading denotes decision regions for class A. Smooth closed contours bound input distributions for classes A and B. Nodes in all nets use hard limiting nonlinearities.

"error function" was defined in terms of the deviations between the calculated and desired outputs. The so-called backpropagation algorithm was used for this purpose (see Rummelhart and McClelland (1990)).

A merit of ANNs is that they can be made to approximate any desired input/output relationship. Suppose an ANN is trained to predict future values of a given acoustical response, using preceding values. If such an ANN is presented with data generated from a different system, then poor prediction of the time series will most likely result. Hence, a monitoring system could be constructed by first training an ANN to predict the acoustical response of an intact object. Then the learning is stopped while continuing to take data from the same object. If after some time the trained ANN fails to predict the ongoing time series, then the conclusion is drawn that the object has changed.

Another way to use the ANN depends on its connection weights. Assume that an ANN has been trained for an intact system as just described. Now instead of halting the learning, the weights of the ANN are saved and the training continues, which should not change the ANN weights very much, provided the structure under test remains the same. However, if a relatively large change in the weights is observed, then this indicates a defect has occurred in the system being monitored.

One of the most straightforward approaches to the use of ANNs for predictive analysis is the time delay neural network (TDNN), which is a feed-forward network that uses delayed samples of the time series as inputs. The output of the TDNN is a sample of the time series more recent than the inputs, such that the TDNN can be said to predict a future sample of the time series using past and present samples. This is the approach that was used in the application illustrated in Fig. 17.29. Steel beams of rectangular cross

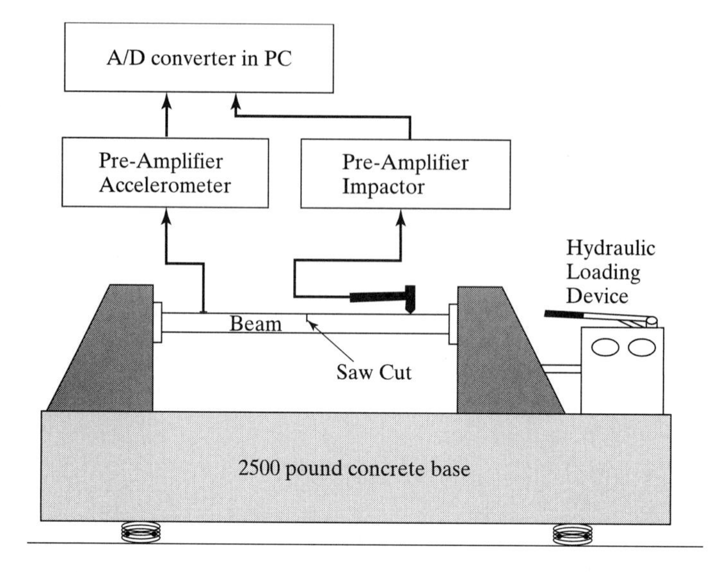

Fig. 17.29 Experimental setup for acoustic monitoring with an artificial neural network. (After Akerberg et al., 1995.)

section with dimensions of 36 inches by 2 inches by 0.75 inches were mounted in a test fixture described by Man et al. (1994). Vibrations were induced by the impact of a hammer at 30 inches from the end of the beam and measured by accelerometers attached 10 inches from the same end of the beam. The hammer was equipped with a transducer to determine the impact level imparted to the structure, this level being used for normalization and calibration purposes. The outputs from the beam and hammer accelerometers were amplified, low-pass filtered to 10 kHz, and sampled at 20 kHz.

Measurements were first made on an intact beam. Then the beam was dismounted and a 0.1 inch slot was cut in the cross section at the center of the beam. The beam was then remounted and the measurements repeated. Next the beam was taken out and the depth of the slot was increased by 0.1 inch, after which the beam was remounted, and so on. Sixty-two measurements were taken, with slots up to 0.7 inch deep. Specifically, four repeat measurements were obtained for each of the slot depths (including 0.0 inch, the intact beam). Thirty additional measurements were available for the beam with a 0.4 inch slot, for a total of thirty-four recordings, which were obtained by mounting and dismounting the beam four times.

The weights were updated after each pass of all the data (referred to as epoch training). Training was halted after 30,000 epochs. The trained neural nets can be used in two ways to determine the similarity of the training and the testing signals by comparing the prediction error of the trained ANN for the training and testing sequences. The approach involves "freezing" the ANN weights. The signal to be tested is then presented to the ANN to obtain a predicted signal. The predicted values are compared to the target signal and an error function called NE is computed. A small NE indicates that the test signal was produced by the same system as originally used to train the ANN. On the other hand, a large NE indicates that a change in the system has occurred.

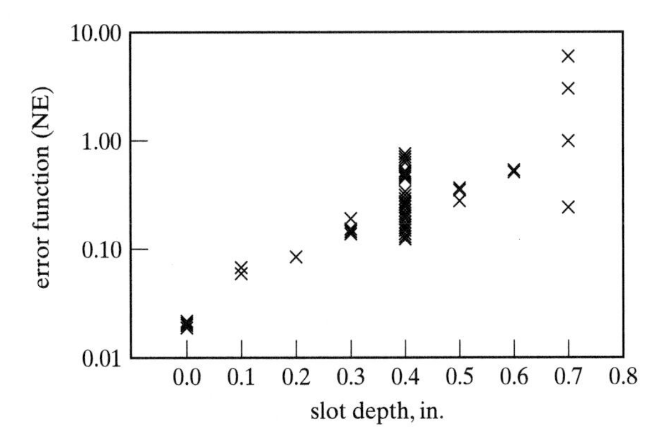

Fig. 17.30 Values of error function, NE, obtained by comparison of signals predicted from vibration testing of slotted steel beams with target signal from training on intact beams. (After Akerberg et al., 1995.)

A TDNN with a 10-8-1 configuration was trained on a single acoustic response from the intact beam for 30,000 epochs. The trained TDNNs were presented with data from beams with slot depths cut in 0.1 inch increments to 0.7 inch. The resulting values of NE are plotted in Fig. 17.30. Note that there are more samples for the beam with the 0.4 inch slot than for the other slot depths. These data have a larger variance than the measurements obtained for other slot depths because the data from the beam with the 0.4 inch slot included data from remounts. The NE measure was a linear function of slot depth. Also, NE for the intact beams was significantly different from all the slotted ones at the 97.5% level of significance.

An advantage of this approach is that no a priori assumptions are made regarding the data. For any given architecture of the ANN, there are limitations on which data sets can be successfully modeled, but the general approach is likely to work for a wide variety of applications. This generality would be especially useful for nonlinear, difficult-to-model systems.

17.6 Information Theory and Model-Based Systems

We have alluded to the fact that in the historical development of engineering systems the mathematical formulation of the principles underlying their operation frequently followed the implementation of successful physical embodiments. A wide variety of mechanisms and structures existed in the engineering repertoire long before Galileo and Newton elucidated the laws of mechanics. Similarly, many operating heat engines were built before the laws of thermodynamics were completely understood. It might be argued that the first law of thermodynamics (conservation of energy) is simply an extension of mechanical principles to the idea that energy in its many forms (mechanical work, heat, chemical, and electrical energy) were equivalent. But the second law was clearly something new: the tendency for entropy to increase. This was first grasped in terms of the availability of heat energy, and only after many years was the concept seen to have a deeper meaning, namely, the increase of disorder during physical processes. We will now expand on this statement.

The second law of thermodynamics arose from the study of the efficiency of heat engines and was first enunciated in 1824 by Carnot, who imagined an ideal engine taking heat from a high temperature source and delivering part of it to a low temperature sink, the remainder being converted into mechanical work. The earliest statement of the second law was that no real engine could have a higher efficiency than the Carnot cycle. Real engines had smaller efficiencies because some of the heat from the source was conducted away to the surroundings by a process described in a mathematical formulation only in 1811 by Fourier, who pointed out that heat flows from higher to lower temperatures at a rate proportional to the temperature gradient, a process that is irreversible. If a hot body is placed in contact with a cold one, then heat flows from the hot one to the cold one until they both eventually reach the same temperature. Heat never flows from a cold body to a hot one. In 1852, Kelvin restated the second law as the tendency of a closed system to come to thermal equilibrium. The irreversible heat loss during a process in a closed system was defined by Clausius to be proportional to an increase in a quantity called entropy. Clausius produced yet another version of the second law wherein the entropy of a closed system always tends towards a maximum.

It was thought at first that the second law of thermodynamics could not be explained on the basis of mechanics. Maxwell was the first to propose that the link could be established through probability. It was Boltzmann who suggested that the entropy of a system was proportional to the logarithm of the probability of its being in a given state. The entropy of a gas is developed in the following way. Suppose that the gas is comprised of N molecules. Each molecule has three components of velocity so that the whole system is in a state defined by the set of values assumed by 3N variables. Boltzmann argued that a quantity proportional to the logarithm of the probability would tend to a maximum. Each possible velocity value has a certain probability defined in terms of the mean value. Presuming that this mean value is determined by the temperature of the ensemble and assuming the velocity distribution obeys Gaussian statistics, it is possible to calculate Boltzmann's quantity and show that it is identical to the entropy of the gas. According to Boltzmann's view, a system will tend to the state of maximum probability. Chaotic states are more probable than ordered states, so the second law became a statement of the tendency of nature to favor disorder over order.

The concept of entropy can be applied to any situation in which probability plays a role. A familiar example involves the shuffling of a deck of cards. A new deck is usually ordered (e.g., by suit and number) so that we can predict the sequence as the cards are dealt. After shuffling, however, the order is unpredictable. Entropy has increased to a maximum. If we now sort the cards into their original sequence, entropy will be lowered.

The concept of entropy was used again in the development of information theory (or as it might be better named, communication theory). Here the issue concerns the transmission of messages along communication channels, such as telephone lines. Communication involves a sender, a receiver, and a channel. Suppose the sender and the receiver have a repertoire of symbols (an alphabet) available for use. A message is then a sequence of such symbols. The smaller the probability of a sequence of symbols being sent, the greater the information content of the message. If the source transmits messages whose statistical probability is low, then it is said to be of low entropy. On the other hand, it has a high entropy if the symbols transmitted are random. The imposition of some structure on the message source (e.g., use of the English language) reduces its entropy. Information theory is used in communication engineering to determine how certain structures permit more efficient encoding of messages.

Information theory, however tells little or nothing about the *meaning* of a transmitted message, Rapoport (1986) discusses what is involved in transmitting a message with meaning, such as "it is raining outside." The message implies that the sender has been to the window and has received optical and perhaps acoustic signals typically associated with rain. These signals will have been processed in his brain and then encoded into words, perhaps spoken into a telephone with A/D conversion and digital transmission. The stream of 0s and 1s travelling along the telephone line carries the message "it is raining outside" to be extracted by the recipient through the reverse process at the receiving end, with meaning extracted by exercise of the receiver's "imagination" (virtual reality?).

It is clear that the processing of such information in the brain is accompanied by the operation of certain rules, which we call "common sense." Recent work in artificial intelligence has demonstrated the importance of common sense in human data processing. In addition to such everyday knowledge, we also possess a deeper theoretical knowledge, built up by learning and scientific inquiry. Such is the knowledge described in this book and believed by acousticians to be of value in understanding our world. This knowledge is contained in mathematical models of acoustical objects. For example, there is extensive literature on the modeling of defects in vibrating beams, much of which has been developed by Dimorogonas and his students. Could such models be incorporated into acoustic monitoring systems? Would it improve on the work using the pattern recognition recounted in Sections 17.4 and 17.5, which used signal processing techniques that do not take into account the nature of the test object?

Man et al. (1996) performed work whose first objective was the development of a mathematical model of a defective test article. A beam was chosen as the test body, since it is a basic component of most structures, and a slot was used as a simulated defect. Because the main purpose of having a model is to program it into the data processing, a more computationally efficient model is desirable. To do this, a perturbation method can be used. The formulae resulting from the perturbation method can be incorporated in real-time data processing. An added advantage of the perturbation method is that the process using it is invertible, so that the defect characteristics can be determined solely from the measured vibration response of the system.

The Euler-Bernoulli theory for vibration of a beam with varying moment of inertia and cross-sectional area leads (see Section 12.10) to the equation

$$\frac{\partial^2}{\partial x^2}\left(EI(x)\frac{\partial^2 v}{\partial x^2}\right) + m\frac{\partial^2 v}{\partial t^2} = 0 \tag{17.6.1}$$

where E is Young's modulus; I(x) is the moment of inertia, in general, a function of the spatial coordinate x; v(x,t) is the displacement; and m is the mass per unit length of the beam. Eq. (17.6.1) is the Euler-Bernoulli equation, and it is essentially the same as Eq. (12.10.7). For a nonslotted uniform beam, the vibration equation may be written in terms of the nth mode shape $V_{n0}(x)$ and the nth modal frequency ω_{n0} of the uniform beam, thus

$$EI_0\frac{\partial^4 V_{n0}(x)}{\partial x^4} - \rho A_0 \omega_{n0}^2 V_{n0}(x) = 0 \tag{17.6.2}$$

where ρ is the mass density and A is the cross-sectional area. The modal frequency of such a beam is determined by the boundary conditions. For example, the nth modal frequency of a cantilever is given by

$$1 + \cosh(\beta_n \ell) \cos(\beta_n \ell) = 0 \tag{17.6.3}$$

where ℓ is the beam length and

$$\beta_{n0}^4 = \frac{\rho A_0 \omega_{n0}^2}{EI_0} \tag{17.6.4}$$

For a slotted beam, Eq. (17.6.1) becomes

$$\frac{\partial^2}{\partial x^2}\left(EI(x) \frac{\partial^2 V_n(x)}{\partial x^2} \right) - \rho A \omega_n^2 V_n(x) = 0 \tag{17.6.5}$$

where ω_n and $V_n(x)$ are the nth modal frequency and the nth mode shape for the slotted beam. The presence of the slot causes the moment of inertia and the cross-sectional area to be reduced at the slot location, thus increasing the local flexibility which will tend to decrease the modal frequency and change the mode shape. On the other hand, the presence of the slot implies that there is mass reduction, and thus the modal frequency tends to increase. The parameters of the slotted beam are shown in Fig. 17.31. The changes in the moment of inertia and the cross-sectional area are directly related to the slot depth, slot width, and location and can be described by

$$I(x) = I_0 - I_1(x) = I_0 - \kappa I_0[u(x - a) - u(x - a - c)] \tag{17.6.6}$$

where $k = \frac{1}{\alpha} - 1$ is a slot depth factor related to the moment of inertia, $\alpha = {}^{I'}\!/_{I_0}$, the ratio of moment of inertia with and without a slot; a is the location of the left slot edge from the left end of the beam; c is the slot width; $u(x-a)$ is the unit step function, $u(x - a) = 1$ for $x > a$ and 0 elsewhere; and

$$A(x) = A_0 - A_1(x) = A_0 - \mu A_0[u(x - a) - u[x - a - c)] \tag{17.6.7}$$

$\mu = \frac{1}{\gamma} - 1$ is a slot depth factor related to cross-sectional area, and $\gamma = {}^{A'}\!/_{A_0}$ is the ratio of cross-sectional area with and without a slot. In the case of small slots, $I_1(x)$ and

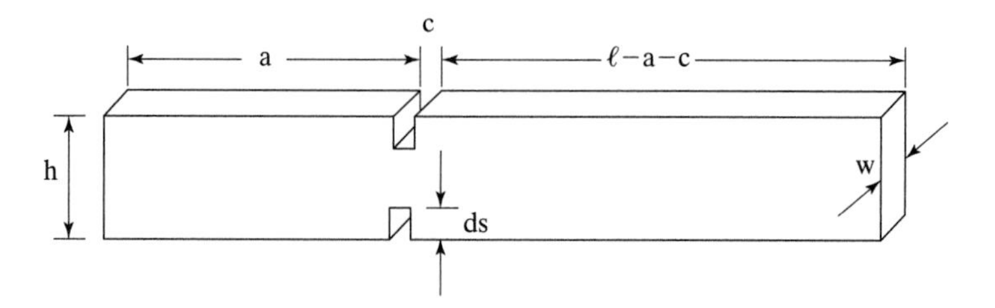

Fig. 17.31 Parameters for a beam with a symmetric slot of width c, and depth ds from top and bottom of the beam. The slot is located at $x = a$ from the left end. The beam is ℓ by h by w.

$A_1(x)$ can be taken as the perturbation introduced by the slot. The parameters κ and μ depend on the relative slot depth. Both of them are zero for uniform beams.

We can assume that the modal parameters have a small perturbation $V_{n1}(x)$ due to the defect given by the form

$$V_n(x) = V_{n0}(x) + V_{n1}(x) = V_{n0}(x) + \sum_{\substack{j=1,\\ j\neq n}}^{N} \varepsilon_{nj} V_{j0}(x) \tag{17.6.8}$$

Since the eigenvalue depends on the square of the frequency, we can assume the perturbation in the nth modal frequency is given by

$$\omega_n^2 = \omega_{n0}^2 - \omega_{n1}^2 \tag{17.6.9}$$

Solving Eq. (17.6.5) using the assumed perturbations (see Man (1996)) yields

$$\frac{1}{N_{mn}} \sum_{\substack{j=1,\\ j\neq n}}^{N} \varepsilon_{nj}(\omega_{j0}^2 - \omega_{n0}^2) N_{mj}\delta_{mj} - \kappa\omega_{n0}^2 A_{nm} + \mu\omega_{n0}^2 B_{nm} + \omega_{n1}^2 \delta_{mn} = 0 \tag{17.6.10}$$

where

$$\int_0^{\ell} V_{m0}(x) V_{n0}(x)\, dx = N_{mn}\delta_{mn} \tag{17.6.11}$$

$$\frac{EI_0}{N_{nn}\rho A_0 \omega_{n0}^2} \int_a^{a+c} \frac{\partial^2 V_{n0}}{\partial x^2} \frac{\partial^2 V_{j0}}{\partial x^2}\, dx = A_{nj} \tag{17.6.12}$$

and

$$\frac{1}{N_{nn}} \int_a^{a+c} V_{n0} V_{j0}\, dx = B_{nj} \tag{17.6.13}$$

Thus, we have for the nth modal frequency of the slotted beam

$$\omega_n^2 = \omega_{n0}^2 - \omega_{n1}^2 = \omega_{n0}^2 - \omega_{n0}^2[\kappa A_{nn} - \mu B_{nn}] \tag{17.6.14}$$

and the nth mode shape is

$$V_n(x) = V_{n0}(x) + \sum_{\substack{m=1\\ m\neq n}}^{N} \left[\frac{\kappa A_{mn}\omega_{m0}^2 - \mu\omega_{n0}^2 B_{mn}}{\omega_{m0}^2 - \omega_{n0}^2}\right] V_{m0}(x) \tag{17.6.15}$$

If the slot depth $ds = 0$ (i.e., $\kappa = \mu = 0$), Eqs. (17.6.14) and (17.6.15) reduce to those of the uniform beam. If the slot width $c = 0$, then $A_{mn} = B_{mn} = 0$, and Eqs. (17.6.14) and (17.6.15) reduce to those of the uniform beam. Eqs. (17.6.14) and (17.6.15) are valid for relatively small slot depth because they include only the first-order perturbation, although no assumption is made on the slot width in their derivation. To validate the model, another special case is considered. When the slot width is the same as the beam length, the "slotted beam" is actually a beam with reduced overall height; in other words, it is another uniform beam. In this case, we have

$$A_{mn} = \delta_{mn} \tag{17.6.16}$$

and

$$B_{mn} = \delta_{mn} \tag{17.6.17}$$

and therefore Eqs. (17.6.14) and (17.6.15) reduce to

$$\omega_n^2 = \omega_{n0}^2[1 - \kappa + \mu] \tag{17.6.18}$$

and

$$V_n(x) = V_{n0}(x) \tag{17.6.19}$$

Eq. (17.6.19) states that the slotted beam with a slot width equal to the beam length has the same mode shape as the original uniform beam, which is definitely true. The effects of changing slot width could be discussed using Eq. (17.6.18). This model will not apply when the slot is large or the beam is cut through $(ds/h \rightarrow 1)$ where h is the beam height, because the fundamental assumption of the perturbation method is that the perturbation terms are very small, which implies $ds/h \ll 1$.

A clamped-clamped steel beam was used to compare the results from an exact model, based on Bernoulli beam theory and the perturbation method along with the appropriate measured data as shown in Fig. 17.32. The beam is 36 inches by 2 inches by 0.75 inches with a slot centrally located. The frequency resolution for the measured data is about 5 Hz. As expected, the perturbation method predicted smaller frequency reductions when compared with those from the previous model. When the slot is about 20% of the beam height, the more exact model predicts a frequency shift with about 7.5% discrepancy from the measured value and the perturbation method has a discrepancy of about 30%.

The perturbation model can also be used to predict the modal frequency for a beam as a function of slot parameters. For example, the third modal frequency for a slotted cantilever (18 inches by 2 inches by 0.75 inches) was calculated for different slot locations and depths, as shown in Fig. 17.33. We can see that at a given location, the larger the slot depth, the larger the frequency reduction. The third modal frequency shift versus slot location resembles the shape of the third mode. The reduction in the cross-sectional area is related to the removal of mass, which therefore tends to increase the modal frequency. The reduction in the moment of inertia is related to a decrease in stiffness, and therefore leads to a reduction of the modal frequency. Close to the clamped end, the effect due to the reduction in the moment of inertia dominates, and the modal frequency then has a large reduction. Close to the free-free end, the effect due to the reduction in the cross-sectional area dominates, and actually results in an increase in the modal frequency. This is equivalent to shortening the beam.

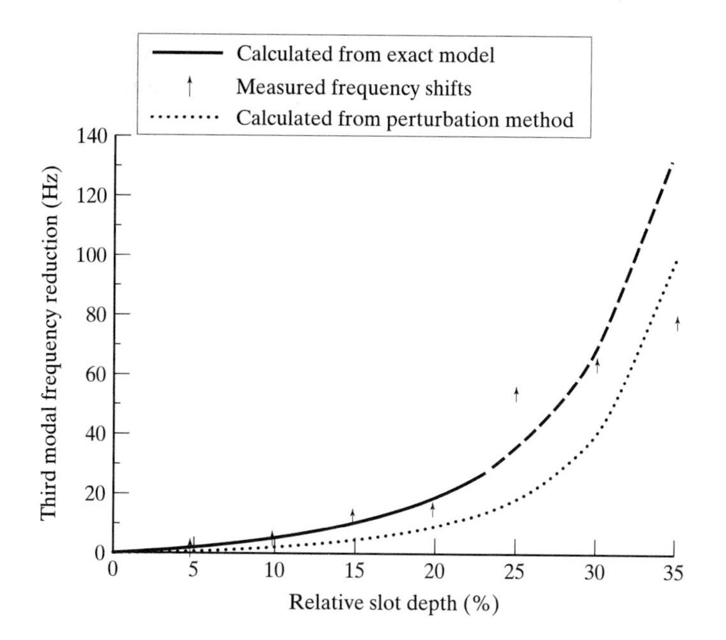

Fig. 17.32 Comparison of results from the exact model and the perturbation method with the measured data. The beam is a clamped-clamped beam of 36 in. by 2 in. by 0.75 in. (After Man et al., 1994.)

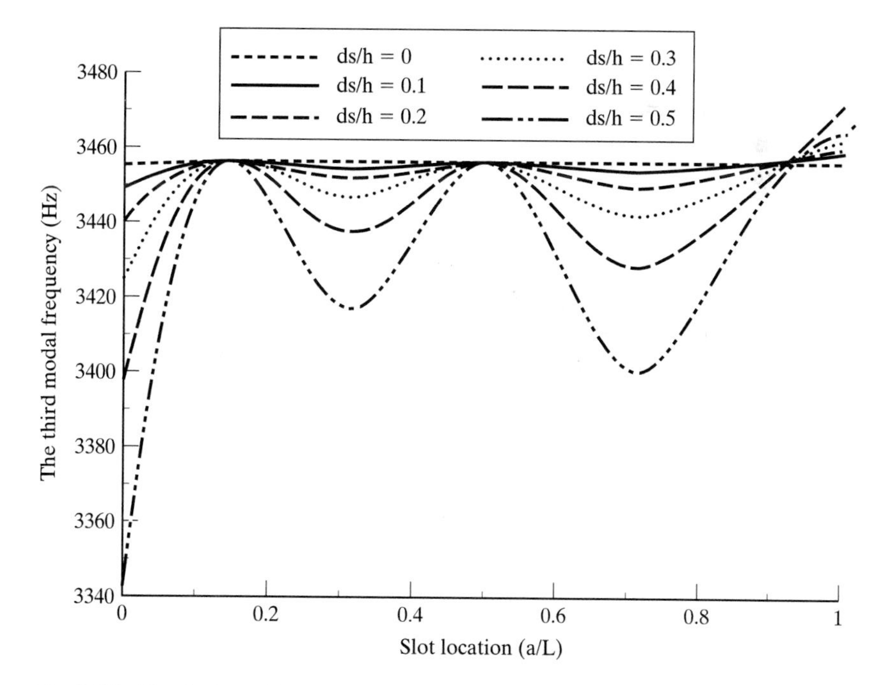

Fig. 17.33 The third modal frequency versus relative slot location (a/L) for six different slot depths (ds/h = 0 to 0.5). Slot width c is constant 0.04 in. The cantilever beam is 18 in. by 2 in. by 0.75 in. and is clamped at x = 0. (After Man et al., 1994.)

For any assumed slot characteristics—including the slot depth, width, and location—the first several modal frequencies and shapes can be computed from Eqs. (17.6.14) and (17.6.15). For example, using a constant slot width, frequency contours are shown for the first, second, and third modes of a cantilever in Figs. 17.34, 17.35, and 17.36, respectively. Each point located on one of these frequency contours represents a combination of slot depth and location, which will produce a resonance for the given mode at the given frequency. Figs. 17.34 through 17.36 illustrate how frequency contours may be used inversely to determine the slot characteristics. Suppose there is a slot in the beam with ds/h = 0.6 and a/L = 0.3, then the first mode contour passing through (0.3,0.6) is at a frequency of 195 Hz; the second mode contour is at a frequency of 1230 Hz; and the third mode contour is at a frequency of 3410 Hz. The situation is illustrated in Fig. 17.37. Thus, in an inverse process, starting with experimental data, if the modal frequencies are known to be at 195 Hz, 1230 Hz, and 3410 Hz, then the slot parameters can be determined by finding the points at which the contours for these frequencies intersect.

Steel beams were tested on the fixture described in Fig. 17.29. The vibration was excited using an impact hammer. The vibration response signals were recorded with an accelerometer. A cantilever beam (18 inches by 2 inches by 0.75 inches) was cut with a slot whose depth was increased in increments from zero to 0.7 inches final total depth. For each slot depth, the cantilever beam was placed on the fixture. The third modal frequency reductions versus the slot depth for the cantilever beam are shown in Fig. 17.38, with a 0.04 inch wide slot located at 5 inches from the clamped end. As we can see, the perturbation method predicted the third modal frequency with acceptable accuracy for

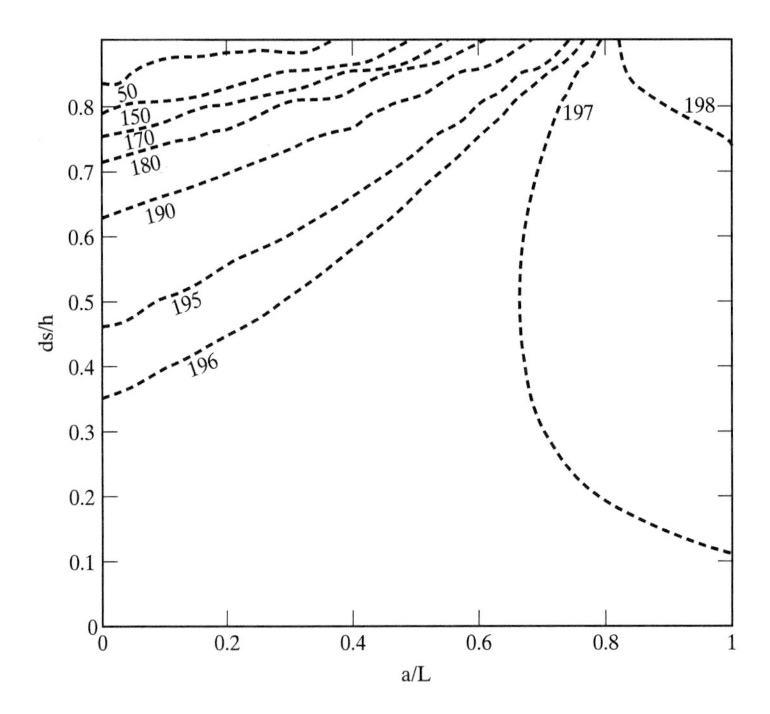

Fig. 17.34 Frequency contours for the first mode of a cantilever. (After Man et al., 1994.)

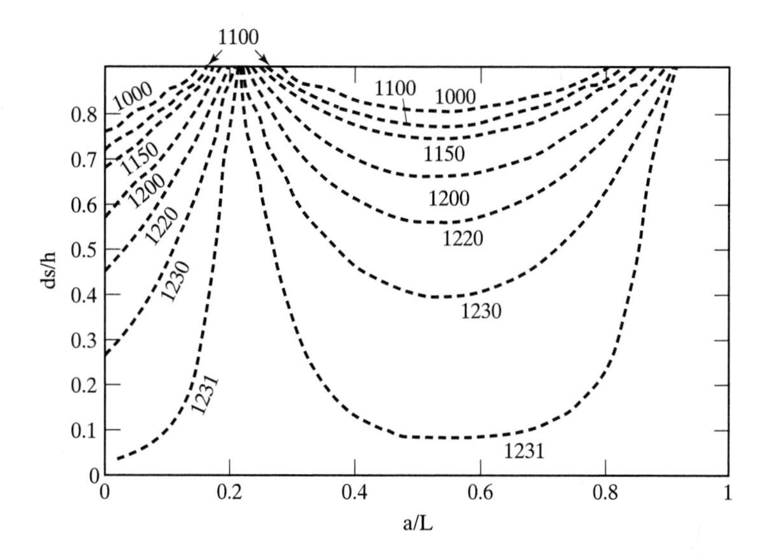

Fig. 17.35 Frequency contours for the second mode of a cantilever. (After Man et al., 1994.)

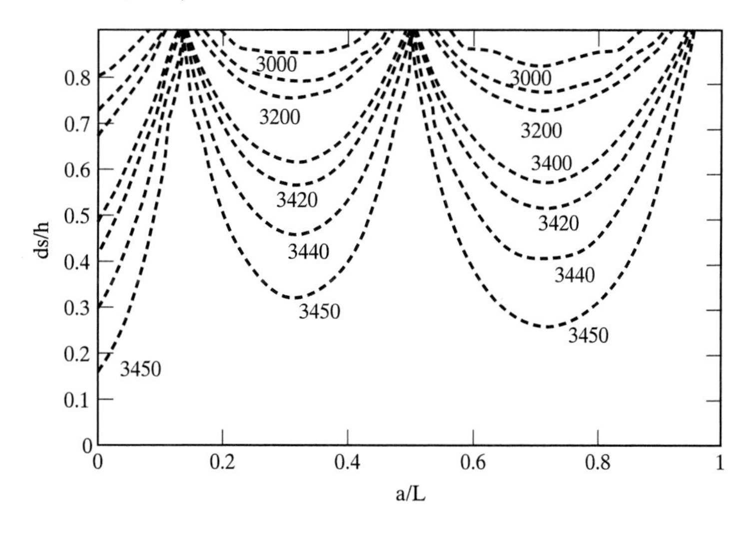

Fig. 17.36 Frequency contours for the third mode of a cantilever. (After Man et al., 1994.)

slots of a depth less than 20% of the beam. The discrepancy between the experimental modal frequency and the calculated values becomes large as the slot depth increases, and thus the applicability of this model is limited to slots of relatively small depth.

Eleven steel beams (seven slotted and four uniform) were used as a test population. As a first exercise, the defective beams were separated from the nondefective ones using the measured modal frequency of each. A defectiveness index (DI) was defined by

$$\mathrm{DI} = \sum_{i=1}^{n} \frac{|f_i^{cal} - f_i^{est}|}{f_i^{est}} \times 100\% \tag{17.6.20}$$

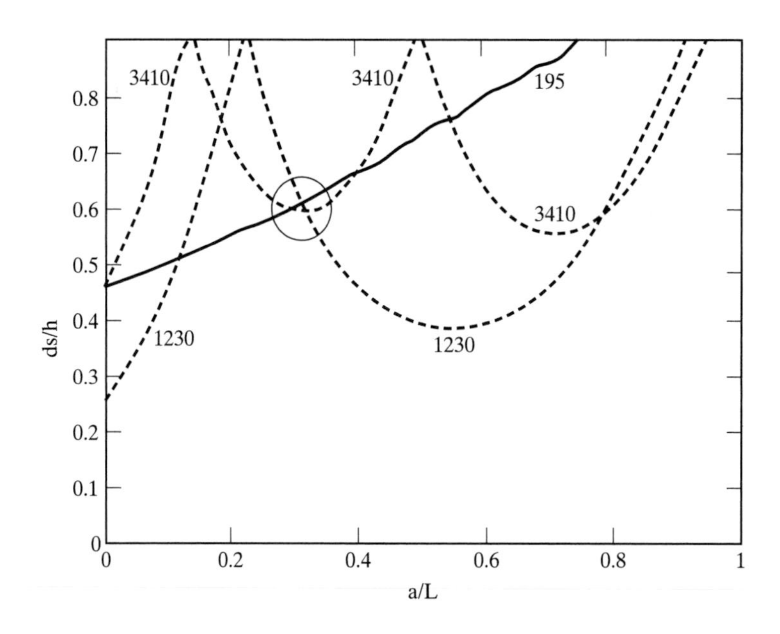

Fig. 17.37 Contours for first, second, and third resonance modes corresponding to ds/h = 0.6 and a/L = 0.3. (After Man et al., 1994.)

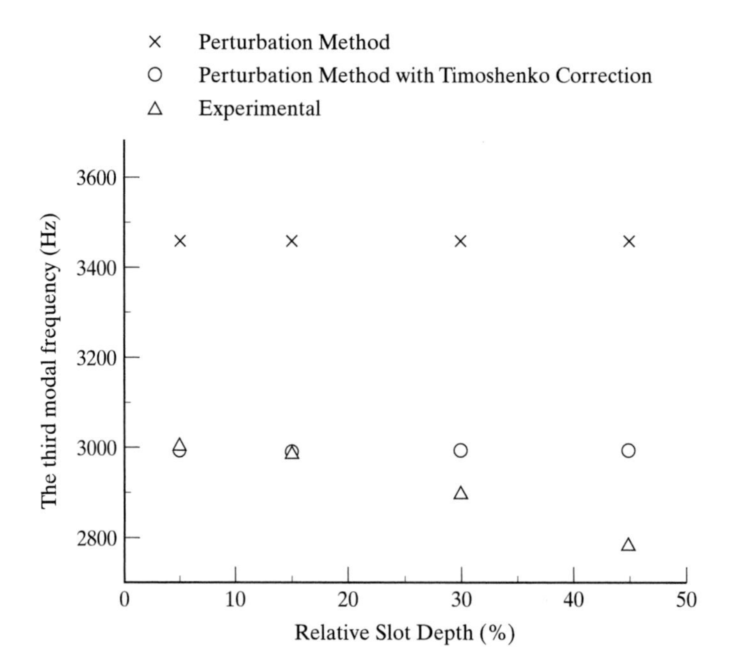

Fig. 17.38 Comparison of experimental and theoretical frequency changes with slot depth in a beam. (After Man et al., 1994.)

Here, f_i^{est} is the experimentally estimated ith modal frequency and f_i^{cal} is the value calculated from Euler-Bernoulli beam theory. Comparatively, the DI values are significantly different for the defective beams and for the nondefective ones.

After separating the defective and nondefective beams, the next objective was to identify the defect size and location. Each defective beam had a slot of the same width, but a different depth and location. Eq. (17.6.16) was used to compute the numerical frequency shifts for several modes. The frequency shift contours were obtained, and thus the size of the slot was assessed. The first defective beam measured had frequency shifts of 18.0 Hz, 19.8 Hz, and 420.5 Hz in the first three modes. The corresponding contours for these three modal frequency shifts were recalled and are shown in Fig. 17.39. The relative slot depth and location were found from Fig. 17.39 as those corresponding to the intersection of the three contours. As the modal frequency was always estimated within a range of uncertainty, the "intersection" was an area of overlap that determines the range of the slot depth and location. For this beam, the slot was determined to have a relative depth of about 0.75 inches and a relative location of about 0.27 inches (i.e., a depth of 1.5 inches and a location at 4.9 inches from the fixed end). The actual slot was 0.7 inches deep at 5.0 inches from the fixed end. The discrepancy in estimation of the slot depth is more than 100%, while it is less than 5% in identification of the slot location. The slot depth is estimated to be much larger than the actual value.

If we had only one modal frequency shift to use, the slot depth and location could be any of an infinite number of possible combinations. If two modal frequency shifts are available, the possible slot location and depth combinations reduce to those at the intersections of the two frequency shift contours. If more than one combination is found from the two frequency contours, then the frequency shift for another mode has to be employed. Using

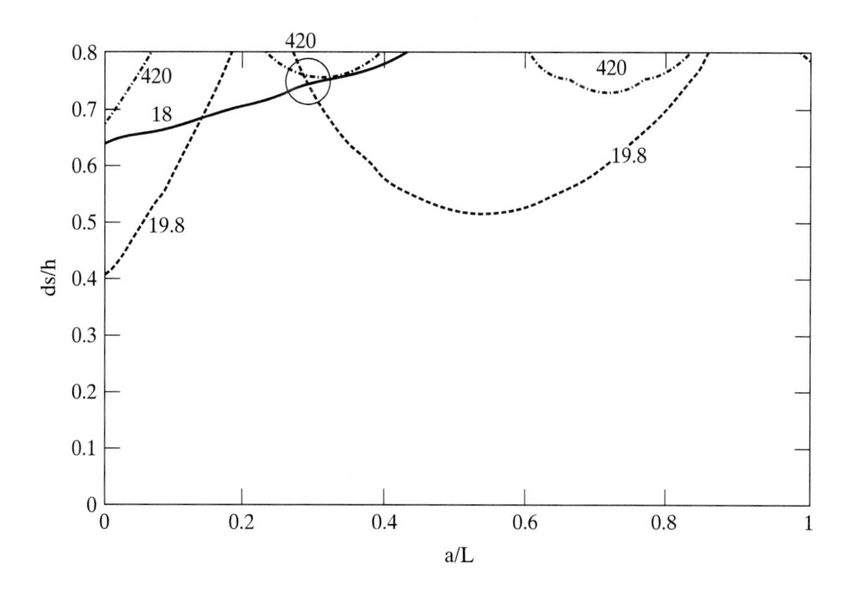

Fig. 17.39 The first three frequency shift contours for a defective beam. Relative slot location versus depth. The solid, dot-dashed, and dashed contours are the frequency shifts in the first, second, and third modes, respectively. (After Man et al., 1994.)

more modes eventually leads to the determination of the slot depth and location. To illustrate this, the first three modal frequency shifts of a clamped-clamped slotted beam were used. They had the frequency shift contours shown in Fig. 17.40 from which three possible combinations of the slot location and depth were found. The slot might be at the relative slot location at 0.15 inches, 0.5 inches, and 0.85 inches with corresponding relative depth of 0.8 inches, 0.85 inches, and 0.85 inches. Of these, the first and the third combinations are virtually the same, as the beam is symmetric about its center.

To determine the real slot depth and location, the fourth modal frequency shift of 12.1 Hz was added, as shown in Fig. 17.41. Only a slot located at the middle of the beam (i.e., the relative location a/L = 0.5, a = 18 in.) with a slot depth 1.6 inches (i.e., the relative slot depth ds/h = 0.8) was left. The actual slot was located at the middle of the beam with a slot depth of 0.7 inches. The discrepancy in the slot depth estimation is

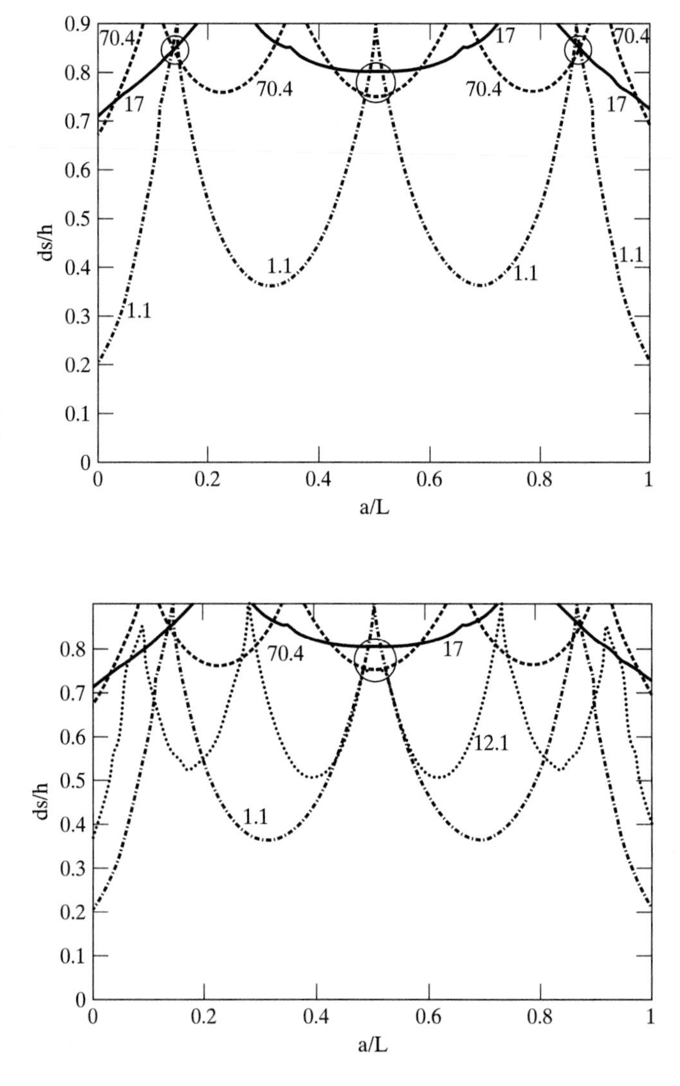

Fig. 17.40 First three modal frequency shifts of a clamped-clamped slotted beam. Three possible combinations of slot location and depth are determined. (After Man et al., 1994.)

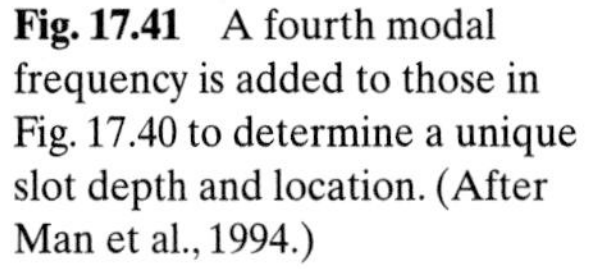

Fig. 17.41 A fourth modal frequency is added to those in Fig. 17.40 to determine a unique slot depth and location. (After Man et al., 1994.)

more than 120% for this large slot ($ds/h = 35\%$). The slot locations were estimated with great accuracy, while the slot depths were always assessed much larger than the actual values because the perturbation method tends to predict smaller frequency shifts than the actual values.

17.7 Systems for Speech and Hearing

17.7.1 Speech Production

We have already discussed the physiological mechanisms of speech production in the human vocal tract and the physiological basis of hearing. From the engineering viewpoint, there are two major concerns: speech synthesis and speech recognition. Speech synthesis is used in articulated computers, appliance outputs, and machine reading of text. Speech recognition is used in talking to computers, as well as in typewriters, toys, televisions, appliances, locks, automobiles, and wristwatches. The deepest level of understanding in this area is required for the related application of machine translation. About 1970, major engineering research programs were started by companies (such as AT&T and IBM) and the U.S. Department of Defense. Progress in the field has been steady but slow because of the considerable challenges. One is the numerous relevant disciplines, including philosophy, linguistics, physiology, psychology, acoustics, signal processing, and chip design. In fact, the production of an utterance involves various levels of information processing as shown in Fig. 17.42. Each of these levels can be studied in depth. We will begin our discussion by reviewing some salient scientific background.

As we saw in Section 16.4, the production of vowels may be represented by

$$p(s) = U(s)T(s)R(s) \tag{17.7.1}$$

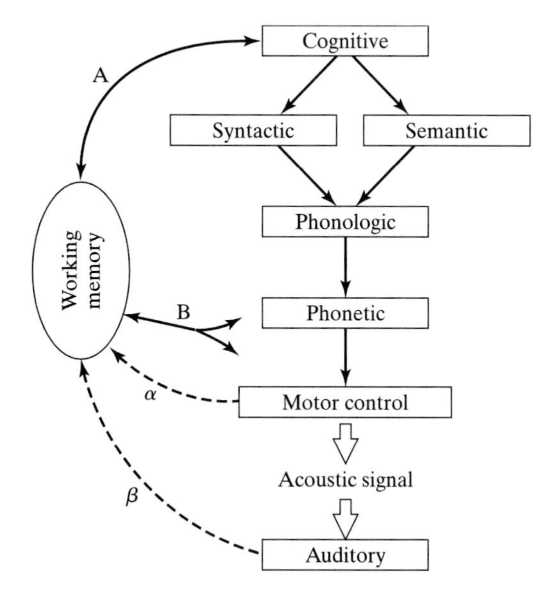

Fig. 17.42 This simplified information-processing model of utterance generation depicts the various levels of processing that contribute to the production of an utterance such as a sentence. The solid lines A and B represent demands on working memory. The broken lines α and β designate feedback channels. (After Kent and Read, 1992.)

where p(s) is the radiated pressure signal, U(s) is the volume velocity source spectrum, T(s) is the transfer function of the vocal tract, and R(s) is a radiation characteristic. There are three main resources in the vocal tract, giving rise to the so-called formants. Fig. 17.43 shows a formant synthesizer, which can be used to generate vowel sounds artificially. The relative values of the formant frequencies are determined by the position of the mouth and tongue. Fig. 17.44 shows the ratios of the first two formants for a number of speakers, and Fig. 17.45 shows a model that can be used to reproduce various consonant classes. Much work in speech research has been devoted to tracking the spectral movement of formants and in following the dynamics of the lips, tongue, velum, and vocal chords during speech production. An acoustic signal can be predicted quite accurately given a description of the vocal tract configuration; conversely, the vocal tract configuration can be estimated from the acoustic signal. Not so well understood is the sequence of events coordinating the movement of the 100 odd muscles that control the vocal tract. Even less understood is the structure of the neural activity, which in turn controls the muscles. It is in this area that linguistic theories have some relevance.

Our knowledge of the speech production process is sufficiently advanced, however, that artificial speech synthesis from text is a practical reality. A schema for doing this is shown in Fig. 17.46.

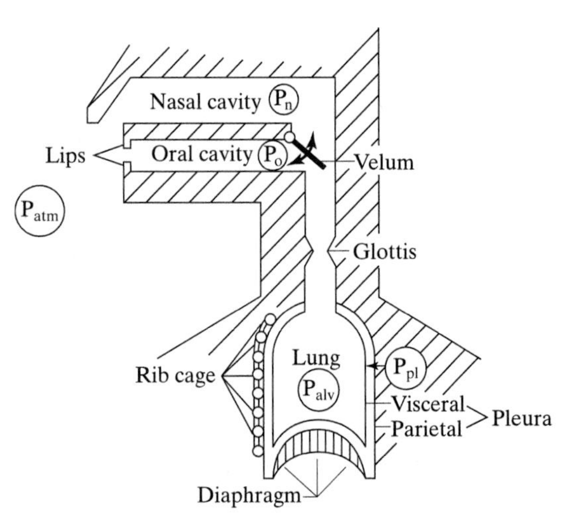

Fig. 17.43 A schematic illustration of the speech production system that shows the sites of measurement of five principal air pressures: alveolar pressure within the alveoli of the lungs, pleural pressure between the pleural linings of the rib cage, oral pressure within the oral cavity, nasal pressure within the nasal cavity, and the reference atmospheric pressure. (Adapted from Peterson and Barney, 1952.)

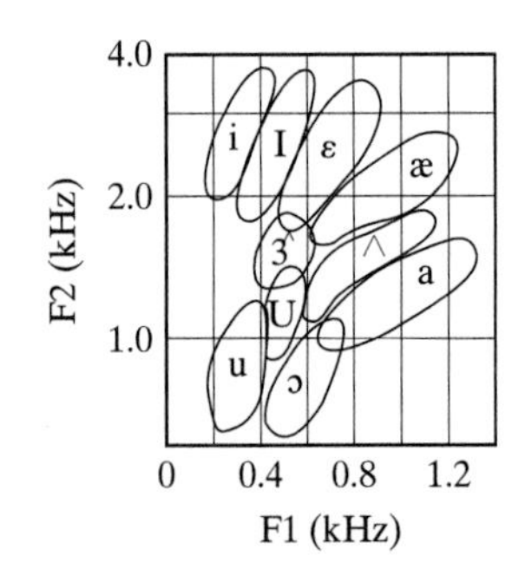

Fig. 17.44 F1/F2 ellipses for the vowels of 76 speakers (men, women, and children). (Adapted from Peterson and Barney, 1952.)

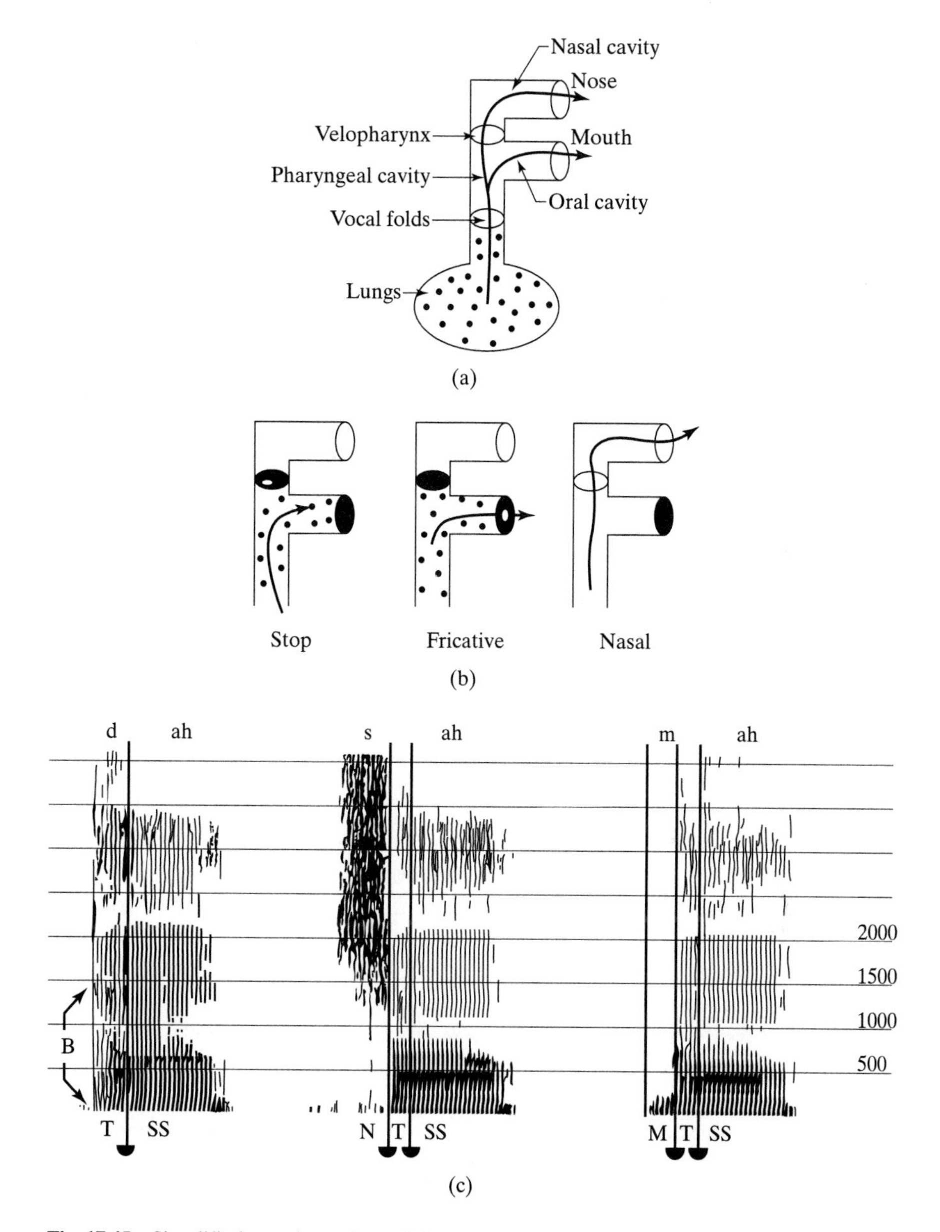

Fig. 17.45 Simplified aerodynamic model of speech (a), adaptations of model for three consonant classes (b), and spectrograms of examples of each consonant type (c). (After Kent and Read, 1992.)

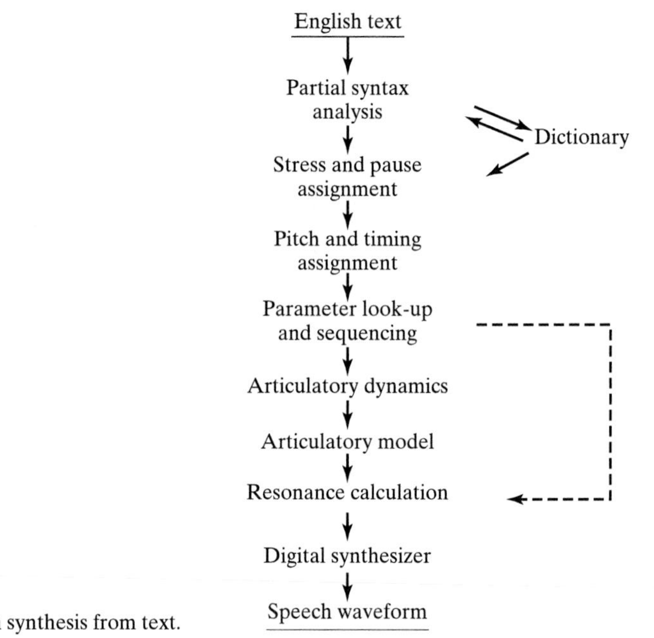

Fig. 17.46 Kent's schema for speech synthesis from text.

17.7.2 Speech Perception

While the speech synthesizer is important, the other important engineering problem is the construction of a speech perceptor. Although the complete physics of perception is not understood, it is possible that progress could be made with the engineering objective and such progress could perhaps help the scientific effort.

The engineering task might be formulated as the problem of converting the continuous acoustic signal into a sequence of words. A major difficulty has always been the uncertainty of the acoustic signal due to variations in the human voice (male, female, and children), the rate of speaking, dialect, etc. The medium through which the speech travels (distance, reverberation, etc.) also contributes to the variability of the received signal. It is commonly assumed that the sound signal is wrapped into a phonetic code relating to sound segments, such as consonants and vowels. But it is becoming clear that this is a rather gross assumption, and that the auditory system encodes speech, even at the level of the auditory nerve, using complex forms of signal processing. Fig. 17.47 illustrates the problem: There are no obvious boundaries demarking the words, consonants, and vowels.

Early speech research concentrated on finding acoustic cues corresponding to the phonetic dimensions of speech; that is, features that could be seen on the sound speech graph (burst frequency, second formant transition, etc.). Research on vowel perception has shown that the relative frequencies of the first two formants is sufficient for perception of most vowels. However, for many acoustic cues, there is no simple correspondence between the cue and the phonetic dimension (e.g., with the same burst frequency, various different consonants would be heard).

Another approach to speech perception involves the application of pattern recognition techniques. We discussed in Section 17.4 how sounds such as those from impacts on railroad wheels can be recognized and classified. A similar approach has been

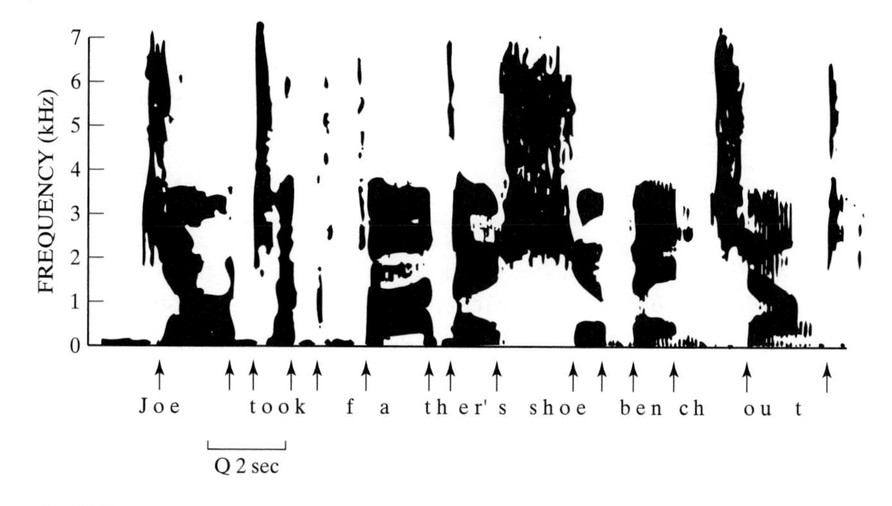

Fig. 17.47 Spectrogram of the sentence "Joe took father's shoe bench out." The arrows at the bottom indicate times where a rapid spectrum change occurs, corresponding to a consonantal event. It is hypothesized that certain phonetic features corresponding to place of articulation, voicing, etc., are signaled by the detailed acoustic properties in the vicinity of these consonantal events. (After Kent and Read, 1992.)

used for speech. The utterance is frequency analyzed as a function of time and the resulting spectrum compared with templates obtained from known utterances. The results are best when the utterances are short (e.g., words) and from the same speaker using careful pronunciation under controlled conditions. Even such a limited capability has applications in "hands busy" activities, such as baggage handling, parcel sorting, or recording microscope data. Neural networks have also proved very effective in recognizing limited utterances. Improvement results when the utterance is divided into syllable-length units that are matched against templates using a technique known as dynamic programming.

The best results seem to have been obtained when the processing combines a good model of the speech apparatus as well as stochastic methods. Raj Reddy and his students at Carnegie Mellon have pioneered other improvements, such as incorporating a model for word variation due to position in the sentence, proceeding and following words, stress by the speaker, rate of speaking, and who is talking. This was done using "Hidden Markov" modeling. Every distinct sound in English is treated as a computer that can generate an array of pronunciations. The Hidden Markov method finds the sound most likely to have produced a given vocal input. Further improvements have resulted from the inclusion of syntactical rules and a priori probabilities of the occurence of sound sequences. If the machine is used for a particular application, then a domain-specific vocabulary is helpful.

The costs of commercial systems increase dramatically with the size of the vocabulary, with requirements for speaker independence, and for processing continuous speech. As computer processing speed improves, speech processing applications will become more economically viable.

But perhaps the biggest breakthrough will occur when speech recognition systems can be backed up by artificial intelligence. Many words are acoustically equivalent (e.g., pair, pear, and pare) but their meaning is inferred by a human listener. Furthermore, a human processes speech, in general, by understanding its meaning. A machine that can match human performance in this way is likely decades away, although the greater the progress, the greater the economic investment in research will become.

17.8 Musical Acoustics

Musical acoustics presents us with several illustrations of different aspects of systems science. Fig. 17.48 illustrates the parts of a musical system. Music in the mind of the player is transmitted to the listener first by the energizing of an instrument and then through some space or auditorium. Musical instruments consist of components that we have discussed earlier in this book: vibrating air columns (the orchestral wind section), vibrating strings (the orchestral string section and the piano), and various vibrating bars, membranes, and shells (the orchestral percussion section). Strings and bars are poorly matched to the surrounding air and require some intervening device (such as a "resonator" or "sounding board") to produce an adequate output. The vibrating element is usually sharply resonant, so that to cover the audible range, either some means

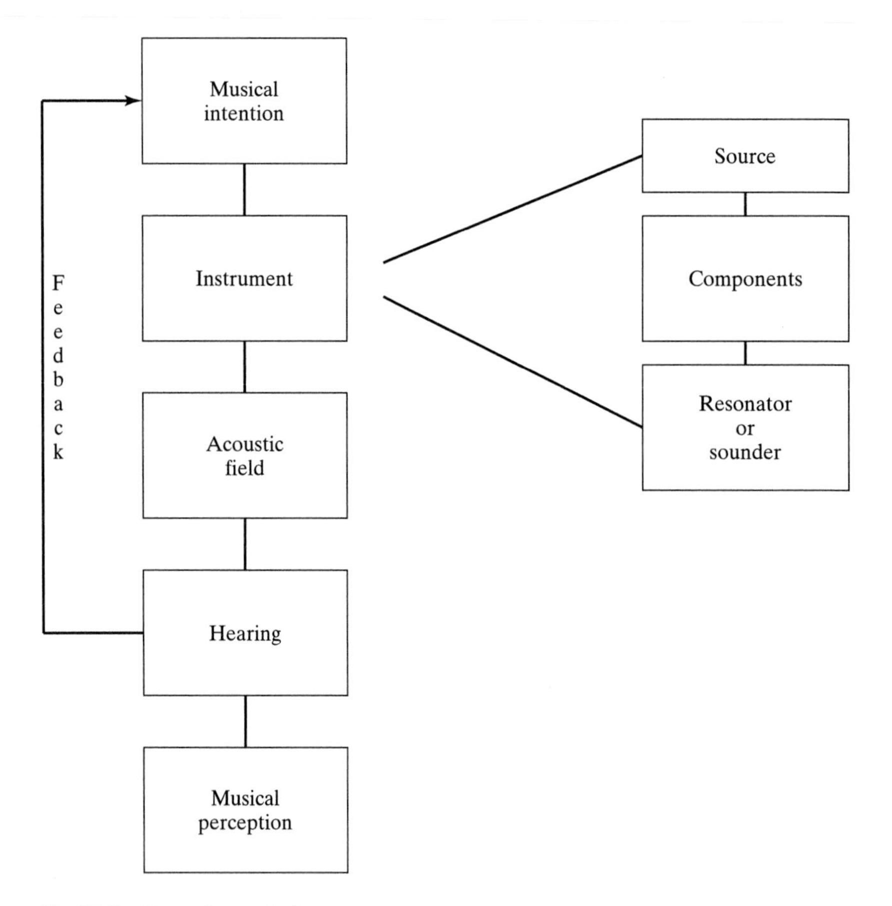

Fig. 17.48 Parts of a musical system.

must be provided to alter its resonant frequency or a number of elements resonant at different frequencies have to be employed. The sounds produced by the instruments are transmitted to the listener in rooms, studios, and concert halls, sometimes with electroacoustic enhancement. Finally, the human auditory system transmits signals to the brain, resulting in the perception of music.

The family of musical instruments has been evolved from simple, naturally occurring elements available to prehistoric people into the complex systems found in the modern orchestra. It is only in the recent past that acoustical science has had any appreciable impact on the design of these instruments, which were originally the work of craftsmen with input from the musicians they served. As with speech processing, the engineer has to wonder at the human abilities involved in such a process and then to determine how to improve it with scientific and engineering methods. Considerable progress has been made in the science of musical acoustics in quantifying the behavior of the instruments viewed as physical systems. Excellent reviews of this knowledge and its extensive literature can be found in Benade (1990) and in Backus (1977). Here, we will cover only the highlights to illustrate the systems viewpoint.

We already discussed some basic ideas of sound propagation in tubes and horns in Chapter 4, including the various possible resources and the input impedance. Benade (1990) showed that the spectrum of the impedance can be influenced by a series of tone holes, as illustrated in Fig. 17.49, which introduce a cutoff frequency given by

$$f_c = 0.11\frac{b}{a}\frac{c}{\sqrt{s(t + 1.5b)}} \tag{17.8.1}$$

The flute and clarinet have essentially conical bores, while most other woodwinds (oboe, English horn, bassoon, and saxophone) are conical. The bassoon has two bends to make it of manageable size, and the English horn has a bell with a sinusoidal profile. Among the brasses, both conical and flaring cross sections are found.

We will now discuss the manner in which the wind instruments are powered. In both woodwinds and brasses, the energy source is air pressure from the musician's lungs, although the control over the air flow is effected differently. In the woodwinds, the control is by means of a short tapered beam, called a reed, whose vibration is governed by feedback from the air column. In some woodwinds (flute and clarinet), there is a single reed, while in others (oboe, English horn, bassoon, and contra bassoon), there is a double reed. In the brasses, the control is effected by the player's lips. The situation is illustrated in Fig. 17.50 for both types. The principle behind pressure-controlled feedback was illustrated in a demonstration conceived of by Benade as shown in Fig. 17.51. In this device, known as the water trumpet, when the water level rises at the throat of the horn, a valve admits more water.

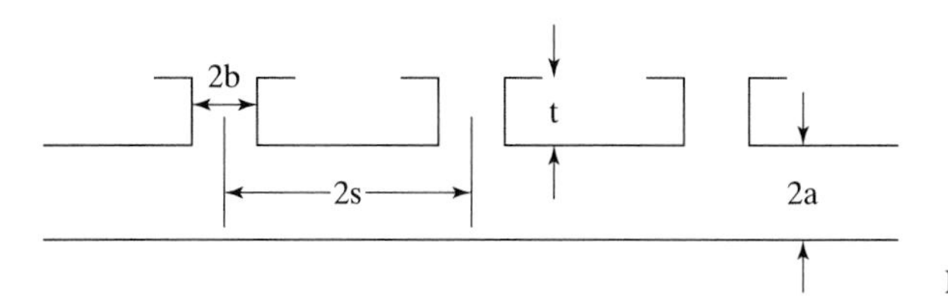

Fig. 17.49 Parameters of tone holes.

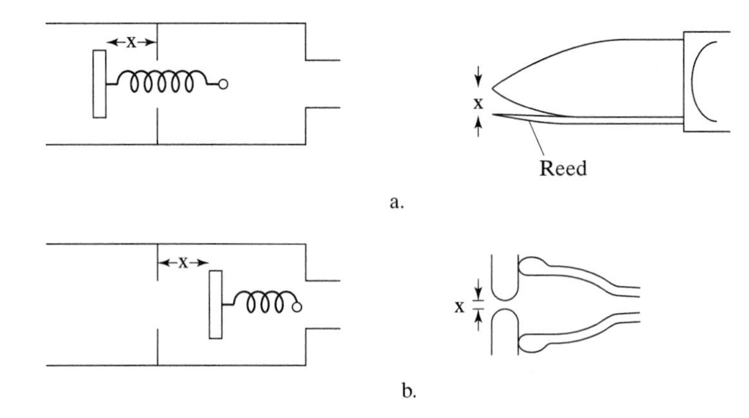

Fig. 17.50 Two types of pressure-controlled valves in wind instruments: a) "inward striking" reed of a clarinet; and b) "outward striking" lips of a brass player. In each case, the opening x depends on $p_o - p$.

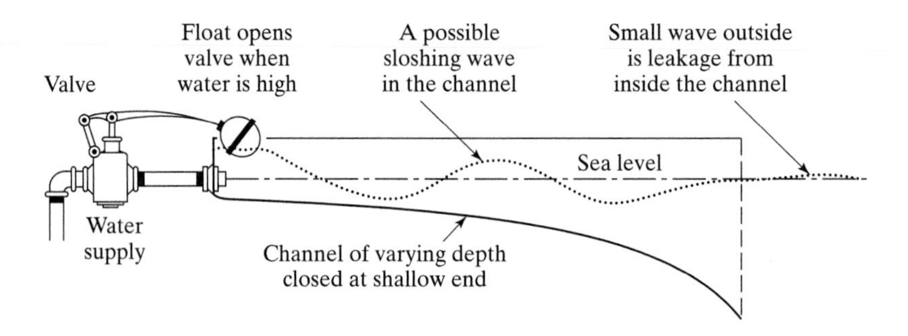

Fig. 17.51 A "water trumpet" illustrating pressure-controlled feedback in a wind instrument. (Adapted from Benade, 1990.)

The percussion instruments are easier to understand and indeed are similar as systems to the impacted objects discussed in Section 17.4. An impact is a forcing function with a wide frequency spectrum, which thus excites most of the modes in the object being struck. The bar percussion instruments (xylophone, glockenspiel, and marimba) are assemblies of beams that are excited into flexural vibration when struck. The xylophone and glockenspiel have metallic bars, whereas the bars of the marimba are of rosewood or fiberglass. The frequency of the various modes are not harmonically related, which is what gives the instruments their characteristic timbre. A vibrating bar is not a very efficient radiator, and in some instruments (e.g., the marimba) resonators are employed to increase the radiation efficiency. The alternative is to use a vibrating shell (i.e., a bell), which has a larger area. Bells have been used since ancient times and have developed differently in the East and in the West. In the West, there are church bells, carillon, and hand bells. The acoustics of bells have been studied since the 1700s and received attention from Rayleigh, who was interested in the "hum" note. Rossing (1982) points out that the various modes of a bell can be set out in a form similar to the periodic table of the elements. None of these modes matches the frequency of the hum note, which is now believed to be a psychoacoustic phenomenon. Much interest was aroused by the discovery of a cache of ancient Chinese bells in the 1990s. The concern was to determine how similar their design might be to the western bell family.

The third major family of instruments can be considered to have originated in the action of a single string stretched between two supports and caused to vibrate, as discussed in Section 12.4. The overtones of a string are harmonic so that the quality of the note emitted depends on the relative amplitude of the harmonics, which in turn is strongly influenced by the method in which the string is excited, whether by percussion, plucking, or bowing. The piano may be regarded as a percussion instrument as regards its excitation, but it may also be classified as a member of the stringed instrument family. Plucked stringed instruments include the harp and the guitar. The violin family is bowed into excitation. Each of these groups of stringed instruments has its own history of development. In each case it is necessary to provide a means of increasing the radiation efficiency of an isolated string. For the harp and the piano, this is done by attaching the strings to a massive and flexible frame. In the violin and guitar family, there is a flexible case that encloses a resonant cavity, a combination that serves the same purpose of increasing the radiation efficiency. A single string between two supports would provide only one note, so that in each case it is necessary to provide either a series of strings (as in the piano and to some extent in the violins), or a means of varying the vibration length of the sting by fingering (as in the guitar). The full range of the audible spectrum is covered in the violin family by the individual members of the family (double bass, bass, cello, viola, and violin). An interesting project in musical acoustics has been the development of a new instrumental family by scaling the dimensions of the violin. The members of this new family have similarities and dissimilarities to the traditional instruments. Helmholtz discovered that bowing produces a saw-tooth-forcing function (slip-stick) and that this then excites a characteristic frequency spectrum. The hourglass shape of the violin is believed to have evolved to permit free movement of the bow. The manufacture of the traditional violins and the production of good bows and strings are arts or technologies all in themselves, handed down from master to student.

We discussed the modes of a rectangular room in Section 10.6. From the viewpoint of this basic theory, it appears that the primary requirements of the music space are the provision of a sound level sufficiently above the noise level at the listening location. The room should be configured so that spatial variations in the transmitted field are minimized and absorption in the room should provide a suitable reverberation time. Beyond these basic requirements, listeners and musicians have other desiderati for a musical room. Beranek (1996) made a study of good concert halls and produced a listing of these other attributes, which include musical intimacy. Beranek ascribes this impression of intimacy to the closeness in arrival time of the first reflected sound and the first direct sound. If the separation between these exceeds 20 ms, the two sounds will not be perceived as coming from the same source. Haas (1951) showed that when the time separation is less than 35 ms, the two sounds are perceived as coming from the direction of the first arrival. The knowledge gained from the research of Beranek and others has allowed the design of halls different from the traditional shoe-box configuration. Electronic sound reinforcement for rooms is now commonplace and permits the transmission of music to far larger audiences than would be feasible without their presence. Another innovation permitted by the advance of electroacoustics has been the recording of music.

In listening to music it is clear that the human auditor has the ability to distinguish pure tones of frequencies differing by more than a certain amount. The musician would say that these sounds differ in "pitch." According to the American National Standards Institute (ANSI) standard of 1994: "Pitch is that attribute of auditory sensation in terms of which sounds may be ordered on a scale extending from low to high. Pitch depends mainly on the frequency content of the sound stimulus, but it also depends on the sound pressure and the waveform of the stimulus." Listeners have no difficulty distinguishing between two sounds of the same pitch: For example, middle C as played on the piano and on the organ, in which case the sounds are said to be of different timbre. The musician need not be concerned as to the physiological mechanism(s) that enables the making of these perceptions. To an engineer having to duplicate the functions of the hearing system, however, they are clearly of some concern, and the field is of considerable research interest. As an aid to the engineering student, at some risk of incurring the displeasure of psychoacousticians, we might say that the ear has the ability to perform signal processing with a similarity to cepstal analysis.

The difficulties presented to the engineer in the design of modern musical systems are probably more challenging than those in speech processing. If speech processing systems are going to require successful artificial intelligence for their implementation, then what will musical systems require? In turn, this requires us to answer the question: What is music? Studies by Manfred Kleins suggest that sounds have an emotional effect on a listener depending on whether they rise quickly or slowly. This hypothesis certainly correlates with the musician's stress on the importance of the "attack" of a note. Eventually there may be a comprehensive theory of the cognitive and emotional aspects of music. What is considered to be good music depends on the listener's personal background and musical education. There are also cultural influences to be considered. Since people first made music a number of social systems for its support have evolved. The invention of the system of musical notation was clearly a major development, as was the organization of ensembles and orchestras. We now have a huge musical repertoire both in notation and on record. Also of interest is the role of the composer. From the viewpoint of Dewey, we would see the activity of the composer in the same category as that of the engineer, namely a creator of human products.

17.9 Evolution of Acoustic Systems

A puzzle arose from nineteenth-century thermodynamics. According to Boltzmann's view, a system will tend to the state of maximum probability. Chaotic states are more probable than ordered states, so the second law became a statement of the tendency of nature to favor disorder over order. But it was about the same time that this interpretation of the second law was gaining currency that Darwin was propounding the theory of evolution, in which very improbable events were supposed to be favored, and where increasingly complex but ordered molecular structures appeared to be better suited to survive.

The answer to the paradox lies in the fact that the thermodynamics of the nineteenth century related only to equilibrium conditions. We now understand that there are three regions of thermodynamics: the equilibrium region, where entropy production, the fluxes (e.g., heat and mass diffusion), and the forces (gradients of thermodynamic variables) are all zero; the near equilibrium region, where the rates are linear functions of the forces, as in Fourier's law of heat conduction; and the far-from-equilibrium region, where the rates are no longer simple linear functions of the rates. Onsager (1931) published results showing that in the linear region systems evolve toward a stationary state in which the entropy production is a minimum subject to the constraints imposed on the system. This result was what might have been anticipated based on equilibrium thermodynamics. However, when efforts were made to extend the results to situations far from equilibrium, it proved impossible to do so. A system far from equilibrium may still evolve toward some steady state, but that state can no longer be characterized in terms of a simple thermodynamic potential. The stability of such a state is no longer to be taken for granted. Certain fluctuations, instead of dying away, may be amplified until they take over the whole system, driving it to a new regime that may be qualitatively quite different from states corresponding to minimum energy production. Phenomena of this kind are well-known in hydrodynamics and fluid flow. Turbulence and the Benard instability are well-known examples. Prigogine (1984) provides further details on the new thermodynamics. Two points relevant to our discussion are first, if there are any systems that do not evolve towards entropy maxima, then it is no longer guaranteed that the universe will end in a state of thermal equilibrium; and second, that the evolution of improbable, highly ordered states is indeed possible in the thermodynamics of systems far from equilibrium.

The evolution of biological systems is a remarkable business covered extensively by many authors. We have referred to the evolution of the hearing system in Section 16.1. As an example of the evolution of acoustical artifacts, consider the sonar system. The earliest sonar systems were active; that is, they emitted a pulse of ultrasound into the water and received an echo from the target, as described in Section 17.2. The target was a submarine, which would then be liable to torpedo attack. A solution to the problem of sonar detection was found by coating the submarine with a sound absorber (air bubbles embedded in rubber). Furthermore, it was possible to determine the source of the sonar "pings" and thus counterattack by torpedoing the transmitter. This in turn led to the world's navies adopting passive sonar techniques (i.e., listening systems).

These adaptations in sonar have parallels in the biological realm of bats and their prey. The similarities between biological evolution and the evolution of artifacts can be explained in terms of systems theory. A biological system is determined by its genome, and the artifact is determined by some specifications that are usually spelled out by a blueprint or set of instructions. In both cases, the system as a whole is generally decomposed into a number of subsystems, which may themselves be further decomposed into smaller subsystems. The mechanism of evolution in both cases consists of changes in the subsystems. In the biological case, it is supposed that changes in the genome are random mutations, while in the engineered artifact the changes will be due to deliberate design changes. Successive generations of biological systems are subject to a process of selection described by Darwin as the survival of the fittest. For the engineered system, the process of selection may depend on the artifact's fitness for the tasks in which it is employed. The factors determining the evolution of engineering artifacts have been detailed by Balsalla (1988), who discusses the relative influence on the process of military, economic, social, and scientific factors. Even at this level there are parallels between biological and engineering evolution. However, the use of systematically acquired knowledge by the engineering designer does not seem to have a parallel in biological evolution, unless we include the developments in the new field of bioengineering as a part of biological evolution.

The point being argued here is that evolution of artifacts (and systems in general) describes the way in which human beings progress. Engineering is a pragmatic activity in which systems are conceived, constructed, evaluated, and redesigned based on prior knowledge of the behavior of both the natural and the artificial worlds. Acoustics is a part of the sum total of human knowledge, and its future progress depends on the continued existence of the systems that have supported its growth in the past: the journals and books; libraries, schools, and universities; government agencies and laboratories; professional societies and businesses; as well as the professional dedication of individual practicing engineers and researchers in many disciplines.

REFERENCES AND FURTHER READING

Akerberg, P.M., B.H. Jansen, and R.D. Finch. 1995. "Neural Net-Based Monitoring of Steel Beams. 98 *J. Acoust. Soc. Am.* 1505–1509.

Backus, J. 1977. *The Acoustical Foundations of Music*. Norton.

Balsalla, G. 1988. *The Evolution of Technology*. Cambridge Univ. Press.

Benade, A.H. 1990. *Foundations of Musical Acoustics*. Dover.

Beranek. L. 1996. *Concert and Opera Halls: How They Sound*. Acoustical Society of America.

Bertalanffy, L. von. 1968. *General System Theory*. George Braziller.

Carpenter, G.A. 1989. "Neural Network Models for Pattern Recognition and Associative Memory." 2 *Neural Networks* 243–257.

Cespedes, E.I. 1993. *Elastography: A Method for Imaging of Biological Tissue Elasticity*. Ph.D. dissertation, University of Houston.

Claerbout, J.F. 1985. *Imaging the Earth's Interior*. Blackwell.

Cook, B.D., and R.E. Werchan. 1971. "Mapping Ultrasonic Fields With Cholesteric Liquid Crystals." 9 *Ultrasonics* 101–102.

Dawkins, R. 1976. *The Selfish Gene*. Oxford University Press.

Dimarogonas, A. 1996. *Vibration for Engineers*. 2d ed. Prentice Hall.

Dimarogonas, A.D., and C.A. Papadopoulos. 1983. "Vibration of Cracked Shafts in Bending." 91 *Sound and Vib.* 583–593.

Ewins, D.J. 1984. *Modal Testing: Theory and Practice*. Research Studies Press.

Fink, L.H., G.L. Blankenship, and J.B. Cruz. 1982. "Systems Engineering," in *Electronics Engineers' Handbook*, edited by D.G. Fink and D. Christiansen. McGraw Hill.

Gabor, D. 1948. "A New Microscopic Principle." 161 *Nature* 777–778.

Gell-Mann, M. 1994. *The Quark and the Jaguar*. W.H. Freeman and Co.

Grossberg, S. 1986. *The Adaptive Brain I: Cognition, Learning, Reinforcement, and Rhythm*. Elsevier.

Grossberg, S. 1986. *The Adaptive Brain II: Vision, Speech, Language, and Motor Control*. Elsevier.

Haas, H. 1951. "Uber den Einfluss Eines Einfachechos auf die Horsamkeit von Sprache." 1 *Acustica* 49–58.

Haran, S., B.H. Jansen, and R.D. Finch. 1989. "Application of an Automated Package of Pattern Recognition Techniques to Acoustic Signature Inspection of Railroad Wheels." 85 *J. Acoust. Soc. Am.* 440–449.

Hebb, D.O. 1949. *The Organization of Behavior*. Wiley.

Hickman, L.A. 1990. *John Dewey's Pragmatic Technology*. Indiana Univ. Press.

Hopfield, J.J. 1982. "Neural Networks and Physical Systems With Emergent Collective Computational Abilities." 79 *Proc. Natl. Acad. Sci. USA* 2554–2558.

Kent, R., and C. Read. 1992. *The Acoustic Analysis of Speech*. Whurr.

Lemons, R.A., and C.F. Quate. 1979. "Acoustic Microscopy." In *Physical Acoustics*, edited by W.P. Mason and R.N. Thurston. Academic Press.

Man, X.C. 1996. *Vibration Monitoring of Beams Using an Analytical Model*. Ph.D. dissertation, University of Houston.

Man, X.C., L. McClure, Z. Wang, R.D. Finch, P. Robin, and B. Jansen. 1994. "Slot depth resolution in vibration signature monitoring of beams using frequency shift." 95 *J. Acoust. Soc. Am.* 2029–2037.

McCulloch, W.S., and W. Pitts. 1943. "A Logical Calculus of the Ideas Imminent in Nervous Activity." 5 *Bull. Math. Biophys.* 115–133.

Minsky, M., and S. Papert. 1969. *Perceptrons: An Introduction to Computational Geometry*. MIT Press.

Nagy, K., D.A. Dousis, and R.D. Finch. 1978. "Detection of Flaws in Railroad Wheels Using Acoustic Signatures." 100 *J. Engrg. For Ind.* 459–465.

Onsager, L. 1931. "Reciprocal relations in irreversible processes I and II." *Phys. Rev.* 37, 405–426 and 38, 2265–2279.

Pohlman, R. 1937. "Uber die Richtende Wirkung des Schallfeldes auf Suspensionen Nicht Kugelformiger Teilchen." 107 *Z. fur Physik* 497–507.

Prigogine, I. 1984. *Order Out of Chaos*. Bantam.

Rapoport, A. 1986. *General System Theory*. Abacus Press.

Robin, P.Y., B.H. Jansen, X.T.C. Man, Z. Wang, and R.D. Finch. 1994. "Vibration Monitoring of Steel Beams by Evaluation of Resonance Frequency Decay Rates." 96(2) *J. Acoust. Soc. Am.* 867–873.

Rosenblatt, R. 1959. *Principles of Neurodynamics*. Spartan Books.

Rossing, T.D. 1982. *The Science of Sound*. Addison-Wesley.

Rummelhart, D.E., and J.M. McClelland. 1990. *Parallel Distributed Processing*. MIT Press.

Sokolov, S.J. 1935. "Ultrasonic Oscillations and Their Applications." 2 *Tech. Phys. USSR* 522.

Wade, G. 1976. *Acoustic Imaging*. Plenum Press.

Wells, P.N. T. 1977. "Biomedical Ultrasonics," Academic Press.

Widrow, B., and M.E. Hoff. 1960. "Adaptive Switching Circuits." IRE WESCON Conv. Record, Part 4, 96–104.

Widrow, B., and S.D. Stearns. 1985. *Adaptive Signal Processing*. Prentice Hall.

PROBLEMS

17.1 Discuss the concept of a system

- **a.** What does the term mean to you?
- **b.** What is the difference between a simple and a complex system?
- **c.** Discuss the difference between order and disorder.
- **d.** Present a taxonomy of the major types of systems.

17.2 Explain what is meant by "block diagram algebra."

- **a.** Show how block diagram algebra can be applied in the analysis of the functioning of a loudspeaker.
- **b.** Show how the concept of a "transfer function" may be applied in the analysis of an acoustic system to be used for ranging.

17.3 State the second law of thermodynamics. How is the second law interpreted in terms of probability theory?

17.4 **a.** What is a switch? Is it linear or nonlinear? What is the essential difference between analog and digital devices?

b. Describe the essential elements of a computer. How could you build a computer using switches?

17.5 **a.** Describe a system that might be used to recognize a certain sound.

b. After decompression, bubbles may occur in the blood stream of divers. Design a system to detect these bubbles.

17.6 Describe the systems used in human speech and hearing. Why do these systems operate in the frequency range that they do?

17.7 What constitutes the essential difference(s) between human beings and apes? How could you represent such differences in terms of systems?

17.8 **a.** What are the possible subsystems in a musical system? Illustrate your answer by reference to musical instruments with which you are familiar.

b. Do you think the computer will replace the instruments of the orchestra? Discuss.

17.9 Draw a systems diagram to represent the Acoustical Society of America. The ASA is said to be a very successful society. Would you venture an opinion as to why this might be so? Could the reason for its success be explained in terms of your system diagram?

17.10 In his book *The Selfish Gene*, (1976) Dawkins introduced the concept of a "meme." Compare and contrast this idea with that of the "schema" introduced by Gell-Mann (1994) in *The Quark and the Jaguar*.

Appendix A

Greek Alphabet

Greek Letter	Greek Name	English Equivalent
Α α	Alpha	a
Β β	Beta	b
Γ γ	Gamma	g
Δ δ	Delta	d
Ε ε, ϵ	Epsilon	e
Ζ ζ	Zeta	z
Η η	Eta	e
Θ θ	Theta	th
Ι ι	Iota	i
Κ κ	Kappa	k
Λ λ	Lambda	l
Μ μ	Mu	m
Ν ν	Nu	n
Ξ ξ	Xi	x
Ο ο	Omicron	o
Π π	Pi	p
Ρ ρ	Rho	r
Σ σ	Sigma	s
Τ τ	Tau	t
Υ υ	Upsilon	u
Φ φ	Phi	ph
Χ χ	Chi	kh
Ψ ψ	Psi	ps
Ω ω	Omega	o

Appendix B

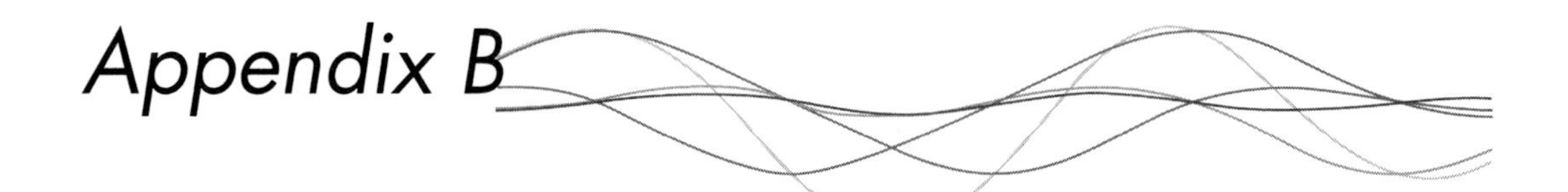

Basic Physical and Mathematical Constants

$\pi = 3.14159$

$e = 2.71828$

Gas constant $= R = 8.32$ J/mol-K

Boltzmann's constant $= k = 1.38 \times 10^{-23}$ J/K

Avogadro's number $=$ No $= 6.02 \times 10^{23}$ molecules/mole

Ratio of specific heats for air (20°C) $= \gamma = 1.40$

Planck's constant $= h = 6.63 \times 10^{-34}$ J-s

Acceleration due to gravity $= g = 9.81$ m/s^2

Atmospheric pressure $= 1.013 \times 10^5$ N/m^2 $= 14.7$ lb/in^2 $= 760$ mm-Hg

Charge on the electron $= e = 1.60 \times 10^{-19}$ C

Permittivity of free space $= \varepsilon_0 = 8.854 \times 10^{-12}$ F/m

Permeability of free space $= \mu_0 = 1.257 \times 10^{-6}$ H/m

Velocity of light in free space $= 3.0 \times 10^8$ m/s

Wavelength of sodium yellow light $= 5892$ A

Appendix C

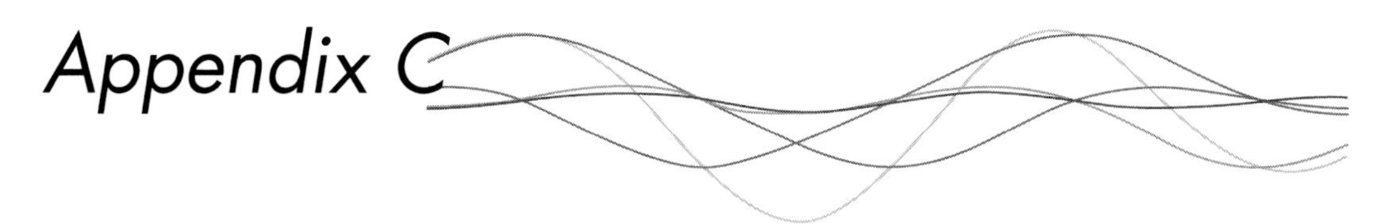

Definitions

Acoustic energy density: acoustic energy per unit volume at a point

Acoustic impedance: ratio of the sound pressure to the volume velocity in a wave

Acoustic intensity: acoustic energy crossing unit area in unit time at a point

Bar: a unit of pressure equal to 10^5 N/m^2 (approximately equal to one atmosphere)

Cycle: a complete repetition of the values of a periodic phenomenon

Decibel: the relative magnitude of two quantities expressed as one twentieth of the logarithm to the base ten of the ratio of the quantities

Frequency: the number of times a periodic phenomenon repeats in unit time

Loudness: the subjective measure of the strength of a sound

Period: the time required for one cycle of a periodic phenomenon

Radiation impedance: the mechanical impedance due to an acoustic wave

Sone: the unit of loudness.

Sound intensity level: $10\log_{10}(I_0/I_r)$, where I_0 is the intensity of the sound being measured and I_r is a reference sound intensity $= 10^{-12}$ W/m^2

Sound pressure level: $20\log_{10}(p_0/p_r)$, where p_0 is the pressure amplitude of the sound being measured and p_r is a reference sound pressure amplitude $= 20$ μPa

Specific acoustic impedance: ratio of the sound pressure to the particle velocity in a wave

Volume velocity: volume flow across an area

Appendix D

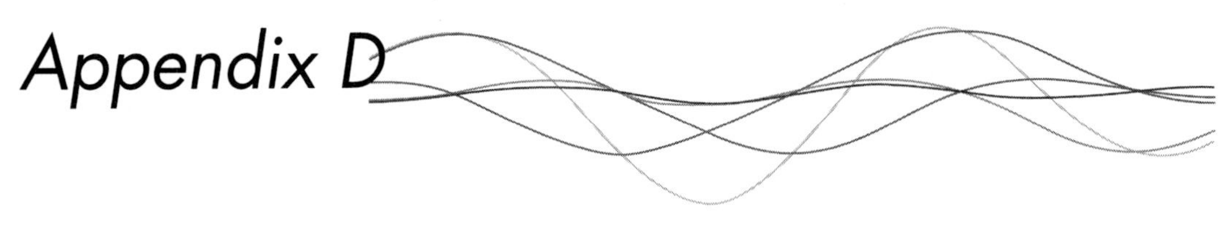

Units

Increasingly, work in acoustics uses the International System of Units, frequently referred to as SI units. But it is also still quite common to find the older British engineering units in use, so it is necessary to be familiar with both.

Quantity	Dimensions	SI Unit Name	Symbol	British Unit	Symbol	Conversion Factor*
Length	L	meter	m	foot	ft	3.281
Mass	M	kilogram	kg	slug	slug	6.854×10^{-2}
Time	T	second	s	second	sec	1
Frequency	T^{-1}	hertz	Hz		sec^{-1}	1
Velocity	LT^{-1}	meter/sec	$m\text{-}s^{-1}$		ft/sec	3.281
Force	MLT^{-2}	newton	N	pound-force	lb	0.2248
Magnetic Flux		weber	Wb			
Pressure	$ML^{-1}T^{-2}$	pascal	Pa		lb/ft^2	2.09×10^{-2}
Volume Velocity	L^3T^{-1}	cu meter/sec	$m^3\text{-}s^{-1}$	cu feet/sec	ft^3/sec	35.32
Energy	ML^2T^{-2}	joule	J		ft-lb	0.7376
Mechanical Impedance	MT^{-1}	mechanical ohm				
Acoustical Impedance	$ML^{-4}T^{-1}$	acoustical ohm				
Electric Charge		coulomb	C			
Electric Current		ampere	A			
Electric Potential		volt	V			
Capacitance		farad	F			
Power	$ML^{-2}T^{-3}$	watt	W	horsepower	hp	1.34×10^{-3}
Inductance		henry	H			

*Multiply a quantity expressed in SI units by the conversion factor to obtain the quantity in British units.

Appendix E

Bessel Functions and Directivity and Impedance Functions for a Piston

	Bessel Functions		Directivity Functions		Piston Impedance Functions	
x	$J_0(x)$	$J_1(x)$	Pressure $2J_1(x)/x$	Intensity $[2J_1(x)/x]$	$R_1(x)$	$X_1(x)$
0.0	1.0000	0.0000	1.0000	1.0000	0.0000	0.0000
0.1	0.9975	0.0499	0.9980	0.9960	0.0020	0.0424
0.2	0.9900	0.0995	0.9950	0.9900	0.0050	0.0847
0.3	0.9776	0.1483	0.9887	0.9775	0.0113	0.1272
0.4	0.9604	0.1960	0.9800	0.9604	0.0200	0.1680
0.5	0.9385	0.2423	0.9692	0.9393	0.0308	0.2120
0.6	0.9120	0.2867	0.9557	0.9133	0.0443	0.2486
0.7	0.8812	0.2390	0.6829	0.4663	0.3171	0.2968
0.8	0.8463	0.3688	0.9220	0.8501	0.0780	0.3253
0.9	0.8075	0.4059	0.9020	0.8136	0.0980	0.3816
1.0	0.7652	0.4401	0.8802	0.7748	0.1198	0.3969
1.1	0.7196	0.4709	0.8562	0.7330	0.1438	0.4296
1.2	0.6711	0.4983	0.8305	0.6897	0.1695	0.4624
1.3	0.6201	0.5220	0.8031	0.6449	0.1969	0.4915
1.4	0.5669	0.5419	0.7741	0.5993	0.2259	0.5207
1.5	0.5118	0.5579	0.7439	0.5533	0.2561	0.5460
1.6	0.4554	0.5699	0.7124	0.5075	0.2876	0.5713
1.7	0.3980	0.5778	0.6798	0.4621	0.3202	0.5924
1.8	0.3400	0.5815	0.6461	0.4175	0.3539	0.6134
1.9	0.2818	0.5811	0.6117	0.3742	0.3883	0.6301
2.0	0.2239	0.5767	0.5767	0.3326	0.4233	0.6468
2.2	0.1104	0.5560	0.5055	0.2555	0.4945	0.6711
2.4	0.0025	0.5202	0.4335	0.1879	0.5665	0.6862
2.6	−0.0968	0.4708	0.3622	0.1312	0.6378	0.6925
2.8	−0.1850	0.4097	0.2926	0.0856	0.7074	0.6903
3.0	−0.2601	0.3391	0.2261	0.0511	0.7739	0.6800
3.2	−0.3202	0.2613	0.1633	0.0267	0.8367	0.6623
3.4	−0.3643	0.1792	0.1054	0.0111	0.8946	0.6381
3.6	−0.3918	0.0955	0.0531	0.0028	0.9469	0.6081

(Continued)

	Bessel Functions		Directivity Functions		Piston Impedance Functions	
x	$J_0(x)$	$J_1(x)$	Pressure $2J_1(x)/x$	Intensity $[2J_1(x)/x]$	$R_1(x)$	$X_1(x)$
3.8	−0.4026	0.0128	0.0067	0.0000	0.9933	0.5733
4.0	−0.3971	−0.0660	(0.0330)	0.0011	1.0330	0.5349
5.0	−0.1776	−0.3276	(0.1310)	0.0172	1.1310	0.3232
6.0	0.1507	−0.2767	(0.0922)	0.0085	1.0922	0.1594
7.0	0.3001	−0.0047	(0.0013)	0.0000	1.0013	0.0989
8.0	0.1716	0.2346	0.0587	0.0034	0.9414	0.1219
9.0	−0.0903	0.2453	0.0545	0.0030	0.9455	0.1663
10.0	−0.2459	0.4350	0.0870	0.0076	0.9130	0.1784
11.0	−0.1712	−0.1768	(0.0321)	0.0010	1.0321	0.1464
12.0	0.0477	−0.2234	(0.0372)	0.0014	1.0372	0.0973

After Kinsler et al., 1982.

Appendix F
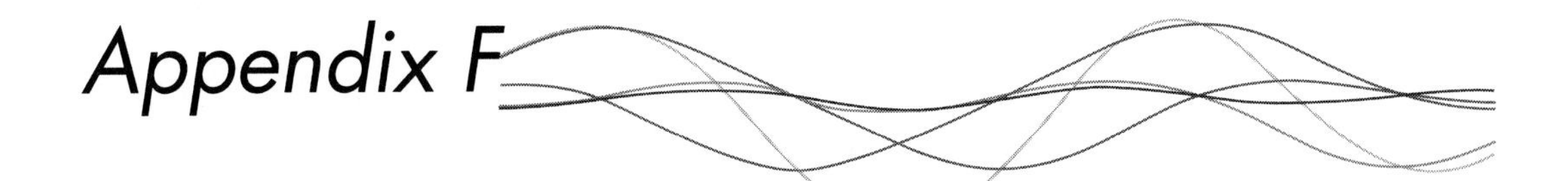

Some Useful Results from Vector Analysis

Green's theorem

$$\int\int\int_v \nabla \cdot \mathbf{A}\, dv = \int\int_s \mathbf{A} \cdot d\mathbf{s}$$

where **A** is a vector, and v is volume bounded by a surface S.

Stoke's theorem

$$\int\int_s (\nabla \times \mathbf{A}) \cdot d\mathbf{S} = \int \mathbf{A} \cdot d\mathbf{l}$$

where **A** is a vector and the integration is carried out over the curve ***l*** and the surface **S**.

Helmholtz's theorem

$$\nabla \times \nabla \times \mathbf{A} = \nabla\nabla \cdot \mathbf{A} - \nabla \cdot \nabla\mathbf{A}$$

Appendix G

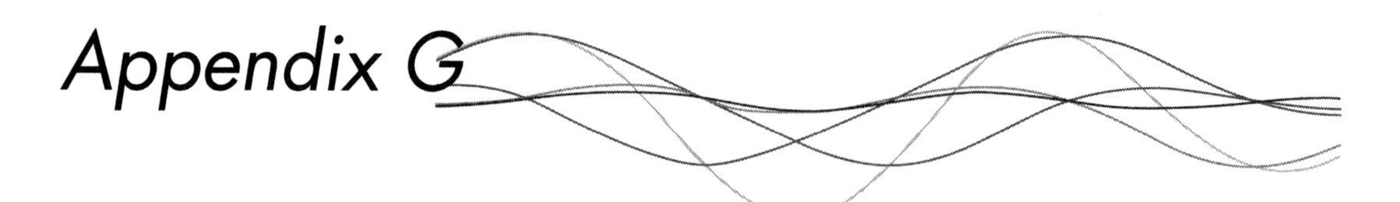

Notes on Matrices

A matrix is a rectangular array of numbers with m rows and n columns. An example is the fare table used by a bus conductor. But instead of being bus fares, the elements of the array can be any real or complex numbers. Many real-world systems can be represented in terms of such arrays. There are many situations in which we want to perform some mathematical operation on the entire set of numbers in the array, in studying the vibrations of multidegree of freedom systems for example. If these mathematical operations obey certain rules, then we call the array a *matrix*.

A matrix can be written as follows:

$$[A] = [a_{ij}] = \begin{bmatrix} a_{11} & a_{12} & a_{13} & a_{1j} & a_{1n} \\ a_{21} & a_{22} & & a_{2j} & a_{2n} \\ & & & & \\ a_{i1} & a_{i2} & & a_{ij} & a_{in} \\ & & & & \\ a_{m1} & a_{m2} & & a_{mj} & a_{mn} \end{bmatrix}$$

Matrix Algebra: If $[A] = [a_{pq}]$ and $[B] = [b_{pq}]$ are matrices of the same order m by n, then

1. $[A] = [B]$ if and only if $a_{pq} = b_{pq}$.
2. The sum [S] and difference [D] are the matrices defined by

$$[S] = [A] + [B] = [a_{pq} + b_{pq}],\ [D] = [A] - [B] = [a_{pq} - b_{pq}]$$

3. The product $[P] = [A]\,[B]$ is defined only when the number of columns in [A] equals the number of rows in [B] and is then given by

$$[P] = [A][B] = [a_{pq}][b_{pq}] = [a_{pr}b_{rq}]$$

where $a_{pr}b_{rq} = \sum_{r=1}^{n} a_{pr}b_{rq}$ by the summation convention.

4. The inverse of a square matrix [A] is a matrix $[A]^{-1}$ such that $[A]\,[A]^{-1} = [1]$ where [1] is the unit matrix.
5. The transpose of a matrix [A] is a matrix $[A]^T$, which is formed from [A] by interchanging its rows and columns. Thus, if $[A] = [a_{pq}]$, then $[A]^T = [a_{qp}]$.

Appendix H

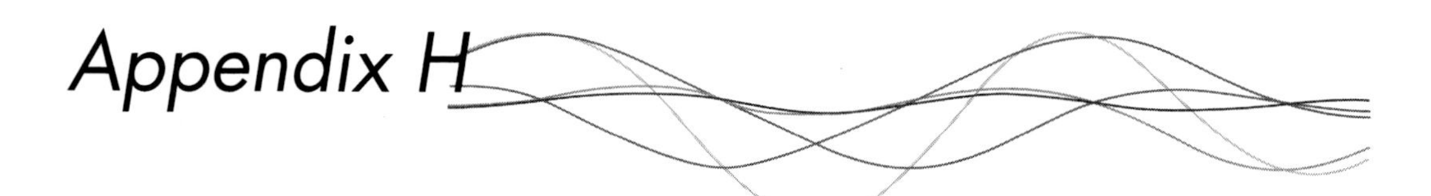

Properties of Solids

Solid	Density (kg/m³)	Young's Modulus (Pa) × 10¹⁰	Shear Modulus (Pa) × 10¹⁰	Bulk Modulus (Pa) × 10¹⁰	Poisson's Ratio	Sound Velocity (m/s)		Characteristic Impedance Pa · s/m × 10⁶	
						Bar	Bulk	Bar	Bulk
Aluminum	2700	7.1	2.6	7.5	0.33	5150	6300	13.9	17.0
Brass	8500	10.4	3.8	13.6	0.37	3500	4700	29.8	40.0
Copper	8900	12.2	4.4	16.0	0.35	3700	5000	33.0	44.5
Iron (cast)	7700	10.5	4.4	8.6	0.28	4480	4350	28.5	33.5
Lead	11300	1.65	0.55	4.2	0.44	1200	2050	13.6	23.2
Nickel	8800	21.0	8.0	19.0	0.31	4900	5850	43.0	51.5
Silver	10500	7.8	2.8	10.5	0.37	2700	3650	28.4	39.0
Steel	7700	19.5	8.3	17.0	0.28	5050	6100	39.0	47.0
Glass (Pyrex)	2300	6.2	2.5	3.9	0.24	5170	5600	12.0	12.9
Quartz (x-cut)	2650	7.9	3.9	3.3	0.33	5750	5750	14.5	15.3
Lucite	1200	0.4	0.14	0.65	0.4	1840	2680	2.15	3.2
Concrete	2600						3100		8.0
Ice	920						3200		2.95
Cork	240						500		0.12
Oak	720						4000		2.9
Pine	450						3500		1.57
Rubber (hard)	1100	0.23	0.1	0.5	0.4	1450	2400	1.6	2.64
Rubber (soft)	950	0.0005		0.1	0.5	70	1050	0.065	1.0
Rubber (rho-c)	1000			0.24			1550		1.55

Appendix I

Properties of Fluids

Liquids

Liquid	Temperature (°C)	Density (kg/m^3)	Isothermal Bulk Modulus (Pa) $\times 10^9$	Ratio of Specific Heats	Sound Velocity (m/s)	Characteristic Impedance (Pa · s/m) $\times 10^6$
Water (fresh)	20	998	2.18	1.004	1481	1.48
Water (sea)	13	1026	2.28	1.01	1500	1.54
Alcohol (ethyl)	20	790			1150	0.91
Castor (oil)	20	950			1540	1.45
Mercury	20	13600	25.3	1.13	1450	19.7
Turpentine	20	870	1.07	1.27	1250	1.11
Glycerin	20	1260			1980	2.5

Gases

Gas (at 1 atm)	Temperature (°C)	Density (kg/m^3)	Ratio of Specific Heats	Sound Velocity (m/s)	Characteristic Impedance (Pa · s/m)
Air	0	1.293	1.402	331.5	429
Air	20	1.21	1.402	343	415
O_2	0	1.43	1.40	317.2	453
CO_2	0	1.98	1.304	258	512
H_2	0	0.090	1.41	1269.5	114
Steam	100	0.6	1.324	404.8	242

Appendix J

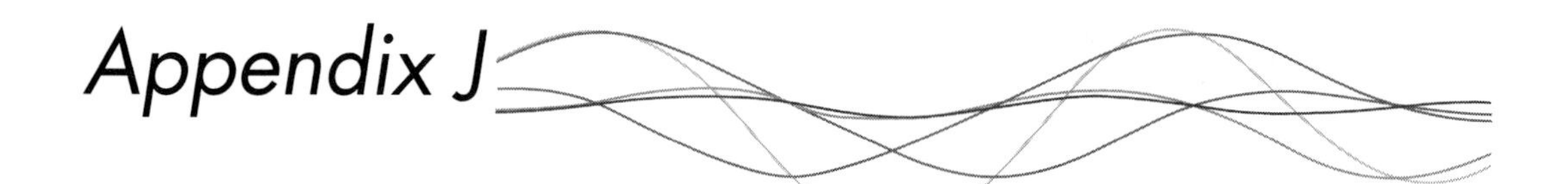

Absorption Coefficients at Various Frequencies

	128 Hz	256 Hz	512 Hz	1024 Hz	2048 Hz
Audience	1 - 2	3.5	4.1	4.9	4.2
Seat	1.4	4.3	4.5	4.5	3.6
Brick wall	0.02	0.02	0.03	0.04	0.05
Fabric curtains	0.05	0.12	0.35	0.45	0.38
Carpet	0.1	0.06	0.24	0.24	0.24
Acoustic tile	0.41	0.48	0.68	0.79	0.75
Acoustic perforated panel (fiber glass enclosed in metal)	0.50	0.98	0.99	0.92	0.82
Poured concrete	0.01	0.01	0.02	0.02	0.02
Wood sheathing	0.10	0.11	0.10	0.08	0.08

Appendix K

Acceptable Ambient Sound Levels for Unoccupied Spaces*

Type of Room	A-weighted Sound Level (dB)
Recording studios	18–28
Concert and recital halls	23–28
Television studios, music rooms	28–33
Theaters	28–33
Private residences	33–38
Conference rooms	33–38
Lecture rooms, classrooms	33–38
Executive offices	33–38
Private offices	38–43
Churches	38–43
Cinemas	38–43
Apartments, hotel rooms	38–43
Courtrooms	43–48

*After Bies and Hansen, 1988.

Type of Room	A-weighted Sound Level (dB)
Open-plan offices, schools	43–48
Libraries	43–48
Lobbies, public areas	43–48
Restaurants	48–53
Public offices (large)	48–53

Recommended Reverberation Times (sec) at 512 Hz for Rooms of Various Sizes

	Volume (thousand cubic feet)		
Room Usage	10	100	1000
Speech	0.70	0.90	1.08
Movie theater	0.93	1.10	1.30
Chamber music	0.98	1.18	1.40
School auditorium	1.03	1.26	1.49
Average music	1.17	1.44	1.71
Church music	1.33	1.67	2.02

Index

Q

R

S